SECOND EDITION

VISUALIZING PSYCHOLOGY

Siri Carpenter, PhD

Karen Huffman

Palomar College

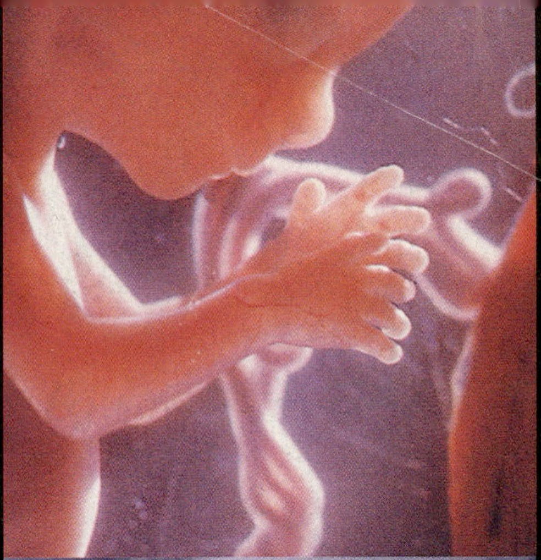

WILEY

In collaboration with

THE NATIONAL GEOGRAPHIC SOCIETY

CREDITS

VICE PRESIDENT AND PUBLISHER Jay O'Callaghan
MANAGING DIRECTOR Helen McInnis
EXECUTIVE EDITOR Christopher Johnson
DIRECTOR OF DEVELOPMENT Barbara Heaney
MANAGER, PRODUCT DEVELOPMENT Nancy Perry
DEVELOPMENT EDITOR Carolyn Smith
ASSISTANT EDITOR Eileen McKeever
EDITORIAL ASSISTANT Sean Boda
EXECUTIVE MARKETING MANAGER Jeffrey Rucker
MARKETING MANAGER Danielle Torio
PRODUCTION MANAGER Micheline Frederick
PRODUCTION EDITIOR Kerry Weinstein
CREATIVE DIRECTOR Harry Nolan
COVER DESIGNER Harry Nolan
INTERIOR DESIGN Vertigo Design
SENIOR PHOTO EDITOR Elle Wagner
PHOTO RESEARCHER Stacy Gold, National Geographic Society
SENIOR ILLUSTRATION EDITOR Sandra Rigby
MEDIA EDITORS Lynn Pearlman, Bridget O'Lavin
PRODUCTION SERVICES Furino Production

Cover Credits: Main image: Courtesy of St. Jude Medical, Inc. Illustration of the Libra® Deep Brain Stimulation System, an investigational device being tested in clinical research studies for major depressive disorder. The Libra system sends mild electrical pulses to a specific target in the brain from a device implanted in the chest. This device was developed by St. Jude Medical Inc.

Film Strip (left to right): Stacy Gold/NG Image Collection; © Adrianna Williams/Corbis; Masterfile; Anne Keiser/NG Image Collection; John Burcham/NG Image Collection

Image opposite title page: Gordon Wilsie/NGS Images Sales

This book was set in Times New Roman by GGS Higher Education Resources, a Divison of PreMedia Global, Inc., printed and bound by Quad/Graphics. The cover was printed by Phoenix Color.

ISBN 978-0-470-41017-2
BRV ISBN 978-0-470-55627-6

Printed in the United States of America

15 14 13 12 11 10

FOREWORD *From the Publisher*

V*isualizing Psychology, Second Edition,* is designed to help your students learn effectively. Created in collaboration with the National Geographic Society and our Wiley Visualizing Consulting Editor, Professor Jan Plass of New York University, *Visualizing Psychology* integrates rich visuals and media with text to direct students' attention to important information. This approach represents complex processes, organizes related pieces of information, and integrates information into clear representations. Beautifully illustrated, *Visualizing Psychology* shows your students what the discipline is all about, its main concepts and applications, while also instilling an appreciation and excitement about the richness of the subject.

Visuals, as used throughout this text, are instructional components that display facts, concepts, processes, or principles. They create the foundation for the text and do more than simply support the written or spoken word. The visuals include diagrams, graphs, maps, photographs, illustrations, schematics, animations, and videos.

Why should a textbook based on visuals be effective? Research shows that we learn better from integrated text and visuals than from either medium separately. Beginners in a subject benefit most from reading about the topic, attending class, and studying well-designed and integrated visuals. A visual, with good accompanying discussion, really can be worth a thousand words!

Well-designed visuals can also improve the efficiency with which information is processed by a learner. The more effectively we process information, the more likely it is that we will learn. This processing of information takes place in our working memory. As we learn, we integrate new information in our working memory with existing knowledge in our long-term memory.

Have you ever read a paragraph or a page in a book, stopped, and said to yourself: "I don't remember one thing I just read?" This may happen when your working memory has been overloaded, and the text you read was not successfully integrated into long-term memory. Visuals don't automatically solve the problem of overload, but well-designed visuals can reduce the number of elements that working memory must process, thus aiding learning.

You, as the instructor, facilitate your student's learning. Well-designed visuals, used in class, can help you in that effort. Here are six methods for using the visuals in *Visualizing Psychology* in classroom instruction.

1. **Assign students to study visuals in addition to reading the text.**
   It is important to make sure your students know that the visuals are just as essential as the text.

2. **Use visuals during class discussions or presentations.**
   By pointing out important information as the students look at the visuals during class discussions, you can help focus students' attention on key elements of the visuals and help them begin to organize the information and develop an integrated model of understanding.

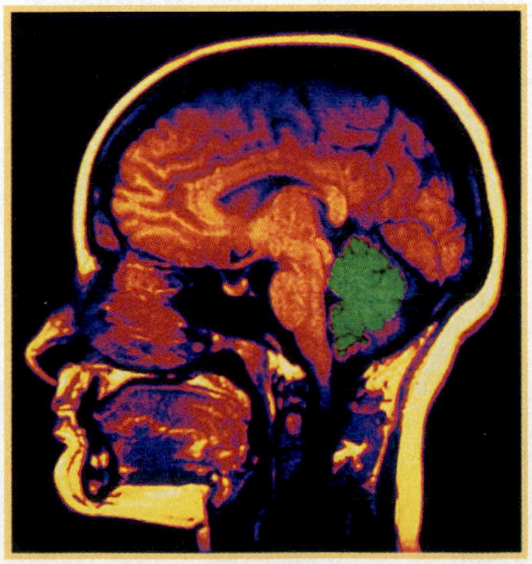

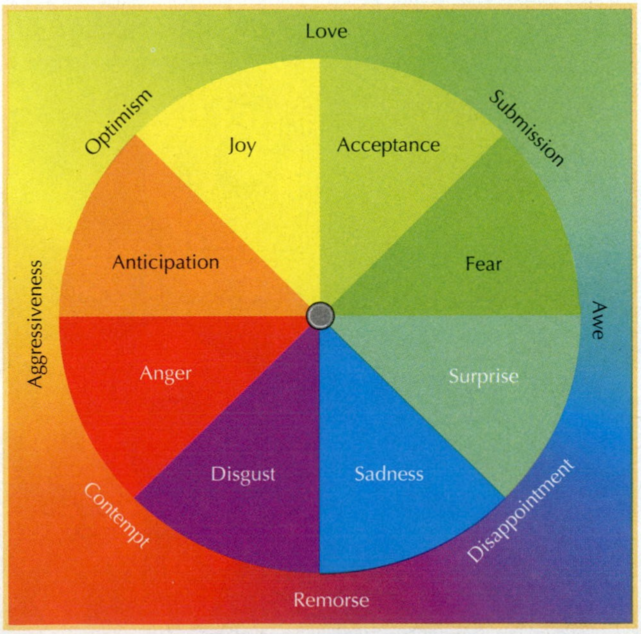

3. **Use visuals to review content knowledge.**
   Students can review key concepts, principles, processes, vocabulary, and relationships displayed visually. Better understanding results when new information in working memory is linked to prior knowledge.

4. **Use visuals for assignments or when assessing learning.**
   Visuals can be used for comprehension activities or assessments. For example, students could be asked to identify examples of concepts portrayed in visuals. Visuals can be very useful for drawing inferences, for predicting, and for problem solving.

5. **Use visuals to situate learning in authentic contexts.**
   Learning is made more meaningful when a learner can apply facts, concepts, and principles to realistic situations or examples. Visuals can provide that realistic context.

6. **Use visuals to encourage collaboration.**
   Collaborative groups often are required to practice interactive processes. These interactive, face-to-face processes provide the information needed to build a verbal mental model. Learners also benefit from collaboration in many instances, such as decision making or problem solving.

Visualizing Psychology not only aids student learning with extraordinary use of visuals, but it also offers an array of remarkable photos, media, and film from the National Geographic Society collections.

National Geographic has also performed an invaluable service in fact-checking *Visualizing Psychology*: they have verified every fact in the book with two outside sources, ensuring the accuracy and currency of the text.

Given all of its strengths and resources, *Visualizing Psychology* will immerse your students in the discipline, its main concepts and applications, while also instilling an appreciation and excitement about the subject area.

Additional information on learning and instructional design is provided electronically, including an *Instructor's Manual* that provides guidelines and suggestions on using the text and visuals most effectively. Other supplementary materials include the Test Bank with visuals used in assessment; PowerPoints; Image Gallery to provide you with the same visuals used in the text; web-based learning materials for homework and assessment including images, video, and media resources from National Geographic.

PREFACE

Do I contradict myself?
Very well then I contradict myself,
(I am large, I contain multitudes).
WALT WHITMAN, 1819–1892

Our own senses tell us—and psychological science confirms—that human beings are extraordinarily diverse. Bold, shy, cheerful, gloomy, analytical, artistic, clever, foolish, conscientious, careless, energetic, lazy, tense, relaxed, trusting, cautious—each person is a unique collection of qualities. We all see our world and respond to it just a little differently from any other person. Yet we also can observe remarkable constants in human thought, feeling, and behavior. Even in our endless variety, we are part of an integrated whole. To paraphrase the poet Walt Whitman, we contain multitudes. *Visualizing Psychology, Second Edition*, invites readers to explore the complexities and nuances of behavior—both human and nonhuman—that make the study of psychology so compelling.

As you might expect, the compelling (and rewarding) nature of psychology has attracted the attention and devotion of literally millions of readers, along with a multitude of psychology books. Why do we need another text? What makes *Visualizing Psychology* unique? Your authors and the editors and publisher of this text all believe that *active learning* and *critical thinking* (two synonymous and inseparable terms) are key ingredients to true understanding and lifelong learning. Therefore, we have developed and incorporated a large set of active learning and critical thinking pedagogical tools that will help you, the reader, personally unlock the fascinating mysteries and excitement of psychology. These tools will also teach you how to apply the wealth of insights and knowledge from psychological science to your everyday life. Best of all, active learning and critical thinking can make your study and mastery of psychology easier and more rewarding.

As the name implies, *Visualizing Psychology* is also unique in its focus on visuals. Based in part on the old saying that a "picture is worth a thousand words," this text covers the basic content of a standard psychology text enhanced by an educationally sound and carefully designed visual art program. For example, each chapter of this text contains a unique *Visualizing* feature and *Process Diagram* that present a key concept or topic and then explore it in detail using a combination of illustrative photos or figures. While reading through the text, be sure to take full advantage of these and other special study tools, including *Learning Objectives, What a Psychologist Sees, Applying Psychology, Psychological Research, Study Organizers, Concept Checks*, and *Summaries*. On pages of this Preface, we provide selected samples of these special features.

Through this premier art program, combined with our strong emphasis on active learning and critical thinking, *Visualizing Psychology* provides readers with a new and innovative approach to the understanding of psychology's major issues, from stereotyping to stem cells. In the context of an engaging visual presentation, we offer solid discussions of critical psychological concepts, ranging from the impact of stress on health to the psychological foundations of prejudice.

This book is intended to serve as a broad overview of the entire field of psychology. Despite its shortened

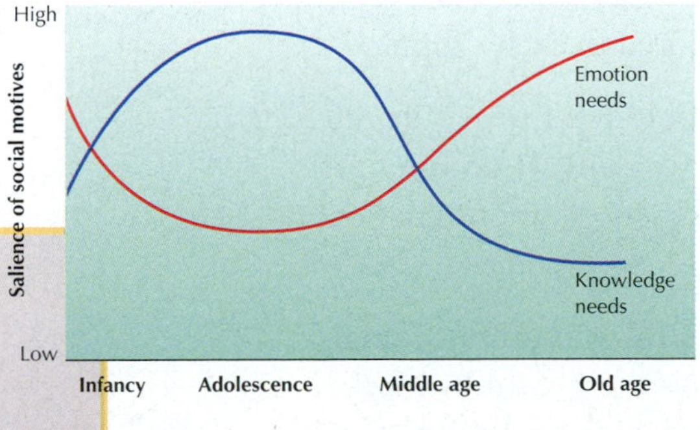

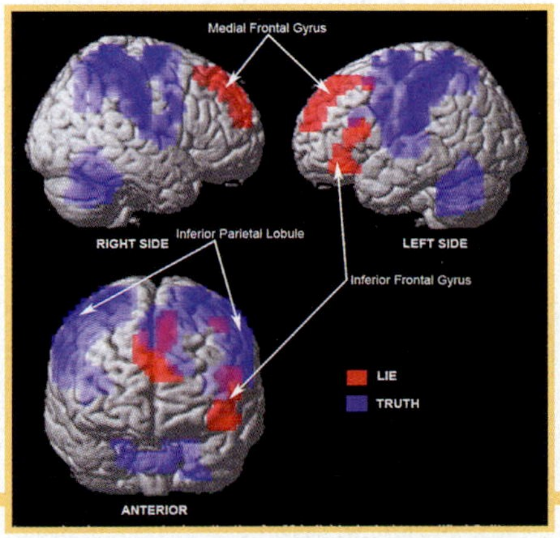

and condensed nature, *Visualizing Psychology*, like most other large survey and general education texts, contains a large number of unfamiliar terms and complex concepts. Do not be dismayed. The language of psychology is new to all but the most seasoned scholars. With a little hard work and concentrated study, you can master this material, and your work will pay off with immediate and unforeseen rewards that can last a lifetime.

As you can see, your authors feel passionate about psychology, and we believe that the study of psychology offers all of us an incomparable window into not only ourselves, but also the world that sustains us. We're eager to share our passion for psychology with you. We also welcome feedback from our readers. Please feel free to contact us at khuffman@palomar.edu.

ORGANIZATION

Poet Walt Whitman describes himself as containing "multitudes," and the field of psychology is similarly multitudinous. However, if you're like most beginning introductory psychology students, you probably think of psychology as primarily the study of abnormal behavior and therapy. You'll be surprised, and hopefully delighted, to discover that our field is much larger and richer than just these two areas. To organize such a diverse and complex field, our book is divided into 15 chapters that are arranged in a somewhat "microscopic/telescoping" fashion. We tend to move from the smallest element of behavior (the neuron and neuroscience) out to the largest (the group, culture, and social psychology). Here is a brief summary of the major topics explored in each chapter:

- **CHAPTER 1** describes psychology's history, its different theoretical perspectives and fundamental questions, and how psychologists go about answering those questions.
- **CHAPTER 2** explains the neural and other biological bases of behavior, and lays the groundwork for further discussions of biological foundations that appear in later chapters.
- **CHAPTER 3** examines interactions among stress, health, and behavior.
- **CHAPTERS 4 THROUGH 8** present aspects of cognition, including sensation, perception, consciousness, learning, memory, thinking, language, and intelligence. These chapters examine both cognition under healthy circumstances and cases where cognition goes awry. Throughout these discussions, we provide examples and exercises that connect basic research on cognition to real-world situations.
- **CHAPTERS 9 AND 10** of *Visualizing Psychology* explore human development across the life span, including physical, cognitive, social, moral, and personality development. We have organized these chapters topically, rather than chronologically, to help students appreciate the trajectory that each facet of development takes over the course of a lifetime.
- **CHAPTERS 11 AND 12** discuss processes and qualities that are integral to our most basic experiences and interactions with one another: motivation, emotion, and personality.
- **CHAPTER 13** addresses six major categories of psychological disorders. But first, we begin by discussing what constitutes "abnormal behavior," how psychological disorders are identified and classified. We also explore how psychological disorders vary across cultures.

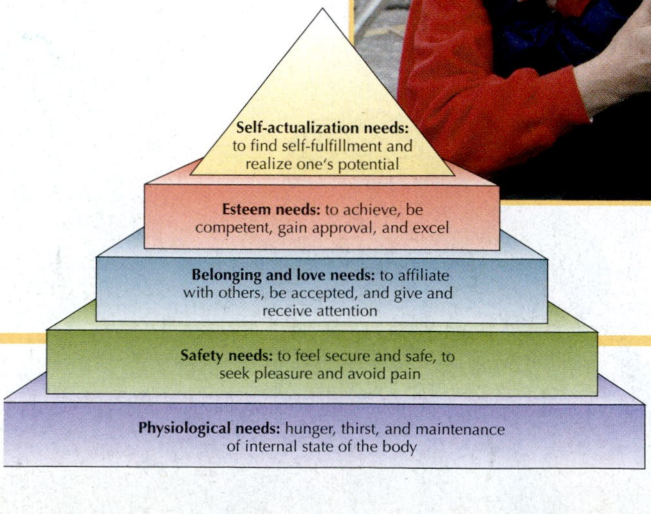
Self-actualization needs: to find self-fulfillment and realize one's potential
Esteem needs: to achieve, be competent, gain approval, and excel
Belonging and love needs: to affiliate with others, be accepted, and give and receive attention
Safety needs: to feel secure and safe, to seek pleasure and avoid pain
Physiological needs: hunger, thirst, and maintenance of internal state of the body

- **CHAPTER 14** describes and evaluates major forms of therapy, organizing the most widely used treatments into three groups: insight therapies, behavior therapies, and biomedical therapies.
- **CHAPTER 15**, social psychology, is in some ways the culmination of all the previous chapters, as there is no aspect of psychology that is irrelevant to how we think about, feel about, and act toward others. In this final chapter, we explore a range of social psychological phenomena, ranging from perceptions of others' intentions, to romantic attractions, to prejudice and discrimination.

NEW TO THIS EDITION

This Second Edition of *Visualizing Psychology* is dedicated to further enhancing the student learning experience through several new and unique features, including:

- **Enhanced visuals.** Throughout the text, photos, figures, diagrams, and other illustrations have been carefully examined and revised to increase their diversity and overall effectiveness as aids to learning.
- **Expanded coverage of important topics**. Throughout, this edition contains new or expanded discussions of topics such as sources of stress, positive reinforcement, mirror neurons, the misinformation effect, divergent thinking, the personal fable, parenting styles, nonverbal communication of emotion, the sharing of delusions on the Internet, and recent findings in support of psychoanalysis.
- **New *Applying Psychology* features.** These application sections help students relate psychological concepts to their own lives and understand how these concepts are applied in various sectors of society, such as the workplace.
- **New *Psychological Research* features**. *Visualizing Psychology* has always emphasized the empirical, scientific nature of psychology. This edition offers expanded descriptions of current research findings, explanations of their significance, and applications.
- **More opportunities for critical thinking**. Each *Applying Psychology* and *Psychological Science* box is accompanied by questions designed to encourage students to critically evaluate the topic of the box within the context of what they have learned in the text. Many figure captions also include critical thinking questions to further enhance student comprehension and critical thinking skills.
- **New study aid.** Carefully developed *Study Organizers* make it easy to compare different aspects of a topic, thus providing students with a useful tool for enhancing their understanding of the topic and preparing for exams. Among the topics treated in this way are the major psychological perspectives, properties of vision and hearing, schedules of reinforcement, stages of language development, parenting styles, and defense mechanisms.

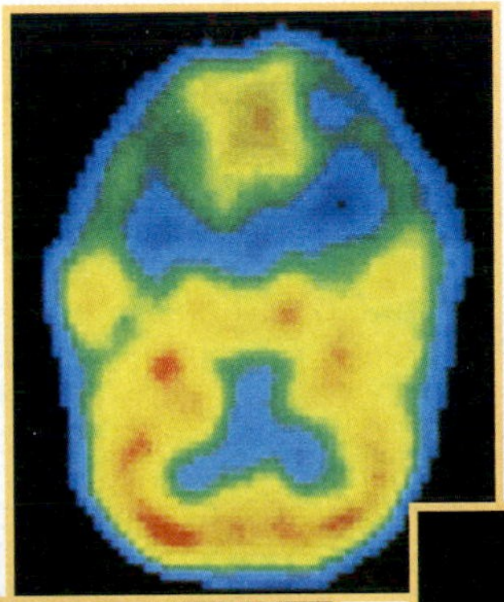

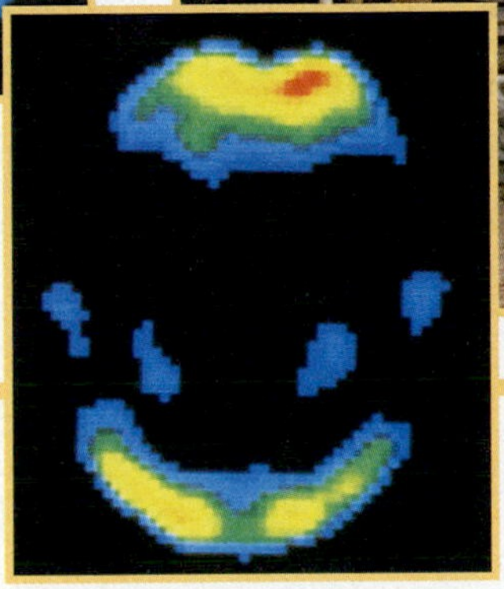

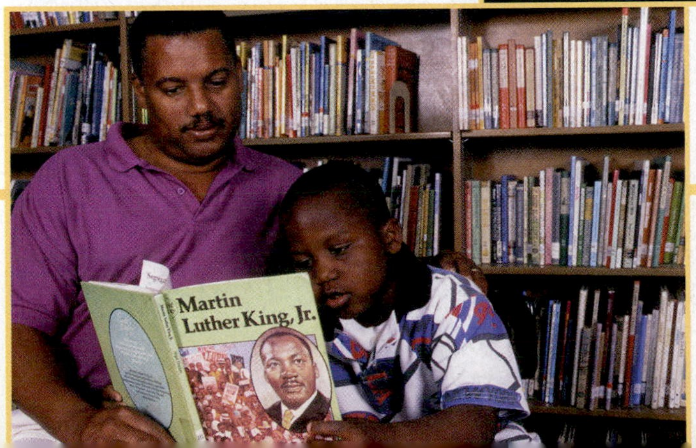

I L L U S T R A T E D

FEATURES THAT HELP STUDENTS VISUALIZE PSYCHOLOGY

A number of pedagogical features using visuals have been developed specifically for *Visualizing Psychology, Second Edition*. Presenting the highly varied and often technical concepts woven throughout psychological science raises challenges for reader and instructor alike. This Illustrated Book Tour provides a guide to the diverse features contributing to the book's pedagogical plan.

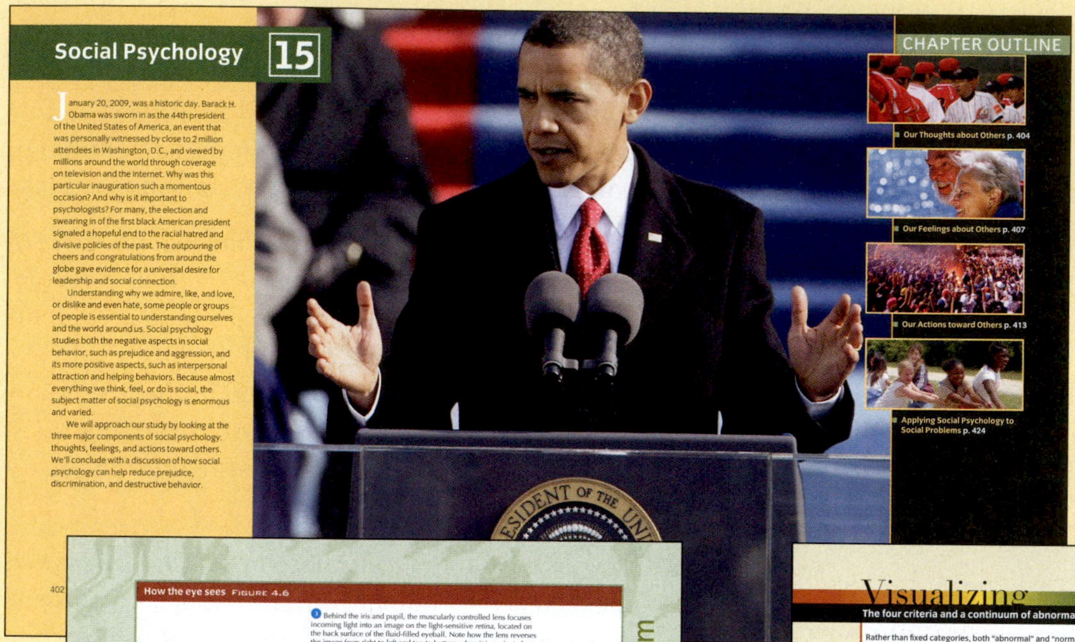

CHAPTER INTRODUCTIONS illustrate certain concepts in the chapter with concise stories that underscore some of psychology's most compelling questions. These narratives are featured alongside striking accompanying photographs. The chapter openers also include **CHAPTER OUTLINES** listing the main sections and features of the chapter.

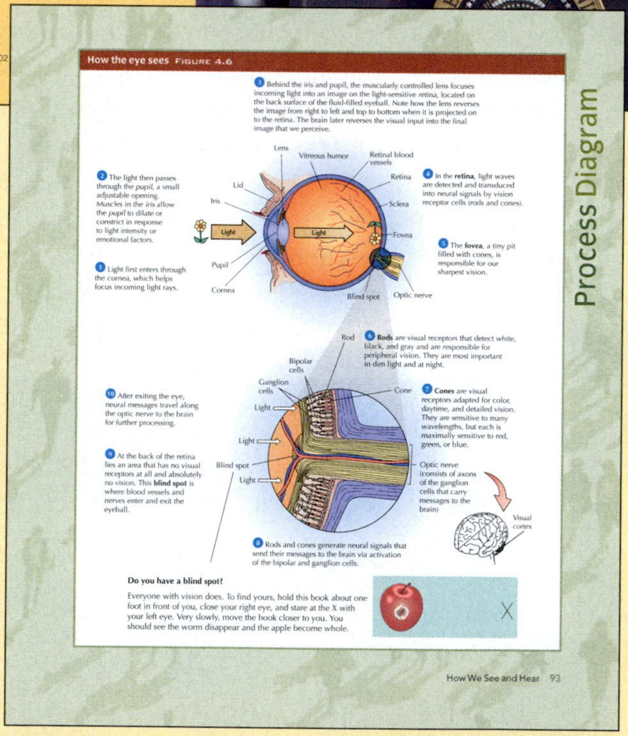

PROCESS DIAGRAMS present a series of figures or a combination of figures and photos that describe and depict a complex process, helping students to observe, follow, and understand the process.

VISUALIZING features are specially designed multipart visual spreads that focus on a key concept or topic in the chapter, exploring it in detail or in broader context using a combination of photos and figures.

BOOK TOUR

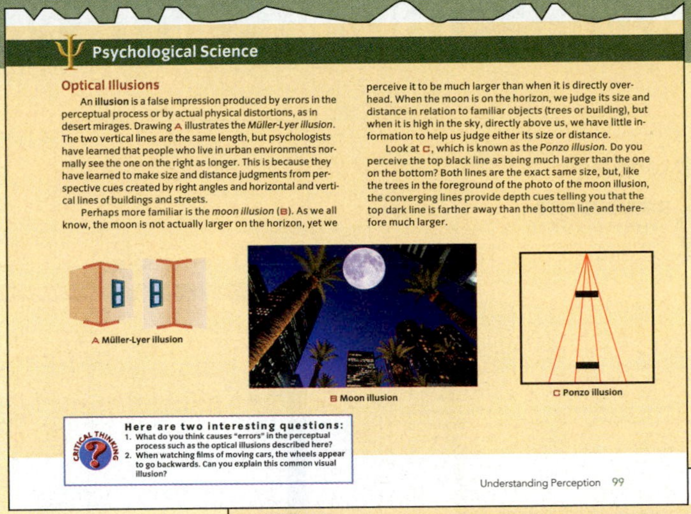

Psychological Science

Optical Illusions

An **illusion** is a false impression produced by errors in the perceptual process or by actual physical distortions, as in desert mirages. Drawing A illustrates the *Müller-Lyer illusion*. The two vertical lines are the same length, but psychologists have learned that people who live in urban environments normally see the one on the right as longer. This is because they have learned to make size and distance judgments from perspective cues created by right angles and horizontal and vertical lines of buildings and streets.

Perhaps more familiar is the *moon illusion* (B). As we all know, the moon is not actually larger on the horizon, yet we perceive it to be much larger than when it is directly overhead. When the moon is on the horizon, we judge its size and distance in relation to familiar objects (trees or building), but when it is high in the sky, directly above us, we have little information to help us judge either its size or distance.

Look at C, which is known as the *Ponzo illusion*. Do you perceive the top black line as being much larger than the one on the bottom? Both lines are the exact same size, but, like the trees in the foreground of the photo of the moon illusion, the converging lines provide depth cues telling you that the top dark line is farther away than the bottom line and therefore much larger.

A Müller-Lyer illusion

B Moon illusion

C Ponzo illusion

CRITICAL THINKING

Here are two interesting questions:
1. What do you think causes "errors" in the perceptual process such as the optical illusions described here?
2. When watching films of moving cars, the wheels appear to go backwards. Can you explain this common visual illusion?

Understanding Perception 99

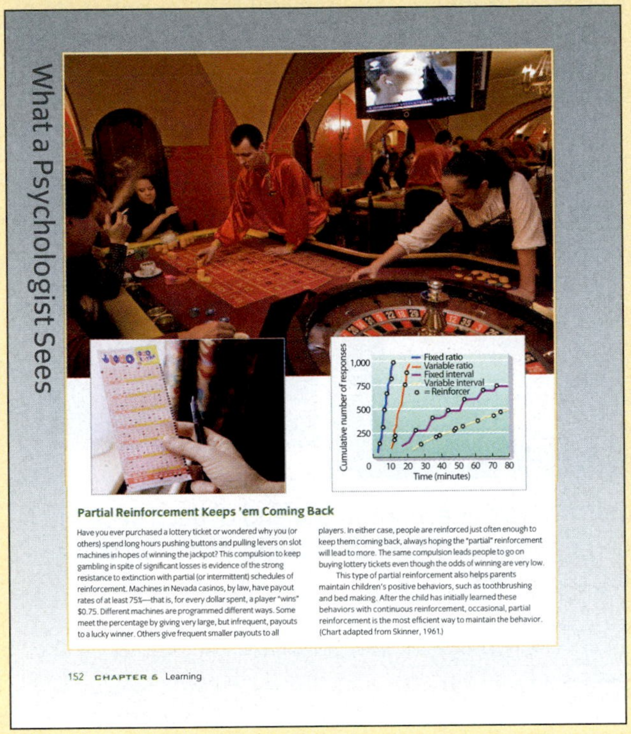

What a Psychologist Sees

Partial Reinforcement Keeps 'em Coming Back

Have you ever purchased a lottery ticket or wondered why you (or others) spend long hours pushing buttons and pulling levers on slot machines in hopes of winning the jackpot? This compulsion to keep gambling in spite of significant losses is evidence of the strong resistance to extinction with partial (or intermittent) schedules of reinforcement. Machines in Nevada casinos, by law, have payout rates of at least 75%—that is, for every dollar spent, a player "wins" $0.75. Different machines are programmed different ways. Some meet the percentage by giving very large, but infrequent, payouts to a lucky winner. Others give frequent smaller payouts to all players. In either case, people are reinforced just often enough to keep them coming back, always hoping the "partial" reinforcement will lead to more. The same compulsion leads people to go on buying lottery tickets even though the odds of winning are very low.

This type of partial reinforcement also helps parents maintain children's positive behaviors, such as toothbrushing and bed making. After the child has initially learned these behaviors with continuous reinforcement, occasional, partial reinforcement is the most efficient way to maintain the behavior. (Chart adapted from Skinner, 1961.)

152 CHAPTER 6 Learning

WHAT A PSYCHOLOGIST SEES features highlight a concept or phenomenon, using photos and figures that would stand out to a professional in the field, and helping students to develop observational skills.

PSYCHOLOGICAL SCIENCE features emphasize the empirical, scientific nature of psychology by presenting expanded descriptions of current research findings, along with explanations of their significance and possible applications.

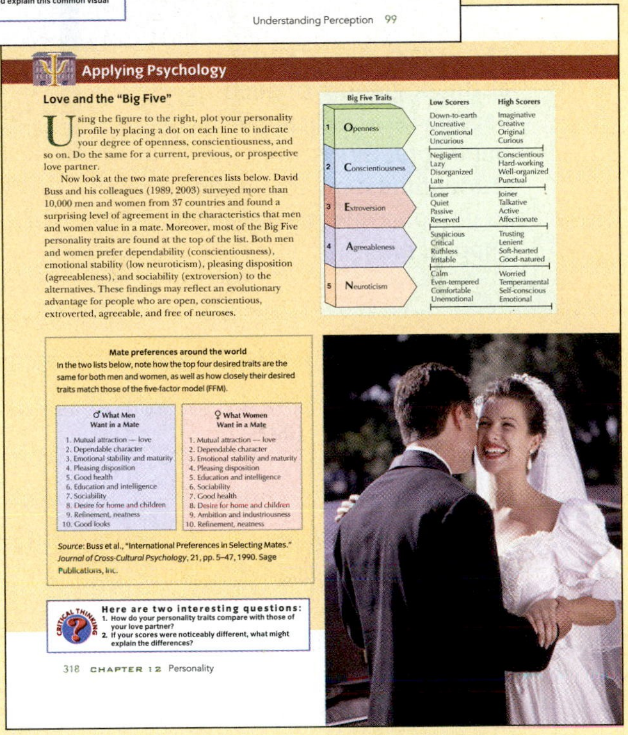

Applying Psychology

Love and the "Big Five"

Using the figure to the right, plot your personality profile by placing a dot on each line to indicate your degree of openness, conscientiousness, and so on. Do the same for a current, previous, or prospective love partner.

Now look at the two mate preferences lists below. David Buss and his colleagues (1989, 2003) surveyed more than 10,000 men and women from 37 countries and found a surprising level of agreement in the characteristics that men and women value in a mate. Moreover, most of the Big Five personality traits are found at the top of the list. Both men and women prefer dependability (conscientiousness), emotional stability (low neuroticism), pleasing disposition (agreeableness), and sociability (extroversion) to the alternatives. These findings may reflect an evolutionary advantage for people who are open, conscientious, extroverted, agreeable, and free of neuroses.

Big Five Traits	Low Scorers	High Scorers
1 Openness	Down-to-earth Uncreative Conventional Uncurious	Imaginative Creative Original Curious
2 Conscientiousness	Negligent Lazy Disorganized Late	Conscientious Hard-working Well-organized Punctual
3 Extroversion	Loner Quiet Passive Reserved	Joiner Talkative Active Affectionate
4 Agreeableness	Suspicious Critical Ruthless Irritable	Trusting Lenient Soft-hearted Good-natured
5 Neuroticism	Calm Even-tempered Comfortable Unemotional	Worried Temperamental Self-conscious Emotional

Mate preferences around the world

In the two lists below, note how the top four desired traits are the same for both men and women, as well as how closely their desired traits match those of the five-factor model (FFM).

♂ What Men Want in a Mate	♀ What Women Want in a Mate
1. Mutual attraction — love	1. Mutual attraction — love
2. Dependable character	2. Dependable character
3. Emotional stability and maturity	3. Emotional stability and maturity
4. Pleasing disposition	4. Pleasing disposition
5. Good health	5. Education and intelligence
6. Education and intelligence	6. Sociability
7. Sociability	7. Good health
8. Desire for home and children	8. Desire for home and children
9. Refinement, neatness	9. Ambition and industriousness
10. Good looks	10. Refinement, neatness

Source: Buss et al., "International Preferences in Selecting Mates." *Journal of Cross-Cultural Psychology*, 21, pp. 5–47, 1990. Sage Publications, Inc.

CRITICAL THINKING

Here are two interesting questions:
1. How do your personality traits compare with those of your love partner?
2. If your scores were noticeably different, what might explain the differences?

318 CHAPTER 12 Personality

APPLYING PSYCHOLOGY sections help students relate psychological concepts to their own lives and understand how these concepts are applied in various sectors of society, such as the workplace.

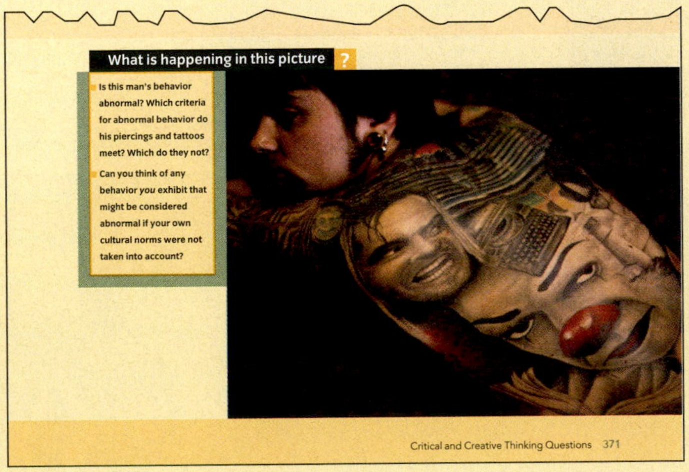

What is happening in this picture ?

- Is this man's behavior abnormal? Which criteria for abnormal behavior do his piercings and tattoos meet? Which do they not?
- Can you think of any behavior *you* exhibit that might be considered abnormal if your own cultural norms were not taken into account?

Critical and Creative Thinking Questions 371

WHAT IS HAPPENING IN THIS PICTURE? is an end-of-chapter feature that presents students with a photograph relevant to chapter topics but that illustrates a situation students are not likely to have encountered previously. The photograph is paired with questions designed to stimulate creative thinking.

PROVIDE STUDENTS WITH PROVEN LEARNING TOOLS

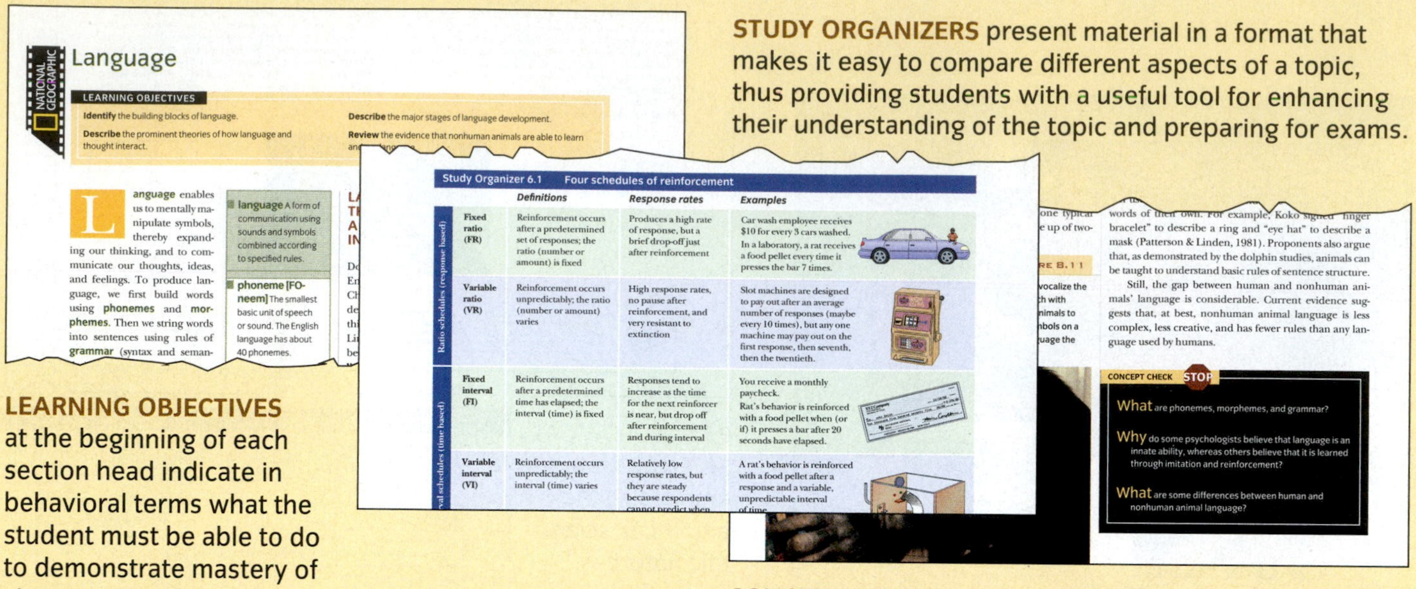

STUDY ORGANIZERS present material in a format that makes it easy to compare different aspects of a topic, thus providing students with a useful tool for enhancing their understanding of the topic and preparing for exams.

LEARNING OBJECTIVES at the beginning of each section head indicate in behavioral terms what the student must be able to do to demonstrate mastery of chapter material.

CONCEPT CHECK questions at the end of each section encourage students to test their comprehension of the learning objectives.

behavioral genetics The study of the relative effects of heredity and environment on behavior and mental processes.

evolutionary psychology A branch of psychology that studies the ways in which natural selection and adaptation can explain behavior and mental processes.

neuroscience An interdisciplinary field studying how biological processes relate to behavioral and mental processes.

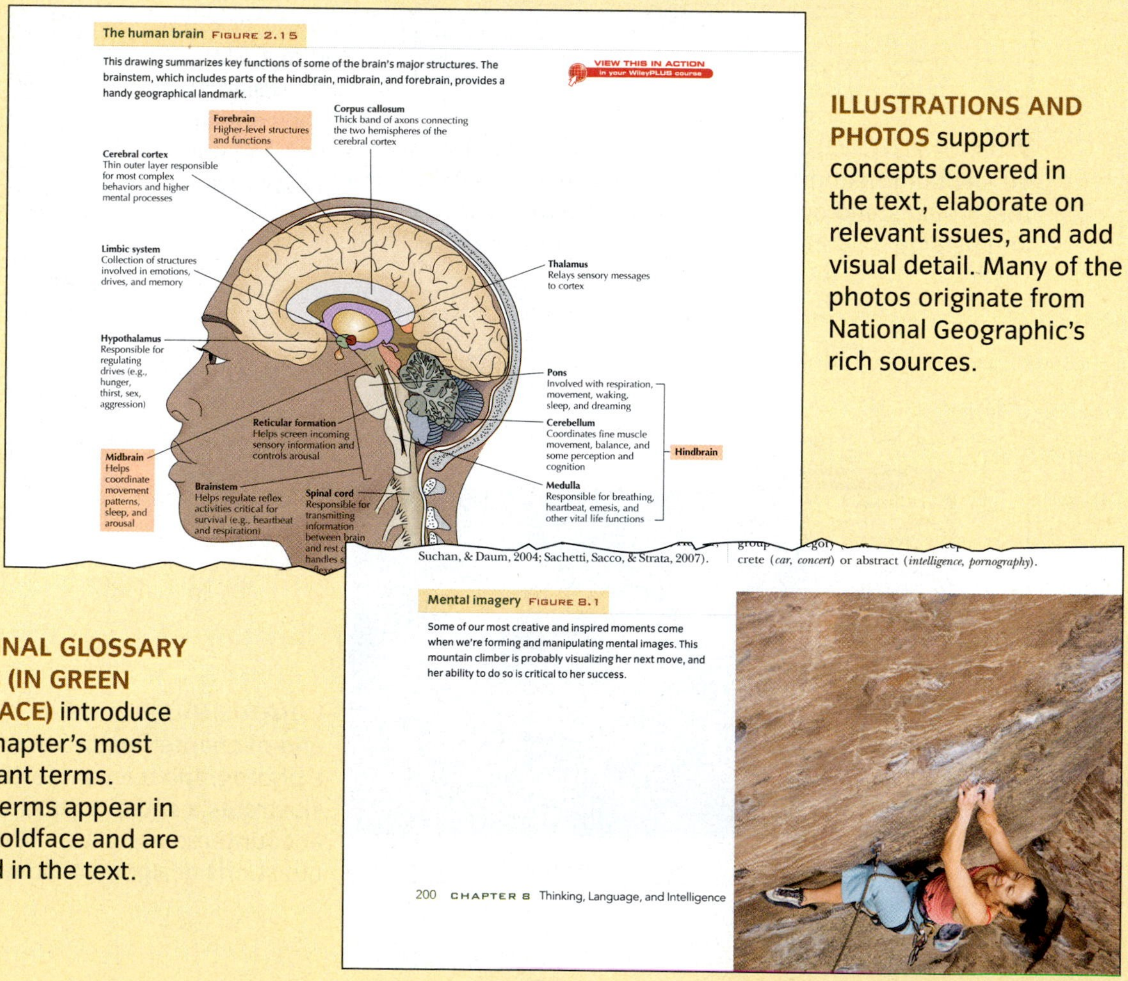

ILLUSTRATIONS AND PHOTOS support concepts covered in the text, elaborate on relevant issues, and add visual detail. Many of the photos originate from National Geographic's rich sources.

MARGINAL GLOSSARY TERMS (IN GREEN BOLDFACE) introduce each chapter's most important terms. Other terms appear in black boldface and are defined in the text.

PROVIDE STUDENTS WITH PROVEN REVIEW TOOLS

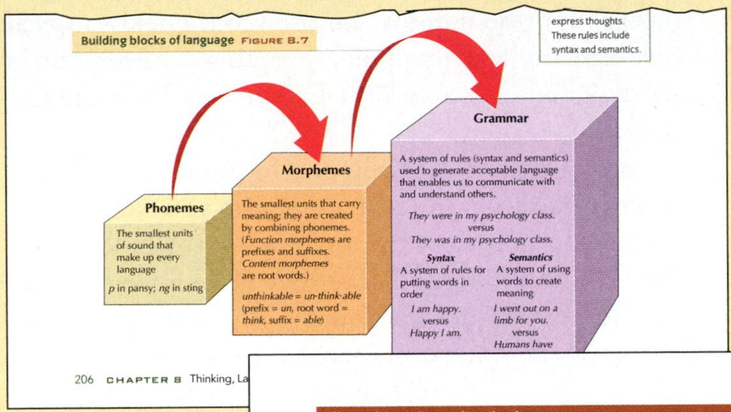

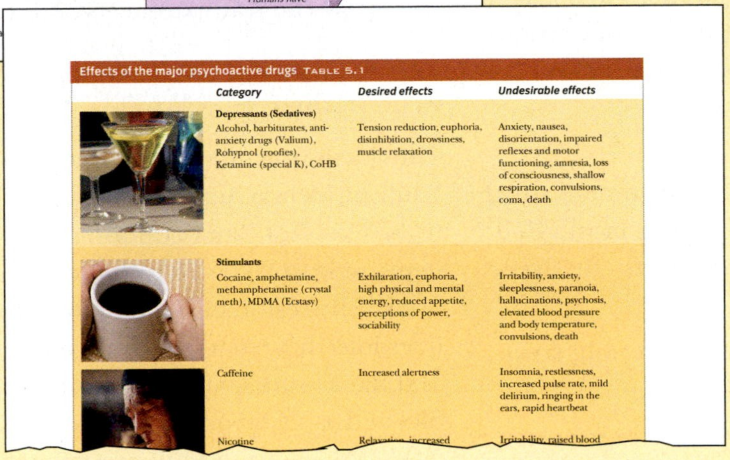

TABLES AND GRAPHS, with data sources cited at the end of the text, summarize and organize important information.

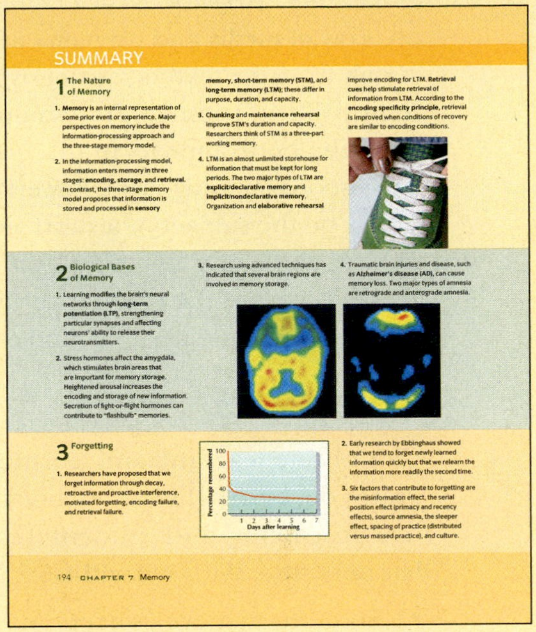

The end-of-chapter **SUMMARY** revisits each learning objective and each marginal glossary term, featured in boldface here, and included in a list of **KEY TERMS**. Students are thus able to study vocabulary words in the context of related concepts. Each portion of the Summary is illustrated with a relevant photo from its respective chapter section.

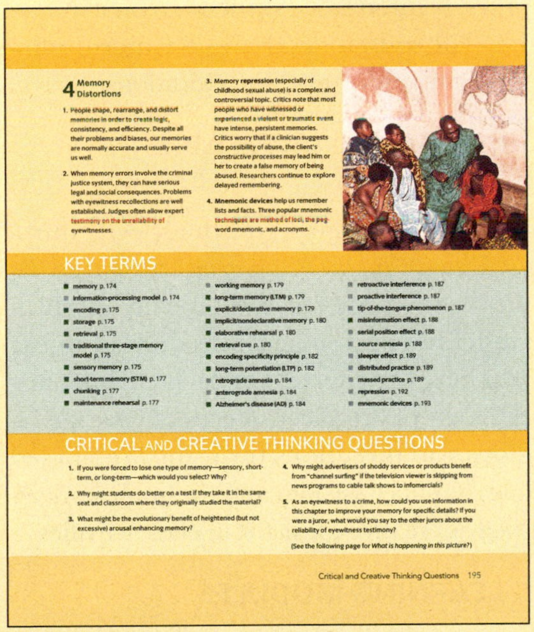

CRITICAL AND CREATIVE THINKING QUESTIONS encourage critical thinking and highlight each chapter's important concepts and applications.

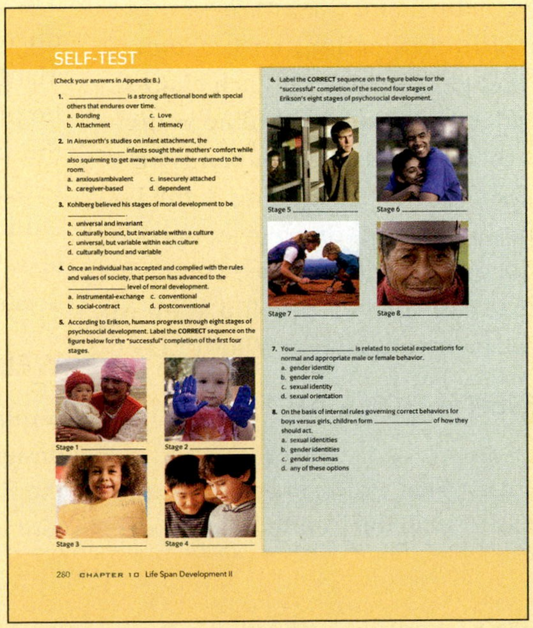

SELF-TESTS at the end of each chapter provide a series of multiple-choice questions, many of them incorporating visuals from the chapter, that review the major concepts.

MEDIA AND SUPPLEMENTS

Visualizing Psychology, Second Edition, is accompanied by an array of media and supplements that incorporate the visuals from the textbook extensively to form a pedagogically cohesive package. For example, a Process Diagram from the book appears in the *Instructor's Manual* with suggestions on using it as a PowerPoint in the classroom; it may be the subject of a short video or an online animation; and it may also appear with questions in the Test Bank, as part of the chapter review, homework assignment, assessment questions, and other online features.

WileyPLUS

This online teaching and learning environment integrates the entire digital textbook with the most effective instructor and student resources to fit every learning style. With **WileyPLUS**:

- Students achieve concept mastery in a rich, structured environment that's available 24/7.
- Instructors personalize and manage their course more effectively with assessment, assignments, grade tracking, and more.

WileyPLUS can be used with or in place of the textbook.

INSTRUCTOR RESOURCES

VIDEOS

A collection of videos, many from the award-winning National Geographic Film Collection, have been selected to accompany and enrich the text. Each chapter includes at least one video clip, available online as digitized streaming video that illustrates and expands on a concept or topic to aid student understanding. A full list of the National Geographic videos is on page xv. Accompanying each of the videos are contextualized commentary and questions that can further develop student understanding. The videos are available in **WileyPLUS**.

POWERPOINT PRESENTATIONS AND IMAGE GALLERY

A complete set of highly visual PowerPoint presentations by Karen Huffman is available online to enhance classroom presentations. Tailored to the text's topical coverage and learning objectives, these presentations are designed to convey key text concepts, illustrated by embedded text art.

Image Gallery All photographs, figures, maps, and other visuals from the text are online and can be used as you wish in the classroom. These online electronic files allow you to easily incorporate them into your PowerPoint presentations as you choose, or to create your own overhead transparencies and handouts.

TEST BANK (AVAILABLE IN WileyPLUS AND ELECTRONIC FORMAT)

The visuals from the textbook are also included in the Test Bank, by Melissa Acevedo of Westchester Community College. The Test Bank contains approximately 1200 test items, at least 25% of which incorporate visuals from the book. The test items include multiple-choice and essay questions that test a variety of comprehension levels. The Test Bank is available in two formats: online in MS Word files and as a Computerized Test Bank on a multiplatform CD-ROM. The easy-to-use test-generation program fully supports graphics, printed tests, student answer sheets, and answer keys. The software's advanced features allow you to create an exam to your exact specifications.

INSTRUCTOR'S MANUAL (AVAILABLE IN ELECTRONIC FORMAT)

The Instructor's Manual begins with the special introduction *Using Visuals in the Classroom,* prepared by Matthew Leavitt of the Arizona State University, in which he provides guidelines and suggestions on how to use the visuals in teaching the course. For each chapter, materials by Lynnel Kiely of the City Colleges of Chicago include suggestions and directions for using Web-based learning modules in the classroom and for homework assignments, as well as creative ideas for in-class activities.

WEB-BASED LEARNING MODULES

A robust suite of multimedia learning resources have been designed for *Visualizing Psychology* focusing on and using the visuals from the book. Delivered via the Web, the content is organized into Tutorial animations. These animations visually support the learning of a difficult concept, process, or theory, many of them built around a specific feature such as a Process Diagram, Visualizing feature, or key visual in the chapter. The animations go beyond the content and visuals presented in the book, providing additional visual examples and descriptive narration.

NATIONAL GEOGRAPHIC SOCIETY VIDEOS

National Geographic videos accompany *Visualizing Psychology, Second Edition.* Below is a brief description of the videos available for each chapter.

Chapter 1 Introduction and Research Methods

1. Among Wild Chimpanzees (3:48) A young Jane Goodall speaks about her work in the wilds of Africa with primates.
2. What Is Psychology? (0:54) What makes us act the way we do? Psychology explores individual differences.

Chapter 2 Neuroscience and Biological Foundations

3. Brain Surgery (4:33) Brain surgery is performed on a young man's tumor while he is awake.
4. Cool Quest (3:59) MRIs map the activity of the brain, exposing "cool" and "uncool" images.
5. Brain Bank (3:08) The Harvard Brain Tissue Resource Center, known as the "Brain Bank," is the largest brain repository in the world.

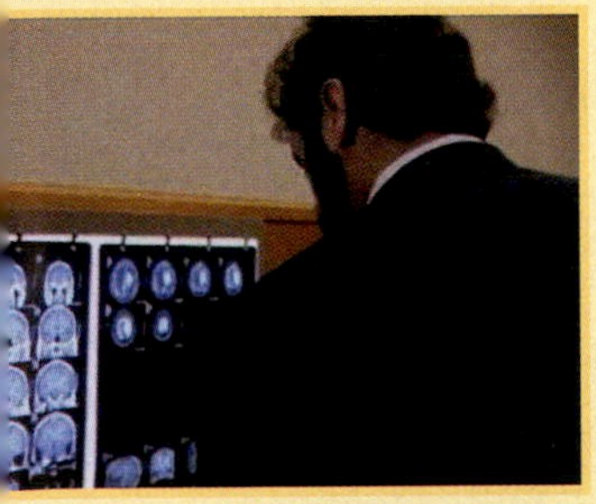

6. Brain Tumor Surgery (3:14) A patient suffering from seizures discovers he has a massive brain tumor near the part of the brain that controls motor activity.
7. MRI (0:38) An actual patient undergoing an MRI, showing the various images that the MRI produces.

Chapter 3 Stress and Health Psychology

8. Science of Stress (3:31) How stress affects the body.

Chapter 4 Sensation and Perception

9. Eye Trick Town (2:35) In Italy, Trompe L'oeil paintings that "trick" the eye into a perception of depth.
10. Camels (1:25) Photography of camels walking in the desert presents an interesting exploration into the relative nature of sensation and perception.

Chapter 5 States of Consciousness

11. Sleep Walking (1:57) The video suggests that during slow wave, non-REM sleep some people's lower part of the brain wakes up while the upper part of the brain responsible for awareness stays asleep.
12. Bali: Trance (3:24) Highlights of a festival in Bali where villagers come close to stabbing themselves while in a trance state.
13. Peyote and the Huichol People (4:17) The Huichol people ingest peyote, a mind-altering drug to enter the spirit world.

Chapter 6 Learning

14. Animal Minds (1:15) Rats are able to learn their way through a maze implying they may have a cognitive map.
15. Thai Monkey (2:11) Monkeys are taught how to retrieve coconuts through an official monkey training school, using both operative conditioning and modeling.

Chapter 7 Memory

16. Taxi Drivers (4:06) Video on the role that the hippocampus plays in consolidating memories, suggesting that there is a structural change to the brains of London taxi drivers.

Chapter 8 Thinking, Language, Intelligence

17. Orangutan Language (3:23) The orangutan language project at the National Zoo provides a stimulating environment where they learn a vocabulary of symbols and construct simple sentences.

Chapter 9 Life Span Development I

18. Feral Children (7:13) A Western researcher studies a young boy found in the jungle of Uganda, Africa, living with wild monkeys, rescued, and raised by a couple.

Chapter 10 Life Span Development II

19. Taboo Childhood (2:04) India allows imprisoned mothers to keep their children with them in jail—a very different value system than our Western view in the rearing of children.

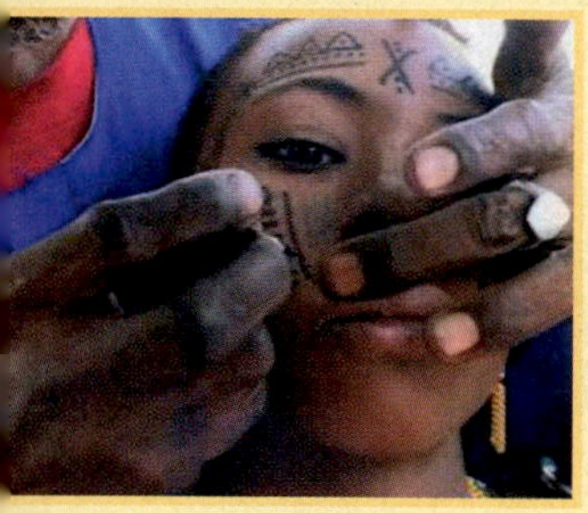

20. Taboo Sexuality: Eunuchs (3:53) Highlights of a group of eunuchs in their struggle to make a life in the Indian culture where they are viewed as social outcasts.
21. Coming of Age Rituals (4:38) The coming-of-age rituals in the Fulani tribe where two young boys have a whipping match, while a girl is given a full face of tattoos.

Chapter 11 Motivation and Emotion

22. Rodeo Clowns (2:28) A window into the world of a dangerous profession where professional rodeo clown bullfighters risk life and limb to entertain.
23. Heidi Howkins (3:22) Howkins's climb of K2, the world's second tallest mountain, risking everything to become the first woman to do so.
24. Fire Fighter Training (3:45) Dedicated smoke-jumpers meeting the challenges season after season, demonstrating not only bravery, but strong motivation.

Chapter 12 Personality

25. Freud (2:31) The life and times of Freud are highlighted, and some of the underlying tenets of psychoanalysis are explained.

Chapter 13 Psychological Disorders

26. Philippines: Exorcism (5:03) The exorcism of a teenage boy and a possible alternative explanation for his behavior are depicted.
27. Mars Desert Research (3:08) The Mars Desert Research Station program simulates conditions of a space mission to Mars, highlighting the psychological aspects of crew selection.
28. Odd Festivals Around the World (3:04) While these festivals in England, Thailand, and Italy may seem odd, they demonstrate the importance of cultural sensitivity.

29. Hindu Festival (5:20) Two brothers and other pilgrims proceed from Kuala Lumpur to the Batu Caves where they pierce their flesh with spikes and hooks, showing the importance of cultural sensitivity.

Chapter 14 Therapy

30. What Is Psychology: Therapy? (2:28) A clinical psychologist and a child psychiatrist are interviewed about the work they perform.
31. Leeches for Curing Illness (2:46) A look at the use of leeches in curing illness—how might this treatment be used in more mainstream medicine?

Chapter 15 Social Psychology

32. Teeth Chiseling (3:48) In Indonesia, a tribal chieftain's wife undergoes teeth chiseling to enhance her beauty—a matter of balance between the soul and body.
33. Leg Stretching (3:31) In China, a doctor performs leg-stretching operations to increase the height of patients, considering the cultural and psychological aspects, and the risks and benefits.
34. The Maroons of Jamaica (1:56) The story of the Maroons of Jamaica, winning independence from the British, and what their lives are like today.

ACKNOWLEDGMENTS

CLASS TESTING AND STUDENT FEEDBACK

To make certain that *Visualizing Psychology, Second Edition,* meets the needs of today's students, we asked several instructors to class test chapters. The feedback that we received from students and instructors confirmed our belief that the visualizing approach taken in this book is highly effective in helping students to learn. We wish to thank the following instructors and their students who provided us with helpful feedback and suggestions:

Sheree Barron
Georgia College & State University

Dale V. Doty
Monroe Community College

William Rick Fry
Youngstown State University

Andy Gauler
Florida Community College at Jacksonville

Bonnie A. Green
East Stroudsburg University

Janice A. Grskovic
Indiana University Northwest

Richard Keen
Converse College

Robin K. Morgan
Indiana University Southwest

Jack A. Palmer
University of Louisiana at Monroe

Marianna Rader
Florida Community College at Jacksonville

Melissa Terlecki
Cabrini College

PROFESSIONAL FEEDBACK

Throughout the process of writing and developing this text and the visual pedagogy, we benefited from the comments and constructive criticism provided by the instructors listed below. We offer our sincere appreciation to these individuals for their helpful review:

Marc W. Barnes
Ivy Tech Community College

Karen Bearce
Mercer County Community College

John Broida
University of Southern Maine

Tracie Burt
Southeast Arkansas College

Barbara Canaday

Richard Cavasina
California University of Pennsylvania

Michelle Caya
Community College of Rhode Island

Diane Cook
Gainesville State College

Curt Dewey
San Antonio College

Dale V. Doty
Monroe Community College

Steve Ellyson
Youngstown State University

Nolen Embry-Bailey
Bluegrass Community and Technical College

Melanie Evans
Eastern Connecticut State University

William Rick Fry
Youngstown State University

Susan Fuhr
Maryville College

Matthew Tyler Giobbi
Mercer County Community College

Betsy Goldenberg
University of Massachusetts, Lowell

Jeffrey Henriques
University of Wisconsin

Kathryn Herbst
Grossmont College

Scott Husband
University of Tampa

Heather Jennings
Mercer County Community College

Richard Keen
Converse College

Dawn Mclin
Jackson State University

Jean Mandernach
University of Nebraska at Kearney

Jan Mendoza
Brooks College/Golden West College

Tara Mitchell
Lock Haven University

Ruby Montemayor
San Antonio College

Robin K. Morgan
Indiana University Southeast

Ronald Mulson
Hudson Valley Community College

Larry Peck
Erie Community College–North

Lori Perez
Community College of Baltimore County

Robin Popp
Chattanooga State Technical Community College

Marianna Rader
Florida Community College at Jacksonville

Christopher Smith
Ivy Tech Community College

Clayton Teem
Gainesville State College

Marianna Torres
San Antonio College

Karina Vargas
San Antonio College

Jameel Walji
San Antonio College

Colin William
Ivy Tech Community College

FOCUS GROUPS AND TELESESSION PARTICIPANTS

A number of professors and students participated in focus groups and telesessions, providing feedback on the text, visuals, and pedagogy. Our thanks to the following instructors for their helpful comments and suggestions.

American Psychological Association Focus Group participants:

Sheree Barron
Georgia College & State University

Joni Caldwell
Union College

Stephen Ray Flora
Youngstown State University

Regan Gurung
University of Wisconsin, Green Bay

Brett Heintz
Delgado Community College

J. Kris Leppien-Christensen
Saddleback College

Mike Majors
Delgado Community College

Debra Murray
Viterbo University

Jack A. Palmer
University of Louisiana at Monroe

Melissa S. Terlecki
Cabrini College

American Psychological Society Focus Group participants:

Jonathan Bates
Hunter College

Michell E. Berman
University of Southern Mississippi

Will Canu
University of Missouri-Rolla

Patricia C. Ellerson
Miami University of Ohio

Renee Engeln-Maddox
Loyola University Chicago

Julie Evey
University of Southern Indiana

Bonnie A. Green
East Stroudsburg University

Janice A. Grskovic
Indiana University Northwest

Keith Happaney
Lehman College

Hector L. Torres
Medial College of Wisconsin

Visualizing format/pedagogy Focus Group and Telesession participants:

Sylvester Allred
Northern Arizona University

David Bastedo
San Bernardino Valley College

Ann Brandt-Williams
Glendale Community College

Natalie Bursztyn
Bakersfield College

Stan Celestian
Glendale Community College

O. Pauline Chow
Harrisburg Area Community College

Diana Clemens-Knott
California State University, Fullerton

Mitchell Colgan
College of Charleston

Linda Crow
Montgomery College

Smruti Desai
Cy-Fair College

Charles Dick
Pasco-Hernando Community College

Donald Glassman
Des Moines Area Community College

Mark Grobner
California State University, Stanislaus

Michael Hackett
Westchester Community College

Gale Haigh
McNeese State University

Roger Hangarter
Indiana University

Michael Harman
North Harris College

Terry Harrison
Arapahoe Community College

Javier Hasbun
University of West Georgia

Hasiotis, Stephen
University of Kansas

Adam Hayashi
Central Florida Community College

Laura Hubbard
University of California, Berkeley

James Hutcheon
Georgia Southern University

Scott Jeffrey
Community College of Baltimore County, Catonsville Campus

Matthew Kapell
Wayne State University

Arnold Karpoff
University of Louisville

Dale Lambert
Tarrant County College NE

Arthur Lee
Roane State Community College

Harvey Liftin
Broward Community College

Walter Little
University at Albany, SUNY

Mary Meiners
San Diego Miramar College

Scott Miller
Penn State University

Jane Murphy
Virginia College Online

Bethany Myers
Wichita State University

Terri Oltman
Westwood College

Keith Prufer
Wichita State University

Ann Somers
University of North Carolina, Greensboro

Donald Thieme
Georgia Perimeter College

Kip Thompson
Ozarks Technical Community College

Judy Voelker
Northern Kentucky University

Arthur Washington
Florida A&M University

Stephen Williams
Glendale Community College

Feranda Williamson
Capella University

SPECIAL THANKS

Our heartfelt thanks also go to the superb editorial and production teams at John Wiley and Sons who guided us through the challenging steps of developing this second edition. We thank in particular: Nancy Perry, Manager, Production Development; this edition would not exist were it not for Nancy's unflagging support, careful eye, and invaluable expertise. We also owe an enormous debt of gratitude to Executive Editor Chris Johnson, who expertly launched and directed our process; Helen McInnis, Managing Director, Wiley Visualizing, who oversaw the concept of the book; Barbara Heaney, Director of Product and Market Development, who provided valuable suggestions for page layout; Micheline Frederick, Production Manager, who stepped in whenever we needed expert advice; Kerry Weinstein, Production Editor, who guided the book through production; Jay O'Callaghan, Vice President and Publisher, who oversaw the entire project; and Jeffrey Rucker, Executive Marketing Manager for Wiley Visualizing, and Danielle Torio, Marketing Manager, who adeptly represent the Visualizing imprint.

In addition, we are deeply indebted to Carolyn Smith, our developmental editor, who contributed long hours of careful and patient editing. She also deserves special acknowledgment for her creative "Study Organizers," which will be a great asset to our readers.

Robin Popp from *Chattanooga State Technical Community College* carefully reviewed, edited, and updated all references for this edition. This type of "backstage" support requires a sharp, professionally trained mind and endless patience—two qualities that are seldom acknowledged (but deeply appreciated) by all authors.

We wish also to acknowledge the contributions of Vertigo Design for the interior design concept, and Harry Nolan, Wiley's Creative Director who gave art direction, refined the design and other elements and the cover. We appreciate the efforts of Elle Wagner in obtaining some of our text photos, and Sandra Rigby for her expertise in managing the illustration program.

Our sincerest thanks are also offered to all who worked on the media and ancillary materials, including Tom Kulesa, Senior Media Editor, and Lynn Pearlman and Bridget O'Lavin, Media Editors, for their expert work in developing the video and electronic components, and a host of others who contributed to the wide assortment of ancillaries.

Stacy Gold, Research Editor and Account Executive at the National Geographic Image Collection, also deserves our thanks for her valuable expertise in selecting NGS photos. Many other individuals at National Geographic offered their expertise and assistance in developing this book: Richard Easby, Supervising Editor, National Geographic School Division; Mimi Dornack, Sales Manager, and Lori Franklin, Assistant Account Executive, National Geographic Image Collection; and Jocelyn Shearer, Jim Burch, and Tracey Stewart of the National Geographic Digital Media TV. We appreciate their contributions and support.

Next, we would like to offer our thanks to all the folks at Furino Production—particularly Jeanine Furino, whose dedication, keen eye for detail, and desire for perfection can be seen throughout this book. The careful and professional approach of Jeanine and her staff was critical to the successful production of this edition.

All the writing, producing, and marketing of this book would be wasted without an energetic and dedicated sales staff. We wish to sincerely thank all the publishing representatives for their tireless efforts and good humor. It's a true pleasure to work with such a remarkable group of people.

From Siri Carpenter: thank you to my husband, Joe Carpenter, for his thoughtful advice and steadfast support throughout the production of this book. My appreciation also to colleagues who provided helpful feedback in one way or another: Tracy Banaszynski, Jennifer Randall Crosby, Brian Detweiler-Bedell, Jerusha Detweiler-Bedell, Meghan Dunn, and Kristi Lemm.

From Karen Huffman: continuing appreciation to my family and students who supported and inspired me. Last, and definitely not least, I thank my beloved husband, Bill Barnard.

CONTENTS *in Brief*

(perception, memory, attention)
Parietal lobe
Temporal lobe
Occipital lobe
Frontal lobe (working memory, temporal integration, abstract thinking, introspection, cognitive flexibility)
Cerebral cortex
Limbic system (emotions, learning, memory consolidation)
Brain stem (arousal)

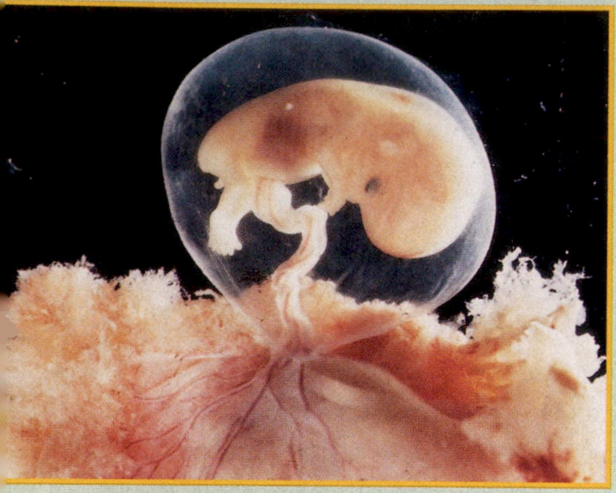

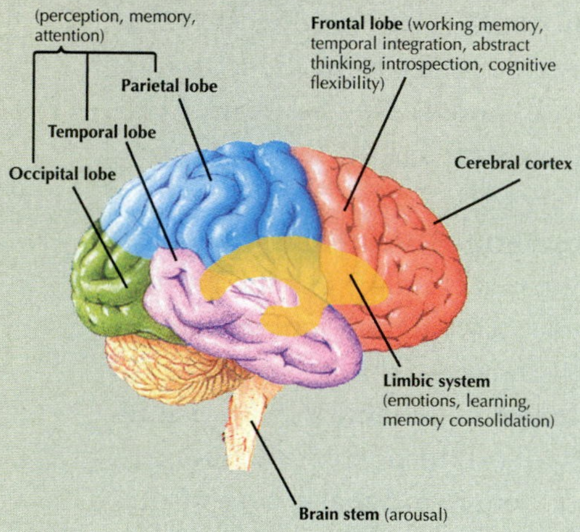

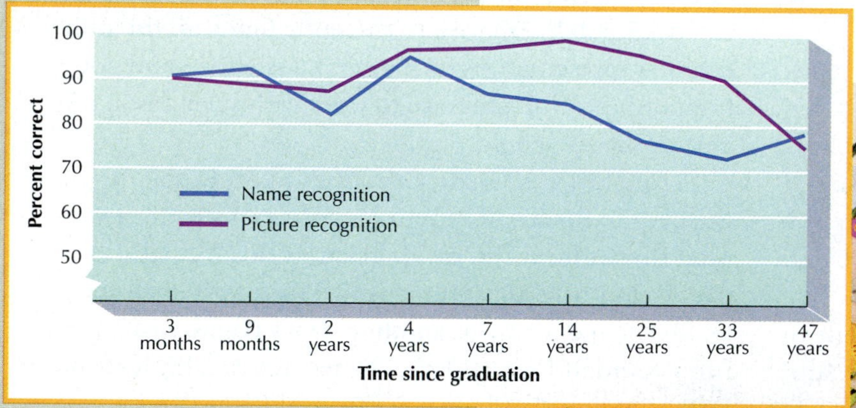

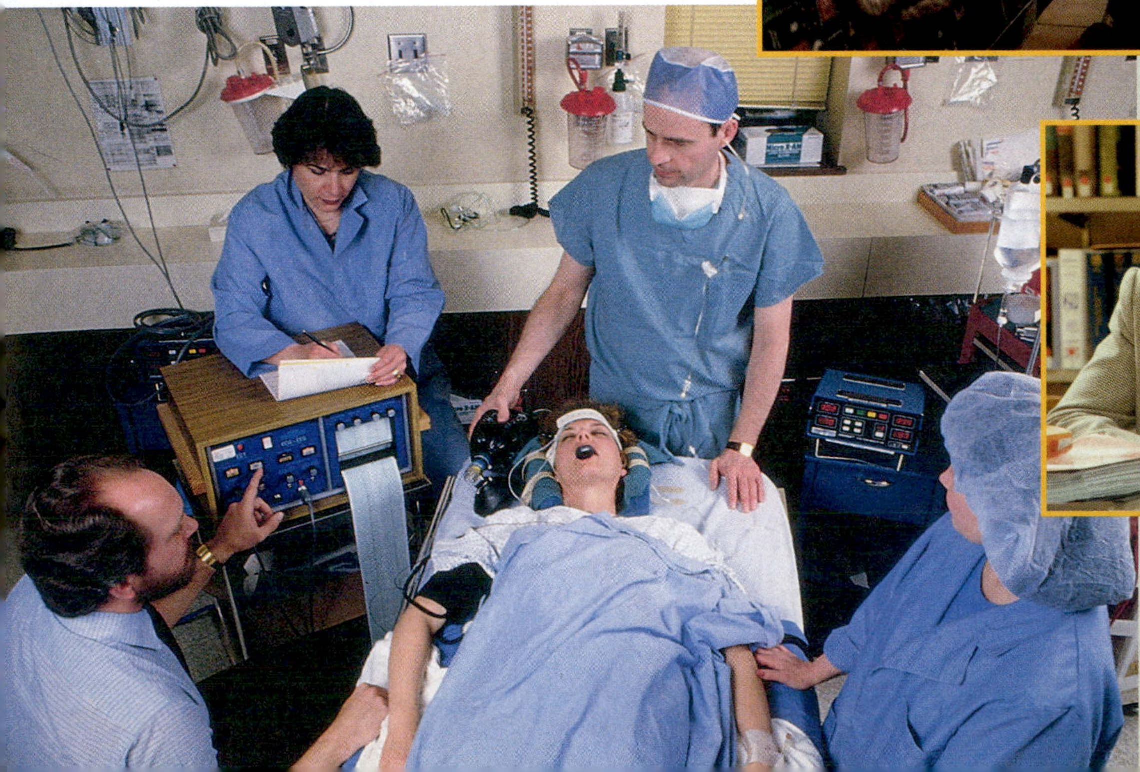

CONTENTS

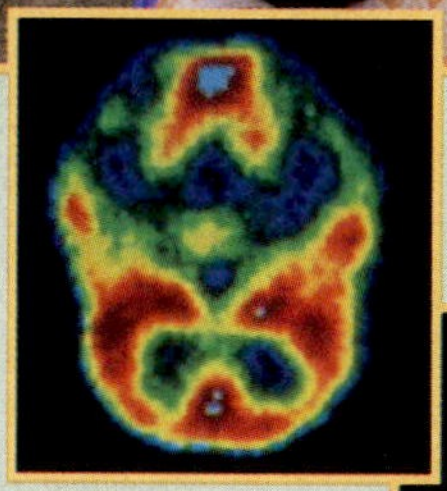

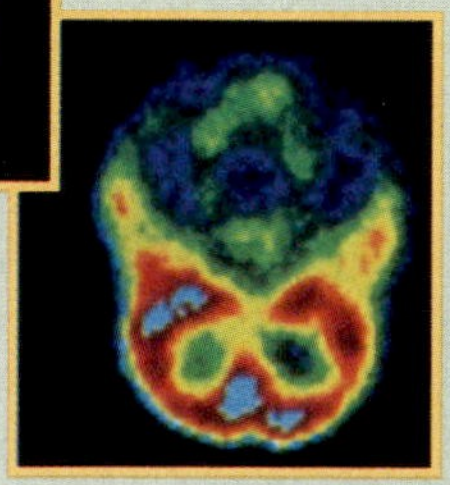

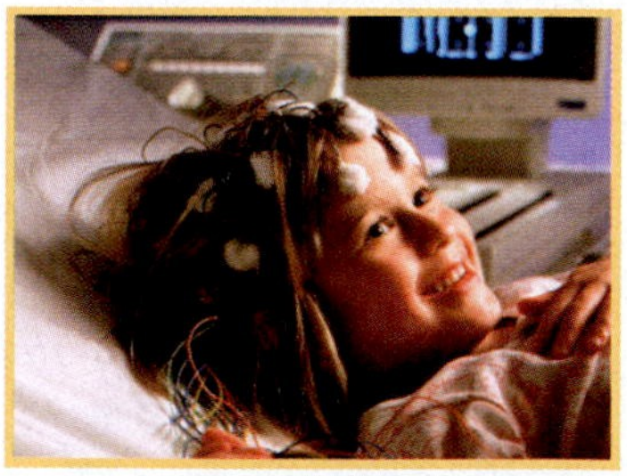

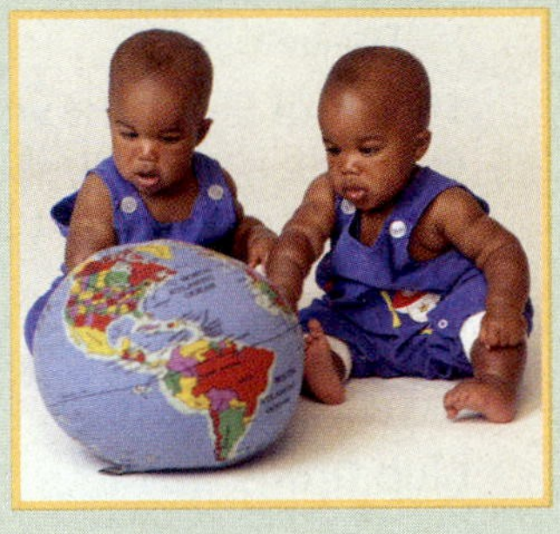

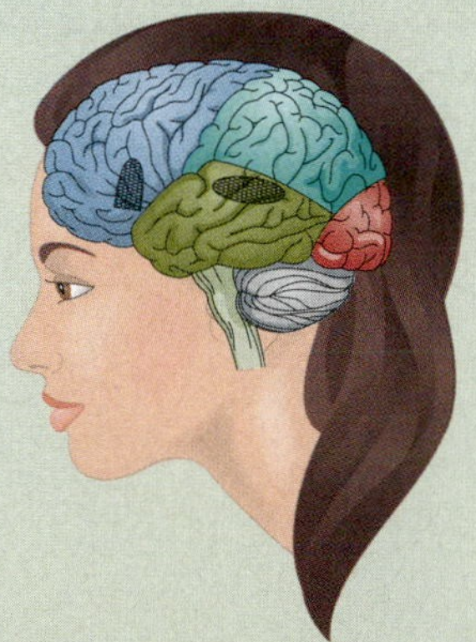

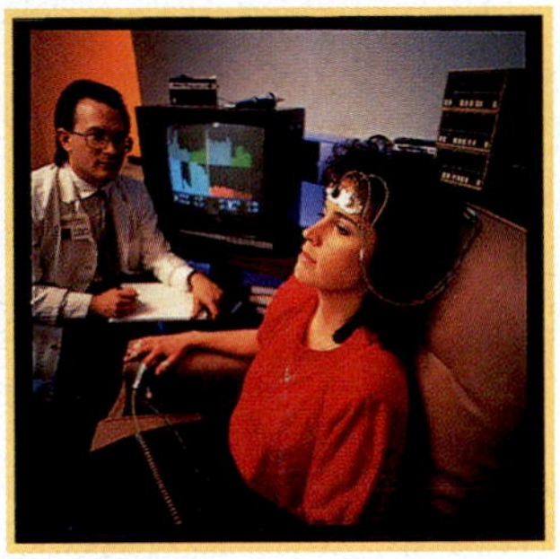

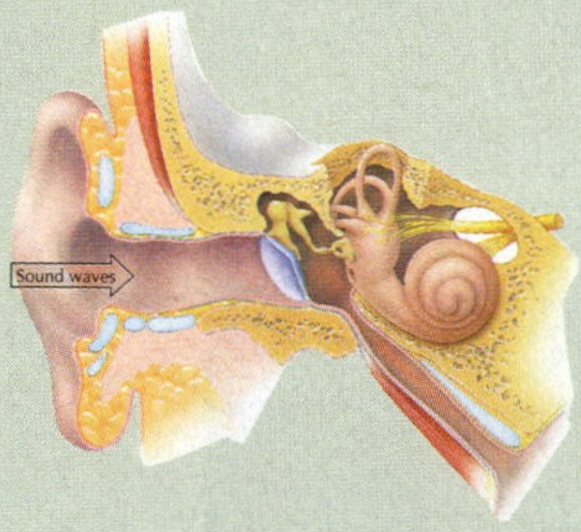
Sound waves

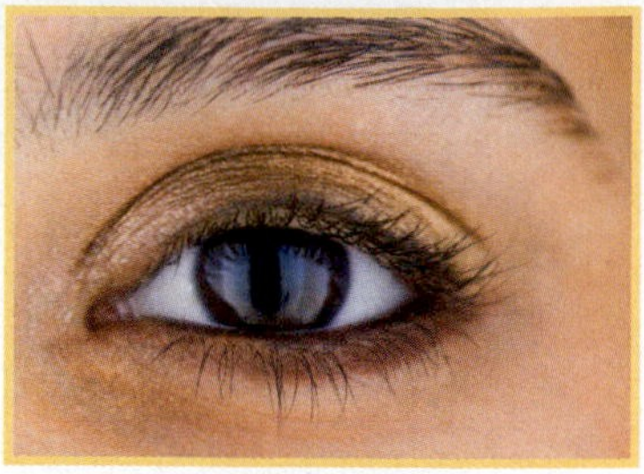

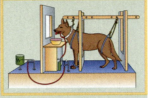

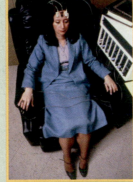

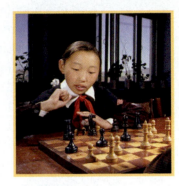

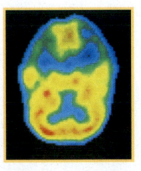

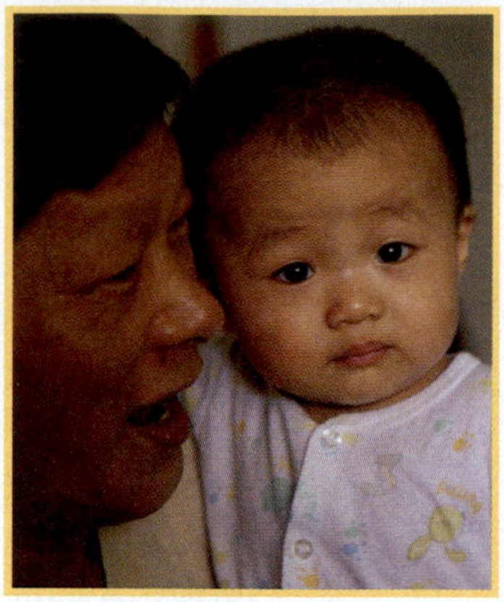

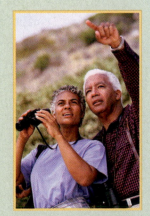

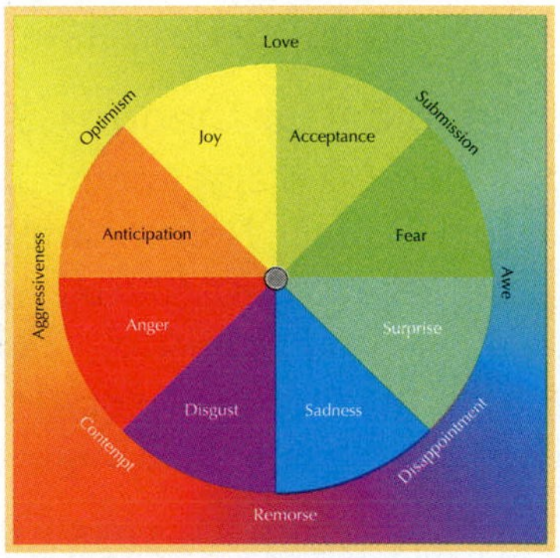

Love
Submission
Awe
Disappointment
Remorse
Contempt
Aggressiveness
Optimism
Joy
Acceptance
Fear
Surprise
Sadness
Disgust
Anger
Anticipation

VISUALIZING FEATURES

Visualizing Features: Multi-part visual presentations that focus on a key concept or topic in the chapter

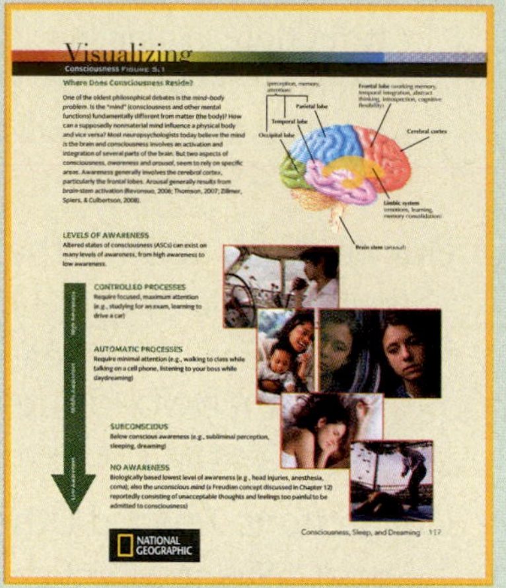

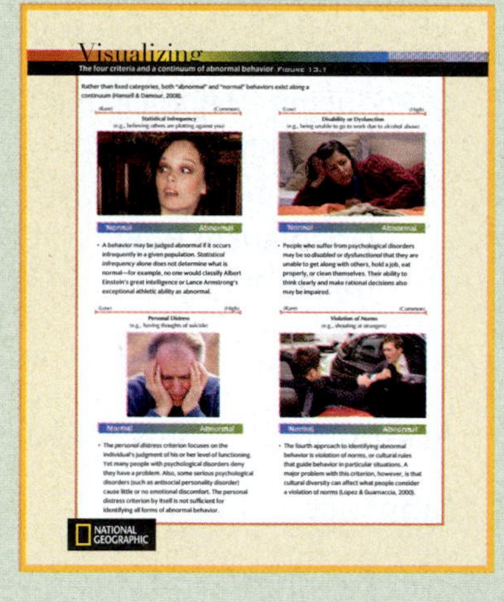

PROCESS DIAGRAMS

Process Diagrams: A series or combination of figures and photos that describe and depict a complex process

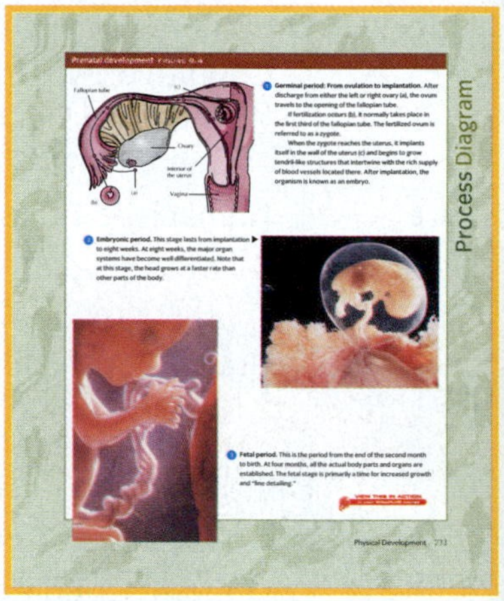

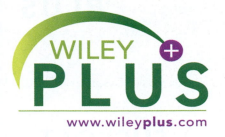

ALL THE HELP, RESOURCES, AND PERSONAL SUPPORT YOU AND YOUR STUDENTS NEED!

2-Minute Tutorials and all of the resources you & your students need to get started
www.wileyplus.com/firstday

Student support from an experienced student user
Ask your local representative for details!

Collaborate with your colleagues, find a mentor, attend virtual and live events, and view resources
www.WhereFacultyConnect.com

Pre-loaded, ready-to-use assignments and presentations
www.wiley.com/college/quickstart

Technical Support 24/7
FAQs, online chat, and phone support
www.wileyplus.com/support

Your *WileyPLUS* Account Manager
Training and implementation support
www.wileyplus.com/accountmanager

MAKE IT YOURS!

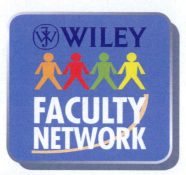

SECOND EDITION

VISUALIZING PSYCHOLOGY

Introduction and Research Methods

1

What might compel a person to willingly dangle hundreds of feet above the ground? What binds mothers—human and nonhuman alike—to their young? How can chronic stress contribute to serious health problems such as cancer and heart disease? What happens when ancient cultural practices collide with the instruments and demands of modern living? All of these questions, and countless more, are the province of psychology.

Psychology is a dynamic field that affects every part of our lives. Our innermost thoughts, our relationships, our politics, our "gut" feelings, and our deliberate decisions are all shaped by a complex psychology that affects us at every level, from the cellular to the cultural. Psychology encompasses not only humankind but our nonhuman compatriots as well—from rats and pigeons to cats and chimps.

Psychologists work in an incredible range of areas, perhaps more than you realize. In addition to studying and treating abnormal behavior, psychologists study sleep, dreaming, stress, health, drugs, personality, sexuality, motivation, emotion, learning, memory, childhood, aging, death, love, conformity, intelligence, creativity, and much more.

We invite you to let us know how your study of psychology (and this text) affects you and your life. What questions have captured your imagination? What have you learned about yourself and your world? Where will you go next in your journeys? You can reach us at siri@nasw.org and khuffman@palomar.edu.

We look forward to hearing from you.

Warmest regards,
Siri Carpenter and Karen Huffman

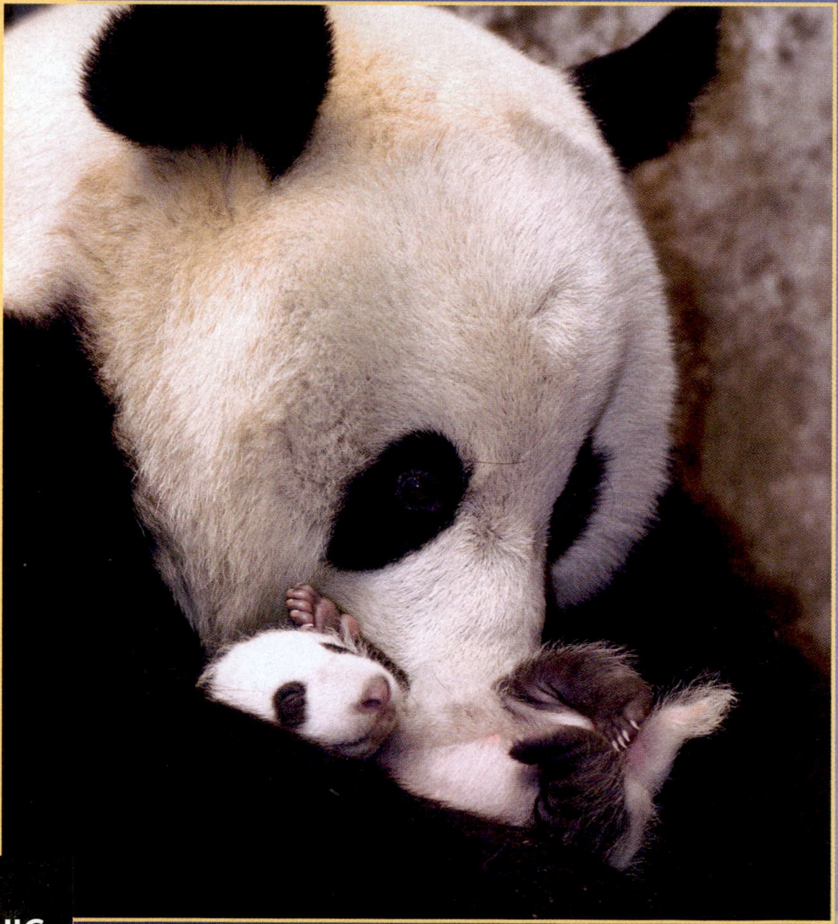

NATIONAL GEOGRAPHIC

CHAPTER OUTLINE

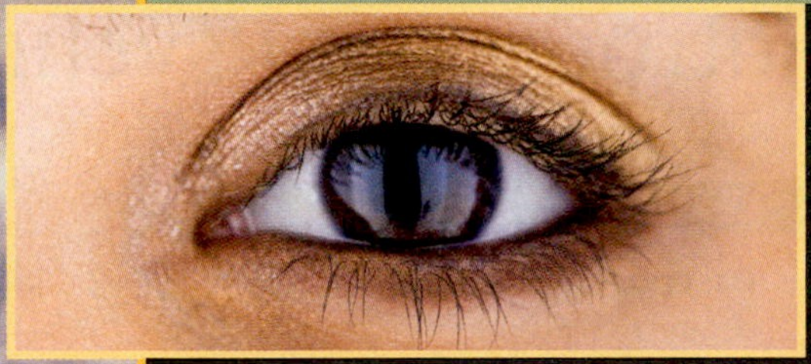

Introducing Psychology

LEARNING OBJECTIVES

Describe how scientific psychology differs from pseudopsychologies.

Review psychology's four main goals.

Identify some of the diverse professional roles that psychologists fill.

psychology The scientific study of behavior and mental processes.

The term **psychology** derives from the roots *psyche*, meaning "mind," and *logos*, meaning "word." Modern psychology is the scientific study of **behavior** and **mental processes**. Behavior is anything we do—from sleeping to rock climbing. Mental processes are our private, internal experiences—thoughts, perceptions, feelings, memories, and dreams.

SCIENCE VERSUS PSEUDOSCIENCE

For many psychologists, the most important part of the definition of psychology is the word *scientific*. Psychology places high value on **empirical evidence** and **critical thinking** (TABLE 1.1).

Be careful not to confuse scientific psychology with **pseudopsychologies**, which only give the appearance of science. (*Pseudo* means "false.") Pseudopsychologies include claims made by psychics (who purport to be able to read thoughts and foretell the future), palmistry (reading people's character from the markings on their palms), psychometry (determining facts about an object by handling it), psychokinesis (moving objects by purely mental means), and astrology (the study of how the positions of the stars and planets influence people's personalities and affairs) (FIGURE 1.1).

empirical evidence Information acquired by direct observation and measurement using systematic scientific methods.

critical thinking The process of objectively evaluating, comparing, analyzing, and synthesizing information.

Test your knowledge of psychology TABLE 1.1

Answer true or false to the following statements:

1. The best way to learn and remember information is to "cram," or study it intensively during one concentrated period.
2. Most brain activity stops during sleep.
3. Advertisers and politicians often use subliminal persuasion to influence our behavior.
4. Punishment is the most effective way to permanently change behavior.
5. Eyewitness testimony is often unreliable.
6. Polygraph ("lie detector") tests can accurately and reliably reveal whether a person is lying.
7. Behaviors that are unusual or that violate social norms indicate a psychological disorder.
8. People with schizophrenia have two or more distinct personalities.
9. Similarity is one of the best predictors of long-term relationships.
10. In an emergency, as the number of bystanders increases, your chance of getting help decreases.

Answers: 1. False (Chapter 1). **2.** False (Chapter 5). **3.** False (Chapter 4). **4.** False (Chapter 6). **5.** True (Chapter 7). **6.** False (Chapter 11). **7.** False (Chapter 13). **8.** False (Chapter 13). **9.** True (Chapter 15). **10.** True (Chapter 15).

"The Amazing Randi" FIGURE 1.1

Do you believe there is such a thing as psychic power? The magician James Randi has dedicated his life to educating the public about fraudulent pseudopsychologists. Along with the prestigious MacArthur Foundation, Randi has offered $1 million to "anyone who proves a genuine psychic power under proper observing conditions" (*About James Randi*, 2002; Randi, 1997). After many years, the money has never been collected. If you would like more information, visit Randi's website at www.randi.org.

Ψ Psychological Science

The Goals of Psychology

Scientific psychology has four basic goals: to *describe, explain, predict,* and *change* behavior and mental processes through the use of scientific methods. Let's consider each within the context of aggressive behavior.

Psychologists usually attempt to describe, or name and classify, particular behaviors by making careful scientific observations. For example, if someone says, "Males are more aggressive than females," does "aggressive" mean angry? Prone to yelling? Likely to throw the first punch? Scientific description requires specificity.

To explain a behavior or mental process, we need to discover and understand its causes. One of the most enduring debates in science has been the **nature-nurture controversy** (Hartwell, 2008; Hudziak, 2008; McCrae, 2004; Rutter, 2007). To what extent are we controlled by biological and genetic factors (the nature side) or by environment and learning (the nurture side)? Today, almost all scientists agree that most psychological and even physical traits reflect an interaction between nature and nurture. For example, research indicates that there are numerous interacting causes or explanations for aggression, including culture, learning, genes, brain damage, and high levels of testosterone (e.g., Juntii, Coats, & Shah, 2008; Kelly et al., 2008; Temcheff et al., 2008).

nature-nurture controversy Ongoing dispute over the relative contributors of nature (heredity) and nurture (environment) to the development of behavior and mental processes.

After describing and explaining a behavior or event, psychologists try to predict the conditions under which that behavior or event is likely to occur. For instance, knowing that alcohol leads to increased aggression (Tremblay, Graham, & Wells, 2008), we might predict that more fights will erupt in places where alcohol is consumed than in those where alcohol isn't consumed.

To psychologists, change means applying psychological knowledge to prevent unwanted outcomes or to bring about desired goals. In almost all cases, change as a goal of psychology is positive. For example, psychologists help people stop addictive behaviors, improve their work environments, become less depressed, improve their family relationships, and so on.

CRITICAL THINKING

Here are two interesting questions:

1. **How would you explain the behavior illustrated in this photo?**
2. **Can you predict the conditions under which such behavior is likely to occur?**

Percentage of psychology degrees awarded by subfield FIGURE 1.2

This is a small sampling of the numerous specialty areas in psychology.

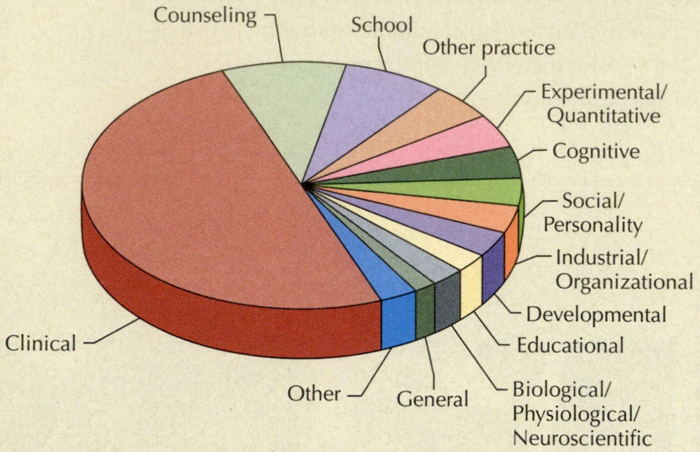

Source: American Psychological Association, 2007.

Many people think of psychologists only as therapists, but there are many psychologists who have no connection with therapy. Instead, they work as researchers, teachers, and consultants in academic, business, industry, and government settings, or in a combination of settings (FIGURE 1.2). For more information about what psychologists do—or how to pursue a career in psychology—check out the websites of the American Psychological Association (APA; www.apa.org) and the Association for Psychological Science (APS; www.psychologicalscience.org).

CONCEPT CHECK STOP

What are the four goals of psychology?

What professional fields might psychologists work in?

Origins of Psychology

LEARNING OBJECTIVES

Describe the different perspectives offered by early psychologists.

Identify a fundamental difference between the psychoanalytic and behavioristic perspectives.

Explore the important role of women and minorities in psychology's history.

Explain the central idea underlying the biopsychosocial model.

Although people have always been interested in human nature, it was not until the first psychological laboratory was founded in 1879 that psychology as a science officially began. As interest in the new field grew, psychologists adopted various perspectives on the "appropriate" topics for psychological research and the "proper" research methods. These diverse viewpoints and subsequent debates molded and shaped modern psychological science.

A BRIEF HISTORY: PSYCHOLOGY'S INTELLECTUAL ROOTS

Wilhelm Wundt (VILL-helm Voont), generally acknowledged as the "father of psychology," established the first psychological laboratory in Germany in 1879, helped train the first scientific psychologists, and wrote one of psychology's most important books, *Principles of Physiological Psychology* (published in 1890).

Wundt and his followers were primarily interested in how we form sensations, images, and feelings. Their chief methodology was termed "introspection," which involved monitoring and reporting on conscious experiences (Goodwin, 2009). Edward Titchener brought Wundt's ideas to the United States. Titchener's approach, now known as **structuralism**, sought to identify the basic building blocks, or structures, of mental life through introspection and then to determine how these elements combine to form the whole of experience. Because introspection could not be used to study animals, children, or more complex mental disorders, structuralism failed as a working psychological approach. Although short-lived, structuralism established a model for studying mental processes scientifically.

Structuralism's intellectual successor, **functionalism**, studied how the mind functions to adapt humans and other animals to their environment. Functionalism was strongly influenced by Charles Darwin's theory of evolution (Segerstrale, 2000). William James was the leading force in the functionalist school (FIGURE 1.3). Although functionalism also eventually declined, it expanded the scope of psychology to include research on emotions and observable behaviors, initiated the psychological testing movement, and influenced modern education and industry.

During the late 1800s and early 1900s, while functionalism was prominent in the United States, the **psychoanalytic school** was forming in Europe (Gay, 2000). Its founder, Austrian physician Sigmund Freud, believed that many psychological problems are caused by conflicts between "acceptable" behavior and "unacceptable" unconscious sexual or aggressive motives (Chapter 12). The theory provided a basis for a system of therapy known as *psychoanalysis* (Chapter 14).

Freud's nonscientific approach and emphasis on sexual and aggressive impulses have long been controversial, and today there are few strictly Freudian psychoanalysts left. But the broad features of his theory have profoundly affected psychotherapy, psychiatry, and modern psychodynamic psychologists, who focus on the importance of unconscious processes and unresolved past conflicts.

In the early 1900s, another major perspective appeared that dramatically shaped the course of psychology. Unlike earlier approaches, the **behavioral perspective** emphasizes objective, observable environmental influences on overt behavior. Behaviorism's founder, John B. Watson (1913), rejected the practice of introspection and the influence of unconscious forces. Instead, Watson adopted Russian physiologist Ivan Pavlov's concept of *conditioning* (Chapter 6) to explain behavior as a result of observable stimuli (in the environment) and observable responses (behavioral actions).

Most early behaviorist research was focused on learning; nonhuman animals were ideal subjects for this research. Onc of the most well-known behaviorists, B. F. Skinner, was convinced that behaviorist approaches could be used to "shape" human behavior. Therapeutic techniques rooted in the behavioristic perspective have been most successful in treating observable behavioral problems, such as phobias and alcoholism (FIGURE 1.4 on the next page).

William James (1842–1910) FIGURE 1.3

William James broadened psychology to include animal behavior, biological processes, and behaviors. His book *Principles of Psychology* (1890) became the leading psychology text.

Although the psychoanalytic and behaviorist perspectives dominated American psychology for some time, in the 1950s a new approach emerged—the **humanistic perspective**, which stressed *free will* (voluntarily chosen behavior) and *self-actualization* (a state of self-fulfillment). According to Carl Rogers and Abraham Maslow (two central humanist figures), all individuals naturally strive to develop and move toward self-actualization. Like psychoanalysis, humanist psychology developed an influential theory of personality and a form of psychotherapy (Chapters 12 and 14). The humanistic approach also led the way to a contemporary research specialty known as **positive psychology**—the scientific study of optimal human functioning (Diener, 2008; Patterson & Joseph, 2007; Seligman, 2003, 2007; Taylor & Sherman, 2008).

B. F. Skinner (1904–1990) FIGURE 1.4

B. F. Skinner was one of the most influential psychologists of the twentieth century. He believed that by using basic learning principles to shape human behavior, we could change what he perceived as the negative course of humankind.

One of the most influential modern approaches, the **cognitive perspective**, recalls psychology's earliest days, in that it emphasizes thoughts, perception, and information processing. Modern cognitive psychologists, however, study how we gather, encode, and store information using a vast array of mental processes. These include perception, memory, imagery, concept formation, problem solving, reasoning, decision making, and language. Many cognitive psychologists use what is called an *information-processing approach*, likening the mind to a computer that sequentially takes in information, processes it, and then produces a response.

During the last few decades, scientists have explored the role of biological factors in almost every area of psychology. Using sophisticated tools and technologies, scientists who take this **neuroscientific/biopsychological perspective** examine behavior through the lens of genetics and biological processes in the brain and other parts of the nervous system.

The **evolutionary perspective** stresses natural selection, adaptation, and evolution of behavior and mental processes. Its proponents argue that natural selection favors behaviors that enhance an organism's reproductive success.

Finally, the **sociocultural perspective** emphasizes social interactions and cultural determinants of behavior and mental processes. Although we are often unaware of their influence, factors such as ethnicity, religion, occupation, and socioeconomic class have an enormous psychological impact on our mental processes and behavior.

Early schools of psychological thought, like structuralism and functionalism, have almost entirely disappeared or have been blended into newer, broader perspectives. Contemporary psychology reflects seven major perspectives: psychoanalytic/psychodynamic, behaviorist, humanist, cognitive, neuroscientific/biopsychological, evolutionary, and sociocultural (STUDY ORGANIZER 1.1). Yet the complex behaviors and mental processes we humans and other animals exhibit require complex explanations. That is why most contemporary psychologists do not adhere to one single intellectual perspective. Instead, a more integrative, unifying theme—the **biopsychosocial model**—has gained wide acceptance. This model views biological processes (e.g., genetics, brain functions, neurotransmitters, and evolution), psychological factors (e.g., learning, thinking, emotion, personality, and motivation), and social forces (e.g., family, school, culture, ethnicity, social class, and politics) as interrelated, inseparable influences. (See *What a Psychologist Sees* on page 10.)

biopsychosocial model
A unifying theme of modern psychology that considers biological, psychological, and social processes.

Study Organizer 1.1 Seven major perspectives in modern psychology

Perspectives		*Major Emphases*
Psychoanalytic/Psychodynamic		Unconscious processes and unresolved past conflicts
Behavioral		Objective, observable environmental influences on overt behavior
Humanistic	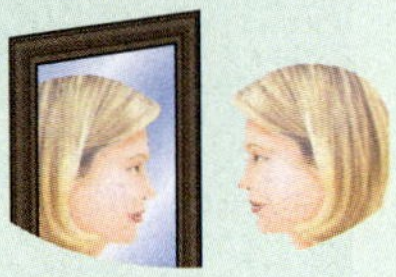	Free-will, self-actualization, and human nature as naturally positive and growth-seeking
Cognitive		Thinking, perceiving, problem solving, memory, language, and information processing
Neuroscientific/Biopsychological	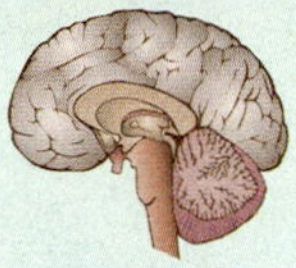	Genetics and biological processes in the brain and other parts of the nervous system
Evolutionary		Natural selection, adaptation, and evolution of behavior and mental processes
Sociocultural	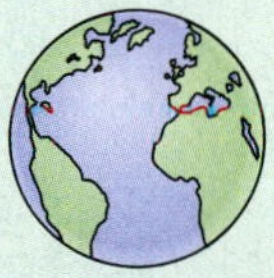	Social interaction and the cultural determinants of behavior and mental processes

THE BIOPSYCHOSOCIAL MODEL

The biopsychosocial model combines and interacts with the seven major perspectives.

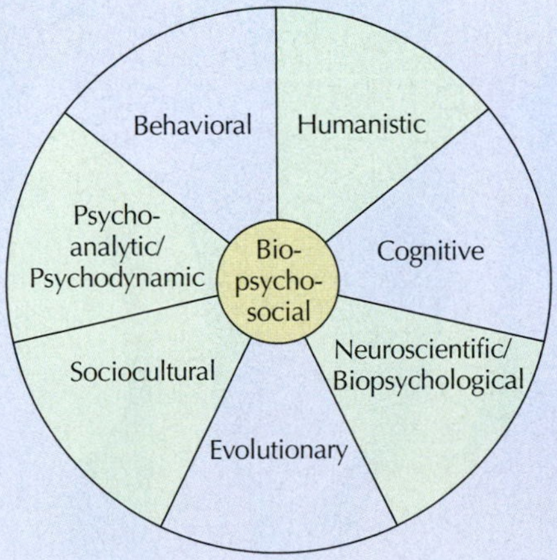

The Biopsychosocial Model

Look at the photo of the child to the right. What might be the cause of her emotional arousal? Now look at the photo below, which shows the child within a broader context. With this "bigger picture" (the child's immediate surroundings, her parents' guiding influence, and her group's enthusiasm for exciting sporting events) in mind, can you better understand why she might be feeling and behaving as she is? The biopsychosocial model recognizes that there is usually no single cause for our behavior or our mental states. For example, our moods and feelings are often influenced by genetics and neurotransmitters (biology), our learned responses and patterns of thinking (psychology), and our socioeconomic status and cultural views of emotion (social).

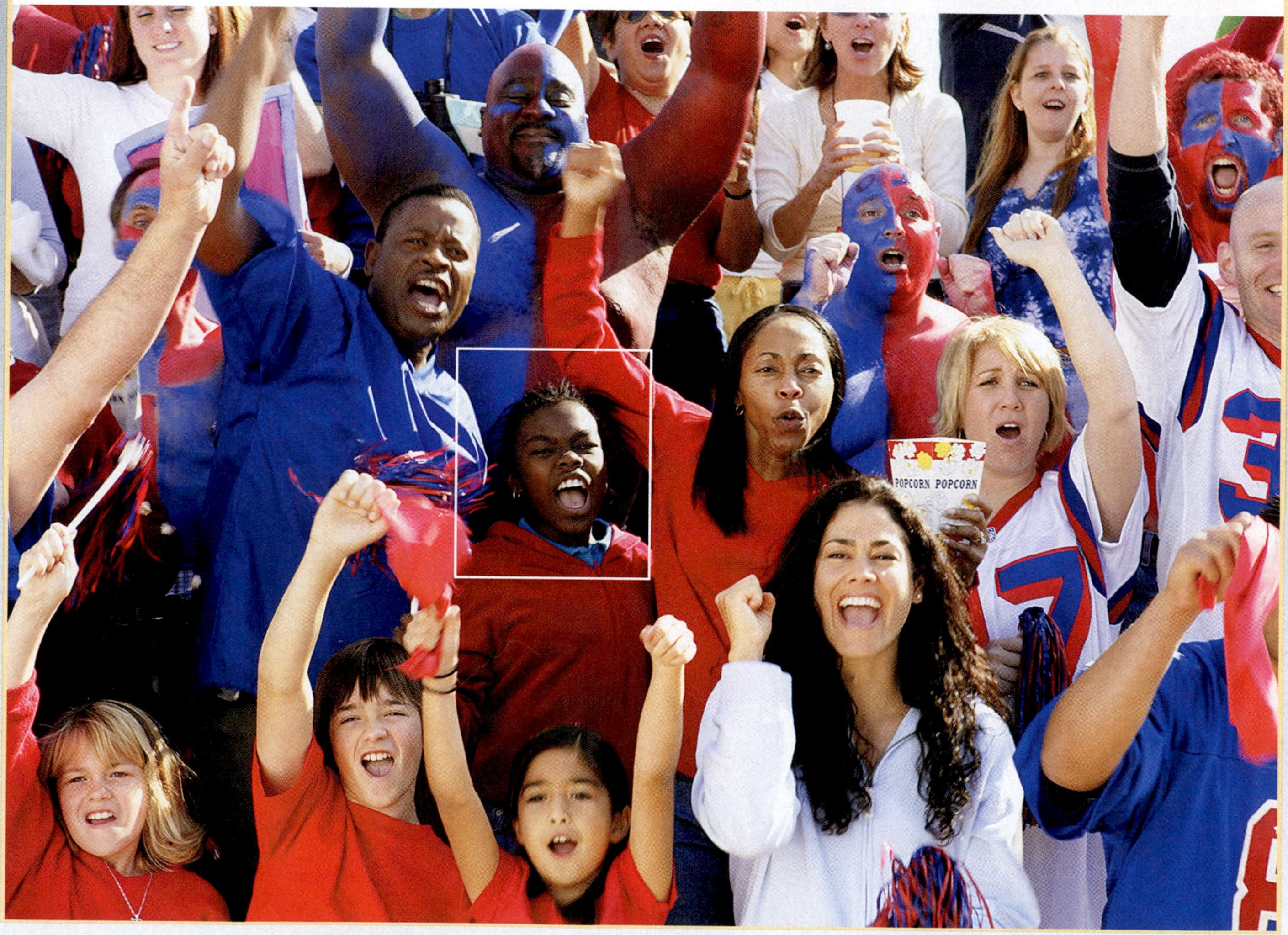

WOMEN AND MINORITIES IN PSYCHOLOGY

During the late 1800s and early 1900s, most colleges and universities provided little opportunity for women and minorities, either as students or as faculty members. Despite these early limitations, both women and minorities have made important contributions to psychology.

One of the first women to be recognized in the field was Mary Calkins (FIGURE 1.5). Calkins performed valuable research on memory, and in 1905 she served as the first female president of the APA. Her achievements are particularly noteworthy, considering the significant discrimination that she overcame. Even after she completed all the requirements for a Ph.D. at Harvard University and was described by William James as his brightest student, the university refused to grant the degree to a woman. The first woman to receive a Ph.D. in psychology was Margaret Floy Washburn (in 1894), who wrote several influential books and served as the second female president of the APA.

Francis Cecil Sumner became the first African American to earn a Ph.D. in psychology. He earned it from Clark University in 1920 and later chaired one of the country's leading psychology departments, at Howard University. In 1971, one of Sumner's students, Kenneth B. Clark, became the first African American to be elected APA president. Clark's research with his wife, Mamie Clark, documented the harmful effects of prejudice and directly influenced the Supreme Court's ultimate ruling against racial segregation in schools (FIGURE 1.6).

Mary Calkins (1863–1930) FIGURE 1.5

Kenneth Clark (1914–2005) and Mamie Clark (1917–1983) FIGURE 1.6

Sumner and Clark, Calkins and Washburn, along with other important minorities and women, made significant and lasting contributions to the developing science of psychology. In recent years, people of color and women have been actively encouraged to pursue graduate degrees in psychology. However, white (non-Hispanic) people still make up the majority of new doctorate recipients in psychology.

CONCEPT CHECK STOP

Which early schools of psychological thought are reflected in modern perspectives?

Why did structuralism decline in popularity?

What modern perspective views biological, psychological, and social forces as interrelated influences on behavior?

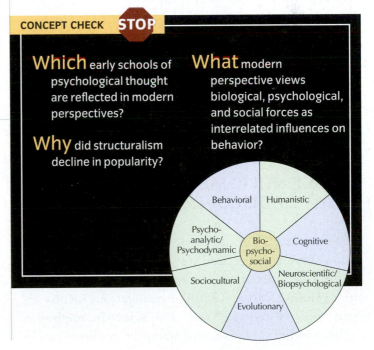

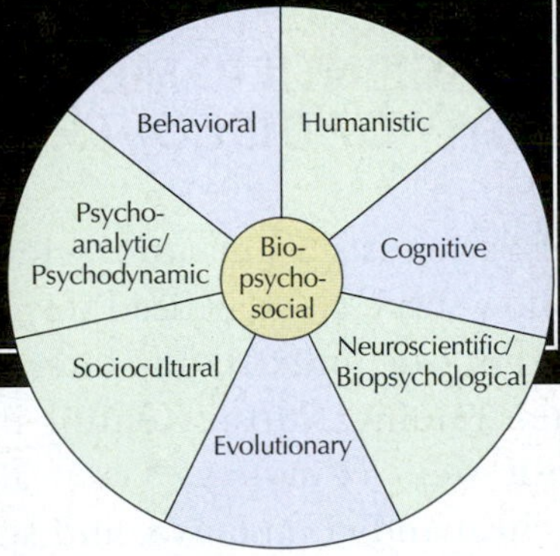

The Science of Psychology

LEARNING OBJECTIVES

Compare the fundamental goals of basic and applied research.

Describe the scientific method.

Identify how psychologists protect the rights of human and nonhuman research participants and psychotherapy clients.

Explain why research with nonhuman animals is valuable and why it is controversial.

In science, research strategies are generally categorized as either basic or applied. **Basic research** is typically conducted in universities or research laboratories by researchers who are interested in advancing general scientific understanding. Basic research meets the first three goals of psychology (description, explanation, and prediction). In contrast, **applied research** is generally conducted outside the laboratory, and it meets the fourth goal of psychology—to change existing real-world problems.

Basic and applied research frequently interact, with one building on the other. For example, after basic research documented a strong relationship between alcohol consumption and increased aggression, applied research led some sports stadium owners to limit the sale of alcohol during the final quarter of football games and the last two innings of baseball games.

basic research Research conducted to advance scientific knowledge rather than for practical application.

applied research Research designed to solve practical problems.

THE SCIENTIFIC METHOD: A WAY OF DISCOVERING

Like scientists in any other scientific field, psychologists follow strict, standardized scientific procedures so that others can understand, interpret, and repeat or test their findings. Most scientific investigations involve six basic steps (FIGURE 1.7). The **scientific method** is cyclical and cumulative, and scientific progress comes from repeatedly challenging and revising existing theories and building new ones. If numerous scientists, using different participants in varied settings, can repeat, or *replicate*, a study's findings, there is much greater scientific confidence in the findings. If the findings cannot be replicated, researchers look for explanations and conduct further studies. When different studies report contradictory findings, researchers often average or combine the results of all such studies and reach conclusions about the overall weight of the evidence, a popular statistical technique called *meta-analysis*.

ETHICAL GUIDELINES: PROTECTING THE RIGHTS OF OTHERS

The two largest professional organizations of psychologists, the American Psychological Association (APA) and the Association for Psychological Science (APS), both recognize the importance of maintaining high ethical standards in research, therapy, and all other areas of professional psychology. The preamble to the APA's publication *Ethical Principles of Psychologists and Code of Conduct* (2002) admonishes psychologists to maintain their competence, to retain objectivity in applying their skills, and to preserve the dignity and best interests of their clients, colleagues, students, research participants, and society.

Process Diagram

The scientific method FIGURE 1.7

VIEW THIS IN ACTION in your WileyPLUS course

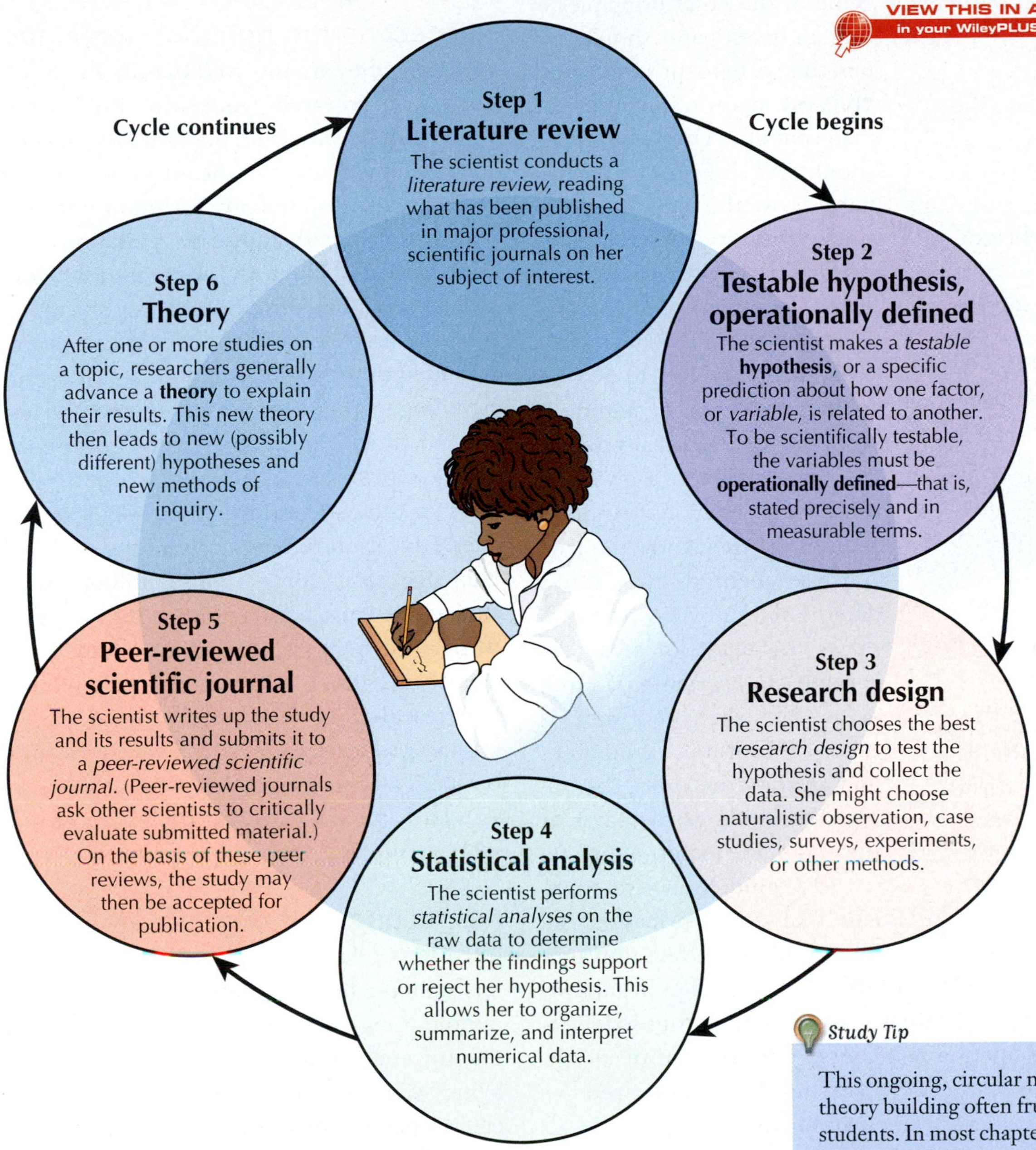

Cycle begins
Step 1
Literature review
The scientist conducts a *literature review*, reading what has been published in major professional, scientific journals on her subject of interest.
Step 2
Testable hypothesis, operationally defined
The scientist makes a *testable* **hypothesis**, or a specific prediction about how one factor, or *variable*, is related to another. To be scientifically testable, the variables must be **operationally defined**—that is, stated precisely and in measurable terms.
Step 3
Research design
The scientist chooses the best *research design* to test the hypothesis and collect the data. She might choose naturalistic observation, case studies, surveys, experiments, or other methods.
Step 4
Statistical analysis
The scientist performs *statistical analyses* on the raw data to determine whether the findings support or reject her hypothesis. This allows her to organize, summarize, and interpret numerical data.
Step 5
Peer-reviewed scientific journal
The scientist writes up the study and its results and submits it to a *peer-reviewed scientific journal*. (Peer-reviewed journals ask other scientists to critically evaluate submitted material.) On the basis of these peer reviews, the study may then be accepted for publication.
Step 6
Theory
After one or more studies on a topic, researchers generally advance a **theory** to explain their results. This new theory then leads to new (possibly different) hypotheses and new methods of inquiry.
Cycle continues

Study Tip

This ongoing, circular nature of theory building often frustrates students. In most chapters you will encounter numerous and sometimes conflicting hypotheses and theories. You will be tempted to ask, "Which theory is right?" But remember that theories are never absolute. Like most aspects of behavior; the "correct" answer is usually an interaction. In most cases, multiple theories contribute to the full understanding of complex concepts.

Respecting the rights of human participants

The APA has developed guidelines regulating research with human participants. One of the chief principles is that an investigator should obtain the participant's **informed consent** before initiating an experiment. The researcher should fully inform the participant as to the nature of the study, including physical risks, discomfort, or unpleasant emotional experiences. The researcher must also explain that participants are free to decline to participate or to withdraw from the research at any time.

informed consent A participant's agreement to take part in a study after being told what to expect.

Participants in a given research study may not respond naturally if they know the true purpose behind the study. Therefore, the APA acknowledges the need for some deception in certain research areas. But when deception is used, important guidelines and restrictions apply, including **debriefing** participants at the end of the experiment.

debriefing Informing participants after a study about the purpose of the study, the nature of the anticipated results, and any deception used.

APA guidelines also stipulate that all information acquired about people during a study must be held confidential and not published in such a way that individuals' privacy is compromised. If research participation is a course requirement or an opportunity for extra credit, the student must be given the choice of an alternative activity of equal value.

Finally, a human subjects committee or institutional review board must first approve all research using human participants conducted at a college, university, or any other reputable institution. Participating in psychological research can be a fascinating way to learn more about the nuts and bolts of the process. Many university psychology departments offer opportunities for students and community members to participate in research. In addition, the Association for Psychological Science (APS) has a website (http://psych.hanover.edu/research/exponnet.html) with links to ongoing studies that need participants.

Respecting the rights of nonhuman animals

Although they are involved in only 7 to 8 percent of psychological research (American Psychological Association, 1984), nonhuman animals—mostly rats and mice—have made significant contributions to almost every area of psychology, including the brain and nervous system, health and stress, sensation and perception, sleep, learning, memory, and emotion. Nonhuman animal research has also produced significant gains for animals themselves—for example, by suggesting more natural environments for zoo animals and more successful breeding techniques for endangered species.

Despite the advantages, using nonhuman animals in psychological research remains controversial (Guidelines for Ethical Conduct, 2008). While debate continues over ethical questions surrounding such research, psychologists take great care in handling research animals. Researchers also actively search for new and better ways to protect them (Appiah, 2008; Guidelines for Ethical Conduct, 2008). In all institutions where nonhuman animal research is conducted, animal care committees ensure proper treatment of research animals, review projects, and set guidelines in accordance with the APA standards for the care and treatment of nonhuman research animals.

Respecting the rights of psychotherapy clients

Like psychological scientists, therapists must maintain the highest of ethical standards; they must also uphold their clients' trust. All personal information and therapy records must be kept confidential, with records being available only to authorized persons and with the client's permission. However, the public's right to safety ethically outweighs the client's right to privacy. Therapists are legally required to break confidentiality if a client threatens violence to him- or herself or to others, if a client is suspected of abusing a child or elderly person, and in other limited situations. In general, however, a counselor's primary obligation is to protect client disclosures (Sue & Sue, 2008).

Any member of the APA who disregards the association's principles for the ethical treatment of human or nonhuman research participants or therapy clients may be censured or expelled from the organization. Clinicians who violate the ethical guidelines for working with clients risk severe sanctions and can permanently lose their licenses to practice. In addition, both researchers and clinicians are held professionally and legally responsible by their institutions as well as by local and state agencies.

CONCEPT CHECK STOP

What is the primary purpose of basic research? Of applied research?

How do scientists generate and refine hypotheses?

What is informed consent? Debriefing? Confidentiality?

Research Methods

LEARNING OBJECTIVES

Explain why only experiments can identify the cause and effect underlying particular patterns of behavior.

Describe the three key types of descriptive research.

Explain what is meant by the axiom "Correlation is not causation."

Identify some important methods used in biological research.

Psychologists draw on four major types of psychological research—experimental, descriptive, correlational, and biological (STUDY ORGANIZER 1.2). All have advantages and disadvantages, and most psychologists use several methods to study a single problem. In fact, when multiple methods lead to similar conclusions, scientists have an especially strong foundation for concluding that one variable does affect another in a particular way.

Study Organizer 1.2 Psychology's four major research methods

	Method	*Purpose*	*Advantages*	*Disadvantages*
	**Experimental** (manipulation and control of variables)	Identify cause and effect (meets psychology's goal of *explanation*)	Allows researchers to have precise control over variables and to identify cause and effect	Ethical concerns, practical limitations, artificiality of lab conditions, uncontrolled variables may confound results, researcher and participant biases
	**Descriptive** (naturalistic observation, surveys, case studies)	Observe, collect, and record data (meets psychology's goal of *description*)	Minimizes artificiality, easier to collect data, allows description of behavior and mental processes as they occur	Little or no control over variables, researcher and participant biases, cannot explain cause and effect
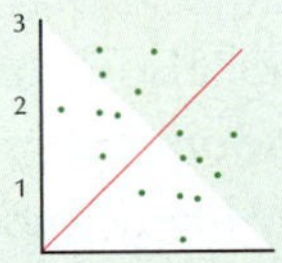	**Correlational** (statistical analyses of relationships between variables)	Identify relationships and assess how well one variable predicts another (meets psychology's goal of *prediction*)	Helps clarify relationships between variables that cannot be examined by other methods and allows prediction	Researchers cannot identify cause and effect
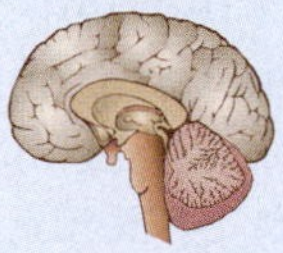	**Biological** (studies of the brain and other parts of the nervous system)	Identify contributing biological factors (meets one or more of psychology's goals)	Shares many or all of the advantages of experimental, descriptive, and correlational research	Shares many or all of the disadvantages of experimental, descriptive, and correlational research

Note that the four methods are not mutually exclusive. Researchers may use two or more methods to explore the same topic.

EXPERIMENTAL RESEARCH: A SEARCH FOR CAUSE AND EFFECT

experiment A carefully controlled scientific procedure that determines whether variables manipulated by the experimenter have a causal effect on other variables.

independent variable (IV) Variable that is manipulated to determine its causal effect on the dependent variable.

dependent variable (DV) Variable that is measured; it is affected by (or dependent on) the independent variable.

The most powerful research method is the **experiment**, in which an experimenter manipulates and controls the variables to determine cause and effect. Only through an experiment can researchers examine a single factor's effect on a particular behavior (Goodwin, 2009). That's because the only way to discover which of many factors has an effect is to experimentally isolate each one. As illustrated in FIGURE 1.8, an experiment has two critical components, an **independent variable (IV)** and a **dependent variable (DV)**, and **experimental** and **control groups**.

The goal of any experiment is to learn how the dependent variable is affected by (depends on) the independent variable. Experiments can also have different *levels* of an independent variable. For example, consider the experiment testing the effect of TV violence on aggression described in Figure 1.8. Two experimental groups could be created, with one group watching two hours of violent programming and the other watching six hours; the control group would watch only nonviolent programming. Then a researcher could relate differences in aggressive behavior (DV) to the *amount* of violent programming viewed (IV).

In experiments, all extraneous variables (such as time of day, lighting conditions, and participants' age and gender) must be held constant across experimental and control groups so that they do not affect the different groups' results.

In addition to the scientific controls mentioned (e.g., operational definitions, a control group, and consistent extraneous variables), a good scientific experiment also protects against potential sources of error from both the researcher and the participants.

Experimenters can unintentionally let their beliefs and expectations affect participants' responses, producing flawed results. For example, imagine what might happen if an experimenter breathed a sigh of relief when a participant gave a response that supported the researcher's hypothesis. One way to prevent such **experimenter bias** from destroying the validity of participants' responses is to establish objective methods for collecting and recording data. For example, an experimenter might use audiotape recordings to present stimuli and computers to record responses.

Another option is to use "blind observers" (neutral helpers) to collect and record the data without knowing what the researcher has predicted. In addition, it is sometimes possible to arrange a **double-blind study**, where neither the observer nor the participant knows which group received the experimental treatment.

Experimenters can also skew their results when they assume that behaviors that are typical in their own culture are typical in all cultures—a bias known as **ethnocentrism**. One way to avoid this problem is to have researchers from two cultures conduct the same study twice, once with their own culture and once with at least one other culture. This kind of **cross-cultural sampling** isolates group differences in behavior that stem from researchers' ethnocentrism.

In addition to potential problems from the researcher, the participants themselves can also introduce error or bias into an experiment. First, **sample bias** can occur if the sample of participants does not accurately reflect the composition of the larger population from which they are drawn. For example, critics suggest that much psychological literature is biased because it primarily uses white participants. One way to minimize sample bias is to randomly select participants who constitute a *representative sample* of the entire population of interest. Assigning participants to experimental groups using a chance, or random system, such as a coin toss, also helps prevent sample bias. This procedure of **random assignment** ensures that each participant is equally likely to be assigned to any particular group.

Bias can also occur when participants are influenced by the experimenter or the experimental conditions. For example, participants may try to present themselves in a good light (the **social desirability response**) or deliberately attempt to mislead the researcher.

Process Diagram

Experimental research design FIGURE 1.8

To test the hypothesis that watching violent television increases aggression, experimenters might randomly assign children to one of two groups: *experimental group* participants, who watch a prearranged number of violent television programs, and *control group* participants, who watch the same amount of television, except the programs that they watch are nonviolent. (*Having at least two groups—a control group and an experimental group—allows the performance of one group to be compared with that of another.*)

Experimenters then observe the children and count how many times—say, within one hour—each child hits, kicks, or punches a large, plastic "Bobo doll" (an operational definition of aggression).

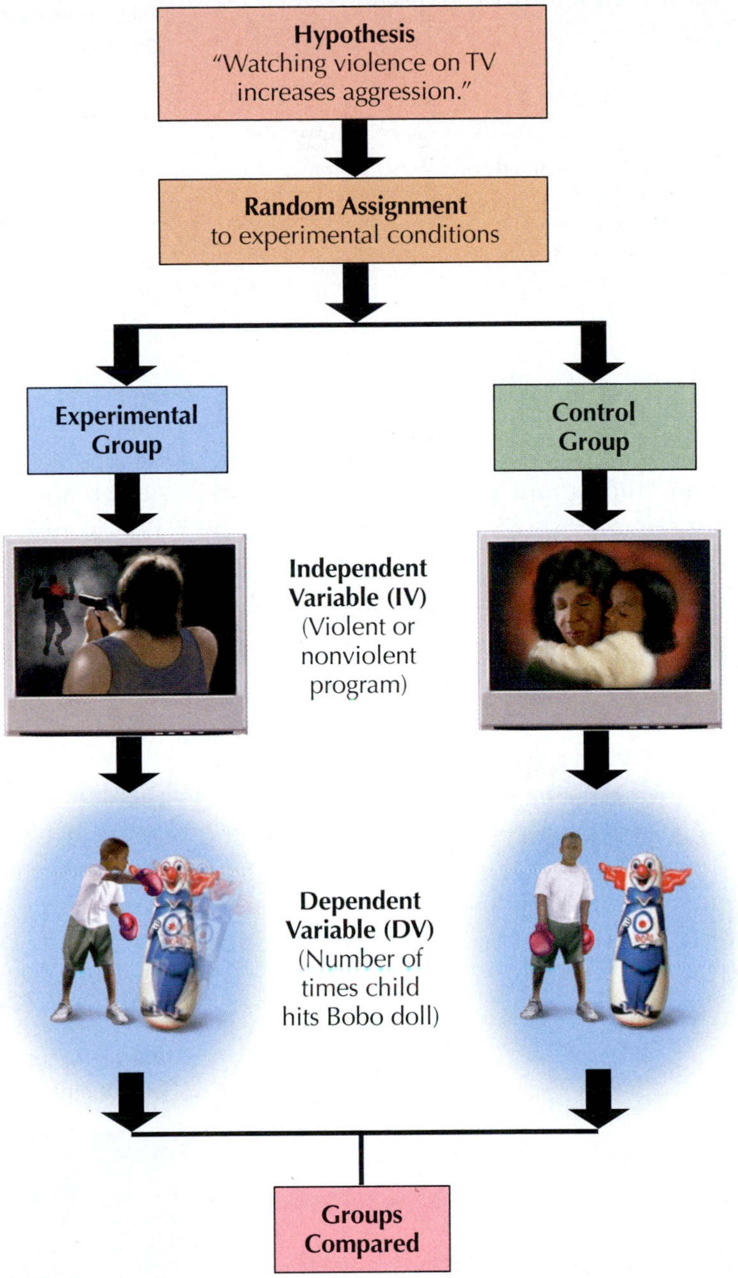

VIEW THIS IN ACTION in your WileyPLUS course

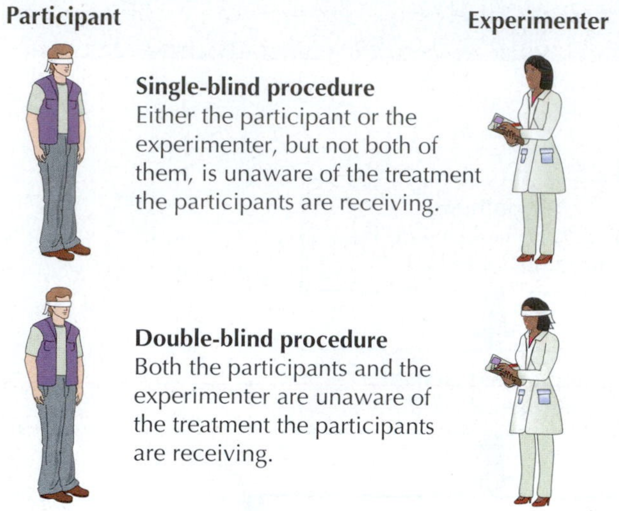
Participant
Experimenter
Single-blind procedure
Either the participant or the experimenter, but not both of them, is unaware of the treatment the participants are receiving.
Double-blind procedure
Both the participants and the experimenter are unaware of the treatment the participants are receiving.

A double-blind experimental design FIGURE 1.9

In a double-blind experiment to test a new drug, both the experimenters administering the drug and the participants are unaware (or "blind") as to who is receiving a **placebo** (a fake pill or injection) and who is receiving the drug itself. Researchers use placebos (thus treating experimental and control group participants exactly the same) because they know that the mere act of taking a pill or receiving an injection can change a participant's condition or responses.

Researchers attempt to control for this type of participant bias by offering anonymous participation and other guarantees of privacy and confidentiality. Single- and double-blind studies and **placebos**, which ensure that participants are unaware of which group they're in, offer additional safeguards (FIGURE 1.9). Finally, one of the most effective (but controversial) ways of preventing participant bias is by deceiving them (temporarily) about the nature of the research project.

DESCRIPTIVE RESEARCH: NATURALISTIC OBSERVATION, SURVEYS, AND CASE STUDIES

Almost everyone observes and describes others in an attempt to understand them, but in conducting **descriptive research**, psychologists do it systematically and scientifically. The key types of descriptive research are naturalistic observation, surveys, and case studies. Most of the problems and safeguards discussed with regard to the experimental method also apply to the nonexperimental methods.

> **descriptive research** Research methods used to observe and record behavior (without producing causal explanations).

When conducting **naturalistic observation**, researchers systematically measure and record participants' behavior, without interfering. Many settings lend themselves to naturalistic observation, from supermarkets to airports to outdoor settings (FIGURE 1.10).

Naturalistic observation FIGURE 1.10

Why is this researcher observing the children's behavior from outside the room?

Teacher Evaluation Team - 9:00
SIPRESS

"Just pretend we're not here, Ms. Robinson..."

The chief advantage of naturalistic observation is that researchers can obtain data about natural behavior, rather than about behavior that is a reaction to an artificial experimental situation. But naturalistic observation can be difficult and time-consuming, and the lack of control by the researcher makes it difficult to conduct observations for behavior that occurs infrequently.

Psychologists use **surveys** (FIGURE 1.11) to measure a variety of psychological behaviors and attitudes. (The survey technique includes tests, questionnaires, polls, and interviews.) One key advantage of surveys is that researchers can gather data from many more people than is possible with other research methods. Unfortunately, most surveys rely on self-reported data, and not all participants are completely honest. Although they can help predict behavior, survey techniques cannot explain causes of behavior.

What if a researcher wants to investigate photophobia (fear of light)? In such a case, it would be difficult to find enough participants to conduct an experiment or to use surveys or naturalistic observation. For rare disorders or phenomena, researchers try to find someone who has the problem and study him or her intensively. Such in-depth studies of a single research participant are called **case studies** (FIGURE 1.12).

Surveys FIGURE 1.11

In conducting surveys, researchers often use questionnaires to gather data from a wide selection of people. Under what conditions would you be willing to participate in such research?

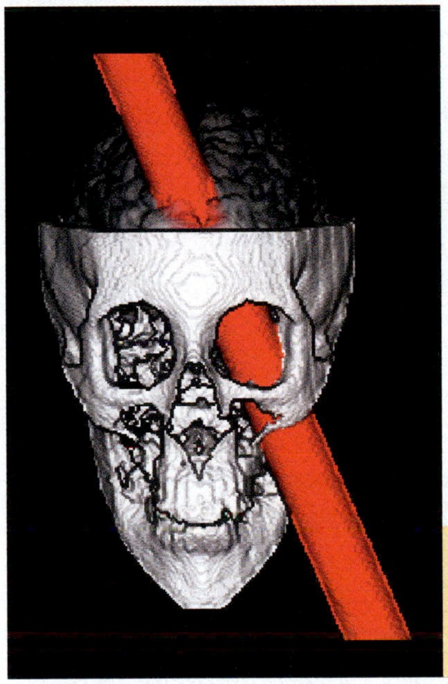

An early case study FIGURE 1.12

In 1848, a railroad foreman named Phineas Gage had a metal rod (13 pounds, 1¼ inches in diameter, and 3½ feet long) blown through the front of his face and brain. Amazingly, Gage was soon up and moving around; he didn't receive medical treatment until 1½ hours later. However, Gage suffered a serious personality transformation. Before the accident, he had been capable, energetic, and well-liked. Afterward, he was described as "fitful, capricious, impatient of advice, obstinate, and lacking in deference to his fellows" (Macmillan, 2000, p. 13).

Gage's injury and "recovery" were carefully documented and recorded by his attending physician, Dr. J. M. Harlow. As his story illustrates, the case study method offers unique advantages. Can you see how this method also poses serious research limits, including lack of generalizability, recorder bias, and inaccurate or biased recall among participants?

CORRELATIONAL RESEARCH: LOOKING FOR RELATIONSHIPS

correlational research Scientific study in which the researcher observes or measures (without directly manipulating) two or more variables to find relationships between them.

When nonexperimental researchers want to determine the degree of relationship (correlation) between two variables, they turn to **correlational research**. As the name implies, when any two variables are "co-related," a change in one is accompanied by a change in the other.

Using the correlational method, researchers measure participants' responses on two or more variables of interest. Next, the researchers analyze their results using a statistical formula that results in a **correlation coefficient**, a numerical value that indicates the degree and direction of the relationship between the two variables. Correlation coefficients are expressed as a number ranging from +1.00 to –1.00. The sign (+ or –) indicates the direction of the correlation, positive or negative (FIGURE 1.13A).

The number (0 to +1.00 or –1.00) associated with a correlation indicates the strength of the relationship. Both +1.00 and –1.00 are the strongest possible relationships. Thus, if you had a correlation of +0.92 or –0.92, you would have a *strong correlation*. By the same token, a correlation of +0.15 or –0.15 would represent a *weak correlation*.

Correlational research is an important research method for psychologists, and understanding correlations can also help us live safer and more productive lives. For example, correlational studies have repeatedly found high correlations between birth defects and a pregnant mother's use of alcohol (Bearer et al., 2004–2005; Gunzerath et al., 2004). This kind of information enables us to reliably predict our relative risks and to make informed decisions (FIGURE 1.13B).

However, people sometimes do not understand that a correlation between two variables does not mean that one variable causes another (FIGURE 1.13C). People sometimes read media reports about correlations between stress and cancer, for example, or between family dynamics and homosexuality, and infer that "stress causes cancer" or that "withdrawn fathers and overly protective mothers cause their sons to become homosexuals." They fail to realize that a third factor, perhaps genetics, may cause greater susceptibility to both of the correlated phenomena. Although correlational studies do sometimes point to possible causes, only the experimental method manipulates the independent variable under controlled conditions and, therefore, can support conclusions about cause and effect.

BIOLOGICAL RESEARCH: TOOLS FOR EXPLORING THE BRAIN AND NERVOUS SYSTEM

biological research Scientific studies of the brain andother parts of the nervous system.

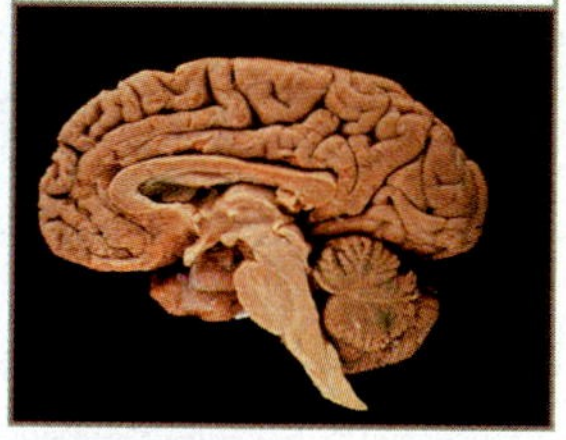

Biological research examines the biological processes that are involved in our feelings, thoughts, and behavior.

The earliest explorers of the brain dissected the brains of deceased humans and conducted experiments on other animals using *lesioning* techniques (systematically destroying brain tissue to study the effects on behavior and mental processes). By the mid-1800s, this research had produced a basic map of the nervous system, including some areas of the brain. Early researchers also relied on clinical observations and case studies of living people who had experienced injuries, diseases, and disorders that affected brain functioning.

Modern researchers still use such methods, but they also employ other techniques to examine biological processes that underlie our behavior (TABLE 1.2 on page 22). Recent advances in brain science have led to various types of brain-imaging scans, which can be used in both clinical and laboratory settings (Benazzi, 2004; Haller et al., 2005). Most of these methods are relatively *noninvasive*—that is, their use does not involve breaking the skin or entering the body.

Visualizing

Correlations FIGURE 1.13

Three types of correlation

A Each dot on these graphs (called *scatterplots*) represents one participant's score on two factors, or variables. ▼

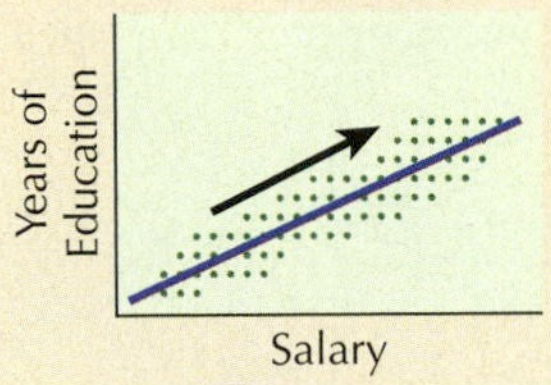

In a positive correlation, the two factors move (or vary) in the same direction.

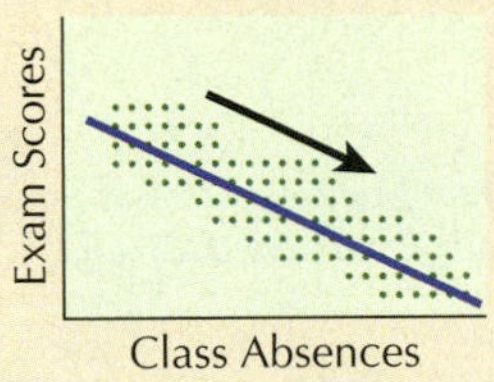

In a negative correlation, the two factors vary in opposite directions—that is, as one factor increases, the other factor decreases.

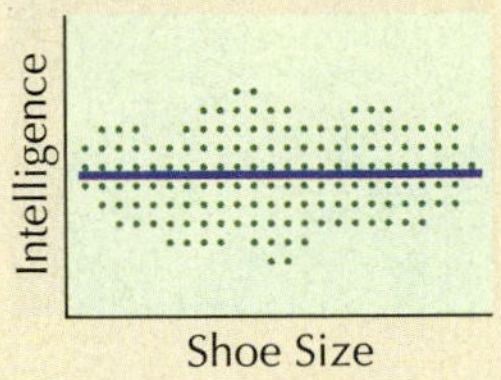

Sometimes there is no relationship between two variables—a zero correlation.

Pregnancy and smoking ▲

B Research shows that cigarette smoking is strongly linked to serious fetal damage. The more the mother smokes, the more the fetus is damaged. Is this a positive or negative correlation?

Correlation is not causation ▲

C Does a high correlation between young children's foot size and their reading speed mean that having small feet causes slower reading? Or that reading quickly causes feet to grow? Obviously not! Instead, both are caused by a third variable—an increase in children's ages.

NATIONAL GEOGRAPHIC

Tools for biological research TABLE 1.2

Each biological method has strengths and weaknesses, but all provide invaluable insights. Findings from these tools are discussed in a number of chapters in this text.

Method	*Description*	*Sample Results*
Brain dissection 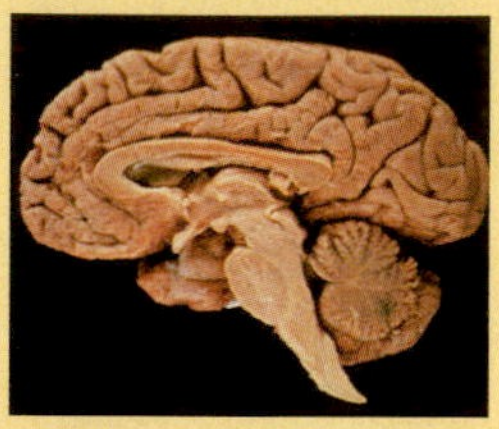 ***Brain dissection.*** *This is an actual photo of a deceased person's brain that has been vertically sliced in half to reveal inner structures.*	Careful cutting and study of a cadaver's brain to reveal structural details.	Brain dissections of Alzheimer's disease victims often show identifiable changes in various parts of the brain (Chapter 7).
Ablation/lesions 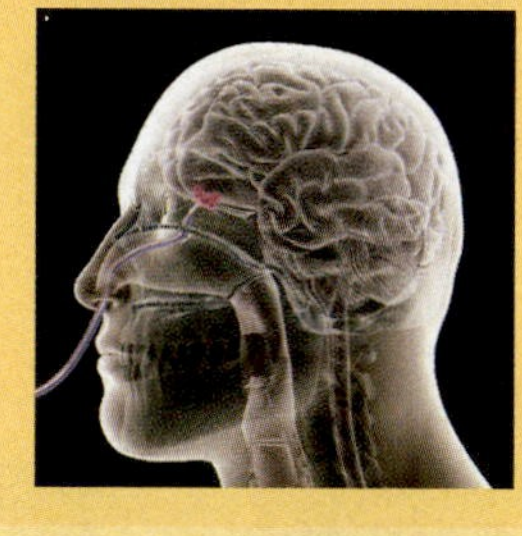	Surgically removing parts of the brain (ablation), or destroying specific areas of the brain (lesioning), is followed by observation for changes in behavior or mental processes.	Lesioning specific parts of the rat's hypothalamus greatly affects its eating behavior (Chapter 11).
Observations/case studies 	Observing and recording changes in personality, behavior, or sensory capacity associated with brain disease or injuries.	Damage to one side of the brain often causes numbness or paralysis on the body's opposite side.
Electrical recordings  ***Electroencephalogram (EEG).*** *Electrical activity throughout the brain sweeps in regular waves across its surface, and the EEG is a readout of this activity. Unfortunately, because the electrodes only record from the surface of the scalp, they provide little precision about the location of the activity.*	Using electrodes attached to the skin or scalp, brain activity is detected and recorded on an electroencephalogram (EEG).	The EEG reveals areas of the brain most active during a particular task or changes in mental states, like sleeping and meditation (Chapter 5); it also traces abnormal brain waves caused by brain malfunctions, like epilepsy or tumors.
Electrical stimulation of the brain (ESB) 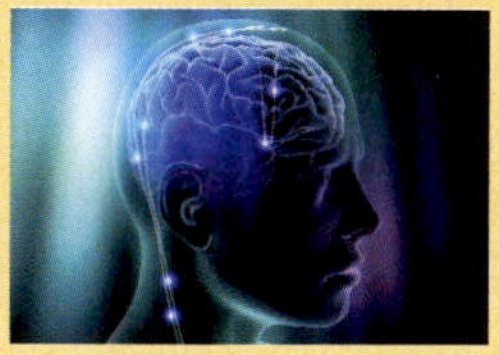	Using an electrode, a weak electric current stimulates specific areas or structures of the brain.	Penfield (1947) mapped the surface of the brain and found that different areas have different functions.

Tools for biological research (continued) TABLE 1.2

Method	Description	Sample Results
CT (computed tomography) scan 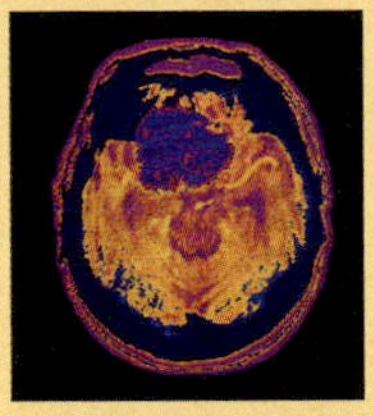 ***CT scans.*** *This CT scan used X-rays to locate a brain tumor, which is the deep purple mass at the top left.*	Computer-created cross sectional X-rays of the brain; least expensive type of imaging and widely used in research.	CT reveals the effects of strokes, injuries, tumors, and other brain disorders.
PET (positron emission tomography) scan 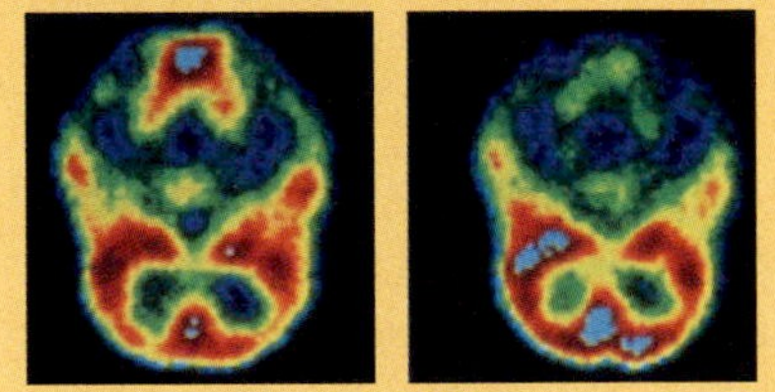 ***PET scans and brain functions.*** *The left scan shows brain activity when the eyes are open, whereas the one on the right is with the eyes closed. Note the increased activity, red and yellow, in the left photo when the eyes are open.*	Radioactive form of glucose is injected into the bloodstream; scanner records glucose levels in active areas of the brain and produces computer-constructed picture of the brain.	PET scans, originally designed to detect abnormalities, are also used to identify brain areas active during ordinary activities (reading, singing, etc.).
MRI (magnetic resonance imaging) scan 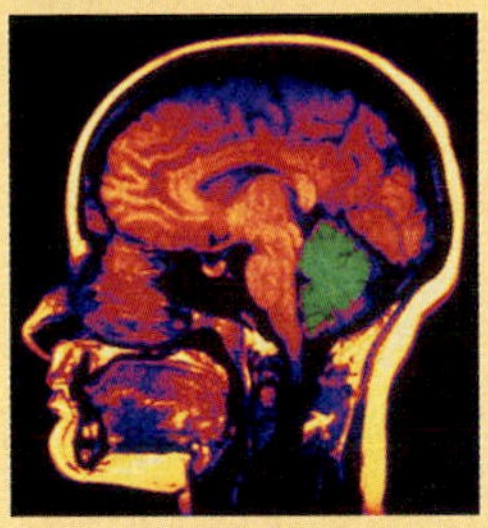 ***Magnetic resonance imaging (MRI).*** *Note the fissures and internal structures of the brain. The throat, nasal airways, and fluid surrounding the brain are dark.*	A high-frequency magnetic field is passed through the brain by means of electromagnets.	The MRI produces high-resolution three-dimensional pictures of the brain useful for identifying abnormalities and mapping brain structures and function.
fMRI (functional magnetic resonance imaging) scan 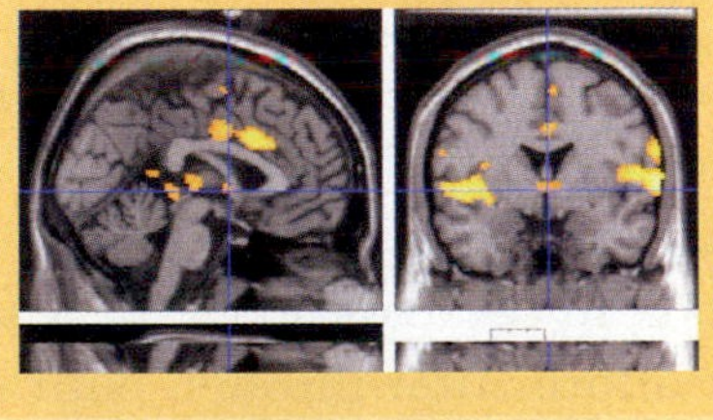	A newer, faster version of the MRI that detects blood flow by picking up magnetic signals from blood that has given up its oxygen to activate brain cells.	The fMRI measures blood flow, which indicates areas of the brain that are active or inactive during ordinary activities or responses (like reading or talking); also shows changes associated with disorders.
TMS (Transcranial magnetic stimulation) 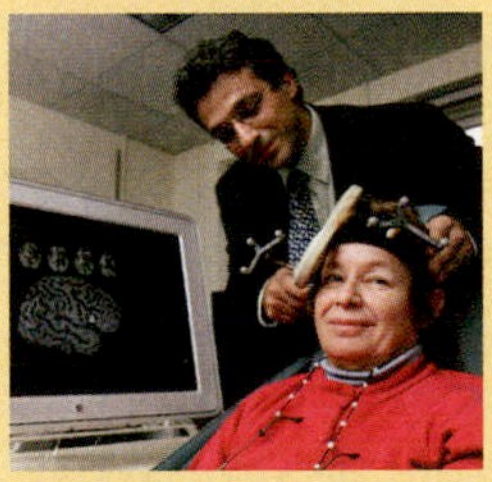	Recent method of brain stimulation that delivers a large current through a wire coil placed on the skull.	Can be used to elicit a motor response or to temporarily inactivate an area and observe the effects; also used to treat depression (Chapter 14).

CONCEPT CHECK STOP

How do psychologists guard against bias?

Which research method involves observing participants' behavior in the real world?

What is the difference between a positive and a negative correlation?

What does it mean to call a procedure "noninvasive"?

Getting the Most from Your Study of Psychology

LEARNING OBJECTIVES

Describe the steps you can take to read more accurately.

Explain how visual features can enhance learning.

Examine your current time-management habits, and identify how you might improve them.

Explain the benefits of distributed study and overlearning.

Summarize the grade-improvement and test-taking strategies that students can use to ensure success in their courses.

In this section, we offer several well-documented techniques that will help you "work smarter"—not just longer or harder (Dickinson, O'Connell, & Dunn, 1996)—so that you can get the most from your study of psychology (or any other subject). Mastering these skills will initially take some time, but you'll save hundreds of hours later on.

FAMILIARIZATION

Have you ever noticed that you can read a paragraph many times and still remember nothing from it? Often you must make a conscious effort to learn. There are a number of ways to actively read (and remember) information in a text. The first step is to familiarize yourself with the general text so that you can take full advantage of its contents. In *Visualizing Psychology*, the Preface, Table of Contents, Glossary (both running through the text of each chapter and at the end of each chapter), References, Name Index, and Subject Index will help give you a bird's-eye view of the rest of the text. In addition, as you scan the book to familiarize yourself with its contents, you should also take note of the many tables, figures, photographs, and special feature boxes, all of which will enhance your understanding of the subject.

ACTIVE READING

The most important tool for college success is the ability to read and master the assigned class text. One of the best ways to read actively is to use the **SQ4R method**, which was developed by Francis Robinson (1970). The initials stand for six steps in effective reading: **S**urvey, **Q**uestion, **R**ead, **R**ecite, **R**eview, and w**R**ite. As you might have guessed, *Visualizing Psychology* was designed to incorporate each of these steps (FIGURE 1.14).

VISUAL LEARNING

Our brains are highly tuned to visual cues as well as verbal cues. Photographs, drawings, and other graphical information help us solidify our understanding, organize and internalize new material, recognize patterns and interrelationships, and think creatively.

In some books, photographs and illustrations merely repeat, visually, concepts that are also stated in words. *Visualizing Psychology* is different. We have explicitly designed the book to take advantage of readers' capacity to process information through both visual and verbal channels. The text is tightly integrated with a rich array of visual features that will help you solidify your understanding of the concepts contained in the text. (The Illustrated Book Tour in the Preface describes these features in detail.) The photographs, drawings, diagrams, and graphs in this book carry their own weight—that is, they serve a specific instructional purpose, above and beyond what is stated in words. They are as essential as the text itself; be sure to pay attention to them.

Process Diagram

Using the SQ4R method FIGURE 1.14

Survey
Before you begin reading, skim and scan through the entire chapter. This *survey* will provide a "big picture" of the content and organization. Also, pay attention to the opening, chapter outline, and Learning Objectives. Knowing what to expect and the main points to look for provides a framework that helps hold the details together as you read.

Question
Based on your survey, what questions do you want answered in the chapter? Write them down. Note that the questions provided in the Learning Objective questions serve as a model for questions you should be asking yourself as you read. This active questioning strategy greatly increases your attention and comprehension.

Read
Do not passively slide your eyes over the words. Actively look for and write down your answers to the questions you formed in the previous step. Read carefully in short, concentrated time periods. Also, note the terms highlighted in **boldface type** and those defined in boxes. These key terms are important to your mastery of the material—and to higher scores on quizzes and exams.

Recite
After you've read one short section, stop and silently recite and summarize in your own words the main points of what you've just read—either orally or on paper. Then compare what you remember with the language of the text. This type of recitation gives you immediate feedback on what you have learned—and which areas you need to immediately reread.

Review
Carefully review and answer the Concept Checks that conclude each major section and the Critical and Creative Thinking Questions at the end of each chapter. Upon finishing the chapter, review your questions and answers from the earlier steps and the end-of-chapter summary. Repeating this review process before each quiz or exam will dramatically improve your exam scores!

wRite
You've been writing in most of the earlier parts of the SQ4R method, so why is "wRite" the final step? Research shows that it's critical to read and understand the material first, and then to go back and do a final written summary—in your own words! Also, it helps to take your text to class and write brief notes in the margins or on paper that tie in your professor's lecture to the text.

Applying Psychology

Improving Your Grade

If you are a student, the general learning tools outlined here will make you more efficient and successful in your courses. In addition, here are several specific strategies for grade improvement and test taking that can further improve your performance:

- *Take good notes.* Effective note taking depends on active listening. Ask yourself, "What is the main idea?" Write down key ideas and supporting details and examples.

- *Understand your professor.* The amount of lecture time spent on various topics is generally a good indication of what the instructor considers important.
- *General test taking.* On multiple-choice exams, carefully read each question and all of the alternative answers before responding. Be careful to respond to all questions, and make sure that you have recorded your answers correctly. Finally, practice your test taking by responding to the Concept Check questions, the Critical and Creative Thinking Questions, and the Self-Test in each chapter.
- *Skills courses.* Improve your reading speed and comprehension and your word-processing/typing skills by taking additional courses designed to develop these specific abilities.
- *Additional resources.* Don't overlook important human resources. Your instructors can provide useful tips. Enlist roommates, classmates, friends, and family members as your conscience-coaches.

CRITICAL THINKING

Here are two interesting questions:

1. What factors might prevent you from reading test questions carefully and responding accurately?
2. How could a friend or roommate help you improve your grades on tests?

TIME MANAGEMENT

If you find that you can't always strike a good balance between work, study, and social activities or that you aren't always good at budgeting your time, here are four basic time-management strategies:

- *Establish a baseline.* Before attempting any changes, simply record your day-to-day activities for one to two weeks. You may be surprised by how you spend your time.
- *Set up a realistic schedule.* Make a daily and weekly "to do" list, including all required activities, basic maintenance tasks (like laundry, cooking, child care, and eating), and a reasonable amount of "down time." Then create a daily schedule of activities that includes time for each of these. To make permanent time-management changes, shape your behavior, starting with small changes and building on them.
- *Reward yourself.* Give yourself immediate, tangible rewards for sticking with your daily schedule.
- *Maximize your time.* Time-management experts, such as Alan Lakein (1998), suggest that you should try to minimize the amount of time you spend worrying and complaining and fiddling around getting ready to study ("fretting and prepping"). Also be alert for hidden "time opportunities"—spare moments that normally go to waste, which you might instead use productively.

DISTRIBUTED STUDY

Spaced practice is a much more efficient way to study and learn than massed practice (Chapter 7). That is, you will learn material more thoroughly if you distribute your study over time, rather than trying to cram all the information in at once.

OVERLEARNING

Many people tend to study new material just to the point where they can recite the information, but they do not attempt to understand it more deeply. For best results, you should know how key terms and concepts are related to one another. You should also be able to generate examples other than the ones in the text. You should repeatedly review the material (by visualizing the phenomena that are described and explained in the text and by rehearsing what you have learned) until the information is firmly locked in place. You will find this confidence-building exercise particularly important if you suffer from test anxiety.

CONCEPT CHECK STOP

What are the elements of active reading?

How can a textbook take advantage of readers' visual learning abilities?

Why is distributed practice more effective than massed practice?

What steps will you take to get the most from your study of psychology?

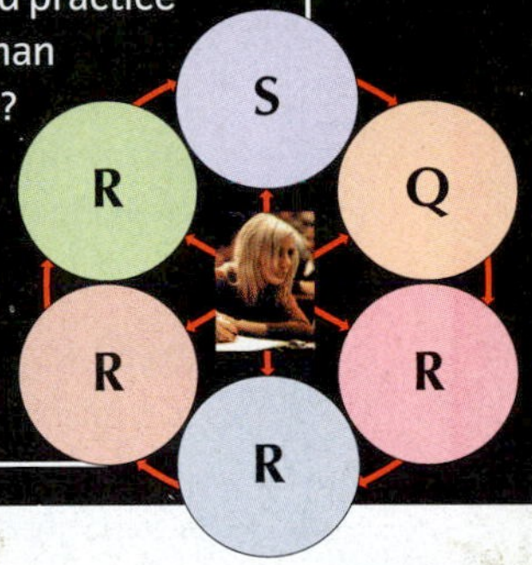

SUMMARY

1 Introducing Psychology

1. **Psychology** is the scientific study of behavior and mental processes. The discipline places high value on **empirical evidence** and **critical thinking**. One of the most enduring debates in science has been the **nature-nurture controversy**.

2. Psychology's four basic goals are to describe, explain, predict, and change behavior and mental processes through the use of scientific methods.

3. Psychologists work as therapists, researchers, teachers, and consultants in a wide range of settings.

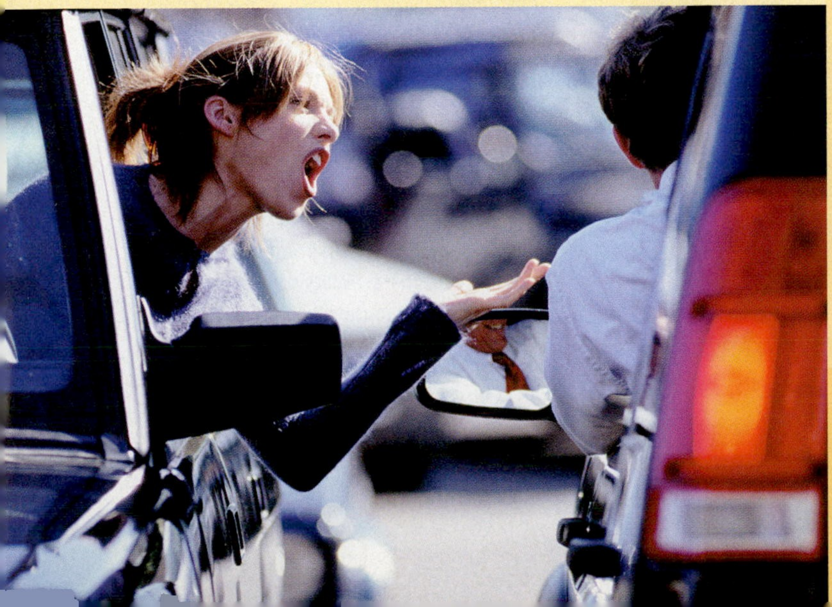

2 Origins of Psychology

1. Wilhelm Wundt, the father of psychology, and his followers, including Edward Titchener, were interested in studying conscious experience. Their approach, structuralism, sought to identify the basic structures of mental life through introspection.

2. The functionalist school, led by William James, studied how the mind functions to adapt humans and other animals to their environment.

3. Contemporary psychology reflects seven major perspectives: psychoanalytic or psychodynamic, behavioral, humanistic, cognitive, neuroscientific or biopsychological, evolutionary, and sociocultural. Most contemporary psychologists embrace a unifying perspective known as the **biopsychosocial model**. This model views biological processes, psychological factors, and social forces as interrelated influences on behavior.

4. Despite early limitations, women and minorities have made important contributions to psychology. Pioneers include Margaret Floy Washburn, Mary Calkins, Francis Cecil Sumner, and Kenneth B. Clark.

3 The Science of Psychology

1. **Basic research** is aimed at advancing general scientific understanding, whereas **applied research** works to address real-world problems.

2. Most scientific investigations involve six basic steps, collectively known as the scientific method. Scientific progress comes from repeatedly challenging and revising existing theories and building new ones.

3. Psychologists must maintain high ethical standards. This includes respecting the rights of both human and nonhuman research participants and of psychotherapy clients.

Informed consent and **debriefing** are critical elements of any research that involves human participants. Researchers and clinicians are held professionally responsible for their actions by the APA, by their institutions, and by local and state agencies.

4 Research Methods

1. Experimental research manipulates and controls variables to determine cause and effect. An **experiment** has two critical components: **independent** and **dependent variables**, and experimental and control groups. A good scientific experiment protects against potential sources of error from both the researcher and the participants.

2. **Descriptive research** involves systematically observing and describing behavior without manipulating variables. The three major types of descriptive research are naturalistic observation, surveys, and case studies.

3. **Correlational research** allows researchers to observe the relationship between two variables. Researchers analyze their results using a correlation coefficient. Correlations can be positive or negative. A correlation between two variables does not necessarily mean that one causes the other.

4. **Biological research** focuses on internal, biological processes that are involved in our feelings, thoughts, and behavior. Recent advances in brain imaging have improved scientists' ability to examine these processes and to do so noninvasively.

5 Getting the Most from Your Study of Psychology

1. Several well-documented techniques will help readers understand and absorb the material in this book most completely. These include familiarization, active reading, visual learning, time management, distributed study, and overlearning.

2. For students, several additional strategies for grade improvement and test taking can further improve course performance. These include more effective note taking, understanding the professor, general test-taking strategies, study skills courses, and other helpful resources.

KEY TERMS

- **psychology** p. 4
- **behavior** p. 4
- **mental processes** p. 4
- **empirical evidence** p. 4
- **critical thinking** p. 4
- **pseudopsychologies** p. 4
- **nature-nurture controversy** p. 5
- **structuralism** p. 7
- **functionalism** p. 7
- **psychoanalytic school** p. 7
- **behavioral perspective** p. 7
- **humanistic perspective** p. 8
- **positive psychology** p. 8
- **cognitive perspective** p. 8
- **neuroscientific/biopsychological perspective** p. 8
- **evolutionary perspective** p. 8
- **sociocultural perspective** p. 8
- **biopsychosocial model** p. 8
- **basic research** p. 12
- **applied research** p. 12
- **scientific method** p. 12
- **hypothesis** p. 13
- **operational definition** p. 13
- **theory** p. 13
- **informed consent** p. 14
- **debriefing** p. 14
- **experiment** p. 16
- **independent variable (IV)** p. 16
- **dependent variable (DV)** p. 16
- **experimenter bias** p. 16
- **double-blind study** p. 16
- **ethnocentrism** p. 16
- **cross-cultural sampling** p. 16
- **sample bias** p. 16
- **random assignment** p. 16
- **social desirability response** p. 16
- **experimental group** p. 17
- **control group** p. 17
- **placebo** p. 18
- **descriptive research** p. 18
- **naturalistic observation** p. 18
- **surveys** p. 19
- **case studies** p. 19
- **correlational research** p. 20
- **correlation coefficient** p. 20
- **biological research** p. 20
- **SQ4R method** p. 24

CRITICAL AND CREATIVE THINKING QUESTIONS

1. Scientific psychologists are among the least likely to believe in psychics, palmistry, astrology, and other paranormal phenomena. Why might that be?
2. Which psychological perspective would most likely be used to study and explain why some animals, such as newly hatched ducks or geese, follow and become attached to (or imprinted on) the first large moving object they see or hear?
3. Why is the scientific method described as a cycle rather than as a simple six-step process?
4. Imagine that a researcher recruited research participants from among her friends and then assigned them to experimental or control groups based on their gender. Why might this be a problem?
5. Which modern methods of examining how the brain influences behavior are noninvasive?
6. What do you think keeps most people from fully employing the active learning strategies and study skills presented in this chapter?

(See the next page for *What is happening in this picture?*)

What is happening in this picture?

Nonhuman animals are sometimes used in psychological research when it would be impractical or unethical to use human participants.

- What research questions might require the use of nonhuman animals? Opinions are sharply divided on the question of whether nonhuman animal research is ethical.
- What safeguards help ensure the proper treatment of these animals?

SELF-TEST

(Check your answers in Appendix B.)

1. In your text, psychology is defined as the ________________.
   a. science of conscious and unconscious forces on behavior
   b. empirical study of the mind
   c. scientific study of the mind
   d. scientific study of behavior and mental processes

2. According to your textbook, the goals of psychology are to ________________.
   a. explore the conscious and unconscious functions of the human mind
   b. understand, compare, and analyze human behavior
   c. improve psychological well-being in all individuals from conception to death
   d. describe, explain, predict, and change behavior and mental processes

3. The father of psychology is ________________.
   a. Sigmund Freud
   b. B. F. Skinner
   c. Wilhelm Wundt
   d. William James

4. The biopsychosocial model is known as a(n) ________________.
   a. integrative, unifying model
   b. concept formation
   c. consolidation model
   d. eclectic conceptualization

5. The term *basic research* is BEST defined as research that ________________.
   a. is basic to one field only
   b. is intended to advance scientific knowledge rather than for practical application
   c. is done to get a grade or a tenured teaching position
   d. solves basic problems encountered by humans and animals in a complex world

6. Identify and label the six steps in the scientific method.

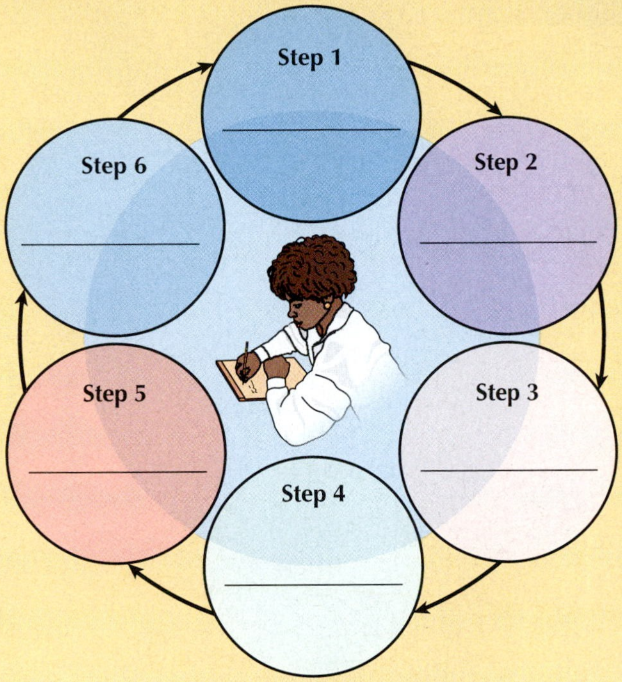

7. According to your text, debriefing is ________________.

a. interviewing subjects after a study to find out what they were thinking during their participation
b. explaining the purpose of the study, anticipated results, and any deception used when the study is over
c. disclosing potential physical and emotional risks, and the nature of the study prior to its beginning
d. interviewing subjects after a study to determine whether any deception used was effective in preventing them from learning the true purpose of the study

8. ________________ are manipulated; ________________ are measured.

a. Dependent variables; independent variables
b. Surveys; experiments
c. Statistics; correlations
d. Independent Variables; Dependent Variables

9. The **BEST** definition of a double-blind study is research in which ________________.

a. nobody knows what they are doing
b. neither the participants in the treatment group nor the control group know which treatment is being given to which group
c. both the researcher and the participants are unaware of who is in the experimental and control groups
d. two control groups (or placebo conditions) must be used

10. In a case study, a researcher is most likely to ________________.

a. interview many research subjects who have a single problem or disorder
b. conduct an in-depth study of a single research participant
c. choose and investigate a single topic
d. use any of these options, which describe different types of case studies

11. In ________________ research, a researcher observes or measures (without directly manipulating) two or more variables to find relationships between them, without inferring a causal relationship.

a. experimental
b. correlational
c. basic
d. applied

12. Label the six steps of the SQ4R study method.

Neuroscience and Biological Foundations

2

In 1996, when cyclist Lance Armstrong was diagnosed with testicular cancer, it seemed uncertain whether he would survive. The cancer had already spread to his lungs and brain. His doctors told him he had only a 40 percent chance of survival. (Later one of them revealed that the real estimate was 3 percent, but they had given him the higher figure to keep him from becoming discouraged!)

After major surgeries and chemotherapy, however, Armstrong overcame the cancer, and by 1998 he was racing again—with remarkable success. He won his first Tour de France race in 1999 and went on to win it an unprecedented six more consecutive times before retiring in 2005. The Tour de France is by far the most prestigious of all cycling competitions. It is also one of the most physically demanding; it runs over three weeks and covers 3000 to 4000 kilometers (1800 to 2500 miles).

What made Armstrong's extraordinary achievements possible? Is he "genetically gifted"? What part of his brain allowed him to monitor limited energy supplies over long days on his bike? Are there special chemicals in his brain—and those of other top athletes—that make it possible to cope with the pain and exhaustion from hours, days, and years of grueling practices and races? In this chapter, we'll look at the biological processes that allow great achievers like Lance Armstrong—as well as the rest of us—to absorb and organize the massive influx of sights, sounds, thoughts, emotions, and memories that compete for the brain's attention.

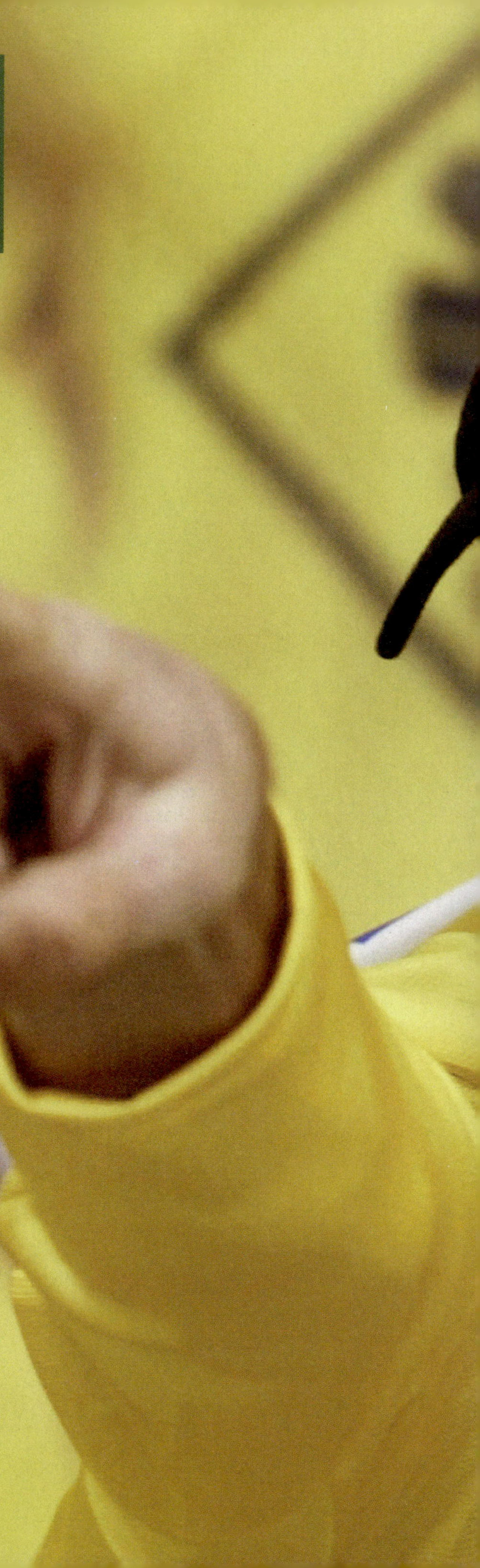

CHAPTER OUTLINE

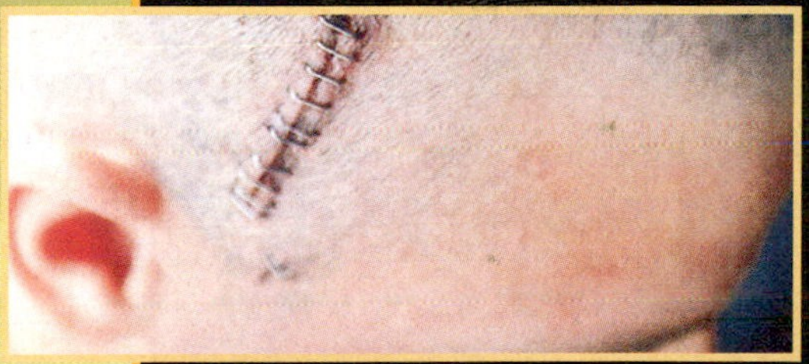

Our Genetic Inheritance

LEARNING OBJECTIVES

Summarize how genetic material passes from one generation to the next.

Review the approaches that scientists take to explore human inheritance.

Describe how natural selection and genetic mutations help explain behavior.

Thousands of years of evolution have contributed to what we are today. Our ancestors foraged for food, fought for survival, and passed on traits that were selected and transmitted down through the generations. How do these transmitted traits affect us today? For answers, psychologists often turn to **behavioral genetics** (how heredity and environment affect us) and **evolutionary psychology** (how the natural process of adapting to our environment affects us).

behavioral genetics The study of the relative effects of heredity and environment on behavior and mental processes.

evolutionary psychology A branch of psychology that studies how natural selection and adaptation help explain behavior and mental processes.

neuroscience An interdisciplinary field studying how biological processes relate to behavioral and mental processes.

BEHAVIORAL GENETICS: IS IT NATURE OR NURTURE?

Ancient cultures, including the Egyptian, Indian, and Chinese, believed the heart was the center of all thoughts and emotions. But we now know that the brain and the rest of the nervous system are the power behind our psychological life and much of our physical being. This chapter introduces you to the field of **neuroscience** and *biopsychology*, the scientific study of the biology of behavior and mental processes. We will discuss genetics and heredity, the nervous system, and the functions of each part of the brain.

At the moment of your conception, your mother and father each contributed to you 23 *chromosomes*. Thousands of *genes* make up each chromosome (FIGURE 2.1). For some traits, such as blood type, a single pair of genes (one from each parent) determines what characteristics you will possess. But most traits are determined by a combination of many genes.

When the two genes for a given trait conflict, the outcome depends on whether the gene is *dominant* or *recessive*. A dominant gene reveals its trait whenever the gene is present. In contrast, the gene for a recessive trait will normally be expressed only if the other gene in the pair is also recessive.

It was once assumed that characteristics such as eye color, hair color, or height were the result of either one dominant gene or two paired recessive genes. But modern geneticists believe that each of these characteristics is *polygenic*, meaning they are controlled by multiple genes. Many

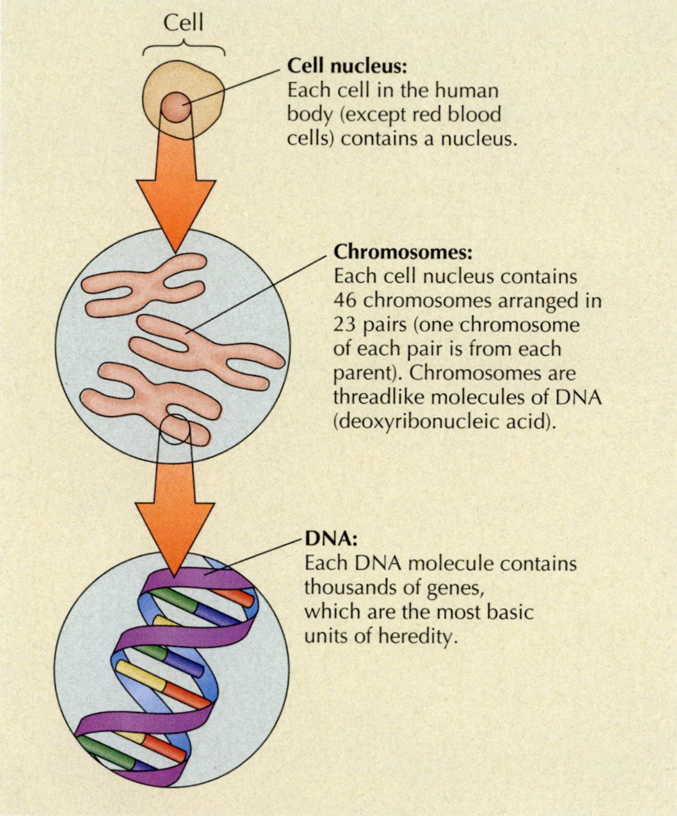

Cell
Cell nucleus:
Each cell in the human body (except red blood cells) contains a nucleus.
Chromosomes:
Each cell nucleus contains 46 chromosomes arranged in 23 pairs (one chromosome of each pair is from each parent). Chromosomes are threadlike molecules of DNA (deoxyribonucleic acid).
DNA:
Each DNA molecule contains thousands of genes, which are the most basic units of heredity.

Hereditary code FIGURE 2.1

Gene–environment interaction FIGURE 2.2

Children who are malnourished may not reach their full potential genetic height or maximum intelligence. Can you see how environmental factors interact with genetic factors to influence many traits?

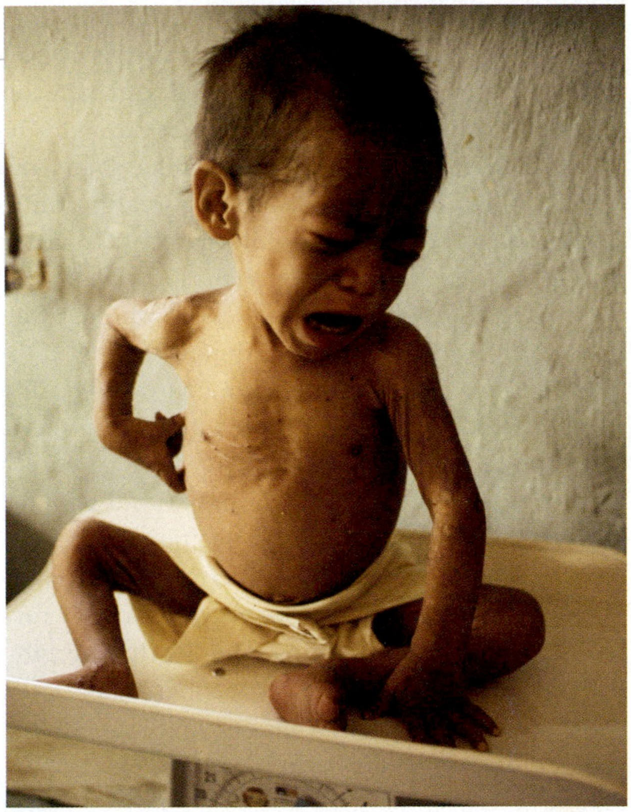

polygenic traits like height or intelligence are also affected by environmental and social factors (FIGURE 2.2). Fortunately, most serious genetic disorders are not transmitted through a dominant gene. Can you understand why?

How do scientists research human inheritance? To determine the influences of heredity or environment on complex traits like aggressiveness, intelligence, or sociability, scientists rely on these indirect methods: twin, family, and adoption studies; studies of genetic abnormalities.

Psychologists are especially interested in the study of twins because they have a uniquely high proportion of shared genes. Identical (*monozygotic*—one egg) twins share 100 percent of the same genes, whereas fraternal (*dizygotic*—two eggs) twins share, on average, 50 percent of their genes, just like any other pair of siblings (FIGURE 2.3).

Identical and fraternal twins FIGURE 2.3

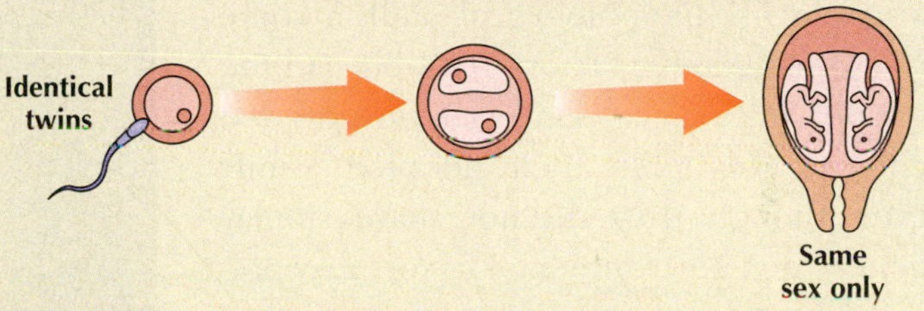

A Monozygotic, or identical, twins develop from a single egg fertilized by a single sperm. They share the same placenta and have the same sex and same genetic makeup.

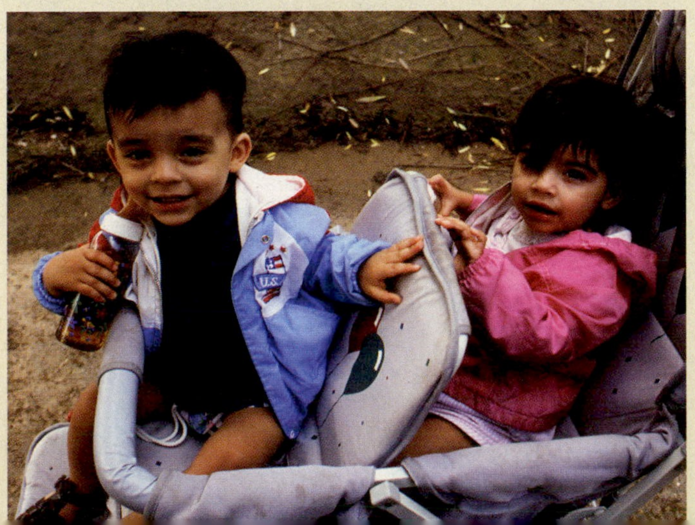

B Fraternal twins are formed when two separate sperm fertilize two separate eggs.

Because both identical and fraternal twins share the same parents and develop in relatively the same environment, they provide a valuable "natural experiment." If heredity influences a trait or behavior to some degree, identical twins should be more alike than fraternal twins. Twin studies have provided a wealth of information on the relative effects of heredity on behavior. For example, studies of intelligence show that identical twins have almost identical IQ scores, whereas fraternal twins are only slightly more similar in their IQ scores than are non-twin siblings (Bouchard, 2004; Plomin, 1999). The difference suggests a genetic influence on intelligence.

Psychologists interested in behavioral genetics can also study entire families. If a specific trait is inherited, blood relatives should show increased trait similarity, compared with unrelated people. Also, closer relatives, like siblings, should be more similar than distant relatives. Family studies have shown that many traits and mental disorders, such as intelligence, sociability, and depression, do indeed run in families.

Studying families with children who have been adopted provides valuable information for researchers. If adopted children are more like their biological family in some trait, then genetic factors probably had the greater influence. Conversely, if adopted children resemble their adopted family even though they do not share similar genes, then environmental factors may predominate.

Finally, research in behavioral genetics explores disorders and diseases that result when genes malfunction. For example, researchers believe that genetic or chromosomal abnormalities are important factors in Alzheimer's disease and schizophrenia.

Findings from these four methods have allowed behavioral geneticists to estimate the *heritability* of various traits. That is, to what degree are individual differences a result of genetic, inherited factors rather than differences in the environment? If genetics contributed nothing to the trait, it would have a heritability estimate of 0 percent. If a trait was completely due to genetics, it would have a heritability estimate of 100 percent. Keep in mind, however, that heritability estimates apply to groups, not individuals (FIGURE 2.4).

As we've seen, behavioral genetics studies help explain the role of heredity (nature) and the environment (nurture) in our individual behavior. To increase our understanding of genetic dispositions, we also need to look at universal behaviors transmitted from our evolutionary past.

EVOLUTIONARY PSYCHOLOGY: DARWIN EXPLAINS BEHAVIOR AND MENTAL PROCESSES

Evolutionary psychology suggests that many behavioral commonalities, from eating to fighting with our ene-

Height and heritability FIGURE 2.4

Height has one of the highest heritability estimates—around 90 percent (Plomin, 1990). However, it's impossible to predict with certainty an individual's height from a heritability estimate. What other factors might have contributed to the difference in height between this mother and daughter?

mies, emerged and remain in human populations because they helped our ancestors (and ourselves) survive. This perspective stems from the writings of Charles Darwin (1859), who suggested that natural forces select traits that are adaptive to the organism's survival. This process of **natural selection** occurs when a particular genetic trait gives a person a reproductive advantage over others. Some people mistakenly believe that natural selection means "survival of the fittest." But what really matters is *reproduction*—the survival of the genome. Because of natural selection, the fastest or otherwise most fit organisms will be most likely to live long enough to pass on their genes to the next generation.

Genetic mutations also help explain behavior. Everyone likely carries at least one gene that has mutated, or changed from the original. Very rarely, a mutated gene will be significant enough to change an individual's behavior. It might cause someone to be more social, more risk taking, more shy, more careful. If the gene then gives the person reproductive advantage, he or she will be more likely to pass on the gene to future generations. However, this mutation doesn't guarantee long-term survival. A well-adapted population can perish if its environment changes.

CONCEPT CHECK STOP

What is the difference between a dominant and a recessive trait?

Why doesn't knowing the heritability of a trait predict what characteristics a given individual will have?

What is natural selection?

How are heredity and evolution linked to human behavior?

Neural Bases of Behavior

NATIONAL GEOGRAPHIC

LEARNING OBJECTIVES

Describe how neurons communicate throughout the body.

Explain the role that neurotransmitters play.

Compare and contrast the functions of neurotransmitters and hormones.

Your brain and the rest of your nervous system essentially consist of **neurons**. Each one is a tiny information-processing system with thousands of connections for receiving and sending electrochemical signals to other neurons. Each human body may have as many as one *trillion* neurons. (Be careful not to confuse the term *neuron* with the term *nerve*. Nerves are large bundles of axons outside the brain and spinal cord.)

Neurons are held in place and supported by **glial cells**. Glial cells surround neurons, perform cleanup tasks, and insulate one neuron from another so their neural messages do not get scrambled. They also play a direct role in nervous system communication (Arriagada et al., 2007; Wieseler-Frank, Maier, & Watkins, 2005; Zillmer, Spiers, & Culbertson, 2008). However, the "star" of the communication show is still the neuron.

No two neurons are alike, but most share three basic features: **dendrites**, the **cell body**,

neuron A nerve cell that receives and conducts electrical impulses from the brain.

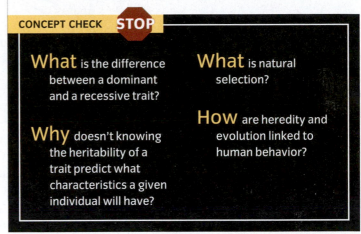

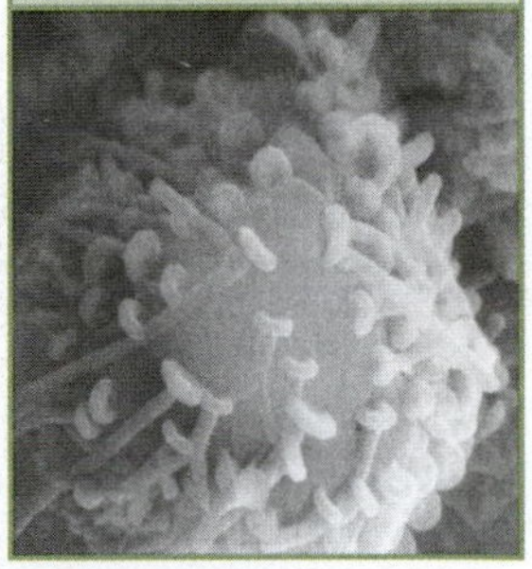

The structure of a neuron FIGURE 2.5

Arrows indicate direction of information flow: dendrites → cell body → axon → terminal buttons of axon.

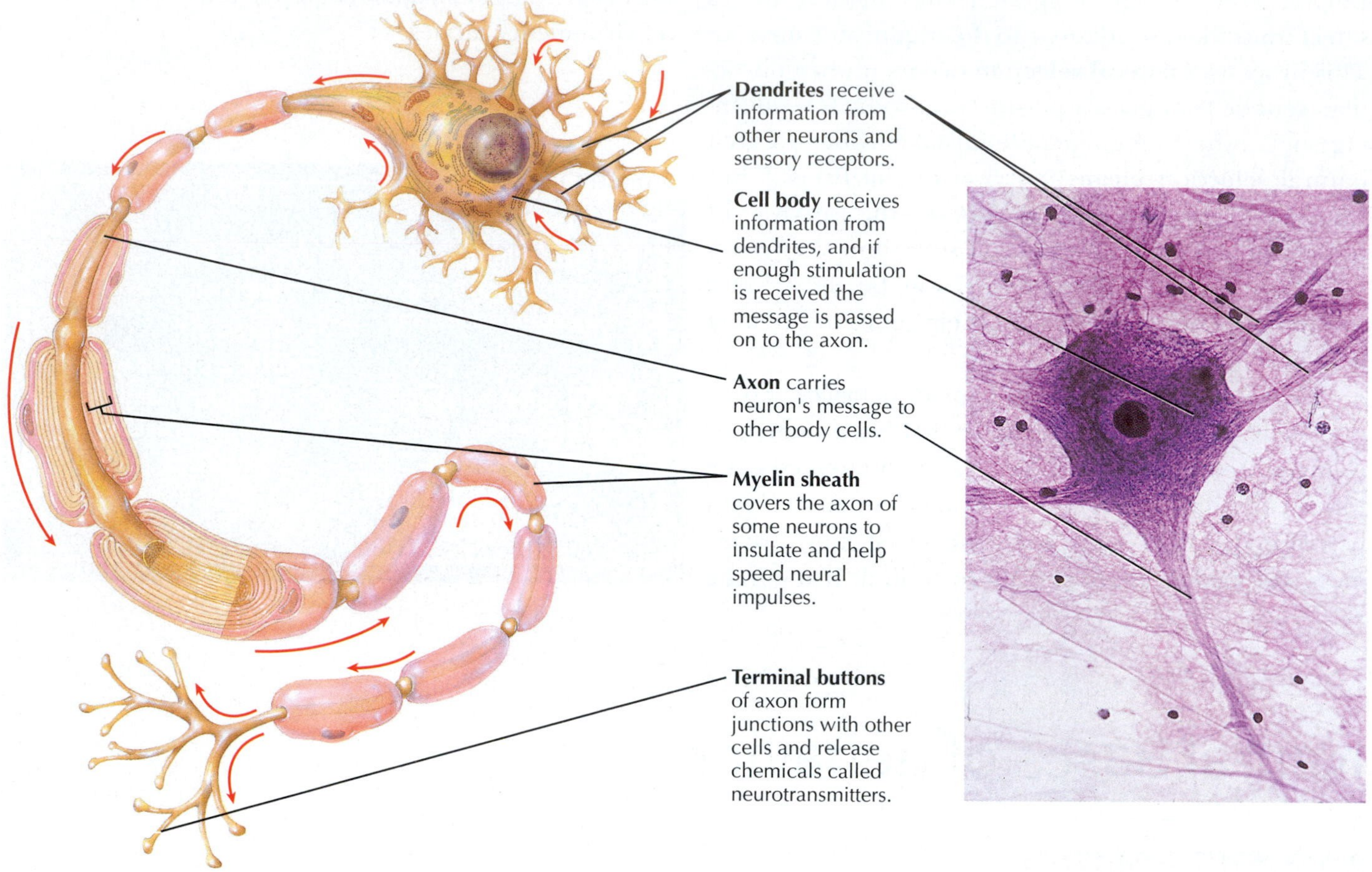

Dendrites receive information from other neurons and sensory receptors.
Cell body receives information from dendrites, and if enough stimulation is received the message is passed on to the axon.
Axon carries neuron's message to other body cells.
Myelin sheath covers the axon of some neurons to insulate and help speed neural impulses.
Terminal buttons of axon form junctions with other cells and release chemicals called neurotransmitters.

and an **axon** (FIGURE 2.5). To remember how information travels through the neuron, think of these three in reverse alphabetical order: *D*endrite → *C*ell *B*ody → *A*xon.

HOW DO NEURONS COMMUNICATE?

A neuron's basic function is to transmit information throughout the nervous system. Neurons "speak" in a type of electrical and chemical language. The process of neural communication begins within the neuron itself, when the dendrites and cell body receive electrical "messages." These messages move along the axon in the form of a neural impulse, or **action potential** (FIGURE 2.6).

Nerve impulses move much more slowly than electricity through a wire. A neural impulse travels along a bare axon at only about 10 meters per second. (Electricity moves at 36 million meters per second.)

Some axons, however, are enveloped in fatty insulation, the **myelin sheath**. This sheath blankets the axon, with the exception of periodic *nodes,* points at which the myelin is very thin or absent. In a myelinated axon, the nerve impulse moves about 10 times faster than in a bare axon because the action potential jumps from node to node rather than traveling along the entire axon.

Communication *within* the neuron (FIGURE 2.6A) is not the same as communication *between* neurons (FIGURE 2.6B on page 40). Within the neuron, messages travel electrically. But messages are transmitted chemically from one neuron to the next. The chemicals that transmit these messages are called **neurotransmitters**.

> **neurotransmitters** Chemicals that neurons release, which affect other neurons.

How neurons communicate: Communication *within* the neuron FIGURE 2.6A

VIEW THIS IN ACTION
in your WileyPLUS course

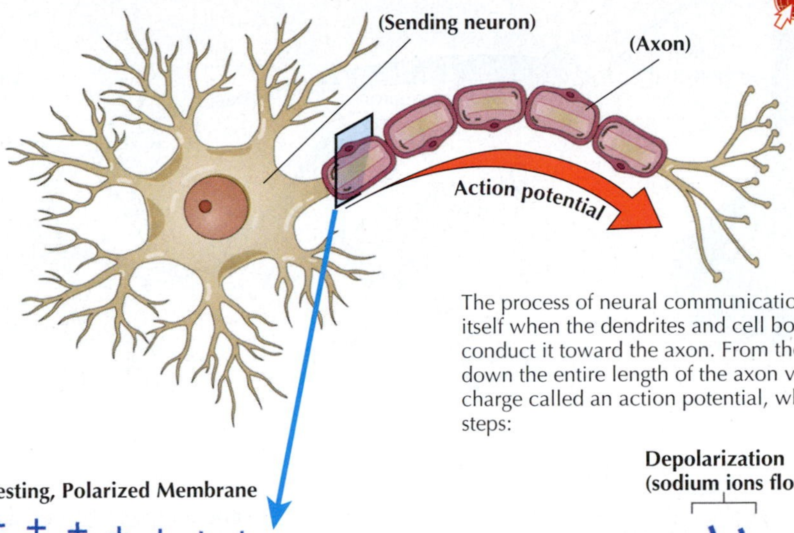

The process of neural communications begins within the neuron itself when the dendrites and cell body receive information and conduct it toward the axon. From there, the information travels down the entire length of the axon via a brief traveling electrical charge called an action potential, which can be described in three steps:

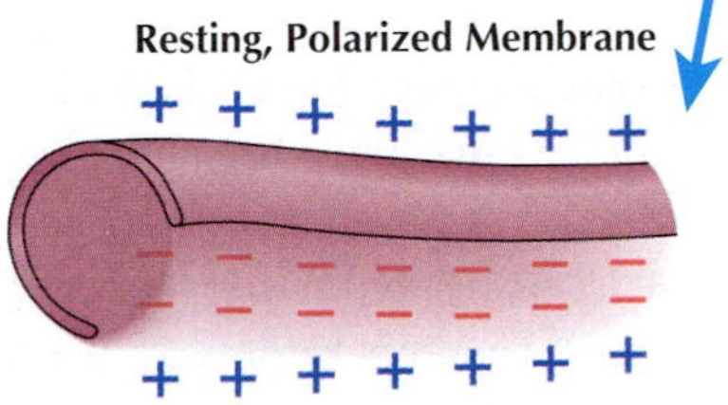

❶ Resting potential
When an axon is not stimulated, it is in a polarized state, called the *resting potential.* "At rest," the fluid inside the axon has more negatively charged ions than the fluid outside. This results from the selective permeability of the axon membrane and a series of mechanisms, called *sodium-potassium pumps,* which pull potassium ions in and pump sodium ions out of the axon. The inside of the axon has a charge of about −70 millivolts relative to the outside.

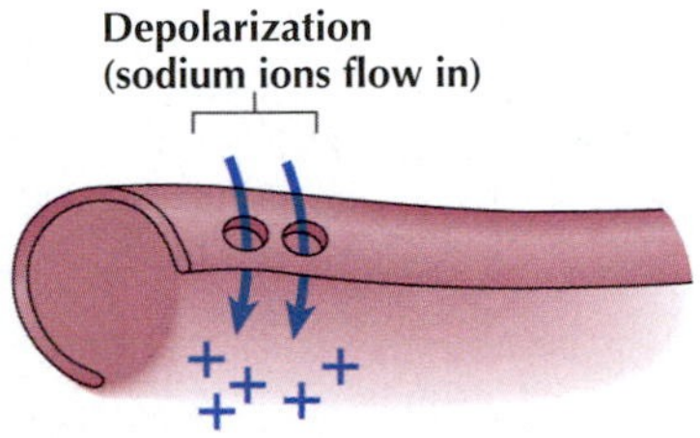

❷ Action potential initiation
When an "at rest" axon membrane is stimulated by a sufficiently strong signal, it produces an *action potential* (or depolarization). This action potential begins when the first part of the axon opens its "gates" and positively charged sodium ions rush through. The additional sodium ions change the previously negative charge inside the axon to a positive charge—thus depolarizing the axon.

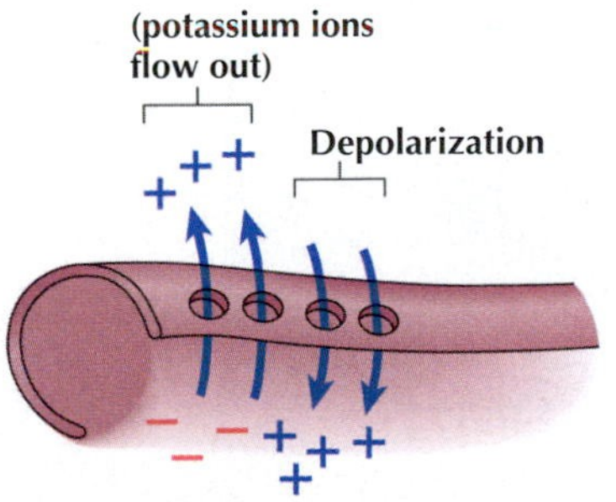

❸ Spreading of action potential and repolarization
The initial depolarization (or action potential) of Step 2 produces a subsequent imbalance of ions in the adjacent axon membrane. This imbalance thus causes the action potential to spread to the next section. Meanwhile, "gates" in the axon membrane of the initially depolarized section open and potassium ions flow out, thus allowing the first section to repolarize and return to its resting potential.

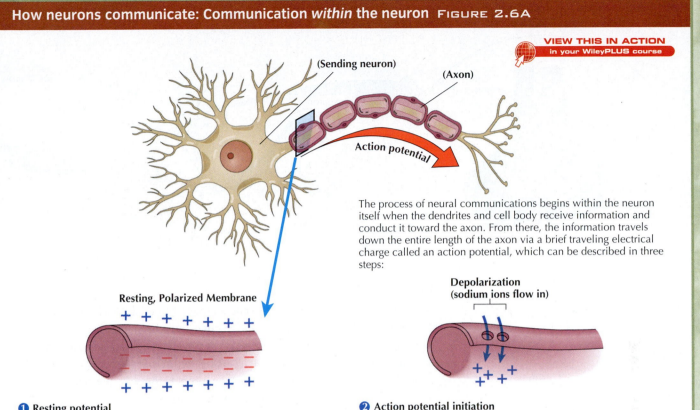

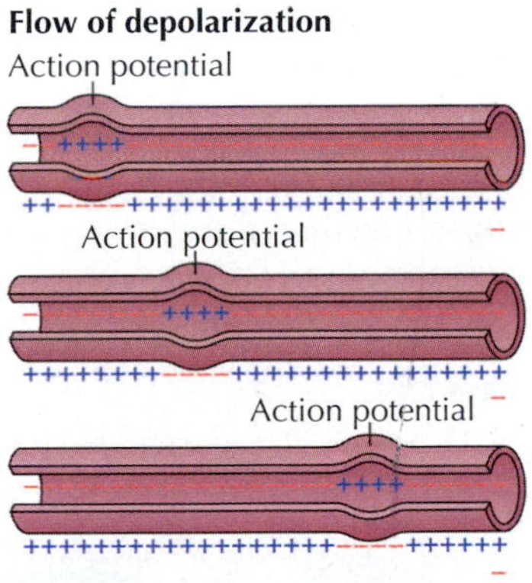

❹ Overall summary
As you can see in the figure above, this sequential process of depolarization, followed by repolarization, transmits the action potential along the entire length of the axon from the cell body to the terminal buttons. This is similar to an audience at an athletic event doing "the wave". One section of fans initially stands up for a brief time (action potential). This section then sits down (resting potential), and the "wave" then spreads to adjacent sections.

(See next page for Figure 2.6B)

Process Diagram

Process Diagram

How neurons communicate: Communication *between* the neurons FIGURE 2.6B

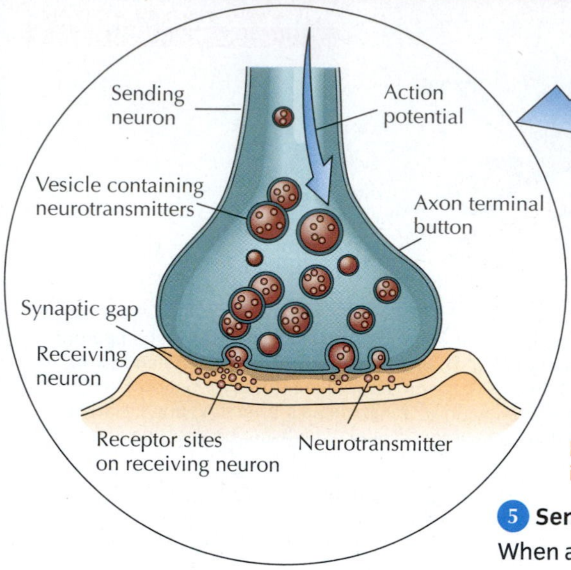

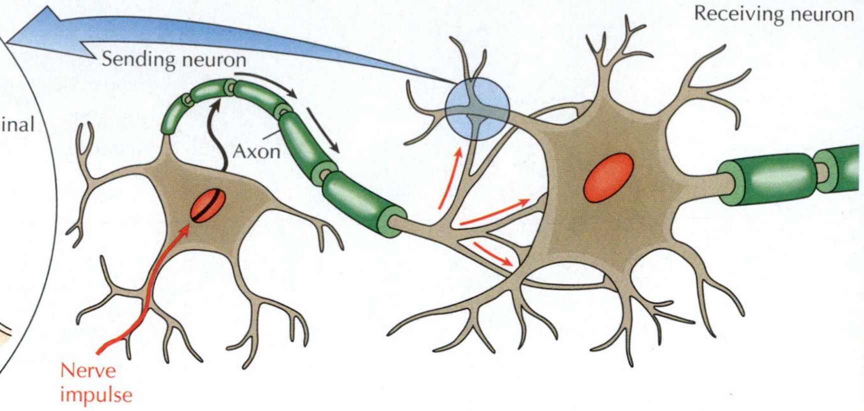

5 Sending a chemical signal

When action potentials reach the branching axon terminals, they trigger the terminal buttons at the axon's end to open and release thousands of neurotransmitters into the *synaptic gap*, the tiny opening between the sending and receiving neuron. These chemicals then move across the synaptic gap and attach to the membranes of the receiving neuron. In this way, they carry the message from the sending neuron to the receiving neuron.

6 Receiving a chemical signal

After a chemical message flows across the synaptic gap, it attaches to specific receiving neurons. It's important to know that each receiving neuron gets multiple neurotransmitter messages. As you can see in this close-up photo, the axon terminals from thousands of other nearby neurons almost completely cover the cell body of the receiving neuron. It's also important to understand that neurotransmitters deliver either excitatory or inhibitory messages, and that the receiving neuron will only produce an action potential and pass along the message if the number of excitatory messages outweigh the inhibitory messages.

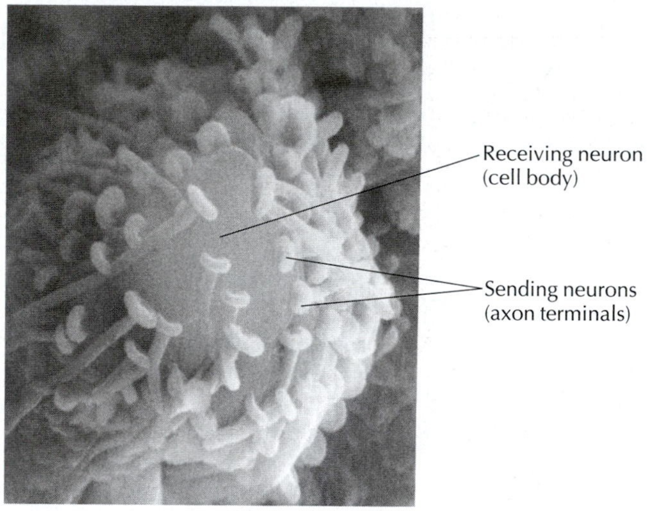

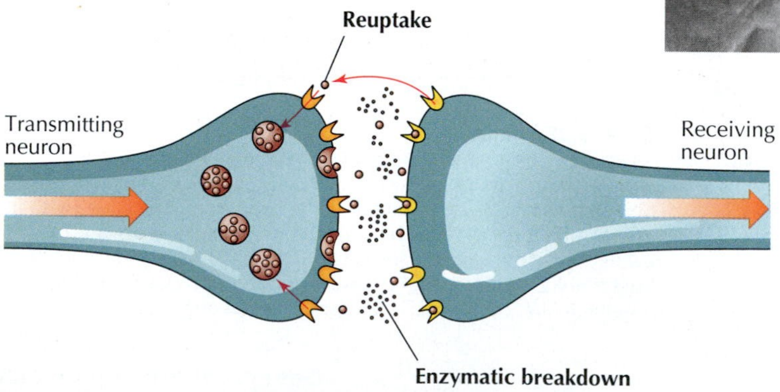

7 Dealing with leftovers

What happens to excess neurotransmitters or to those that do not "fit" into the adjacent receptor sites? The sending neuron normally reabsorbs the excess (called "reuptake") or they are broken down by special enzymes.

What a Psychologist Sees

How Poisons and Drugs Affect Our Brain

Neurotransmitters help explain how poisons and mind-altering drugs affect the brain.

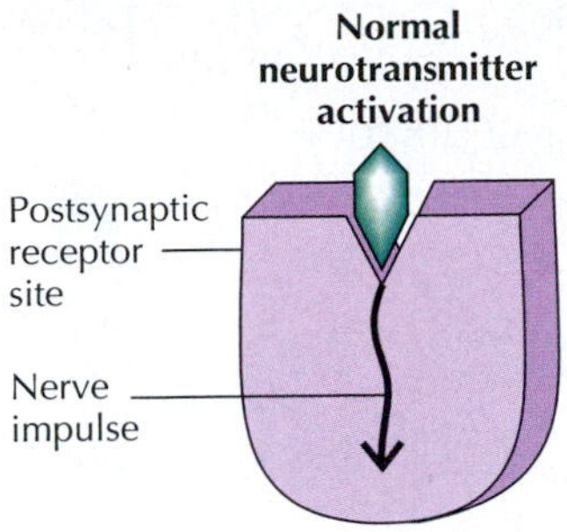

A Like a key fitting into a lock, receptor sites on receiving neurons' dendrites recognize neurotransmitters by their particular shape.

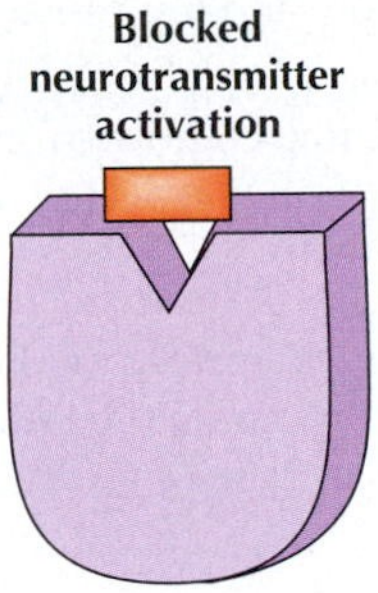

B Molecules without the correct shape don't fit the receptors, so they cannot stimulate the dendrite.

C Some *agonist drugs*, like the poison in the black widow spider or the nicotine in cigarettes, are similar enough in structure to a certain neurotransmitter (in this case, *acetylcholine*) that they mimic its effects on the receiving neuron.

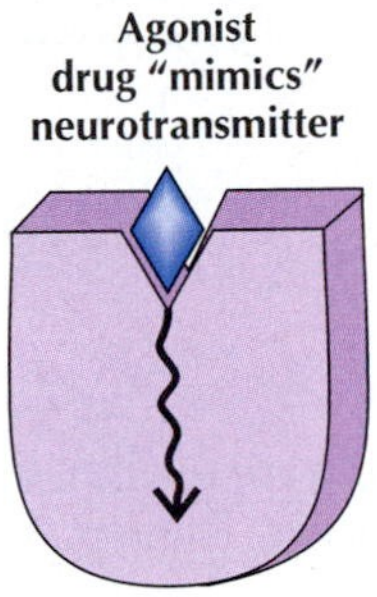

D Some *antagonist drugs* block neurotransmitters like acetylcholine. Because acetylcholine is vital in muscle action, blocking it paralyzes muscles, including those involved in breathing, which can be fatal.

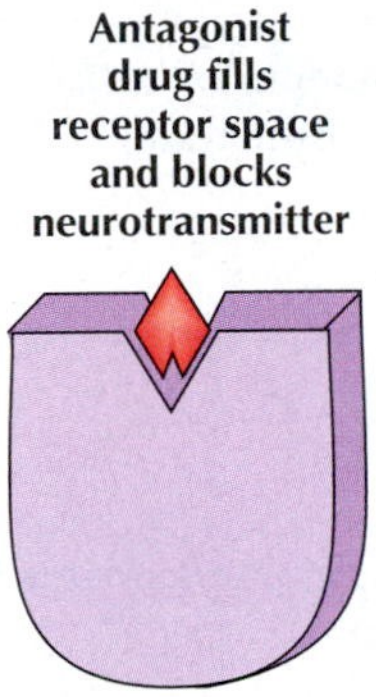

Examples of antagonists to acetylcholine include most snake venom and some poisons, like *botulinum* toxin (Botox®), which is used to treat some medical conditions involving abnormal muscle contraction—as well as for some cosmetic purposes.

Researchers have discovered hundreds of substances that function as neurotransmitters. These substances regulate a wide variety of physiological processes. See *What a Psychologist Sees* for a description of some of these effects.

One of the benefits of studying your brain and its neurotransmitters is that it will help you understand some common medical problems. For example, we know that decreased levels of the neurotransmitter dopamine are associated with Parkinson's disease (PD), whereas excessively high levels of dopamine appear to contribute to some forms of schizophrenia. (STUDY ORGANIZER 2.1 on the next page presents additional examples of the better understood neurotransmitters.)

Perhaps the best-known neurotransmitters are the endogenous opioid peptides, commonly known as **endorphins** (a contraction of *endogenous* [self-produced] and *morphine*). These chemicals mimic the effects of opium-based drugs such as morphine—they elevate mood and reduce pain. They also affect memory, learning, blood pressure, appetite, and sexual activity.

Study Organizer 2.1 How neurotransmitters affect us

Neurotransmitter	*Known or suspected effects*	
Serotonin	Mood, sleep, appetite, sensory perception, temperature regulation, pain suppression, and impulsivity. Low levels associated with depression.	
Acetylcholine (ACh)	Muscle action, cognitive functioning, memory, rapid-eye-movement (REM) sleep, emotion. Suspected role in Alzheimer's disease.	
Dopamine (DA)	Movement, attention, memory, learning, and emotion. Excess DA associated with schizophrenia, too little with Parkinson's disease. Also plays a role in addiction and the reward system.	
Norepinephrine (NE) (or noradrenaline)	Learning, memory, dreaming, emotion, waking from sleep, eating, alertness, wakefulness, reactions to stress. Low levels of NE associated with depression, high levels with agitated, manic states.	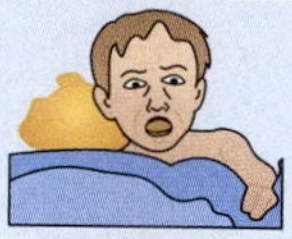
Epinephrine (or adrenaline)	Emotional arousal, memory storage, and metabolism of glucose necessary for energy release.	
Gamma aminobutyric acid (GABA)	Neural inhibition in the central nervous system. Tranquilizing drugs, like Valium, increase GABA's inhibitory effects and thereby decrease anxiety.	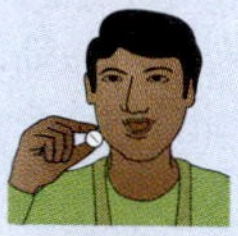
Endorphins	Mood, pain, memory, and learning.	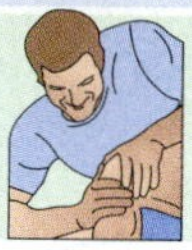

The endocrine system
FIGURE 2.7

This figure shows the major endocrine glands, along with some internal organs to help you locate the glands.

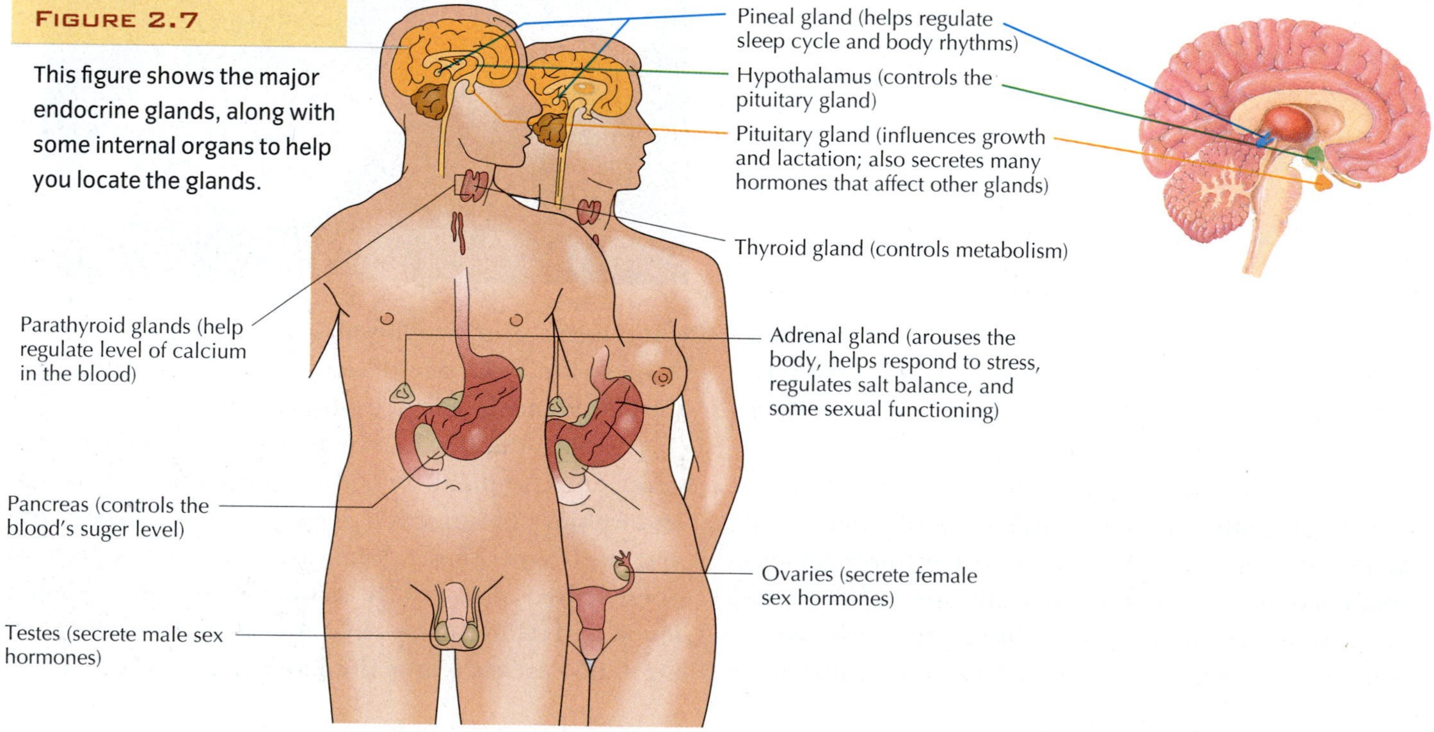

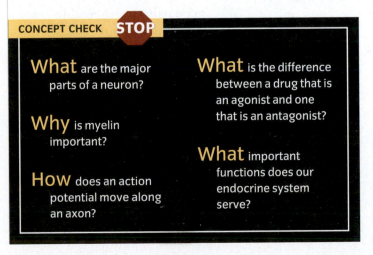

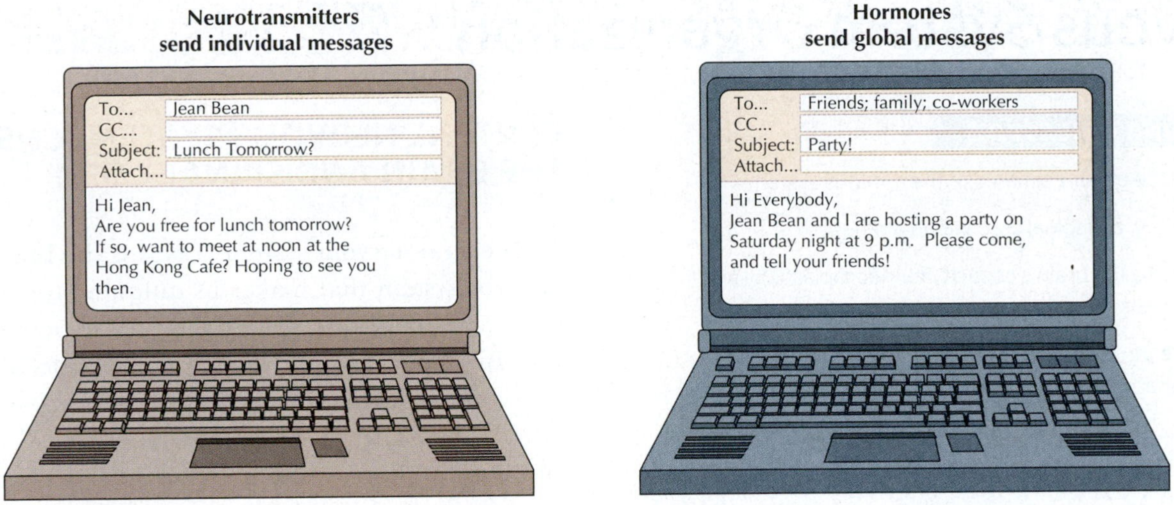
Neurotransmitters send individual messages
To... Jean Bean
CC...
Subject: Lunch Tomorrow?
Attach...
Hi Jean,
Are you free for lunch tomorrow?
If so, want to meet at noon at the Hong Kong Cafe? Hoping to see you then.
Hormones send global messages
To... Friends; family; co-workers
CC...
Subject: Party!
Attach...
Hi Everybody,
Jean Bean and I are hosting a party on Saturday night at 9 p.m. Please come, and tell your friends!

Why do we need two communication systems? FIGURE 2.8

You can think of neurotransmitters as individual e-mails that you send to particular people. Neurotransmitters deliver messages to specific receptors, which other neurons nearby probably don't "overhear." Hormones, in contrast, are like a global e-mail message that you send to everyone in your address book. Endocrine glands release hormones directly into the bloodstream, which travel throughout the body, carrying messages to any cell that will listen. Hormones also function like your global e-mail recipients forwarding your message to yet more people. For example, a small part of the brain called the hypothalamus releases hormones that signal the pituitary (another small brain structure), which stimulates or inhibits the release of other hormones.

HORMONES: A GLOBAL COMMUNICATION SYSTEM

We've just seen how the nervous system uses neurotransmitters to transmit messages throughout the body. A second type of communication system also exists. This second system is made up of a network of glands, called the **endocrine system** (FIGURE 2.7). Rather than neurotransmitters, this system uses **hormones** to carry its messages. (See FIGURE 2.8.)

hormones Chemicals manufactured by endocrine glands and circulated in the bloodstream to produce bodily changes or maintain normal bodily function.

Your endocrine system has several important functions. It helps regulate long-term bodily processes, such as growth and sexual characteristics. It also maintains ongoing bodily processes (such as digestion and elimination). Finally, hormones control the body's response to emergencies. In times of crisis, the hypothalamus sends messages through two pathways—the neural system and the endocrine system (primarily the pituitary). The pituitary sends hormonal messages to the adrenal glands (located right above the kidneys). The adrenal glands then release **cortisol**, a "stress hormone" that boosts energy and blood sugar levels, *epinephrine* (commonly called adrenaline), and *norepinephrine*. (Remember that these same chemicals also serve as neurotransmitters.)

CONCEPT CHECK STOP

What are the major parts of a neuron?

Why is myelin important?

How does an action potential move along an axon?

What is the difference between a drug that is an agonist and one that is an antagonist?

What important functions does our endocrine system serve?

Nervous System Organization

LEARNING OBJECTIVES

Identify the major elements of the nervous system.

Explain how the spinal cord initiates reflexes.

Explain why the brain's capacity for neuroplasticity and neurogenesis are important.

Describe the opposing roles of the sympathetic and parasympathetic nervous systems.

Have you heard the expression "Information is power?" Nowhere is this truer than in the human body. Without information, we could not survive. Neurons within our nervous system must take in sensory information from the outside world and then decide what to do with the information. Just as the circulatory system handles blood, which conveys chemicals and oxygen, our nervous system uses chemicals and electrical processes that convey information.

The nervous system is divided and subdivided into several branches (**FIGURE 2.9**). One main part of our nervous system includes the brain and a bundle of nerves that form the *spinal cord*. Because this system is located in the center of your body (within your skull and spine), it is called the **central nervous system (CNS)**. The CNS is primarily responsible for processing and organizing information.

The second major part of your nervous system includes all the nerves outside the brain and spinal cord. This **peripheral nervous system (PNS)** carries messages (action potentials) to and from the central nervous system to the periphery of the body. Now, let's take a closer look at the CNS and the PNS.

central nervous system (CNS) The brain and spinal cord.

peripheral nervous system (PNS) All nerves and neurons connecting the CNS to the rest of the body.

CENTRAL NERVOUS SYSTEM (CNS): THE BRAIN AND SPINAL CORD

The central nervous system (CNS) is the branch of the nervous system that makes us unique. Most other animals can smell, run, see, and hear far better than we can. But thanks to our CNS, we can process information and adapt to our environment in ways that no other animal can. Unfortunately, our CNS is also incredibly fragile. Unlike neurons in the PNS that can regenerate and require less protection, serious damage to neurons in the CNS is usually permanent. However, the brain may not be as "hard wired" as we once believed.

Scientists long believed that after the first two or three years of life, humans and most animals are unable to repair or replace damaged neurons in the brain or spinal cord. We now know that the brain is capable of lifelong **neuroplasticity** and **neurogenesis**.

Neuroplasticity

Rather than being a fixed, solid organ, the brain is capable of changing its structure and function as a result of usage and experience (Deller et al., 2006; Romero et al., 2008; Rossignol et al., 2008). This "rewiring" is what makes our brains so wonderfully adaptive. For example, it makes it possible for us to learn a new sport or a foreign language.

Remarkably, this rewiring has even helped "remodel" the brain following strokes. For example, psychologist Edward Taub and his colleagues (2002, 2004, 2007) have had success working with stroke patients (**FIGURE 2.10**).

neuroplasticity The brain's ability to reorganize and change its structure and function through the life span.

neurogenesis The division and differentiation of nonneuronal cells to produce neurons.

Neurogenesis

Our brains continually replace lost cells with new cells that originate deep within the brain and migrate to become part of its circuitry. The source of

The nervous system FIGURE 2.9

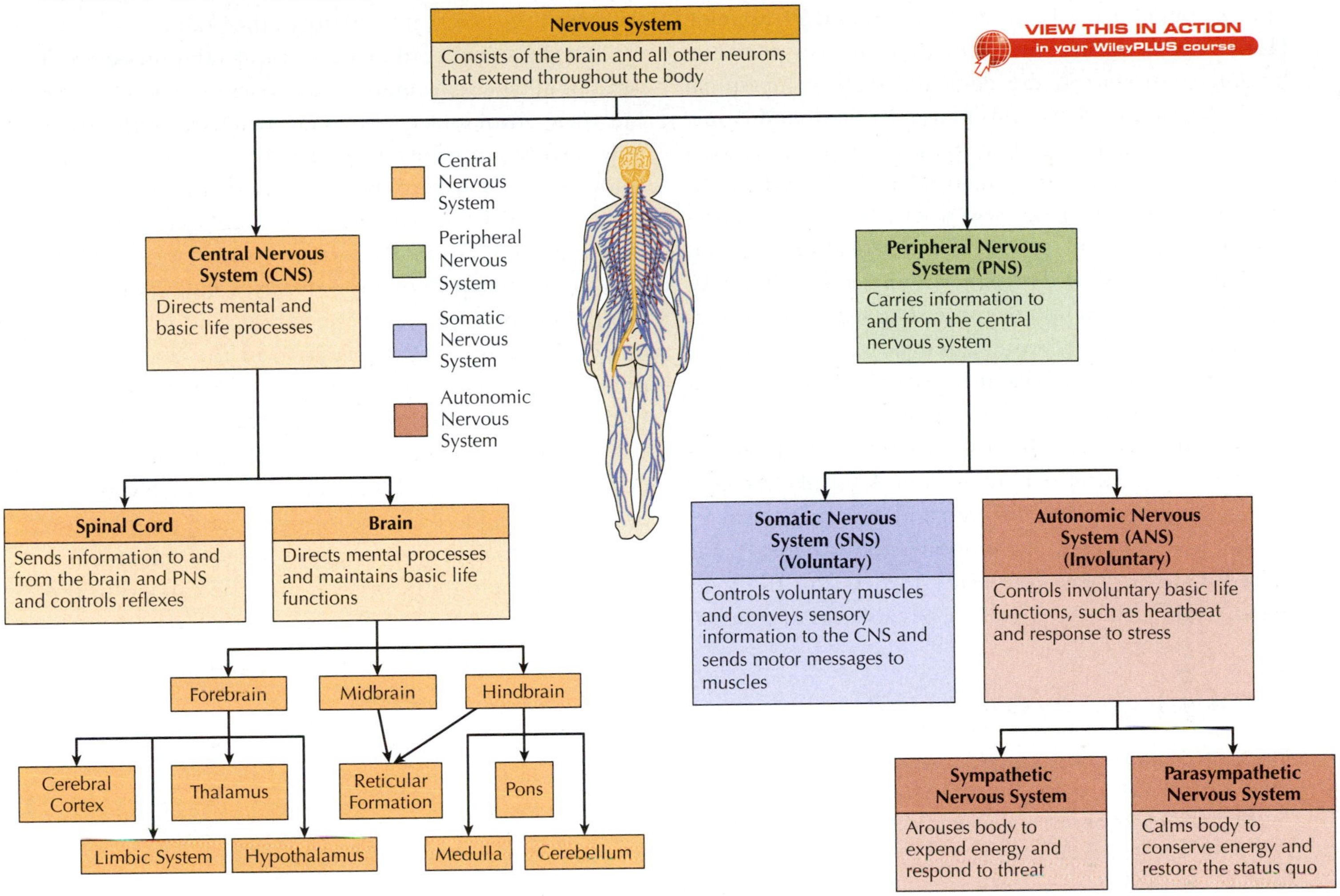

these newly created cells is neural **stem cells**—rare, immature cells that can grow and develop into any type of cell. Their fate depends on the chemical signals they receive (Abbott, 2004; Kim, 2004). Stem cells have been used for bone marrow transplants, and clinical trials using stem cells to repopulate or replace cells devastated by injury or disease have helped patients suffering from strokes, Alzheimer's, Parkinson's, epilepsy, stress, and depression (Chang et al., 2005; Fleischmann & Welz, 2008; Hampton, 2006, 2007; Leri, Anversa, & Frishman, 2007).

stem cells
Precursor (immature) cells that give birth to new specialized cells; a stem cell holds all the information it needs to make bone, blood, brain—any part of a human body—and can also copy itself to maintain a stock of stem cells.

A breakthrough in neuroscience FIGURE 2.10

By immobilizing the unaffected arm or leg and requiring rigorous and repetitive exercise of the affected limb, psychologist Edward Taub and colleagues "recruit" stroke patients' intact brain cells to take over for damaged cells. The therapy has restored function in some patients as long as 21 years after their strokes.

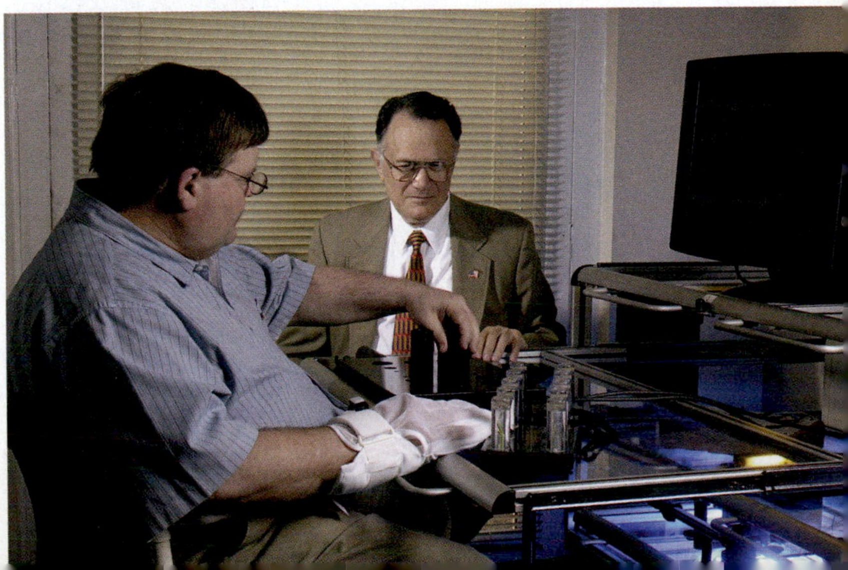

Does this mean that people paralyzed from spinal cord injuries might be able to walk again? At this point, neurogenesis in the brain and spinal cord is minimal. However, one possible bridge might be to transplant embryonic stem cells in the damaged area of the spinal cord. Researchers have transplanted mouse embryonic stem cells into a damaged rat spinal cord (Jones, Anderson, & Galvin, 2003; McDonald et al., 1999). When the damaged spinal cord was viewed several weeks later, the implanted cells had survived and spread throughout the injured spinal cord area. More important, the transplant rats also showed some movement in previously paralyzed parts of their bodies. Medical researchers have also begun human trials using nerve grafts to repair damaged spinal cords (Lopez, 2002; Saltus, 2000).

Now that we have discussed the remarkable adaptability of the central nervous system, let's take a closer look at its components: the brain and the spinal cord.

Because of its central importance for psychology and behavior, we'll discuss the brain in detail in the next major section. But the spinal cord is also important. Beginning at the base of the brain and continuing down the back, the spinal cord carries vital information from the rest of the body into and out of the brain.

But the spinal cord doesn't simply relay messages. It can also initiate some automatic behaviors on its own. We call these involuntary, automatic behaviors **reflexes** or **reflex arcs** because the response to the incoming stimuli is automatically "reflected" back (FIGURE 2.11).

We're all born with numerous reflexes, many of which fade over time. But even as adults, we still blink in response to a puff of air in our eyes, gag when something touches the back of the throat, and urinate and defecate in response to pressure in the bladder and rectum. Reflexes even influence our sexual responses. Certain stimuli, such as the stroking of the genitals, can lead to arousal and the reflexive muscle contractions of orgasm in both men and women. However, in order to have the passion, thoughts, and emotion we normally associate with sex, the sensory information from the stroking and orgasm must be carried to the brain.

The workings of the spinal cord FIGURE 2.11

Reflexes provide an evolutionary advantage. If messages had to travel all the way to the brain before they could be acted upon, an animal (human or nonhuman) might be fatally wounded in the meantime. Why do you think we have reflexes like the knee jerk?

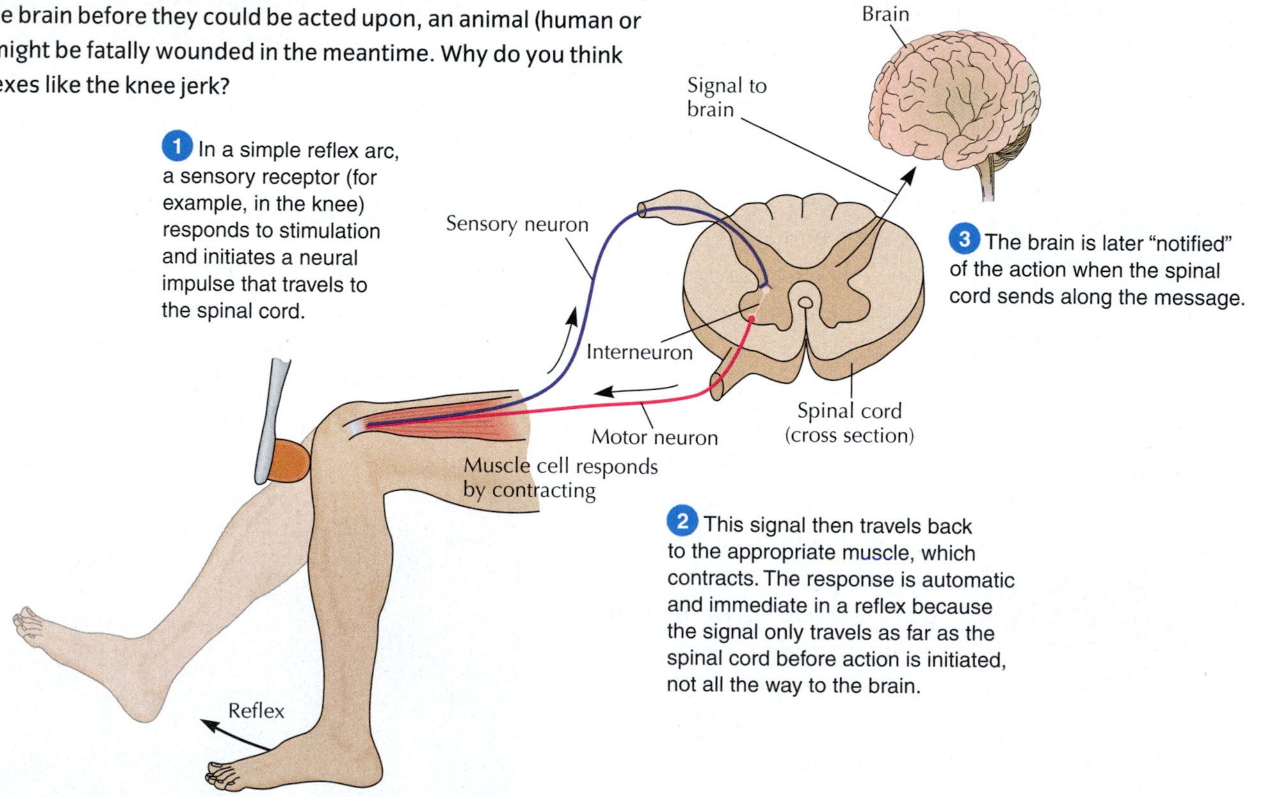
Brain
Signal to brain
1 In a simple reflex arc, a sensory receptor (for example, in the knee) responds to stimulation and initiates a neural impulse that travels to the spinal cord.
Sensory neuron
3 The brain is later "notified" of the action when the spinal cord sends along the message.
Interneuron
Spinal cord (cross section)
Motor neuron
Muscle cell responds by contracting
2 This signal then travels back to the appropriate muscle, which contracts. The response is automatic and immediate in a reflex because the signal only travels as far as the spinal cord before action is initiated, not all the way to the brain.
Reflex

Applying Psychology

Testing for Reflexes

If you have a newborn or young infant in your home, you can easily (and safely) test for these simple reflexes. (Note: Most infant reflexes disappear within the first year of life. If they reappear in later life, it generally indicates damage to the central nervous system.)

> **CRITICAL THINKING**
>
> **Here are two interesting questions:**
> 1. What might happen if infants lacked these reflexes?
> 2. Can you imagine why most infant reflexes disappear within the first year?

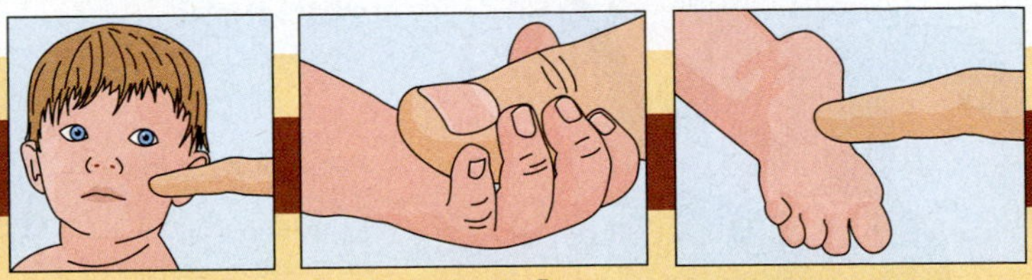

Rooting Reflex | Grasping Reflex | Babinski Reflex

A Rooting reflex. Lightly stroke the infant's cheek or side of the mouth, and watch how he or she will automatically (reflexively) turn toward the stimulation and attempt to suck.

B Grasping reflex. Place your finger in the infant's palm and note the automatic grasp.

C Babinski reflex. Lightly stroke the sole of the infant's foot, and the toes will fan out and the foot will twist inward.

PERIPHERAL NERVOUS SYSTEM (PNS): CONNECTING THE CNS TO THE REST OF THE BODY

> **somatic nervous system (SNS)** Subdivision of the peripheral nervous system (PNS). The SNS connects the sensory receptors and controls the skeletal muscles.

The chief function of the peripheral nervous system (PNS) is to carry information to and from the central nervous system. It links the brain and spinal cord to the body's sense receptors, muscles, and glands.

The PNS is subdivided into the somatic nervous system and the autonomic nervous system. In a kind of "two-way street," the **somatic nervous system (SNS)** (also called the skeletal nervous system) first carries sensory information to the CNS, and then carries messages from the CNS to skeletal muscles (FIGURE 2.12).

The other subdivision of the PNS is the **autonomic nervous system (ANS)**. The ANS is responsible for involuntary tasks, such as heart rate,

> **autonomic nervous system (ANS)** Subdivision of the peripheral nervous system (PNS) that controls involuntary functions. It includes the *sympathetic* nervous system and the *parasympathetic* nervous system.

Sensory and motor neurons FIGURE 2.12

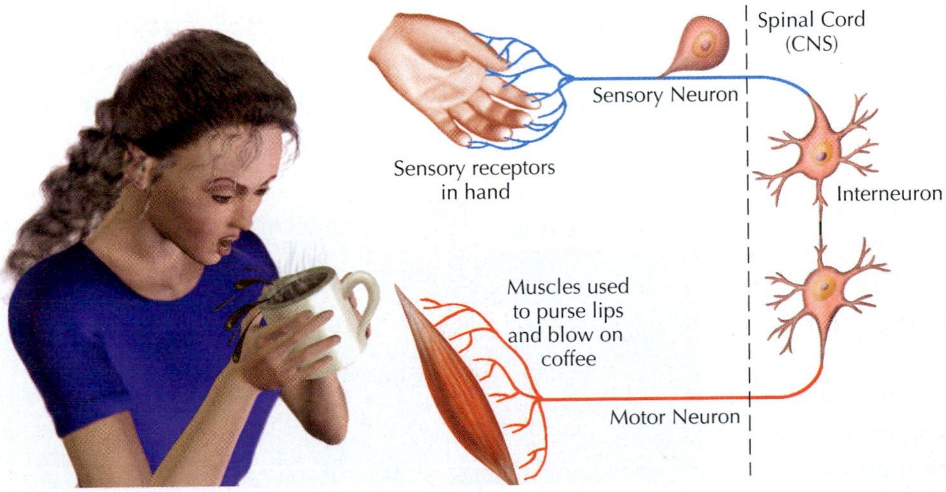

In order for you to be able to function, your brain must communicate with your body. This is the job of the somatic nervous system, which receives sensory information, sends it to the brain, and allows the brain to direct the body to act. Messages (action potentials) within the nervous system can cross the synapse in only one direction. Sensory neurons carry messages to the CNS. Motor neurons carry messages away from the CNS. Interneurons internally communicate and intervene between the sensory inputs and the motor outputs. Most of the neurons in the brain are interneurons.

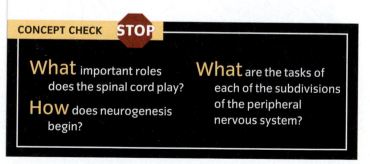

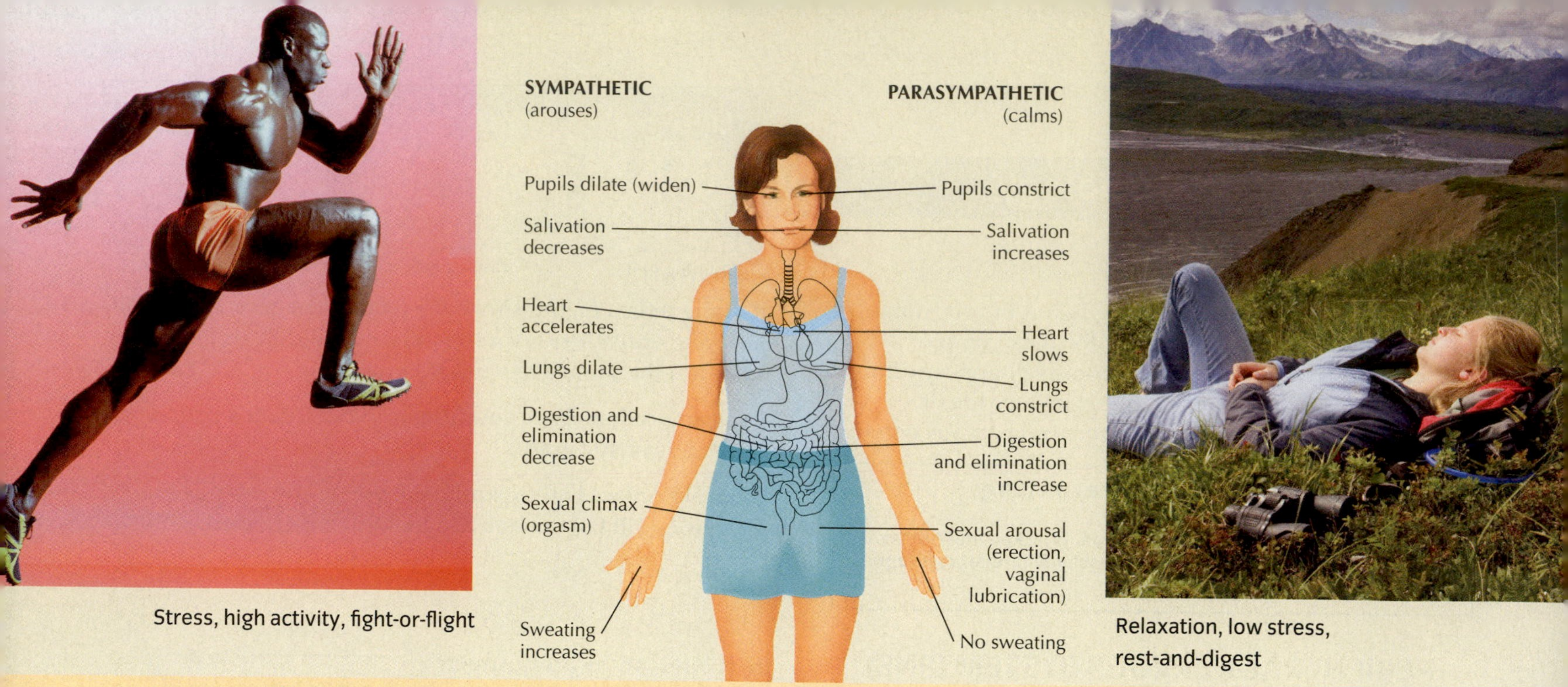

Actions of the autonomic nervous system (ANS) FIGURE 2.13

digestion, pupil dilation, and breathing. Like an automatic pilot, the ANS can sometimes be consciously overridden. But as its name implies, the autonomic system normally operates on its own (autonomously).

The autonomic nervous system is further divided into two branches, the sympathetic and parasympathetic, which tend to work in opposition to each other to regulate the functioning of such target organs as the heart, the intestines, and the lungs (FIGURE 2.13). Like two children on a teeter-totter, one will be up while the other is down, but they essentially balance each other out.

During stressful times, either mental or physical, the **sympathetic nervous system** mobilizes bodily resources to respond to the stressor. This emergency response is often called the "fight-or-flight" response. If you noticed a dangerous snake coiled and ready to strike, your sympathetic nervous system would increase your heart rate, respiration, and blood pressure, stop your digestive and eliminative processes, and release hormones, such as cortisol, into the bloodstream. The net result of sympathetic activation is to get more oxygenated blood and energy to the skeletal muscles, thus allowing you to cope with the stress—to "fight or flee."

In contrast to the sympathetic nervous system, the **parasympathetic nervous system** is responsible for returning your body to its normal functioning by slowing your heart rate, lowering your blood pressure, and increasing your digestive and eliminative processes.

The sympathetic nervous system provides an adaptive, evolutionary advantage. At the beginning of human evolution, when we faced a dangerous bear or an aggressive human intruder, there were only two reasonable responses—fight or flight. This automatic mobilization of bodily resources can still be critical even in modern times. However, less life-threatening events, such as traffic jams, also activate our sympathetic nervous system. Our bodies still respond to these sources of stress with sympathetic arousal. As the next chapter discusses, ongoing sympathetic system response to such chronic, daily stress can become detrimental to our health. *What a Psychologist Sees* describes a familiar example of the interaction between the sympathetic and parasympathetic nervous systems.

CONCEPT CHECK **STOP**

What important roles does the spinal cord play?

How does neurogenesis begin?

What are the tasks of each of the subdivisions of the peripheral nervous system?

Sexual Arousal

The complexities of sexual interaction—and in particular, the difficulties that couples sometimes have in achieving sexual arousal or orgasm—illustrate the balancing act between the sympathetic and parasympathetic nervous systems.

Parasympathetic dominance: Sexual arousal and excitement require that the body be relaxed enough to allow increased blood flow to the genitals—in other words, the nervous system must be in *parasympathetic dominance*. Parasympathetic nerves carry messages from the central nervous system directly to the sexual organs, allowing for a localized response (increased blood flow and genital arousal).

Sympathetic dominance: When a person experiences strong emotions, such as anger, anxiety, or fear, the body shifts to *sympathetic dominance*, which can disrupt the sexual response patterns that occur under parasympathetic dominance. During sympathetic dominance, impulses from the central nervous system are sent to transfer stations outside their end destinations; the transfer stations quickly relay the messages all around the body. Blood flow to the genitals and organs decreases as the body readies for a "fight or flight." As a result, the person is unable (or less likely) to become sexually aroused. Any number of circumstances—for example, performance anxiety, fear of unwanted pregnancy or disease, or tensions between partners—can trigger sympathetic dominance.

A Tour Through the Brain

NATIONAL GEOGRAPHIC

LEARNING OBJECTIVES

Identify the major structures of the hindbrain, midbrain, and forebrain, and of the cerebral cortex.

Summarize the major roles of the lobes of the cerebral cortex.

Describe what scientists have learned from split-brain research.

Explain why it's a mistake to believe that the right brain is usually "neglected."

Describe some examples of localization of function in the brain.

We begin our exploration of the brain at the lower end, where the spinal cord joins the base of the brain, and move upward toward the skull. As we move from bottom to top, "lower," basic processes, like breathing, generally give way to more complex mental processes (FIGURE 2.14 and FIGURE 2.15).

LOWER-LEVEL BRAIN STRUCTURES: THE HINDBRAIN, MIDBRAIN, AND PARTS OF THE FOREBRAIN

Brain size and complexity vary significantly from species to species. For example, fish and reptiles have smaller, less complex brains than do cats and dogs. The most complex brains belong to whales, dolphins, and higher primates such as chimps, gorillas, and humans. The billions of neurons that make up the human brain control much of what we think, feel, and do. Certain brain structures are specialized to perform certain tasks, a process known as **localization of function**. However, most parts of the brain perform integrating, overlapping functions.

localization of function Specialization of various parts of the brain for particular functions.

The hindbrain

You are deep in REM sleep as you begin your last dream of the night. Your vital signs increase as the dream gets more exciting. But then your sleep—and your dream—is shattered by a buzzing alarm clock. All your automatic behaviors and survival responses in this scenario are controlled or influenced by parts of the hindbrain. The **hindbrain** includes the medulla, pons, and cerebellum.

The **medulla** is essentially an extension of the spinal cord, with many nerve fibers passing through it carrying information to and from the brain. It also controls many essential automatic bodily functions, such as respiration and heart rate.

The **pons** is involved in respiration, movement, sleeping, waking, and dreaming (among other things). It also contains axons that cross from one side of the brain to the other (*pons* is Latin for "bridge").

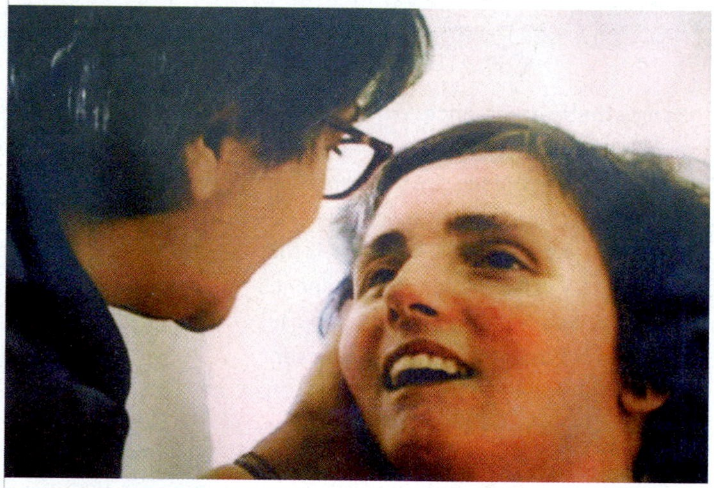

Damage to the brain FIGURE 2.14

The 2005 debate over Terri Schiavo was largely about whether her husband should be allowed to remove her feeding tube because she was still able to move and breathe on her own and showed some reflexive responses. Terri's parents and others believed that these lower-level brain functions were sufficient proof of life. Advocates on the other side felt that once the cerebral cortex ceases functioning, the "person" is dead and there is no ethical reason to keep the body alive. What do you think?

The human brain FIGURE 2.15

This drawing summarizes key functions of some of the brain's major structures. The brainstem, which includes parts of the hindbrain, midbrain, and forebrain, provides a handy geographical landmark.

VIEW THIS IN ACTION in your WileyPLUS course

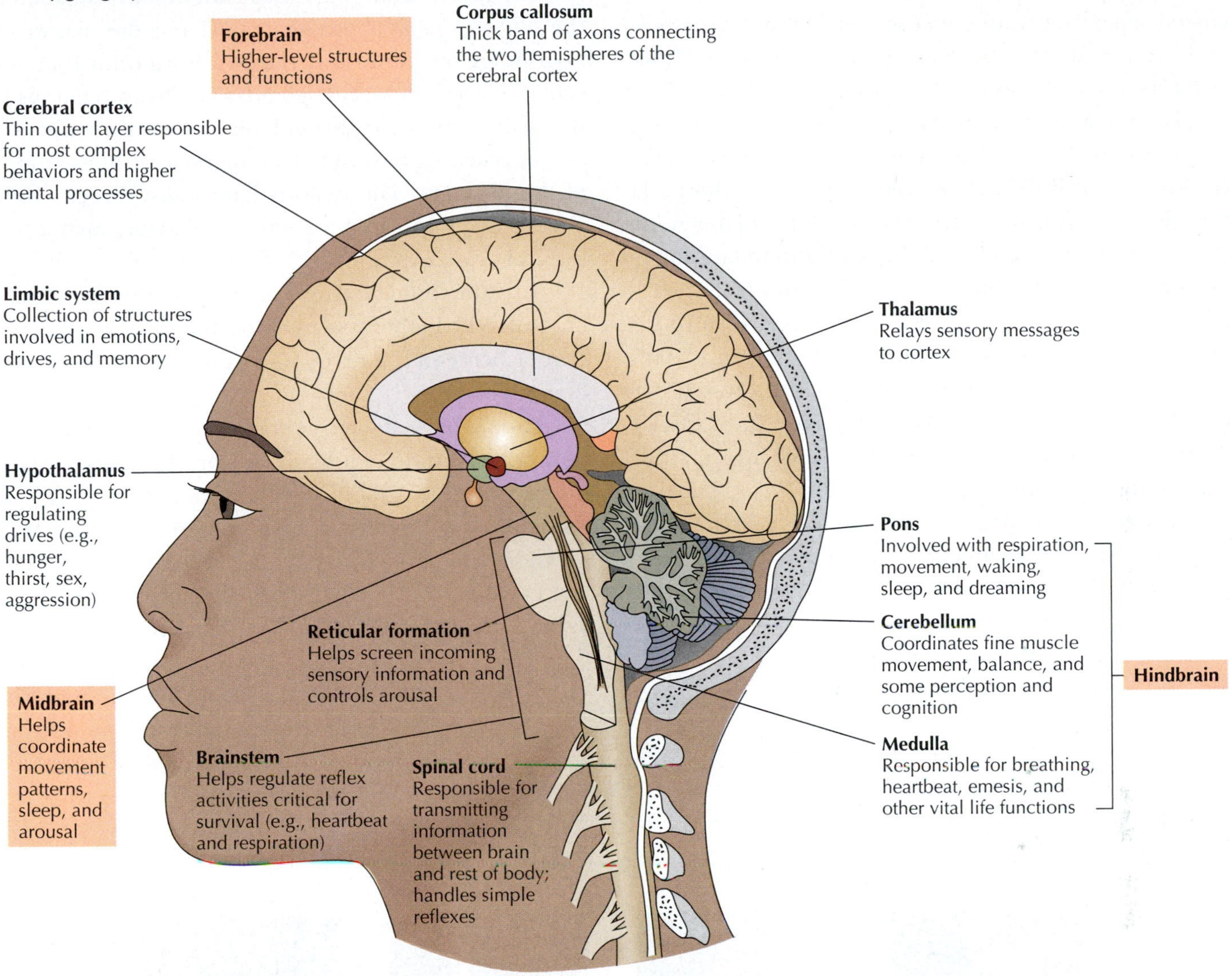

Forebrain
Higher-level structures and functions
Corpus callosum
Thick band of axons connecting the two hemispheres of the cerebral cortex
Cerebral cortex
Thin outer layer responsible for most complex behaviors and higher mental processes
Limbic system
Collection of structures involved in emotions, drives, and memory
Thalamus
Relays sensory messages to cortex
Hypothalamus
Responsible for regulating drives (e.g., hunger, thirst, sex, aggression)
Pons
Involved with respiration, movement, waking, sleep, and dreaming
Reticular formation
Helps screen incoming sensory information and controls arousal
Cerebellum
Coordinates fine muscle movement, balance, and some perception and cognition
Hindbrain
Midbrain
Helps coordinate movement patterns, sleep, and arousal
Medulla
Responsible for breathing, heartbeat, emesis, and other vital life functions
Brainstem
Helps regulate reflex activities critical for survival (e.g., heartbeat and respiration)
Spinal cord
Responsible for transmitting information between brain and rest of body; handles simple reflexes

The cauliflower-shaped **cerebellum** ("little brain" in Latin) is, evolutionarily, a very old structure. It coordinates fine muscle movement and balance. Researchers using functional magnetic resonance imaging (fMRI) have also shown that parts of the cerebellum are active during perceptual, cognitive, and language tasks (Picard et al., Rönnberg et al., 2004; Thompson, 2005; Woodruff-Pak & Disterhoft, 2008).

The midbrain The **midbrain** helps us orient our eye and body movements to visual and auditory stimuli, and works with the pons to help control sleep and level of arousal. It also contains a small structure involved with the neurotransmitter dopamine, which deteriorates in Parkinson's disease.

Running through the core of the hindbrain, midbrain, and brainstem is the **reticular formation** (RF). This diffuse, finger-shaped network of neurons filters incoming sensory information and alerts the higher brain centers to important events. Without your reticular formation, you would not be alert or perhaps even conscious.

The forebrain The **forebrain** is the largest and most prominent part of the human brain. It includes the thalamus, hypothalamus, limbic system, and cerebral cortex (FIGURE 2.16). The first three structures are located near the top of the brainstem. The cerebral cortex (discussed separately in the next section) is wrapped above and around them. (*Cerebrum* is Latin for "brain," and *cortex* is Latin for "covering" or "bark.")

The **thalamus** integrates input from the senses, and it may also be involved in learning and memory (Bailey & Mair, 2005; Ridley et al., 2005). Think of the thalamus as an air traffic control center that receives information from all aircraft and directs them to landing or takeoff areas. The thalamus receives input from nearly all sensory systems and directs the information to the appropriate cortical areas. Because the thalamus is the brain's major sensory relay center to the cerebral cortex, damage or abnormalities might cause the cortex to misinterpret or not receive vital sensory information. Interestingly, brain-imaging research links thalamus abnormalities to schizophrenia, a serious psychological disorder involving problems with sensory filtering and perception (Byne et al., 2008; Clinton & Meador-Woodruff, 2004; Preuss et al., 2005).

Beneath the thalamus lies the kidney bean-sized **hypothalamus** (*hypo-* means "under"). It has been called the "master control center" for emotions and many basic motives such as hunger, thirst, sex, and aggression (Hinton et al., 2004; Williams et al., 2004; Zillmer, Spiers, & Culbertson, 2008). It regulates the body's internal environment, including temperature control, which it accomplishes by regulating the endocrine system. Hanging down from the hypothalamus, the *pituitary gland* is usually considered the master endocrine gland because it releases hormones that activate the other endocrine glands. The hypothalamus influences the pituitary through direct neural connections and by releasing its own hormones into the blood supply of the pituitary. The hypothalamus also directly influences some important aspects of behavior, such as eating and drinking patterns.

An interconnected group of forebrain structures, known as the **limbic system**, is located roughly along the border between the cerebral cortex and the lower-level brain structures.

The limbic system is generally responsible for emotions, drives, and memory. However, the major focus of interest in the limbic system, and particularly the **amygdala**, has been its production and regulation of aggression and fear (Asghar et al., 2008; Carlson, 2008; LeDoux, 1998, 2002, 2007). Another well-known function of the limbic system is its role in pleasure or reward (Dackis & O'Brien, 2001; Olds & Milner, 1954). Even though limbic system structures and neurotransmitters are instrumental in emotional behavior, the cerebral cortex also tempers emotion in humans.

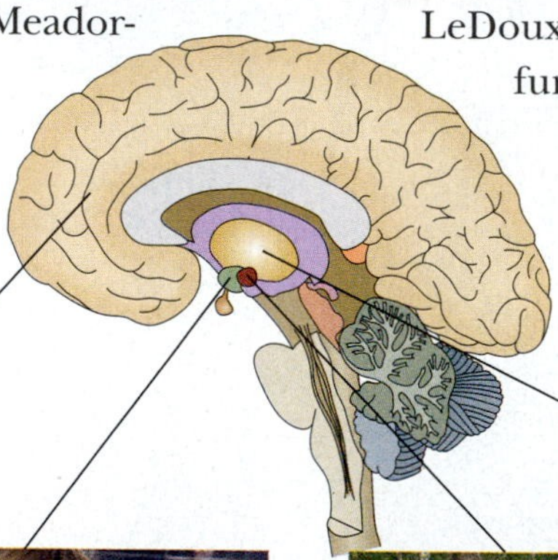

Cerebral cortex Governs higher mental processes

Hypothalamus Controls basic drives, such as hunger

Limbic system Involved in emotions, drives, and memory

Thalamus Integrates input from the senses

Structures of the forebrain FIGURE 2.16

THE CEREBRAL CORTEX: THE CENTER OF "HIGHER" PROCESSING

cerebral cortex Thin surface layer on the cerebral hemispheres that regulates most complex behavior, including receiving sensations, motor control, and higher mental processes.

The gray, wrinkled **cerebral cortex** is responsible for most complex behaviors and higher mental processes. It plays such a vital role that many consider it the essence of life. In fact, physicians may declare a person legally dead when the cortex dies, even when the lower-level brain structures and the rest of the body are fully functioning.

Although the cerebral cortex is only about one-eighth of an inch thick, it's made up of approximately 30 billion neurons and nine times as many glial cells. It contains numerous "wrinkles" called *convolutions* (think of a crumpled-up newspaper), which allow it to fit in the restricted space of the skull.

The full cerebral cortex and the two cerebral hemispheres beneath it closely resemble an oversized walnut. The division, or *fissure,* down the center marks the left and right *hemispheres* of the brain, which make up about 80 percent of the brain's weight. They are mostly filled with axon connections between the cortex and the other brain structures. Each hemisphere controls the opposite side of the body.

The cerebral hemispheres are divided into eight distinct areas or lobes—four in each hemisphere (FIGURE 2.17). Like the lower-level brain structures, each lobe specializes in somewhat different tasks—another example of localization of function. However, some functions overlap between lobes.

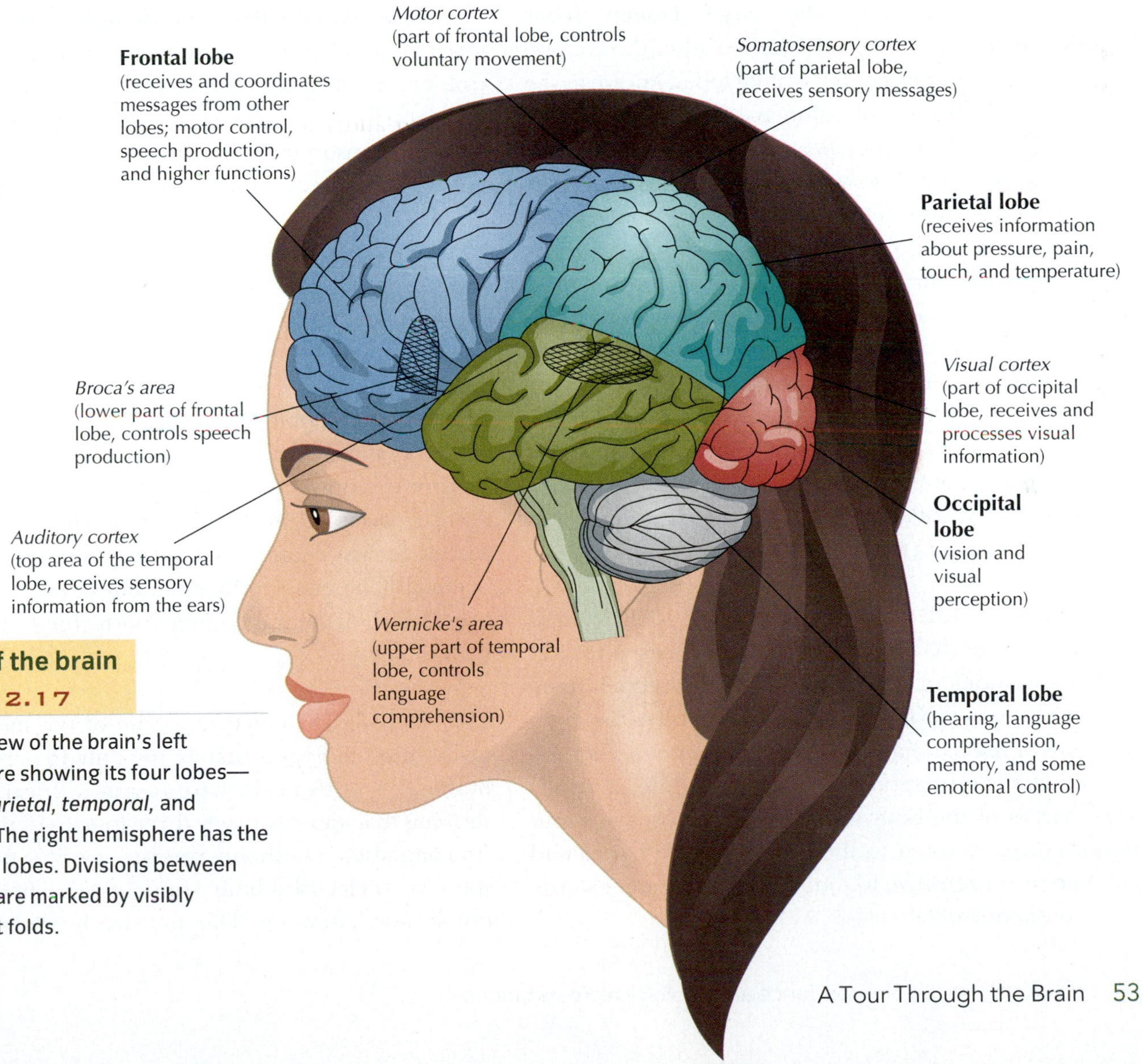

Lobes of the brain
FIGURE 2.17

This is a view of the brain's left hemisphere showing its four lobes—*frontal, parietal, temporal,* and *occipital*. The right hemisphere has the same four lobes. Divisions between the lobes are marked by visibly prominent folds.

Accidental proof of specialized brain functions? FIGURE 2.18

In 1998, construction worker Travis Bogumill was accidentally shot with a nail gun near the rear of his right frontal lobe. Remarkably, Bogumill experienced only an impaired ability to perform complex mathematical problems. This case supports experimental research showing that the frontal lobes and short-term memory are responsible for reasoning, problem solving, mathematical calculation, and thinking about future rewards or actions (Carlson, 2008; Hill, 2004; Neubauer et al., 2004).

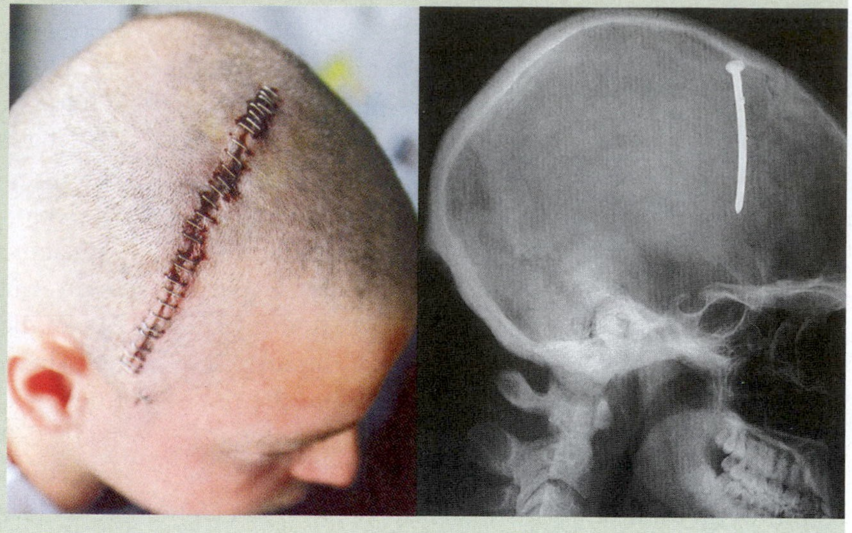

The frontal lobes

The large **frontal lobes** coordinate messages received from the other three lobes. An area at the very back of the frontal lobes, known as the *motor cortex,* instigates all voluntary movement. In the lower left frontal lobe lies *Broca's area.* In 1865, French physician Paul Broca discovered that damage to this area causes difficulty in speech, but not language comprehension. This type of impaired language ability is known as *Broca's aphasia.* Finally, the frontal lobes control most higher functions that distinguish humans from other animals, such as thinking, personality, emotion, and memory. Abnormalities in the frontal lobes are often observed in patients with schizophrenia (Chapter 13). And as the case of Phineas Gage (Chapter 1) and other research indicate, damage to the frontal lobe affects motivation, drives, creativity, self-awareness, initiative, reasoning, and emotional behavior (FIGURE 2.18).

The parietal lobes, the temporal lobes, and the occipital lobes

The **parietal lobes** interpret bodily sensations including pressure, pain, touch, temperature, and location of body parts. A band of tissue on the front of the parietal lobe, called the *somatosensory cortex,* receives information about touch in different body areas. Areas of the body with more somatosensory and motor cortex devoted to them (such as the hands and face) are most sensitive to touch and have the most precise motor control (FIGURE 2.19).

The **temporal lobes** are responsible for hearing, language comprehension, memory, and some emotional control. The *auditory cortex* (which processes sound) is located at the top front of each temporal lobe. This area processes incoming sensory information and sends it to the parietal lobes, where it is combined with other sensory information.

An area of the left temporal lobe, *Wernicke's area,* is involved in language comprehension. About a decade after Broca's discovery, German neurologist Carl Wernicke noted that patients with damage in this area could not understand what they read or heard, but they could speak quickly and easily. However, their speech was often unintelligible because it contained made-up words, sound substitutions, and word substitutions. This syndrome is now referred to as *Wernicke's aphasia.*

The **occipital lobes** are responsible, among other things, for vision and visual perception. Damage to the occipital lobe can produce blindness, even though the eyes and their neural connection to the brain are perfectly healthy.

The association areas

One of the most popular myths in psychology is that we use only 10 percent of our brain. This myth might have begun with early research showing that approximately three-fourths of the cortex is "uncommitted" (with no precise, specific function responsive to electrical brain stimulation). These areas are not dormant, however. They are clearly involved in interpreting, integrating, and acting on information processed by other parts of the brain. They are called **association areas** because they associate, or connect, various areas and functions of the brain. The association areas in the frontal lobe, for example, help in decision making and planning. Similarly, the association area right in front of the motor cortex is involved in the planning of voluntary movement.

TWO BRAINS IN ONE? A HOUSE DIVIDED

We mentioned earlier that the brain's left and right cerebral hemispheres control opposite sides of the body. Each hemisphere also has separate areas of specialization. (This is another example of *localization* of function, yet it is technically referred to as *lateralization.*)

Early researchers believed that the right hemisphere was "subordinate" or "nondominant" to the left, with few special functions or abilities. In the 1960s, landmark research with *split-brain* patients began to change this view.

The primary connection between the two cerebral hemispheres is a thick, ribbon-like band of nerve fibers under the cortex called the **corpus callosum**. In some rare cases of severe epilepsy, when other forms of treatment have failed, surgeons cut the corpus callosum to stop the spread of epileptic seizures from one hemisphere to the other. Because this operation cuts the only direct communication link between the two hemispheres, it reveals what each half of the brain can do in isolation from the other. The resulting research has profoundly improved our understanding of how the two halves of the brain function.

Body representation of the motor cortex and somatosensory cortex FIGURE 2.19

This drawing represents a vertical cross section taken from the left hemisphere's motor cortex and right hemisphere's somatosensory cortex. If body areas were truly proportional to the amount of tissue on the motor and somatosensory cortices, our bodies would look like the oddly shaped human figures draped around the outside edge of the cortex.

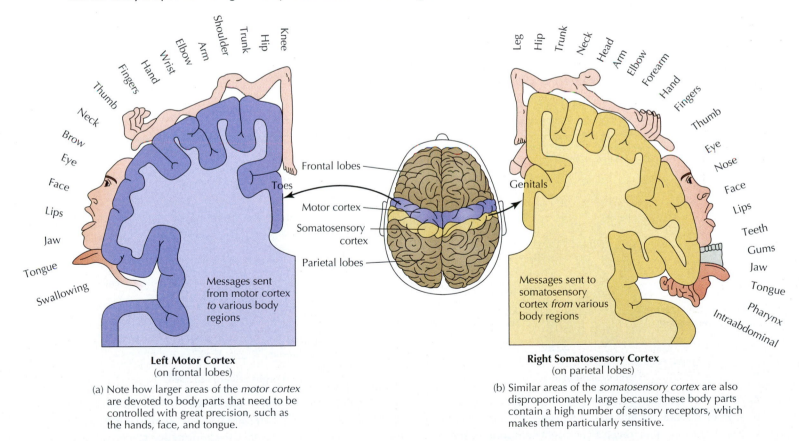

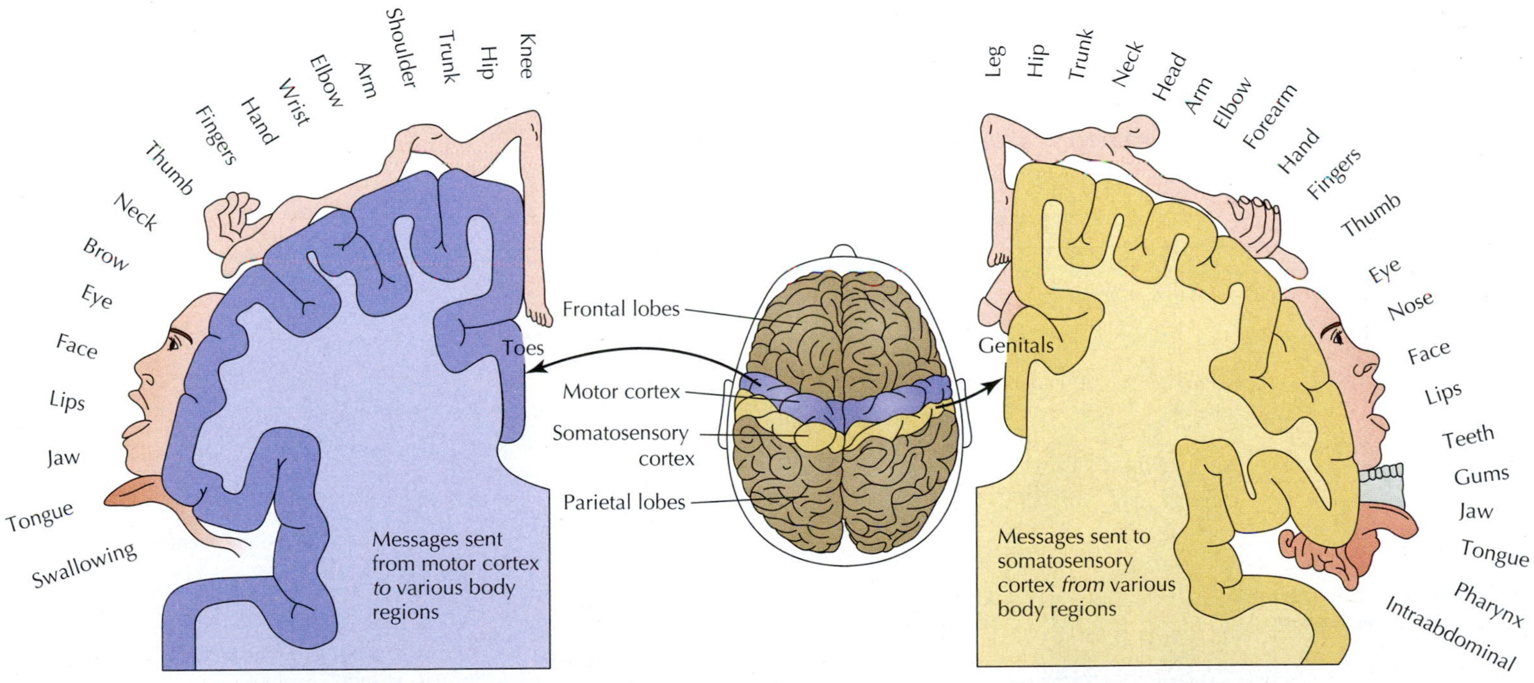

Left Motor Cortex
(on frontal lobes)

(a) Note how larger areas of the *motor cortex* are devoted to body parts that need to be controlled with great precision, such as the hands, face, and tongue.

Right Somatosensory Cortex
(on parietal lobes)

(b) Similar areas of the *somatosensory cortex* are also disproportionately large because these body parts contain a high number of sensory receptors, which makes them particularly sensitive.

Visualizing

Split-brain research FIGURE 2.20

Experiments on split-brain patients often present visual information to only the patient's left or right hemisphere, which leads to some intriguing results. For example,

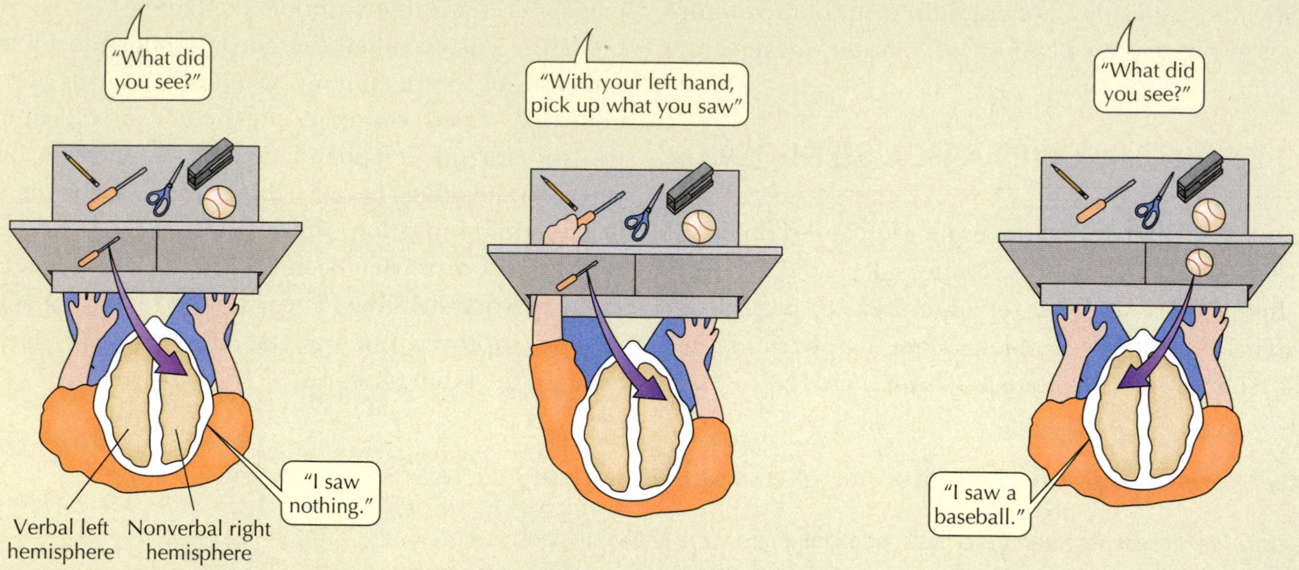

A When a split-brain patient is asked to stare straight ahead while a photo of a screwdriver is flashed only to the right hemisphere, he will report that he "saw nothing."

B However, when asked to pick up with his left hand what he saw, he can reach through and touch the items hidden behind the screen and easily pick up the screwdriver.

C When the left hemisphere receives an image of a baseball, the split-brain patient can easily name it.

Assuming you have an intact, nonsevered corpus callosum, if the same photos were presented to you in the same way, you could easily name both the screwdriver and the baseball. Can you explain why? The answers lie in our somewhat confusing visual wiring system:

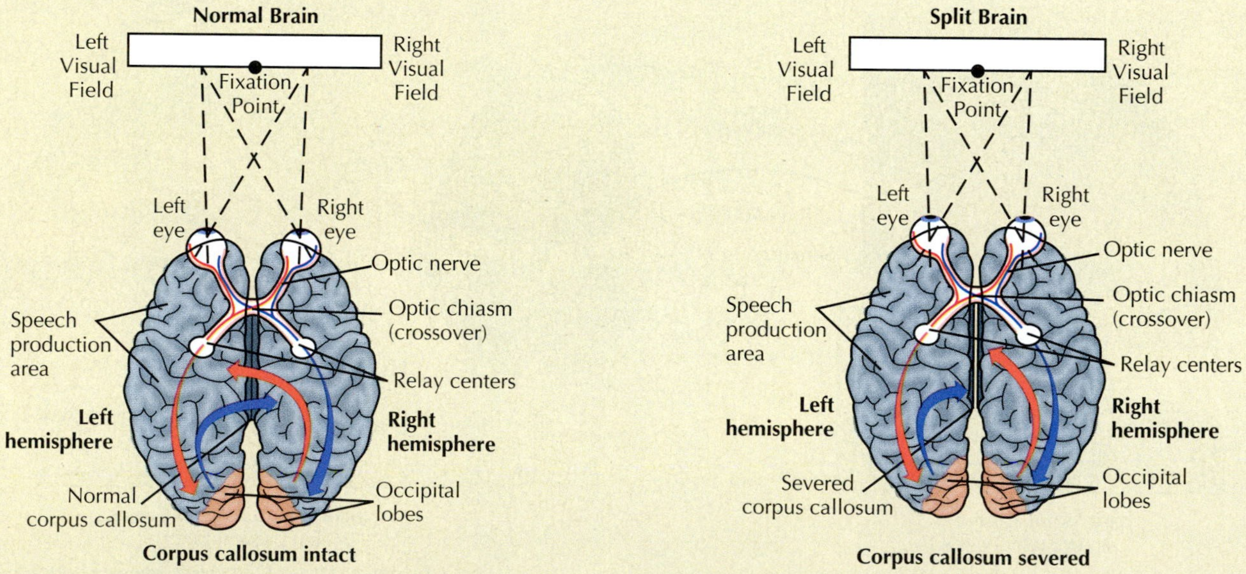

D As you can see, our eyes connect to our brains in such a way that, when we look straight ahead, information from the left visual field (the blue line) travels to our right hemisphere, and information from the right visual field (the red line) travels to our left hemisphere. The messages received by either hemisphere are then quickly sent to the other across the corpus callosum.

E When the corpus callosum is severed, and information is presented only to the right hemisphere, a split-brain patient cannot verbalize what he sees because the information cannot travel to the opposite (verbal) hemisphere.

If you met and talked with a split-brain patient, you probably wouldn't even know he or she had had the operation. The subtle changes in split-brain patients normally appear only with specialized testing (FIGURE 2.20).

Dozens of studies on split-brain patients, and newer research on people whose brains are intact, have documented several differences between the two brain hemispheres (FIGURE 2.21). In one study, researchers reported that different aspects of personality appear in the different hemispheres. For example, in one patient, the right hemisphere seemed more disturbed by childhood memories of being bullied than did the left (Schiffer et al., 1998).

Interestingly, left and right brain specialization is not usually reversed in left-handed people. About 68 percent of left-handers and 97 percent of right-handers have their major language areas on the left hemisphere. This suggests that even though the right side of the brain is dominant for movement in left-handers, other skills are often localized in the same brain areas as for right-handers.

What about the popular conception of the "neglected right brain?" Courses and books directed at "right-brain thinking" often promise to increase your intuition, creativity, and artistic abilities by waking up your "neglected" and "underused" right brain (e.g., Bragdon & Gamon, 1999; Edwards, 1999). This myth of the neglected right brain arose from popularized accounts of split-brain patients and exaggerated claims and unwarranted conclusions about differences between the left and right hemispheres. The fact is that the two hemispheres work together in a coordinated, integrated way, with each making important contributions.

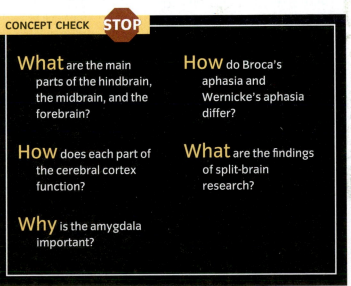

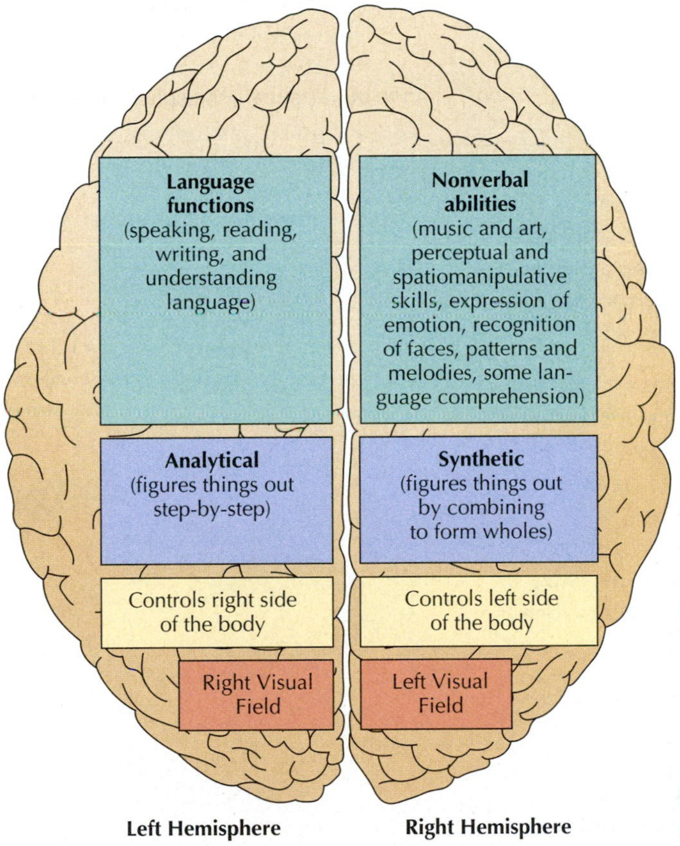

Functions of the left and right hemispheres

FIGURE 2.21

The left hemisphere specializes in verbal and analytical functions. The right hemisphere focuses on nonverbal abilities, such as spatiomanipulative skills, art and musical abilities, and visual recognition tasks. Keep in mind that both hemispheres are activated when we perform almost any task or respond to any stimuli. If your brain had only one hemisphere, would you want it to be the right hemisphere or the left?

CONCEPT CHECK STOP

What are the main parts of the hindbrain, the midbrain, and the forebrain?

How does each part of the cerebral cortex function?

Why is the amygdala important?

How do Broca's aphasia and Wernicke's aphasia differ?

What are the findings of split-brain research?

SUMMARY

1 Our Genetic Inheritance

1. **Neuroscience** studies how biological processes relate to behavioral and mental processes.
2. Genes (dominant or recessive) hold the code for inherited traits. Scientists use **behavioral genetics** methods to determine the relative influences of heredity and environment (heritability) on complex traits.
3. **Evolutionary psychology** suggests that many behavioral commonalities emerged and remain in human populations through natural selection, because they helped ensure our "genetic survival."

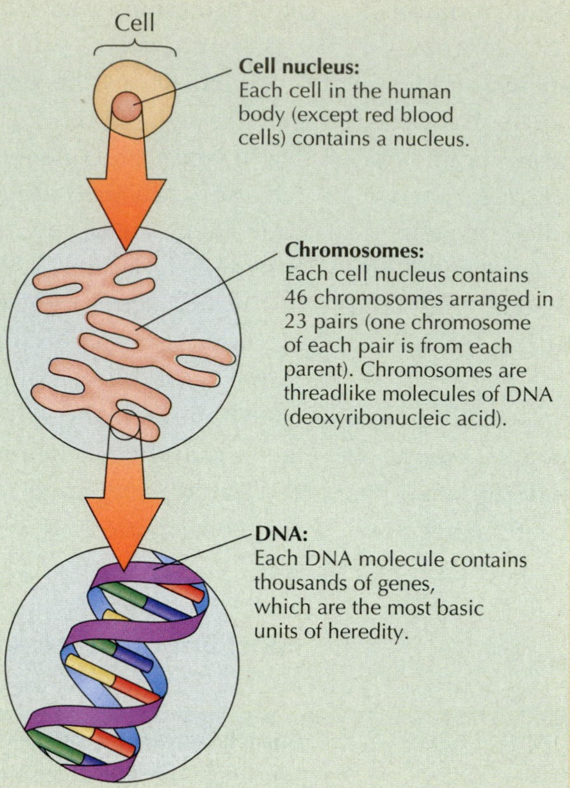

Cell
Cell nucleus:
Each cell in the human body (except red blood cells) contains a nucleus.
Chromosomes:
Each cell nucleus contains 46 chromosomes arranged in 23 pairs (one chromosome of each pair is from each parent). Chromosomes are threadlike molecules of DNA (deoxyribonucleic acid).
DNA:
Each DNA molecule contains thousands of genes, which are the most basic units of heredity.

3 Nervous System Organization

1. The **central nervous system (CNS)** includes the brain and spinal cord. The CNS allows us to process information and adapt to our environment in ways that no other animal can. The spinal cord transmits information between the brain and the rest of the body, and initiates involuntary reflexes. Although the CNS is very fragile, recent research shows that the brain is capable of lifelong **neuroplasticity** and **neurogenesis**. Neurogenesis is made possible by **stem cells**.

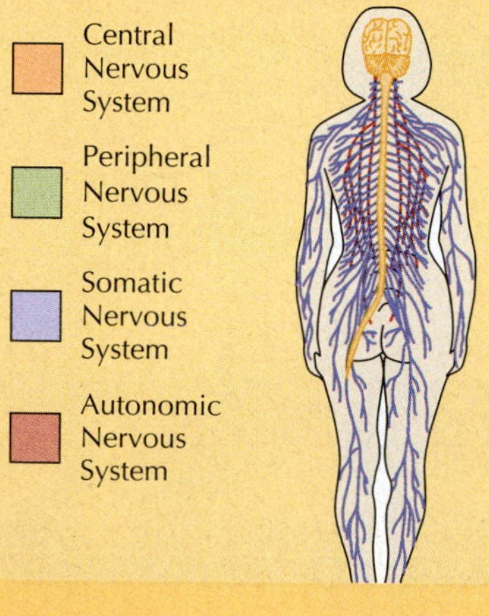

Central Nervous System
Peripheral Nervous System
Somatic Nervous System
Autonomic Nervous System

2. The **peripheral nervous system (PNS)** includes all the nerves outside the brain and spinal cord. It links the brain and spinal cord to the body's sense receptors, muscles, and glands. The PNS is subdivided into the **somatic nervous system (SNS)** and the **autonomic nervous system (ANS)**.
3. The ANS includes the sympathetic nervous system and the parasympathetic nervous system. The sympathetic nervous system mobilizes the body's "fight-or-flight" response. The parasympathetic nervous system returns the body to its normal functioning.

2 Neural Bases of Behavior

1. **Neurons**, supported by glial cells, receive and send electrochemical signals to other neurons and to the rest of the body. Their major components are dendrites, a cell body, and an axon.
2. Within a neuron, a neural impulse, or action potential, moves along the axon. An action potential is an all-or-none event.
3. Neurons communicate with each other using **neurotransmitters**, which are released at the synapse and attach to the receiving neuron. Neurons receive input from many synapses, some excitatory and some inhibitory. Hundreds of different neurotransmitters regulate a wide variety of physiological processes. Many poisons and drugs act by mimicking or interfering with neurotransmitters.
4. The endocrine system uses **hormones** to broadcast messages throughout the body. The system regulates long-term bodily processes, maintains ongoing bodily processes, and controls the body's response to emergencies.

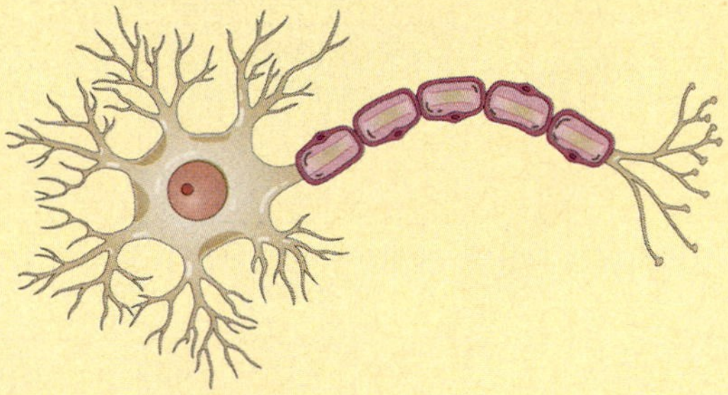

4 A Tour Through the Brain

1. The brain is divided into the hindbrain, the midbrain, and the forebrain. The brainstem includes parts of each of these. Certain brain structures are specialized to perform certain tasks **(localization of function)**.

2. The hindbrain (including the medulla, pons, and cerebellum) controls automatic behaviors and survival responses.

3. The midbrain helps us orient our eye and body movements, helps control sleep and arousal, and is involved with the neurotransmitter dopamine. The reticular formation runs through the core of the hindbrain, midbrain, and brainstem.

4. Forebrain structures (including the thalamus, hypothalamus, and limbic system) integrate input from the senses, control basic motives, regulate the body's internal environment, and regulate emotions, learning, and memory.

5. The **cerebral cortex**, part of the forebrain, governs most higher processing and complex behaviors. It is divided into two hemispheres, each controlling the opposite side of the body. The corpus callosum links the hemispheres. Each hemisphere is divided into frontal, parietal, temporal, and occipital lobes. Each lobe specializes in somewhat different tasks, but a large part of the cortex is devoted to integrating actions performed by different brain regions.

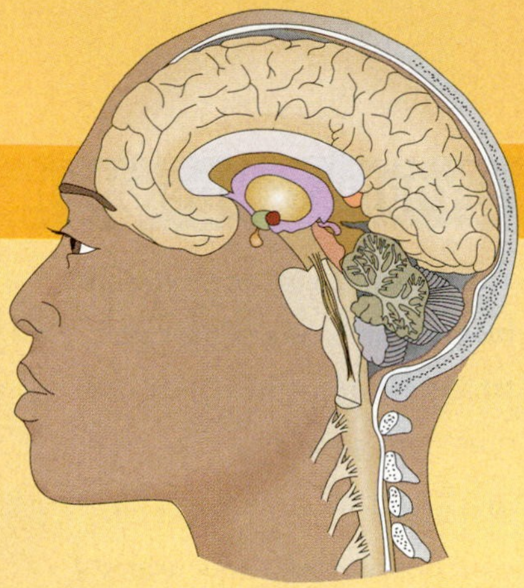

6. Split-brain research shows that each hemisphere performs somewhat different functions, although they work in close communication.

KEY TERMS

- **behavioral genetics** p. 34
- **evolutionary psychology** p. 34
- **neuroscience** p. 34
- **natural selection** p. 37
- **neuron** p. 37
- **glial cells** p. 37
- **dendrites** p. 37
- **cell body** p. 37
- **axon** p. 38
- **action potential** p. 38
- **myelin sheath** p. 38
- **neurotransmitters** p. 38
- **endorphins** p. 41
- **endocrine system** p. 43
- **hormones** p. 43
- **cortisol** p. 43
- **central nervous system (CNS)** p. 44
- **peripheral nervous system (PNS)** p. 44
- **neuroplasticity** p. 44
- **neurogenesis** p. 44
- **stem cells** p. 45
- **relexes/reflex arcs** p. 46
- **somatic nervous system (SNS)** p. 47
- **autonomic nervous system (ANS)** p. 47
- **sympathetic nervous system** p. 48
- **parasympathetic nervous system** p. 48
- **localization of function** p. 50
- **hindbrain** p. 50
- **medulla** p. 50
- **pons** p. 50
- **cerebellum** p. 51
- **midbrain** p. 51
- **reticular formation** p. 51
- **forebrain** p. 52
- **thalamus** p. 52
- **hypothalamus** p. 52
- **limbic system** p. 52
- **amygdala** p. 52
- **cerebral cortex** p. 53
- **frontal lobes** p. 54
- **parietal lobes** p. 54
- **temporal lobes** p. 54
- **occipital lobes** p. 54
- **association areas** p. 55
- **corpus callosum** p. 55

CRITICAL AND CREATIVE THINKING QUESTIONS

1. Imagine that scientists were able to identify specific genes linked to serious criminal behavior, and it was possible to remove or redesign these genes. Would you be in favor of this type of gene manipulation? Why or why not?

2. Why is it valuable for scientists to understand how neurotransmitters work at a molecular level?

3. What are some everyday examples of neuroplasticity—that is, of how the brain is changed and shaped by experience?

(See page 60 for *What is happening in this picture?*)

What is happening in this picture ?

The ability to curl one's tongue lengthwise is one of the few traits that depend on only one dominant gene.

- What does it mean if both of your parents are "noncurlers"?

Unlike tongue-curling ability, most traits are polygenic, meaning that they are controlled by more than one gene.

- Can you imagine why humans and other animals have evolved to possess such "complex traits"?

SELF-TEST

(Check your answers in Appendix B.)

1. Behavioral genetics is the study of ________________.
   a. the relative effects of behavior and genetics on survival
   b. the relative effects of heredity and environment on behavior and mental processes
   c. the relative effects of genetics on natural selection
   d. how genetics affects correct behavior

2. Evolutionary psychology studies ________________.
   a. the ways in which humans adapted their behavior to survive and evolve
   b. the ways in which humankind's behavior has changed over the millennia
   c. the ways in which humans can evolve to change behavior
   d. the ways in which natural selection and adaptation can explain behavior and mental processes

3. This is a measure of the degree to which a characteristic is related to genetic, inherited factors.
   a. heritability
   b. inheritance
   c. the biological ratio
   d. the genome statistic

4. The term ________________ refers to the evolutionary concept that those with adaptive genetic traits will live and reproduce.
   a. natural selection
   b. evolution
   c. survival of the fittest
   d. all of these options

5. Label the following parts of a neuron, the cell of the nervous system responsible for receiving and transmitting electrochemical information:
   a. dendrites
   b. cell body
   c. axon
   d. myelin sheath
   e. terminal buttons of axon

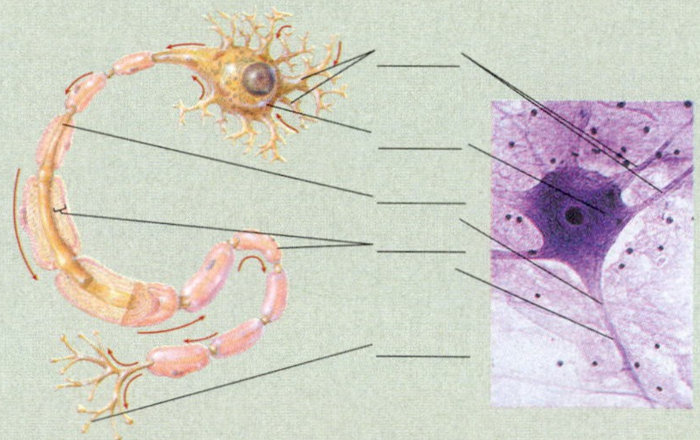

6. Your textbook's definition of an **action potential** is ________________.
   a. the likelihood that a neuron will take action when stimulated
   b. the tendency for a neuron to be potentiated by neurotransmitters
   c. a neural impulse that carries information along the axon of a neuron
   d. the firing of a nerve, either toward or away from the brain

7. Too much of this neurotransmitter may be related to schizophrenia, whereas too little of this neurotransmitter may be related to Parkinson's disease.
   a. acetylcholine
   b. dopamine
   c. norepinephrine
   d. serotonin

8. Chemicals that are manufactured by endocrine glands and circulated in the bloodstream to change or maintain bodily functions are called ________________.
   a. vasopressors
   b. gonadotropins
   c. hormones
   d. steroids

9. Label the main glands of the endocrine system:
   a. pineal
   b. pituitary
   c. adrenal--
   d. thyroid
   e. thymus

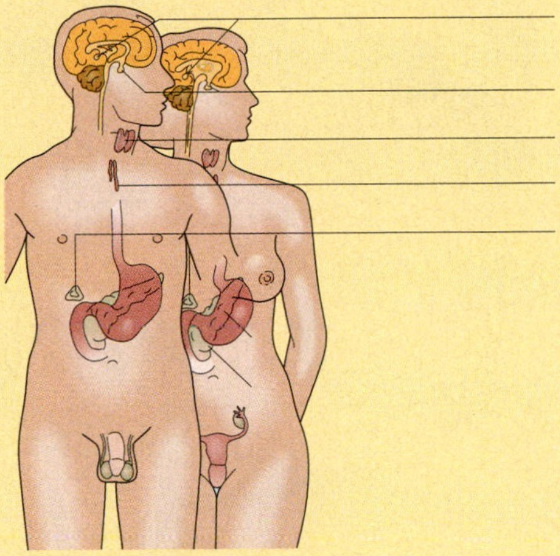

10. The central nervous system ________________.
    a. consists of the brain and spinal cord
    b. is the most important nervous system
    c. includes the automatic and other nervous systems
    d. all of these options

11. The peripheral nervous system ________________.
    a. is composed of the spinal cord and peripheral nerves
    b. is less important than the central nervous system
    c. is contained within the skull and spinal column
    d. includes all the nerves and neurons outside the brain and spinal cord

12. The ________________ nervous system is responsible for fight or flight, whereas the ________________ nervous system is responsible for maintaining calm.
    a. central; peripheral
    b. parasympathetic; sympathetic
    c. sympathetic; parasympathetic
    d. autonomic; somatic

13. Label the following structures/areas of the brain:
    a. forebrain
    b. midbrain
    c. hindbrain
    d. thalamus
    e. hypothalamus
    f. cerebral cortex

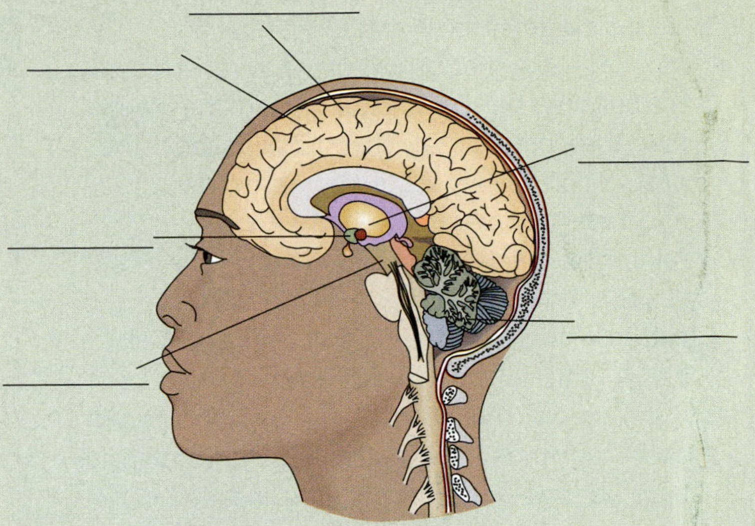

14. Label the four lobes of the brain:
    a. frontal lobe
    b. parietal lobe
    c. temporal lobe
    d. occipital lobe

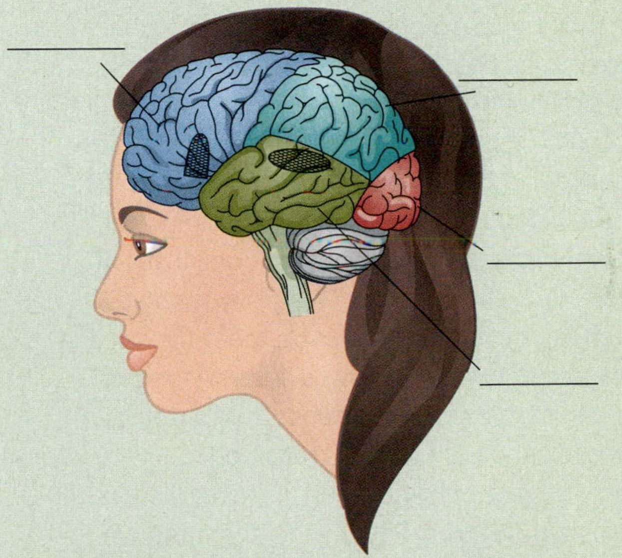

15. Identify the following functions of the brain as either left or right hemisphere.
    a. language function, ________________ hemisphere
    b. nonverbal abilities, ________________ hemisphere
    c. analytic skills, ________________ hemisphere
    d. control of the left side of the body, ________________ hemisphere

Stress and Health Psychology

3

Where were you on September 11, 2001? Even now, you can probably recall in crystalline detail the events of that Tuesday morning—the suddenness of the terrorist attacks, the buildings burning and tumbling to the ground, even the incongruously cheerful skies above the destruction. Many people who were directly affected—and even some who were not—remember the day and its aftermath as the most stressful time in their lives. How does such extreme stress affect people's health and well-being, both immediately and down the road? Is there any way that people can cope with stress of such magnitude? What about life's more mundane aggravations—rude drivers, long lines at the grocery store, nagging decisions about which job to take or college to attend? Do these also take a toll on our well-being? If so, what can we do about it?

Throughout most of history, people have understood that emotions and thoughts affect physical health. However, in the late 1800s, after discovering physiological causes for infectious diseases such as typhoid and syphilis, scientists began to focus primarily on the physiological causes of disease. Today, the major causes of death have shifted from contagious diseases (such as pneumonia, influenza, tuberculosis, and measles) to noncontagious diseases (such as cancer, cardiovascular disease, and chronic lung disease), and the focus has returned to psychological behaviors and lifestyles (Leventhal et al., 2008; Straub, 2007). In this chapter, we explore how biological, psychological, and social factors (the *biopsychosocial model*) affect illness as well as health and well-being.

NATIONAL GEOGRAPHIC

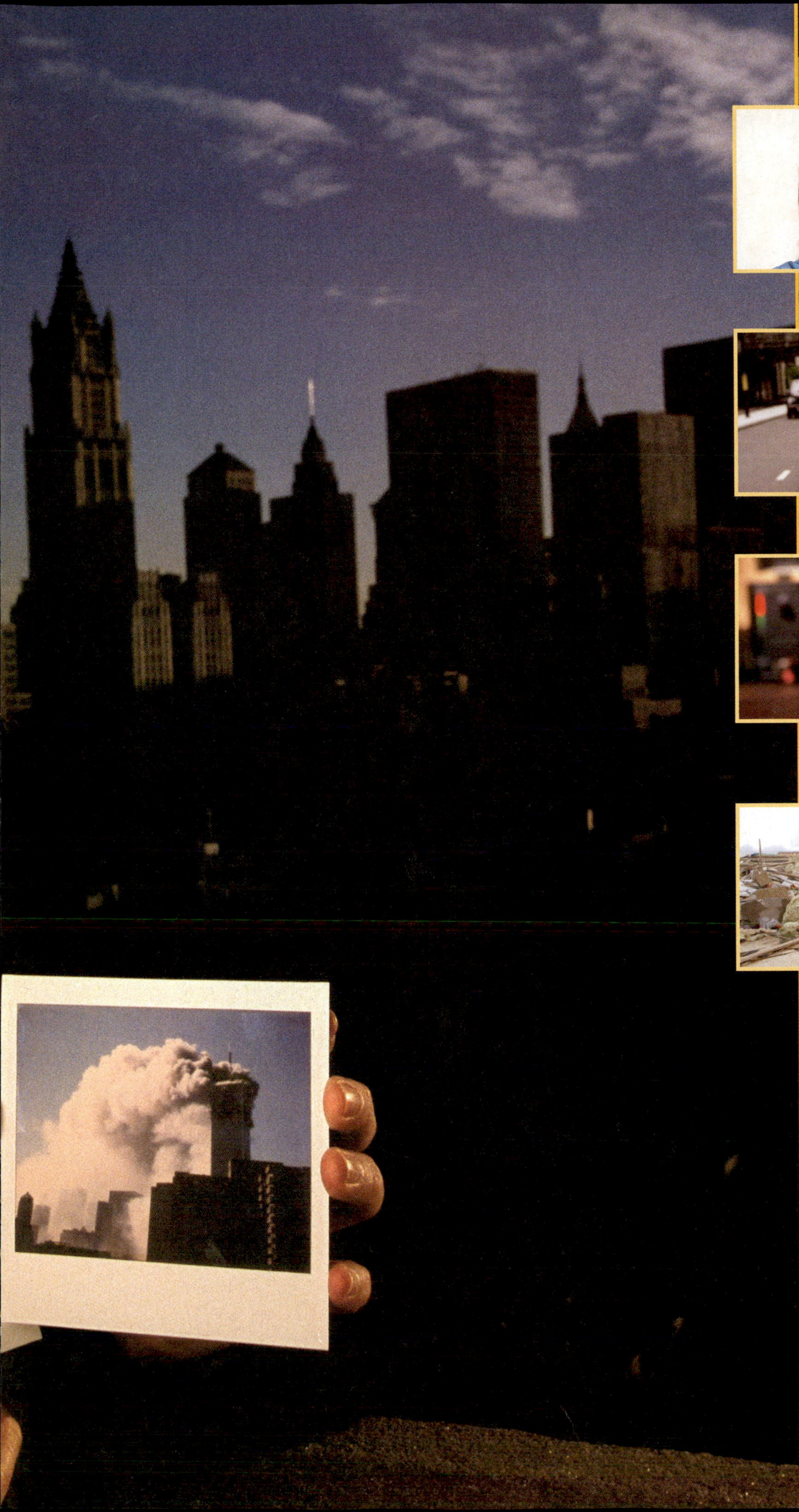

CHAPTER OUTLINE

Understanding Stress

LEARNING OBJECTIVES

Describe some common sources of stress.

Explain how the body responds to stress, immediately and over the long term.

Review the three phases of the general adaptation syndrome (GAS).

Anything that places a demand on the body can cause **stress**. The trigger that prompts the stressful reaction is called a **stressor**. Stress reactions can occur in response to either internal cognitive stimuli, external stimuli, or environmental stimuli (Sarafino, 2008; Straub, 2007).

Pleasant or beneficial stress, such as moderate exercise, is called **eustress**. Stress that is unpleasant or objectionable, as from chronic illness, is called **distress** (Selye, 1974). The total absence of stress would mean the total absence of stimulation, which would eventually lead to death. Because health psychology has been chiefly concerned with the negative effects of stress, we will adhere to convention and use the word "stress" to refer primarily to harmful or unpleasant stress.

stress The body's nonspecific response to any demand made on it; physical and mental arousal to situations or events that we perceive as threatening or challenging.

SOURCES OF STRESS

Although stress is pervasive in all our lives, psychological science has focused on seven major sources (FIGURE 3.1). For example, early stress researchers Thomas Holmes and Richard Rahe (1967) believed that any **life change** that required some adjustment in behavior or lifestyle could cause some degree of stress. They also believed that exposure to numerous stressful events within a short period could have a direct detrimental effect on health.

To investigate the relationship between change and stress, Holmes and Rahe created a Social Readjustment Rating Scale (SRRS) that asked people to check off all the life events they had experienced in the previous year (TABLE 3.1).

The SRRS scale is an easy and popular way to measure stress. Cross-cultural studies have shown that most people rank the magnitude of stressful events in similar ways (De Coteau, Hope, & Anderson, 2003; Scully, Tosi, & Banning, 2000). But the SRRS is not foolproof. For example, it only shows a correlation between stress and illness; it does not prove that stress actually causes illnesses. Moreover, not all stressful situations are **cataclysmic events**, such as a terrorist attack, or single events like a death or a birth. **Chronic stressors**, such as a bad marriage, poor working conditions, or an intolerable political climate, can be significant too. Even the stress of low-frequency noise is associated with measurable hormonal and cardiac changes (Waye et al., 2002). Our social lives can also be chronically stressful because making and maintaining friendships involves considerable thought and energy (Sias et al., 2004).

Given the recent global economic meltdown, one of our most pressing concerns is **job stress**, which includes unemployment, keeping or changing jobs, job performance, etc. (Moore, Grunberg, & Greenberg, 2004). The most stressful jobs are those that make great demands on performance and concentration but allow little creativity or opportunity for advancement (Smith et al., 2008; Straub, 2007).

Seven major sources of stress FIGURE 3.1

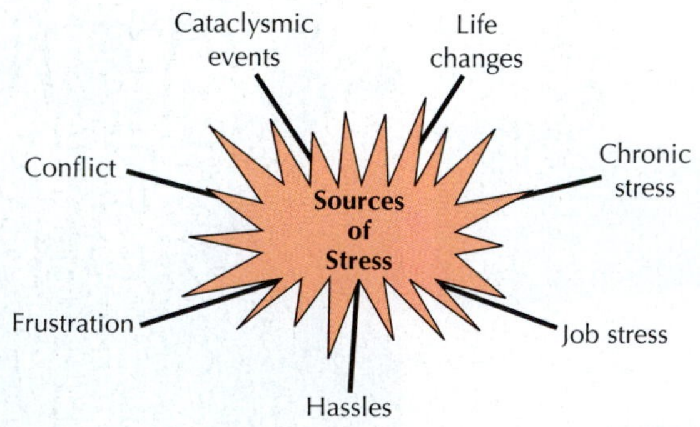

Measuring life changes TABLE 3.1

Social Readjustment Rating Scale

To score yourself on this scale, add up the "life change units" for all life events you have experienced during the last year and compare your score with the following standards: 0–150 = No significant problems; 150–199 = Mild life crisis (33 percent chance of illness); 200–299 = Moderate life crisis (50 percent chance of illness); 300 and above = Major life crisis (80 percent chance of illness).

Life events	Life change units	Life events	Life change units
Death of spouse	100	Son or daughter leaving home	29
Divorce	73	Trouble with in-laws	29
Marital separation	65	Outstanding personal achievement	28
Jail term	63	Spouse begins or stops work	26
Death of a close family member	63	Begin or end school	26
Personal injury or illness	53	Change in living conditions	25
Marriage	50	Revision of personal habits	24
Fired at work	47	Trouble with boss	23
Marital reconciliation	45	Change in work hours or conditions	20
Retirement	45	Change in residence	20
Change in health of family member	44	Change in schools	20
Pregnancy	40	Change in recreation	19
Sex difficulties	39	Change in church activities	19
Gain of a new family member	39	Change in social activities	18
Business readjustment	39	Mortgage or loan for lesser purchase (car, major appliance)	17
Change in financial state	38	Change in sleeping habits	16
Death of a close friend	37	Change in number of family get-togethers	15
Change to different line of work	36	Change in eating habits	15
Change in number of arguments with spouse	35	Vacation	13
Mortgage or loan for major purchase	31	Christmas	12
Foreclosure on mortgage or loan	30	Minor violations of the law	11
Change in responsibilities at work	29		

Source: Reprinted from *Journal of Psychosomatic Research*, Vol. III; Holmes and Rahe: "The Social Readjustment Rating Scale," 213–218, 1967, with permission from Elsevier.

Stress at work can also cause serious stress at home, not only for the worker but for other family members as well. In our private lives, divorce, child and spousal abuse, alcoholism, and money problems can place severe stress on all members of a family (Aboa-Éboulé, 2008; Abou-Éboulé et al. 2007; Ort-Gomér, 2007).

In addition to chronic stressors, the minor **hassles** of daily living can pile up and become a major source of stress. We all share many hassles, such as time pressures and financial concerns. But our reactions to these hassles vary. Persistent hassles can lead to a form of physical, mental, and emotional exhaustion known as **burnout** (Sarafino, 2008; Tümkaya, 2007). (FIGURE 3.2).

Some authorities believe that hassles can be more significant than major life events in creating stress (Kraaij, Arensman, & Spinhoven, 2002; Kubiak et al., 2008). Divorce is extremely stressful, but it may be so because of the increased number of hassles—a change in finances, childcare arrangements, longer working hours, and so on.

Is nursing a stressful career? FIGURE 3.2

Over time, some people in chronically stressful professions who think of their job as a "calling" become emotionally drained and disillusioned and feel a loss of personal accomplishment—they "burn out." Burnout can cause more work absences, less productivity, and increased risk for physical problems. What other occupations might pose an especially high risk for burnout?

Study Organizer 3.1 Types of Conflict

Type	*Description*	*Example*
Approach–approach Conflict	Forced choice between two or more favorable alternatives. Either choice will have positive results; the requirement to choose is the source of stress.	You must choose between two jobs: one that will be inherently interesting and one that will look impressive on your résumé.
Avoidance–avoidance Conflict	Forced choice between two or more unpleasant alternatives that will lead to negative results no matter which choice is made.	You must choose between missing class and missing an important job interview.
Approach–avoidance Conflict	Forced choice between alternatives that will have both desirable and undesirable results, which generally leads to a great deal of ambivalence.	You want to spend more time in a close relationship, but that means you won't be able to see your old friends as much.

In the book (and film) *Sophie's Choice*, Sophie and her two children are sent to a German concentration camp. A soldier demands that Sophie give up either her daughter or her son, or else both children will be killed. Obviously, both alternatives will have tragic results. What kind of conflict does this example illustrate?

Like hassles, **frustration** can cause stress. And the more motivated we are, the more frustrated we are when our goals are blocked.

Finally, stress can arise when we experience **conflict**—that is, when we are forced to make a choice between at least two incompatible alternatives. There are three basic types of conflict as shown in STUDY ORGANIZER 3.1.

The longer any conflict exists or the more important the decision, the more stress a person will experience. Generally, approach–approach conflicts are the easiest to resolve and produce the least stress. Avoidance–avoidance conflicts, on the other hand, are usually the most difficult because all choices lead to unpleasant results.

HOW STRESS AFFECTS THE BODY

When mentally or physically stressed, your body undergoes several physiological changes. The **sympathetic nervous system**, or SAM system, the **HPA Axis**, and the **general adaptation syndrome (GAS)** control the most significant of these changes (FIGURES 3.3 and 3.4 on pages 67 and 68).

general adaptation syndome (GAS) Selye's three-stage (alarm, resistance, exhaustion) reaction to chronic stress.

Cortisol, a key element of the HPA Axis, plays a critical role in the long-term effects of stress. Prolonged

Stress—an interrelated system FIGURE 3.3

Under stress, the sympathetic nervous system prepares us for immediate action—to "fight or flee." Parts of the brain and endocrine system then kick in to maintain our arousal. How does this happen?

1. The **SAM system** (short for *Sympatho-Adreno-Medullary*) provides an initial, rapid-acting stress response thanks to cooperation between the sympathetic nervous system and the adrenal medulla.

2. The **HPA Axis** (short for the Hypothalamic-Pituitary-Adrenocortical system) responds more slowly but lasts longer. It also helps restore the body to its baseline state, *homeostasis*.

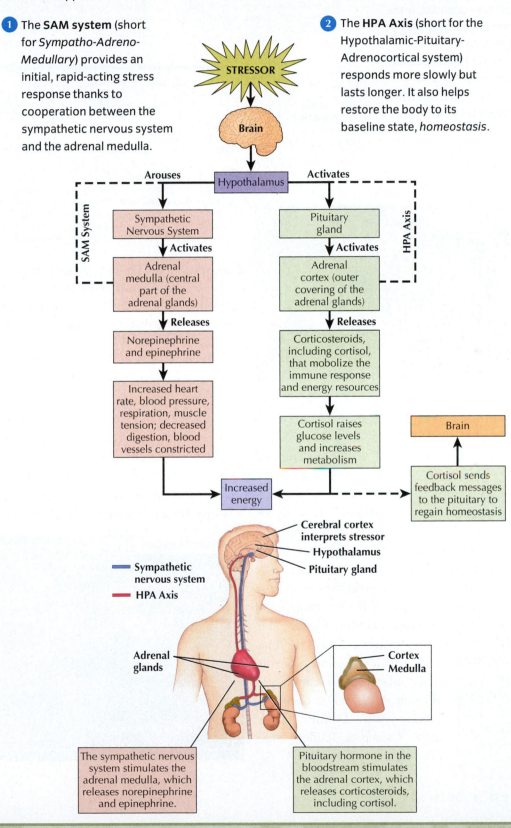

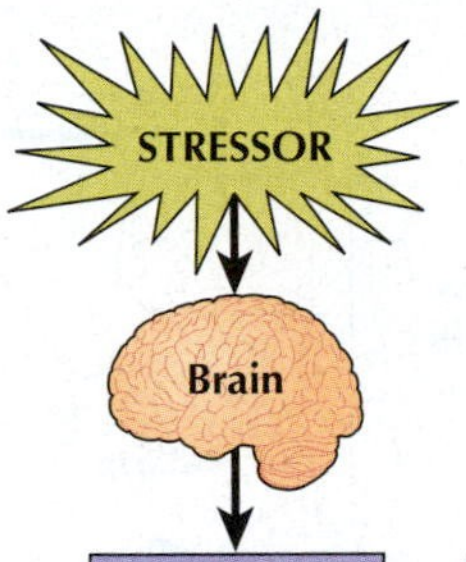

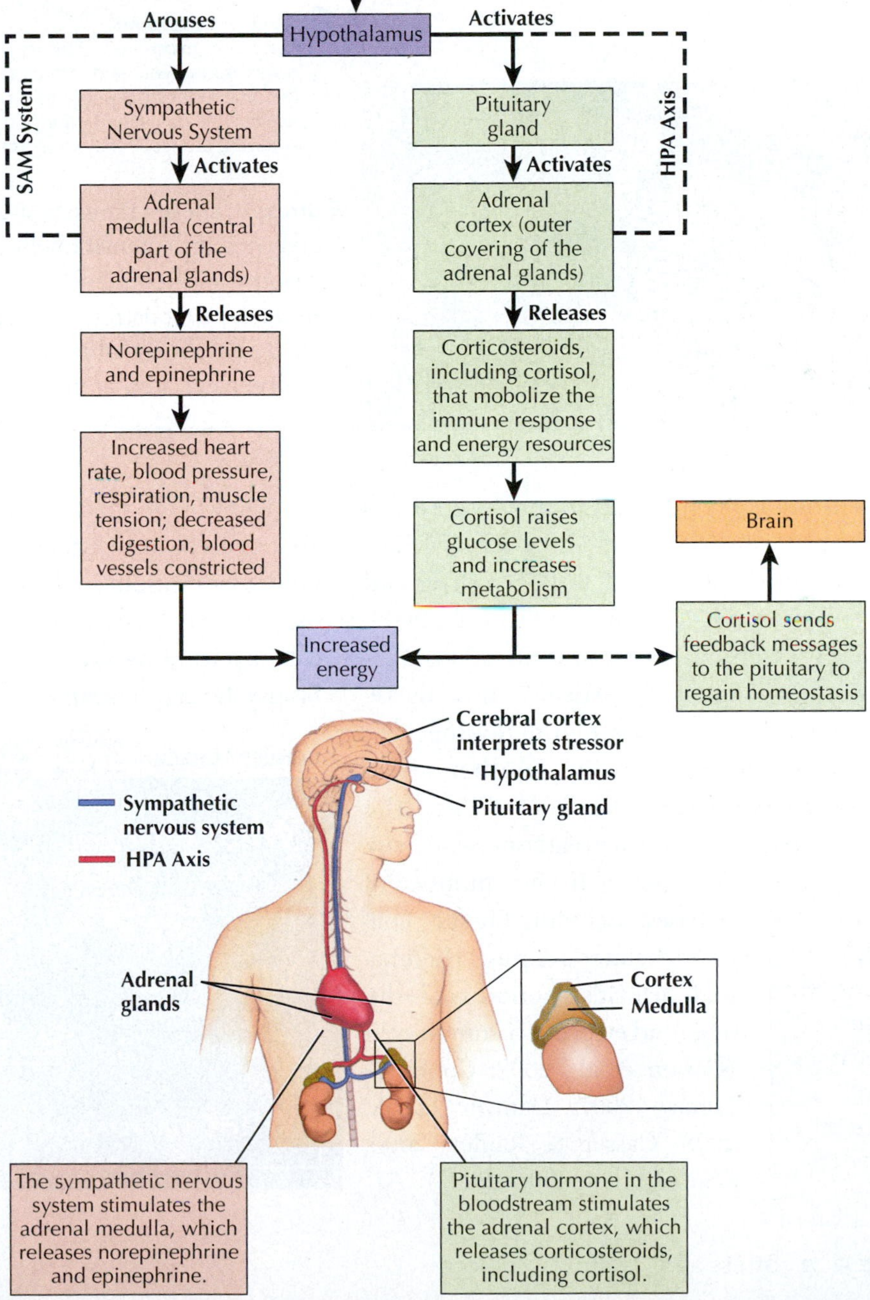

Process Diagram

The general adaptation syndrome (GAS) FIGURE 3.4

Note how the three stages of this syndrome (*alarm, resistance,* and *exhaustion*) focus on the biological response to stress—particularly the "wear and tear" on the body with prolonged stress. As a critical thinker, can you see how the alarm stage corresponds to the SAM system, whereas the resistance and exhaustion stages are part of the HPA axis? ▶

1 Alarm Reaction
In the initial *alarm reaction,* your body experiences a temporary state of shock, and your resistance to illness and stress falls below normal limits.

2 Stage of Resistance
If the stressor remains, your body attempts to endure the stressor, and enters the *resistance phase.* Physiological arousal remains higher than normal, and there is a sudden outpouring of hormones. Selye maintained that one outcome of this stage for some people is the development of *diseases of adaption,* including asthma, ulcers, and high blood pressure.

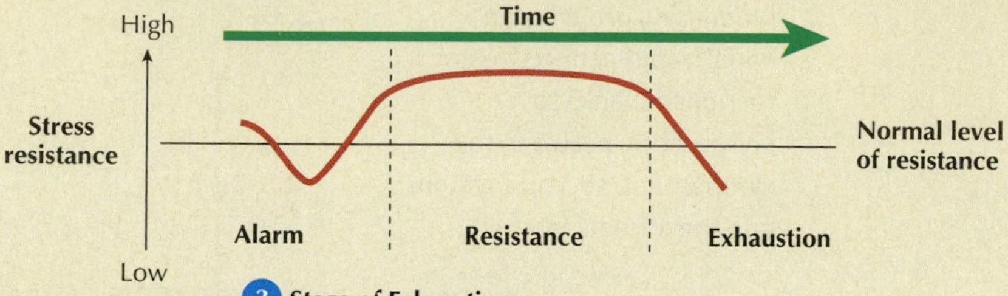

3 Stage of Exhaustion
Long-term exposure to the stressor eventually depletes your body's reserves, and you enter the *exhaustion phase.* In this phase, you become more susceptible to serious illnesses, and possibly irreversible damage to your body. Unless a way of relieving stress is found, the result may be complete collapse and death.

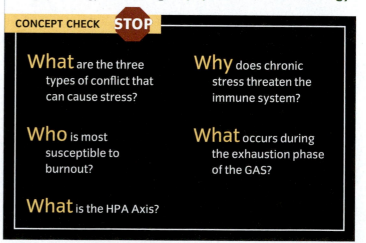

◀ **Stress in ancient times** As shown in these ancient cave drawings, the automatic fight-or-flight response was adaptive and necessary for early human survival. However, in modern society it occurs as a response to ongoing situations where we often cannot fight or flee. This repeated arousal could be detrimental to our health.

elevation of cortisol has been linked to increased levels of depression, posttraumatic stress disorder (PTSD), memory problems, unemployment, and even drug and alcohol abuse (Ayers et al., 2007; Bremner et al., 2004; Johnson, Delahanty, & Pinna, 2008; Sarafino, 2008). Perhaps most important, increased cortisol is directly related to impairment of immune system functioning.

The discovery of the relationship between stress and the immune system is very important. When the immune system is impaired, we are at greatly increased risk of suffering from a number of diseases, including bursitis, colitis, Alzheimer's disease, rheumatoid arthritis, periodontal disease, and even the common cold (Cohen et al., 2002; Cohen & Lemay, 2007; Dantzer et al., 2008; Gasser & Raulet, 2006; Segerstrom & Miller, 2004).

psychoneuroimmunology [sye-koh-NEW-roh-IM-you-NOLL-oh-gee] The interdisciplinary field that studies the effects of psychological factors on the immune system.

Knowledge that psychological factors have considerable control over infectious diseases has upset the long-held assumption in biology and medicine that these diseases are "strictly physical." The clinical and theoretical implications are so important that a new field of biopsychology has emerged: **psychoneuroimmunology**.

CONCEPT CHECK STOP

What are the three types of conflict that can cause stress?

Who is most susceptible to burnout?

What is the HPA Axis?

Why does chronic stress threaten the immune system?

What occurs during the exhaustion phase of the GAS?

Stress and Illness

LEARNING OBJECTIVES

Explain why an immune system compromised by stress might be more vulnerable to cancer growth.

Define the personality patterns that can influence how we respond to stress.

Describe the key symptoms of posttraumatic stress disorder (PTSD).

Explain how biological and psychological factors can jointly influence the development of gastric ulcers.

As we've just seen, stress has dramatic effects on our bodies. This section explores how stress is related to four serious illnesses—cancer, cardiovascular disorders, posttraumatic stress disorder, and gastric ulcers. Next we'll briefly discuss each of these.

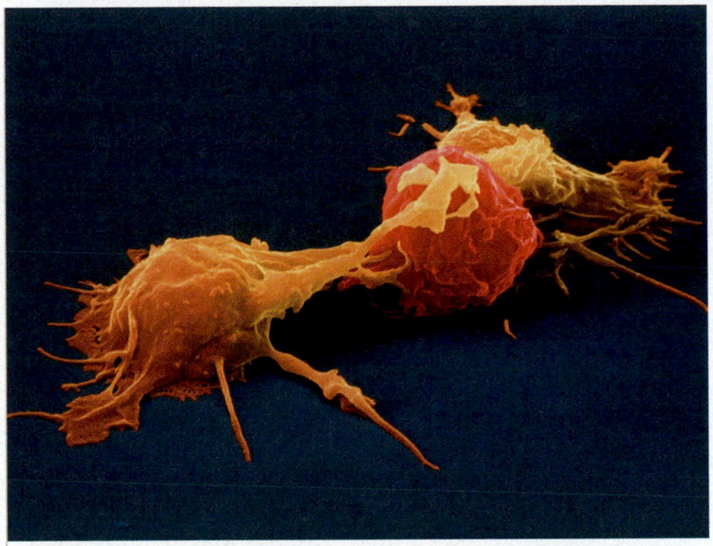

The immune system in action FIGURE 3.5

Stress can compromise the immune system, but the actions of a healthy immune system are shown here. The round red structures are leukemia cells. Note how the yellow killer cells are attacking and destroying the cancer cells.

CANCER: A VARIETY OF CAUSES—EVEN STRESS

Cancer is among the leading causes of death for adults in the United States. It occurs when a particular type of primitive body cell begins rapidly dividing and then forms a tumor that invades healthy tissue. Unless destroyed or removed, the tumor eventually damages organs and causes death. More than 100 types of cancer have been identified. They appear to be caused by an interaction between environmental factors and inherited predispositions.

In a healthy person, whenever cancer cells start to multiply, the immune system checks the uncontrolled growth by attacking the abnormal cells (FIGURE 3.5).

As mentioned earlier, stress causes the adrenal glands to release hormones that suppress the immune system. The compromised immune system is then less able to resist infection and cancer development (Ben-Eliyahu, Page, & Schleifer, 2007; Kemeny, 2007).

The good news is that we can substantially reduce our risk of developing cancer by making changes that reduce our stress levels and enhance our immune systems. For example, researchers have found that interrupting people's sleep decreased the number of *natural killer cells* (a type of immune system cell) by 28 percent (Irwin et al., 1994). Fortunately, these researchers also found that a normal night's sleep after the deprivation returned the killer cells to their normal levels.

CARDIOVASCULAR DISORDERS: THE LEADING CAUSE OF DEATH IN THE UNITED STATES

Cardiovascular disorders cause over half of all deaths in the United States (American Heart Association, 2008). Understandably, health psychologists are concerned because stress is a major contributor to these deaths. **Heart disease** is a general term for all disorders that eventually affect the heart muscle and lead to heart failure. *Coronary heart disease* occurs when the walls of the coronary arteries thicken, reducing or blocking the blood supply to the heart. Symptoms of such disease include *angina* (chest pain due to insufficient blood supply to the heart) and *heart attack* (death of heart muscle tissue). Controllable factors that contribute to heart disease include stress, smoking, certain personality characteristics, obesity, a high-fat diet, and lack of exercise (Aboa-Éboulé, 2008; Ayers et al., 2007; Sarafino, 2008).

When the body is stressed, the autonomic nervous system releases epinephrine and cortisol into the bloodstream. These hormones increase heart rate and release fat and glucose from the body's stores to give muscles a readily available source of energy.

If no physical "fight-or-flight" action is taken (and this is most likely the case in our modern lives), the fat that was released into the bloodstream is not burned as fuel. Instead, it may adhere to the walls of blood vessels. These fatty deposits are a major cause of blood supply blockage, which causes heart attacks.

The effects of stress on heart disease may be amplified if a person tends to be hard driving, competitive, ambitious, impatient, and hostile—a **Type A personality**. The antithesis of the Type A personality is the **Type B personality**, which is characterized by a laid-back, relaxed attitude toward life. (See *What a Psychologist Sees* for more about the Type A personality.)

Type A personality Behavior characteristics that include intense ambition, competition, exaggerated time urgency, and a cynical, hostile outlook.

Type B personality Behavior characteristics consistent with a calm, patient, relaxed attitude.

Have you ever wondered why some people survive in the face of great stress (personal tragedies, demanding jobs, and even a poor home life) while others do not? Suzanne Kobasa was among the first to study this question (Kobasa, 1979; Maddi et al., 2006; Vogt et al., 2008). Examining male executives with high levels of stress, she found that some people are more resistant to stress than others because of a personality factor called **hardiness**, a resilient type of optimism that comes from three distinctive attitudes: *commitment, control,* and *challenge.*

First, hardy people feel a strong sense of commitment to both their work and their personal life. They also make intentional commitments to purposeful activity and problem solving. Second, these people see themselves as being in control of their lives, rather than as victims of their circumstances. Finally, hardy people look at change as a challenge, an opportunity for growth and improvement—not as a threat (FIGURE 3.6).

The important lesson from this research is that hardiness is a learned behavior, not something based on luck or genetics. If you're not one of the hardy souls, you can develop the trait.

Of course, Type A personality and lack of hardiness are not the only controllable risk factors associated with heart disease. *Smoking, obesity,* and *lack of exercise* are also very important factors. Smoking restricts blood circulation, and obesity stresses the heart by causing it to pump more blood to the excess body tissue. A high-fat diet, especially one that is high in cholesterol, contributes to the fatty deposits that clog blood vessels. Lack of exercise contributes to weight gain and prevents the body from obtaining important exercise benefits, including strengthened heart muscle, increased heart efficiency, and the release of neurotransmitters, such as serotonin, that alleviate stress and promote well-being.

Hardiness in action FIGURE 3.6

Based on his smile and cheery wave, it looks like this patient may be one of those lucky "hardy souls." Can you see how an optimistic attitude toward commitment, control, and challenge might help him cope and recuperate from his serious injuries?

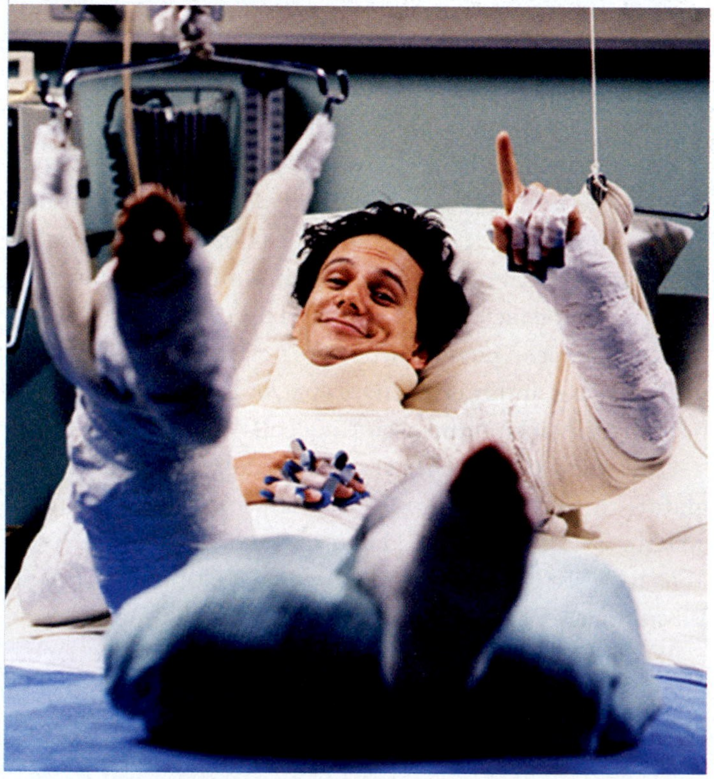

What a Psychologist Sees

Type A Personality and Hostility

Among Type A characteristics, *hostility* is the strongest predictor of heart disease (Krantz & McCeney, 2002; Mittag & Maurischat, 2004). In particular, the constant stress associated with cynical hostility—constantly being "on watch" for problems—translates physiologically into higher blood pressure and heart rate and production of stress-related hormones. In addition, people who are hostile, suspicious, argumentative, and competitive tend to have more interpersonal conflicts. This can heighten autonomic activation, leading to increased risk of cardiovascular disease (Boyle et al., 2004; Bunde & Suls, 2006; Eaker et al., 2007).

Health psychologists have developed two types of behavior modification for people with Type A personalities. The *shotgun approach* aims to eliminate or modify all Type A behaviors. The major criticism of this approach is that it eliminates not only undesirable Type A behaviors but also desirable traits, such as ambition. In contrast, the *target behavior approach* focuses on only those Type A behaviors that are likely to cause heart disease—namely, cynical hostility.

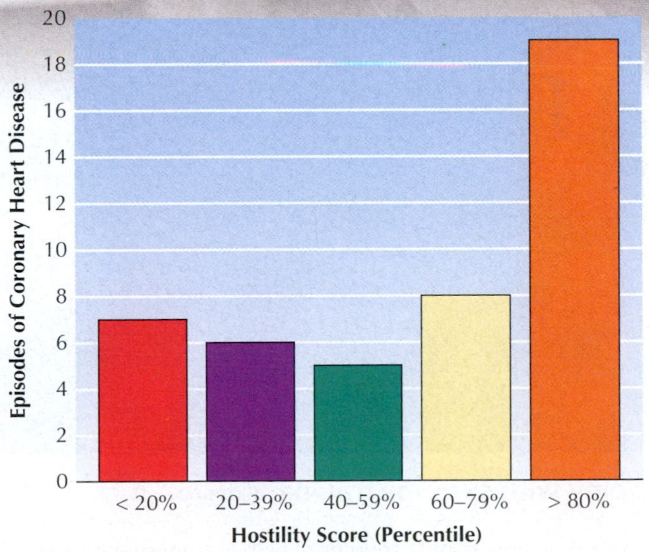

Source: Niaura et al., *Health Psychology*, 2002.

posttraumatic stress disorder (PTSD) An anxiety disorder following exposure to a life-threatening or other extreme event that evoked great horror or helplessness. It is characterized by flashbacks, nightmares, and impaired functioning.

POSTTRAUMATIC STRESS DISORDER (PTSD): A DISEASE OF MODERN TIMES?

One of the most powerful examples of the effects of severe stress is **posttraumatic stress disorder (PTSD)**. Children as well as adults can experience the symptoms of PTSD, which include feelings of terror and helplessness during the trauma and recurrent flashbacks, nightmares, impaired concentration, and emotional numbing afterward. These symptoms may continue for months or years after the event. Some victims of PTSD turn to alcohol and other drugs, which often compound the problem (Kaysen et al., 2008; Sullivan & Holt, 2008).

Coping with extreme trauma FIGURE 3.7

Can you see how finding sources of help for those who have experienced trauma is important for both practical and psychological reasons?

During the Industrial Revolution, workers who survived horrific railroad accidents sometimes developed a condition very similar to PTSD. It was called "railway spine" because experts thought the problem resulted from a twisting or concussion of the spine. Later, doctors working with combat veterans referred to the disorder as "shell shock" because they believed it was a response to the physical concussion caused by exploding artillery. Today, we know that PTSD is caused by any exposure to extraordinary stress (FIGURE 3.7).

PTSD's essential feature is *severe anxiety* (a state of constant or recurring alarm and fearfulness) that develops after experiencing a traumatic event, learning about a violent or unexpected death of a family member, or even being a witness or bystander to violence (American Psychiatric Association, 2002).

According to the Facts for Health website (www.factsforhealth.org), approximately 10 percent of Americans have had or will have PTSD at some point in their lives. TABLE 3.2 summarizes the primary symptoms of PTSD and offers five important tips for coping with traumatic events.

GASTRIC ULCERS: ARE THEY CAUSED BY STRESS?

Beginning in the 1950s, psychologists reported strong evidence that stress can lead to **ulcers**—painful lesions to the lining of the stomach and upper part of the small intestine. Correlational studies have found that people who live in stressful situations develop ulcers more often than people who don't. And numerous experiments with laboratory animals have shown that stressors, such as shock or confinement to a very small space for a few hours, can produce ulcers in some laboratory animals (Andrade & Graeff, 2001; Bhattacharya & Muruganandam, 2003; Gabry et al., 2002; Landeira-Fernandez, 2004).

The relationship between stress and ulcers seemed well established until researchers reported a bacterium

Identifying PTSD and coping with crisis TABLE 3.2

Primary symptoms of posttraumatic stress disorder (PTSD)	Five important tips for coping with crisis
• Re-experiencing the event through vivid memories or flashbacks • Feeling "emotionally numb" • Feeling overwhelmed by what would normally be considered everyday situations • Diminished interest in performing normal tasks or pursuing usual interests • Crying uncontrollably • Isolating oneself from family and friends and avoiding social situations • Relying increasingly on alcohol or drugs to get through the day • Feeling extremely moody, irritable, angry, suspicious, or frightened • Having difficulty falling or staying asleep, sleeping too much, and experiencing nightmares • Feeling guilty about surviving the event or being unable to solve the problem, change the event, or prevent the disaster • Feeling fear and sense of doom about the future	1. Recognize your feelings about the situation and talk to others about your fears. Know that these feelings are a normal response to an abnormal situation. 2. Be willing to listen to family and friends who have been affected and encourage them to seek counseling if necessary. 3. Be patient with people. Tempers are short in times of crisis, and others may be feeling as much stress as you. 4. Recognize normal crisis reactions, such as sleep disturbances and nightmares, withdrawal, reverting to childhood behaviors, and trouble focusing on work or school. 5. Take time with your children, spouse, life partner, friends, and coworkers to do something you enjoy. *Source:* American Counseling Association, 2006, and adapted from Pomponio, 2002.

(*Helicobacter pylori* or *H. pylori*) associated with ulcers. Because many people prefer medical explanations (like bacteria or viruses) to psychological ones, the idea that stress causes ulcers has been largely abandoned.

Most ulcer patients have the *H. pylori* bacterium in their stomachs, and it clearly damages the stomach wall. Antibiotic treatment helps many patients. However, approximately 75 percent of normal control subjects' stomachs also have the bacterium. This suggests that the bacterium may cause the ulcer, but only in people whose systems are compromised by stress. Behavior modification and other psychological treatments, working alongside antibiotics, can help ease ulcers. Finally, studies of the amygdala (a part of the brain involved in emotional response) show that it plays an important role in gastric ulcer formation (Aou, 2006; Tanaka et al., 1998). Apparently, stressful situations and direct stimulation of the amygdala cause an increase in stress hormones and hydrochloric acid and a decrease in blood flow in the stomach walls. This combination leaves the stomach more vulnerable to attack by the *H. pylori* bacteria.

In sum, it appears that *H. pylori*, increased hydrochloric acid, increased stress hormones, and decreased blood flow lead to the formation of gastric ulcers. Once again, we see how biological, psychological, and social forces influence one another (the biopsychosocial model). For now, the psychosomatic explanation for ulcers is back in business (Overmier & Murison, 2000).

CONCEPT CHECK **STOP**

How does stress suppress the immune system?

Which aspect of the Type A personality is most strongly related to heart disease?

What are the components of psychological hardiness?

What is the primary feature of PTSD?

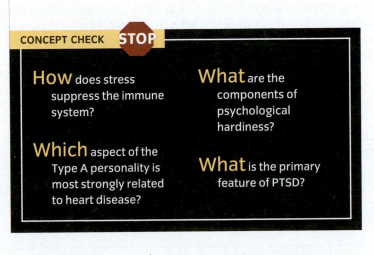

Health Psychology in Action

LEARNING OBJECTIVES

Explain what health psychologists do.

Review the challenges in preventing teen smoking.

Describe what the term binge drinking means.

Describe some psychologically based treatments for chronic pain.

Health psychology is the study of how biological, psychological, and social factors affect health and illness.

In this section, we consider the psychological components of two major health risks—tobacco and alcohol. We also explore the psychological factors that affect chronic pain.

health psychology The study of how biological, psychological, and social factors affect health and illness.

TOBACCO: HAZARDOUS TO YOUR HEALTH

Most people know that smoking is bad for their health and that the more they smoke, the more at risk they are. *Tobacco* use endangers both smokers and those who breathe secondhand smoke, so it is not surprising that most health psychologists and medical professionals are concerned with preventing smoking and getting those who already smoke to stop.

The first puff on a cigarette is rarely pleasant. So why do people start smoking? The answer is complex. First, smoking usually starts when people are young. A survey of U.S. middle schools, grades 6 to 8, found that one in eight students were experimenting with some form of tobacco, such as cigarettes, cigars, and chewing tobacco. Peer pressure and imitation of role models (such as celebrities) are particularly strong factors in young people's decision to begin smoking.

Psychological Science

What Does a Health Psychologist Do?

Health psychologists are interested in how people's lifestyles and activities, emotional reactions, ways of interpreting events, and personality characteristics influence their physical health and well-being.

As researchers, they are particularly interested in the relationship between stress and the immune system.

As practitioners, health psychologists can work as independent clinicians or as consultants with physicians, physical and occupational therapists, and other health care workers. The goal of health psychologists is to reduce psychological distress and unhealthy behaviors. They also help patients and families make critical decisions and prepare them psychologically for surgery or other treatment.

Health psychologists also educate the public about health maintenance. They provide information about the effects of stress, smoking, alcohol, and lack of exercise, as well as other health issues. In addition, health psychologists help people cope with chronic problems, such as pain, diabetes, and high blood pressure, as well as unhealthful behaviors, such as anger expression and lack of assertiveness.

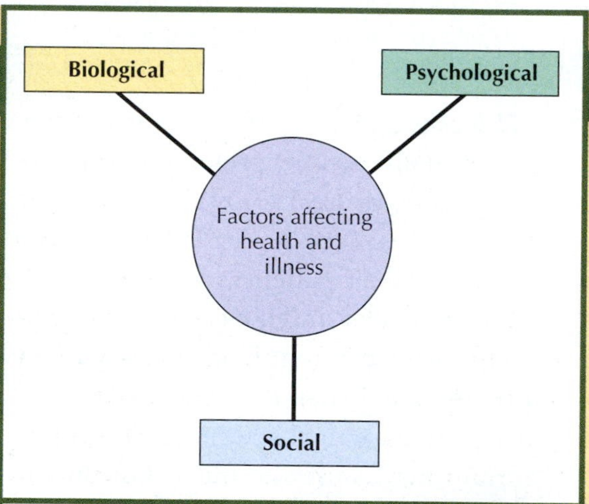

CRITICAL THINKING

Here are two interesting questions:

1. How might the health psychologist in this photo help a patient coping with a serious illness?
2. Would you like to be a health psychologist? Why or why not?

Applying Psychology

Preventing Teen Smoking

For adolescents, smoking's long-term health disadvantages seem irrelevant compared with its short-term social rewards and the addictive, reinforcing properties of nicotine. Therefore, as shown in this photo, many **smoking prevention programs** focus on more immediate problems with smoking. Films and discussion groups also educate teens about peer pressure and the media's influence on smoking, as well as help them hone decision-making and coping skills.

Unfortunately, these psychosocial programs can be complicated, controversial, and expensive (Hatsukami, 2008; Pierce, 2007; Vijgen et al., 2008). To have even a modest effect, these programs must begin early and continue for many years. To reduce the health risk and help fight peer pressure, many schools ban smoking in college buildings and offer smoke-free dormitories. The rising cost of cigarettes—now more than $4 per pack in most states when taxes are included—may also deter many young people from smoking. For a person who smokes a pack of cigarettes a day, the annual cost is nearly $1500.

ELLER
Smoking can
cause impotence
in young men.

CRITICAL THINKING Here are two interesting questions:

1. Why do you think teenagers tend to disregard the health risks associated with smoking, even when they are aware of those risks?
2. Do you think the cost of cigarettes is a sufficient deterrent to teenage smoking? Can you think of a more effective deterrent?

Second, nicotine is addictive. So once a person begins to smoke, there is a biological need to continue. Nicotine addiction appears to be very similar to heroin, cocaine, and alcohol addiction (Brody et al., 2004). When we inhale tobacco smoke, the nicotine quickly increases the release of acetylcholine and norepinephrine in our brain. These *neurotransmitters* (see Chapter 2) increase alertness, concentration, memory, and feelings of pleasure. Nicotine also stimulates the release of dopamine, the neurotransmitter that is most closely related to reward centers in the brain (Fehr et al., 2008; Yang et al., 2008).

Finally, smokers learn to associate smoking with pleasant things, such as good food, friends, and sex. (Ironically, as shown in FIGURE 3.8, anti-smoking laws may actually create stronger bonds among smokers and enhance their resolve to keep smoking.)

In addition to these social and psychological associations, smokers also learn to associate the act of smoking with the "high" that nicotine gives them. When smokers are deprived of cigarettes, they go through an unpleasant *physical withdrawal.* Nicotine relieves the withdrawal symptoms, so smoking is rewarded. Is it any surprise that many scientists believe that the best way to reduce the number of smokers is to stop people from ever taking that first puff?

Smokers unite? FIGURE 3.8

Anti-smoking laws may have made quitting smoking even more difficult for some. Being forced to gather together outside to smoke may forge stronger social bonds among smokers and strengthen their resolve to "suffer the tyranny" of anti-smoking laws. In addition, having to wait longer for their next nicotine dose increases the severity of withdrawal symptoms (Palfai et al., 2000). In effect, the smoker gets repeated previews of just how unpleasant quitting would be. Does this mean anti-smoking laws should be abolished?

Some people find that the easiest way for them to stop smoking is to suddenly and completely stop. However, the success rate for this "cold turkey" approach is extremely low. Even with medical aids, such as patches, gum, or pills, it is still very difficult to quit. Any program designed to help smokers break their habit must combat the social rewards of smoking as well as the physical addiction to nicotine (Koop, Richmond, & Steinfeld, 2004).

Some **smoking cessation programs** combine nicotine replacement therapy with cognitive and behavioral techniques. This helps smokers identify stimuli or situations that make them feel like smoking and then change or avoid them (Brandon et al., 2000). Smokers can also refocus their attention on something other than smoking or remind themselves of the benefits of not smoking (Taylor et al., 2000). Behaviorally, they might cope with the urge to smoke by chewing gum, exercising, or chewing on a toothpick after a meal instead of lighting a cigarette.

ALCOHOL: A PERSONAL AND SOCIAL HEALTH PROBLEM

The American Medical Association considers *alcohol* to be the most dangerous and physically damaging of all drugs (American Medical Association, 2003). After tobacco, it is the leading cause of premature death in America and most European countries (Abadinsky, 2008; Cohen et al., 2004; Maisto, Galizio, & Conners, 2008). Drinking alcohol may also cause serious brain damage (Crews et al., 2004). Alcohol also seems to increase aggression, which helps explain why it's a major factor in most murders, suicides, spousal assaults, incidents of child abuse, and accidental deaths in the United States (Levinthal, 2008; Sebre et al., 2004; Sher, Grekin, & Williams, 2005).

Most people are now aware of the major risks of drinking alcohol and driving—heavy fines, loss of driver's license, serious injuries, jail time, and even death. But the effects of drinking alcohol itself can also be fatal.

binge drinking
Occurs when a man consumes five or more drinks, or a woman consumes four or more drinks in about two hours.

Because alcohol depresses neural activity throughout the brain, if blood levels of alcohol rise to a certain level, the brain's respiratory center stops functioning and the person dies. This is why **binge drinking** is so dangerous (FIGURE 3.9).

Binge drinking is of particular concern on college campuses. A national survey of college students found that between 1993 and 2001, approximately 44 percent of college students were binge drinkers (Wechsler et al., 2002). This survey also found that white students are significantly more likely to drink heavily (50.2 percent) than are students of other ethnicities.

Alcohol is a serious health problem for all college-age students. Research shows that every year:

- 1400 college students die from alcohol-related causes,
- 500,000 students suffer nonfatal, alcohol-related injuries,
- 1.2–1.5 percent of students attempt suicide because of alcohol or other drug use,
- 400,000 students have unprotected sex, and more than 100,000 are too intoxicated to know whether they consented to sexual intercourse (Task Force of the National Advisory, 2002).

College administrators, however, are increasingly aware of the problems of binge drinking and other types of alcohol abuse. They are developing policies and pro-

Binge drinking FIGURE 3.9

What's wrong with this picture? Unfortunately, many college students believe that heavy drinking is harmless fun and a natural part of college life. But excessive alcohol consumption, including binge drinking, carries serious health consequences and can even be fatal.

Applying Psychology

Do You Have an Alcohol Problem?

In our society, drinking alcohol (within limits) is generally considered normal and appropriate. However, many people abuse alcohol. If you'd like to make a quick check of your own drinking behavior, place a mark next to each of the symptoms that describes your current drinking behavior.

Seven Signs of Alcohol Dependence Syndrome

_____ • Drinking increases, sometimes to the point of almost continuous daily consumption.

_____ • Drinking is given higher priority than other activities, in spite of its negative consequences.

_____ • More and more alcohol is required to produce behavioral, subjective, and metabolic changes; large amounts of alcohol can be tolerated.

_____ • Even short periods of abstinence bring on withdrawal symptoms, such as sweatiness, trembling, and nausea.

_____ • Withdrawal symptoms are relieved or avoided by further drinking, especially in the morning.

_____ • You are aware of a craving for alcohol and have little control over the quantity and frequency of alcohol you consume.

_____ • If you begin drinking again after a period of abstinence, you rapidly return to the previous high level of consumption and other behavioral patterns.

Source: World Health Organization, 2008.

CRITICAL THINKING

Here are two interesting questions:

1. Why is it important to determine whether you have an alcohol problem?
2. What effects might an individual's alcohol dependence have on his or her family, community, and society?

grams that go beyond traditional educational programs to also include the physical, social, legal, and economic environment on college campuses and the surrounding communities (Kapner, 2004).

CHRONIC PAIN: AN ONGOING THREAT TO HEALTH

chronic pain
Continuous or recurrent pain lasting six months or longer.

Normally, pain is essential to our survival and well-being. It alerts us to dangerous or harmful situations and forces us to rest and recover from injury (Watkins & Maier, 2000). In contrast, **chronic pain**—the kind that comes with a chronic disease or continues long past the healing of a wound—does not serve a useful function.

Chronic pain is a serious problem with no simple solution. Although exercise increases the body's supply of *endorphins,* which help block the perception of pain, patients with chronic pain tend to decrease their activity and exercise.

Psychological factors such as anxiety frequently intensify pain and increase the related anguish and disability (Bieber et al., 2008; Keefe, Abernathy, & Campbell, 2005; McGuire, Hogan, & Morrison, 2008). Even well-meaning family members can exacerbate pain just by asking about it, because talking about pain focuses attention on it and increases its intensity (Roth et al., 2007; Sullivan, 2008).

Health psychologists often use psychologically oriented treatments to treat chronic pain. For example, a health psychologist may begin a **behavior modification** program for both the patient with chronic pain and his or her family. Such programs often involve establishing an individualized pain management plan that incorporates daily exercise and relaxation techniques. Health psychologists also monitor each patient's adherence to the plan and provide rewards for following through with the treatment program.

Biofeedback for chronic pain FIGURE 3.10

Biofeedback using an EMG is most helpful when the pain involves extreme muscle tension, such as with tension headache and lower back pain.

In **biofeedback** (also called *neurofeedback*) for chronic pain, an electromyograph (EMG) measures muscle tension by recording electrical activity in the skin. When a patient is sufficiently relaxed, the machine signals with a tone or light. This feedback teaches patients self-regulation skills that help control their pain (FIGURE 3.10). Biofeedback is sometimes as effective as more expensive and lengthier forms of pain treatment (Hammond, 2007; Monastra, 2008).

Finally, special **relaxation techniques**—like the breathing and muscle relaxation exercises that are taught in childbirth classes—can help divert attention from pain, reducing discomfort (Astin, 2004).

CONCEPT CHECK STOP

Why is it so difficult for people to quit smoking?

What are some key indicators of problem drinking?

How do behavior modification programs for chronic pain treatment work?

Why might relaxation techniques ease chronic pain, even without eliminating the pain itself?

Health and Stress Management

LEARNING OBJECTIVES

Compare emotion-focused and problem-focused forms of coping.

Explain the role that interpretation plays in shaping our responses to stressors.

Review some major resources for combating stress.

COPING WITH STRESS

Because we can't escape stress, we need to learn how to effectively cope with it. Simply defined, **coping** is an attempt to manage stress in some effective way. It is not one single act but a process that allows us to deal with various stressors (FIGURE 3.11).

Our level of stress generally depends on both our interpretation of and our reaction to stressors. **Emotion-focused forms of coping** are emotional or cognitive strategies that help us manage a stressful situation. For example, suppose you were refused a highly desirable job. You might reappraise the situation and decide that the job wasn't the right match for you or that you weren't really qualified or ready for it.

emotion-focused forms of coping Managing one's emotional reactions to a stressor.

When faced with unavoidable stress, people commonly use **defense mechanisms**. That is, they unconsciously distort reality to protect their egos and to avoid anxiety. For instance, fantasizing about what you will do on your next vacation can help relieve stress. Sometimes, however, defense mechanisms can be destructive. For example, people often fabricate excuses for not attaining particular goals. Such *rationalization* keeps us from seeing a situation more clearly and realistically and can prevent us from developing valuable skills or qualities.

Emotion-focused forms of coping that are accurate reappraisals of stressful situations and that do not distort reality may alleviate stress in some situations (Giacobbi, Foore, & Weinberg, 2004; Patterson, Holm, & Gurney, 2004). Many times, however, it is necessary and more effective to

Cognitive appraisal and coping FIGURE 3.11

Research suggests that our emotional response to an event depends largely on how we interpret the event.

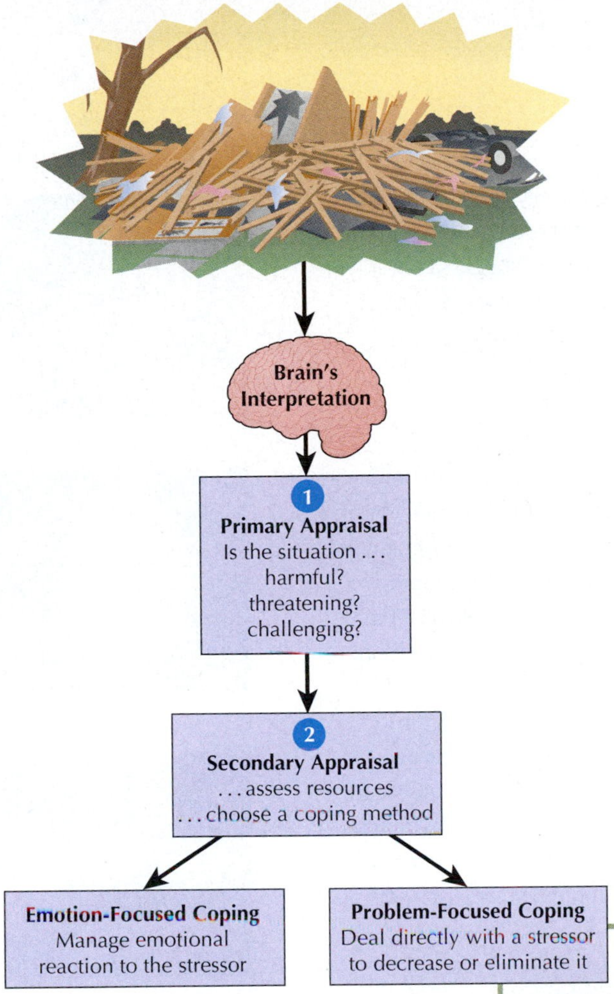

Brain's Interpretation
1 Primary Appraisal
Is the situation . . .
harmful?
threatening?
challenging?
2 Secondary Appraisal
. . . assess resources
. . . choose a coping method
Emotion-Focused Coping
Manage emotional reaction to the stressor
Problem-Focused Coping
Deal directly with a stressor to decrease or eliminate it

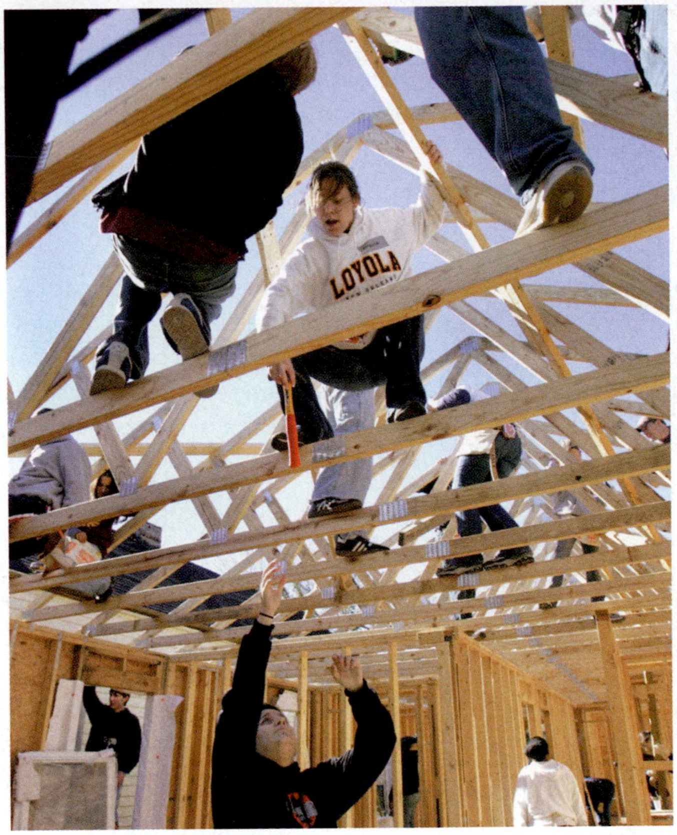

LOYOLA

People often combine *emotion-focused* and *problem-focused* coping strategies to resolve complex stressors or to respond to a stressful situation that is in flux. In some situations, an emotion-focused strategy can allow people to step back from an especially overwhelming problem. Then they can reappraise the situation and use the problem-solving approach to look for solutions. Can you see how each form of coping is represented in these two photos?

Process Diagram

Resources for healthy living TABLE 3.3

Health and exercise	Exercising and keeping fit helps minimize anxiety, depression, and tension, which are associated with stress. Exercise also helps relieve muscle tension; improves cardiovascular efficiency; and increases strength, flexibility, and stamina.	
Positive beliefs	A positive self-image and attitude can be especially significant coping resources. Even temporarily raising self-esteem reduces the amount of anxiety caused by stressful events. Also, hope can sustain a person in the face of severe odds, as is often documented in news reports of people who have triumphed over seemingly unbeatable circumstances.	
Social skills	People who acquire social skills (such as knowing appropriate behaviors for certain situations, having conversation starters up their sleeves, and expressing themselves well) suffer less anxiety than people who do not. In fact, people who lack social skills are more at risk for developing illness than those who have them. Social skills not only help us interact with others but also communicate our needs and desires, enlist help when we need it, and decrease hostility in tense situations.	
Social support	Having the support of others helps offset the stressful effects of divorce, the loss of a loved one, chronic illness, pregnancy, physical abuse, job loss, and work overload. When we are faced with stressful circumstances, our friends and family often help us take care of our health, listen, hold our hands, make us feel important, and provide stability to offset the changes in our lives.	

use **problem-focused forms of coping**, which deal directly with the situation or the stressor to eventually decrease or eliminate it (Bond & Bunce, 2000). These direct coping strategies include:

- identifying the stressful problem,
- generating possible solutions,
- selecting the appropriate solution,
- applying the solution to the problem, thus eliminating the stress.

problem-focused forms of coping Dealing directly with a stressor to decrease or eliminate it.

RESOURCES FOR HEALTHY LIVING: FROM GOOD HEALTH TO MONEY

A person's ability to cope effectively depends on the stressor itself—its complexity, intensity, and duration—and on the type of coping strategy used. It also depends on available resources. Eight important resources for healthy living and stress management are health and exercise, positive beliefs, social skills, social support, control, material resources, relaxation, and sense of humor. These resources are described in TABLE 3.3.

Control	Believing that you are the "master of your own destiny" is an important resource for effective coping. People with an **external locus of control** feel powerless to change their circumstances and are less likely to make healthy changes, follow treatment programs, or positively cope with a situation. Conversely, people with an **internal locus of control** believe that they are in charge of their own destinies and are therefore able to adopt more positive coping strategies.	
Material resources	Money increases the number of options available for eliminating sources of stress or reducing the effects of stress. When faced with the minor hassles of everyday living, or when faced with chronic stressors or major catastrophes, people with money and the skills to effectively use it generally fare better and experience less stress than people without money.	
Relaxation	There are a variety of relaxation techniques. Biofeedback is often used in the treatment of chronic pain, but it is also useful in teaching people to relax and manage their stress. **Progressive relaxation** helps reduce or relieve the muscular tension commonly associated with stress. Using this technique, patients first tense and then relax specific muscles, such as those in the neck, shoulders, and arms. This technique teaches people to recognize the difference between tense and relaxed muscles.	
Sense of humor	Research shows that humor is one of the best ways to reduce stress. The ability to laugh at oneself, and at life's inevitable ups and downs, allows us to relax and gain a broader perspective. In short: "Don't sweat the small stuff."	

external locus of control Belief that chance or outside forces beyond one's control determine one's fate.

internal locus of control Belief that one controls one's own fate.

CONCEPT CHECK STOP

What is a defense mechanism?

Why would it sometimes be useful for people to combine coping strategies?

Why is exercise important for counteracting stress?

How can social skills protect us against stress?

SUMMARY

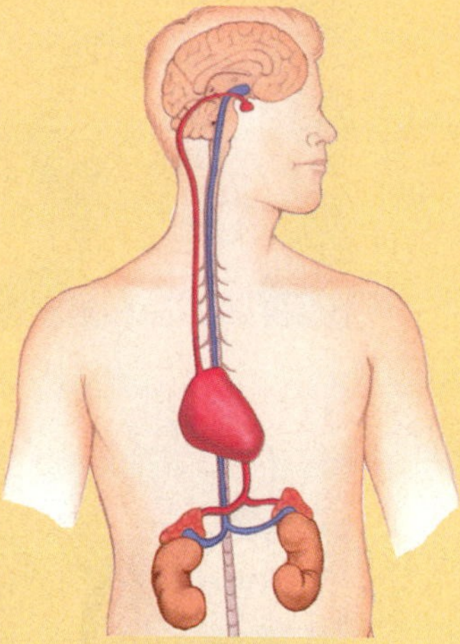

1 Understanding Stress

1. **Stress** is the body's nonspecific response to any demand placed on it. Pleasant or beneficial stress is called eustress. Stress that is unpleasant or objectionable is called distress. Life changes, chronic stressors, minor hassles, frustrations, and conflict all can cause stress. There are three basic types of conflict: **approach–approach conflict, avoidance–avoidance conflict,** and **approach–avoidance conflict.**

2. The sympathetic nervous system and the HPA Axis control significant physiological responses to stress. The sympathetic nervous system prepares us for immediate action; the HPA Axis responds more slowly but lasts longer. Researchers' understanding that psychological factors influence many diseases has spawned a new field of biopsychology called **psychoneuroimmunology**.

3. The general adaptation syndrome (GAS) describes the body's three-stage reaction to stress. The phases are the initial alarm reaction, the resistance phase, and the exhaustion phase (if resistance to stress is not successful).

2 Stress and Illness

1. Stress makes the immune system less able to resist infection and cancer development.

2. Increased stress hormones can cause fat to adhere to blood vessel walls, increasing the risk of heart attack. Having a **Type A personality** amplifies the risk of stress-related heart disease. The **Type B personality** is its antithesis.

3. Exposure to extraordinary stress can cause **posttraumatic stress disorder (PTSD)**, a type of severe anxiety that increases the likelihood of substance abuse problems.

4. Stress increases the risk of developing gastric ulcers among people who have the *H. pylori* bacterium in their stomachs.

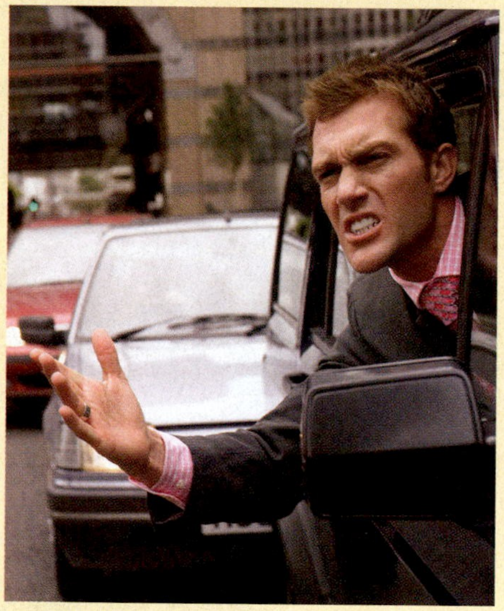

3 Health Psychology in Action

1. **Health psychology**, the study of how biological, psychological, and social factors affect health and illness, is a growing field in psychology.

2. Most approaches to helping people quit smoking include cognitive and behavioral techniques to aid smokers in their withdrawal from nicotine, along with nicotine replacement therapy.

3. Alcohol is a leading cause of premature death and may cause serious brain damage. Alcohol also seems to increase aggression. Alcohol, particularly **binge drinking**, is of major concern on college campuses. Although many college students believe that heavy drinking is a harmless part of college life, college administrators are becoming increasingly aware of the problems of binge drinking and other types of alcohol abuse.

4. **Chronic pain** is a serious problem with no simple solution. Psychological factors frequently intensify pain and increase related problems. To treat chronic pain, health psychologists often focus their efforts on psychologically oriented treatments, including behavior modification, biofeedback, and relaxation techniques.

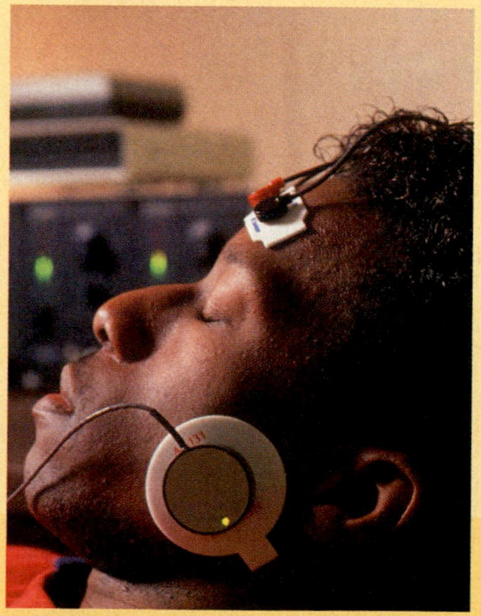

4 Health and Stress Management

1. Our level of stress generally depends on both our interpretation of and our reaction to stressors. **Emotion-focused forms of coping** are emotional or cognitive strategies that change how we view a stressful situation. Some emotion-focused defense mechanisms, such as rationalization, can be destructive because they prevent us from developing valuable skills or qualities. It is often more effective to use **problem-focused forms of coping** strategies. People often combine problem-focused and emotion-focused coping strategies to resolve complex stressors or to respond to a stressful situation that is in flux.
2. Eight important resources for healthy living and stress management are health and exercise, positive beliefs, social skills, social support, control, material resources, relaxation, and a sense of humor. A more specific look at control tells us that believing that one controls one's own fate—**an internal locus of control**—is a healthy way to manage stress. Having an **external locus of control** means believing that chance or outside forces beyond one's control determine one's fate.

KEY TERMS

- **stress** p. 64
- **stressor** p. 64
- **eustress** p. 64
- **distress** p. 64
- **life change** p. 64
- **cataclysmic event** p. 64
- **chronic stressors** p. 64
- **job stress** p. 64
- **hassles** p. 65
- **burnout** p. 65
- **frustration** p. 66
- **conflict** p. 66
- **approach–approach conflict** p. 66
- **avoidance–avoidance conflict** p. 66
- **approach–avoidance conflict** p. 66
- **sympathetic nervous system** p. 66
- **HPA Axis** p. 66
- **general adaptation syndrome (GAS)** p. 66
- **psychoneuroimmunology** p. 68
- **cancer** p. 69
- **heart disease** p. 69
- **Type A personality** p. 70
- **Type B personality** p. 70
- **hardiness** p. 70
- **posttraumatic stress disorder (PTSD)** p. 72
- **ulcers** p. 72
- **health psychology** p. 74
- **smoking prevention programs** p. 75
- **smoking cessation programs** p. 76
- **binge drinking** p. 76
- **chronic pain** p. 77
- **behavior modification** p. 77
- **biofeedback** p. 78
- **relaxation techniques** p. 78
- **coping** p. 78
- **emotion-focused forms of coping** p. 78
- **defense mechanisms** p. 78
- **problem-focused forms of coping** p. 80
- **external locus of control** p. 81
- **internal locus of control** p. 81
- **progressive relaxation** p. 81

CRITICAL AND CREATIVE THINKING QUESTIONS

1. What are the major sources of stress in your life?
2. If you were experiencing a period of high stress, what would you do to avoid illness?
3. Why are smoking-prevention efforts often aimed at adolescents?
4. What are the benefits of involving a person's family in behavior modification programs to treat chronic pain?
5. Which forms of coping do you most often draw on?
6. Which of the resources for stress management seem most important to you?

(See page 84 for *What is happening in this picture?*)

What is happening in this picture ?

Anti-smoking laws, which have become increasingly common since the 1990s, have forced smokers to go outside to smoke.

- **What psychological effect of these laws do cigarette companies exploit in this advertisement?**

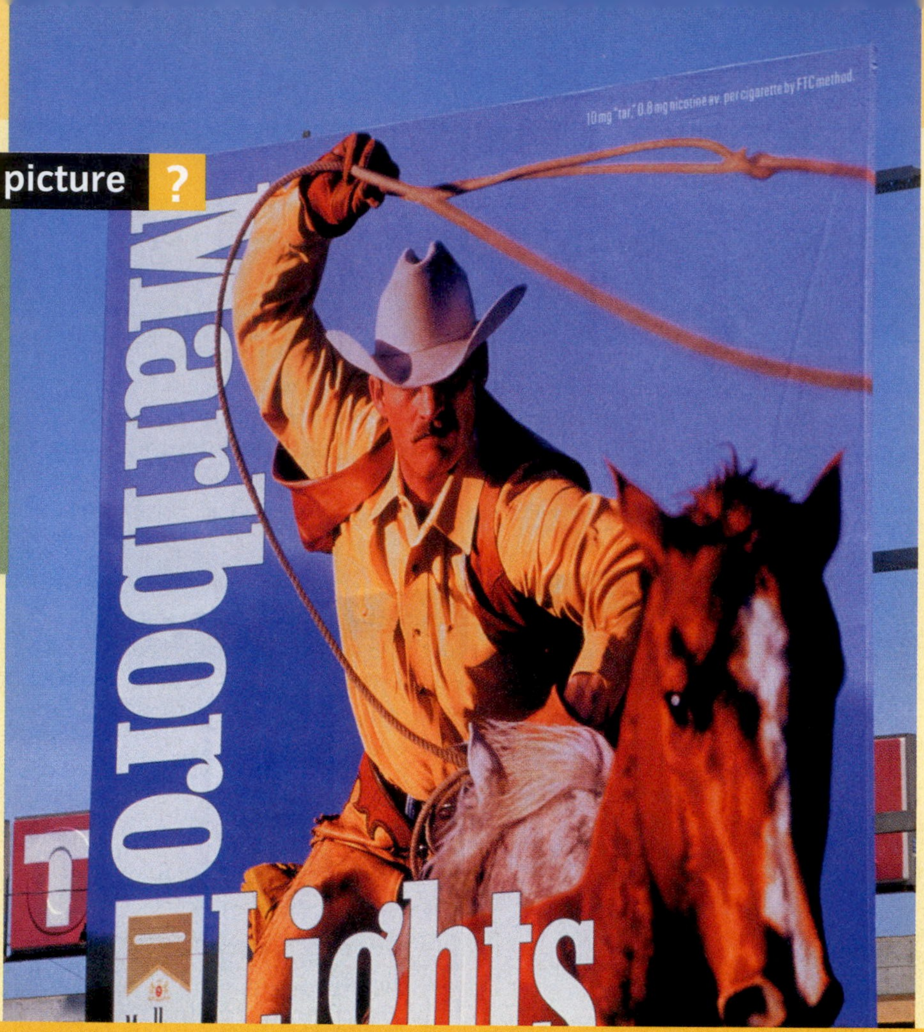

SELF-TEST

(Check your answers in Appendix B.)

1. In your text, the physical and mental arousal to situations that we perceive as threatening or challenging is called ________________.

a. distress
b. eustress
c. stress
d. all of these options

2. ________________ is one of the largest sources of **CHRONIC** stress for adults.

a. Birth
b. Work
c. Christmas
d. Moving

3. A state of physical, emotional, and mental exhaustion attributable to long-term involvement in emotionally demanding situations is called ________________.

a. primary conflict
b. technostress
c. burnout
d. secondary conflict

4. In an *approach–approach conflict,* a person must choose between two or more goals that will lead to ________________, whereas in an *avoidance–avoidance conflict,* a person must choose between two or more goals that will lead to ________________.

a. less conflict; more conflict
b. frustration; hostility
c. a desirable result; an undesirable result
d. effective coping; ineffective coping

5. Label the *HPA Axis*, which enables us to deal with chronic stressors:

a. hypothalamus
b. pituitary
c. adrenal cortex

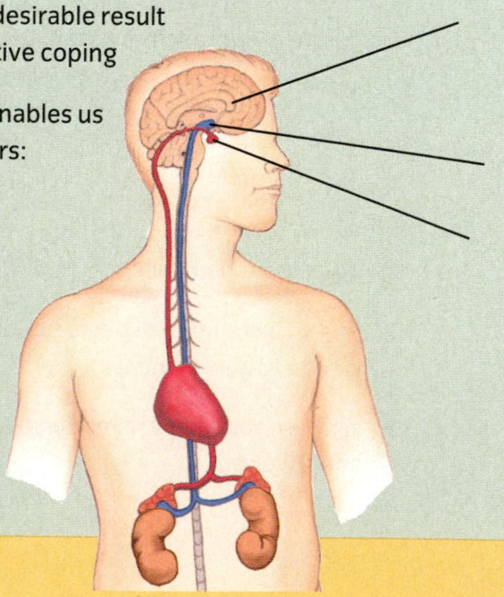

6. ______________ is the field that studies the effects of psychological factors on the immune system.
a. Psychosomatology
b. Neurobiology
c. Psychoneuroimmunology
d. Biopsychology

7. Research suggests that one particular *Type A* characteristic, illustrated in this photo, is **MOST** associated with heart disease. What characteristic is this?
a. intense ambition
b. impatience
c. cynical hostility
d. time urgency

8. People who experience flashbacks, nightmares, and impaired functioning following a life-threatening or other horrifying event are ______________.
a. suffering from a psychosomatic illness
b. experiencing posttraumatic stress disorder
c. having a nervous breakdown
d. weaker than people who take such events in stride

9. Which of the following is **TRUE** of *health psychology*?
a. It studies the relationship between psychological behavior and physical health.
b. It studies the relationship between psychological behavior and illness.
c. It emphasizes wellness and the prevention of illness.
d. All of these statements are true.

10. The effects of nicotine are related to the release of ______________ in the brain.
a. acetylcholine
b. norepinephrine
c. dopamine
d. all of these options

11. At several parties in the past couple of weeks, Arash consumed five drinks in a row and his girlfriend consumed four drinks in a row. This means that ______________ met the definition for binge drinking.
a. Arash
b. his girlfriend
c. both Arash and his girlfriend
d. neither Arash nor his girlfriend

12. Any pain that continues or recurs for six months or more is considered to be ______________.
a. life-threatening
b. chronic
c. psychosomatic
d. caused by an incompetent physician

13. Biofeedback using ______________ measures muscle tension and provides feedback regarding a patient's level of relaxation.
a. EMG
b. EKG
c. EGG
d. EEG

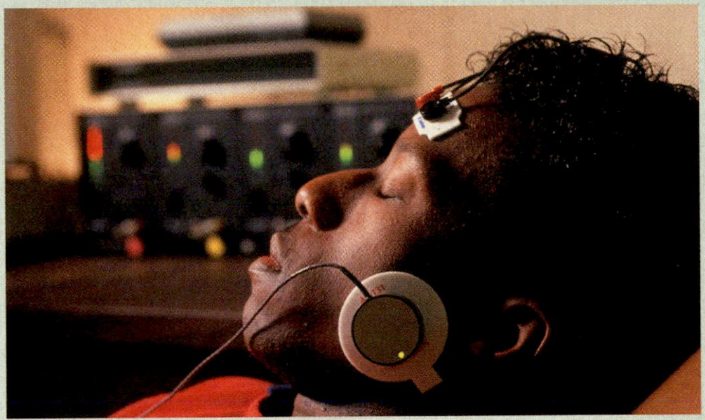

14. *Emotion-focused forms of coping* are based on changing your ______________ when faced with stressful situations.
a. feelings
b. perceptions
c. strategies
d. all of these options

15. Research suggests that people with higher ______________ have less psychological stress than those with higher ______________.
a. external locus of control; internal locus of control
b. internal locus of control; external locus of control
c. emotion-focused coping styles; problem-focused coping styles
d. problem-focused coping styles; emotion-focused coping styles

Sensation and Perception

4

Imagine that your visual field were suddenly inverted and reversed, so that things you expected to be on your right would be on your left, and things you expected to be above your head would be below your head. You would certainly have trouble getting around. Do you think you could ever adapt to this distorted world?

To answer that question, psychologist George Stratton (1896) wore special lenses for eight days. For the first few days, Stratton had a great deal of difficulty navigating in this environment and coping with everyday tasks. But by the third day, his experience had begun to change. He noted:

> *Walking through the narrow spaces between pieces of furniture required much less care than hitherto. I could watch my hands as they wrote, without hesitating or becoming embarrassed thereby.*

By the fifth day, Stratton had almost completely adjusted to his strange perceptual environment. His expectations of how the world should be arranged had changed. As Stratton's experiment shows, we are able to adapt even our most basic perceptions by retraining our brains to adapt to unfamiliar physical sensations, creating a newly coherent world. This chapter focuses on two separate, but inseparable, aspects of how we experience the world: sensation and perception. The boundary between these two processes is not precise. **Sensation** is the process of receiving, translating, and transmitting raw sensory data from the external and internal environments to the brain. **Perception** is the "higher level" process of selecting, organizing, and interpreting sensory data into useful mental representations of the world.

CHAPTER OUTLINE

Understanding Sensation

LEARNING OBJECTIVES

Describe how raw sensory stimuli are converted to signals in the brain.

Explain how the study of thresholds helps to explain sensation.

Describe why adapting to sensory stimuli provides an evolutionary advantage.

Identify the factors that govern pain perception.

When presented with a high-pitched tone, a musician reported, "It looks like fireworks tinged with a pink-red hue. The color feels rough and unpleasant, and it has an ugly taste—rather like that of a briny pickle" (Luria, 1968). This musician was describing a rare condition known as **synesthesia**, which means "mixing of the senses." People with synesthesia routinely blend their sensory experiences. They may "see" temperatures, "hear" colors, or "taste" shapes. To appreciate how extraordinary synesthesia is, we must first understand the basic processes of normal, nonblended sensations. For example, how do we turn light and sound waves from the environment into something our brain can comprehend? To do this, we must have both a means of detecting stimuli and a means of converting them into a language the brain can understand.

PROCESSING: DETECTION AND CONVERSION

Our eyes, ears, skin, and other sense organs all contain special cells called receptors, which receive and process sensory information from the environment. For each sense, these specialized cells respond to a distinct stimulus, such as sound waves or odor molecules. During the process of **transduction**, the receptors convert the stimulus into neural impulses, which are sent to the brain. For example, in hearing, tiny receptor cells in the inner ear convert mechanical vibrations from sound waves into electrochemical signals. These signals are carried by neurons to the brain, where specific sensory receptors detect and interpret the information. How does our brain differentiate between sensations, such as sounds and smells? Through a process known as **coding**, different physical stimuli are interpreted as distinct sensations because their neural impulses travel by different routes and arrive at different parts of the brain (FIGURE 4.1).

transduction Process by which a physical stimulus is converted into neural impulses.

coding Process that converts a particular sensory input into a specific sensation.

We also have structures that purposefully reduce the amount of sensory information we receive. Without this natural filtering of stimuli we would constantly hear blood rushing through our veins and feel our clothes brushing against our skin. Some level of filtering is needed so the brain is not overwhelmed with unnecessary information.

All species have evolved selective receptors that suppress or amplify information for survival (FIGURE 4.2). In the process of **sensory reduction**, we filter incoming sensations and analyze the sensations that

sensory reduction Filtering and analyzing incoming sensations before sending a neural message to the cortex.

Sensory processing within the brain
FIGURE 4.1

Neural impulses travel from the sensory receptors to various parts of the brain.

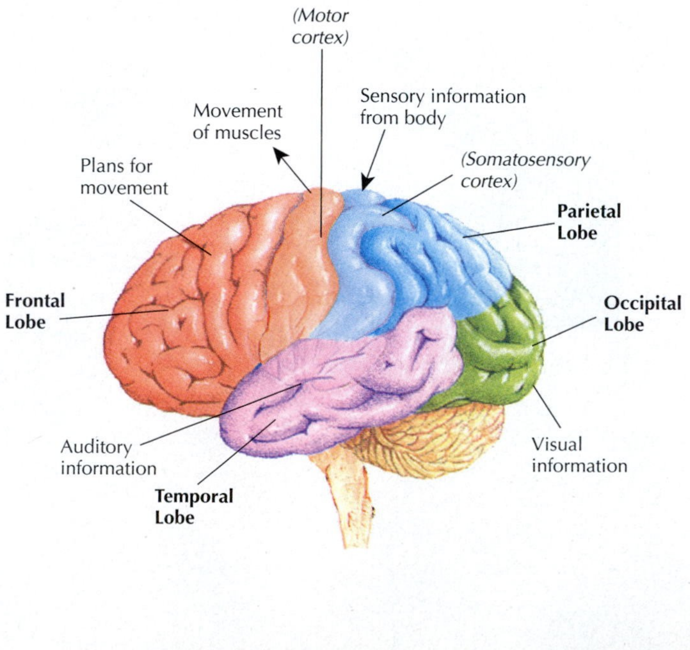

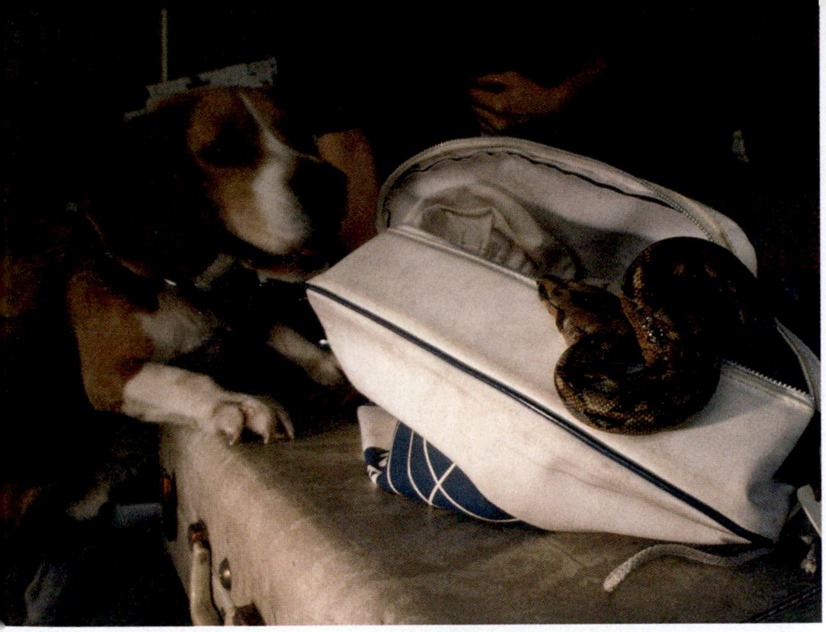

Dogs' sense of smell FIGURE 4.2

Many animals possess extraordinary sensory abilities. For example, dogs' sense of smell is far more sensitive than humans'. For this reason, specially trained dogs provide invaluable help in sniffing out dangerous plants, animals, drugs, and explosives, tracking criminals, and assisting in search-and-rescue operations. Some researchers believe dogs can even detect chemical signs of certain illnesses, such as diabetes or cancer (Akers & Denbow, 2008).

the body does register before a neural impulse is finally sent to the various parts of the brain. Humans, for example, cannot sense ultraviolet light, microwaves, the ultrasonic sound of a dog whistle, or infrared heat patterns from warm-blooded animals, as some other animals can. However, we can see a candle burning 30 miles away on a dark, clear night, hear a watch tick at 20 feet under quiet conditions, smell one drop of perfume in a six-room apartment, and taste 1 teaspoon of sugar dissolved in 2 gallons of water (FIGURE 4.3).

ADAPTATION: WEAKENING THE RESPONSE

Imagine that friends have invited you to come visit their beautiful new baby. As they greet you at the door, you are overwhelmed by the odor of a wet diaper. Why don't your friends do something about that smell? The answer lies in the previously mentioned sensory reduction and **sensory adaptation**. When a constant stimulus is presented for a length of time, sensation often fades or disappears. In sensory adaptation, receptors higher up in the sensory system get "tired" and actually fire less frequently.

sensory adaptation
Repeated or constant stimulation decreases the number of sensory messages sent to the brain, which causes decreased sensation.

Sensory adaptation makes sense from an evolutionary perspective. We can't afford to waste attention and time on unchanging, normally unimportant stimuli. "Turning down the volume" on repetitive information helps the brain cope with an overwhelming amount of sensory stimuli and allows time to pay attention to change. Sometimes, however, adaptation can be dangerous, as when people stop paying attention to a gas leak in the kitchen.

Measuring the senses
FIGURE 4.3

How do we know that humans can hear a watch ticking at 20 feet or smell one drop of perfume in a six-room apartment? The answer comes from research in **psychophysics** which studies the link between the physical characteristics of stimuli and our sensory experience of them. Researchers study how the strength or intensity of a stimulus affects an observer. Consider this example:

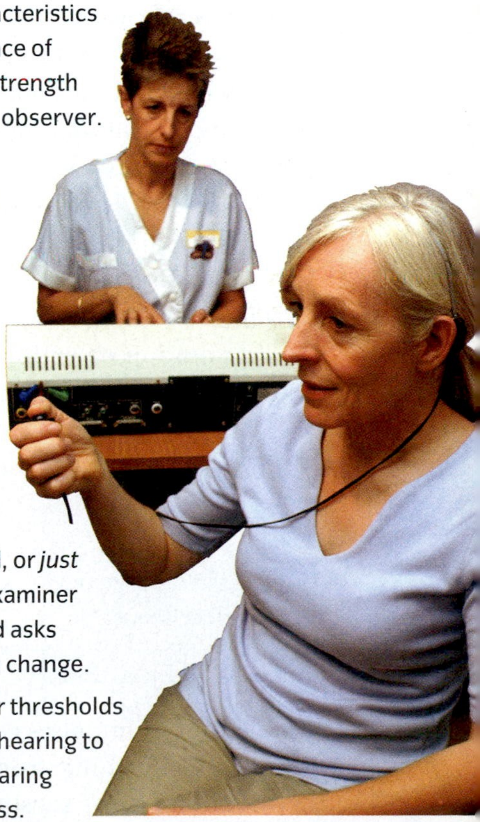

- To test for hearing loss, a hearing specialist uses a tone generator to produce sounds of differing pitches and intensities.
- You listen with earphones and indicate the earliest point at which you hear a tone. This is your **absolute threshold**, or the smallest amount of a stimulus that an observer can reliably detect.
- To test your **difference threshold**, or *just noticeable difference (JND)*, the examiner gradually changes the volume and asks you to respond when you notice a change.
- The examiner then compares your thresholds with those of people with normal hearing to determine whether you have a hearing loss and, if so, the extent of the loss.

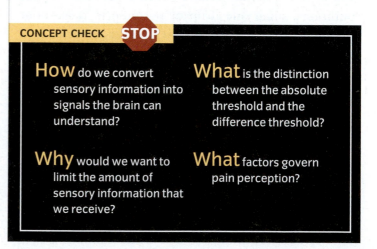

Sidelined? FIGURE 4.4

The body's ability to inhibit pain perception sometimes makes it possible for athletes to "play through" painful injuries. Do you think the potential for lasting damage is too high a price to pay for a medal or trophy?

Although some senses, like smell and touch, adapt quickly, we never completely adapt to visual stimuli or to extremely intense stimuli, such as the odor of ammonia or the pain of a bad burn. From an evolutionary perspective, these limitations on sensory adaptation aid survival, for example, by reminding us to avoid strong odors and heat or to take care of that burn.

If we don't adapt to pain, how do athletes keep playing despite painful injuries? In certain situations, the body releases natural painkillers called *endorphins* (see Chapter 2), which inhibit pain perception (FIGURE 4.4).

In addition to endorphin release, one of the most accepted explanations of pain perception is the **gate-control theory**, first proposed by Ronald Melzack and Patrick Wall (1965). According to this theory, the experience of pain depends partly on whether the neural message gets past a "gatekeeper" in the spinal cord. Normally, the gate is kept shut, either by impulses coming down from the brain or by messages coming from large-diameter nerve fibers that conduct most sensory signals, such as touch and pressure. However, when body tissue is damaged, impulses from smaller pain fibers open the gate.

gate-control theory Theory that pain sensations are processed and altered by mechanisms within the spinal cord.

According to the gate-control theory, massaging an injury or scratching an itch can temporarily relieve discomfort because pressure on large-diameter neurons interferes with pain signals. Messages from the brain can also control the pain gate, explaining how athletes and soldiers can carry on despite excruciating pain. When we are soothed by endorphins or distracted by competition or fear, our experience of pain can be greatly diminished. On the other hand, when we get anxious or dwell on our pain, we can intensify it (Roth et al., 2007; Sullivan, 2008; Sullivan, Tripp, & Santor, 1998). Ironically, well-meaning friends who ask chronic pain sufferers about their pain may unintentionally reinforce and increase it (Jolliffe & Nicholas, 2004).

Research also suggests that the pain gate may be chemically controlled, that a neurotransmitter called *substance P* opens the pain gate, and that endorphins close it (Bianchi et al., 2008; Cesaro & Ollat, 1997; Liu, Mantyh, & Basbaum, 1997). Other research (Melzack, 1999; Vertosick, 2000) finds that when normal sensory input is disrupted, the brain can generate pain and other sensations on its own. Amputees sometimes continue to feel pain (and itching or tickling) long after a limb has been amputated. This *phantom limb pain* occurs because nerve cells send conflicting messages to the brain. The brain interprets this "static" as pain because it arises in the area of the spinal cord responsible for pain signaling. When amputees are fitted with prosthetic limbs and begin using them, phantom pain generally disappears (Crawford, 2008; Gracely, Farrell, & Grant, 2002).

CONCEPT CHECK STOP

How do we convert sensory information into signals the brain can understand?

What is the distinction between the absolute threshold and the difference threshold?

Why would we want to limit the amount of sensory information that we receive?

What factors govern pain perception?

How We See and Hear

LEARNING OBJECTIVES

Identify the three major characteristics of light and sound waves.

Explain how the eye captures and focuses light energy and how it converts it into neural signals.

Describe the path that sound waves take in the ear.

Summarize the two theories that explain how we distinguish among different pitches.

Even the most complex visual and auditory experiences depend on our basic ability to detect light and sound. Both light and sound move in waves, similar to the movement of waves in the ocean (FIGURE 4.5).

WAVES OF LIGHT AND SOUND

Light waves are a form of electromagnetic energy, and different types of waves on the *electromagnetic spectrum* have different wavelengths.

In contrast to light waves, which are particles of electromagnetic energy, sound waves are produced by air molecules moving in a particular wave pattern. This occurs when an impact or vibrating objects, such as vocal cords or guitar strings, cause a sudden change in air pressure.

Differences between light and sound waves FIGURE 4.5

A Watching a fireworks show is only one of many ways you've probably learned that light travels faster than sound. But light also travels differently from sound, which must pass through a physical material to be heard. The speed of light is always 300 million meters (186,000 miles) per second no matter what it passes through. Sound travels through air at 344 meters (1100 ft) per second at 70°F. But if you were observing fireworks under water, you'd notice much less of a gap between the burst of light and the arrival of sound—in water sound will travel at a speed of about 1500 meters (5000 ft) per second, which is about five times its speed in air.

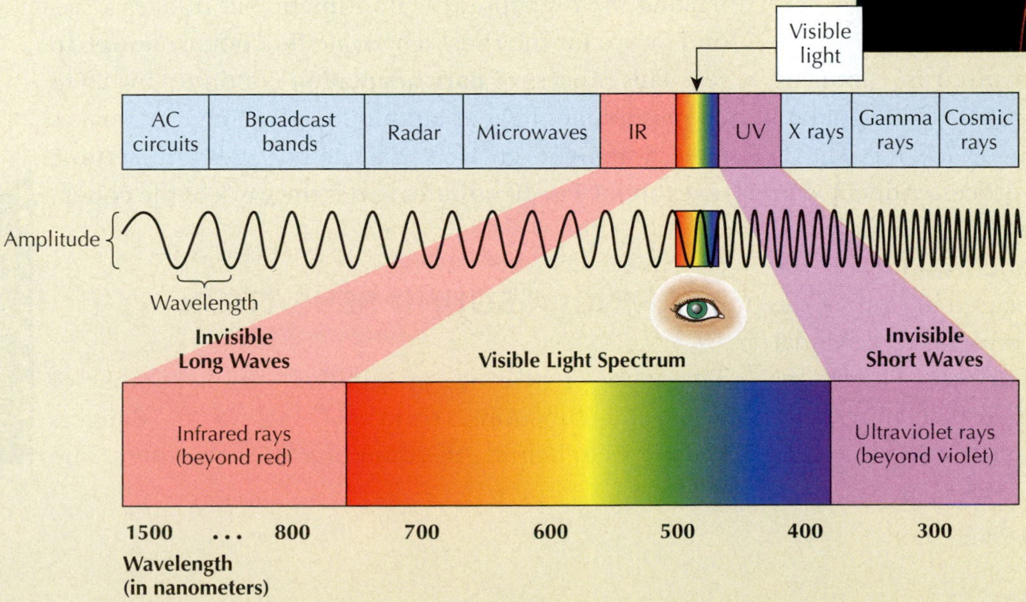

B The human eye can see only visible light, a small part of the full spectrum. Visible light with a short wavelength is perceived as blue, visible light with a medium wavelength is green or yellow, and visible light with a long wavelength is red.

Study Organizer 4.1 Properties of vision and hearing

Physical Properties	**Wavelength:** The distance between successive peaks.	**Wave amplitude:** The height from peak to trough.	**Range of wave-lengths:** the mixture of waves.
	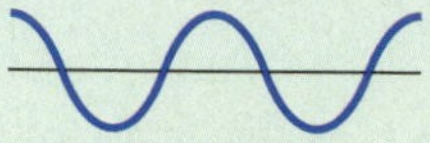 *Long wavelength/low frequency* 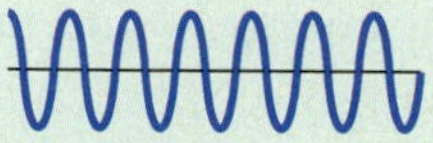 *Short wavelength/high frequency*	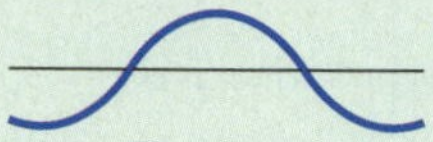 *Low amplitude/low intensity* 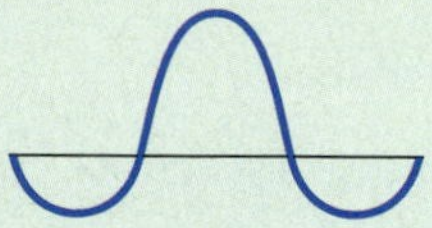 *High amplitude/high intensity*	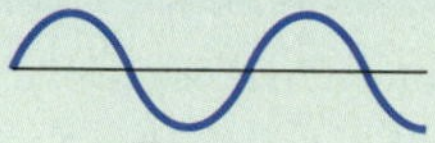 *Low range/low complexity* 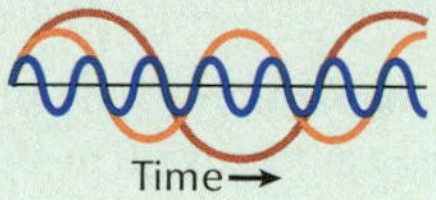 *High range/high complexity*
VISION (Light waves)	**Hue:** Short wavelengths produce higher frequency and bluish colors; long wavelengths produce lower frequency and reddish colors.	**Brightness:** Great amplitude produces more intensity and bright colors; small amplitude produces less intensity and dim colors.	**Saturation:** Wider range produces more complex color; narrow range produces less complex color.
AUDITION (Sound waves)	**Pitch:** Shorter wavelengths produce higher frequency and high-pitched sounds; long wavelengths produce lower frequency and low-pitched sounds.	**Loudness:** Great amplitude produces louder (more intense) sounds; small amplitude produces soft sounds.	**Timbre:** Wider range produces more complex sound with a mix of multiple frequencies. Narrower range produces less complex sound with one or a few frequencies.

Both light waves and sound waves vary in **wavelength**, **amplitude** (height), and **range**—each with a distinct effect on vision and hearing, or audition, as shown in STUDY ORGANIZER 4.1.

VISION: THE EYES HAVE IT

Several structures in the eye are involved in capturing and focusing light and converting it into neural signals to be interpreted by the brain, as shown in FIGURE 4.6.

Thoroughly understanding the processes detailed in Figure 4.6 gives us clues for understanding some visual peculiarities. For example, small abnormalities in the eye sometimes cause images to be focused in front of the retina (**nearsightedness**, also called myopia) or behind it (**farsightedness**, or hyperopia). Corrective lenses or laser surgery can correct most such visual acuity problems. During middle age, most people's lenses lose elasticity and the ability to accommodate for near vision, a condition known as presbyopia that can normally be treated with corrective lenses.

If you walk into a dark movie theater on a sunny afternoon, you will at first be blinded. This is because in bright light, the pigment inside the rods is bleached, making them temporarily nonfunctional. It takes a second or two for the rods to become functional enough to see. This process of **dark adaptation** continues for 20 to 30 minutes. Light adaptation, the adjustment that takes place when you go from darkness to a bright setting, takes about 7 to 10 minutes and is the work of the cones.

HEARING: A SOUND SENSATION

The sense of hearing, or **audition**, has a number of important functions, from alerting us to dangers around us to helping us communicate with others. The

How the eye sees Figure 4.6

Process Diagram

3 Behind the iris and pupil, the muscularly controlled lens focuses incoming light into an image on the light-sensitive *retina*, located on the back surface of the fluid-filled eyeball. Note how the lens reverses the image from right to left and top to bottom when it is projected on to the retina. The brain later reverses the visual input into the final image that we perceive.

2 The light then passes through the *pupil*, a small adjustable opening. Muscles in the *iris* allow the *pupil* to dilate or constrict in response to light intensity or emotional factors.

1 Light first enters through the *cornea,* which helps focus incoming light rays.

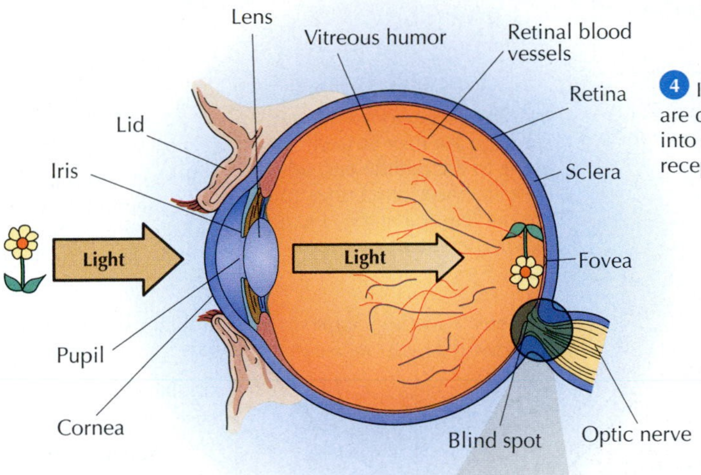

4 In the **retina**, light waves are detected and transduced into neural signals by vision receptor cells (rods and cones).

5 The **fovea**, a tiny pit filled with cones, is responsible for our sharpest vision.

6 **Rods** are visual receptors that detect white, black, and gray and are responsible for peripheral vision. They are most important in dim light and at night.

7 **Cones** are visual receptors adapted for color, daytime, and detailed vision. They are sensitive to many wavelengths, but each is maximally sensitive to red, green, or blue.

10 After exiting the eye, neural messages travel along the optic nerve to the brain for further processing.

9 At the back of the retina lies an area that has no visual receptors at all and absolutely no vision. This **blind spot** is where blood vessels and nerves enter and exit the eyeball.

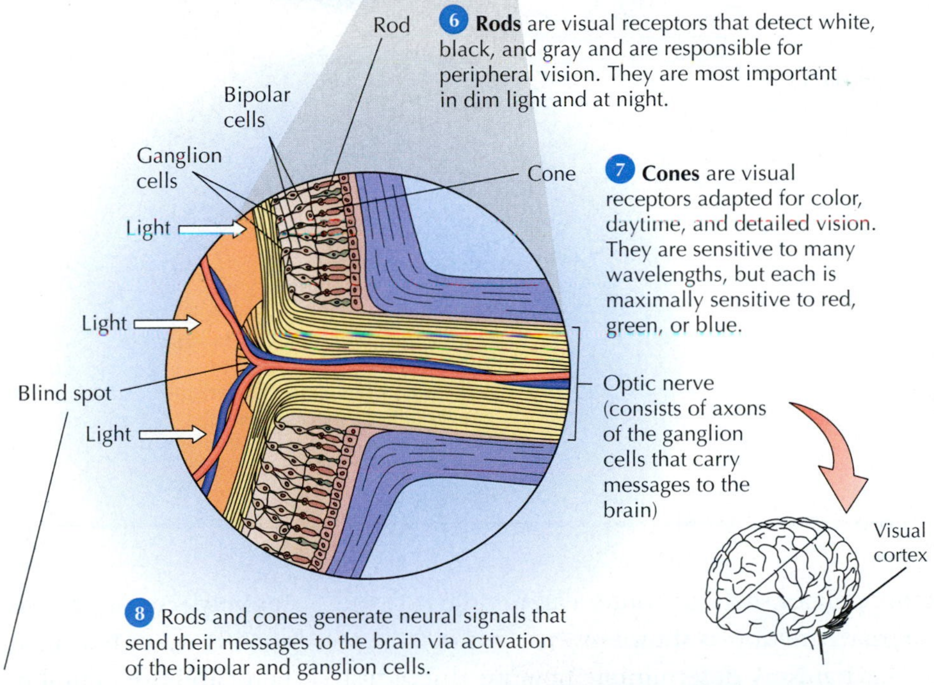

8 Rods and cones generate neural signals that send their messages to the brain via activation of the bipolar and ganglion cells.

Do you have a blind spot?

Everyone with vision does. To find yours, hold this book about one foot in front of you, close your right eye, and stare at the X with your left eye. Very slowly, move the book closer to you. You should see the worm disappear and the apple become whole.

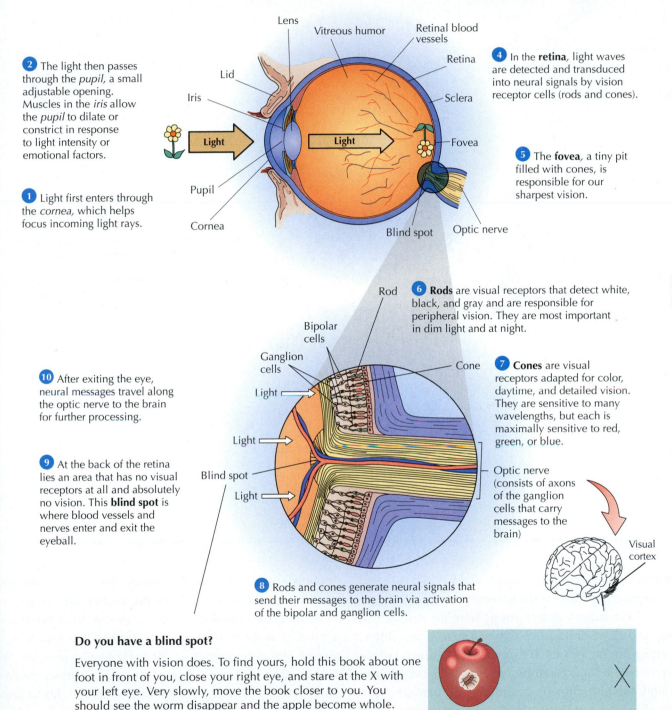

Process Diagram

How the ear hears Figure 4.7

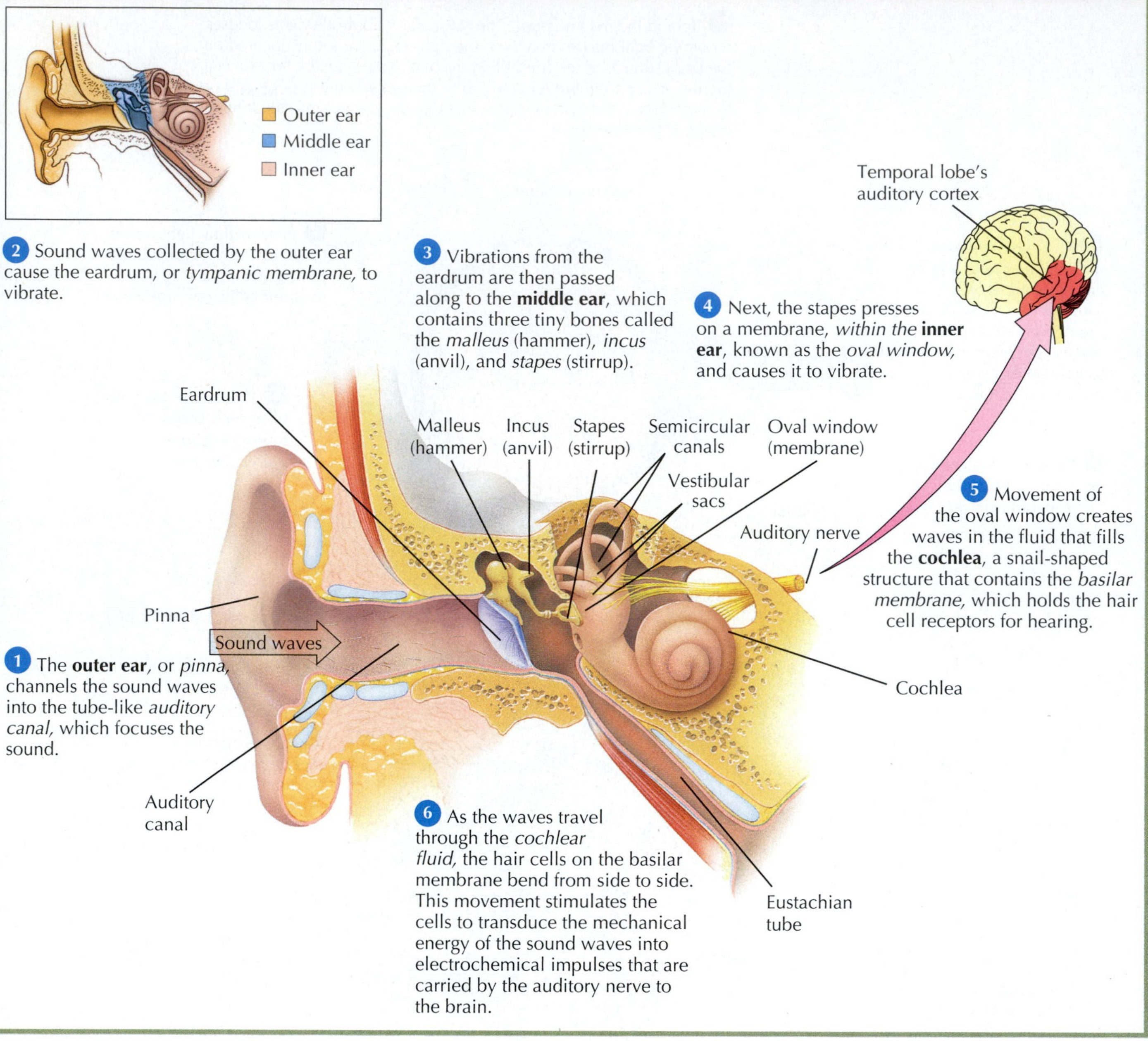
Outer ear
Middle ear
Inner ear
1 The **outer ear**, or *pinna*, channels the sound waves into the tube-like *auditory canal*, which focuses the sound.
2 Sound waves collected by the outer ear cause the eardrum, or *tympanic membrane*, to vibrate.
3 Vibrations from the eardrum are then passed along to the **middle ear**, which contains three tiny bones called the *malleus* (hammer), *incus* (anvil), and *stapes* (stirrup).
4 Next, the stapes presses on a membrane, *within the* **inner ear**, known as the *oval window*, and causes it to vibrate.
5 Movement of the oval window creates waves in the fluid that fills the **cochlea**, a snail-shaped structure that contains the *basilar membrane*, which holds the hair cell receptors for hearing.
6 As the waves travel through the *cochlear fluid*, the hair cells on the basilar membrane bend from side to side. This movement stimulates the cells to transduce the mechanical energy of the sound waves into electrochemical impulses that are carried by the auditory nerve to the brain.
Temporal lobe's auditory cortex
Eardrum
Malleus (hammer)
Incus (anvil)
Stapes (stirrup)
Semicircular canals
Oval window (membrane)
Vestibular sacs
Auditory nerve
Pinna
Sound waves
Cochlea
Auditory canal
Eustachian tube

ear has three major sections (outer ear, middle ear, and inner ear) that function as shown in Figure 4.7.

The mechanisms determining how we distinguish among sounds of different pitches (low to high) differ, depending on the sounds' **frequency**. According to **place theory**, different high-frequency sound waves (which produce high-pitched sounds) maximally stimulate the hair cells at different locations along the basilar membrane (Figure 4.7). Hearing for low-pitched sounds works differently. According to **frequency theory**, low-pitched sounds cause hair cells along the basilar membrane to bend and fire neural messages (action potentials) at the same rate as the frequency of that sound. For example, a sound with a frequency of 90 hertz would produce 90 action potentials per second in the auditory nerve. Interestingly, as we age, we tend to lose our ability to hear high-pitched sounds, while still being able to hear low-pitched sounds (Figure 4.8).

Whether we detect a sound as soft or loud depends on its intensity. Waves with high peaks and low valleys produce loud sounds; those that have relatively low peaks and shallow valleys produce soft sounds. The relative loudness or softness of sounds is measured on a scale of *decibels* (**FIGURE 4.9**).

Hearing loss can stem from two major causes: (1) **conduction deafness**, or middle-ear deafness, which results from problems with the mechanical system that conducts sound waves to the inner ear, and (2) **nerve deafness**, or inner-ear deafness, which involves damage to the cochlea, hair cells, or auditory nerve (Figure 4.9).

Although most conduction deafness is temporary, damage to the auditory nerve or receptor cells is almost always irreversible. The only treatment for nerve deafness is a small electronic device called a **cochlear implant**. If the auditory nerve is intact, the implant bypasses hair cells to stimulate the nerve. Currently, cochlear implants produce only a crude approximation of hearing, but the technology is improving. It's best to protect your sense of hearing by avoiding exceptionally loud noises, wearing earplugs when such situations cannot be avoided, and paying attention to bodily warnings of possible hearing loss, including a change in your normal hearing threshold and *tinnitus*, a whistling or ringing sensation in your ears.

For whom the bell tolls FIGURE 4.8

Stealthy teenagers now have a biological advantage over their teachers: a cell phone ringtone that sounds at 17 kilohertz—too high for adult ears to detect. The ringtone is an ironic offshoot of another device using the same sound frequency. That invention, dubbed the Mosquito, was designed to help shopkeepers annoy and deter loitering teens.

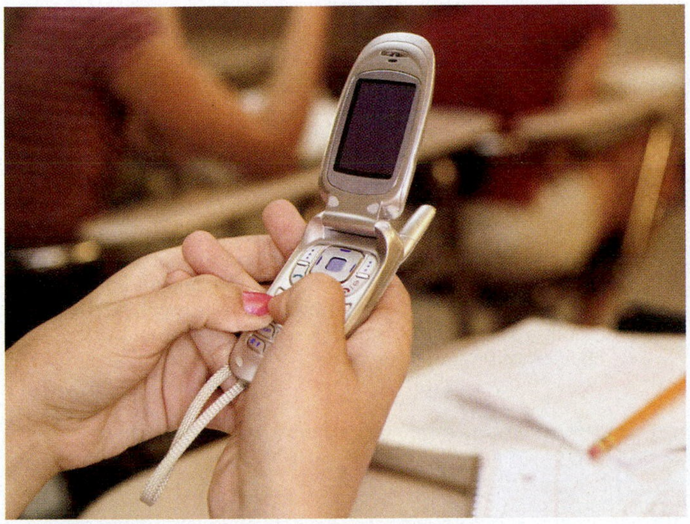

How loud is too loud? FIGURE 4.9

The loudness of a sound is measured in *decibels,* and the higher a sound's decibel reading, the more damaging it is to the ear. Chronic exposure to loud noise, such as loud music or heavy traffic—or brief exposure to really loud sounds, such as a stereo at full blast, a jackhammer, or a jet engine—can cause permanent nerve deafness. Disease and biological changes associated with aging can also cause nerve deafness.

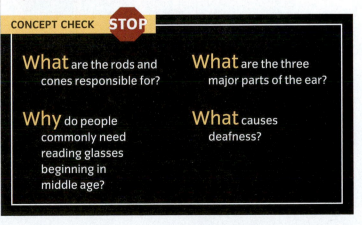

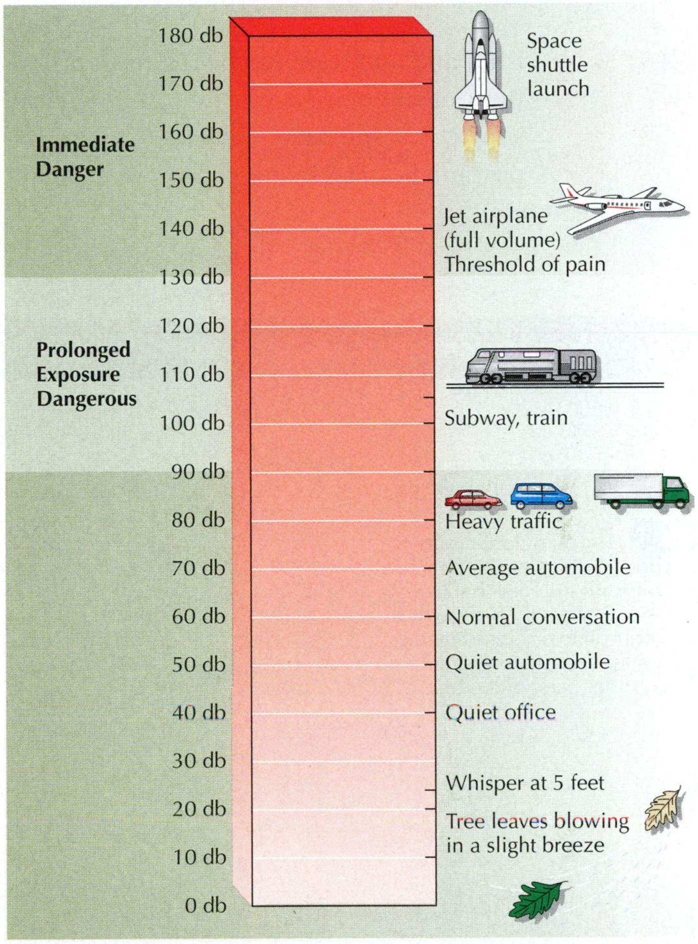

CONCEPT CHECK STOP

What are the rods and cones responsible for?

Why do people commonly need reading glasses beginning in middle age?

What are the three major parts of the ear?

What causes deafness?

Our Other Senses

LEARNING OBJECTIVES

Explain the importance of smell and taste to survival.

Describe how the information contained in odor molecules reaches the brain.

Identify the locations of receptors for the body senses.

Explain the role of our vestibular and kinesthetic senses.

Vision and audition may be the most prominent of our senses, but the others—taste, smell, and the body senses—are also important for gathering information about our environment.

SMELL AND TASTE: SENSING CHEMICALS

Smell and taste are sometimes referred to as the *chemical senses* because they both involve chemoreceptors that are sensitive to certain chemical molecules. Smell and

Process Diagram

How the nose smells FIGURE 4.10

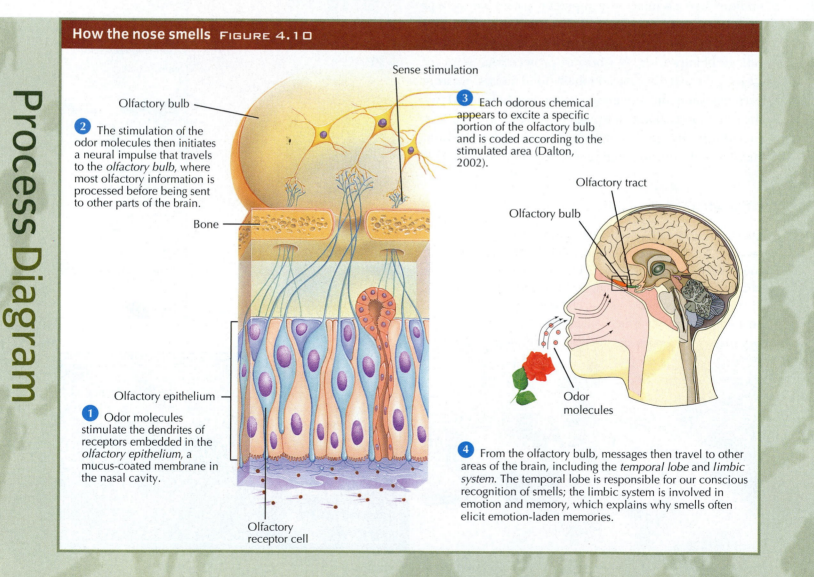

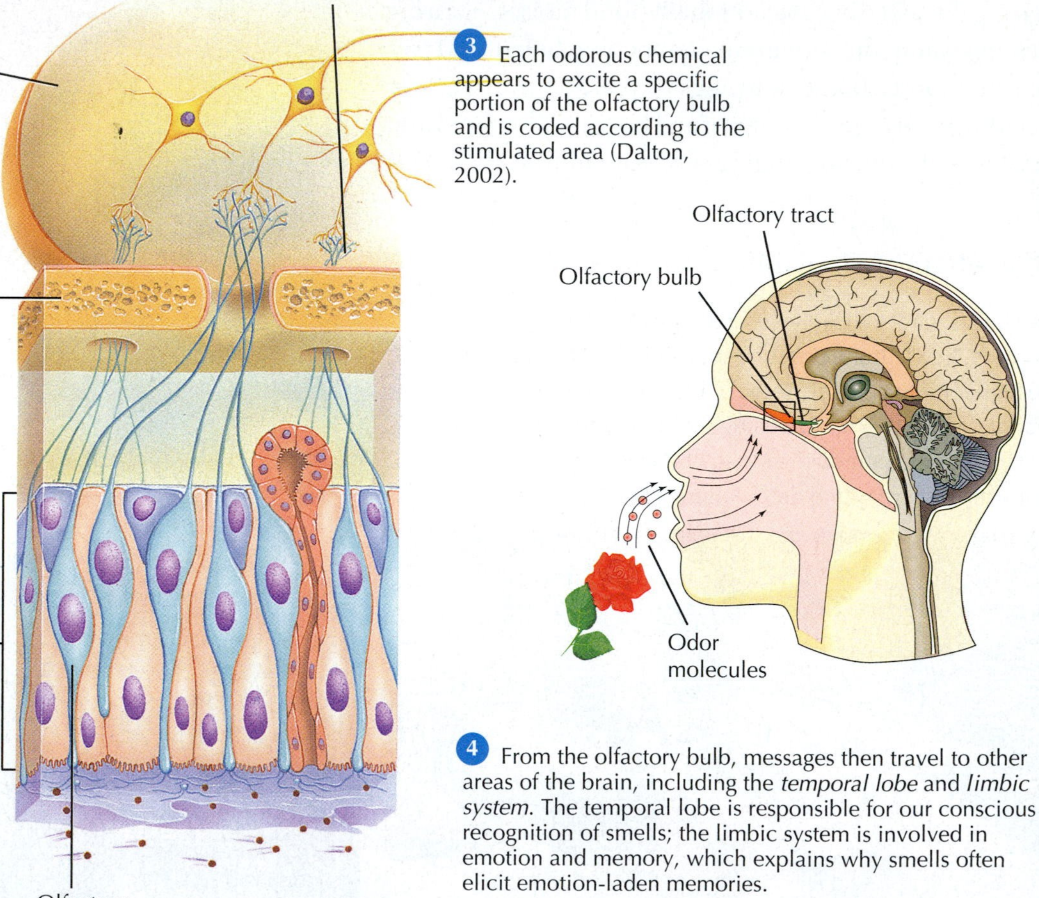

1 Odor molecules stimulate the dendrites of receptors embedded in the *olfactory epithelium*, a mucus-coated membrane in the nasal cavity.

2 The stimulation of the odor molecules then initiates a neural impulse that travels to the *olfactory bulb*, where most olfactory information is processed before being sent to other parts of the brain.

3 Each odorous chemical appears to excite a specific portion of the olfactory bulb and is coded according to the stimulated area (Dalton, 2002).

4 From the olfactory bulb, messages then travel to other areas of the brain, including the *temporal lobe* and *limbic system*. The temporal lobe is responsible for our conscious recognition of smells; the limbic system is involved in emotion and memory, which explains why smells often elicit emotion-laden memories.

taste receptors are located near each other and often interact so closely that we have difficulty separating the sensations.

Our sense of smell, **olfaction**, is remarkably useful and sensitive. We possess more than 1,000 types of olfactory receptors, allowing us to detect more than 10,000 distinct smells (FIGURE 4.10). The nose is more sensitive to smoke than any electronic detector, and—through practice—blind people can quickly recognize others by their unique odors.

Some research on **pheromones**—compounds found in natural body scents that may affect the behavior of others, including their sexual behavior—supports the idea that these chemical odors increase sexual behaviors in humans (Savic, Berglund, & Lunstrom, 2007; Thornhill et al., 2003). However, other findings question the results (Hays, 2003), suggesting that human sexuality is far more complex than that of other animals—and more so than perfume advertisements would have you believe.

Today, the sense of taste, **gustation**, may be the least critical of our senses. In the past, however, it probably contributed to our survival. The major function of taste, aided by smell, is to help us avoid eating or drinking harmful substances. Because many plants that taste bitter contain toxic chemicals, an animal is more likely to survive if it avoids bitter-tasting plants (Cooper et al., 2002; Kardong, 2008; Skelhorn et al., 2008). Humans and other animals have a preference for sweet foods, which are generally nonpoisonous and are good sources of energy. Children's taste buds are replaced more quickly than adults', so they often dislike foods with strong or unusual tastes. Many food and taste preferences are learned from childhood experiences and cultural influences, so that one person's delicacy can be a source of revulsion to others.

When we take away the sense of smell, there are five distinct tastes: sweet, sour, salty, bitter, and umami. Umami means "delicious" or "savory" and refers to sensitivity to an amino acid called glutamate (Chandrashekar et al., 2006; McCabe & Rolls, 2007). Glutamate is found in meats, meat broths, and monosodium glutamate (MSG).

Taste receptors respond differentially to food molecules of different shapes. The major taste receptors (*taste buds*) are clustered on our tongues within little bumps called *papillae* (FIGURE 4.11).

Taste sensation FIGURE 4.11

When we eat and drink, liquids and dissolved foods flow over bumps on our tongue called *papillae* and into the pores to the *taste buds*, which contain the receptors for taste.

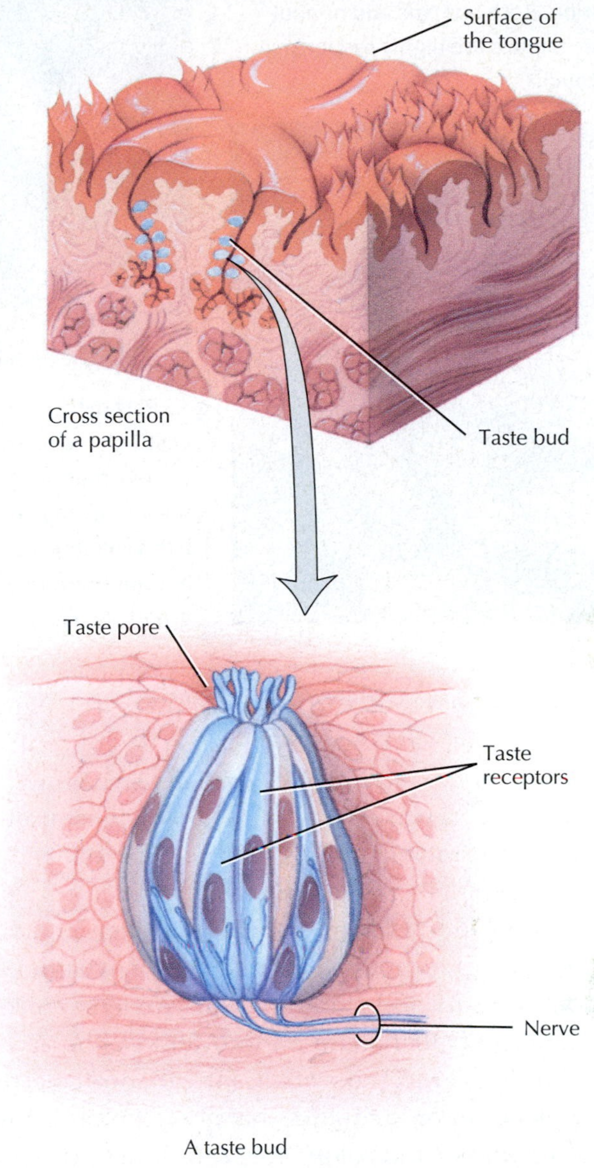

THE BODY SENSES: MORE THAN JUST TOUCH

The senses that tell the brain how the body is oriented, where and how the body is moving, and what it touches or is touched by are called the body senses. They include the **skin senses**, the **vestibular sense**, and **kinesthesia**.

The body senses FIGURE 4.12

A The skin senses are vital. Not only do they protect our internal organs, but they provide our brains with basic attachment and survival information. Both humans and nonhuman animals are highly responsive to touch.

B Part of the "thrill" of amusement park rides comes from our vestibular sense becoming confused. The vestibular sense is used by the eye muscles to maintain visual fixation and sometimes by the body to change body orientation. We can become dizzy or nauseated if the vestibular sense becomes "confused" by boat, airplane, or automobile motion. Children between ages 2 and 12 years have the greatest susceptibility to motion sickness.

C Without her finely tuned kinesthetic sense to provide information about her bodily posture, orientation, and movement, U.S. gymnast Nastia Liukin would have been on her way to the hospital rather than the Olympics.

Our skin is sensitive to touch (or pressure), temperature, and pain (FIGURE 4.12A). The concentration and depth of the receptors for each of these stimuli vary. For example, touch receptors are most concentrated on the face and fingers and least so in the back and legs. Some receptors respond to more than one type of stimulation—for example, itching, tickling, and vibrating sensations seem to be produced by light stimulation of both pressure and pain receptors. We have more skin receptors for cold than for warmth, and we don't seem to have any "hot" receptors at all. Instead, our cold receptors detect not only coolness but also extreme temperatures—both hot and cold (Craig & Bushnell, 1994).

The vestibular sense is responsible for balance—it informs the brain of how the body, and particularly the head, is oriented with respect to gravity and three-dimensional space. When the head moves, liquid in the *semicircular canals*, located in the inner ear, moves and bends hair cell receptors. At the end of the semicircular canals are the *vestibular sacs*, which contain hair cells sensitive to the specific angle of the head—straight up and down or tilted. Information from the semicircular canals and the *vestibular sacs* is converted to neural impulses that are then carried to the appropriate section of the brain (FIGURE 4.12B).

Kinesthesia is the sense that provides the brain with information about bodily posture, orientation, and movement. Kinesthetic receptors are found throughout the muscles, joints, and tendons of the body. They tell the brain which muscles are being contracted or relaxed, how our body weight is distributed, where our arms and legs are in relation to the rest of our body, and so on (FIGURE 4.12C).

CONCEPT CHECK **STOP**

Why are our senses of smell and taste called the chemical senses?

What are the three basic skin sensations?

How do the vestibular and kinesthetic senses differ?

Understanding Perception

LEARNING OBJECTIVES

Document the relationship between selective attention, feature detectors, and habituation.

Summarize the factors involved in perceptual interpretation.

Describe the limitations of subliminal perception.

We are ready to move from sensation and the major senses to perception, the process of selecting, organizing, and interpreting incoming sensations into useful mental representations of the world.

Normally, our perceptions agree with our sensations. When they do not, the result is an *illusion*. See *Psychological Science: Optical Illusions* for more information about how illusions provide psychologists with a tool for studying the normal process of perception.

Psychological Science

Optical Illusions

An **illusion** is a false impression produced by errors in the perceptual process or by actual physical distortions, as in desert mirages. Drawing A illustrates the *Müller-Lyer illusion*. The two vertical lines are the same length, but psychologists have learned that people who live in urban environments normally see the one on the right as longer. This is because they have learned to make size and distance judgments from perspective cues created by right angles and horizontal and vertical lines of buildings and streets.

Perhaps more familiar is the *moon illusion* (B). As we all know, the moon is not actually larger on the horizon, yet we perceive it to be much larger than when it is directly overhead. When the moon is on the horizon, we judge its size and distance in relation to familiar objects (trees or building), but when it is high in the sky, directly above us, we have little information to help us judge either its size or distance.

Look at C, which is known as the *Ponzo illusion*. Do you perceive the top black line as being much larger than the one on the bottom? Both lines are the exact same size, but, like the trees in the foreground of the photo of the moon illusion, the converging lines provide depth cues telling you that the top dark line is farther away than the bottom line and therefore much larger.

A Müller-Lyer illusion

B Moon illusion

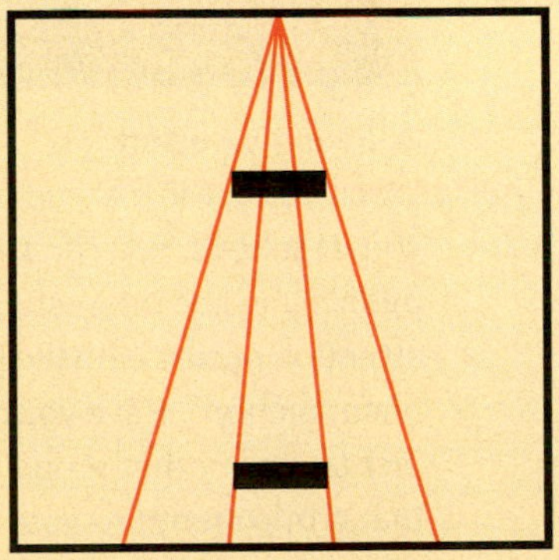

C Ponzo illusion

CRITICAL THINKING

Here are two interesting questions:

1. What do you think causes "errors" in the perceptual process such as the optical illusions described here?
2. When watching films of moving cars, the wheels appear to go backwards. Can you explain this common visual illusion?

SELECTION: EXTRACTING IMPORTANT MESSAGES

In almost every situation, we confront more sensory information than we can reasonably pay attention to. Three major factors help us focus on some stimuli and ignore others: **selective attention**, **feature detectors**, and **habituation** (FIGURE 4.13).

selective attention Filtering out and attending only to important sensory messages.

Certain basic mechanisms for perceptual selection are built into the brain. For example, through the process of selective attention (see FIGURE 4.13A), the brain picks out the information that is important to us and discards the rest (Folk & Remington, 1998; Kramer et al., 2000).

feature detectors Specialized brain cells that respond only to certain sensory information.

habituation Tendency of the brain to ignore environmental factors that remain constant.

In humans and other animals, the brain contains specialized cells, called feature detectors, that respond only to certain sensory information (see FIGURE 4.13B). For example, humans have feature detectors in the temporal and occipital lobes that respond maximally to faces. Interestingly, people with a condition called prosopagnosia (*prospon* means "face" and *agnosia* means "failure to know") can recognize that they are looking at a face, but they cannot say whose face is reflected in a mirror, even if it is their own or that of a friend or relative.

Visualizing

Selection FIGURE 4.13

A Selective attention
When you are in a group of people, surrounded by various conversations, you can still select and attend to the voices of people you find interesting. Another example of selective attention occurs with the well-known "cocktail party phenomenon." Have you noticed how you can suddenly pick up on another group's conversation if someone in that group mentions your name?

B Feature detectors
Cats possess cells, known as feature detectors, that respond to specific lines and angles (Hubel & Wiesel, 1965, 1979). Researchers found that kittens reared in a vertical world fail to develop their innate ability to detect horizontal lines or objects. On the other hand, kittens restricted to only horizontal lines cannot detect vertical lines. A certain amount of interaction with the environment is apparently necessary for feature detector cells to develop normally (Blakemore & Cooper, 1970).

C Habituation
These two girls' brains may "choose to ignore" their painful braces. (Sensory adaptation may have also occurred. Over time the girls' pressure sensors send fewer messages to the brain.)

NATIONAL GEOGRAPHIC

Other examples of the brain's ability to filter experience are evidenced by habituation. Apparently, the brain is "pre-wired" to pay more attention to changes in the environment than to stimuli that remain constant. For example, when braces are first applied or when they are tightened, they can be very painful. After a while, however, awareness of the pain diminishes (see FIGURE 4.13C).

As advertisers and political operatives well know, people tend to automatically select stimuli that are intense, novel, moving, contrasting, and repetitious. For sheer volume of sales (or votes), the question of whether you like the ad is irrelevant. If it gets your attention, that's all that matters.

ORGANIZATION: FORM, CONSTANCY, DEPTH, AND COLOR

Raw sensory data are like the parts of a watch—they must be assembled in a meaningful way before they are useful. We organize sensory data in terms of form, constancy, depth, and color.

Form perception

Gestalt psychologists were among the first to study how the brain organizes sensory impressions into a **gestalt**—a German word meaning "form" or "whole." They emphasized the importance of organization and patterning in enabling us to perceive the whole stimulus rather than perceiving its discrete parts as separate entities. The Gestaltists proposed several laws of organization that specify how people perceive form (FIGURE 4.14).

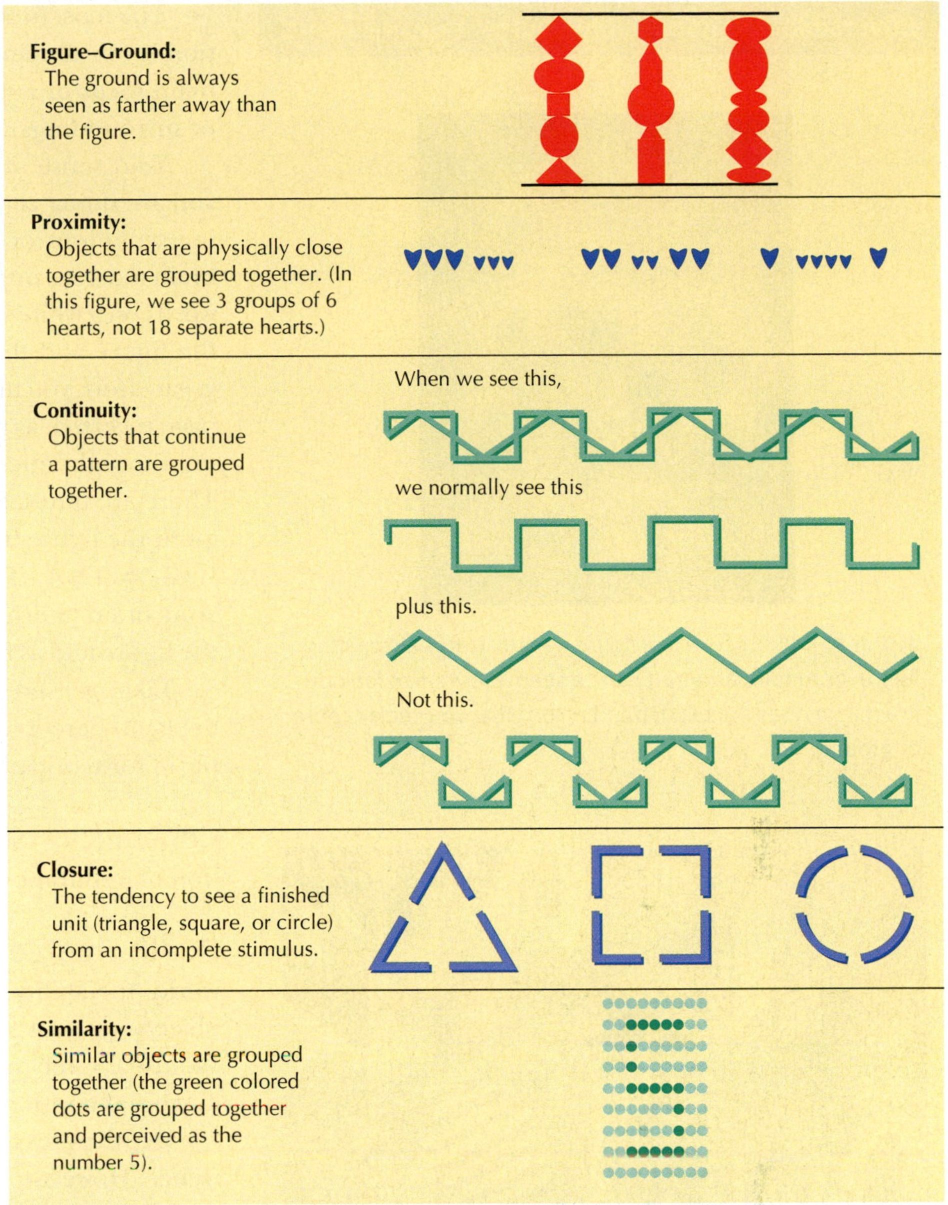

Figure–Ground:
The ground is always seen as farther away than the figure.
Proximity:
Objects that are physically close together are grouped together. (In this figure, we see 3 groups of 6 hearts, not 18 separate hearts.)
Continuity:
Objects that continue a pattern are grouped together.
When we see this,
we normally see this
plus this.
Not this.
Closure:
The tendency to see a finished unit (triangle, square, or circle) from an incomplete stimulus.
Similarity:
Similar objects are grouped together (the green colored dots are grouped together and perceived as the number 5).

Gestalt principles of organization FIGURE 4.14

Gestalt principles are based on the notion that we all share a natural tendency to force patterns onto whatever we see. Although the examples of the Gestalt principles in this figure are all visual, each principle applies to other modes of perception as well. For example, the Gestalt principle of *contiguity* cannot be shown because it involves nearness in time, not visual nearness. You also may have experienced *aural figure and ground effects* at a movie or a concert when there was a conversation going on close by and you couldn't sort out what sounds were the background and what you wanted to be your focus.

Another good example of a Gestalt principle not shown in this figure is *visual closure*, which happens every time you watch television. The picture on the TV screen appears to be a solid image, but it's really a very fast stream of small dots being illuminated one by one, "painting" tiny horizontal lines down the screen one line at a time. Your brain closes the momentary blank gaps on the screen.

VIEW THIS IN ACTION in your WileyPLUS course

Reversible and impossible figures FIGURE 4.15

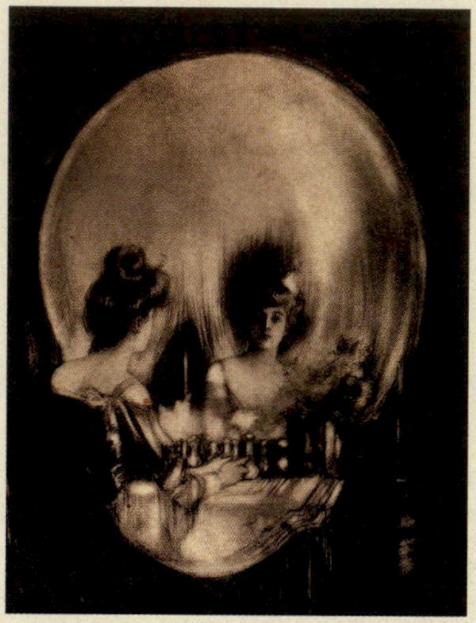

A This so-called *reversible figure* demonstrates alternating figure–ground relations. It can be seen as a woman looking in a mirror or as a skull, depending on what you see as figure or ground.

B When you first glance at the famous painting by the Dutch artist M. C. Escher, you detect specific features of the stimuli and judge them as sensible figures. But as you try to sort and organize the different elements into a stable, well-organized whole, you realize they don't add up—they're impossible. There is no one-to-one correspondence between your actual sensory input and your final perception.

The most fundamental Gestalt principle of organization is our tendency to distinguish between figure (our main focus of attention) and ground (the background or surroundings).

Your sense of figure and ground is at work in what you are doing right now—reading. Your brain is receiving sensations of black lines and white paper, but your brain is organizing these sensations into black letters and words on a white background. You perceive the letters as the figure and the white as the ground. If you make a great effort, you might be able to force yourself to see the page reversed, as though a black background were showing through letter-shaped holes in a white foreground. There are times, however, when it is very hard to distinguish the figure from the ground as can be seen in FIGURE 4.15A. This is known as a "reversible figure." Your brain alternates between seeing the light areas as the figure and as the ground.

Like *reversible figures*, **impossible figures** help us understand perceptual principles—in this case, the principle of form organization (FIGURE 4.15B).

Constancy See *What a Psychologist Sees* for a discussion of four important perceptual constancies.

Depth perception In our three-dimensional world, the ability to perceive the depth and distance of objects—as well as their height and width—is essential. We usually rely most heavily on vision to perceive distance and depth.

Depth perception is learned primarily through experience. However, research using an apparatus called the *visual cliff* (FIGURE 4.16 on page 104) suggests that some depth perception is inborn.

One mechanism by which we perceive depth is the interaction of both eyes to produce **binocular cues**; the other involves **monocular cues**, which work with each eye separately.

One of the most important binocular cues for depth perception comes from **retinal disparity** (FIGURE 4.17 on page 104). Because our eyes are about two and one-half inches apart, each retina receives a slightly different view of the world. (Watch what happens when you point at a distant object, closing one eye and then the other.) When both eyes are open, the brain fuses the different images into one, an effect known as *stereoscopic vision*.

What a Psychologist Sees

Four Perceptual Constancies

As noted earlier with sensory adaptation and habituation, we are particularly alert to change in our environment. However, we also are attuned to consistencies. Without this **perceptual constancy**, our world would be totally chaotic. We learn about perceptual constancies through prior experience and learning.

There are four basic perceptual constancies:

1. *Size Constancy* Our retinal image of the father and daughter in the foreground is that they are much larger than the trees and mountains behind them. Thanks to size constancy, however, we readily perceive them as people of normal size. ▶

2. *Shape Constancy* As the coin is rotated, it changes shape, but we still perceive it as the same coin because of shape constancy. ▼

Want another example of size and shape constancy?

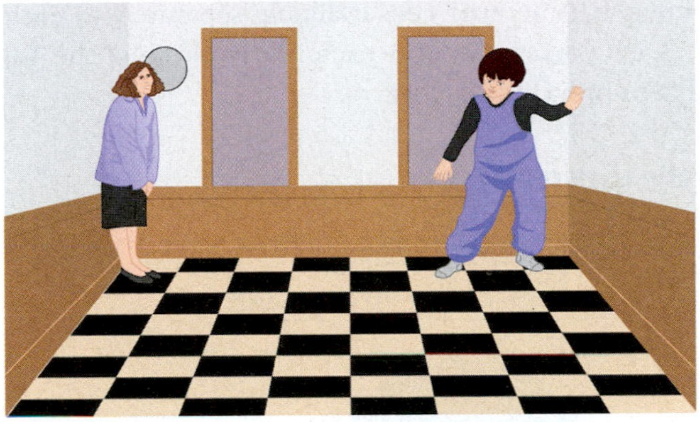

The young boy on the right appears to be much larger than the woman on the left. The illusion is so strong that when a person walks from the left corner to the right, the observer perceives the person to be "growing," even though that is not possible. How can this be? ▶

This so-called *Ames room illusion* is based on the unusual construction of the room, and our perceptual constancies have falsely filled in the wrong details. To the viewer, peering through the peephole, the room appears to be a normal cubic-shaped room. But the true shape is trapezoidal: the walls are slanted, and the floor and ceiling are at an incline. Because our brains mistakenly assume the two people are the same distance away, we compensate for the apparent size difference by making the person on the left appear much smaller. ▶

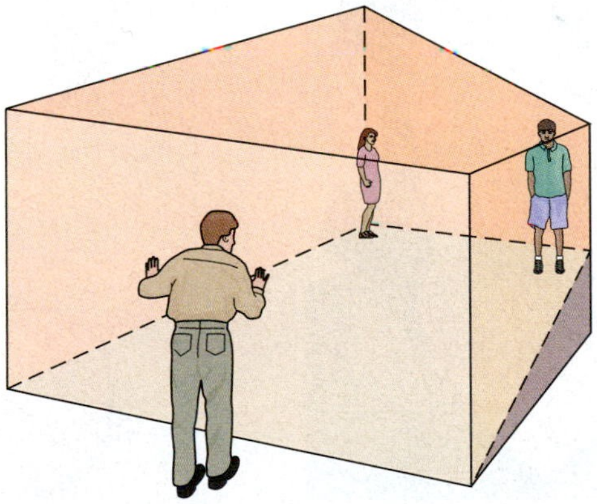

Several Ames room sets were used in *The Lord of the Rings* film series to make the heights of the hobbits appear correct when standing next to Gandalf.

◀ **3.** *Color Constancy* and **4.** *Brightness Constancy* We perceive the dog's fur in this photo as having a relatively constant hue (or color) and brightness despite the fact that the wavelength of light reaching our retinas may vary as the light changes.

Visual cliff FIGURE 4.16

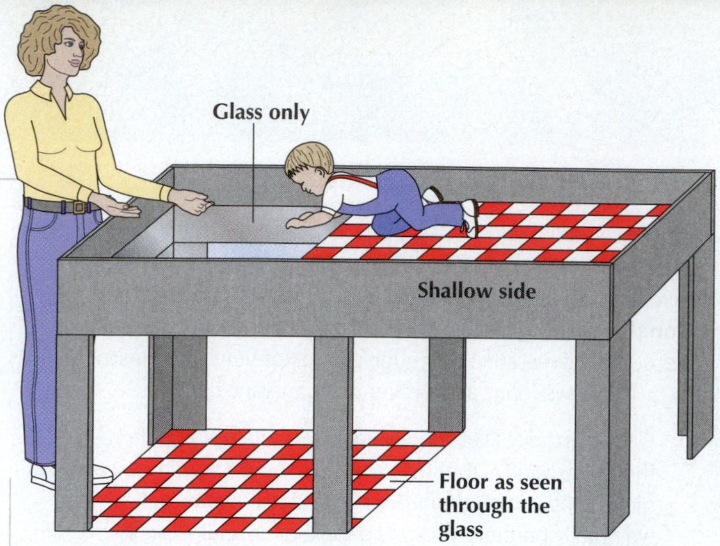

Crawling infants hesitate or refuse to move to the "deep end" of the visual cliff (Gibson & Walk, 1960), indicating that they perceive the difference in depth. (The same is true for baby animals that walk almost immediately after birth.) Even two-month-old infants show a change in heart rate when placed on the deep versus shallow side of the visual cliff (Banks & Salapatek, 1983).

As we move closer to an object, a second binocular cue, **convergence**, helps us judge depth. The closer the object, the more our eyes are turned inward. The resulting amount of eye-muscle strain helps the brain interpret distance.

The binocular (two eyes) cues of retinal disparity and convergence are inadequate in judging distances longer than a football field. Luckily, we have several monocular (one eye) cues available separately to each eye. See if you can identify each in the photo of the Taj Mahal in India, in FIGURE 4.18.

Two additional monocular cues are **accommodation** of the lens (discussed earlier) and **motion parallax**. In accommodation, muscles that adjust the shape of the lens as it focuses on an object send neural impulses to the brain, which interprets the signal to perceive distance. Motion parallax refers to the fact that when we are moving, close objects appear to whiz by whereas farther objects seem to move more slowly or remain stationary. This effect can easily be seen when traveling by car or train.

Retinal disparity FIGURE 4.17

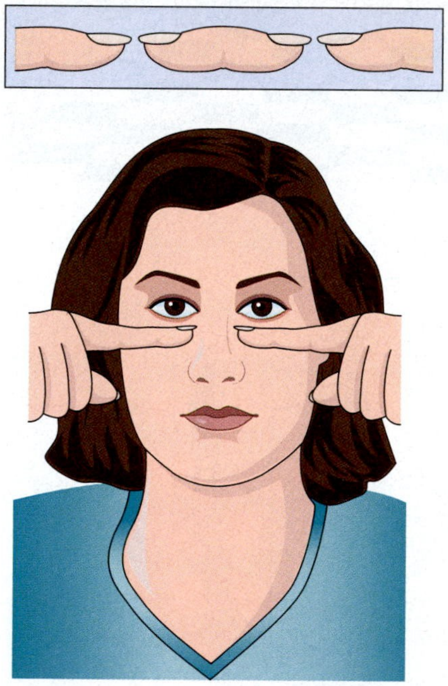

A Stare at your two index fingers a few inches in front of your eyes with their tips an inch apart. Do you see the "floating finger"? Move it farther away and the "finger" will shrink. Move it closer and it will enlarge.

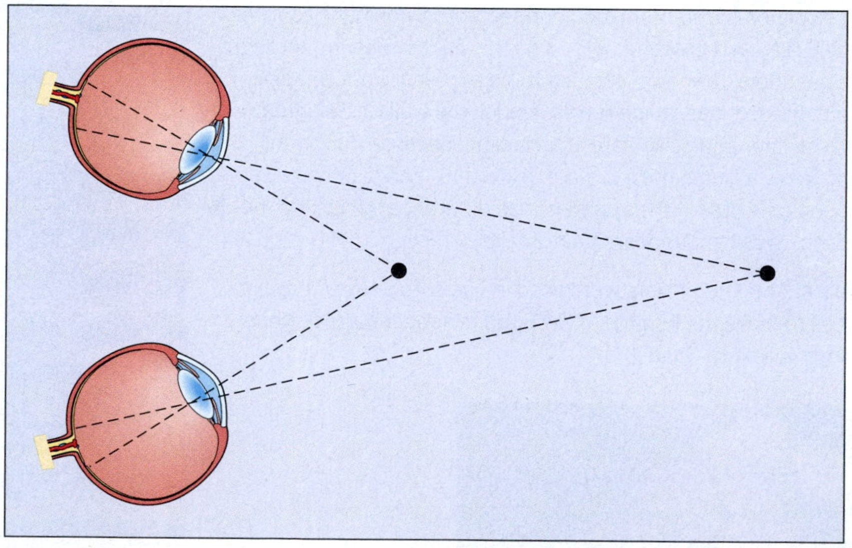

B Because of retinal disparity, objects at different distances (such as the "floating finger") project their images on different parts of the retina. Far objects project on the retinal area near the nose, whereas near objects project farther out, closer to the ears.

Monocular cues FIGURE 4.18

Linear perspective Parallel lines converge, or angle toward one another, as they recede into the distance.

Interposition Objects that obscure or overlap other objects are perceived as closer.

Relative size Close objects cast a larger retinal image than distant objects.

Texture gradient Nearby objects have a coarser and more distinct texture than distant ones.

Aerial perspective Distant objects appear hazy and blurred compared to close objects because of intervening atmospheric dust or haze.

Light and shadow Brighter objects are perceived as being closer than distant objects.

Relative height Objects positioned higher in our field of vision are perceived as farther away.

Color perception Our color vision is as remarkable as our ability to perceive depth and distance. Humans may be able to discriminate among seven million different hues, and research conducted in many cultures suggests that we all seem to see essentially the same colored world (Davies, 1998). Furthermore, studies of infants old enough to focus and move their eyes show that they are able to see color nearly as well as adults (Knoblauch, Vital-Durand, & Barbur, 2000; Werner & Wooten, 1979).

Although we know color is produced by different wavelengths of light, the actual way in which we perceive color is a matter of scientific debate. Traditionally, there have been two theories of color vision, the trichromatic (three-color theory) and the opponent-process theory. The **trichromatic theory** (from the Greek word *tri*—meaning "three," and *chroma*—meaning "color") was first proposed by Thomas Young in the early nineteenth century and was later refined by Herman von Helmholtz and others. Apparently, we have three "color systems," as they called them—one system that is maximally sensitive to red, another maximally sensitive to green, and another maximally sensitive to blue (Young, 1802). The proponents of this theory demonstrated that mixing lights of these three colors could yield the full spectrum of colors we perceive. Unfortunately this theory has its flaws. One is that it doesn't explain *color after effects,* a phenomenon you can experience in the *Applying Psychology* box on page 106.

trichromatic theory Theory that color perception results from mixing three distinct color systems—red, green, and blue.

The **opponent-process theory**, proposed by Ewald Hering later in the nineteenth century, also suggested the three color systems, but he suggested that each system is sensitive to two opposing colors—blue and yellow, red and green, black and white—in an "on-off" fashion. In other words, each color receptor responds either to blue or yellow or to red or green, with the black-or-white systems responding to differences in brightness levels. This

opponent-process theory Theory that color perception is based on three systems of color receptors, each of which responds in an on-off fashion to opposite-color stimuli: blue-yellow, red-green, and black-white.

Applying Psychology

Color Aftereffects

Try staring at the dot in the middle of this color-distorted American flag for 60 seconds. Then stare at a plain sheet of white paper. You should get interesting color aftereffects—red in place of green, blue in place of yellow, and white in place of black: a "genuine" American flag.

What happened? This is a good example of the *opponent-process theory*. As you stared at the green, black, and yellow colors, the neural systems that process those colors became fatigued. Then when you looked at the plain white paper, which reflects all wavelengths, a reverse *opponent process* occurred. Each fatigued receptor responded with its opposing red, white, and blue colors!

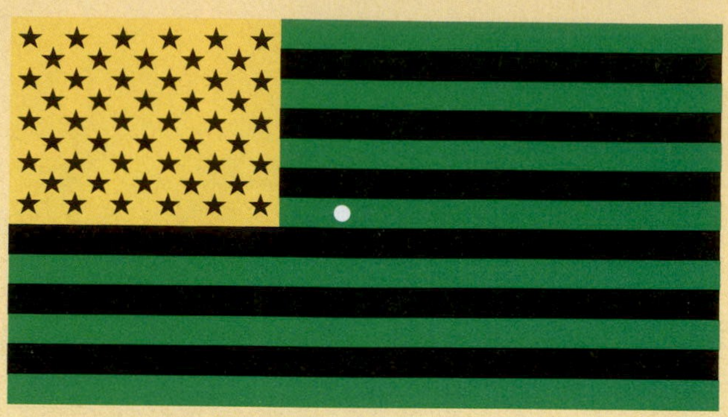

CRITICAL THINKING

Here's an interesting question:

1. In what kinds of situations do you think color aftereffects are likely to occur?

theory makes a lot of sense because when different-colored lights are combined, people are unable to see reddish greens and bluish yellows. In fact, when red and green lights or blue and yellow lights are mixed in equal amounts, we see white.

In 1964, research by George Wald and colleague Paul Brown showed we do have three kinds of cones in the retina and confirmed the central proposition of the trichromatic theory. At nearly the same time, R. L. DeValois (1965) was studying electrophysiological recording of cells in the optic nerve and optic pathways to the brain. He discovered that cells respond to color in an opponent-fashion in the thalamus. The findings reconciled the once-competing trichromatic and opponent-process theories. Now, we know that color is processed in a trichromatic fashion at the level of the cones, in the retina, and in an opponent fashion at the level of the optic nerve and the thalamus, in the brain.

Color-deficient vision Most people perceive three different colors—red, green, and blue—and are called *trichromats*. However, a small percentage of the population has a genetic deficiency in either the red-green system, the blue-yellow system, or both. Those who perceive only two colors are called *dichromats*. People who are sensitive to only the black-white system are called *monochromats*, and they are totally color blind. If you'd like to test yourself for red-green color blindness, see FIGURE 4.19.

INTERPRETATION: EXPLAINING OUR PERCEPTIONS

After selectively sorting through incoming sensory information and organizing it, the brain uses this information to explain and make judgments about the external

NATIONAL GEOGRAPHIC

Color-deficient vision FIGURE 4.19

Are you color blind? People who suffer red-green deficiency have trouble perceiving the number within this design. Although we commonly use the term *color blindness*, most problems are color confusion rather than color blindness. Furthermore, most people who have some color blindness are not even aware of it.

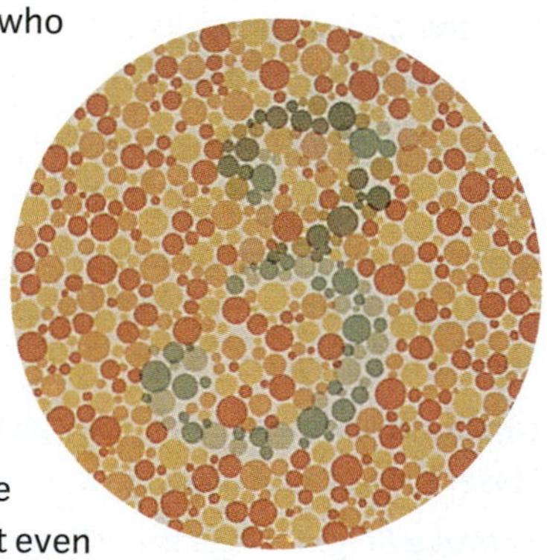

world. This final stage of perception—interpretation—is influenced by several factors, including perceptual adaptation, perceptual set, frame of reference, and bottom-up or top-down processing.

Do you remember the upside-down photo and the discussion about the George Stratton (1896) experiment in the chapter opener? Stratton's experiment illustrates the critical role that **perceptual adaptation** plays in how we interpret the information that our brains gather. Without his ability to adapt his perceptions to a skewed environment, Stratton would not have been able to function. His brain's ability to "retrain" itself to his new surroundings allowed him to create coherence out of what would otherwise have been chaos.

Our previous experiences, assumptions, and expectations also affect how we interpret and perceive the world, by creating a **perceptual set**, or a readiness to perceive in a particular manner, based on expectations—in other words, we largely see what we expect to see (FIGURE 4.20). In one study involving participants who were members of a Jewish organization (Erdelyi & Applebaum, 1973), collections of symbols were briefly flashed on a screen. When a swastika was at the center,

Perceptual set? FIGURE 4.20

When you look at this drawing, do you see a young woman looking back over her shoulder, or an older woman with her chin buried in a fur collar? It may depend on your age. Younger students tend to first see a young woman, and older students first see an older woman. Although basic sensory input stays the same, your brain's attempt to interpret ambiguous stimuli creates a type of perceptual dance, shifting from one interpretation to another (Gaetz et al., 1998).

Ψ Psychological Science

Subliminal Perception

Is the public under siege by sneaky advertisers and politicians lobbing **subliminal** (literally, "below the threshold") messages that can undermine our intentions? Can you lose weight, stop smoking, or relieve stress by listening to "subliminal tapes" promising to solve your problems without your having to pay attention?

It is, in fact, possible to perceive something without conscious awareness (Aarts, 2007; Boccato et al., 2008; Cleeremans & Sarrazin, 2007). For example, in one study the experimenter very briefly flashed one of two pictures subliminally (either a happy or an angry face) followed by a neutral face. The experimenter found this subliminal presentation evoked matching unconscious facial expressions in the participants' own facial muscles (Dimberg, Thunberg, & Elmehed, 2000).

Despite this and much other evidence that *subliminal perception* occurs, that does not mean that such processes lead to *subliminal persuasion*. Subliminal stimuli are basically weak stimuli. At best, they have a modest (if any) effect on consumer behavior and absolutely no effect on citizens' voting behavior (Begg, Needham, & Bookbinder, 1993; Dijksterhuis, Aarts, & Smith, 2005; Karremans, Stroebe, & Claus, 2006). As for subliminal self-help tapes, you're better off with old-fashioned, conscious methods of self-improvement.

CRITICAL THINKING

Here are two interesting questions:

1. Why do you think humans developed the capacity for subliminal perception?
2. What role might subliminal perception play in everyday life today?

Bottom-up, top-down FIGURE 4.21

When first learning to read, you used bottom-up processing. You initially learned that certain arrangements of lines and "squiggles" represented specific letters. You later realized that these letters make up words.

Now, yuor aiblity to raed uisng top-dwon prcessoing mkaes it psosible to unedrstnad thsi sntenece desipte its mnay mssipllengis.

the Jewish participants were less likely to recognize and remember the other symbols. Can you see how the life experiences of the Jewish subjects led them to create a perceptual set for the swastika?

How we perceive people, objects, or situations is also affected by the **frame of reference**, or context. An elephant is perceived as much larger when it is next to a mouse than when it stands next to a giraffe.

Finally, recall that we began this chapter by discussing how we receive sensory information (sensation) and work our way upward to the top levels of perceptual processing (perception). Psychologists refer to this type of information processing as **bottom-up processing** (Mulckhuyse et al., 2008; Prouix, 2007). In contrast, **top-down processing** begins with "higher," "top"-level processing involving thoughts, previous experiences, expectations, language, and cultural background and works down to the sensory level (Schuett et al., 2008; Zhaoping & Guyader, 2007) (FIGURE 4.21).

Science and ESP

So far in this chapter, we have talked about sensations provided by our eyes, ears, nose, mouth, and skin. What about a so-called sixth sense? Can some people perceive things that cannot be perceived with the usual sensory channels, by using **extrasensory perception (ESP)**? People who claim to have ESP profess to be able to read other people's minds (telepathy), perceive objects or events that are inaccessible to their normal senses (clairvoyance), predict the future (precognition), or move or affect objects without touching them (psychokinesis).

Scientific investigations of ESP began in the early 1900s, and some work on the subject continues today. The most important criticism of both experimental and casual claims of ESP is their lack of stability and replicability—a core requirement for scientific acceptance (Hyman, 1996). A meta-analysis of 30 studies using strong scientific controls reported absolutely no evidence of ESP (Milton & Wiseman, 1999, 2001; Valeo & Beyerstein, 2008).

So why do so many people believe in ESP? One reason is that, as mentioned earlier in the chapter, our motivations and interests often influence our perceptions, driving us to selectively attend to things we want to see or hear. In addition, the subject of extrasensory perception (ESP) often generates strong emotional responses. When individuals feel strongly about an issue, they sometimes fail to recognize the faulty reasoning underlying their beliefs.

Belief in ESP is particularly associated with illogical or noncritical thinking. For example, people often fall victim to the *fallacy of positive instances,* noting and remembering events that confirm personal expectations and beliefs and ignoring nonsupportive evidence. Other times, people fail to recognize chance occurrences for what they are. Finally, human information processing often biases us to notice and remember the most vivid information—such as a detailed (and spooky) anecdote or a heartfelt personal testimonial.

CONCEPT CHECK STOP

Why do we experience perceptual illusions?

What are the processes that allow us to pay attention to some stimuli in our environments and ignore others?

What kinds of cues do we use to perceive depth and distance?

What factors entice some people to believe in ESP?

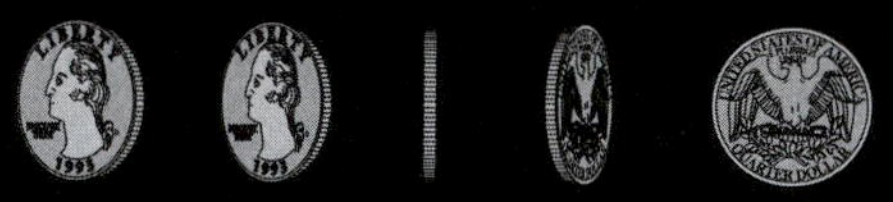

SUMMARY

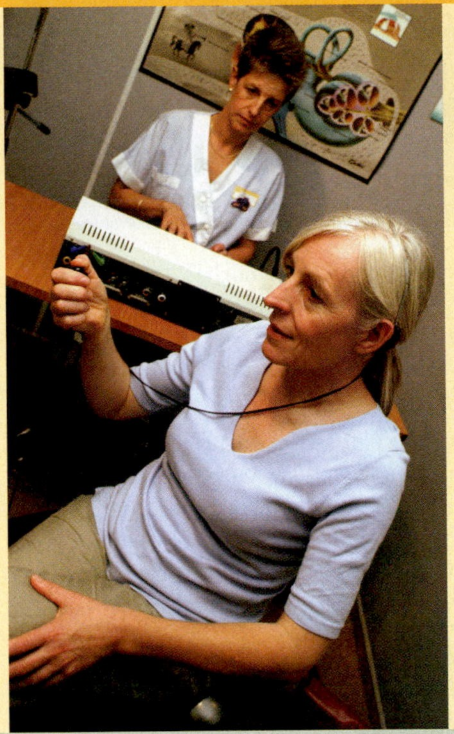

1 Understanding Sensation

1. **Sensation** is the process by which we detect stimuli and convert them into neural signals (**transduction**). During **coding**, the neural impulses generated by different physical stimuli travel by separate routes and arrive at different parts of the brain. In **sensory reduction**, we filter and analyze incoming sensations.

2. In **sensory adaptation**, sensory receptors fire less frequently with repeated stimulation, so that over time, sensation decreases.

3. The absolute threshold is the smallest amount of a stimulus needed to detect a stimulus, and the difference threshold, or just noticeable difference (JND), is the smallest change in stimulus intensity that a person can detect.

4. According to the **gate-control theory**, our experience of pain depends partly on whether the neural message gets past a "gatekeeper" in the spinal cord, which researchers believe is chemically controlled.

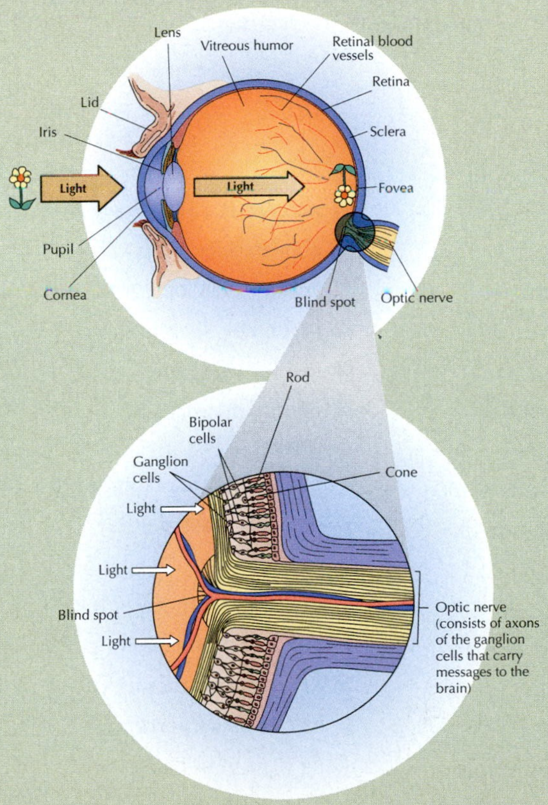

2 How We See and Hear

1. Light and sound move in waves. Light waves are a form of electromagnetic energy, and sound waves are produced when air molecules move in a particular wave pattern. Both light waves and sound waves vary in length, height, and range.

2. Light enters the eye at the front of the eyeball. The cornea protects the eye and helps focus light rays. The iris provides the eye's color, and muscles in the iris dilate or constrict the pupil. The lens further focuses light, adjusting to allow focusing on objects at different distances. At the back of the eye, incoming light waves reach the retina, which contains light-sensitive rods and cones. A network of neurons in the retina transmits neural information to the brain.

3. The outer ear gathers sound waves; the middle ear amplifies and concentrates the sounds; and the inner ear changes sounds' mechanical energy into neural impulses. Sounds' frequency and intensity determine how we distinguish among sounds of different pitches and loudness, respectively.

SUMMARY

3 Our Other Senses

1. Smell and taste, sometimes called the chemical senses, involve chemoreceptors that are sensitive to certain chemical molecules. In olfaction, odor molecules stimulate receptors in the olfactory epithelium, in the nose. The resulting neural impulse travels to the olfactory bulb, where the information is processed before being sent elsewhere in the brain. Our sense of taste (gustation) involves five tastes: sweet, sour, salty, bitter, and umami (which means "savory" or "delicious"). The taste buds are clustered on our tongues within the papillae.

2. The body senses tell the brain how the body is oriented, where and how it is moving, and what it touches. They include the skin senses, the vestibular sense, and kinesthesia.

4 Understanding Perception

1. **Perception** is the process of selecting, organizing, and interpreting incoming sensations into useful mental representations of the world. **Selective attention** allows us to filter out unimportant sensory messages. **Feature detectors** are specialized cells that respond only to certain sensory information. **Habituation** is the tendency to ignore stimuli that remain constant. People tend to automatically select stimuli that are intense, novel, moving, contrasting, and repetitious.

2. To be useful, sensory data must be assembled in a meaningful way. We organize sensory data in terms of form, constancy, depth, and color. Traditionally there have been two theories of color vision: the **trichromatic theory** and the **opponent-process theory**.

3. Perceptual adaptation, perceptual set, frame of reference, and bottom-up versus top-down processing affect our interpretation of what we sense and perceive. Subliminal stimuli, although perceivable, have a modest effect on behavior.

KEY TERMS

- sensation p. 86
- perception p. 86
- synesthesia p. 88
- transduction p. 88
- coding p. 88
- sensory reduction p. 88
- sensory adaptation p. 89
- psychophysics p. 89
- absolute threshold p. 89
- difference threshold p. 89
- gate-control theory p. 90
- wavelength p. 92
- amplitude p. 92
- range p. 92
- nearsightedness p. 92
- farsightedness p. 92
- dark adaptation p. 92
- audition p. 92
- retina p. 93
- fovea p. 93
- rods p. 93
- cones p. 93
- blind spot p. 93
- outer ear p. 94
- middle ear p. 94
- inner ear p. 94
- cochlea p. 94
- frequency p. 94
- place theory p. 94
- frequency theory p. 94
- conduction deafness p. 95
- nerve deafness p. 95
- cochlear implant p. 95
- olfaction p. 97
- pheromones p. 97
- gustation p. 97
- skin senses p. 97
- vestibular sense p. 97
- kinesthesia p. 97
- selective attention p. 100
- feature detectors p. 100
- habituation p. 100
- gestalt p. 101
- impossible figures p. 102
- depth perception p. 102
- binocular cues p. 102
- monocular cues p. 102
- retinal disparity p. 102
- perceptual constancy p. 103
- convergence p. 104
- accommodation p. 104
- motion parallax p. 104
- trichromatic theory p. 105
- opponent-process theory p. 105
- perceptual adaptation p. 107
- perceptual set p. 107
- subliminal p. 107
- frame of reference p. 108
- bottom-up processing p. 108
- top-down processing p. 108
- extrasensory perception (ESP) p. 108

CRITICAL AND CREATIVE THINKING QUESTIONS

1. Sensation and perception are closely linked. What is the central distinction between the two?
2. If we sensed and attended equally to each stimulus in the world, the amount of information would be overwhelming. What sensory and perceptual processes help us lessen the din?
3. If we don't adapt to pain, why is it that people can sometimes "tune out" painful injuries?
4. What senses would likely be impaired if a person were somehow missing all of the apparatus of the ear (including the outer, middle, and inner ear)?
5. Can you explain how your own perceptual sets might create prejudice or discrimination?

What is happening in this picture ?

This man willingly endures what would normally be excruciating pain.

- What psychological and biological factors might make this possible for him?
- Do you think this man would feel more pain, or less, if his friends and family members were frequently and solicitously asking how he was feeling?

SELF-TEST

(Check your answers in Appendix B.)

1. Transduction is the process of converting _______________.
   a. neural impulses into mental representations of the world
   b. receptors into transmitters
   c. a physical stimulus into neural impulses
   d. receptors into neural impulses

2. Sensory reduction refers to the process of _______________.
   a. reducing your dependence on a single sensory system
   b. decreasing the number of sensory receptors that are stimulated
   c. filtering and analyzing incoming sensations before sending a neural message to the cortex
   d. reducing environmental sensations by physically preventing your sensory organs from seeing, hearing, etc.

3. HOW THE EYE SEES: Identify the parts of the eye, placing the appropriate label on the figure below:

cornea	rod
iris	cone
pupil	fovea
lens	blind spot
retina	

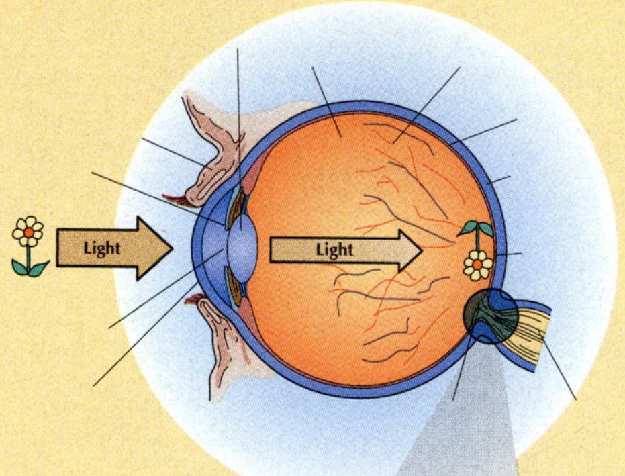

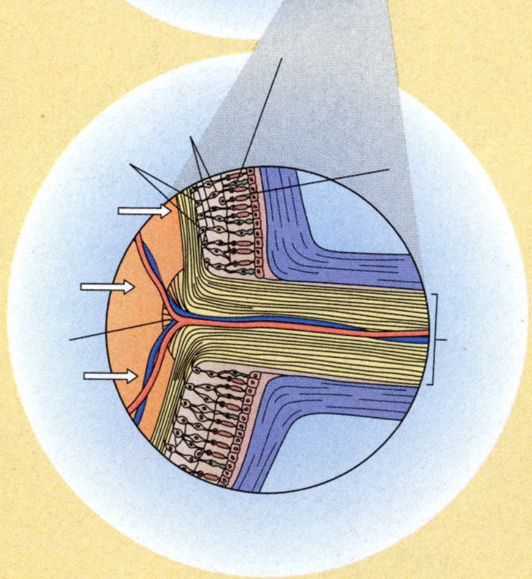

4. A visual acuity problem that occurs when the cornea and lens focus an image in front of the retina is called _______________.
   a. farsightedness
   b. hyperopia
   c. myopia
   d. presbyopia

5. HOW THE EAR HEARS: Identify the parts of the ear, placing the appropriate label on the figure below:

pinna	stapes
tympanic membrane	oval window
malleus	cochlea
incus	

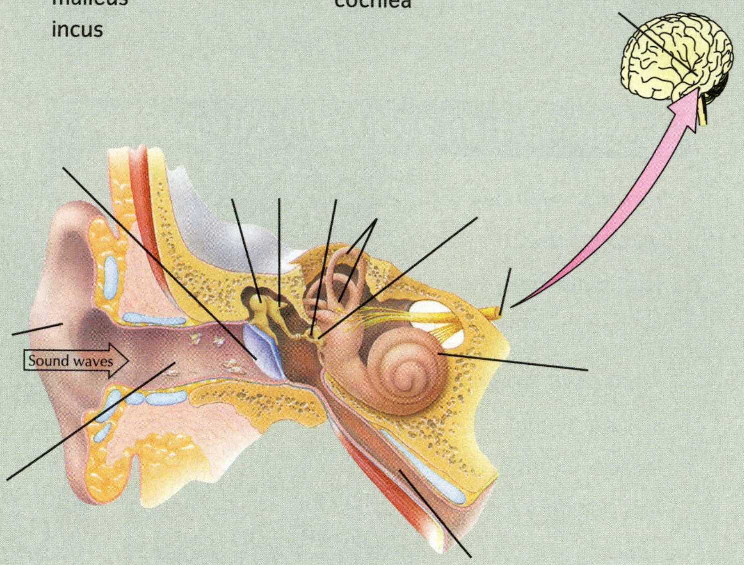

6. Chronic exposure to loud noise can cause permanent _______________.
   a. auditory illusions
   b. auditory hallucinations
   c. nerve deafness
   d. conduction deafness

7. Most information related to smell is processed in the _______________.
   a. nasal cavity
   b. temporal lobe
   c. olfactory bulb
   d. parietal lobe

8. Most of our taste receptors are found on the _______________.
   a. olfactory bulb
   b. gustatory cells
   c. taste buds
   d. frenulum

9. Identify which of these photos, 1 or 2, illustrates:

a. vestibular sense: photo ______

b. kinesthetic sense: photo ______

1

2

10. When the brain is sorting out and attending only to the most important messages from the senses, it is engaged in the process of ________________.

a. sensory adaptation
b. sensory habituation
c. selective attention
d. selective sorting

11. In a(n) ________________, the discrepancy between figure and ground is too vague and you may have difficulty perceiving which is figure and which is ground.

a. illusion
b. reversible figure
c. optical illusion
d. hallucination

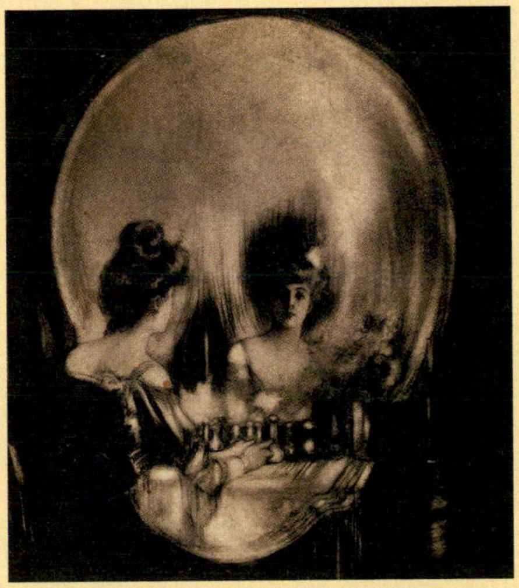

12. The tendency for the environment to be perceived as remaining the same even with changes in sensory input is called ________________.

a. perceptual constancy
b. the constancy of expectation
c. an illusory correlation
d. Gestalt's primary principle

13. The theory of color vision proposed by Thomas Young that says color perception results from mixing three distinct color systems is called the ________________.

a. tricolor theory
b. trichromatic theory
c. tripigment theory
d. opponent-process theory

14. The illustration below is an example of ________________.

a. top-down processing
b. frame of reference
c. subliminal persuasion
d. perceptual adaptation

15. A readiness to perceive in a particular manner is known as ________________.

a. sensory adaptation
b. perceptual set
c. habituation
d. frame of reference

States of Consciousness

5

In 1995 Jean-Dominique Bauby, suffered a stroke that left him with a condition known as "locked-in syndrome." He was unable to speak or move his hands. He could communicate only by blinking his left eye. Yet Bauby remained alert—so much so that he was able to dictate a memoir, *The Diving Bell & the Butterfly*, by "winking" one letter at a time.

What if Bauby had not been able to open or move his eye? If he couldn't communicate, would he still be "conscious"? What exactly is consciousness? Is it simple awareness? How can we study our consciousness when the only tool of discovery is the object itself?

When psychology first became a scientific discipline, it defined itself as "the study of human consciousness." But the nebulousness of this definition led to dissatisfaction within the field. One group, the behaviorists, believed that behavior, not consciousness, was the proper focus of the new science.

More recently, psychology has renewed its original interest in consciousness. In addition, technological advances allow study of brain activity during alternate states of consciousness (ASCs): sleep and dreaming, chemically induced changes from psychoactive drugs, daydreaming, fantasies, hypnosis, fasting, meditation, and even the so-called runner's high.

In this chapter, we begin with the definition and description of consciousness. Then we examine how consciousness changes because of circadian rhythms, sleep, and dreams. We also look at psychoactive drugs and their effects. Finally, we explore alternative routes to altered consciousness, such as meditation and hypnosis.

NATIONAL GEOGRAPHIC

CHAPTER OUTLINE

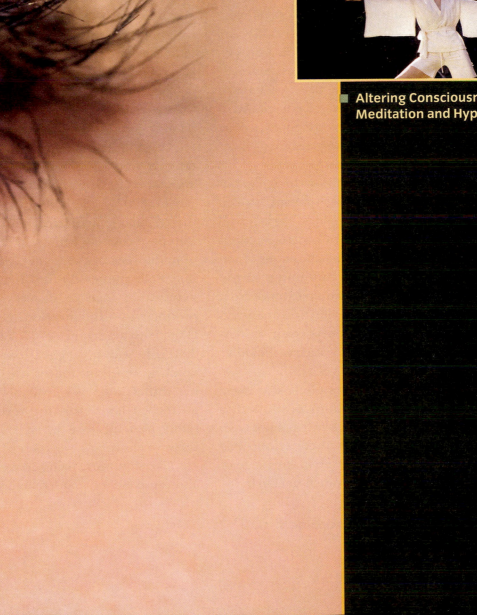

Consciousness, Sleep, and Dreaming

LEARNING OBJECTIVES

Explain the difference between controlled and automatic processes.

Describe the effects of sleep deprivation and disruption of circadian rhythms.

Review the stages of sleep.

Compare and contrast the theories of why we sleep and dream.

Summarize the types of sleep disorders.

William James, the first American psychologist, likened **consciousness** to a stream that's constantly changing yet always the same. It meanders and flows, sometimes where the person wills and sometimes not. However, through the process of selective attention (Chapter 4), we can control our consciousness by deliberate concentration and full attention. For example, at the present moment you are, hopefully, awake and concentrating on the words on this page. At times, however, your control may weaken, and your stream of consciousness may drift to thoughts of a computer you want to buy, a job, or an attractive classmate.

consciousness An organism's awareness of its own self and surroundings (Damasio, 1999).

altered state of consciousness (ASC) Mental states found during sleep, dreaming, psychoactive drug use, hypnosis, and so on.

In addition to meandering and flowing, your "stream of consciousness" also varies in depth. Consciousness is not an all-or-nothing phenomenon—conscious or unconscious. Instead, it exists along a continuum. As you can see in FIGURE 5.1, this continuum extends from high awareness and sharp, focused alertness at one extreme, to middle levels of awareness such as daydreaming, to nonconsciousness and coma at the other extreme.

Two of the more common states of consciousness are sleep and dreaming. You may think of yourself as being unconscious while you sleep, but that's not true. Rather, you are in an **altered state of consciousness (ASC)**. In this chapter, you will learn about this state of consciousness and why we spend so much time in it.

To understand sleep and dreaming, we need to first explore the topic of **circadian rhythms**. Most animals have adapted to our planet's cycle of days and nights by developing a pattern of bodily functions that wax and wane over each 24-hour period. Our alertness, core body temperature, moods, learning efficiency, blood pressure, metabolism, and pulse rate all follow these circadian rhythms (Leglise, 2008; Oishi et al., 2007; Sack et al., 2007). Usually, these activities reach their peak during the day and their low point at night (see FIGURE 5.2 on page p. 118).

circadian [ser-KAY-dee-an] **rhythms** Biological changes that occur on a 24-hour cycle (in Latin, *circa* means about, and *dies* means day).

Disruptions in circadian rhythms cause increased fatigue, decreased concentration, sleep disorders, and other health problems (James, Cermakian, & Boivin, 2007; Lader 2007; Salvatore et al., 2008). Although many physicians, nurses, police, and others who have rotating work schedules (about 20 percent of employees in the United States) manage to function well, studies do find that shift work and sleep deprivation lead to decreased concentration and productivity—and increased accidents (Dembe et al., 2006; Papadelis et al., 2007; Yegneswaran & Shapiro, 2007). Research suggests that shifting from days to evenings to nights makes it easier to adjust to rotating shift schedules—probably because it's easier to go to bed later than the reverse. Productivity and safety also increase when shifts are rotated every three weeks versus every week.

Like shift work, flying across several time zones can also cause fatigue and irritability—symptoms of jet lag.

Visualizing

Consciousness FIGURE 5.1

Where Does Consciousness Reside?

One of the oldest philosophical debates is the *mind–body problem*. Is the "mind" (consciousness and other mental functions) fundamentally different from matter (the body)? How can a supposedly nonmaterial mind influence a physical body and vice versa? Most neuropsychologists today believe the mind *is* the brain and *consciousness* involves an activation and integration of several parts of the brain. But two aspects of consciousness, *awareness* and *arousal*, seem to rely on specific areas. Awareness generally involves the *cerebral cortex*, particularly the frontal lobes. Arousal generally results from *brain-stem* activation (Revonsuo, 2006; Thomson, 2007; Zillmer, Spiers, & Culbertson, 2008).

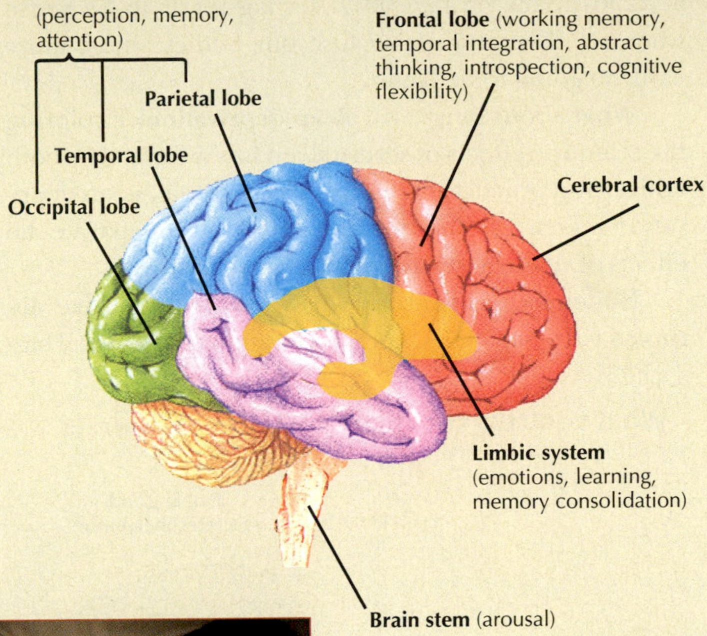

LEVELS OF AWARENESS

Altered states of consciousness (ASCs) can exist on many levels of awareness, from high awareness to low awareness.

High Awareness — Middle Awareness — Low Awareness

CONTROLLED PROCESSES

Require focused, maximum attention (e.g., studying for an exam, learning to drive a car)

AUTOMATIC PROCESSES

Require minimal attention (e.g., walking to class while talking on a cell phone, listening to your boss while daydreaming)

SUBCONSCIOUS

Below conscious awareness (e.g., subliminal perception, sleeping, dreaming)

NO AWARENESS

Biologically based lowest level of awareness (e.g., head injuries, anesthesia, coma); also the *unconscious mind* (a Freudian concept discussed in Chapter 12) reportedly consisting of unacceptable thoughts and feelings too painful to be admitted to consciousness)

Jet lag correlates with decreased alertness, mental agility, and efficiency, as well as exacerbation of psychiatric disorders (Dawson, 2004; Leglise, 2008; Morgenthalar et al., 2007; Sack et al., 2007). Jet lag tends to be worse when we fly eastward because our bodies adjust more easily to going to bed later.

What about long-term **sleep deprivation**? Exploring the scientific effects of severe sleep loss is limited by both ethical and practical concerns. For example, sleep deprivation increases stress, making it difficult to separate the effects of sleep deprivation from those of stress.

Nonetheless, researchers have learned that, like disrupted circadian cycles, sleep deprivation poses several hazards. These include reduced cognitive and motor performance, irritability and other mood alterations, decreased self-esteem, and increased *cortisol* levels (a sign of stress) (Dembe et al., 2006; Mirescu et al., 2006; Papadelis et al., 2007; Sack et al., 2007; Yegneswaran & Shapiro, 2007).

The consequences of such impairments are wide-ranging, from threatening students' school performance to endangering physical health. Perhaps the most frightening danger is that lapses in attention among sleep-deprived pilots, physicians, truck drivers, and other workers cause serious accidents and cost thousands of lives each year (Dembe et al., 2006; de Pinho et al., 2006; Paice et al., 2002; Yegneswaran & Shapiro, 2007).

What controls circadian rhythms? FIGURE 5.2

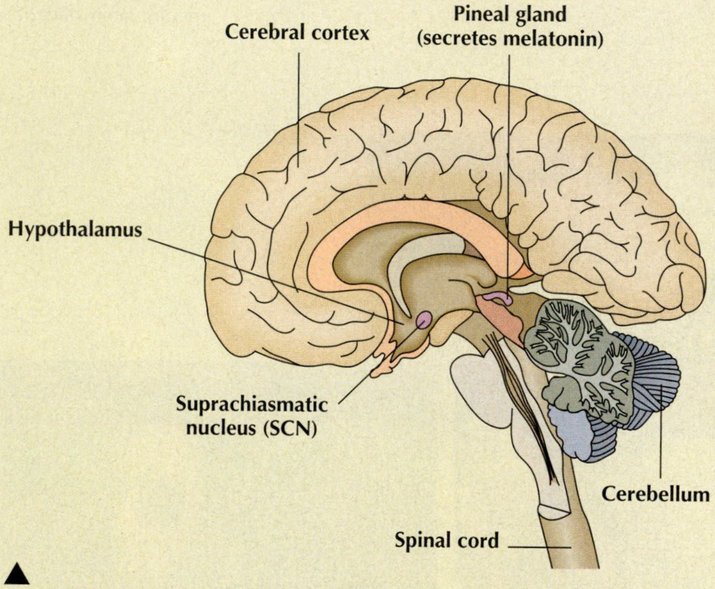

A Circadian rhythms are regulated by a part of the hypothalamus called the *suprachiasmatic nucleus* (SCN).

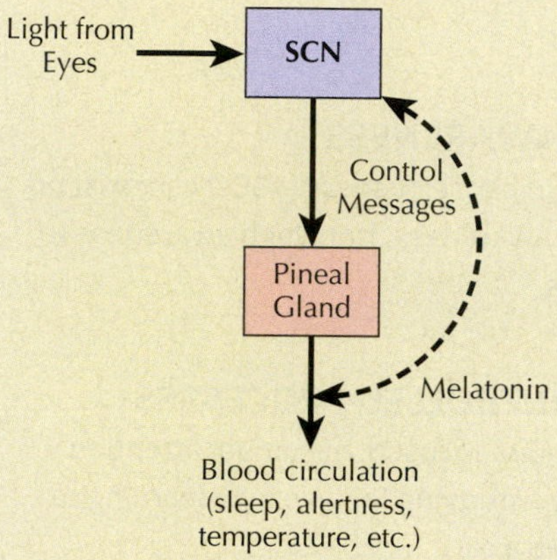

B The SCN receives information about light and darkness from the eyes and then sends control messages to the *pineal gland*, which releases *melatonin*—a hormone thought to influence sleep, alertness, and body temperature. Like other feedback loops in the body, the level of melatonin in the blood is sensed by the SCN, which then can modify the output of the pineal to maintain the "desired" level.

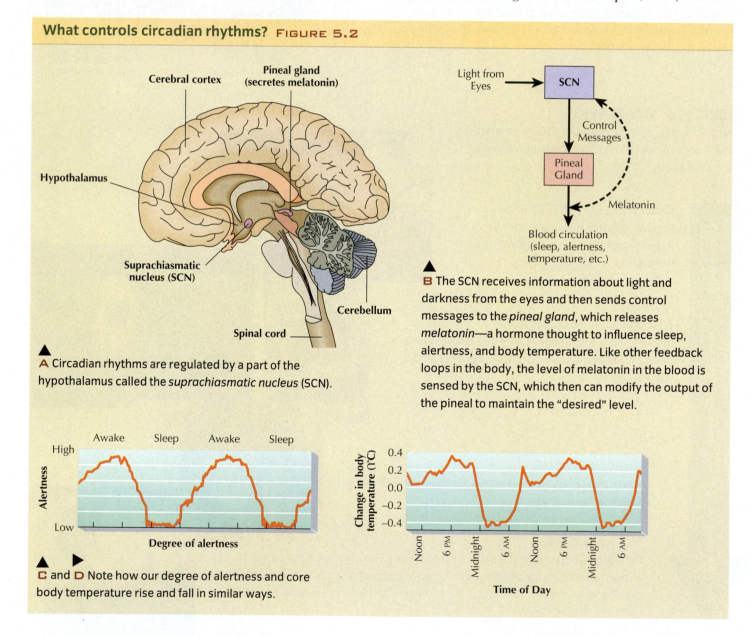

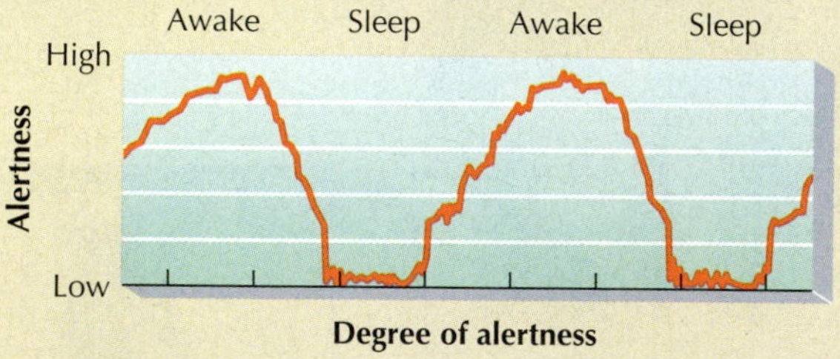

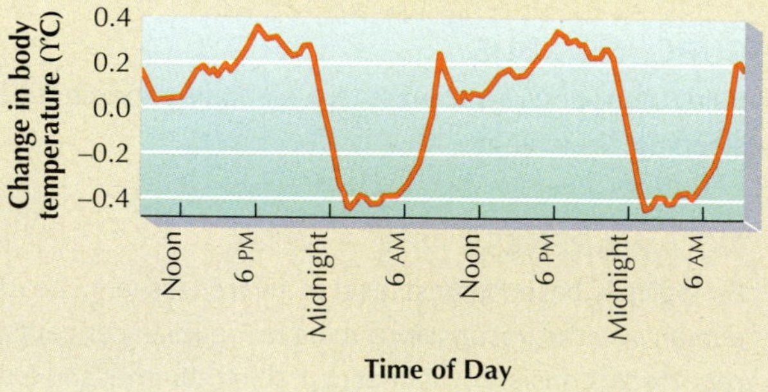

C and D Note how our degree of alertness and core body temperature rise and fall in similar ways.

Applying Psychology

Are You Sleep Deprived?

Take the following test to determine whether you are sleep deprived.

Part 1 Set up a small mirror next to the text and see if you can trace the black star pictured here, using your nondominant hand while watching your hand in the mirror. The task is difficult, and sleep-deprived people typically make many errors. If you are not sleep deprived, it may be difficult to trace the star, but you'll probably trace it accurately

Part 2 Give yourself one point each time you answer yes to the following questions:

Do you often fall asleep . . .
watching TV?
during boring meetings or lectures or in warm rooms?
after heavy meals or after a small amount of alcohol?
while relaxing after dinner?
within five minutes of getting into bed?

In the morning, do you generally . . .
need an alarm clock to wake up at the right time?
struggle to get out of bed?
hit the snooze bar several times to get more sleep?

During the day, do you . . .
feel tired, irritable, and stressed out?
have trouble concentrating and remembering?
feel slow when it comes to critical thinking, problem solving, and being creative?
feel drowsy while driving?
need a nap to get through the day?
have dark circles around your eyes?

If you answered yes to three or more items, you probably are not getting enough sleep.

Source: From POWER SLEEP by James B. Maas and M. L. Wherry, copyright © 1997 by James Maas. Used by permission of Villard Books, a division of Random House, Inc.

CRITICAL THINKING

Here are two interesting questions:

1. If you are sleep deprived, what steps could you take to get more sleep?
2. If you feel drowsy while driving, have you considered the possible consequences?

STAGES OF SLEEP: HOW SCIENTISTS STUDY SLEEP

Surveys and interviews can provide some information about the nature of sleep, but researchers in sleep laboratories use a number of sophisticated instruments to study physiological changes during sleep.

Imagine that you are a participant in a sleep experiment. When you arrive at the sleep lab, you are assigned one of several bedrooms. The researcher hooks you up to various physiological recording devices (FIGURE 5.3A on the next page). You will probably need a night or two to adapt to the equipment before the researchers can begin to monitor your typical night's sleep. At first, you enter a relaxed, *presleep* state. As you continue relaxing, your brain's electrical activity slows even further. In the course of about an hour, you progress through four distinct stages of sleep (Stages 1 through 4), each progressively deeper (FIGURE 5.3B). Then the sequence begins to reverse itself. Although we don't necessarily go through all sleep stages in this sequence, during the course of a night, people usually complete four to five cycles of light to deep sleep and back, each lasting about 90 minutes.

Visualizing

The scientific study of sleep and dreaming FIGURE 5.3

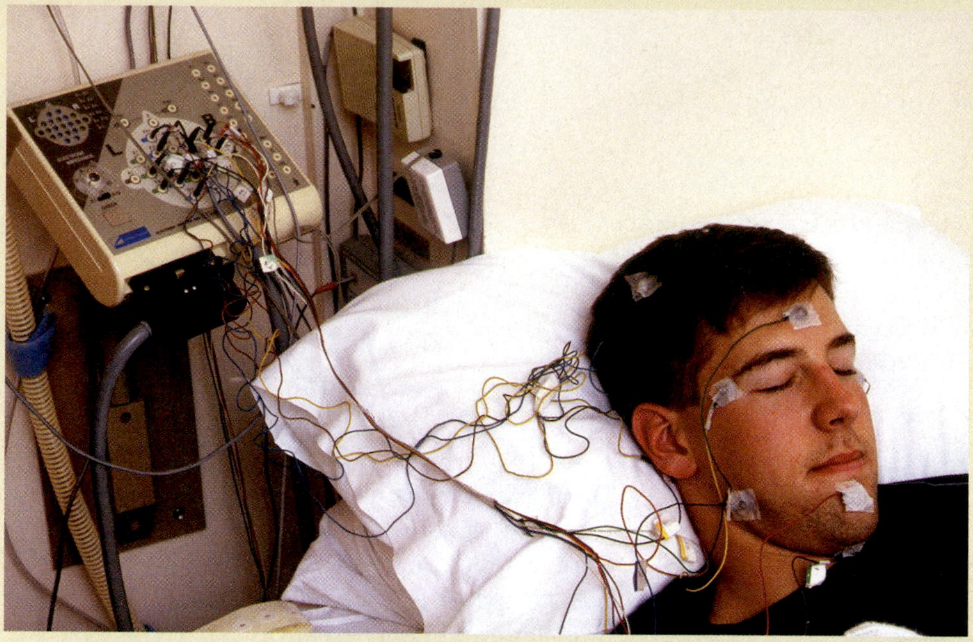

◀ A Sleep research participants wear electrodes on their heads and bodies to measure the brain's and body's responses during the sleep cycle.

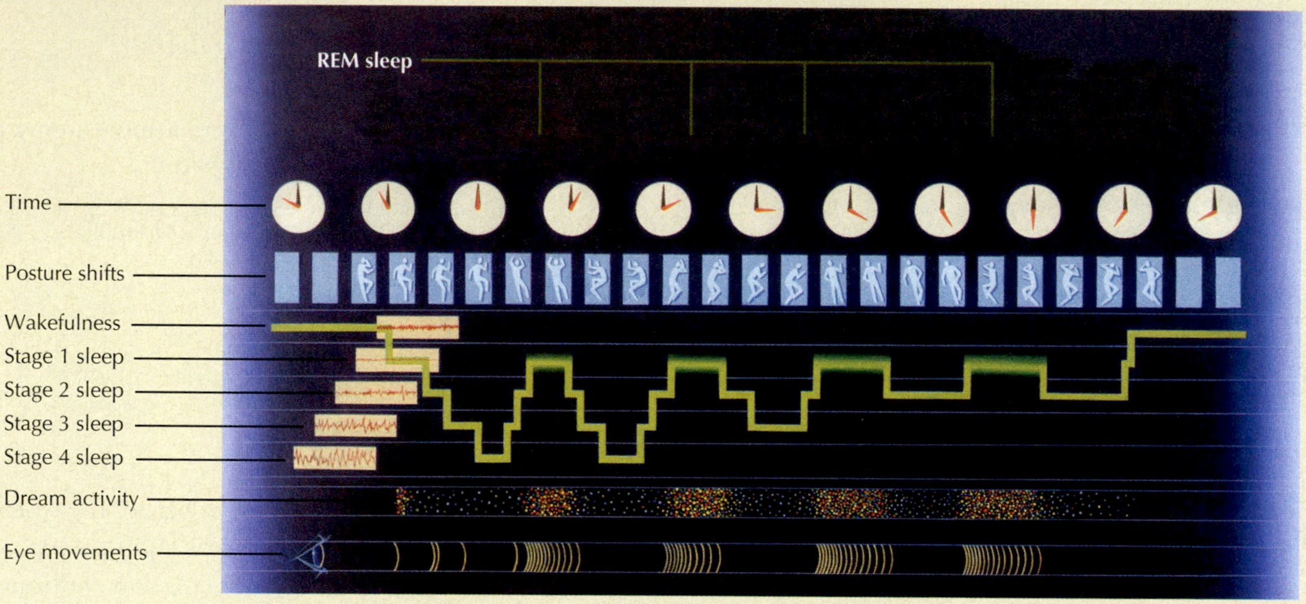

▲

B An **electroencephalogram (EEG)** records brain wave changes by means of small electrodes on the scalp. Other electrodes measure muscle activity and eye movements. The stages of sleep, defined by telltale changes in brain waves, are indicated by the green stepped lines. The compact brain waves of alertness gradually lengthen as we drift into Stages 1–4. By the end of Stage 4, a change in body position generally occurs and heart rate, blood pressure, and respiratory rates all decrease. The sleeper then reverses through Stages 3 and 2 before entering the first REM period of the night. Although the brain and body are giving many signs of active arousal, the musculature is deeply relaxed and unresponsive. Sleepers awakened from REM sleep often report vivid, bizarre dreams, indicated in the figure by red and yellow dots. Those awakened from Stages 1–4 sleep often have more peaceful thoughts (indicated by muted dots). The heavy green line on the graph showing all four stages of sleep indicates the approximate time spent in REM sleep. Note how the length of the REM period increases as the night progresses.

REM and NREM sleep

Figure 5.3B also shows an interesting phenomenon that occurs at the end of the first sleep cycle (and subsequent cycles). You reverse back through Stages 3 and 2. Then your scalp recordings abruptly display a pattern of small-amplitude, fast-wave activity, similar to an awake, vigilant person's brain waves. Your breathing and pulse rates become fast and irregular, and your genitals likely show signs of arousal. Yet your musculature is deeply relaxed and unresponsive. Because of these contradictory qualities, this stage is sometimes referred to as *paradoxical sleep*.

During this stage of paradoxical sleep, rapid eye movements occur under your closed eyelids. Researchers refer to this stage as **rapid-eye-movement (REM) sleep**. When awakened from REM sleep, people almost always report dreaming. Because REM sleep is so different from the other periods of sleep, Stages 1 through 4 are often collectively referred to as **non-rapid-eye-movement (NREM) sleep**. Dreaming also occurs during NREM sleep, but less frequently, and the dreams usually contain a simple experience, such as "I dreamed of a house" (Hobson, 2002; Squier & Domhoff, 1998).

Scientists believe that REM sleep is important for learning and consolidating new memories (Marshall & Born, 2007; Massicotte-Marquez et al., 2008; Silvestri & Root, 2008). Further evidence of the importance of REM sleep for complex brain functions comes from the fact that the amount of REM sleep increases after periods of stress or intense learning and that fetuses, infants, and young children spend a large percentage of their sleep time in this stage. In addition, REM sleep occurs only in mammals of higher intelligence (Rechtschaffen & Siegel, 2000).

REM sleep meets an important biological need. When deprived of REM sleep, most people "catch up" later by spending more time than usual in REM sleep (Dement & Vaughan, 1999).

NREM sleep may be even more important to our biological functioning than REM sleep. When people are deprived of *total* sleep, they spend more time in NREM sleep during their first uninterrupted night of sleep (Borbely, 1982). Also, it is only after our need for NREM sleep has been satisfied each night that we begin to devote more time to REM sleep. Further, studies show that adults who sleep five or fewer hours each night spend less time in REM sleep than do those who sleep nine or more hours. Similarly, infants get much more sleep and have a higher percentage of REM sleep (close to 50 percent of total daily sleep during the first six months of life) than do adults (22 percent of total daily sleep at age 19, declining to about 14 percent in old age) (FIGURE 5.4).

The effect of aging on the sleep cycle FIGURE 5.4

Have you ever noticed how much babies sleep and how little sleep older people seem to need? Our biological sleep requirements change throughout our lifetimes. The pie charts in this figure show the relative amounts of REM sleep (dark blue), non-REM sleep (medium blue), and awake time (light blue) that the average person experiences as an infant, as an adult, and as an elderly person. An infant sleeps 14 hours and spends 40 percent of that time in REM. An adult sleeps about 7.5 hours, with 20 percent of that in REM. The average 70-year-old sleeps only about six hours, with 14 percent of that in REM.

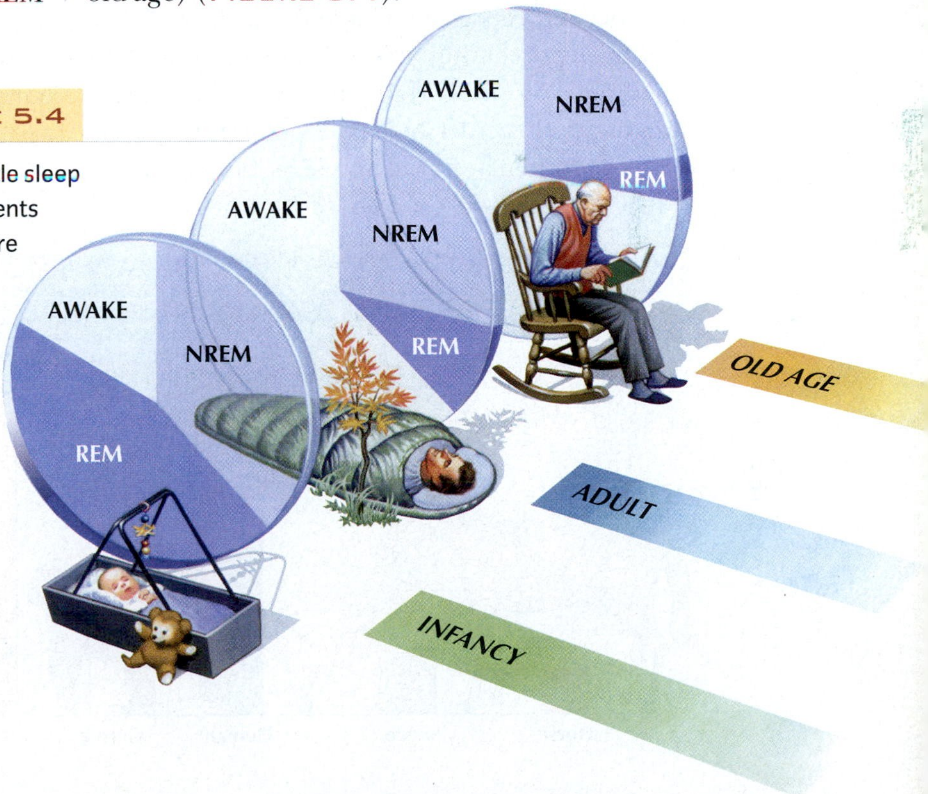

Why do we sleep? There are two prominent theories about why we sleep. The **evolutionary/circadian theory** emphasizes the relationship of sleep to basic circadian rhythms. Sleep keeps animals still when predators are active (Siegel, 2008) (FIGURE 5.5).

In contrast, the **repair/restoration theory** suggests that sleep helps us recuperate—physically, emotionally, and intellectually—from depleting daily activities (Maas, 1999).

It may be that both theories of sleep are correct—that sleep initially served to conserve energy and keep us out of trouble and that, over time, it has evolved to allow for repair and restoration.

evolutionary/circadian theory As a part of circadian rhythms, sleep evolved to conserve energy and to serve as protection from predators.

repair/restoration theory Sleep serves a recuperative function, allowing organisms to repair or replenish key factors.

Why do we dream? The question of why we dream—and whether dreams carry special meaning or information—has fascinated and perplexed psychologists at least as much as the question of why we sleep. Let's look at three theories of why we dream.

One of the oldest and most scientifically controversial explanations for why we dream is Freud's **psychoanalytic view**. Freud proposed that unacceptable desires, which are normally repressed, rise to the surface of consciousness during dreaming. We avoid anxiety, Freud believed, by disguising our forbidden unconscious needs (what Freud called the dream's latent content) as symbols (manifest content). For example, a journey is supposed to symbolize death; horseback riding and dancing would symbolize sexual intercourse; and a gun might represent a penis.

Most modern scientific research does not support Freud's view (Domhoff, 2004; Dufresne, 2007). Critics also say that Freud's theory is highly subjective and that the symbols can be interpreted according to the particular analyst's view or training.

In contrast to Freud, a biological view called the **activation–synthesis hypothesis** suggests that dreams are a by-product of random stimulation of brain cells during REM sleep (Hobson, 1999, 2005). Alan Hobson and Robert McCarley (1977) proposed that specific neurons in the brain stem fire spontaneously during REM sleep and that the cortex struggles to "synthesize" or make sense out of this random stimulation by manufacturing dreams. This is *not* to say that dreams are totally meaningless. Hobson (1988, 2005) suggests that even if dreams begin with essentially random brain activity, your individual personality, motivations, memories, and life experiences guide how your brain constructs the dream.

Finally, some researchers support the **cognitive view** that dreams are simply another type of information processing. That is, our dreams help us to periodically sift and sort our everyday experiences and thoughts. For example, some research reports strong similarities between dream content and waking thoughts, fears, and concerns (Domhoff, 2005, 2007; Erlacher & Schredl, 2004).

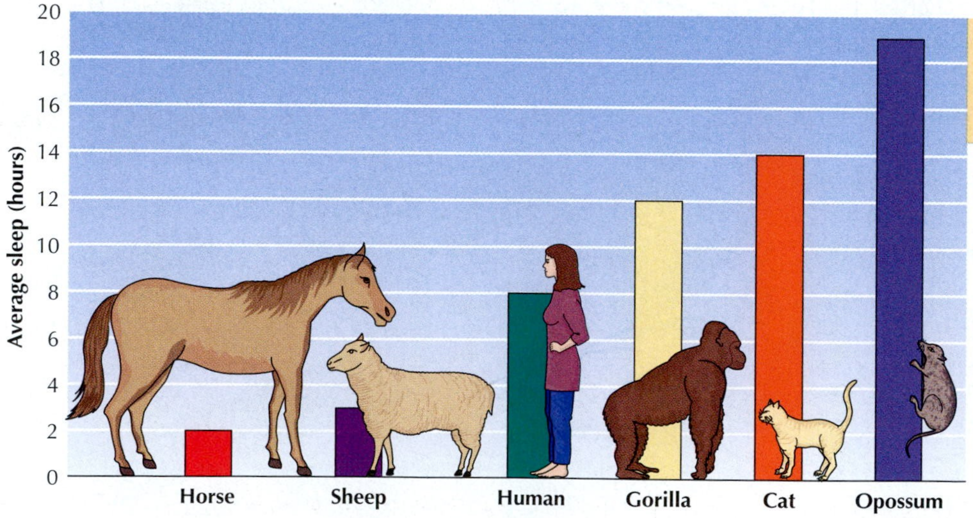

Average daily hours of sleep for different mammals
FIGURE 5.5

According to the evolutionary/circadian theory, differences in diet and number of predators affect different species' sleep habits.

SLEEP DISORDERS: WHEN SLEEP BECOMES A PROBLEM

An estimated two-thirds of American adults suffer from sleep problems, and about 25 percent of children under age 5 have a sleep disturbance (Lader, Cardinali, & Pandi-Perumal, 2006; National Sleep Foundation, 2007; Wilson & Nutt, 2008). Each year, Americans spend more than $98 million on nonprescription sleep medications, and about half that amount on caffeine tablets for daytime use. One in five adults is so sleepy during the day that it interferes with daily activities, and 20 percent of all drivers have fallen asleep for a few seconds (a phenomenon called microsleep) at the wheel.

Mental health professionals divide sleep disorders into two major diagnostic categories: dyssomnias, which describe problems sleeping, and parasomnias, which describe abnormal sleep disturbances.

The most common **dyssomnia** is **insomnia**. Although it's normal to have trouble sleeping before an exciting event, as many as one in ten people has persistent difficulty falling asleep or staying asleep, or wakes up too early. Nearly everybody has insomnia at some time in their life (Pearson, Johnson, & Nahin, 2006; Riemann & Volderholzer, 2003; Wilson & Nutt, 2008). A telltale sign of insomnia is that the person feels poorly rested the next day. Most people with serious insomnia have other medical or psychological disorders as well (Riemann & Voderholzer, 2003; Taylor, Lichstein, & Durrence, 2003).

dyssomnias Problems in the amount, timing, and quality of sleep, include insomnia, sleep apnea, and narcolepsy.

Unfortunately, nonprescription insomnia pills generally don't work. Prescription tranquilizers and barbiturates do help people sleep, but they decrease Stage 4 and REM sleep, seriously affecting sleep quality. In the short term, drugs such as Ambien, Dalmane, Xanax, Halcion, and Lunesta may be helpful in treating sleep problems related to anxiety and acute, stressful situations. However, chronic users run the risk of psychological and physical drug dependence (Leonard, 2003; McKim, 2002).

Narcolepsy is another serious dyssomnia, characterized by sudden and irresistible onsets of sleep during normal waking hours. Narcolepsy afflicts about 1 person in 2,000 and generally runs in families (Billiard, 2007; Pedrazzoli et al., 2007; Siegel, 2000). During an attack, REM-like sleep suddenly intrudes into the waking state of consciousness. Victims may experience sudden, incapacitating attacks of muscle weakness or paralysis (known as cataplexy). Such people may fall asleep while walking, talking, or driving a car. Although long naps each day and stimulant or antidepressant drugs may help reduce the frequency of attacks, both the causes and cure of narcolepsy are still unknown (FIGURE 5.6).

Perhaps the most serious dyssomnia is **sleep apnea**. People with sleep apnea may fail to breathe for a minute or longer and then wake up gasping for breath. When they do breathe during their sleep, they often snore. Although people with sleep apnea are often unaware of it, repeated awakenings result in insomnia and leave the person feeling tired and sleepy during the

Narcolepsy FIGURE 5.6

William Dement and his colleagues at Stanford University's Sleep Disorders Center have bred a group of narcoleptic dogs, which has increased our understanding of the genetics of this disorder. Research on these specially bred dogs has found degenerated neurons in certain areas of the brain (Siegel, 2000). Whether human narcolepsy results from similar degeneration is a question for future research.

day. Sleep apnea seems to result from blocked upper airway passages or from the brain's ceasing to send signals to the diaphragm, thus causing breathing to stop. This disorder can lead to high blood pressure, stroke, and heart attack (Billiard, 2007; Hartenbaum et al., 2006; National Sleep Foundation, 2007; McNicholas & Javaheri, 2007).

Treatment for sleep apnea depends partly on its severity. If the problem occurs only when you're sleeping on your back, sewing tennis balls to the back of your pajama top may help remind you to sleep on your side. Because obstruction of the breathing passages is related to obesity and heavy alcohol use (Christensen, 2000), dieting and alcohol restriction are often recommended. For others, surgery, dental appliances that reposition the tongue, or machines that provide a stream of air to keep the airway open may be the answer (FIGURE 5.7).

Recent findings suggest that even "simple" snoring (without the breathing stoppage characteristic of sleep apnea) can lead to heart disease and possible death (Stone & Redline, 2006). Although occasional mild snoring remains somewhat normal, chronic snoring is a possible "warning sign that should prompt people to seek help" (Christensen, 2000, p. 172).

Sleep apnea FIGURE 5.7

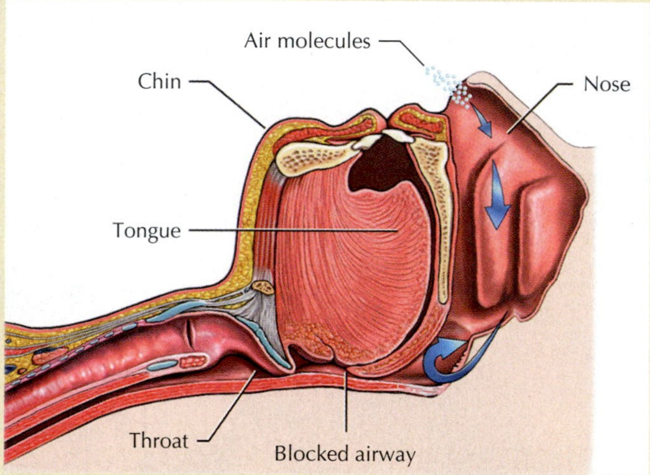

A During sleep apnea, airways are blocked, causing breathing to be severely restricted. To treat this disorder, researchers and doctors have created equipment to promote respiration.

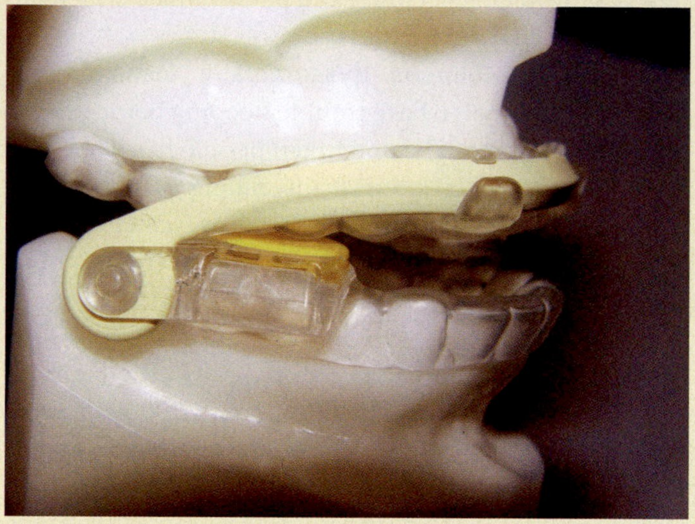

B For some patients, there is help from dental devices that reposition the tongue and open the airway.

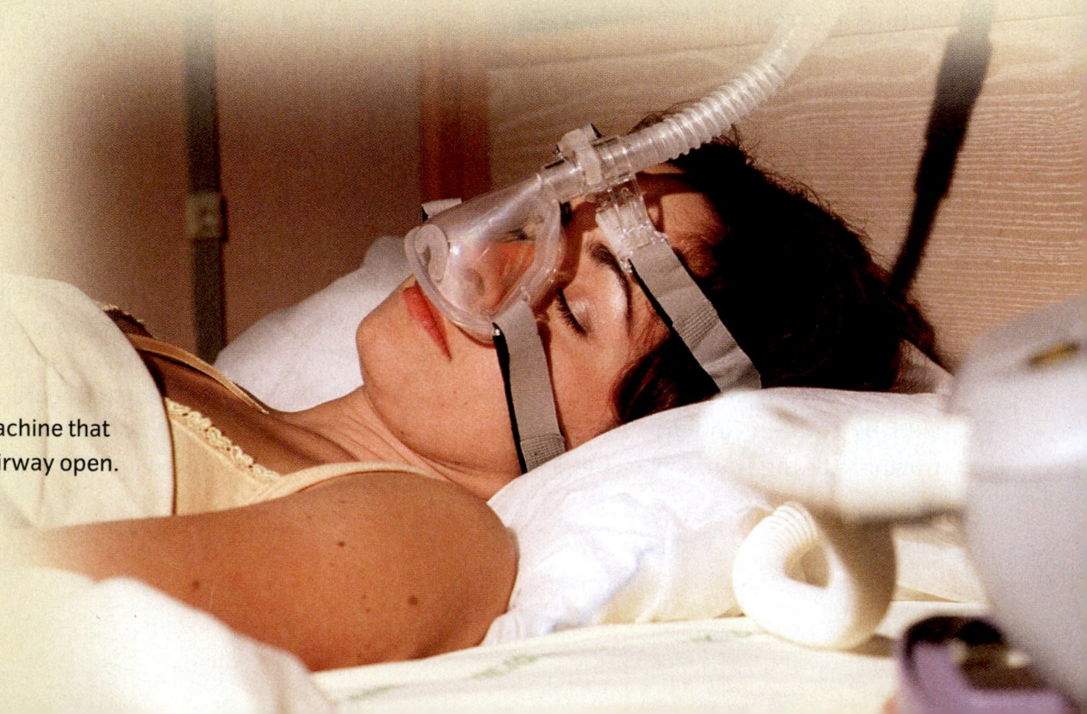

C Another treatment for sleep apnea is a machine that provides a steady supply of air to keep the airway open.

Nightmare or night terrors? FIGURE 5.8

Nightmares, or bad dreams, occur toward the end of the sleep cycle, during REM sleep. Less common but more frightening are night terrors, which occur early in the cycle, during Stage 3 or Stage 4 of NREM sleep. Like the child in this photo, the sleeper may sit bolt upright, screaming and sweating, walk around, and talk incoherently, and the person may be almost impossible to awaken.

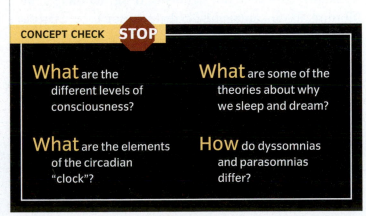

The second major category of sleep disorders, **parasomnias**, includes abnormal sleep disturbances such as **nightmares** and **night terrors** (FIGURE 5.8).

> **parasomnias** Abnormal disturbances occurring during sleep, including nightmares, night terrors, sleepwalking, and sleep talking.

Sleepwalking, which tends to accompany night terrors, usually occurs during NREM sleep. (Recall that large muscles are paralyzed during REM sleep, which explains why sleepwalking normally occurs during NREM sleep.) **Sleep talking** can occur during any stage of sleep, but appears to arise most commonly during NREM sleep. It can consist of single indistinct words or long, articulate sentences. It is even possible to engage some sleep talkers in a limited conversation.

Nightmares, night terrors, sleepwalking, and sleep talking are all more common among young children, but they can also occur in adults, usually during times of stress or major life events (Billiard, 2007; Hobson & Silvestri, 1999). Patience and soothing reassurance at the time of the sleep disruption are usually the only treatment recommended for both children and adults.

A recent large-scale study reported that behavior therapy had good success in treating some sleep problems (Constantino et al., 2007; Smith et al., 2005). You can use similar techniques in your own life. For example:

- When you're having a hard time going to sleep, don't keep checking the clock and worrying about your loss of sleep. Instead, remove all TVs, stereos, and books, and limit the use of the bedroom to sleep.
- Work off tension through exercise (but not too close to bedtime).
- Avoid stimulants such as caffeine and nicotine.
- Avoid late meals and heavy drinking.
- Follow the same presleep routine every evening. It might include listening to music, writing in a diary, or meditating.
- Use *progressive muscle relaxation*. Alternately tense and relax various muscle groups.
- Practice yoga or deep breathing, or take a warm bath to help you relax.

If you need additional help, try the relaxation techniques from the Better Sleep Council, a nonprofit education organization in Burtonsville, Maryland. Also, check out these websites: www.sleepfoundation.org and www.stanford.edu/~dement.

CONCEPT CHECK STOP

What are the different levels of consciousness?

What are some of the theories about why we sleep and dream?

What are the elements of the circadian "clock"?

How do dyssomnias and parasomnias differ?

Psychoactive Drugs

LEARNING OBJECTIVES

Explain the difference between psychological and physical drug dependence.

Summarize the differences among the four major types of psychoactive drugs.

Compare how different psychoactive drugs affect the nervous system.

Have you noticed how difficult it is to have a logical, nonemotional discussion about drugs? In our society, where the most popular **psychoactive drugs** are caffeine, tobacco, and ethyl alcohol, people often become defensive when these drugs are grouped with illicit drugs such as marijuana and cocaine. Similarly, marijuana users are disturbed that their drug of choice is grouped with "hard" drugs like heroin. Most scientists believe that there are good and bad uses of all drugs. How drug use differs from drug abuse and how chemical alterations in consciousness affect a person, psychologically and physically, are important topics in psychology.

psychoactive drugs Chemicals that change conscious awareness, mood, or perception.

Psychoactive drugs influence the nervous system in a variety of ways. Alcohol, for example, has a diffuse effect on neural membranes throughout the nervous system. Most psychoactive drugs, however, act in a more specific way: by either enhancing a particular neurotransmitter's effect (an **agonistic** drug action) or inhibiting it (an **antagonistic** drug action) (FIGURES 5.9 and 5.10).

Is drug abuse the same as drug addiction? The term **drug abuse** generally refers to drug taking that causes emotional or physical harm to oneself or others. The drug consumption is also typically compulsive, frequent, and intense. **Addiction** is a broad term referring to a condition in which a person feels compelled to use a specific drug. People now use the term to describe almost any type of compulsive activity, from working to surfing the Internet (Coombs, 2004). In fact, recent research has shown that risky trading in financial markets can create a high that is indistinguishable from those experienced by drug addicts (Zweig, 2007).

Cocaine: An agonist drug in action FIGURE 5.9

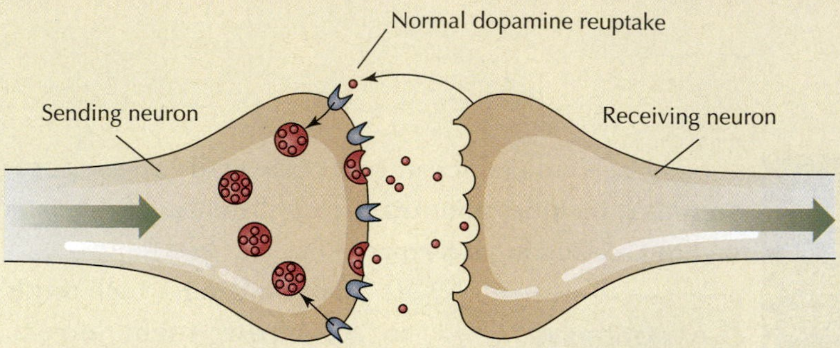

A After releasing neurotransmitter into the synapse, the sending neuron normally reabsorbs (or reuptakes) excess neurotransmitter back into the terminal buttons.

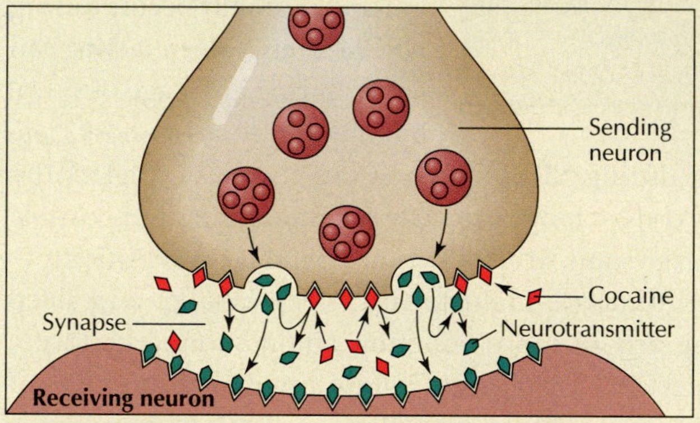

B However, if cocaine is present in the synapse, it will block the reuptake of dopamine, serotonin, and norepinephrine. This blockage intensifies the normal mood-altering effect of these three mood- and energy-activating neurotransmitters.

Process Diagram

How agonistic and antagonistic drugs produce their psychoactive effect FIGURE 5.10

Most psychoactive drugs produce their mood-, energy-, and perception-altering effects by changing the body's supply of neurotransmitters. They can alter synthesis, storage, and release of neurotransmiters (Step 1). Psychoactive drugs can also alter the effect of neurotransmittters on the receiving site of the receptor neuron (Step 2). After neurotransmitters carry their messages across the synapse, the sending neuron normally deactivates the excess, or leftover, neurotransmitter (Step 3).

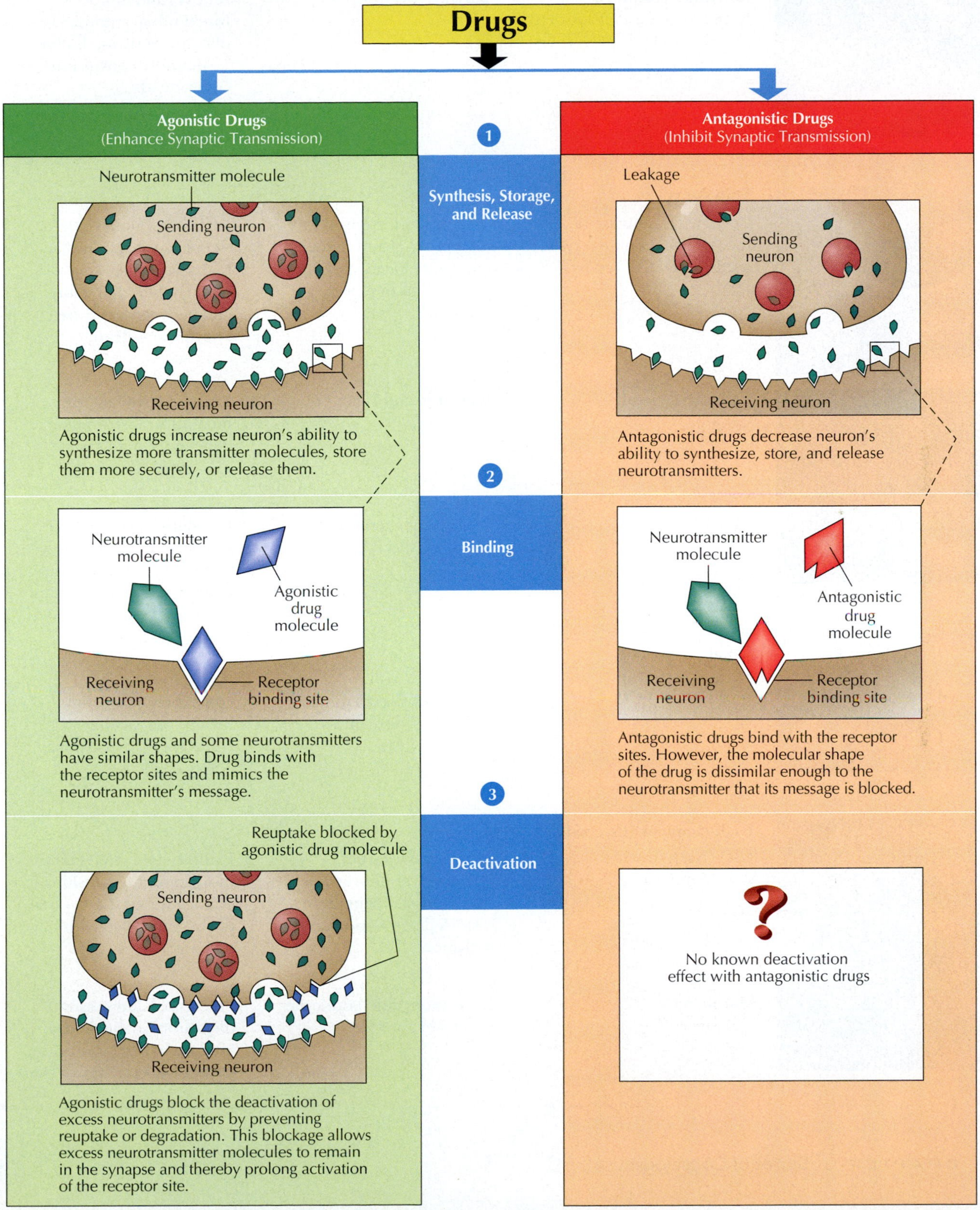

How agonistic and antagonistic drugs produce their psychoactive effect FIGURE 5.10

Most psychoactive drugs produce their mood-, energy-, and perception-altering effects by changing the body's supply of neurotransmiters. They can alter synthesis, storage, and release of neurotransmiters (Step 1). Psychoactive drugs can also alter the effect of neurotransmittters on the receiving site of the receptor neuron (Step 2). After neurotransmitters carry their messages across the synapse, the sending neuron normally deactivates the excess, or leftover, neurotransmitter (Step 3).

Drugs

Agonistic Drugs (Enhance Synaptic Transmission)

Antagonistic Drugs (Inhibit Synaptic Transmission)

1 Synthesis, Storage, and Release

Neurotransmitter molecule
Sending neuron
Receiving neuron

Leakage
Sending neuron
Receiving neuron

Agonistic drugs increase neuron's ability to synthesize more transmitter molecules, store them more securely, or release them.

Antagonistic drugs decrease neuron's ability to synthesize, store, and release neurotransmitters.

2 Binding

Neurotransmitter molecule
Agonistic drug molecule
Receiving neuron
Receptor binding site

Neurotransmitter molecule
Antagonistic drug molecule
Receiving neuron
Receptor binding site

Agonistic drugs and some neurotransmitters have similar shapes. Drug binds with the receptor sites and mimics the neurotransmitter's message.

Antagonistic drugs bind with the receptor sites. However, the molecular shape of the drug is dissimilar enough to the neurotransmitter that its message is blocked.

3 Deactivation

Reuptake blocked by agonistic drug molecule
Sending neuron
Receiving neuron

?
No known deactivation effect with antagonistic drugs

Agonistic drugs block the deactivation of excess neurotransmitters by preventing reuptake or degradation. This blockage allows excess neurotransmitter molecules to remain in the synapse and thereby prolong activation of the receptor site.

Effects of the major psychoactive drugs TABLE 5.1

	Category	*Desired effects*	*Undesirable effects*
	**Depressants (Sedatives)** Alcohol, barbiturates, anti-anxiety drugs (Valium), Rohypnol (roofies), Ketamine (special K), CoHB	Tension reduction, euphoria, disinhibition, drowsiness, muscle relaxation	Anxiety, nausea, disorientation, impaired reflexes and motor functioning, amnesia, loss of consciousness, shallow respiration, convulsions, coma, death
	**Stimulants** Cocaine, amphetamine, methamphetamine (crystal meth), MDMA (Ecstasy)	Exhilaration, euphoria, high physical and mental energy, reduced appetite, perceptions of power, sociability	Irritability, anxiety, sleeplessness, paranoia, hallucinations, psychosis, elevated blood pressure and body temperature, convulsions, death
	Caffeine	Increased alertness	Insomnia, restlessness, increased pulse rate, mild delirium, ringing in the ears, rapid heartbeat
	Nicotine	Relaxation, increased alertness, sociability	Irritability, raised blood pressure, stomach pains, vomiting, dizziness, cancer, heart disease, emphysema
	**Opiates (Narcotics)** Morphine, heroin, codeine	Euphoria, "rush" of pleasure, pain relief, prevention of withdrawal discomfort	Nausea, vomiting, constipation, painful withdrawal, shallow respiration, convulsions, coma, death
	**Hallucinogens (Psychedelics)** LSD (lysergic acid diethylamide)	Heightened aesthetic responses, euphoria, mild delusions, hallucinations, distorted perceptions and sensations	Panic, nausea, longer and more extreme delusions, hallucinations, perceptual distortions ("bad trips"), psychosis
	Marijuana	Relaxation, mild euphoria, increased appetite	Perceptual and sensory distortions, hallucinations, fatigue, lack of motivation, paranoia, possible psychosis

For the sake of clarity, many researchers use the term **psychological dependence** to refer to the mental desire or craving to achieve a drug's effects. They use the term **physical dependence** to refer to changes in bodily processes that make a drug necessary for minimum daily functioning. Physical dependence appears most clearly when the drug is withheld and the user undergoes painful **withdrawal** reactions, including physical pain and intense cravings. After repeated use of a drug, many of the body's physiological processes adjust to higher and higher levels of the drug, producing a decreased sensitivity called **tolerance**.

Tolerance leads many users to escalate their drug use and to experiment with other drugs in an attempt to re-create the original pleasurable altered state. Sometimes, using one drug increases tolerance for another. This is known as **cross-tolerance**. Developing tolerance or cross-tolerance does not prevent drugs from seriously damaging the brain, heart, liver, and other organs.

Psychological dependence is no less damaging than physical dependence. The craving in psychological dependence can be strong enough to keep the user in a constant drug-induced state—and to lure an "addict" back to a drug habit long after he or she has overcome physical dependence.

PSYCHOACTIVE DRUGS: FOUR CATEGORIES

For convenience, psychologists divide psychoactive drugs into four broad categories: depressants, stimulants, opiates, and hallucinogens. TABLE 5.1 provides examples of each and their effects.

Depressants (sometimes called "downers") act on the central nervous system to suppress or slow bodily processes and to reduce overall responsiveness. Because tolerance and dependence (both physical and psychological) are rapidly acquired with these drugs, there is strong potential for abuse.

Although alcohol is primarily a depressant, at low doses it has stimulating effects, thus explaining its reputation as a "party drug." As consumption increases, symptoms of drunkenness appear (FIGURE 5.11).

Alcohol should not be combined with any other drug, but combining alcohol and barbiturates—both depressants—is particularly dangerous. Together, they can relax the diaphragm muscles to such a degree that the person literally suffocates.

Depressants suppress central nervous system activity, whereas **stimulants** (uppers) increase the overall

Alcohol's effect on the body and behavior FIGURE 5.11

Alcohol's effects are determined primarily by the amount that reaches the brain. Because the liver breaks down alcohol at the rate of about one ounce per hour, the number of drinks and the speed of consumption are both very important. People can die after drinking large amounts of alcohol in a short period of time (Chapter 3). In addition, men's bodies are more efficient at breaking down alcohol. Even after accounting for differences in size and muscle-to-fat ratio, women have a higher blood alcohol level than men following equal doses of alcohol.

Number of drinks[a] in two hours	*Blood alcohol (content (%)[b]*	*Effect*
(2)	0.05	Relaxed state; increased sociability
(3)	0.08	Everyday stress lessened
(4)	0.10	Movements and speech become clumsy
(7)	0.20	Very drunk; loud and difficult to understand; emotions unstable
(12)	0.40	Difficult to wake up; incapable of voluntary action
(15)	0.50	Coma and/or death

[a]A drink refers to one 12-ounce beer, a 4-ounce glass of wine, or a 1.25-ounce shot of hard liquor.
[b]In America, the legal blood alcohol level for "drunk driving" varies from 0.05 to 0.12.

activity and responsiveness of the central nervous system. Even legal stimulants can lead to serious problems. For example, the U.S. Public Health Service considers cigarette smoking the single most preventable cause of death and disease in the United States. Researchers have found that nicotine activates the same brain areas (nucleus accumbens) as cocaine—a dangerous stimulant well known for its addictive potential (Champtiaux, Kalivas, & Bardo, 2006; McQuown et al., 2007; Zanetti, Picciotto, & Zoli, 2007). Nicotine's effects (relaxation, increased alertness, diminished pain and appetite) are so powerfully reinforcing that some people continue to smoke even after having a cancerous lung removed.

Opiates (or narcotics), which are derived from the opium poppy, are used medically to relieve pain (Kuhn, Swartzwelder, & Wilson, 2003). They mimic the brain's natural endorphins (Chapter 2), which numb pain and elevate mood. This creates a dangerous pathway to drug abuse. After repeated flooding with artificial opiates, the brain eventually reduces or stops the production of its own opiates. If the user later attempts to stop, the brain lacks both the artificial and normal level of painkilling chemicals, and withdrawal becomes excruciatingly painful (FIGURE 5.12). Interestingly, when opiates are used medically to relieve intense pain, they are very seldom habit-forming. However, when taken recreationally, they are strongly addictive (Fields, 2007; Levinthal, 2008).

So far, we have discussed three of the four types of psychoactive drugs: depressants, stimulants, and opiates. One of the most intriguing alterations of consciousness comes from **hallucinogens**, drugs that produce sensory or perceptual distortions, including visual, auditory, and kinesthetic hallucinations. Some cultures have used hallucinogens for religious purposes, as a way to experience "other realities" or to communicate with the supernatural. In Western societies, most people use hallucinogens for their reported "mind-expanding" potential.

Hallucinogens are commonly referred to as psychedelics (from the Greek for "mind manifesting"). They include mescaline (derived from the peyote cactus), psilocybin (derived from mushrooms), phencyclidine (chemically derived), and LSD (lysergic acid diethylamide, derived from ergot, a rye mold).

LSD, or acid, is a synthetic substance that produces dramatic alterations in sensation and perception. LSD use by high school and college students has been increasing (Connolly, 2000; Hedges & Burchfield, 2006; Yacoubian, Green, & Peters, 2003). LSD can be an extremely dangerous drug. Bad LSD trips can be terrifying and may lead to accidents, deaths, or suicide. Dangerous flashbacks may unpredictably recur long after the initial

How opiates create physical dependence FIGURE 5.12

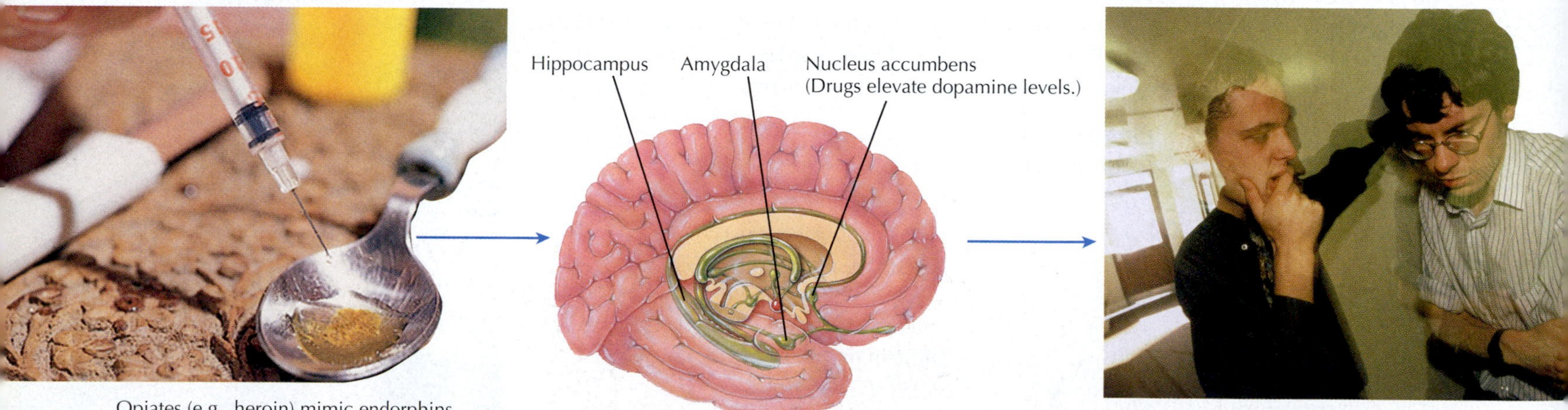

Opiates (e.g., heroin) mimic endorphins, which elicit euphoria and pain relief.

These key brain areas are associated with reward, pleasure, and addiction.

Absence of the drug triggers withdrawal symptoms (e.g., intense pain and cravings).

ingestion. They can be brought on by stress, fatigue, marijuana use, illness, or occasionally by the individual's intentional effort (Abadinsky, 2008).

Marijuana is also classified as a hallucinogen even though it has some properties of a depressant (it induces drowsiness and lethargy) and some of a narcotic (it acts as a weak painkiller). In low doses, marijuana produces mild euphoria; moderate doses lead to an intensification of sensory experiences and the illusion that time is passing slowly. High doses may produce hallucinations, delusions, and distortions of body image (Köfalvi, 2008; Ksir, Hart, & Ray, 2008). The active ingredient in marijuana (cannabis) is THC, or tetrahydrocannabinol, which attaches to receptors that are abundant throughout the brain.

Some research has found marijuana to be therapeutic in the treatment of glaucoma (an eye disease), in alleviating the nausea and vomiting associated with chemotherapy, and with other health problems (Darmani & Crim, 2005; Fogarty et al., 2007; Köfalvi, 2008).

Chronic marijuana use can lead to throat and respiratory disorders, impaired lung functioning and immune response, declines in testosterone levels, reduced sperm count, and disruption of the menstrual cycle and ovulation (Levinthal, 2008; Murphy, 2006; Nahas et al., 2002; Rossato & Pagano, 2008; Roth et al., 2004). While some research supports the popular belief that marijuana serves as a "gateway" to other illegal drugs, other studies find little or no connection (Ksir, Hart, & Ray, 2008; Sabet, 2007; Tarter et al., 2006).

Marijuana also can be habit-forming, but few users experience the intense cravings associated with cocaine or opiates. Withdrawal symptoms are mild because the drug dissolves in the body's fat and leaves the body very slowly, which explains why a marijuana user can test positive for days or weeks after the last use.

CLUB DRUGS

As you may know from television or newspapers, psychoactive drugs like Rohypnol (the date rape drug) and MDMA (3-4 methylenedioxymethamphetamine, commonly known as Ecstasy) are fast becoming some of our nation's most popular drugs of abuse, especially at all-night dance parties. (See **FIGURE 5.13** on the next page.) Other "club" drugs, like GHB (gamma-hydroxybutyrate), ketamine (Special K), methamphetamine (crystal meth), and LSD, are also gaining in popularity (Abadinsky, 2008; Weaver & Schnoll, 2008).

Although these drugs can produce desirable effects (e.g., Ecstasy's feeling of great empathy and connectedness with others), almost all psychoactive drugs may cause serious health problems—in some cases, even death (National Institute on Drug Abuse, 2005). Club drugs (like most psychoactive drugs) affect motor coordination, perceptual skills, and reaction time necessary for safe driving. As with all illicit drugs, there are no truth-in-packaging laws to protect club drug buyers from unscrupulous practices. Sellers often substitute unknown, cheaper, and possibly even more dangerous substances for the ones they claim to be selling.

Club drug use may lead to risky sexual behaviors and an increased risk of contracting AIDS (acquired immunodeficiency syndrome) and other sexually transmitted diseases. Some drugs, such as Rohypnol, are odorless, colorless, and tasteless, and they can easily be added to beverages by individuals who want to intoxicate or sedate others, so the dangers of club drug use go far beyond the drug itself (Fernández et al., 2005; National Institute on Drug Abuse, 2005). For more information on the specific dangers and effects of club drugs, check out www.drugabuse.gov/ClubAlert/Clubdrugalert.html.

CONCEPT CHECK STOP

What are the four types of psychoactive drugs?

What are the differences between psychological and physical drug dependence?

How do psychoactive drugs affect nervous system functioning?

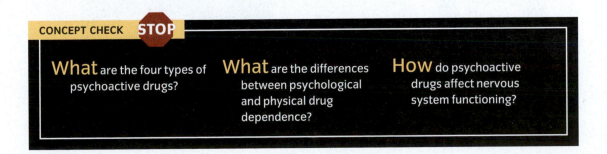

Visualizing

Your brain on "club" drugs FIGURE 5.13

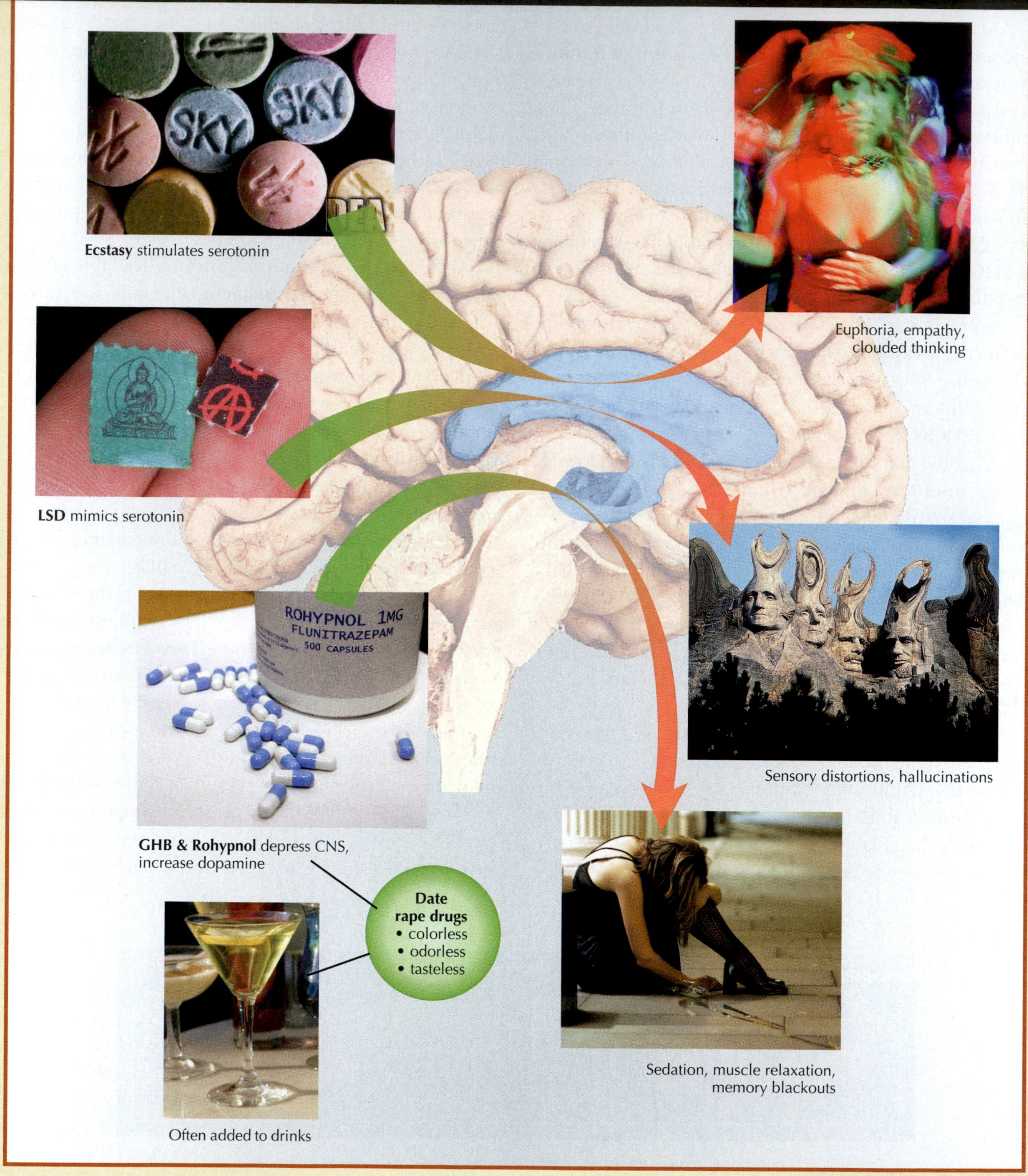

Ecstasy stimulates serotonin

LSD mimics serotonin

GHB & Rohypnol depress CNS, increase dopamine

Often added to drinks

Euphoria, empathy, clouded thinking

Sensory distortions, hallucinations

Sedation, muscle relaxation, memory blackouts

NATIONAL GEOGRAPHIC

Altering Consciousness Through Meditation and Hypnosis

LEARNING OBJECTIVES

Summarize some major forms of meditation.

Explain some of the uses of hypnosis and some of the myths about hypnosis.

As we have seen, factors such as sleep, dreaming, and psychoactive drug use can create altered states of consciousness. Changes in consciousness can also be achieved by means of meditation and hypnosis.

MEDITATION: A HEALTHY "HIGH"

"Suddenly, with a roar like that of a waterfall, I felt a stream of liquid light entering my brain through the spinal cord . . . I experienced a rocking sensation and then felt myself slipping out of my body, entirely enveloped in a halo of light. I felt the point of consciousness that was myself growing wider, surrounded by waves of light" (Krishna, 1999, pp. 4–5).

This is how spiritual leader Gopi Krishna described his experience with **meditation**. Although most people in the beginning stages of meditation report a simpler, mellow type of relaxation followed by a mild euphoria, some advanced meditators report experiences of profound rapture and joy or strong hallucinations.

meditation A group of techniques designed to refocus attention, block out all distractions, and produce an alternate state of consciousness.

The highest functions of consciousness occur in the frontal lobe, particularly in the cerebral cortex. Scientists are seeing increasing evidence that the altered state of consciousness one experiences during meditation occurs when one purposely changes how the prefrontal cortex (the area immediately behind your eyes) functions. Typically, your prefrontal cortex is balancing your working memory, temporal integration, and higher-order thinking, among other tasks. Scientists theorize, based on brain imaging, that when you focus on a single object, emotion, or word, you diminish the amount of brain cells that must be devoted to these multiple tasks, and they instead become involved in the singular focus of your meditation. This narrow focus allows other areas of the brain to be affected, and since the neurons devoted to time have changed focus, you experience a sense of timelessness and mild euphoria (Aftanas & Golosheikim, 2003; Castillo, 2003; Harrison, 2005) (FIGURE 5.14). Evidence shows that meditation, and yoga as well, can help

How meditation alters consciousness FIGURE 5.14

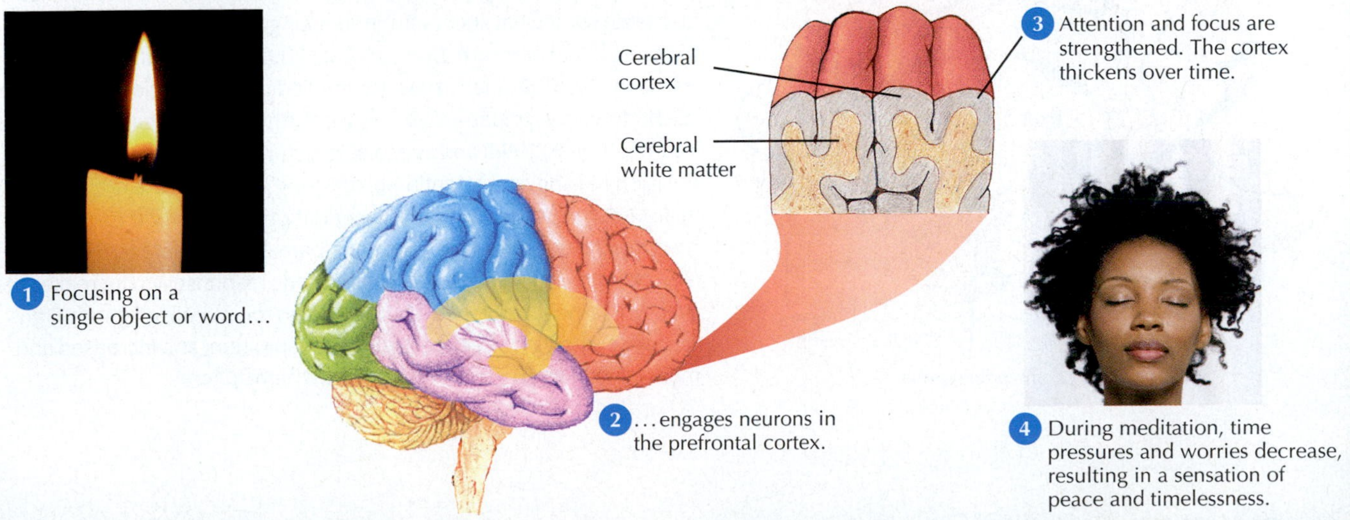

1 Focusing on a single object or word...
2 ...engages neurons in the prefrontal cortex.
Cerebral cortex
Cerebral white matter
3 Attention and focus are strengthened. The cortex thickens over time.
4 During meditation, time pressures and worries decrease, resulting in a sensation of peace and timelessness.

Process Diagram

Meditation and the Brain

Some meditation techniques, such as t'ai chi and hatha yoga, involve body movements and postures, while in other techniques the meditator remains motionless, chanting or focusing on a single point.

Researchers have recently found that a wider area of the brain responds to sensory stimuli during meditation, suggesting that meditation enhances the coordination between the brain hemispheres (see graphic, right; and Lyubimov, 1992). Researchers have also found that those who meditate use a larger portion of their brain, and that faster and more powerful gamma waves exist in individuals who meditate regularly than in those who do not (Lutz et al., 2004). The increased coordination associated with more powerful gamma waves correlates with improvements in focus, memory, learning, and consciousness.

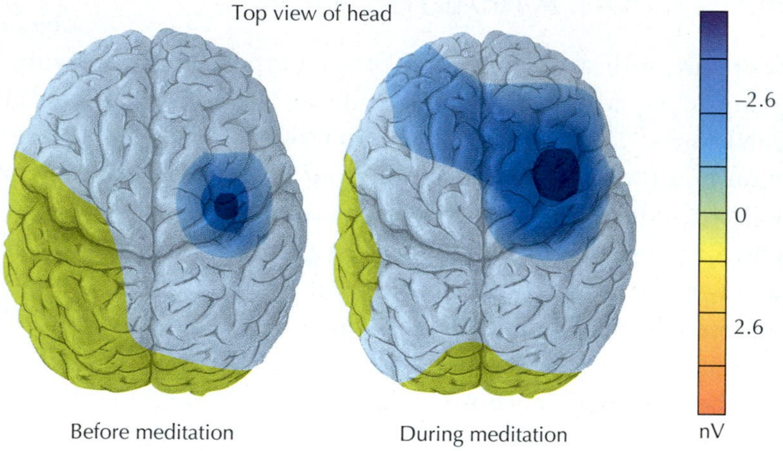

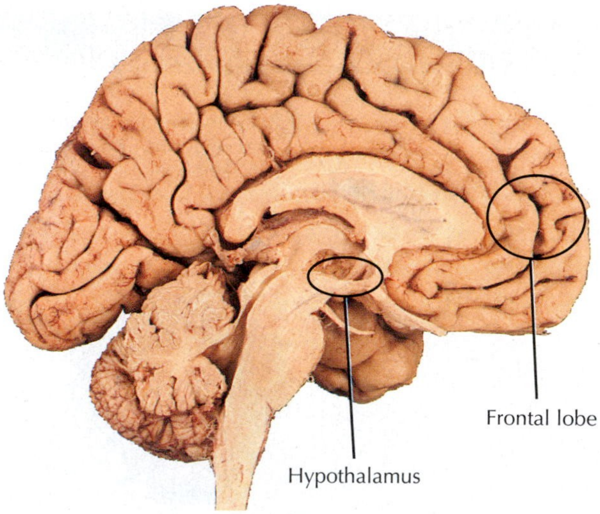

Research has verified that meditation can produce dramatic changes in basic physiological processes, including heart rate, oxygen consumption, sweat gland activity, and brain activity. Meditation has also been somewhat successful in reducing anxiety and stress and lowering blood pressure (Carlson et al., 2007; Evans et al., 2008; Harrison, 2005). Studies have even implied that meditation can change the body's parasympathetic response (Sathyaprabha et al., 2008; Young & Taylor, 1998) and increase structural support for the sensory, decision-making, and attention-processing centers of the brain (Lazar et al., 2005). It seems that during meditation, the part of the brain that is responsible for both sympathetic and parasympathetic responses, the *hypothalamus*, diminishes the sympathetic response and increases the parasympathetic response. Shutting down the so-called fight-or-flight response, allows for deep rest, slower respiration, and increased and more coordinated use of the brain's two hemispheres.

individuals cope better with stress (see Chapter 3 and *What a Psychologist Sees*).

HYPNOSIS: USES AND MYTHS

Relax . . . your eyelids are so very heavy . . . your muscles are becoming more and more relaxed . . . your breathing is becoming deeper and deeper . . . relax . . . your eyes are closing . . . let go . . . relax.

Hypnotists use suggestions like these to begin **hypnosis**. Once hypnotized, some people can be convinced that they are standing at the edge of the ocean listening to the sound of the waves and feeling the ocean mist on their faces. Invited to eat a delicious apple that is actually an onion, the hypnotized person may relish the flavor. Told they are watching a very funny or sad movie, hypnotized people may begin to laugh or cry at their self-created visions.

hypnosis A trance-like state of heightened suggestibility, deep relaxation, and intense focus.

From the 1700s to modern times, entertainers and quacks have used (and abused) hypnosis (TABLE 5.2); but physicians, dentists, and therapists have also long

Hypnosis myths and facts TABLE 5.2

Myth	*Fact*
Forced hypnosis: *People can be involuntarily hypnotized or hypnotically "brainwashed."*	Hypnosis requires a willing, conscious choice to relinquish control of one's consciousness to someone else. The best potential subjects are those who are able to focus attention, are open to new experiences, and are capable of imaginative involvement or fantasy (Carvalho et al., 2008; Hutchinson-Phillips, Gow, & Jamieson 2007; Wickramasekera, 2008).
Unethical behavior: *Hypnosis can make people behave immorally or take dangerous risks against their will.*	Hypnotized people retain awareness and control of their behavior, and they can refuse to comply with the hypnotist's suggestions (Kirsch & Braffman, 2001; Kirsch, Mazzoni, & Montgomery, 2006).
Faking: *Hypnosis participants are "faking it," playing along with the hypnotist.*	Although most participants are not consciously faking hypnosis, some researchers believe the effects result from a blend of conformity, relaxation, obedience, suggestion, and role playing (Fassler et al., 2008; Lynn, 2007; Orne, 2006). Other theorists believe that hypnotic effects result from a special altered state of consciousness (Bob, 2008; Bowers & Woody, 1996; Hilgard, 1978, 1992; Naisch, 2007). A group of "unified" theorists suggests that hypnosis is a combination of both relaxation/role playing and a unique alternate state of consciousness.
Superhuman strength: *Hypnotized people can perform acts of superhuman strength.*	When nonhypnotized people are simply asked to try their hardest on tests of physical strength, they generally can do anything that a hypnotized person can (Orne, 2006).
Exceptional memory: *Under hypnosis, people can recall things they otherwise could not.*	Although the heightened relaxation and focus that hypnosis engenders improves recall for some information, people also are more willing to guess (Stafford & Lynn, 2002; Wagstaff et al., 2007; Wickramasekera, 2008). Because memory is normally filled with fabrication and distortion (Chapter 7), hypnosis generally increases the potential for error.

employed it as a respected clinical tool. Modern scientific research has removed much of the mystery surrounding hypnosis. A number of features characterize the hypnotic state (Jamieson & Hasegawa, 2007; Jensen et al., 2008; Nash & Barnier, 2008):

- Narrowed, highly focused attention (ability to "tune out" competing sensory stimuli)
- Increased use of imagination and hallucinations
- A passive and receptive attitude
- Decreased responsiveness to pain
- Heightened suggestibility, or a willingness to respond to proposed changes in perception ("This onion is an apple")

Today, even with available anesthetics, hypnosis is occasionally used in surgery and for the treatment of chronic pain and severe burns (Jensen et al., 2008; Nash & Barnier, 2008). Hypnosis has found its best use in medical areas, such as dentistry and childbirth, in which patients have a high degree of anxiety, fear, and misinformation. Because tension and anxiety strongly affect pain, any technique that helps the patient relax is medically useful.

In psychotherapy, hypnosis can help patients relax, remember painful memories, and reduce anxiety. Despite the many myths about hypnosis, it has been used with modest success in the treatment of phobias and in helping people to lose weight, stop smoking, and improve study habits (Amundson & Nuttgens, 2008; Golden, 2006; Manning, 2007).

CONCEPT CHECK STOP

What are some major features of meditation?

Who might benefit from hypnosis?

Why can't people be hypnotized against their will?

SUMMARY

1 Consciousness, Sleep, and Dreaming

1. **Consciousness**, an organism's awareness of its own self and surroundings, exists along a continuum, from high awareness to unconsciousness and coma. Sleep is a particular **altered state of consciousness (ASC)**.

2. Controlled processes demand focused attention and generally interfere with other ongoing activities. Automatic processes require minimal attention and generally do not interfere with other ongoing activities.

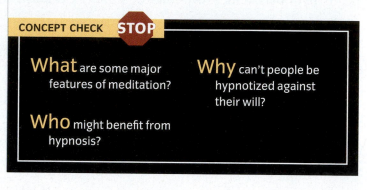

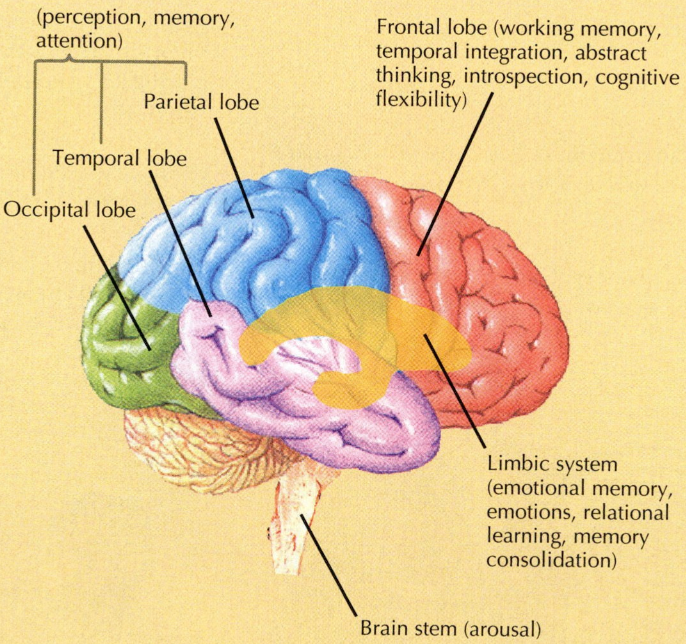

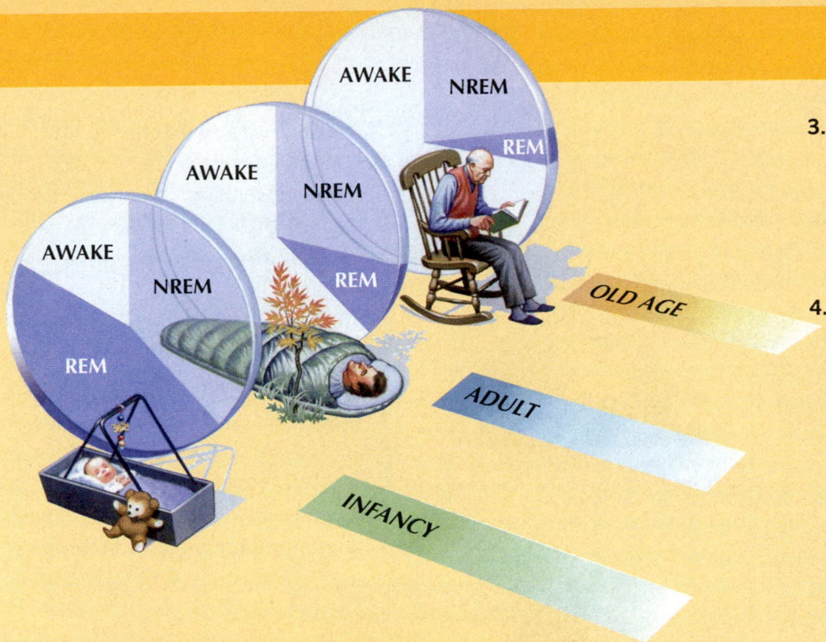

3. Many physiological functions follow 24-hour **circadian rhythms**. Disruptions in circadian rhythms, as well as long-term sleep deprivation, cause increased fatigue, cognitive and mood disruptions, and other health problems.

4. The electroencephalogram (EEG) detects and records electrical changes in the nerve cells of the cerebral cortex. People progress through four distinct stages of non-REM (NREM) sleep, with periods of rapid-eye-movement (REM) sleep occurring at the end of each sleep cycle. Both REM and NREM sleep are important for our biological functioning.

5. The **evolutionary/circadian theory** proposes that sleep evolved to conserve energy and as protection from predators. The **repair/restoration theory** suggests that sleep helps us recuperate from the day's events. Three major theories for why we dream are Freud's psychoanalytic view, the activation–synthesis hypothesis, and the cognitive or information-processing view.

6. **Dyssomnias** are problems in the amount, timing, and quality of sleep; they include insomnia, sleep apnea, and narcolepsy. **Parasomnias** are abnormal disturbances occurring during sleep; they include nightmares, night terrors, sleepwalking, and sleep talking. Although drugs are the most common method of treating sleep disorders, a recent large-scale study reported that behavior therapy had good success in treating some sleep problems.

2 Psychoactive Drugs

1. **Psychoactive drugs** influence the nervous system in a variety of ways. Alcohol affects neural membranes throughout the entire nervous system. Most psychoactive drugs act in a more specific way, by either enhancing a particular neurotransmitter's effect (an *agonistic* drug action) or inhibiting it (an *antagonistic* drug action). Drugs can interfere with neurotransmission at any of four stages: production or synthesis; storage and release; reception; or removal.

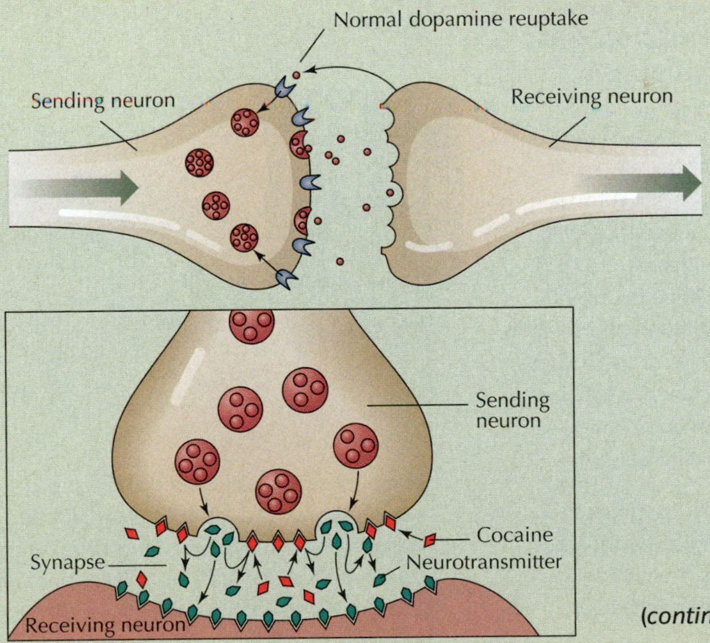

(*continued on page 138*)

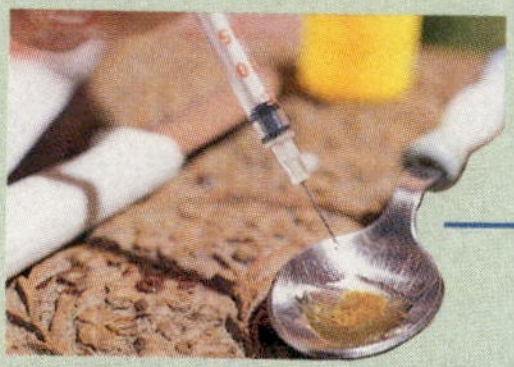

Opiates (e.g., heroin) mimic endorphins, which elicit euphoria and pain relief.

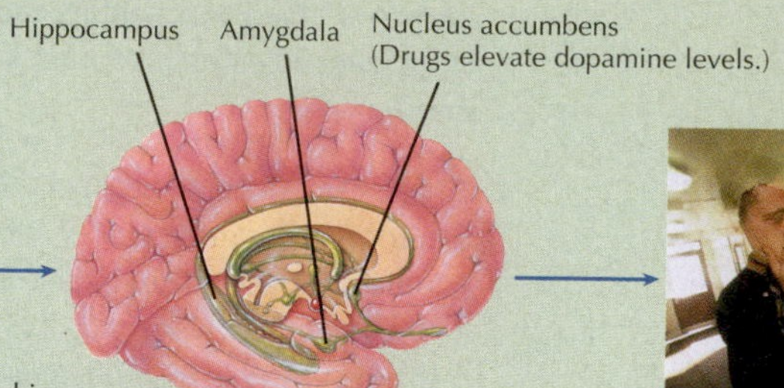

These key brain areas are associated with reward, pleasure, and addiction.

Absence of the drug triggers withdrawal symptoms (e.g., intense pain and cravings).

2. The term **drug abuse** refers to drug-taking behavior that causes emotional or physical harm to oneself or others. Addiction refers to a condition in which a person feels compelled to use a specific drug. Psychological dependence refers to the mental desire or craving to achieve a drug's effects. Physical dependence refers to biological changes that make a drug necessary for minimum daily functioning. Repeated use of a drug can produce decreased sensitivity, or tolerance. Sometimes, using one drug increases tolerance for another (cross-tolerance).

3. Psychologists divide psychoactive drugs into four categories: depressants (such as alcohol, barbiturates, Rohypnol, and Ketamine), stimulants (such as caffeine, nicotine, cocaine, amphetamine, methamphetamine, and Ecstasy), opiates (such as morphine, heroin, and codeine), and hallucinogens (such as marijuana and LSD). Almost all psychoactive drugs may cause serious health problems and, in some cases, even death.

3 Altering Consciousness Through Meditation and Hypnosis

1. The term **meditation** refers to techniques designed to refocus attention, block out distractions, and produce an alternate state of consciousness. Some followers believe that meditation offers a more enlightened form of consciousness, and researchers have verified that it can produce dramatic changes in basic physiological processes.

1. Focusing on a single object or word...

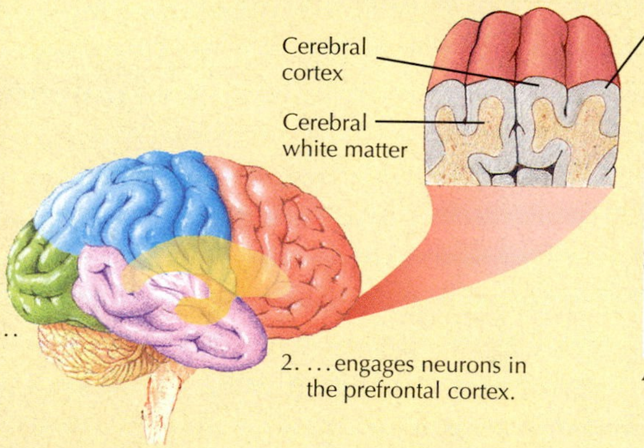

2. ...engages neurons in the prefrontal cortex.

3. Attention and focus are strengthened. The cortex thickens over time.

4. During meditation, time pressures and worries decrease, resulting in a sensation of peace and timelessness.

2. Modern research has removed the mystery surrounding **hypnosis**, a trancelike state of heightened suggestibility, deep relaxation, and intense focus. It is used in surgery and medicine, and it is especially useful in medical areas in which patients have a high degree of anxiety, fear, and misinformation. In psychotherapy, hypnosis can help patients relax, remember painful memories, and reduce anxiety. It is also used to help with a number of behavioral issues, such as efforts to quit smoking, lose weight, or overcome phobias.

KEY TERMS

- consciousness p. 116
- altered state of consciousness (ASC) p. 116
- circadian rhythms p. 116
- controlled processes p. 117
- automatic processes p. 117
- sleep deprivation p. 118
- electroencephalogram (EEG) p. 120
- rapid-eye-movement (REM) sleep p. 121
- non-rapid-eye-movement (NREM) sleep p. 121
- evolutionary/circadian theory p. 122
- repair/restoration theory p. 122
- psychoanalytic view p. 122
- activation–synthesis hypothesis p. 122
- cognitive view p. 122
- dyssomnias p. 123
- insomnia p. 123
- narcolepsy p. 123
- sleep apnea p. 123
- parasomnias p. 125
- nightmares p. 125
- night terrors p. 125
- sleepwalking p. 125
- sleep talking p. 125
- psychoactive drugs p. 126
- agonistic p. 126
- antagonistic p. 126
- drug abuse p. 126
- addiction p. 126
- psychological dependence p. 129
- physical dependence p. 129
- withdrawal p. 129
- tolerance p. 129
- cross-tolerance p. 129
- depressants p. 129
- stimulants p. 129
- opiates p. 130
- hallucinogens p. 130
- meditation p. 133
- hypnosis p. 135

CRITICAL AND CREATIVE THINKING QUESTIONS

1. Do you believe that people have an unconscious mind? If so, how does it affect thoughts, feelings, and behavior?
2. Which of the three main theories of dreams do you most agree with, and why?
3. Do you think marijuana use should be legal? Why or why not?
4. Why might hypnosis help treat people who suffer from chronic pain?

What is happening in this picture?

During NREM sleep, a cat will often sleep in an upright position. With the onset of REM sleep, the cat rolls over on its side.

- **Can you explain why?**
- **Why might the muscle "paralysis" of paradoxical (REM) sleep serve an important adaptive function?**

SELF-TEST

(Check your answers in Appendix B.)

1. *Consciousness* is defined in this text as ______________.
a. ordinary and extraordinary wakefulness
b. an organism's awareness of its own self and surroundings
c. mental representations of the world in the here and now
d. any mental state that requires thinking and processing of sensory stimuli

2. Mental activities that require focused attention are called ______________.
a. thinking processes
b. controlled processes
c. alert states of consciousness
d. conscious awareness

3. *Automatic processes* require ______________ attention.
a. internal
b. unconscious
c. minimal
d. delta wave

4. *Circadian rhythms* are ______________.
a. patterns that repeat themselves on a twice-daily schedule
b. physical and mental changes associated with the cycle of the moon
c. rhythmical processes in your brain
d. biological changes that occur on a 24-hour cycle

5. Identify the main areas of the brain involved in the operation of circadian rhythms.
a. hypothalamus
b. pineal gland
c. suprachiasmatic nucleus

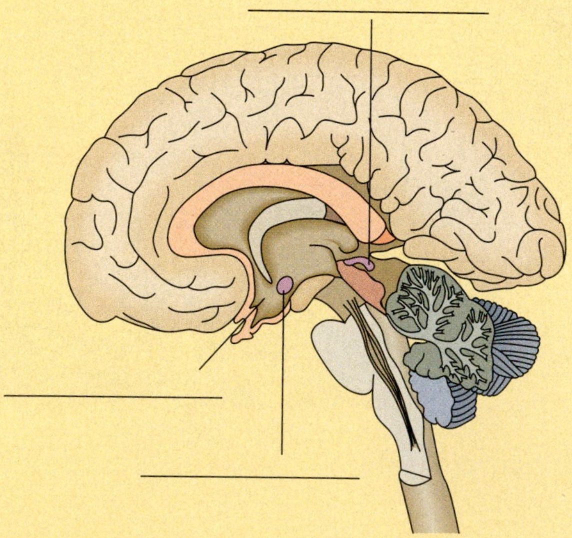

6. This figure displays brain wave changes tracked by means of small electrodes on the scalp. This research tool is called a(n) ______________.
a. EKG
b. PET scan
c. EEG recording
d. EMG

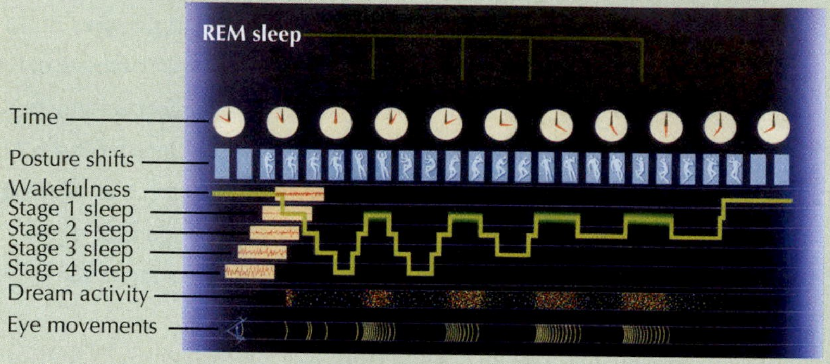

7. The sleep stage marked by irregular breathing, eye movements, high-frequency brain waves, and dreaming is called ______________ sleep.
a. beta
b. hypnologic
c. REM
d. transitional

8. Which of the following people is clearly experiencing insomnia?
a. Joan frequently cannot fall asleep the night before a final exam.
b. Cliff regularly sleeps less than eight hours per night.
c. Consuela persistently has difficulty falling or staying asleep.
d. All of these persons are clearly experiencing insomnia.

9. Dieting, surgery, dental appliances, and tennis balls are all recommended treatments for ______________, a dyssomnia.
a. insomnia
b. parasomnia
c. nightmares
d. sleep apnea

10. *Psychoactive drugs* ________________.
a. change conscious awareness, mood, or perception
b. are addictive, mind altering, and dangerous to your health
c. are illegal unless prescribed by a medical doctor
d. all of these options

11. Match the four major categories of psychoactive drugs with the correct photo:
a. depressants, photo ________________
b. stimulants, photo ________________
c. opiates, photo ________________
d. hallucinogens, photo ________________

12. *Depressants* include all of the following **EXCEPT** ________________.
a. downers such as sedatives, barbiturates, antianxiety drugs
b. alcohol
c. tobacco
d. Rohypnol

13. Altered states of consciousness can be achieved in which of the following ways?
a. during sleep and dreaming
b. via chemical channels
c. through hypnosis and meditation
d. all of these options

14. ________________ is a group of techniques designed to refocus attention and produce an alternate state of consciousness.
a. Hypnosis
b. Scientology
c. Parapsychology
d. Meditation

15. ________________ is an alternate state of heightened suggestibility characterized by deep relaxation and intense focus.
a. Meditation
b. Amphetamine psychosis
c. Hypnosis
d. Daydreaming

Learning

6

On June 7, 1998, James Byrd, a disabled 49-year-old African American, was walking home along Martin Luther King Boulevard in Jasper, Texas. Three young men offered him a ride, but instead, they chained Mr. Byrd by his ankles to the back of their pickup and dragged him along an old country road until his head and right arm were ripped from his body.

What brought about this grisly murder? Byrd's murderers—two of whom were sentenced to death and the third given life imprisonment without the possibility of parole—were "white supremacists." Although white supremacists form only a tiny fraction of the American population, hate crimes are a serious and growing problem around the world. Where does such hatred come from? Is prejudice learned?

We usually think of learning as classroom activities, such as math and reading, or as motor skills, like riding a bike or playing the piano. Psychologists define learning more broadly, as *a relatively permanent change in behavior or mental processes because of practice or experience*. This relative permanence applies not only to useful behaviors (using a spoon or writing great novels) but also to bad habits, racism, and hatred. The good news is that what is learned can be unlearned—through retraining, counseling, and self-reflection.

In this chapter, we discuss several types of conditioning, the most basic form of learning. Then we look at social-cognitive learning and the biological factors involved in learning. Finally, we explore how learning theories and concepts touch on everyday life.

NATIONAL GEOGRAPHIC

CHAPTER OUTLINE

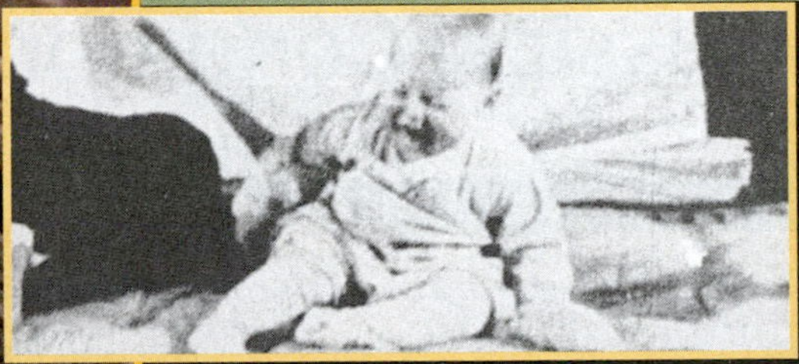

Classical Conditioning

LEARNING OBJECTIVES

Describe Pavlov and Watson's contributions to our understanding of learning.

Explain how stimulus generalization and discrimination affect learning.

Describe the processes of extinction and spontaneous recovery.

Identify an example of higher-order conditioning.

Among the earliest forms of learning to be studied scientifically is *conditioning*. We discuss *classical conditioning*, made famous by Pavlov's dogs, in this section, and a different form of conditioning, known as *operant conditioning*, in the next section.

THE BEGINNINGS OF CLASSICAL CONDITIONING

Why does your mouth water when you see a large slice of chocolate cake or a juicy steak? The answer to this question was accidentally discovered in the laboratory of Russian physiologist Ivan Pavlov (1849–1936). Pavlov's work focused on the role of saliva in digestion, and one of his experiments involved measuring salivary responses in dogs, using a tube attached to the dogs' salivary glands.

One of Pavlov's students noticed that many dogs began salivating at the sight of the food or the food dish, the smell of the food, or even the sight of the person who delivered the food long before receiving the actual food. This "unscheduled" salivation was intriguing. Pavlov recognized that an involuntary reflex (salivation) that occurred before the appropriate stimulus (food) was presented could not be inborn and biological. It had to have been acquired through experience—through **learning**.

learning A relatively permanent change in behavior or mental processes because of practice or experience.

Excited by their accidental discovery, Pavlov and his students conducted several experiments. Their most basic method involved sounding a tone on a tuning fork just before food was placed in the dogs' mouths. After several pairings of tone and food, the dogs would salivate on hearing the tone alone. Pavlov and others went on to show that many things can be conditioned stimuli for salivation if they are paired with food—the ticking of a metronome, a buzzer, a light, and even the sight of a circle or triangle drawn on a card.

The type of learning that Pavlov described came to be known as **classical conditioning** (FIGURE 6.1). To understand classical conditioning, you first need to realize that **conditioning** is just another word for learning. You also need to know that some responses are inborn and don't require conditioning. For example, the inborn salivary reflex consists of an **unconditioned stimulus (UCS)** and an **unconditioned response (UCR)**. That is, the UCS (food) elicits the UCR (salivation) without previous conditioning.

classical conditioning Learning that occurs when a neutral stimulus (NS) becomes paired (associated) with an unconditioned stimulus (UCS) to elicit a conditioned response (CR).

conditioning The process of learning associations between environmental stimuli and behavioral responses.

Before conditioning occurs, a **neutral stimulus (NS)** does not naturally elicit a relevant or consistent response. For example, as shown in Figure 6.1, Pavlov's dogs did not naturally salivate when a tone sounded. Similarly, as the figure shows, the sight of a cardboard box (neutral stimulus) doesn't naturally make a person hungry for a slice of pizza.

Process Diagram

Pavlov's classical conditioning FIGURE 6.1

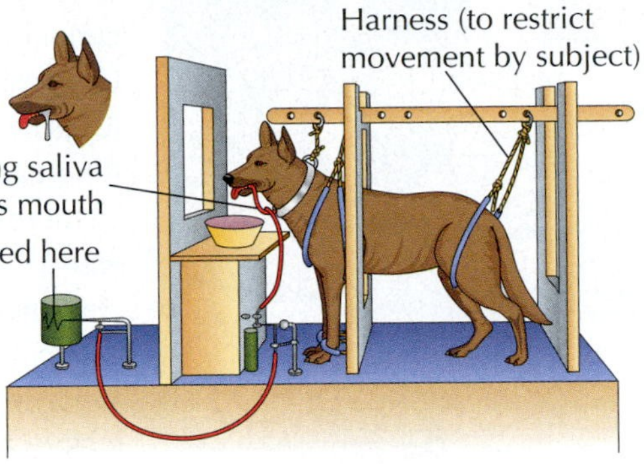

Pavlov Example | **Modern-day Example**

Step 1
Before conditioning
The neutral stimulus (NS) produces no relevant response. The unconditioned (unlearned) *stimulus* (UCS) elicits the unconditioned *response* (UCR).

Step 2
During conditioning
The neutral stimulus (NS) is repeatedly paired with the unconditioned (unlearned) *stimulus* (UCS) to produce the unconditioned *response* (UCR).

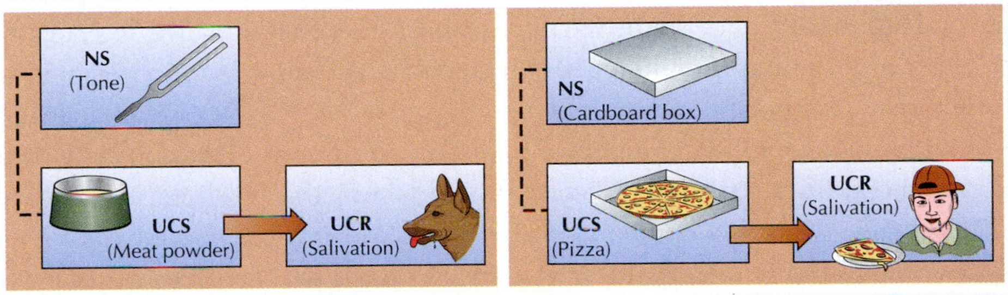

Step 3
After conditioning
The neutral stimulus (NS) has become a conditioned (learned) stimulus (CS). This CS now produces a conditioned (learned) *response* (CR), which is usually similar to the previously unconditioned (unlearned) response (UCR).

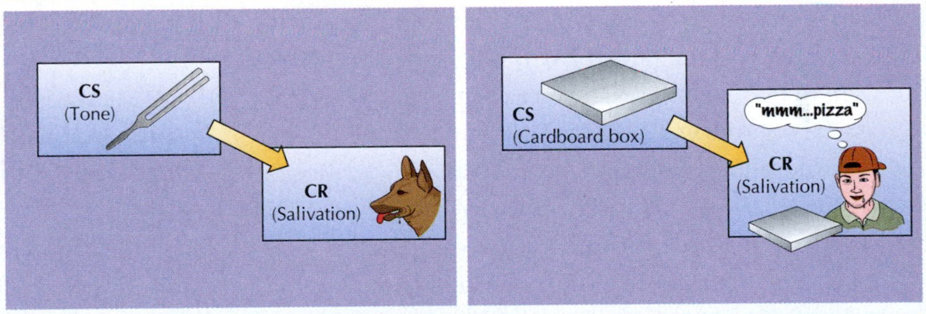

Summary
An originally neutral stimulus (NS) becomes a conditioned stimulus (CS), which elicits a conditioned response (CR).

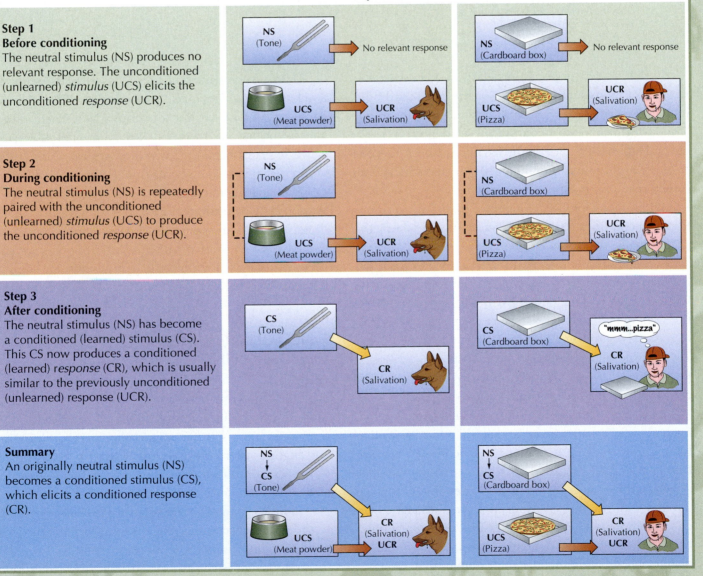

Conditioning and the case of Little Albert FIGURE 6.2

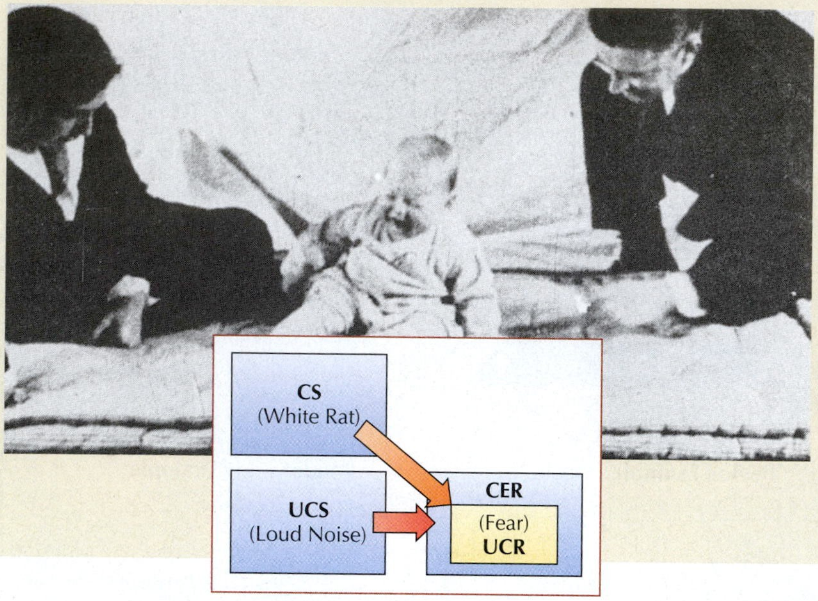

In the famous "Little Albert" study, a healthy 11-month-old child was first allowed to play with a white laboratory rat. Like most infants, Albert was curious and reached for the rat, showing no fear. Using the fact that infants are naturally frightened (UCR) by loud noises (UCS), Watson stood behind Albert and again put the rat (NS) near him. When the infant reached for the rat, Watson banged a steel bar with a hammer. The loud noise frightened Albert and made him cry. The white rat (NS) was paired with the loud noise (UCS) only seven times before the white rat alone produced a *conditioned emotional response* (CER) in Albert, fear of the rat.

Pavlov's discovery was that learning occurs when a neutral stimulus such as a tone (or the cardboard box), is regularly paired with an unconditioned stimulus (food, or the smell of pizza in the cardboard box). The neutral stimulus (tone, or the cardboard box) then becomes a **conditioned stimulus (CS)**, which elicits a **conditioned response (CR)**—salivation.

What does a salivating dog have to do with your life? Classical conditioning is the most fundamental way that all animals, including humans, learn most new responses, emotions, and attitudes. Your love for your parents (or significant other), the hatred and racism that led to the murder of James Byrd, and your drooling at the sight of chocolate cake or pizza are largely the result of classical conditioning.

In a famous experiment, John Watson and Rosalie Rayner (1920) demonstrated how the emotion of fear could be classically conditioned (FIGURE 6.2).

Watson and Rayner's experiment could not be performed today because it violates several ethical guidelines for scientific research (Chapter 1). Moreover, Watson and Rayner ended their experiment without extinguishing (removing) Albert's fear, although they knew that it could endure for a long period. Watson and Rayner also have been criticized because they did not measure Albert's fear objectively. Their subjective evaluation raises doubt about the degree of fear conditioned (Paul & Blumenthal, 1989).

Despite such criticisms, John Watson made important and lasting contributions to psychology. Unlike other psychologists of his time, Watson emphasized strictly observable behaviors, and he founded the school of *behaviorism*, which explains behavior as a result of observable stimuli and observable responses. Watson's study of Little Albert also has had legendary significance for many psychologists. Watson showed us that many of our likes, dislikes, prejudices, and fears are examples of a **conditioned emotional response (CER)**. Watson's research in producing Little Albert's fears also led to powerful clinical tools for eliminating extreme, irrational fears known as *phobias* (Chapter 13).

FINE-TUNING CLASSICAL CONDITIONING

Next, we outline six important principles of classical conditioning: stimulus generalization, stimulus discrimination, extinction, spontaneous recovery, reconditioning, and higher-order conditioning (FIGURE 6.3). These processes complicate classical conditioning, but they also underscore its fundamental nature.

Stimulus generalization occurs when an event similar to the originally conditioned stimulus triggers the same conditioned response. The more the stimulus resembles the conditioned stimulus, the stronger the conditioned

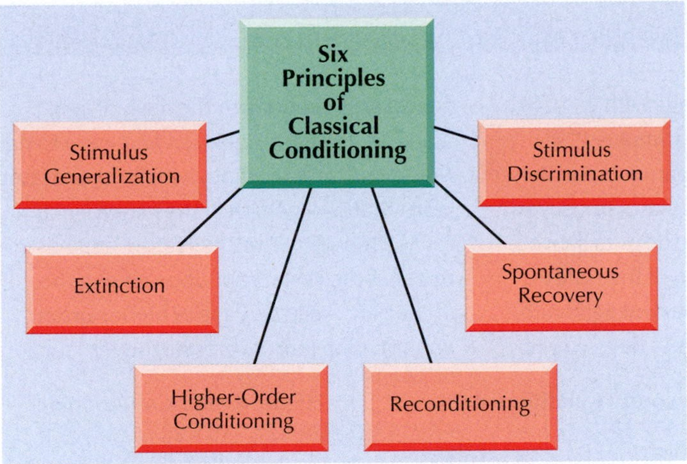

"I don't care if she is a tape dispenser. I love her."

Which of the six basic principles of classical conditioning explain(s) this cartoon?

Answer: Stimulus generalization

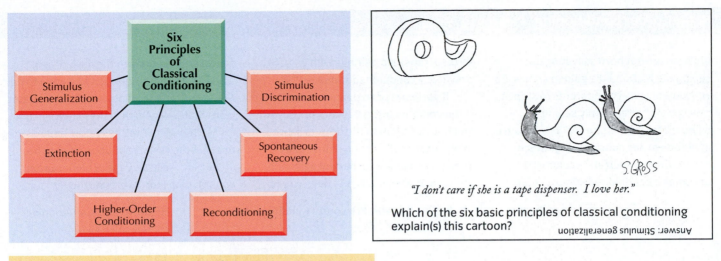

Six principles of classical conditioning FIGURE 6.3

response (Hovland, 1937). For example, after first conditioning dogs to salivate at the sound of low-pitched tones, Pavlov later demonstrated that the dogs would also salivate in response to higher-pitched tones. Similarly, after conditioning, the infant in Watson and Rayner's experiment ("Little Albert") feared not only rats but also a rabbit, a dog, and a bearded Santa Claus mask.

Eventually, through the process of **stimulus discrimination** (a term that refers to a learned response to a specific stimulus, but not to other similar stimuli), Albert presumably learned to recognize differences between rats and other stimuli. As a result, he probably overcame his fear of Santa Claus, even if he remained afraid of white rats. Similarly, Pavlov's dogs learned to distinguish between the tone that signaled food and those that did not.

Most behaviors that are learned through classical conditioning can be weakened or suppressed through **extinction**. Extinction occurs when the unconditioned stimulus (UCS) is repeatedly withheld whenever the conditioned stimulus (CS) is presented, which gradually weakens the previous association. When Pavlov repeatedly sounded the tone without presenting food, the dogs' salivation gradually declined. Similarly, if you have a classically conditioned fear of cats and later start to work as a veterinary assistant, your fear will gradually diminish.

However, extinction is not unlearning—it does not "erase" the learned connection between the stimulus and the response (Bouton, 1994). In fact, on occasion an extinguished response may "spontaneously" reappear (FIGURE 6.4). This **spontaneous recovery** helps

Extinction and spontaneous recovery FIGURE 6.4

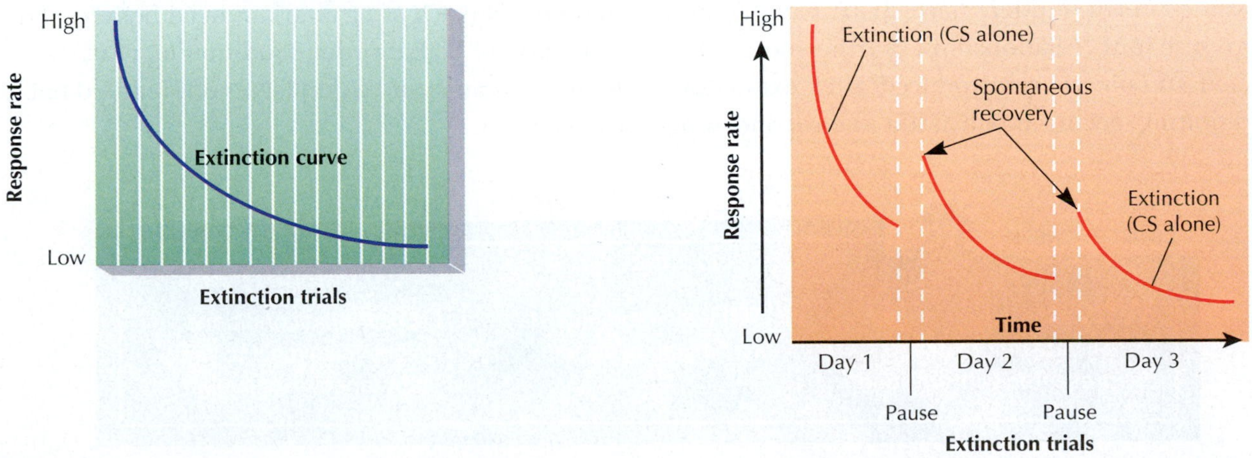

The more often the UCS is withheld whenever the CS is presented, the lower an individual's response rate to the UCS, until extinction occurs.

But an extinguished response may spontaneously reappear, which can trigger a rush of feelings and emotions as though the UCS were present.

Higher-order conditioning FIGURE 6.5

Children are not born salivating upon seeing the McDonald's golden arches. So why do they beg their parents to stop at "Mickey D's" after simply seeing a billboard for the restaurant? It is because of **higher-order conditioning**, which occurs when a neutral stimulus (NS) becomes a conditioned stimulus (CS) through repeated pairings with a previously conditioned stimulus (CS).

If you wanted to demonstrate higher-order conditioning in Pavlov's dogs, you would first condition the dogs to salivate in response to the sound of the tone **A**. Then you would pair a flash of light with the tone **B**. Eventually, the dogs would salivate in response to the flash of light alone **C**. Similarly, children first learn to pair McDonald's restaurants with food and later learn that two golden arches are a symbol for McDonald's. Their salivation and begging to eat at the restaurant upon seeing the arches are a classic case of higher-order conditioning (and successful advertising)

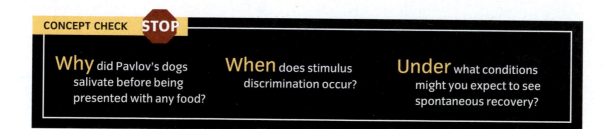

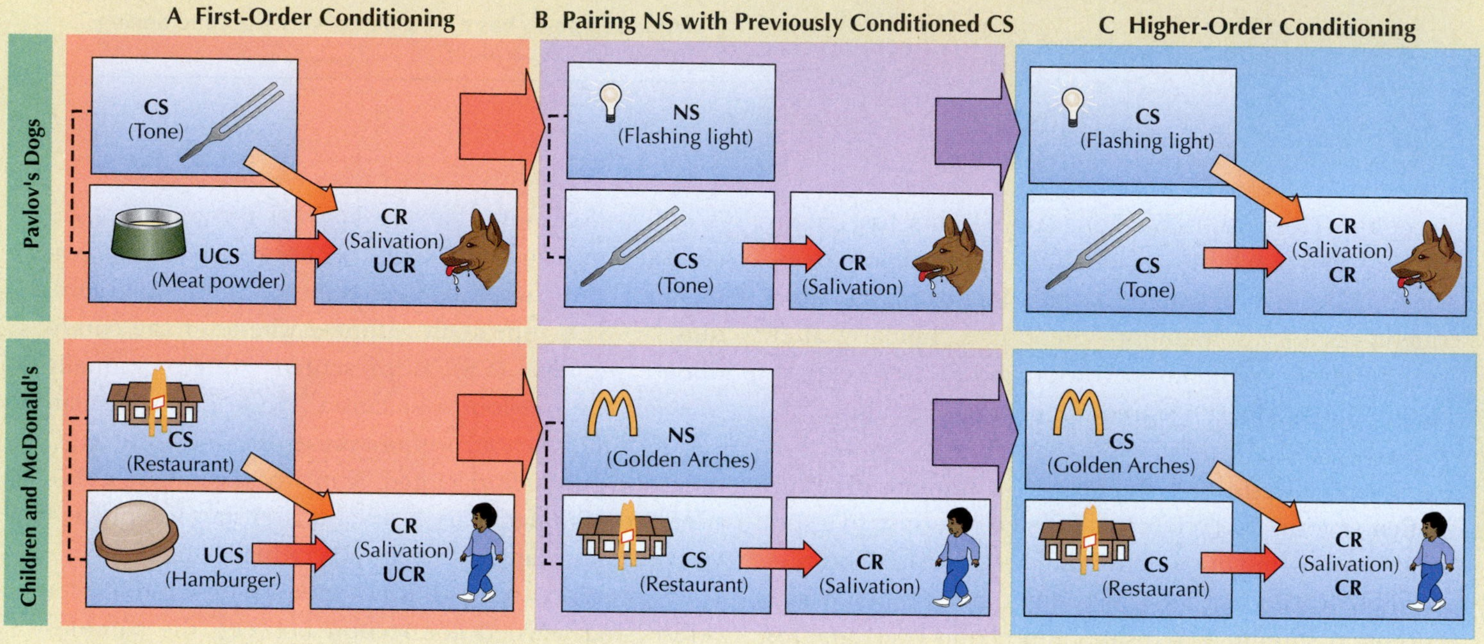

explain why you might suddenly feel excited at the sight of a former girlfriend or boyfriend, even though years have passed (and extinction has occurred). It also explains why couples who've recently broken up sometimes misinterpret a sudden flare-up of feelings and return to unhappy relationships. Furthermore, if a conditioned stimulus is reintroduced after extinction, the conditioning occurs much faster the second time around—a phenomenon known as **reconditioning**. Both spontaneous recovery and reconditioning help underscore why it can be so difficult for us to break bad habits (such as eating too much chocolate cake) or internalize new beliefs (such as egalitarian racial beliefs). The phenomenon of **higher-order conditioning** (FIGURE 6.5) further expands and complicates our learned habits and associations.

CONCEPT CHECK STOP

Why did Pavlov's dogs salivate before being presented with any food?

When does stimulus discrimination occur?

Under what conditions might you expect to see spontaneous recovery?

Operant Conditioning

LEARNING OBJECTIVES

Explain how reinforcement and punishment influence behavior.

Describe Thorndike's and Skinner's contributions to operant conditioning research.

Identify examples of primary and secondary reinforcers.

Explain how different schedules of reinforcement affect behavior.

Describe the negative side effects of punishment.

Consequences are the heart of **operant conditioning**. In classical conditioning, consequences are irrelevant—Pavlov's dogs still got to eat whether they salivated or not. But in operant conditioning, the organism performs a behavior (an **operant**) that produces either reinforcement or punishment. **Reinforcement** strengthens the response, making it more likely to recur. **Punishment** weakens the response, making it less likely to recur.

operant conditioning Learning in which voluntary responses are controlled by their consequences (also known as instrumental or Skinnerian conditioning).

reinforcement A consequence that strengthens a response and makes it more likely to recur.

punishment A consequence that weakens a response and makes it less likely to recur.

Classical and operant conditioning also differ in another important way. In classical conditioning, the organism's response is generally passive and involuntary. In operant conditioning, the organism's response is generally active and voluntary. The learner "operates" on the environment and produces consequences that influence whether the behavior will be repeated. (If your friends smile and laugh when you tell a joke, you are likely to joke more. If they frown, groan, or ridicule you, you are likely to joke less.)

THE BEGINNINGS OF OPERANT CONDITIONING

Edward Thorndike (1874–1949), a pioneer of operant conditioning, determined that the frequency of a behavior is modified by its consequences. He developed the **law of effect** (Thorndike, 1911), a first step in understanding how consequences can modify active, voluntary behaviors (FIGURE 6.6).

law of effect Thorndike's rule that the probability of an action being repeated is strengthened when followed by a pleasant or satisfying consequence.

B. F. Skinner (1904–1990) extended Thorndike's law of effect to more complex behaviors. He emphasized that reinforcement and punishment always occur after the behavior of interest has occurred. In addition, Skinner cautioned that the only way to know how we have influenced someone's behavior is to check whether it increases or decreases. Sometimes, he noted, we think we're reinforcing or punishing, when we're actually doing the opposite. For example, a professor may think she is encouraging shy students to talk by repeatedly praising them each time they speak up in class. But what if shy students are embarrassed by this attention? If so, they may decrease the number of times they talk in class.

Thorndike box FIGURE 6.6

In one famous experiment, Thorndike put a cat inside a specially built puzzle box. When the cat stepped on a pedal inside the box (at first, through trial-and-error), the door opened and the cat could get out to eat. With each additional success, the cat's actions became more purposeful, and it soon learned to open the door immediately (from Thorndike, 1898).

REINFORCEMENT: STRENGTHENING A RESPONSE

Reinforcers, which strengthen a response, can be a powerful tool in all parts of our lives (FIGURE 6.7). Psychologists group them into two types, primary and secondary. **Primary reinforcers** satisfy an intrinsic, unlearned biological need (food, water, sex). **Secondary reinforcers** are not intrinsic; the value of this reinforcer is learned (money, praise, attention). Each type of reinforcer can produce **positive reinforcement** or **negative reinforcement** (TABLE 6.1), depending on whether certain stimuli are added or taken away.

It's easy to confuse negative reinforcement with punishment, but the two concepts are actually completely opposite. Reinforcement (either positive or negative) strengthens a behavior, whereas punishment weakens a behavior. If the terminology seems confusing, it may help to think of positive and negative reinforcement in the mathematical sense (that is, in terms of something being added [+] or taken away [–]) rather than in terms of good and bad.

When you make yourself do some necessary but tedious task—say, paying bills—before letting yourself go to the movies, you are not only using positive reinforcement. You are also using the **Premack principle**, named after psychologist David Premack. Recognizing that you love to go to movies, you intuitively tie your less-desirable, low-frequency activities (paying bills) to your high-frequency behavior (going to the movies).

Should employers use reinforcement?
FIGURE 6.7

Just ask the employees of Google Inc. They can eat for free at any of the company's gourmet cafeterias, make use of free transportation from area train stations, get their car washed and the oil changed at company expense, do laundry for free in company washers and dryers, use on-site workout rooms, get free checkups from company doctors, and much more. No wonder Google employees are among the most loyal in the nation.

How reinforcement strengthens and increases behaviors TABLE 6.1

	Positive reinforcement *Adds to (+) and strengthens behavior*	*Negative reinforcement* *Takes away (–) and strengthens behavior*
Primary reinforcers Satisfy an unlearned biological need	You hug your baby and he smiles at you. The "addition" of his smile strengthens the likelihood that you will hug him again.	Your baby is crying, so you hug him and he stops crying. The "removal" of crying strengthens the likelihood that you will hug him again when he cries.
	You do a favor for a friend and she buys you lunch in return.	You take an aspirin for your headache, which takes away the pain.
Secondary reinforcers Value is learned, not intrinsic or biological	You increase profits and receive $200 as a bonus.	After high sales, your boss says you won't have to work on weekends.
	You study hard and receive a good grade on your psychology exam.	Professor says you won't have to take the final exam because you did so well on your unit exam.

What are the best circumstances for using reinforcement? It depends on the desired outcome. To make this decision, you need to understand various **schedules of reinforcement** (Terry, 2003). This term refers to the rate or interval at which responses are reinforced. Although there are numerous schedules of reinforcement, the most important distinction is whether they are continuous or partial. When Skinner was training his animals, he found that learning was most rapid if the response was reinforced every time it occurred—a procedure called **continuous reinforcement**.

As you have probably noticed, real life seldom provides continuous reinforcement. Yet your behavior persists because your efforts are occasionally rewarded. Most everyday behavior is rewarded on a **partial (or intermittent) schedule of reinforcement**, which involves reinforcing only some responses, not all (Sangha et al., 2002).

Once a task is well learned, it is important to move to a partial schedule of reinforcement. Why? Because under partial schedules, behavior is more resistant to extinction. There are four partial schedules of reinforcement: **fixed ratio** (FR), **variable ratio** (VR), **fixed interval** (FI), and **variable interval** (VI). The type of partial schedule selected depends on the type of behavior being studied and on the speed of learning desired (Kazdin, 2008; Neuringer, Deiss, & Olson, 2000; Rothstein, Jensen, & Neuringer, 2008). A fixed ratio leads to the highest overall response rate, but each of the four types of partial schedules has different advantages and disadvantages (see Study Organizer 6.1).

Study Organizer 6.1 Four schedules of reinforcement

		Definitions	*Response rates*	*Examples*
Ratio schedules (response based)	**Fixed ratio (FR)**	Reinforcement occurs after a predetermined set of responses; the ratio (number or amount) is fixed	Produces a high rate of response, but a brief drop-off just after reinforcement	Car wash employee receives $10 for every 3 cars washed. In a laboratory, a rat receives a food pellet every time it presses the bar 7 times. 
	Variable ratio (VR)	Reinforcement occurs unpredictably; the ratio (number or amount) varies	High response rates, no pause after reinforcement, and very resistant to extinction	Slot machines are designed to pay out after an average number of responses (maybe every 10 times), but any one machine may pay out on the first response, then seventh, then the twentieth. 
Interval schedules (time based)	**Fixed interval (FI)**	Reinforcement occurs after a predetermined time has elapsed; the interval (time) is fixed	Responses tend to increase as the time for the next reinforcer is near, but drop off after reinforcement and during interval	You receive a monthly paycheck. Rat's behavior is reinforced with a food pellet when (or if) it presses a bar after 20 seconds have elapsed. 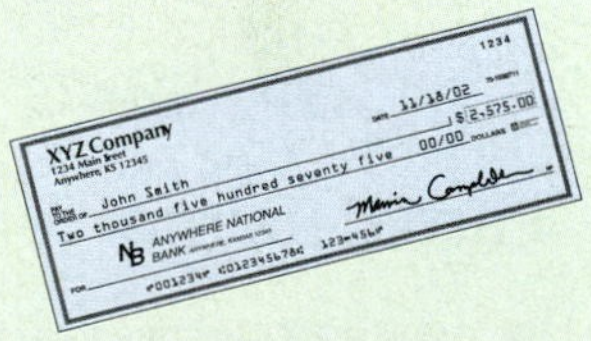
	Variable interval (VI)	Reinforcement occurs unpredictably; the interval (time) varies	Relatively low response rates, but they are steady because respondents cannot predict when reward will come	A rat's behavior is reinforced with a food pellet after a response and a variable, unpredictable interval of time. In a class with pop quizzes, you study at a slow but steady rate because you can't anticipate the next quiz. 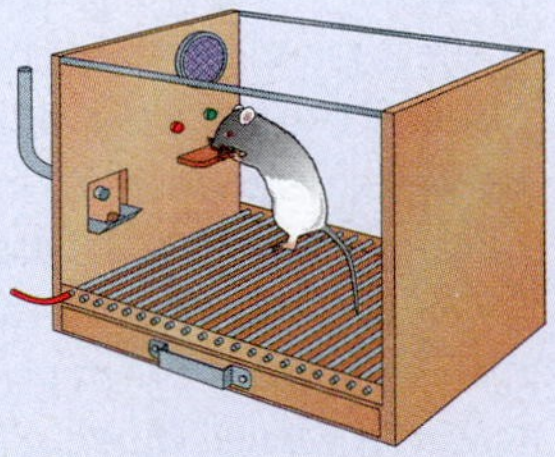

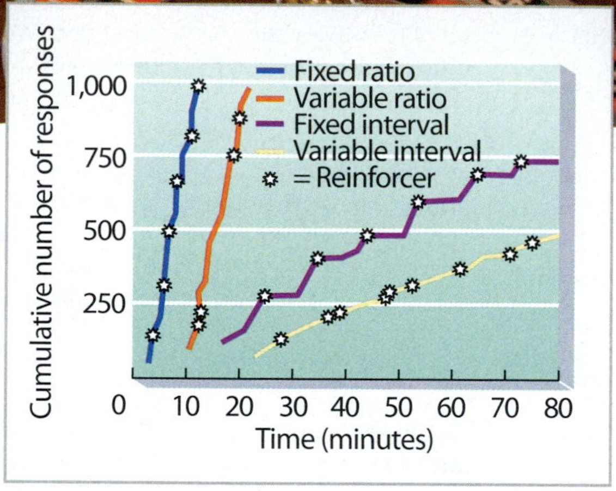

Partial Reinforcement Keeps 'em Coming Back

Have you ever purchased a lottery ticket or wondered why you (or others) spend long hours pushing buttons and pulling levers on slot machines in hopes of winning the jackpot? This compulsion to keep gambling in spite of significant losses is evidence of the strong resistance to extinction with partial (or intermittent) schedules of reinforcement. Machines in Nevada casinos, by law, have payout rates of at least 75%—that is, for every dollar spent, a player "wins" $0.75. Different machines are programmed different ways. Some meet the percentage by giving very large, but infrequent, payouts to a lucky winner. Others give frequent smaller payouts to all players. In either case, people are reinforced just often enough to keep them coming back, always hoping the "partial" reinforcement will lead to more. The same compulsion leads people to go on buying lottery tickets even though the odds of winning are very low.

This type of partial reinforcement also helps parents maintain children's positive behaviors, such as toothbrushing and bed making. After the child has initially learned these behaviors with continuous reinforcement, occasional, partial reinforcement is the most efficient way to maintain the behavior. (Chart adapted from Skinner, 1961.)

How does she do it? FIGURE 6.8

Momoko, a female monkey, is famous in Japan for her water-skiing, deep-sea diving, and other amazing abilities. Can you describe how her animal trainers used the successive steps of shaping to teach her these skills? First, they reinforced Momoko (with a small food treat) for standing or sitting on the water ski. Then they reinforced her each time she put her hands on the pole. Next, they slowly dragged the water ski on dry land and reinforced her for staying upright and holding the pole. Then they took Momoko to a shallow and calm part of the ocean and reinforced her for staying upright and holding the pole as the ski moved in the water. Finally, they took her into the deep part of the ocean.

Partial reinforcement is described further in *What a Psychologist Sees.* Each of the four schedules of partial reinforcement is important for *maintaining* behavior. But how would you teach someone to play the piano or to speak a foreign language? For new and complex behaviors such as these, which aren't likely to occur naturally, **shaping** is an especially valuable tool. Skinner believed that shaping explains a variety of abilities that each of us possesses, from eating with a fork to playing a musical instrument, to driving a stick-shift car. Parents, athletic coaches, teachers, and animal trainers all use shaping techniques (FIGURE 6.8).

> **shaping**
> Reinforcement by a series of successively improved steps leading to desired response.

PUNISHMENT: WEAKENING A RESPONSE

Unlike reinforcement, punishment *decreases* the strength of a response. As with reinforcement, there are two kinds of punishment—positive and negative (Miltenberger, 2008; Skinner, 1953).

Positive punishment is the addition (+) of a stimulus that decreases (or weakens) the likelihood of the response occurring again. **Negative punishment** is the taking away (–) of a reinforcing stimulus, which decreases (or weakens) the likelihood of the response occurring again (TABLE 6.2). (To check your understanding of the principles of reinforcement and punishment, see FIGURE 6.9 on the next page.)

Punishment is a tricky business, and it isn't always intentional. Any process that adds or takes away something and causes a behavior to decrease is punishment. Thus, if parents ignore all the A's on their child's report card ("taking away" encouraging comments) and ask repeated questions about the B's and C's, they may unintentionally be punishing the child's excellent grade achievement and weakening the likelihood of future A's. Similarly, dog owners who yell at or spank their dogs for finally coming to them ("adding" verbal

How punishment weakens and decreases behaviors TABLE 6.2

Positive punishment ***Adds stimulus (+) and weakens the behavior***	***Negative punishment*** ***Takes stimulus away (–) and weakens the behavior***
You must run four extra laps in your gym class because you were late.	You're excluded from gym class because you were late.
A parent adds chores following a child's poor report card.	A parent takes away a teen's cell phone following a poor report card.
Your boss complains about your performance.	Your boss reduces your expense account after a poor performance.

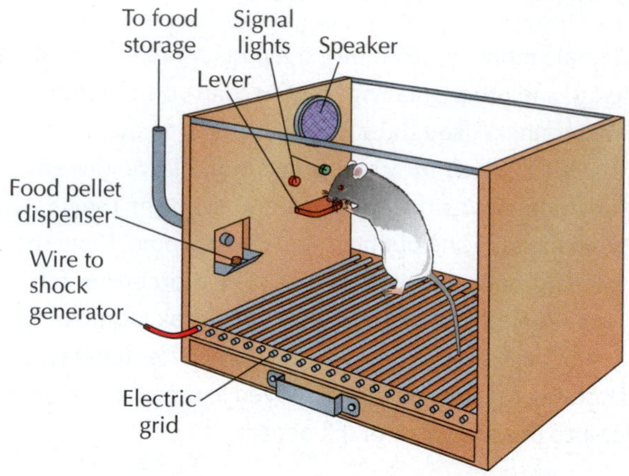

The Skinner box application FIGURE 6.9

To test his behavioral theories, Skinner used an animal, usually a pigeon or a rat, and an apparatus that has come to be called a *Skinner Box*. In Skinner's basic experimental design, an animal such as a rat received a food pellet each time it pushed a lever, and the number of responses was recorded. Note in this drawing that an electric grid on the cage floor could be used to deliver small electric shocks.

Test yourself

Is this positive reinforcement, negative reinforcement, positive punishment, or negative punishment? Fill in the name of the appropriate learning principle in the spaces provided in each box.

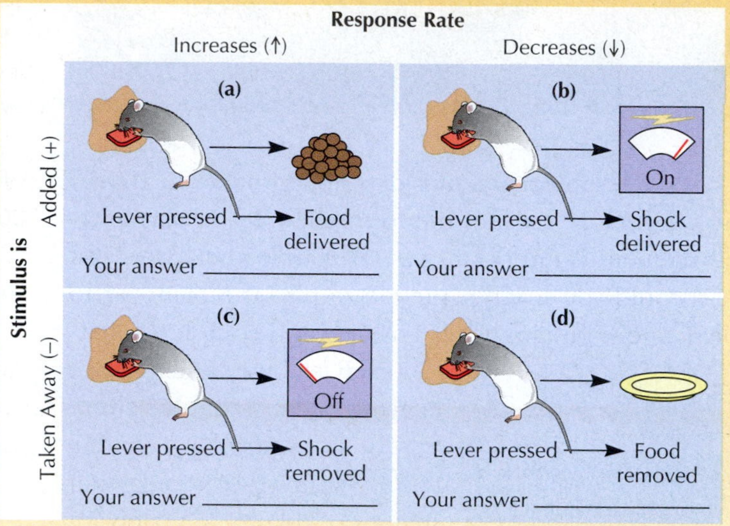

Answers: (a) positive reinforcement, (b) positive punishment, (c) negative reinforcement, (d) negative punishment

or physical aggression) after being called several times are actually punishing the desired behavior—coming when called.

Punishment plays an unavoidable role in our social world. Dangerous criminals must be stopped and possibly removed from society. Parents must stop their teenagers from drinking and driving. Teachers must stop disruptive students in the classroom and bullies on the playground. Yet punishment can be problematic (Borrego et al., 2007; Leary et al., 2008; Loxton et al., 2008).

To be effective, punishment should be immediate and consistent. However, in the real world, this is extremely hard to do. Police officers cannot stop every driver every time they speed. To make matters worse, when punishment is not immediate, during the delay the behavior is likely to be reinforced on a partial schedule, which makes it highly resistant to extinction. Think about gambling. It should be a punishing situation—gamblers usually lose far more than they win. However, the fact that they occasionally win keeps gamblers "hanging in there."

Even if punishment immediately follows the misbehavior, the recipient may learn what not to do but not necessarily what he or she should do. It's much more efficient to teach someone by giving him or her clear examples of correct behavior, rather than simply punishing the incorrect behavior. Finally, punishment can have serious side effects.

Psychological Science

Side effects of punishment

1. **Passive aggressiveness.** For the recipient, punishment often leads to frustration, anger, and eventually aggression. But most of us have learned from experience that retaliatory aggression toward a punisher (especially one who is bigger and more powerful) is usually followed by more punishment. We therefore tend to control our impulse toward open aggression and instead resort to more subtle techniques, such as showing up late or forgetting to mail a letter for someone. This is known as *passive aggressiveness* (Girardi et al., 2007; Johnson, 2008).

2. **Avoidance behavior.** No one likes to be punished, so we naturally try to avoid the punisher. If every time you come home a parent or spouse starts yelling at you, you will delay coming home or find another place to go.
3. **Modeling.** Have you ever seen a parent spank or hit his or her child for hitting another child? Ironically, the punishing parent may unintentionally serve as a "model" for the same behavior he or she is attempting to stop.
4. **Learned helplessness.** Why do some people stay in abusive relationships? Research shows that if you repeatedly fail in your attempts to control your environment, you acquire a general sense of powerlessness or *learned helplessness* and you may make no further attempts to escape (Bargai, Ben-Shakhar, & Salev, 2007; Diaz-Berciano et al., 2008; Kim, 2008; Shea, 2008).
5. **Temporary suppression.** Punishment generally suppresses the behavior only temporarily, while the punisher is nearby.
6. **Increased aggression.** Because punishment often produces a decrease in undesired behavior, at least for the moment, the punisher is in effect rewarded for applying punishment. Thus, a vicious circle may be established in which both the punisher and recipient are reinforced for inappropriate behavior—the punisher for punishing and the recipient for being fearful and submissive. This side effect partially explains the escalation of violence in family abuse and bullying (Anderson, Buckley, & Carnagey, 2008; Fang & Corso, 2007). In addition to fear and submissiveness, the recipient also might become depressed and/or respond with his or her own form of aggression.

CRITICAL THINKING

Here are two interesting questions:

1. Why do you think drivers quickly slow down when they see a police car, and then quickly resume their previous speed once the police officer is out of sight?
2. Given all the problems associated with punishment, why is it so often used?

CONCEPT CHECK STOP

What is the difference between negative reinforcement and punishment?

What are the law of effect and the Premack principle?

Why does a fixed ratio lead to the highest rate of response, compared with other schedules of reinforcement?

When is shaping most useful in learning?

Cognitive-Social Learning

LEARNING OBJECTIVES

Explain how insight and latent learning differ from the principles of classical and operant conditioning.

Summarize the evidence that animals (including humans) form internal, cognitive maps.

Identify Bandura's necessary conditions for observational learning.

So far, we have examined learning processes that involve associations between a stimulus and an observable behavior. Although some behaviorists believe that almost all learning can be explained in such stimulus–response terms, other psychologists feel that there is more to learning than can be explained solely by operant and classical conditioning. **Cognitive-social theory** (also called cognitive-social learning or cognitive-behavioral theory) incorporates the general concepts of conditioning, but rather than relying on a simple S–R (stimulus and response) model, this theory emphasizes the interpretation or thinking that occurs within the organism: S–O–R (stimulus–organism–response). According to this view, animals (including humans) have attitudes, beliefs, expectations, motivations, and emotions that affect learning. Furthermore, both human and nonhuman animals are social creatures that are capable of learning new behaviors through observation and imitation of others. We begin with a look at the *cognitive* part of cognitive-social theory, followed by an examination of the *social* aspects of learning.

cognitive-social theory A perspective that emphasizes the roles of thinking and social learning in behavior.

INSIGHT AND LATENT LEARNING: WHERE ARE THE REINFORCERS?

NATIONAL GEOGRAPHIC

Early behaviorists likened the mind to a "black box" whose workings could not be observed directly, but German psychologist Wolfgang Köhler wanted to look inside the box. He believed that there was more to learning—especially learning to solve a complex problem—than responding to stimuli in a trial-and-error fashion. In one of a series of experiments, he placed a banana just outside the reach of a caged chimpanzee (FIGURE 6.10). To reach the banana, the chimp had to use a stick to extend its reach. Köhler noticed that the chimp

Is this Insight? FIGURE 6.10

Grande, one of Köhler's chimps, has just solved the problem of how to get the banana. (Also, the chimp in the foreground is engaged in observational learning, our next topic.

insight Sudden understanding of a problem that implies the solution.

did not solve this problem in a random trial-and-error fashion but, instead, seemed to sit and think about the situation for a while. Then, in a flash of **insight**, the chimp picked up the stick and maneuvered the banana to within its grasp (Köhler, 1925).

Another of Köhler's chimps, an intelligent fellow named Sultan, was put in a similar situation. This time there were two sticks available to him, and the banana was placed even farther away, too far to reach with a single stick. Sultan seemingly lost interest in the banana, but he continued to play with the sticks. When he later discovered that the two sticks could be interlocked, he instantly used the now longer stick to pull the banana to within his grasp. Köhler designated this type of learning **insight learning** because some internal mental event that he could only describe as "insight" went on between the presentation of the banana and the use of the stick to retrieve it.

Like Köhler, Edward C. Tolman (1898–1956) believed that previous researchers underestimated animals' cognitive processes and cognitive learning. He noted that, when allowed to roam aimlessly in an experimental maze with no food reward at the end, rats seemed to develop a **cognitive map**, or mental representation of the maze.

To test the idea of cognitive learning, Tolman allowed one group of rats to aimlessly explore a maze with no reinforcement. A second group was reinforced with food whenever they reached the end of the maze. The third group was not rewarded during the first 10 days of the trial, but starting on day 11 they found food at the end of the maze. As expected from simple operant conditioning, the first and third groups were slow to learn the maze, whereas the second group, which had reinforcement, showed fast, steady improvement. However, when the third group started receiving reinforcement (on the 11th day), their learning quickly caught up to the group that had been reinforced every time (Tolman & Honzik, 1930). This showed that the nonreinforced rats had been thinking and building cognitive maps of the area during their aimless wandering and that their **latent learning** only showed up when there was a reason to display it (the food reward).

latent learning Hidden learning that exists without behavioral signs.

Cognitive maps and latent learning are not limited to rats. For example, a chipmunk will pay little attention to a new log in its territory (after initially checking it for food). When a predator comes along, however, the chipmunk heads directly for and hides beneath the log. Recent experiments provide additional clear evidence of latent learning and the existence of internal, cognitive maps in both human and nonhuman animals (Lahav & Mioduser, 2008; McNaughton et al., 2006) (FIGURE 6.11).

Is this learning? FIGURE 6.11

This child often rides through her neighborhood for fun, without a specific destination. Could she tell her mother where the nearest mailbox is located? What type of cognitive maps have you created?

OBSERVATIONAL LEARNING: WHAT WE SEE IS WHAT WE DO

observational learning Learning new behavior or information by watching others (also known as social learning or modeling).

In addition to classical and operant conditioning and cognitive processes (such as insight and latent learning), we also learn many things through **observational learning**. From birth to death, observational learning is very important to our biological, psychological, and social survival (the *biopsychosocial model*). Watching others helps us avoid dangerous stimuli in our environment, teaches us how to think and feel, and shows us how to act and interact socially.

Some of the most compelling examples of observational learning come from the work of Albert Bandura and his colleagues (Bandura, 2003; Bandura, Ross, & Ross, 1961; Bandura & Walters, 1963; Huesmann & Kirwil, 2007). In several experiments, children watched an adult kick, punch, and shout at an inflated Bobo doll. Later, children who had seen the aggressive adult were much more aggressive with the Bobo doll than children who had not seen the aggression. In other words, "monkey see, monkey do" (FIGURE 6.12).

According to Bandura, observational learning requires at least four separate processes: attention, retention, reproduction, and reinforcement (FIGURE 6.13).

Bandura's classic Bobo doll studies
FIGURE 6.12

Bandura's "Bobo doll" study is considered a classic in psychology. It proved that children will imitate models they observe. Why is this new or important? Are there circumstances where observational learning (for example, gleaned from television) could have positive effects?

CONCEPT CHECK STOP

Why were Köhler's studies of insight in chimpanzees important?

What is a cognitive map?

Why was Bandura's work on modeling of aggressiveness important?

Four key factors in observational learning FIGURE 6.13

1. ATTENTION

Observational learning requires attention. This is why teachers insist on having students watch their demonstrations.

2. RETENTION

To learn new behaviors, we need to carefully note and remember the model's directions and demonstrations.

3. REPRODUCTION

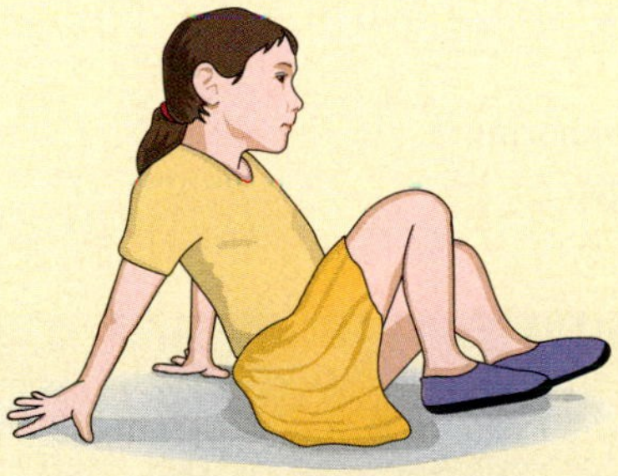

Observational learning cannot occur if we lack the motivation or motor skills necessary to imitate the model.

4. REINFORCEMENT

We are more likely to repeat a modeled behavior if the model is reinforced for the behavior.

The Biology of Learning

LEARNING OBJECTIVES

Explain how an animal's environment might affect learning and behavior.

Identify an example of biological preparedness.

Describe how instinctive drift constrains learning.

So far in this chapter, we have considered how external forces—from reinforcing events to our observations of others—affect learning. But we also know that for changes in behavior to persist over time, lasting biological changes must occur within the organism. In this section, we will examine neurological and evolutionary influences on learning.

NEUROSCIENCE AND LEARNING: THE ADAPTIVE BRAIN

Each time we learn something, either consciously or unconsciously, that experience creates new synaptic connections and alterations in a wide network of brain structures, including the cortex, cerebellum, hypothalamus, thalamus, and amygdala (Culbertson, 2008; May et al., 2007; Mohler et al., 2008; Romero, 2008).

Evidence that experience changes brain structure first emerged in the 1960s from studies of animals raised in enriched versus deprived environments. Compared with rats raised in a stimulus-poor environment, those raised in a colorful, stimulating "rat Disneyland" had a thicker cortex, increased nerve growth factor (NGF), more fully developed synapses, more dendritic branching, and improved performance on many tests of learning and memory (Gresack, Kerr, & Frick, 2007; Lores-Arnaiz et al., 2007; Pham et al., 2002; Rosenzweig & Bennett, 1996).

Admittedly, it is a big leap from rats to humans, but research suggests that the human brain also responds to environmental conditions (**FIGURE 6.14**). For example, older adults who are exposed to stimulating environments generally perform better on intellectual and perceptual tasks than those who are in restricted environments (Schaie, 1994, 2008).

Daily enrichment FIGURE 6.14

For humans and nonhuman animals alike, environmental conditions play an important role in enabling learning. How might a classroom rich with stimulating toys, games, and books foster intellectual development in young children?

MIRROR NEURONS AND IMITATION

Recent research has identified another neurological influence on learning processes, particularly imitation. Using fMRIs and other brain-imaging techniques, researchers have identified specific mirror neurons believed to be responsible for human empathy and imitation (Ahlsén, 2008; Fogassi et al., 2005; Hurley, 2008; Jacob, 2008). These neurons are found in several key areas of the brain, and they help us identify with what others are feeling and to imitate their actions. When we see another person in pain, one reason we empathize and "share their pain" is that our mirror neurons are firing. Similarly, if we watch others smile, our mirror neurons make it harder for us to frown.

Mirror neurons were first discovered by neuroscientists who implanted wires in the brains of monkeys to monitor areas involved in planning and carrying out movement (Ferrari, Rozzi, & Fogassi, 2005; Rizzolatti et al., 2002, 2006). When these monkeys moved and grasped an object, specific neurons fired, but they also fired when the monkeys simply observed another monkey performing the same or similar tasks.

Mirror neurons in humans also fire when we perform a movement or watch someone else perform. Have you noticed how spectators at an athletic event sometimes slightly move their arms or legs in synchrony with the athletes, or how newborns tend to imitate adult facial expressions? Mirror neurons may be the underlying biological mechanism for this imitation, as well as for the infants' copying of lip and tongue movements necessary for speech. They also might help explain the emotional deficits of children and adults with autism or schizophrenia who often misunderstand the verbal and nonverbal cues of others (Arbib & Mundhenk, 2005; Dapretto et al., 2006; Martineau et al., 2008).

EVOLUTION AND LEARNING: BIOLOGICAL PREPAREDNESS AND INSTINCTIVE DRIFT

Humans and other animals are born with various innate reflexes and instincts. Although these biological tendencies help ensure evolutionary survival, they are inherently inflexible. Only through learning are we able to react to important environmental cues—such as spoken words and written symbols—that our innate reflexes and instincts do not address. Thus, from an evolutionary perspective, learning is an adaptation that enables organisms to survive and prosper in a constantly changing world.

Because animals can be operantly conditioned to perform a variety of novel behaviors (like waterskiing), learning theorists initially believed that the fundamental laws of conditioning would apply to almost all species and all behaviors. However, researchers have identified several biological constraints that limit the generality of conditioning principles. These include biological preparedness and instinctive drift.

Years ago, a young woman named Rebecca unsuspectingly bit into a Butterfinger candy bar filled with small, wiggling maggots. Horrified, she ran gagging and screaming to the bathroom. Many years later, Rebecca still feels nauseated when she sees a Butterfinger candy bar (but, fortunately, she doesn't feel similarly nauseated by the sight of her boyfriend, who bought her the candy).

Rebecca's graphic (and true!) story illustrates an important evolutionary process. When a food or drink is associated with nausea or vomiting, that particular food or drink can become a conditioned stimulus (CS) that triggers a conditioned **taste aversion**. Like other classically conditioned responses, taste aversions develop involuntarily (FIGURE 6.15 on the next page).

Can you see why this automatic response would be adaptive? If one of our cave-dwelling ancestors became ill after eating a new plant, it would increase his or her chances for survival if he or she immediately developed an aversion to that plant—but not to other family members who might have been present at the time. Similarly, people tend to develop phobias of snakes, darkness, spiders, and heights more easily than of guns, knives, and electrical outlets. We apparently inherit a built-in (innate) readiness to form associations between certain stimuli and responses. This is known as **biological preparedness**.

Laboratory experiments have provided general support for both taste aversion and

biological preparedness
Built-in (innate) readiness to form associations between certain stimuli and responses.

Taste aversion in the wild FIGURE 6.15

In applied research, Garcia and his colleagues used classical conditioning to teach coyotes not to eat sheep (Gustavson & Garcia, 1974). The researchers began by lacing freshly killed sheep with a chemical that caused extreme nausea and vomiting in the coyotes that ate the tainted meat. The conditioning worked so well that the coyotes would run away from the mere sight and smell of sheep. This taste aversion developed involuntarily. This research has since been applied many times in the wild and in the laboratory with coyotes and other animals (Aubert & Dantzer 2005; Domjan, 2005; Workman & Reader, 2008).

biological preparedness. For example, Garcia and his colleagues (1966) produced taste aversion in lab rats by pairing flavored water (NS) and a drug (UCS) that produced gastrointestinal distress (UCR). After being conditioned and then recovering from the illness, the rats refused to drink the flavored water (CS) because of the conditioned taste aversion. Remarkably, however, Garcia discovered that only certain neutral stimuli could produce the nausea. Pairings of a noise (NS) or a shock (NS) with the nausea-producing drug (UCS) produced no taste aversion. Garcia suggested that when we are sick to our stomachs, we have a natural, evolutionary tendency to attribute it to food or drink. Being biologically prepared to quickly associate nausea with food or drink is adaptive because it helps us avoid that or similar food or drink in the future (Domjan, 2005; Garcia, 2003; Kardong, 2008).

Just as Garcia couldn't produce noise-nausea associations, other researchers have found that an animal's natural behavior pattern can interfere with operant conditioning. For example, the Brelands (1961) tried to teach a chicken to play baseball. Through shaping and reinforcement, the chicken first learned to pull a loop that activated a swinging bat and then learned to actually hit the ball. But instead of running to first base, it would chase the ball as if it were food. Regardless of the lack of reinforcement for chasing the ball, the chicken's natural behavior took precedence. This biological constraint is known as **instinctive drift**.

instinctive drift The tendency of some conditioned responses to shift (or drift) back toward innate response pattern.

CONCEPT CHECK STOP

What changes occur in rats' brains when they are raised in an enriched (versus deprived) environment?

Why is taste aversion evolutionarily adaptive?

Why don't the fundamental laws of conditioning apply to all species and all behaviors?

Conditioning and Learning in Everyday Life

LEARNING OBJECTIVES

Explain how classical conditioning applies to prejudice, marketing, medical treatments, and phobias.

Explain the role of operant conditioning in prejudice, biofeedback, and superstition.

Describe how cognitive-social learning applies to prejudice and to media influences on behavior.

In this section, we discuss several everyday applications for classical conditioning, operant conditioning, and cognitive-social learning.

CLASSICAL CONDITIONING: FROM PREJUDICE TO PHOBIAS

One common—and very negative—instance of classical conditioning is prejudice. In a classic study in the 1930s, Kenneth Clark and Mamie P. Clark (1939) found that given a choice, both black and white children preferred white dolls to black dolls. When asked which doll was good and which was bad, both groups of children responded that the white doll was good and nice and that the black doll was bad, dirty, and ugly.

The Clarks reasoned that the children had learned to associate inferior qualities with darker skin and more positive qualities with lighter skin. You may think that this 1930s study no longer applies, but follow-up research in the late 1980s found that 65 percent of the African American children and 74 percent of the white children still preferred the white doll (Powell-Hopson & Hopson, 1988) (FIGURE 6.16).

Classical conditioning is also a primary tool for marketing and advertising professionals, filmmakers and musicians, medical practitioners, psychotherapists, and politicians who want to influence our purchases, motivations, emotions, health behavior, and votes.

Instances of classical conditioning are also found in the medical field. For example, for alcohol-addicted patients, some hospitals pair the smell and taste of alcohol with a nausea-producing drug. Afterward, just the smell or taste of alcohol makes the person sick. Some patients, but not all, have found this treatment successful (Chapter 14).

Finally, researchers have found that classically conditioned emotional responses explain most everyday fears and even most *phobias*, which are exaggerated and irrational fears of a specific object or situation (Cal et al., 2006; Field, 2006; Ressler & Davis, 2003; Stein & Matsunaga, 2006). The good news is that extreme fears—for example, of heights, spiders, or public places—can be effectively treated with *behavior therapy* (Chapter 14).

Conditioned race prejudice FIGURE 6.16

The Clarks' findings played a pivotal role in the famous *Brown v. Board of Education of Topeka* decision in 1954, which ruled that segregation of public facilities was unconstitutional. One effect of segregation was to condition students to associate dark skin with inferior qualities. Can you explain how classical conditioning would create this type of negative association?

Applying Psychology

Classical Conditioning as a Marketing Tool

Marketers have mastered numerous classical conditioning principles. For example, TV commercials, magazine ads, and business promotions often pair a company's products or logo (the neutral stimulus/NS) with pleasant images, such as attractive models and celebrities (the conditioned stimulus/CS), which, through higher-order conditioning, automatically trigger favorable responses (the conditioned response/CR). Advertisers hope that after repeated viewings, the previously neutral stimulus (the company's products or logo) will become a conditioned stimulus that elicits favorable responses (CR)—we buy the advertised products.

Researchers caution that these ads also help produce visual stimuli that trigger conditioned responses, such as urges to smoke, overeat, and drink alcohol (Kazdin, 2008; Martin et al., 2002; Tirodkar & Jain, 2003; Wakefield et al., 2003).

To appreciate the influences of classical conditioning on your own life, try this:

- Look through a popular magazine and find several advertisements. What images are used as the unconditioned stimulus (UCS) or conditioned stimulus (CS)? Note how you react to these images.
- While watching a movie or a favorite TV show, note what sounds and images are used as conditioned stimuli (CS). (*Hint:* certain types of music are used to set the stage for happy stories, sad events, and fearful situations.) What are your conditioned emotional responses (CERs)?

CRITICAL THINKING

Here are two interesting questions:

1. In what way is the ad pictured here an application of psychology?
2. Can you think of other situations in which sounds and images are used as conditioned sitimuli?

OPERANT CONDITIONING: PREJUDICE, BIOFEEDBACK, AND SUPERSTITION

Just as people can learn prejudice through classical conditioning, we also can learn prejudice through operant conditioning. Demeaning others gains attention and sometimes approval from others, as well as increases one's self-esteem (at the expense of the victim), positively reinforcing prejudice and discrimination (Fein & Spencer, 1997; Kassin, Fein, & Markus, 2008) (FIGURE 6.17A). People also may have a single negative (punishing) experience with a specific member of a group, which they then generalize and apply to all members of the group (Vidmar, 1997)—an example of stimulus generalization.

But what explains behavior that is as extreme as that of the men who murdered James Byrd? Why would people do anything that they know could bring the death penalty? Punishment does weaken and suppress behavior, but as mentioned before, to be effective it must be consistent and immediate. Unfortunately, this seldom happens. Instead, when a criminal gets away with one or more crimes, that criminal behavior is put on a partial (intermittent) schedule of reinforcement, making it more likely to be repeated and to become more resistant to *extinction.*

Biofeedback is another example of operant conditioning in action (FIGURE 6.17B). Something is added (feedback) that increases the likelihood that the behavior will be repeated—positive reinforcement. The biofeedback itself is a secondary reinforcer because of the learned value of the relief from pain or other aversive stimuli (primary reinforcer). Finally, biofeedback involves shaping. A person using biofeedback watches a monitor screen (or other instrument) that provides graphs or numbers indicating some bodily state. Like a mirror, the biofeedback reflects back the results of the

Visualizing

Operant conditioning in everyday life FIGURE 6.17

A Prejudice and Discrimination ▶
What reinforcement might these boys receive for teasing this girl? Can you see how operant conditioning, beginning at an early age, can promote unkind or prejudiced behavior?

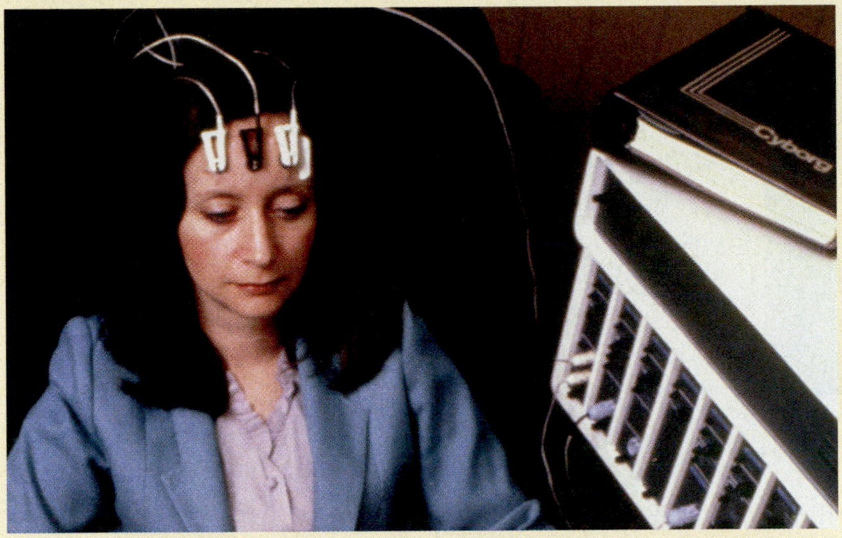

▲ **B Biofeedback**
In biofeedback, internal bodily processes (like blood pressure, muscle tension, or brain wave patterns) are electrically recorded and reported back to the patient. This information helps the person gain control over processes that are normally involuntary. Researchers have successfully used biofeedback (usually in conjunction with other techniques, such as behavior modification) to treat hypertension, anxiety, epilepsy, urinary incontinence, cognitive functioning, chronic pain, and headache (Andrasik, 2006; Bohm-Starke et al., 2007; Hammond, 2007; Kazdin, 2008; Moss, 2004; Stetter & Kupper, 2002; Tatrow, Blanchard, & Silverman, 2003).

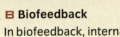

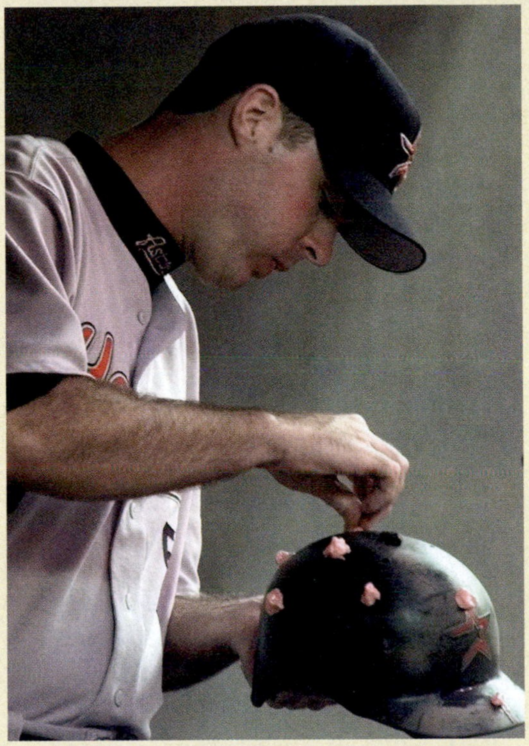

C Superstition ▶
Like Skinner's pigeons, we humans also believe in many superstitions that may have developed from accidental reinforcement—from wearing "something old" during one's wedding to knocking on wood for good fortune to performing some bizarre ritual before every athletic competition.

various strategies that the participant uses to gain control. Through trial and error, the participant gets progressively better at making the desired changes.

Even accidental reinforcement can exert a powerful effect, sometimes causing superstitious behavior (FIGURE 6.17C). In a fascinating experiment, B. F. Skinner (1948, 1992) set the feeding mechanism in the cages of eight pigeons to release food once every 15 seconds. No matter what the birds did, they were reinforced at 15-second intervals. Six of the pigeons acquired behaviors that they repeated over and over, even though the behaviors were not necessary to receive the food. For example,

John William "Bill" King FIGURE 6.18

King was sentenced to death for the murder of James Byrd. How might King and his two accomplices have learned some of their hatred and prejudice through observation and modeling?

one pigeon kept turning in counterclockwise circles, and another kept making jerking movements with its head. Why did this happen? Although Skinner was not using the food to reinforce any particular behavior, the pigeons associated the food with whatever behavior they were engaged in when the food was originally dropped into the cage.

COGNITIVE-SOCIAL LEARNING: WE SEE, WE DO?

We use cognitive-social learning in many ways in our everyday lives, yet two of the most powerful areas of learning are frequently overlooked—prejudice and media influences. One of James Byrd's murderers, Bill King, had numerous tattoos that proudly proclaimed his various prejudices (FIGURE 6.18). King's family and friends insist that he was a pleasant and quiet man until he began serving an eight-year prison sentence for burglary (Galloway, 1999). What did he learn about prejudice during his prison sentence? Did he model his killing of Byrd after his uncle's well-known killing of a gay traveling salesman a number of years earlier? Or did he learn his prejudices during his numerous years of active membership with the Ku Klux Klan?

The media also propagate some forms of prejudice. When children watch television, go to the movies, and read books and magazines that portray minorities and women in demeaning and stereotypical roles, they learn to expect these behaviors and to accept them as "natural." Exposure of this kind initiates and reinforces the learning of prejudice (Dill & Thill, 2007; Kassin, Fein, & Markus, 2008; Neto & Furnham, 2005).

Both children and adults may be learning other destructive behaviors through observational learning and the media, as well. Correlational evidence from more than 50 studies indicates that observing violent behavior is related to later desensitization and the performance of violent behavior (Anderson, Buckley, & Carnagey, 2008; Coyne, Archer, & Eslea, 2004; Kronenberger et al.,

Video games and aggression FIGURE 6.19

Researchers hypothesize that video games are more likely to model aggressive behavior because, unlike TV and other media, they are interactive, engrossing, and require the player to identify with the aggressor. What forms of learning might be involved in this phenomenon?

2005). In addition, more than 100 experimental studies have also shown a causal link between observing violence and later performing it (Primavera & Herron, 1996).

Researchers are just beginning to study how video games (and virtual reality games) affect behavior (FIGURE 6.19). For example, studies have found that students who play more violent video games in junior high and high school also engage in more aggressive behaviors (Anderson & Bushman, 2001; Bartholow & Anderson, 2002; Boyle et al., 2007; Wei, 2007). Furthermore, psychologists Craig Anderson and Karen Dill (2000) first assigned 210 students to play either a violent or a nonviolent video game and later allowed them to punish their opponents with a loud sound blast. The researchers found that those who played the violent game punished their opponents not only longer but also with greater intensity.

CONCEPT CHECK STOP

How do classical conditioning, operant conditioning, and cognitive-social learning all contribute to prejudice?

How can "accidental reinforcement" affect behavior?

Why might video games (and virtual reality games) foster aggressiveness?

SUMMARY

1 Classical Conditioning

1. **Learning** is a relatively permanent change in behavior or mental processes as a result of practice or experience. Pavlov discovered a fundamental form of **conditioning** (learning) called **classical conditioning**, in which a neutral stimulus becomes associated with an unconditioned stimulus to elicit a conditioned response. In the "Little Albert" experiment, Watson and Rayner demonstrated how many of our likes, dislikes, prejudices, and fears are conditioned emotional responses.

2. Stimulus generalization occurs when an event similar to the originally conditioned stimulus triggers the same conditioned response. With experience, animals learn to distinguish between an original conditioned stimulus and similar stimuli—stimulus discrimination. Most learned behaviors can be weakened through extinction. Extinction is a gradual weakening or suppression of a previously conditioned response. However, if a conditioned stimulus is reintroduced after extinction, an extinguished response may spontaneously recover.

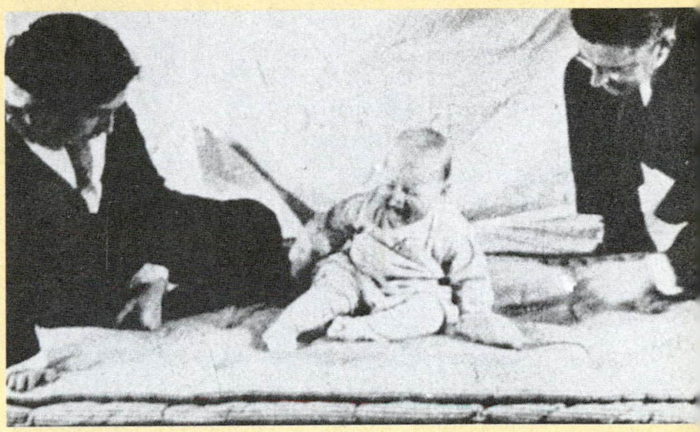

2 Operant Conditioning

1. In **operant conditioning**, an organism performs a behavior that produces either reinforcement or punishment. **Reinforcement** strengthens the response, while **punishment** weakens the response.

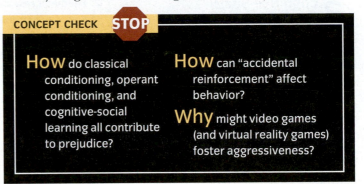

2. Thorndike developed the **law of effect**, a first step in understanding how consequences can modify voluntary behaviors. Skinner extended Thorndike's law of effect to more complex behaviors.

3. Reinforcers can be either primary or secondary, and reinforcement can be either positive or negative.

4. "Schedules of reinforcement" refers to the rate or interval at which responses are reinforced. Most behavior is rewarded on one of four partial schedules of reinforcement: fixed ratio, variable ratio, fixed interval, or variable interval.

5. Organisms learn complex behaviors through **shaping**, in which reinforcement is delivered for successive approximations of the desired response.

6. Punishment weakens the response. To be effective, punishment must be immediate and consistent. Even then, the recipient may learn only what not to do. Punishment can have serious side effects: increased aggression, avoidance behavior, passive aggressiveness, and learned helplessness.

3 Cognitive-Social Learning

1. **Cognitive-social theory** emphasizes cognitive and social aspects of learning. Köhler discovered that animals sometimes learn through sudden insight, rather than through trial and error. Tolman provided evidence of **latent learning** and internal cognitive maps.
2. Bandura's research found that children who watched an adult behave aggressively toward an inflated Bobo doll were later more aggressive than those who had not seen the aggression. According to Bandura, **observational learning** requires attention, retention, motor reproduction, and reinforcement.

4 The Biology of Learning

1. Learning creates structural changes in the brain. Early evidence for such changes came from research on animals raised in enriched versus deprived environments.
2. Learning is an evolutionary adaptation that enables organisms to survive and prosper in a constantly changing world. Researchers have identified biological constraints that limit the generality of conditioning principles: **biological preparedness** and **instinctive drift**.

5 Conditioning and Learning in Everyday Life

1. Prejudice is a negative instance of classical conditioning. Influential research by Clark and Clark indicated that children learn to associate inferior qualities with darker skin and more positive qualities with lighter skin. Classical conditioning is a primary tool for advertisers and others who want to manipulate beliefs, emotions, or behavior. Classical conditioning is also used in medicine and in the development and treatment of phobias.
2. Operant conditioning plays a role in the development of prejudice. Operant conditioning also underlies biofeedback. Even accidental reinforcement can cause superstitious behavior.
3. Cognitive-social learning plays a role in many areas of everyday life, including the development of prejudice and media influences on our behavior.

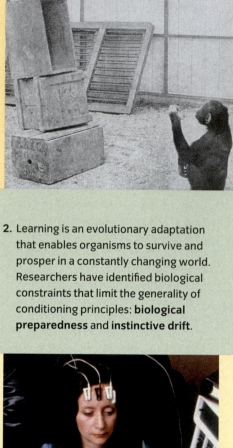

KEY TERMS

- learning p. 144
- classical conditioning p. 144
- conditioning p. 144
- unconditioned stimulus (UCS) p. 144
- unconditioned response (UCR) p. 144
- neutral stimulus (NS) p. 144
- conditioned stimulus (CS) p. 146
- conditioned response (CR) p. 146
- conditioned emotional response (CER) p. 146
- stimulus generalization p. 146
- stimulus discrimination p. 147
- extinction p. 147
- spontaneous recovery p. 147
- reconditioning p. 148
- higher-order conditioning p. 148
- operant conditioning p. 149
- reinforcement p. 149
- punishment p. 149
- law of effect p. 149
- primary reinforcer p. 150
- secondary reinforcer p. 150
- positive reinforcement p. 150
- negative reinforcement p. 150
- Premack principle p. 150
- schedules of reinforcement p. 151
- continuous reinforcement p. 151
- partial (or intermittent) schedule of reinforcement p. 151
- fixed ratio (FR) p. 151
- variable ratio (VR) p. 151
- fixed interval (FI) p. 151
- variable interval (VI) p. 151
- shaping p. 153
- positive punishment p. 153
- negative punishment p. 153
- cognitive-social theory p. 156
- insight p. 157
- insight learning p. 157
- cognitive map p. 157
- latent learning p. 157
- observational learning p. 158
- taste aversion p. 161
- biological preparedness p. 161
- instinctive drift p. 162
- biofeedback p. 164

CRITICAL AND CREATIVE THINKING QUESTIONS

1. How might Watson and Rayner, who conducted the famous "Little Albert" study, have designed a more ethical study of conditioned emotional responses?
2. What are some examples of ways in which observational learning has benefited you in your life? Are there instances in which observational learning has worked to your disadvantage?
3. Do you have any taste aversions? How would you use information in this chapter to remove them?
4. Most classical conditioning is involuntary. Considering this, is it ethical for politicians and advertisers to use classical conditioning to influence our thoughts and behavior? Why or why not?

What is happening in this picture?

For political candidates, kissing babies seems to be as critical as pulling together a winning platform and airing persuasive campaign ads.

- What principle of learning explains why politicians kiss babies?
- What other stimuli or symbols might elicit similar effects?

SELF-TEST

(Check your answers in Appendix B.)

1. ______________ *conditioning* occurs when a neutral stimulus becomes associated with an unconditioned stimulus to elicit a conditioned response.
   a. Reflex
   b. Instinctive
   c. Classical
   d. Basic
2. The process of learning associations between environmental stimuli and behavioral responses is known as ______________.
   a. maturation
   b. contiguity learning
   c. conditioning
   d. latent learning
3. In John Watson's demonstration of classical conditioning with Little Albert, the unconditioned **STIMULUS** was ______________.
   a. symptoms of fear
   b. a rat
   c. a bath towel
   d. a loud noise

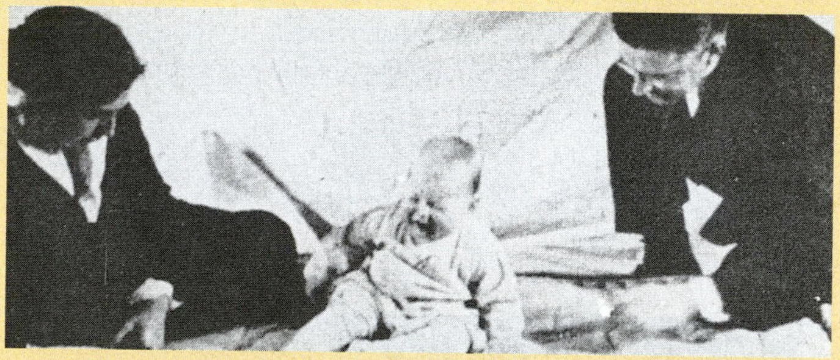

4. Which of the following is an example of the use of classical conditioning in everyday life?
   a. treating alcoholism with a drug that causes nausea when alcohol is consumed
   b. the use of seductive women to sell cars to men
   c. politicians associating themselves with home, family, babies, and the American flag
   d. all of these options
5. *Extinction* ______________.
   a. is a gradual weakening or suppression of a previously conditioned behavior
   b. occurs when a CS is repeatedly presented without the UCS
   c. is a weakening of the association between the CS and the UCS
   d. all of these options
6. When a neutral stimulus is paired with a previously conditioned stimulus to become a conditioned stimulus as well, this is called ______________ conditioning.
   a. operant
   b. classical
   c. higher-order
   d. secondary
7. Learning in which voluntary responses are controlled by their consequences is called ______________.
   a. self-efficacy
   b. classical conditioning
   c. operant conditioning
   d. all of these options
8. Write the number of the illustration after the correct label:
   a. Thorndike box: law of effect, illustration ______________
   b. Skinner box: reinforcement, illustration ______________

1.

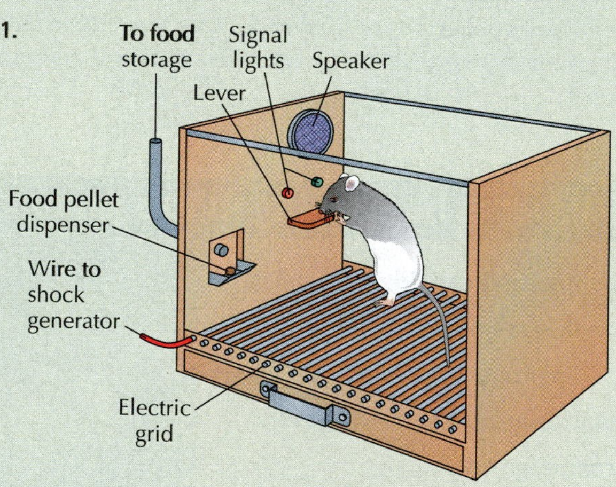

2.

9. The addition of a(n) ______________ stimulus results in *positive reinforcement,* whereas the subtraction of a(n) ______________ stimulus results in *negative reinforcement*.
   a. desirable; painful or annoying
   b. primary; secondary
   c. operant; classical
   d. higher order; lower order

10. The chimpanzee in Köhler's *insight* experiment ______________.
   a. used trial and error to reach a banana placed just out of reach
   b. turned its back on the banana out of frustration
   c. sat for a while, then used a stick to bring the banana within reach
   d. didn't like bananas

11. Learning new behavior or information by watching others is known as ______________.
   a. social learning
   b. observational learning
   c. modeling
   d. all of the above

12. Garcia and his colleagues taught coyotes to avoid sheep by pairing a nausea-inducing drug with freshly killed sheep eaten by the coyotes. This is an example of ______________.
   a. classical conditioning
   b. operant conditioning
   c. positive punishment
   d. negative punishment

13. Mamie has an EMG attached to her forehead. When tension rises, a computer says, "That's too high." When it drops to normal or lower levels, it says, "That's very good." Mamie is using ______________ to decrease her tension headaches.
   a. primary reinforcement
   b. computerized reinforcement
   c. biofeedback
   d. electromyography monitoring

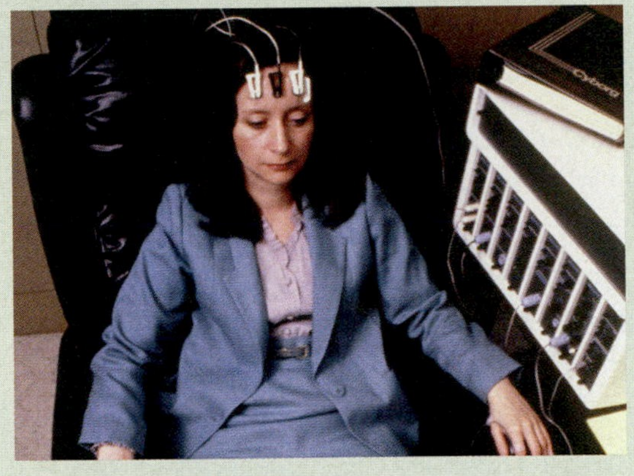

14. According to this text, research has shown that exposure to media portrayals of demeaning and stereotypical roles for minorities and women ______________.
   a. increases critical thinking about minorities and women
   b. initiates and reinforces the learning of prejudice
   c. increases empathy for minorities and women
   d. decreases a child's own stereotypical gender-role behavior

15. Researchers suggest that violent video games may increase aggression because ______________.
   a. they are interactive
   b. they are engrossing to the player
   c. video players identify with the aggressor in the game
   d. all of these options

Memory

7

When Elizabeth was 14, her mother drowned in their pool. As she grew older, the details surrounding her mother's death became increasingly vague. Decades later, a relative told Elizabeth that she had been the one to find the body. Despite her initial shock, the memories slowly started coming back.

> *I could see myself, a thin, dark-haired girl, looking into the flickering blue-and-white pool. My mother, dressed in her nightgown, is floating face down. I start screaming. I remember the police cars, their lights flashing, and the stretcher with the clean, white blanket tucked in around the edges of the body. The memory had been there all along, but I just couldn't reach it. (Loftus & Ketcham, 1994, p. 45)*

Compare Elizabeth's story with this one:

When H. M. was 27, he underwent brain surgery to correct his epileptic seizures. Although the operation improved his medical problem, something was tragically wrong with his long-term memory. H. M. lived another 55 years believing he was still 27 and not recognizing the people who care for him daily. Each time he met his caregivers, read a book, or ate a meal, it was as if for the first time (Carey, 2008; Corkin, 2002). Henry Gustav Molaison, known as H. M. to protect his privacy, died in December 2008.

How could a child forget finding her mother's body? What would it be like to be H. M. and unable to form new memories? How can we remember our second grade teacher's name but forget the name of someone we just met? In this chapter, you'll discover answers to these and other important questions about memory.

NATIONAL GEOGRAPHIC

CHAPTER OUTLINE

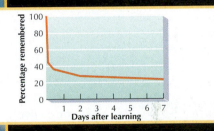

The Nature of Memory

LEARNING OBJECTIVES

Review the principles of the four major memory models.

Describe the different purposes of sensory memory, short-term memory, and long-term memory.

Identify ways of extending the duration and capacity of short-term memory.

Describe the various types of long-term memory.

Explain how organization, elaborative rehearsal, and retrieval cues improve long-term memory.

Memory allows us to learn from our experiences and to adapt to ever-changing environments—without it, we would have no past or future. Yet our memories are also highly fallible. Although some people think of **memory** as a gigantic library or an automatic tape recorder, our memories are not exact recordings of events. Instead, memory is a *constructive process* through which we actively organize and shape information. As you might expect, this construction often leads to serious errors and biases, which we'll discuss throughout this chapter.

memory
An internal record or representation of some prior event or experience.

MEMORY MODELS: A BRIEF OVERVIEW

Over the years, psychologists have developed numerous models for memory. In this section, we first briefly discuss two of the major approaches, and then we'll explore the second (three-stage) model in some depth.

Information-Processing Approach

According to the **information-processing model** (FIGURE 7.1).

Process Diagram

Information-processing model FIGURE 7.1

Memory processes have been likened to a computer's information-processing system. Data are entered on a keyboard and *encoded* in a way that the computer can understand and use. Information is then *stored* on a disk, hard drive, or memory stick, and later *retrieved* and brought to the computer screen for viewing.

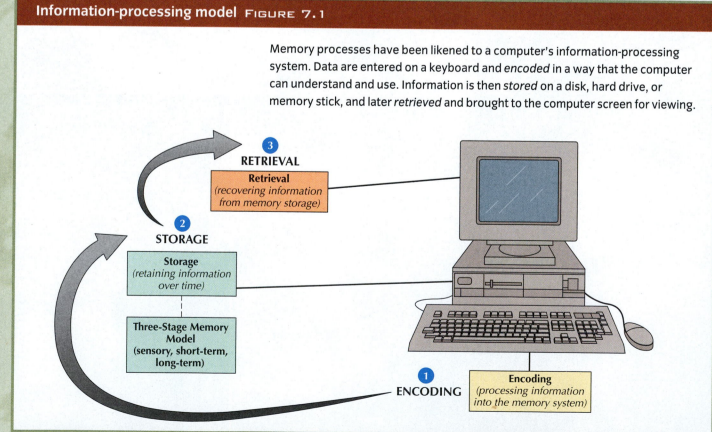

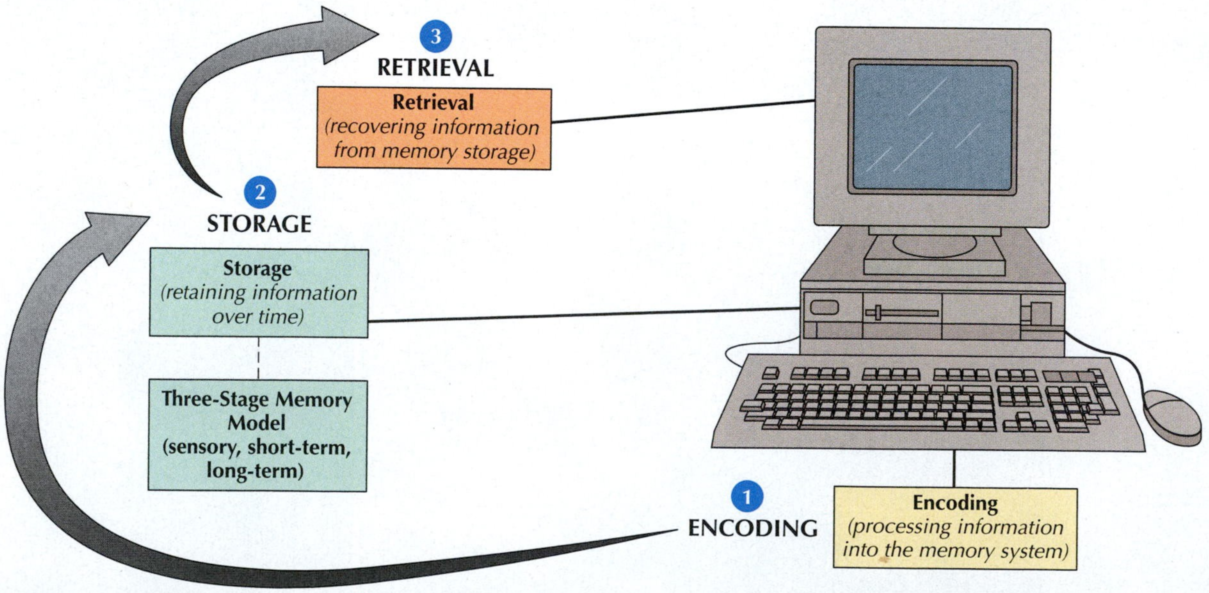

encoding Processing information into the memory system.

storage Retaining information over time.

retrieval Recovering information from memory storage.

the barrage of information that we encounter every day goes through three basic operations: **encoding**, **storage**, and **retrieval**. (This *storage* stage also involves *information processing* in three interacting memory systems—sensory, short-term, and long-term—which is generally considered a separate model, known as the *three-stage memory model.*)

The Three-Stage Model Since the late 1960s, one of the most widely used models in memory research has been the **traditional three-stage memory model** (Atkinson & Shiffrin, 1968). According to this model, three different storage "boxes," or memory stages (*sensory, short-term,* and *long-term*), hold and process information. Each stage has a different purpose, duration, and capacity (FIGURE 7.2). The three-stage memory model remains the leading paradigm in memory research because it offers a convenient way to organize the major findings. Let's discuss the model in more detail.

SENSORY MEMORY

Everything we see, hear, touch, taste, and smell first enters our **sensory memory**. Information remains in sensory memory just long enough to locate relevant bits of information and transfer them to the next stage of memory. For

sensory memory This first memory stage holds sensory information. It has a relatively large capacity, but duration is only a few seconds.

The traditional three-stage memory model FIGURE 7.2

Each "box" represents a separate memory system that differs in purpose, duration, and capacity. When information is not transferred from sensory memory or short-term memory, it is assumed to be lost. Information stored in long-term memory can be retrieved and sent back to short-term memory for use.

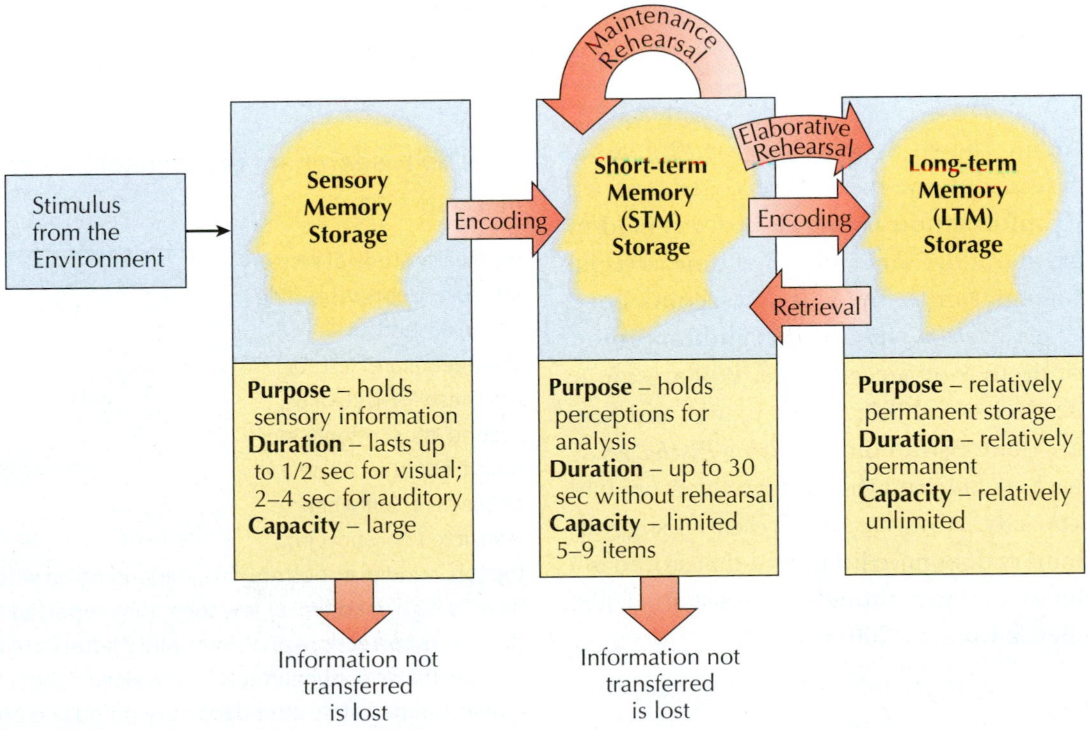

Applying Psychology

Demonstrating Iconic and Echoic Memory

As pointed out in the text, visual and auditory information remains in sensory memory for a very short time. Here are some easy ways to demonstrate this phenomenon.

◀ **Auditory memory**

Auditory, or *echoic memory*, works similarly. Think back to times when someone asked you a question while you were deeply absorbed in a task. Did you ask "What?" and then immediately find you could answer them without hearing their repeated response? Now you know why. A weaker "echo" (echoic memory) of auditory information can last up to four seconds.

Visual memory

For a simple demonstration of the duration of visual, or *iconic memory*, swing a flashlight in a dark room. Because the image, or icon, lingers for a fraction of a second after the flashlight is moved, you see the light as a continuous stream, as in this photo, rather than as a succession of individual points. ▼

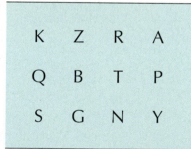

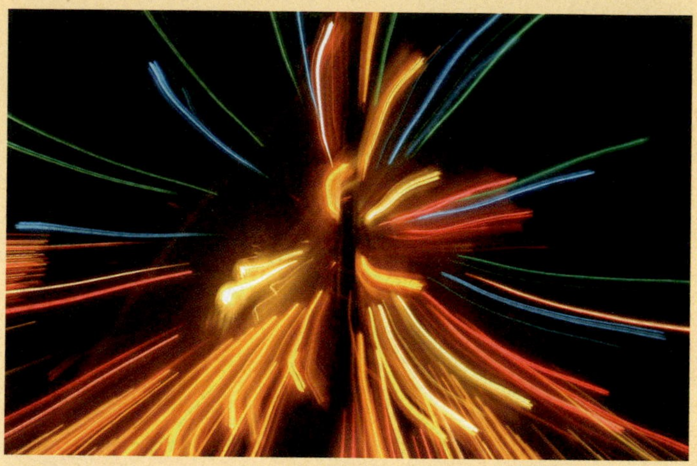

CRITICAL THINKING

Here are two interesting questions:

1. **What do you think would happen if we did not possess iconic or echoic memory?**
2. **What might happen if visual or auditory sensations lingered not for seconds but for minutes?**

visual information, known as *iconic memory,* the visual image (icon) lasts about one-half of a second (FIGURE 7.3). Auditory information (what we hear) is held in sensory memory about the same length of time as visual information, one-quarter to one-half of a second, but a weaker "echo," or *echoic memory,* of this auditory information can last up to four seconds (Lu, Williamson, & Kaufman, 1992; Neisser, 1967). Both iconic and sensory memory are demonstrated in *Applying Psychology.*

Early researchers believed that sensory memory had an unlimited capacity. However, later research suggests that sensory memory does have limits and that stored images are fuzzier than once thought (Goldstein, 2008; Grondin, Ouellet, & Roussel, 2004).

How do researchers test sensory memory?

FIGURE 7.3

K	Z	R	A
Q	B	T	P
S	G	N	Y

In an early study of sensory memory, George Sperling (1960) flashed an arrangement of letters like these for 1/20 of a second. Most people, he found, could recall only 4 or 5 letters. But when instructed to report just the top, middle, or bottom row, depending on whether they heard a high, medium, or low tone, they reported almost all the letters correctly. Apparently, all 12 letters are held in sensory memory right after they are viewed, but only those that are immediately attended to are noted and processed.

short-term memory (STM) This second memory stage temporarily stores sensory information and decides whether to send it on to long-term memory (LTM). Its capacity is limited to five to nine items, and its duration is about 30 seconds.

SHORT-TERM MEMORY

The second stage of memory processing, **short-term memory (STM)**, temporarily stores and processes the sensory stimuli. If the information is important, STM organizes and sends this information along to relatively permanent storage, called *long-term memory (LTM)*. Otherwise, it decays and is lost.

The *capacity* and *duration* of STM are limited (Best, 1999; Kareev, 2000). To extend the *capacity* of STM, you can use a technique called **chunking** (Boucher & Dienes, 2003; Miller, 1956). Have you noticed that credit card, Social Security, and telephone numbers are all grouped into three or four units separated by hyphens? This is because it's easier to remember numbers in "chunks" rather than as a string of single digits, as demonstrated in *What a Psychologist Sees.*

chunking The act of grouping separate pieces of information into a single unit (or chunk).

You can extend the *duration* of your STM almost indefinitely by consciously "juggling" the information, a process called **maintenance rehearsal**. You are using maintenance rehearsal when you look up a phone number and repeat it

maintenance rehearsal Repeating information to maintain it in short-term memory (STM).

What a Psychologist Sees

Chunking in Chess

What do you see when you observe the arrangement of pieces on a chessboard? To the inexpert eye, a chess game in progress looks like little more than a random assembly of black and white game pieces. Accordingly, novice chess players can remember the positions of only a few pieces when a chess game is underway. But expert players generally remember all the positions. To the experts, the scattered pieces form meaningful patterns—classic arrangements that recur often. Just as you group the letters of this sentence into meaningful words and remember them long enough to understand the meaning of the sentence, expert chess players group the chess pieces into easily recalled patterns (or chunks) (Huffman, Matthews, & Gagne, 2001; Waters & Gobet, 2008).

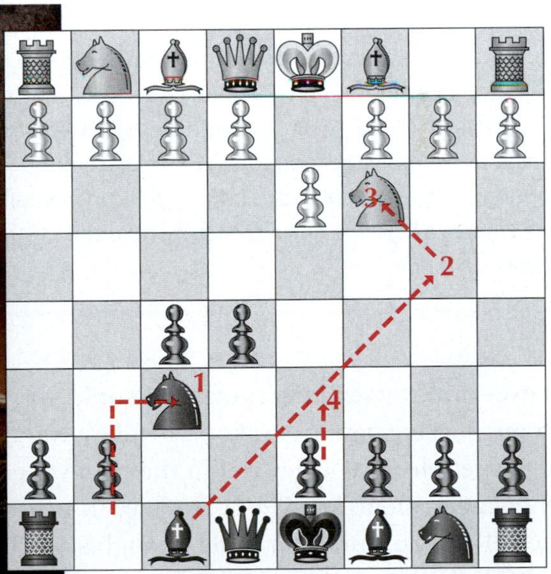

Source: Chess moves from 1986 Kasparov–Karpov game.

Process Diagram

Working memory as a central executive FIGURE 7.4

The *central executive* supervises and coordinates two subsystems, the *phonological rehearsal loop* and the *visuospatial sketchpad*, while also sending and retrieving information to and from LTM. Picture yourself as a food server in a busy restaurant, and a couple has just given you a complicated food order. When you mentally rehearse the food order (the phonological loop) and combine it with a mental picture of the layout of plates on the customers' table (the visuospatial sketchpad), you're using your central executive.

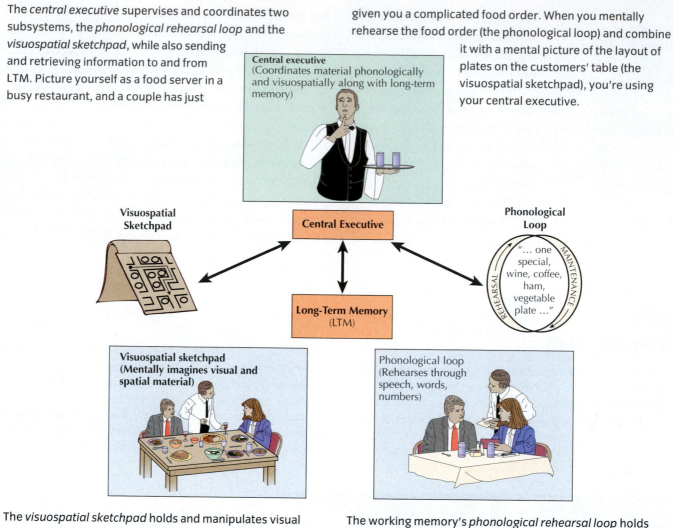

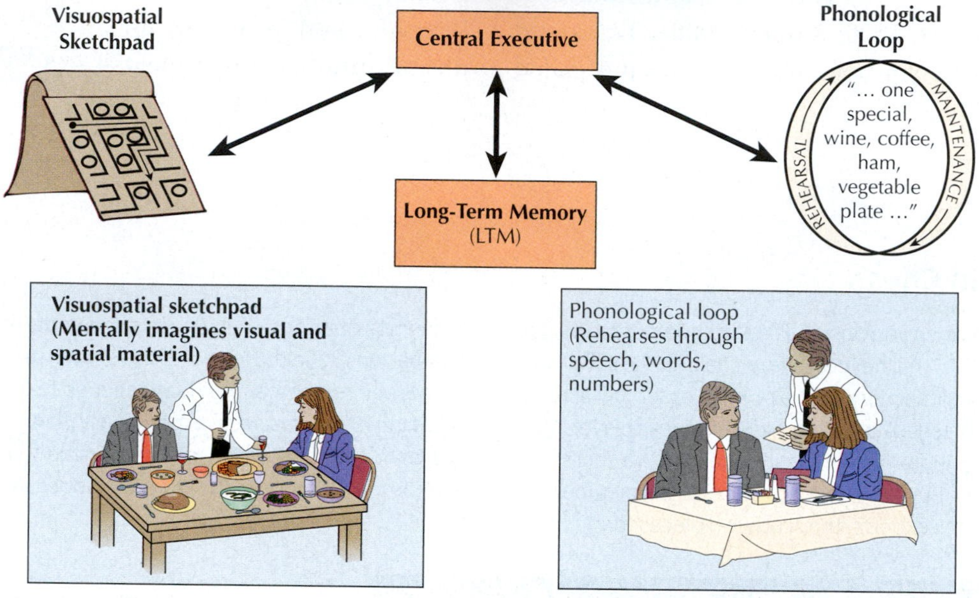

The *visuospatial sketchpad* holds and manipulates visual images and spatial information (Baddeley & Jarrold, 2007; Lehnert & Zimmer, 2008). Again, imagine yourself as the food server who's delivering the food to the same customers. Using your mind's visuospatial sketchpad, you can mentally visualize where to fit all the entrees, side dishes, and dinnerware on their table.

The working memory's *phonological rehearsal loop* holds and manipulates verbal (phonological) information (Dasi et al., 2008; Jonides et al., 2008). Your phonological loop allows you to subvocally repeat all your customers' specific requests while you write a brief description on your order pad.

over and over until you dial the number. People who are good at remembering names also know to take advantage of maintenance rehearsal. They repeat the name of each person they meet, aloud or silently, to keep it active in STM. They also make sure that other thoughts (such as their plans for what to say next) don't intrude.

Short-term memory as a "working memory"

As you can see in FIGURE 7.4, short-term memory is more than just a passive, temporary "holding area." Most current researchers (Baddeley, 1992, 2007; Jonides et al., 2008) realize that active processing of information also occurs in STM. Because the short-term memory is active, or working, we think of STM as a three-part **working memory**. Let's look at each of the three working parts of short-term memory.

LONG-TERM MEMORY

Think back to the opening story of H. M. Although his surgery was successful in stopping the severe epileptic seizures, it also apparently destroyed the mechanism that transfers information from short-term to long-term memory. (That is why he could not remember the people he saw every day or even recognize his own face.)

Once information is transferred from STM, it is organized and integrated with other information in **long-term memory (LTM)**. LTM serves as a storehouse for information that must be kept for long periods. When we need the information, it is sent back to STM for our use. Compared with sensory memory and short-term memory, long-term memory has relatively unlimited *capacity* and *duration* (Klatzky, 1984). But, just as with any other possession, the better we label and arrange our memories, the more readily we'll be able to retrieve them.

> **long-term memory (LTM)** This third memory stage stores information for long periods. Its capacity is limitless; its duration is relatively permanent.

How do we store the vast amount of information that we collect over a lifetime? Several types of LTM exist (see STUDY ORGANIZER 7.1). The two major systems are *explicit/declarative memory* and *implicit/nondeclarative memory*.

Explicit/declarative memory refers to intentional learning or conscious knowledge. If asked to remember your phone number or your mother's name, you can state (*declare*) the answers directly (*explicitly*). Explicit/declarative memory can be further subdivided into two parts. *Semantic* memory is memory for general knowledge, rules, events, facts, and specific information. It is our mental encyclopedia. In contrast, *episodic* memory is like a mental diary. It records the major events (*episodes*) in our lives. Some of our

> **explicit/declarative memory** The subsystem within long-term memory that consciously stores facts, information, and personal life experiences.

Study Organizer 7.1 Types of long-term memory

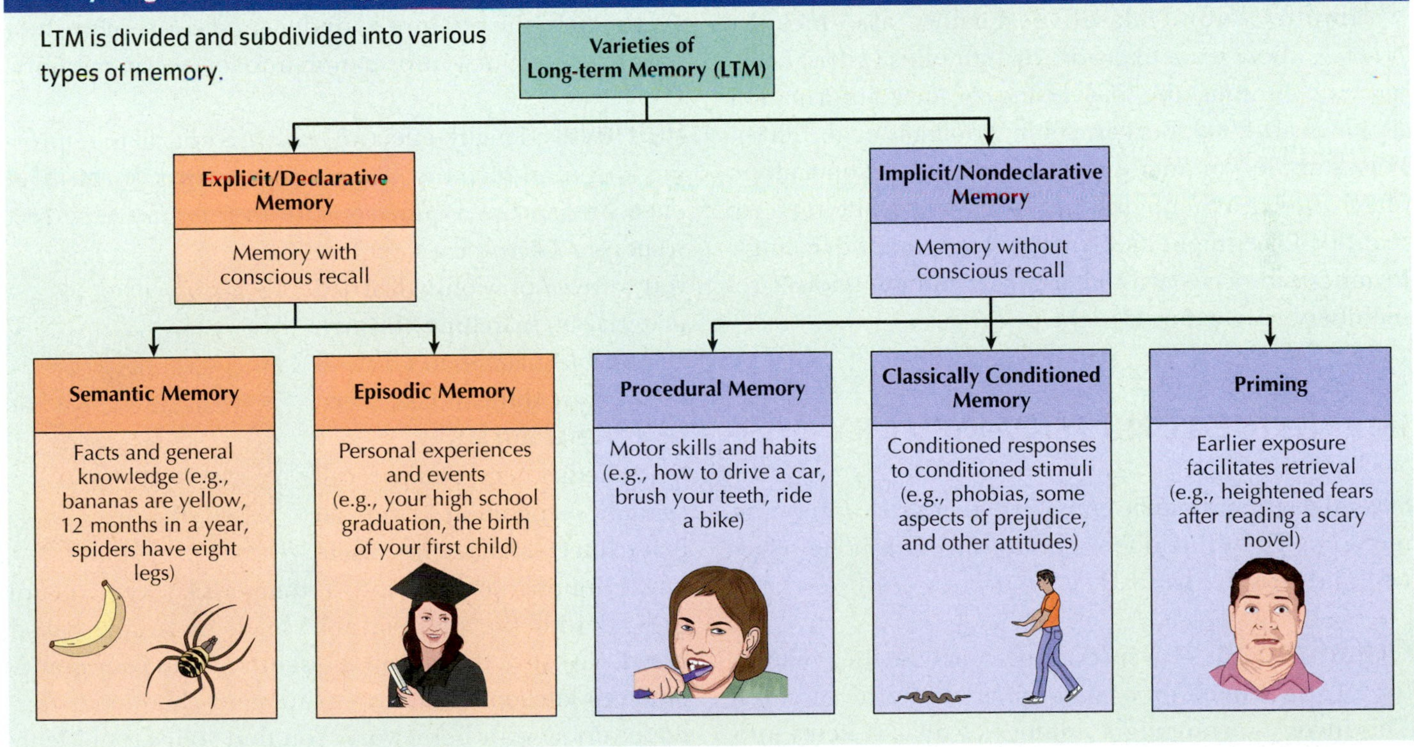

LTM is divided and subdivided into various types of memory.
Varieties of Long-term Memory (LTM)
Explicit/Declarative Memory
Memory with conscious recall
Implicit/Nondeclarative Memory
Memory without conscious recall
Semantic Memory
Facts and general knowledge (e.g., bananas are yellow, 12 months in a year, spiders have eight legs)
Episodic Memory
Personal experiences and events (e.g., your high school graduation, the birth of your first child)
Procedural Memory
Motor skills and habits (e.g., how to drive a car, brush your teeth, ride a bike)
Classically Conditioned Memory
Conditioned responses to conditioned stimuli (e.g., phobias, some aspects of prejudice, and other attitudes)
Priming
Earlier exposure facilitates retrieval (e.g., heightened fears after reading a scary novel)

episodic memories are short-lived, whereas others can last a lifetime.

Have you ever wondered why most adults can recall almost nothing of the years before age 3? Research suggests that a concept of "self," sufficient language development, and growth of the frontal lobes of the cortex (along with other structures) may be necessary before these early events can be encoded and retrieved many years later (Leichtman, 2006; Morris, 2007; Prigatano & Gray, 2008; Suzuki & Amaral, 2004; Wang, 2008).

Implicit/nondeclarative memory refers to unintentional learning or unconscious knowledge. Try telling someone else how you tie your shoelaces without demonstrating the actual behavior. Because your memory of this skill is unconscious and hard to describe (*declare*) in words, this type of memory is sometimes referred to as *nondeclarative.*

> **implicit/nondeclarative memory** The subsystem within long-term memory that consists of unconscious procedural skills, simple classically conditioned responses (Chapter 6), and priming.

Implicit/nondeclarative memory consists of *procedural* motor skills, like tying your shoes or riding a bike, as well as *classically conditioned memory* responses, such as fears or taste aversions.

Implicit/nondeclarative memory also includes *priming*, where prior exposure to a stimulus (*prime*) facilitates or inhibits the processing of new information (Amir et al., 2008; Becker, 2008; Woollams et al., 2008). For example, you might feel nervous being home alone after reading a Stephen King novel, and watching a romantic movie might kindle your own romantic feelings. Priming can occur even when we do not consciously remember being exposed to the prime.

IMPROVING LONG-TERM MEMORY

Several processes can be employed to improve long-term memory. These include *organization, rehearsal* or *repetition*, and effective *retrieval.*

Organization

To successfully encode information for LTM, we need to *organize* material into hierarchies. This involves arranging a number of related items into broad categories that are further divided and subdivided. (This organization strategy for LTM is similar to the strategy of chunking material in STM.) For instance, by grouping small subsets of ideas together (as subheadings under larger, main headings, and within diagrams, tables, and so on), we hope to make the material in this book more understandable and *memorable.*

Admittedly, organization takes time and work. But you'll be happy to know that some memory organization and filing is done automatically while you sleep (Mograss, Guillem, & Godbout, 2008; Siccoli et al., 2008; Wagner et al., 2003). (Unfortunately, despite claims to the contrary, research shows that we can't recruit our sleeping hours to memorize new material, such as a foreign language.)

Rehearsal

Like organization, *rehearsal* also improves encoding for both STM and LTM. If you need to hold information in STM for longer than 30 seconds, you can simply keep repeating it (maintenance rehearsal). But storage in LTM requires *deeper levels of processing* called **elaborative rehearsal** (FIGURE 7.5A).

> **elaborative rehearsal** The process of linking new information to previously stored material.

The immediate goal of elaborative rehearsal is to *understand*—not to memorize. Fortunately, this attempt to understand is one of the best ways to encode new information into long-term memory.

Retrieval

Finally, effective *retrieval* is critical to improving long-term memory. There are two types of **retrieval cues**. *Specific* cues require you only to *recognize* the correct response. *General* cues require you to *recall* previously learned material by searching through all possible matches in LTM—a much more difficult task (FIGURE 7.5B and C).

> **retrieval cue** A clue or prompt that helps stimulate recall and retrieval of a stored piece of information from long-term memory.

Whether cues require recall or only recognition is not all that matters. Imagine that while house hunting, you walk into a stranger's kitchen and are greeted with the unmistakable smell of freshly baked bread. Instantly, the aroma transports you to your grandmother's kitchen, where you spent many childhood afternoons doing your homework. You find yourself suddenly

Visualizing

Rehearsal and retrieval FIGURE 7.5

A Elaborative rehearsal

To improve elaborative rehearsal, you can expand (or elaborate on) the information to try to understand it better, actively explore and question new information, and search for meaningfulness. For example, a student might compare what she reads in an anthropology or history textbook with what she knows about world geography. Can you see how this might deepen her understanding and memory for the subject?

B Retrieval cues

Can you *recall* from memory the names of the eight planets in our solar system? Why is it so much easier to *recognize* the names if you're provided with the first three letters of each planet's name: Mer-, Ven-, Ear-, Mar-, Jup-, Sat-, Ura-, Nep-? *Recall*, like an essay question, requires you to retrieve previously learned material with only general, nonspecific cues. In contrast, *recognition* tasks, as in a multiple-choice question, offer specific cues that only require you to identify (recognize) the correct response. (Note that in 2006, Pluto was officially declassified as a planet, but this finding is still being debated.)

C Longevity and recognition memory

Both name recognition and picture recognition for high school classmates remain high even many years after graduation, whereas recall memory would be expected to drop significantly over time.

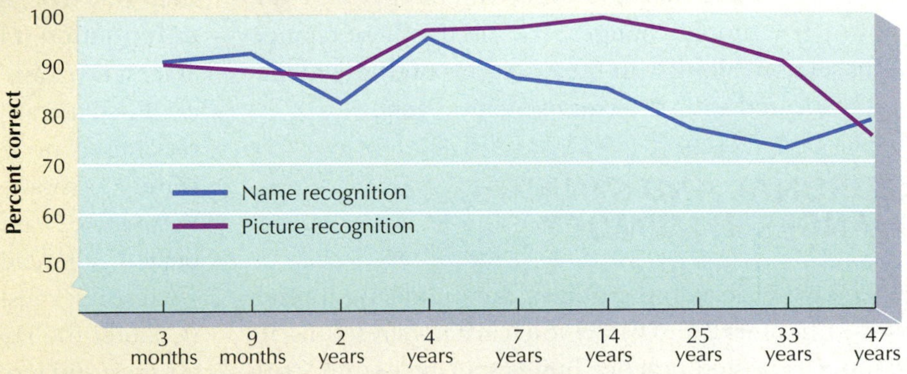

NATIONAL GEOGRAPHIC

thinking of the mental shortcuts your grandmother taught you to help you learn your multiplication tables. You hadn't thought about these little tricks for years, but somehow a whiff of baking bread brought them back to you. Why?

In this imagined episode, you have stumbled upon the **encoding specificity principle** (Tulving & Thompson, 1973). One important contextual cue for retrieval is *location*. In a clever study, Godden and Baddeley (1975) had underwater divers learn a list of 40 words either on land or underwater. The divers had better recall for lists that they had encoded underwater if they were also underwater at the time of retrieval; similarly, lists that were encoded above water were better recalled above water.

encoding specificity principle Retrieval of information is improved when the conditions of recovery are similar to the conditions that existed when the information was encoded.

People also remember information better if their moods during learning and retrieval match (Kenealy, 1997; Nouchi & Hyodo, 2007). This phenomenon, called *mood congruence*, occurs because a given mood tends to evoke memories that are consistent with that mood. When you're sad (or happy or angry), you're more likely to remember events and circumstances from other times when you were sad (or happy or angry).

Finally, as generations of coffee-guzzling university students have discovered, if you learn something while under the influence of a drug, such as caffeine, you will remember it more easily when you take that drug again than at other times (Rezayof et al., 2008; Zarrindast et al., 2005, 2007). This is called *state-dependent retrieval*.

CONCEPT CHECK STOP

What happens to information after it leaves sensory memory?

What is working memory?

What is the difference between recognition and recall?

How do contextual cues such as location, mood, and physiological state affect memory retrieval?

NATIONAL GEOGRAPHIC

Biological Bases of Memory

LEARNING OBJECTIVES

Describe two kinds of biological changes that occur when we learn something new.

Identify some brain areas involved in memory.

Explain how injury and disease can affect memory.

A number of biological changes occur when we learn something new. Among them are neuronal and synaptic changes and hormonal changes. We discuss these changes in this section, along with the questions of where memories are located and what causes memory loss.

NEURONAL AND SYNAPTIC CHANGES IN MEMORY

We know that learning modifies the brain's neural networks (Chapters 2 and 6). As you learn to play tennis, for example, repeated practice builds specific neural "pathways" that make it easier and easier for you to get the ball over the net. This **long-term potentiation (LTP)** happens in at least two ways.

long-term potentiation (LTP) Long-lasting increase in neural excitability believed to be a biological mechanism for learning and memory.

First, as early research with rats raised in "enriched" environments showed (Rosenzweig, Bennett, & Diamond, 1972), repeated stimulation of a synapse can strengthen the synapse by causing the dendrites to grow more spines (Chapter 6). This results in more synapses, more receptor sites, and more sensitivity.

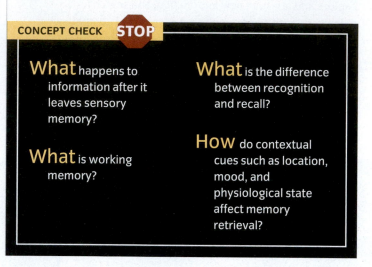

Second, learning affects a particular neuron's ability to release its neurotransmitters. This is demonstrated in research with *Aplysia* (sea slugs), which can be classically conditioned to reflexively withdraw their gills when squirted with water (FIGURE 7.6).

Further evidence comes from research with genetically engineered "smart mice," which have extra receptors for a neurotransmitter named NMDA (N-methyl-d-aspartate). These mice performed significantly better on memory tasks than did normal mice (Tang et al., 2001; Tsien, 2000).

Although it is difficult to generalize from sea slugs and mice, research on long-term potentiation (LTP) in humans has been widely supportive (Berger et al., 2008; Tecchio et al., 2008; Wixted, 2004).

How does a sea slug learn and remember?
FIGURE 7.6

After repeated squirting with water, followed by a mild shock, the sea slug, *Aplysia*, releases more neurotransmitters at certain synapses. These synapses then become more efficient at transmitting signals that allow the slug to withdraw its gills when squirted. As a critical thinker, can you explain why this ability might provide an evolutionary advantage?

HORMONAL CHANGES AND MEMORY

When stressed or excited, we naturally produce hormones that arouse the body, such as *epinephrine* and *cortisol* (Chapter 3). These hormones in turn affect the amygdala (a brain structure involved in emotion), which then stimulates the hippocampus and cerebral cortex (parts of the brain that are important for memory storage). Research has shown that direct injections of epinephrine or cortisol or electrical stimulation of the amygdala will increase the encoding and storage of new information (Hamilton & Gotlib, 2008; Jackson, 2008; van Stegeren, 2008). However, prolonged or excessive stress (and increased levels of cortisol) can interfere with memory (Al'absi, Hugdahl, & Lovallo, 2002; Heffelfinger & Newcomer, 2001; McAllister-Williams & Rugg, 2002).

The powerful effect of hormones on memory also can be seen in what are known as *flashbulb memories*—vivid images of circumstances associated with surprising or strongly emotional events (Brown & Kulik, 1977). In such situations, we secrete fight-or-flight hormones when we initially hear of the event and then replay the event in our minds again and again, which makes for stronger memories. Despite their intensity, flashbulb memories are not as accurate as you might think (Cubelli & Della Sala, 2008; Talarico & Rubin, 2007). For example, when asked how he heard the news of the September 11, 2001, attacks, President George W. Bush's answers contained numerous inconsistencies (Greenberg, 2004). Thus, not even flashbulb memories are immune to alteration.

WHERE ARE MEMORIES LOCATED?

Early memory researchers believed that memory was *localized*, or stored in a particular brain area. Later research suggests that, in fact, memory tends to be localized not in a single area but in many separate areas throughout the brain.

Today, research techniques are so advanced that we can experimentally induce and measure memory-related brain changes as they occur—on-the-spot reporting! For example, James Brewer and his colleagues (1998) used functional magnetic resonance imaging (fMRI) to locate areas of the brain responsible for encoding memories of

Psychological Science

Brain and Memory Formation

Damage to any one of these areas can affect encoding, storage, and retrieval of memories.

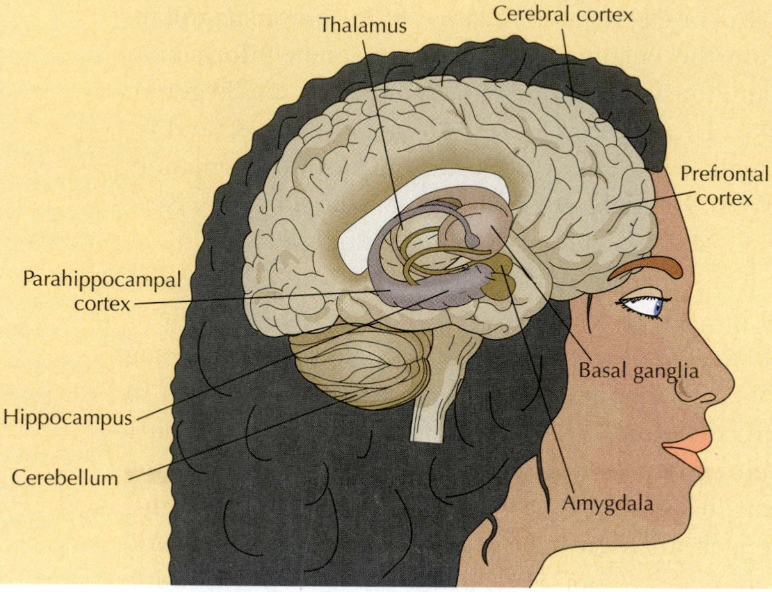

Amygdala	Emotional memory (Gerber et al., 2008; Hamilton & Gotlib, 2008; van Stegeren, 2008)
Basal Ganglia and Cerebellum	Creation and storage of the basic memory trace and implicit (nondeclarative) memories (such as skills, habits, and simple classical conditions responses) (Chiricozzi et al., 2008; Gluck, 2008; Thompson, 2005)
Hippocampal Formation (hippocampus and surrounding area)	Memory recognition; implicit, explicit, spatial, episodic memory; declarative long-term memory; sequences of events (Hamilton & Gotlib, 2008; Yoo et al., 2007)
Thalamus	Formation of new memories and spatial and working memory (Hart & Kraut, 2007; Hofer et al., 2007; Ponzi, 2008)
Cortex	Encoding of explicit (declarative) memories; storage of episodic and semantic memories; skill learning; printing; working memory (Davidson et al., 2008; Dougal et al., 2007; Thompson, 2005)

CRITICAL THINKING

Here are two interesting questions:

1. **What effect might damage to the amygdala have on a person's relationships with others?**
2. **What effect do you think damage to the thalamus might have on a person's day-to-day functioning?**

pictures. They showed 96 pictures of indoor and outdoor scenes to participants while scanning their brains, and then later tested participants' ability to recall the pictures. Brewer and his colleagues identified the *right prefrontal cortex* and the *parahippocampal cortex* as being the most active during the encoding of the pictures. These are only two of several brain regions involved in memory storage.

BIOLOGICAL CAUSES OF MEMORY LOSS: INJURY AND DISEASE

The leading cause of neurological disorders—including memory loss—among Americans between the ages of 15 and 25 is *traumatic brain injury*. These injuries most commonly result from car accidents, falls, blows, and gunshot wounds.

Traumatic brain injury happens when the skull suddenly collides with another object. The compression, twisting, and distortion of the brain inside the skull all cause serious and sometimes permanent damage to the brain. The frontal and temporal lobes often take the heaviest hit because they directly impact against the bony ridges inside the skull.

Loss of memory as a result of brain injury is called *amnesia*. Two major types of amnesia are **retrograde amnesia** and **anterograde amnesia** (FIGURE 7.7). Usually, retrograde amnesia is temporary. Unfortunately, anterograde amnesia is most often permanent, but patients show surprising abilities to learn and remember implicit/nondeclarative tasks (such as procedural motor skills).

Like traumatic brain injuries, disease can alter the physiology of the brain and nervous system, affecting memory processes. For example, **Alzheimer's disease (AD)** is a progressive mental deteriora-

Alzheimer's disease (AD) Progressive mental deterioration characterized by severe memory loss.

Two types of amnesia FIGURE 7.7

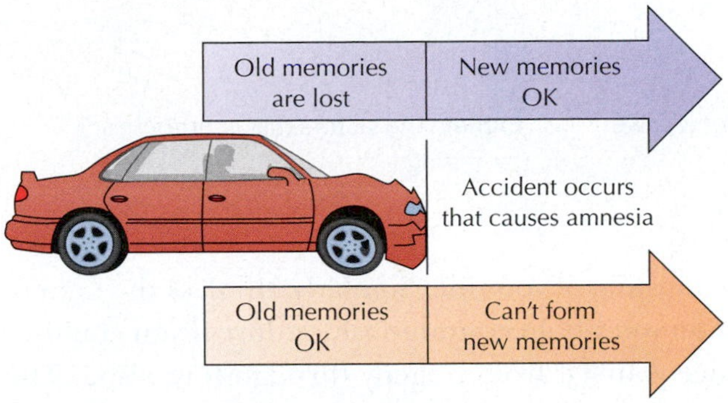

A Retrograde amnesia The person loses memories of events that occurred *before* the accident, yet has no trouble remembering things that happened afterward (old, "retro" memories are lost).

B Anterograde amnesia The person cannot form new memories for events that occur *after* the accident. Anterograde amnesia also may result from a surgical injury or from diseases such as chronic alcoholism.

tion that occurs most commonly in old age (FIGURE 7.8). The most noticeable early symptoms are disturbances in memory, which become progressively worse until, in the final stages, the person fails to recognize loved ones, needs total nursing care, and ultimately dies.

Alzheimer's does not attack all types of memory equally. A hallmark of the disease is an extreme decrease in *explicit/declarative memory* (Haley, 2005; Libon et al., 2007; Satler et al., 2007). Alzheimer's patients fail to recall facts, information, and personal life experiences, yet they still retain some *implicit/nondeclarative* memories, such as simple classically conditioned responses and procedural tasks, like brushing their teeth.

Brain autopsies of people with Alzheimer's show unusual *tangles* (structures formed from degenerating cell bodies) and *plaques* (structures formed from degenerating axons and dendrites). Hereditary Alzheimer's generally strikes its victims between the ages of 45 and 55. Some experts believe that the cause of Alzheimer's is primarily genetic, but others think that genetic makeup may make some people more susceptible to environmental triggers (Diamond & Amso, 2008; Ertekin-Taner, 2007; Persson et al., 2008; Vickers et al., 2000; Weiner, 2008).

CONCEPT CHECK STOP

Why would animals raised in enriched environments develop different neuronal connections from those raised in deprived environments?

How are hormones involved in memory?

What is the difference between retrograde and anterograde amnesia?

The effect of Alzheimer's disease on the brain FIGURE 7.8

A Normal brain
Note the high amount of red and yellow color (signs of brain activity) in the positron emission tomography scans of the normal brain.

B Brain of an Alzheimer's disease patient
The reduced activity in the brain of the Alzheimer's disease patient is evident. The loss is most significant in the temporal and parietal lobes, which indicates that these areas are particularly important for storing memories.

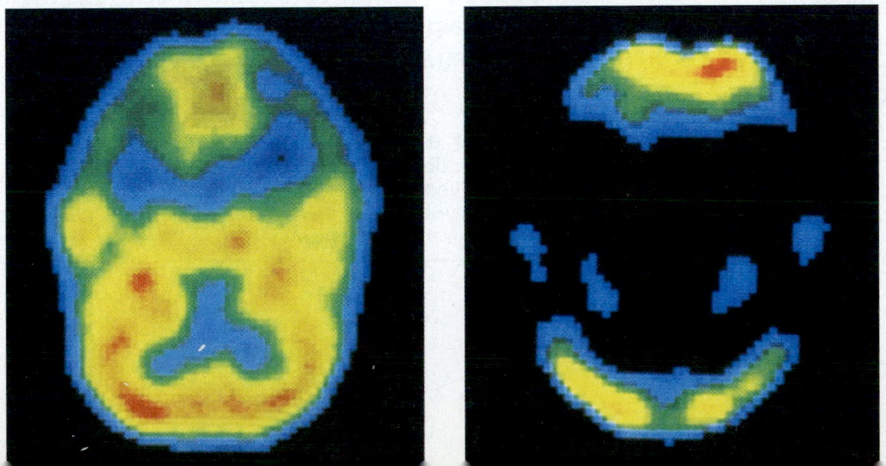

Forgetting

LEARNING OBJECTIVES

Describe Ebbinghaus's research on learning and forgetting.

Review the five key theories of why we forget.

Explain the factors that contribute to forgetting.

Psychologists have developed several theories to explain forgetting and have identified a number of factors that can interfere with the process of forming memories. We discuss some of these theories and factors in this section.

THEORIES OF FORGETTING

If you couldn't forget, your mind would be filled with meaningless data, such as what you ate for breakfast every morning of your life. Similarly, think of the incredible pain you would continuously endure if you couldn't distance yourself from tragedy through forgetting. The ability to forget is essential to the proper functioning of memory. But what about those times when forgetting is an inconvenience or even dangerous?

There are five major theories that explain why forgetting occurs (FIGURE 7.9): decay, interference, motivated forgetting, encoding failure, and retrieval failure. Each theory focuses on a different stage of the memory process or a particular type of problem in processing information.

Psychological Science

How Quickly We Forget

Hermann Ebbinghaus first introduced the experimental study of learning and forgetting in 1885. Using himself as a subject, he calculated how long it took to learn a list of three-letter *nonsense syllables* such as *SIB* and *RAL*. He found that one hour after he knew a list perfectly, he remembered only 44 percent of the syllables. A day later, he recalled 35 percent, and a week later, only 21 percent. This figure shows his famous "forgetting curve."

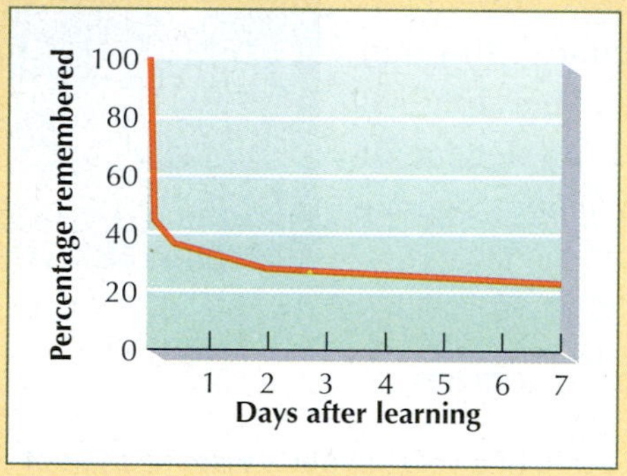

Depressing as these findings may seem, keep in mind that meaningful material is much more memorable than nonsense syllables. Even so, we all forget some of what we have learned.

On a more cheerful note, after some time passed and he had forgotten the list, Ebbinghaus found that *relearning* a list took less time than the initial learning did. This research suggests that we often retain some memory for things that we have learned, even when we seem to have forgotten them completely.

More recent research based on Ebbinghaus's discoveries has found that there is an ideal time to practice something you have learned. Practicing too soon is a waste of time, and if you practice too late you will already have forgotten what you learned. The ideal time to practice is when you are about to forget.

Polish psychologist Piotr Wozniak used this insight to create a software program called SuperMemo. The program can be used to predict the future state of an individual's memory and help the person schedule reviews of learned information at the optimal time. So far the program has been applied mainly to language learning, helping users retain huge amounts of vocabulary. But Wozniak hopes that someday programs like SuperMemo will tell people when to wake and when to exercise, help them remember what they have read and whom they have met, and remind them of their goals (Wolf, 2008).

CRITICAL THINKING

Here are two interesting questions:
1. How do you think Ebbinghaus's findings might be applied to learning to play a musical instrument?
2. Can you think of any disadvantages of depending on a program like SuperMemo?

Why we forget: five key theories Figure 7.9

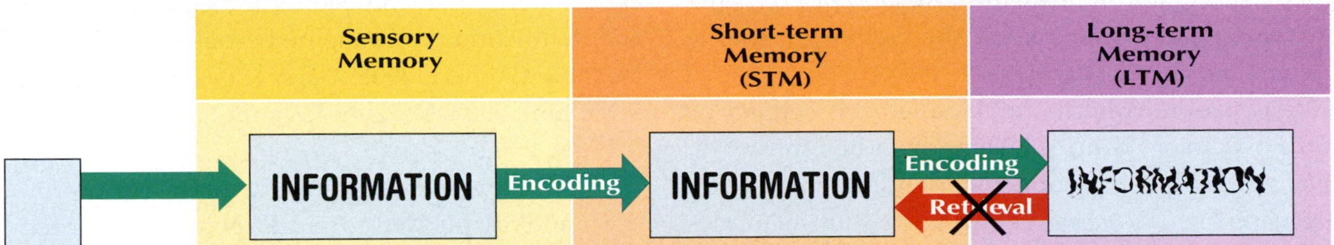

A Decay Theory. Because memory is processed and stored in a physical form—for example, in a network of neurons—connections between neurons probably deteriorate over time. We know that skills and memory degrade if they go unused. In other words, "use it or lose it." However, conclusive experimental support for decay theory is difficult to obtain. How can you prove that a previously stored memory does not exist?

B Interference Theory. Forgetting is caused by two competing memories. Interference is particularly strong among memories with similar qualities. **Retroactive interference** occurs when new information (such as one's new phone number) interferes with old information (such as one's former number). In contrast, **proactive interference** occurs when old information interferes with new information. Calling a new partner by an old partner's name is proactive interference.

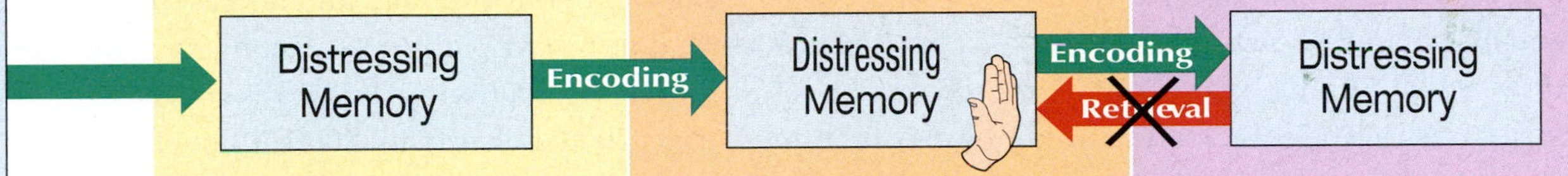

C Motivated Forgetting Theory. We forget some information—such as a dental appointment or an embarrassing remark we once made—for a reason. According to Freudian theory, people forget unpleasant or anxiety-producing information either consciously (suppression) or unconsciously *(repression)*. (See Chapter 13.)

D Encoding Failure Theory. Our sensory memory receives the information and passes it to STM. But during STM, we may decide there is no need to remember the precise details, so we do not fully encode it and pass it on for proper storage in LTM. For example, few people can correctly recognize the details on the U.S. penny (the location of the mint date, the direction Lincoln's head is facing, and so on), despite having seen it many times.

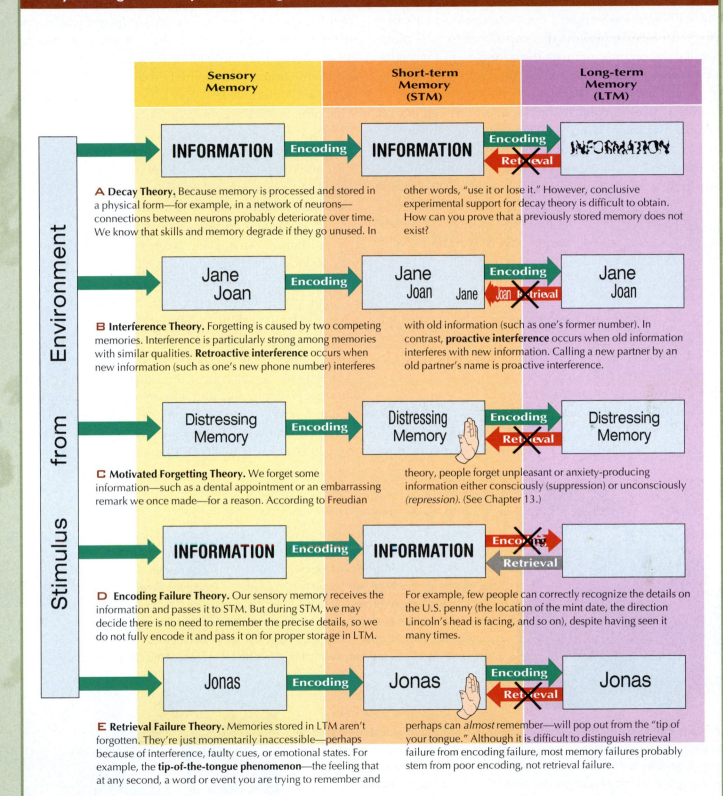

E Retrieval Failure Theory. Memories stored in LTM aren't forgotten. They're just momentarily inaccessible—perhaps because of interference, faulty cues, or emotional states. For example, the **tip-of-the-tongue phenomenon**—the feeling that at any second, a word or event you are trying to remember and perhaps can *almost* remember—will pop out from the "tip of your tongue." Although it is difficult to distinguish retrieval failure from encoding failure, most memory failures probably stem from poor encoding, not retrieval failure.

Process Diagram

FACTORS INVOLVED IN FORGETTING

Since Ebbinghaus's original research, scientists have discovered numerous factors that contribute to forgetting. Six of the most important are the *misinformation effect,* the *serial position effect, source amnesia,* the *sleeper effect, spacing of practice,* and *culture* (FIGURE 7.10).

Many people (who haven't studied this chapter or taken a psychology course) believe that when they're recalling an event, they're remembering it as if it were an "instant replay." However, as you know, our memories are highly fallible and filled with personal "constructions" that we create during encoding and storage. Research on the **misinformation effect** shows that information that occurs after an *event* may further alter and revise those constructions. For example, in one study subjects watched a film of a car driving through the countryside, and were then asked to estimate how fast the car was going when it passed the barn. Although there was no actual barn in the film, subjects were six times more likely to report having seen one than those who were not asked about a barn (Loftus, 1982). Other experiments have created false memories by showing subjects doctored photos of themselves taking a completely fictitious hot-air balloon ride, or by asking subjects to simply imagine an event, such as having a nurse remove a skin sample from their finger. In these and similar cases, a large number of subjects later believed that misleading information was correct and that fictitious or imagined events actually occurred (Allan & Gabbert, 2008; Garry & Gerrie, 2005; Mazzoni & Memon, 2003; Mazzoni & Vannucci, 2007; Pérez-Mata & Diges, 2007).

misinformation effect Distortion of a memory by misleading post-event information.

When study participants are given lists of words to learn and are allowed to recall them in any order they choose, they remember the words at the beginning *(primacy effect)* and the end of the list *(recency effect)* better than those in the middle, which are quite often forgotten (Azizian & Polich, 2008; Healy et al., 2008). This effect, including both the primacy effect and recency effect, is known as the **serial position effect**.

The reasons for this effect are complex, but they do have interesting real-life implications. For example, a potential employer's memory for you might be enhanced if you are either the first or last candidate interviewed.

Each day we read, hear, and process an enormous amount of information, and it's easy to confuse "who said what to whom" and in what context. Forgetting the true source of a memory is known as **source amnesia** (Klieder et al., 2008; Leichtman, 2006; Mitchell et al., 2005).

Fibs or false recall? FIGURE 7.10

◀ **A** Three months after the terrorist attack on the World Trade Center on September 11, 2001, President George Bush was asked how he heard about the first attack. He replied that he was "sitting outside the classroom waiting to go in, and I saw an airplane hit the tower—the TV was obviously on, and I used to fly myself. . . ." However, at the exact moment of the attack on September 11, no one saw the first plane crashing into the north tower on live TV—footage of the first crash only surfaced the next day. Furthermore, there are photos, like this one, showing that the president first learned of the attack when Chief of Staff Andrew Card whispered the news to him while he sat in front of the class. Was this a deliberate lie?

B What about Senator Hillary Clinton's claim that she faced sniper fire on a trip to Bosnia? She repeated this story several times during the 2008 presidential campaign, only to be embarrassed later when videotape and photos surfaced showing that she was in fact greeted with flowers and hugs from a little girl. ▶

Can you use one or more of the six factors that contribute to forgetting to help explain these misstatements?

When we first hear something from an unreliable source, we tend to disregard that information in favor of a more reliable source. However, as the source of the information is forgotten (source amnesia), the unreliable information is no longer discounted. This is called the **sleeper effect** (Appel & Richter, 2007; Kumkale & Albarracin, 2004; Nabi, Moyer-Gusé, & Byrne, 2007).

If we try to memorize too much information at once (as when students "cram" before an exam), we're not likely to learn and remember as much as we would with more distributed study (Donovan & Radosevich, 1999). **Distributed practice** refers to spacing your learning periods, with rest periods between sessions. Cramming is called **massed practice** because the time spent learning is massed into long, unbroken intervals.

Finally, as illustrated in FIGURE 7.11, cultural factors can play a role in how well people remember what they have learned.

Culture and memory FIGURE 7.11

In many societies, tribal leaders pass down vital information through orally related stories. As a result, children living in these cultures have better memories for information that is related through stories than do other children. Can you think of other ways in which culture might influence memory?

CONCEPT CHECK STOP

How does previous learning affect relearning?

What is the difference between proactive and retroactive interference?

Why do we sometimes fall prey to those who offer erroneous information, even when we know it is suspect?

Which is more effective, distributed practice or massed practice?

Memory Distortions

LEARNING OBJECTIVES

Explain why our memories sometimes become distorted.

Describe the dangers of relying on eyewitness testimony.

Summarize the controversy over repressed memories.

One of my first memories would date, if it were true, from my second year. I can still see, most clearly, the following scene, in which I believed until I was about fifteen. I was sitting in my pram, which my nurse was pushing in the Champs-Élysées, when a man tried to kidnap me. I was held in by the strap fastened round me while my nurse bravely tried to stand between the thief and me. She received various scratches, and I can still see vaguely those on her face. Then a crowd gathered, a policeman with a short cloak and a white baton came up, and the man took to his heels. I can still see the whole scene, and can even place it near the tube station. When I was about fifteen, my parents received a letter from my former nurse saying that she had been converted to the Salvation Army. She wanted to confess her past faults, and in particular to return the watch she had been given as a reward on this occasion. She had made up the whole story, faking the scratches. I, therefore, must have heard, as a child, the account of this story, which my parents believed, and projected it into the past in the form of a visual memory, which was a memory of a memory, but false (Piaget, 1962, pp. 187–188).

Applying Psychology

A Memory Test

Carefully read through all the words in the following list.

Bed	Drowse
Awake	Nurse
Tired	Sick
Dream	Lawyer
Wake	Medicine
Snooze	Health
Snore	Hospital
Rest	Dentist
Blanket	Physician
Doze	Patient
Slumber	Stethoscope
Nap	Curse
Peace	Clinic
Yawn	Surgeon

Now cover the list and write down all the words you remember.

Number of correctly recalled words:

21 to 28 words = excellent memory

16 to 20 words = better than most

12 to 15 words = average

8 to 11 words = below average

7 or fewer words = you might need a nap

How did you do? Do you have a good or excellent memory? Did you recall seeing the words *sleep* and *doctor?* Look back over the list. These words are not there. However, over 65 percent of students commonly report seeing these words. Why? As mentioned in the introduction to this chapter, memory is not a faithful duplicate of an event; it is a *constructive process.* We actively shape and build on information as it is encoded and retrieved.

CRITICAL THINKING

Here are two interesting questions:

1. **Did you remember the words *bed* and *surgeon*? If so, why?**
2. **Can you see how this illustrates the serial position effect?**

This is a self-reported childhood memory of Jean Piaget, a brilliant and world-famous cognitive and developmental psychologist. Why did Piaget create such a strange and elaborate memory for something that never happened?

There are several reasons why we shape, rearrange, and distort our memories. One of the most common is our need for *logic* and *consistency*. When we're initially forming new memories or sorting through the old ones, we fill in missing pieces, make "corrections," and rearrange information to make it logical and consistent with our previous experiences. If Piaget's beloved nurse said someone attempted to kidnap him, it was only logical for the boy to "remember" the event.

We also shape and construct our memories for the sake of *efficiency*. We summarize, augment, and tie new information in with related memories in LTM. Similarly, when we need to retrieve the stored information, we leave out seemingly unimportant elements or misremember the details.

Despite all their problems and biases, our memories are normally quite accurate and serve us well in most situations. Our memories have evolved to encode, store, and retrieve vital information. Even while sleeping we process and store important memories. However, when faced with tasks like remembering precise details in a scholarly text, the faces and names of potential clients, or where we left our house keys, our brains are simply not as well equipped.

MEMORY AND EYEWITNESS TESTIMONY

When our memory errors involve the criminal justice system, they may lead to wrongful judgments of guilt or innocence and even life-or-death decisions.

In the past, one of the best forms of trial evidence a lawyer could have was an *eyewitness*—"I was there; I saw it with my own eyes." Unfortunately for lawyers, research has identified several problems with eyewitness testimony (Loftus, 2000, 2001, 2007; Ran, 2007; Rubinstein, 2008; Sharps et al., 2007; Yarmey, 2004). In fact, researchers have demonstrated that it is relatively easy to create false memories (Allan & Gabbert, 2008; Howes, 2007; Loftus & Cahill, 2007; Pérez-Mata & Diges, 2007).

Do you recall our earlier discussion of the misinformation effect and how the experimenters created a false memory of a red barn (page 188)? Participants were first asked to watch a film of a car driving through the countryside. Later, those who were asked to estimate how fast the car was going when it passed the barn (actually nonexistent) were six times as likely to later report that they had seen a barn in the film than participants who hadn't been asked about a barn (Loftus, 1982).

Problems with eyewitness recollections are so well established and important that judges now allow expert testimony on the unreliability of eyewitness testimony and routinely instruct jurors on its limits (Benton et al., 2007; Rubinstein, 2008). If you serve as a member of a jury or listen to accounts of crimes in the news, remind yourself of these problems. Also, keep in mind that research participants in eyewitness studies generally report their inaccurate memories with great self-assurance and strong conviction (Migueles & Garcia-Bajos, 1999). Eyewitnesses to an actual crime may similarly identify—with equally high confidence—an innocent person as the perpetrator (FIGURE 7.12).

Like eyewitness testimony, false memories can have serious legal and social implications. Do you recall the opening story of Elizabeth, who suddenly remembered finding her mother's drowned body decades after it had happened? Elizabeth's recovery of these gruesome childhood memories, although painful, initially brought great relief. It also seemed to explain why she had always been fascinated by the topic of memory.

How often are eyewitnesses mistaken? FIGURE 7.12

In one experiment, participants watched people committing a staged crime. Only an hour later, 20 percent of the eyewitnesses identified innocent people from mug shots, and a week later, 8 percent identified innocent people in a lineup (Brown, Deffenbacher, & Sturgill, 1977). What memory processes might have contributed to the eyewitnesses' errors?

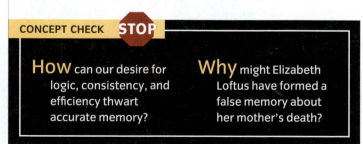

Elizabeth Loftus FIGURE 7.13

Then, her brother called to say there had been a mistake! The relative who told Elizabeth that she had been the one to discover her mother's body later remembered—and other relatives confirmed—that it had actually been Aunt Pearl, not Elizabeth Loftus. Loftus, an expert on memory distortions, had unknowingly created her own *false memory.* (FIGURE 7.13)

REPRESSED MEMORIES

Creating false memories may be somewhat common, but can we recover true memories that are buried in childhood? **Repression** is the supposed unconscious coping mechanism by which we prevent anxiety-provoking thoughts from reaching consciousness. According to some research, repressed memories are *actively* and *consciously* "forgotten" in an effort to avoid the pain of their retrieval (Anderson et al., 2004). Others suggest that some memories are so painful that they exist only in an *unconscious* corner of the brain, making them inaccessible to the individual (Karon & Widener, 1998). In these cases, therapy would be necessary to unlock the hidden memories (Davies, 1996).

This is a complex and controversial topic in psychology. No one doubts that some memories are forgotten and later recovered. What is questioned is the concept of *repressed memories* of painful experiences (especially childhood sexual abuse) and their storage in the unconscious mind (Goodman et al., 2003; Kihlstrom, 2004; Loftus & Cahill, 2007).

Critics suggest that most people who have witnessed or experienced a violent crime, or are adult survivors of childhood sexual abuse, have intense, persistent memories. They have trouble *forgetting,* not remembering. Some critics also wonder whether therapists sometimes inadvertently create false memories in their clients during therapy. Some worry that if a clinician even suggests the possibility of abuse, the client's own *constructive processes* may lead him or her to create a false memory. The client might start to incorporate portrayals of abuse from movies and books into his or her own memory, forgetting their original sources and eventually coming to see them as reliable.

This is not to say that all psychotherapy clients who recover memories of sexual abuse (or other painful incidents) have invented those memories. Indeed, the repressed memory debate has grown increasingly bitter, and research on both sides is hotly contested. The stakes are high because some lawsuits and criminal prosecutions of sexual abuse are sometimes based on recovered memories of childhood sexual abuse. As researchers continue exploring the mechanisms underlying delayed remembering, we must be careful not to ridicule or condemn people who recover true memories of abuse. In the same spirit, we must protect the innocent from wrongful accusations that come from false memories. We look forward to a time when we can justly balance the interests of the victim with those of the accused.

CONCEPT CHECK STOP

How can our desire for logic, consistency, and efficiency thwart accurate memory?

Why might Elizabeth Loftus have formed a false memory about her mother's death?

Applying Psychology

Mnemonic Devices

As you review the key points from this chapter, think about how you might exploit basic principles of memory, using them to your own advantage. One additional "trick" for giving your memory a boost is to use **mnemonic devices** to encode items in a special way. (But be warned—you may get more "bang for your buck" using the well-researched principles discussed throughout this chapter.)

Three popular mnemonic techniques are the following.

A Method of loci Greek and Roman orators developed the *method of loci* to keep track of the many parts of their long speeches. Orators would imagine the parts of their speeches attached to places in a courtyard. For example, if an opening point in a speech was the concept of *justice,* they might visualize a courtroom placed in the first corner of their garden. As they mentally walked around their garden during their speech, they would encounter, in order, each of the points to be made. ▶

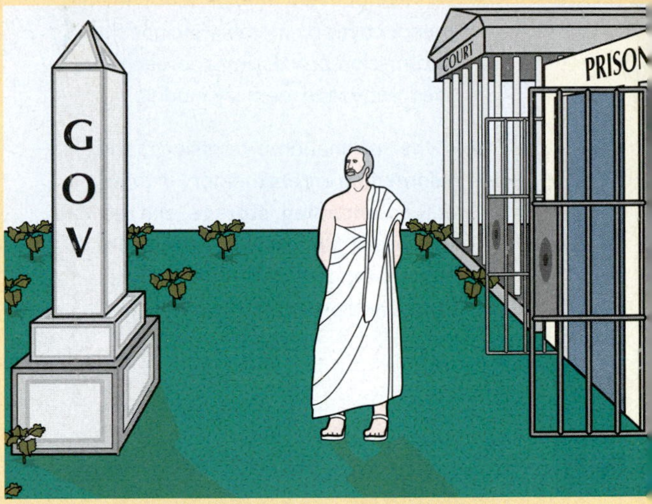

Grocery list

One is a bun.

Two is a shoe.

Three is a tree.

◀ **B Peg words** To use the *peg-word* mnemonic, you first need to memorize a set of 10 images that you can use as "pegs" on which to hang ideas. For example, if you learn 10 items that rhyme with the numbers they stand for, you can then use the images as pegs to hold the items of any list. Try it with items you might want to buy on your next trip to the grocery store: milk, eggs, and bread.

C Acronyms To use the acronym method, create a new code word from the first letters of the items you want to remember. For example, to recall the names of the Great Lakes, think of *HOMES* on a *great lake* (*H*uron, *O*ntario, *M*ichigan, *E*rie, *S*uperior). Visualizing homes on each lake also helps you remember your code word "homes." ▶

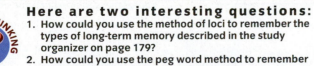

CRITICAL THINKING

Here are two interesting questions:

1. How could you use the method of loci to remember the types of long-term memory described in the study organizer on page 179?
2. How could you use the peg word method to remember the first six words in the list on page 190?

SUMMARY

1 The Nature of Memory

1. **Memory** is an internal representation of some prior event or experience. Major perspectives on memory include the information-processing approach and the three-stage memory model.

2. In the information-processing model, information enters memory in three stages: **encoding, storage**, and **retrieval.** In contrast, the three-stage memory model proposes that information is stored and processed in **sensory memory**, **short-term memory (STM)**, and **long-term memory (LTM)**; these differ in purpose, duration, and capacity.

3. **Chunking** and **maintenance rehearsal** improve STM's duration and capacity. Researchers think of STM as a three-part working memory.

4. LTM is an almost unlimited storehouse for information that must be kept for long periods. The two major types of LTM are **explicit/declarative memory** and **implicit/nondeclarative memory**. Organization and **elaborative rehearsal** improve encoding for LTM. **Retrieval cues** help stimulate retrieval of information from LTM. According to the **encoding specificity principle**, retrieval is improved when conditions of recovery are similar to encoding conditions.

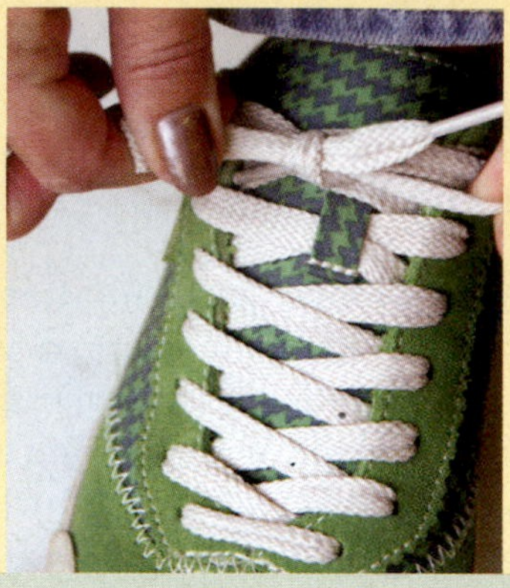

2 Biological Bases of Memory

1. Learning modifies the brain's neural networks through **long-term potentiation (LTP)**, strengthening particular synapses and affecting neurons' ability to release their neurotransmitters.

2. Stress hormones affect the amygdala, which stimulates brain areas that are important for memory storage. Heightened arousal increases the encoding and storage of new information. Secretion of fight-or-flight hormones can contribute to "flashbulb" memories.

3. Research using advanced techniques has indicated that several brain regions are involved in memory storage.

4. Traumatic brain injuries and disease, such as **Alzheimer's disease (AD)**, can cause memory loss. Two major types of amnesia are retrograde and anterograde amnesia.

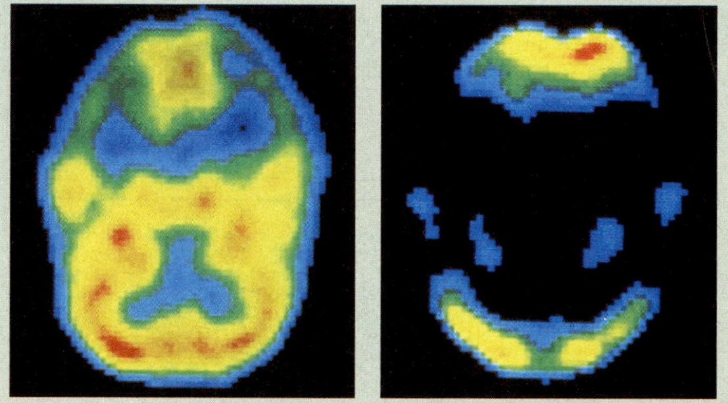

3 Forgetting

1. Researchers have proposed that we forget information through decay, retroactive and proactive interference, motivated forgetting, encoding failure, and retrieval failure.

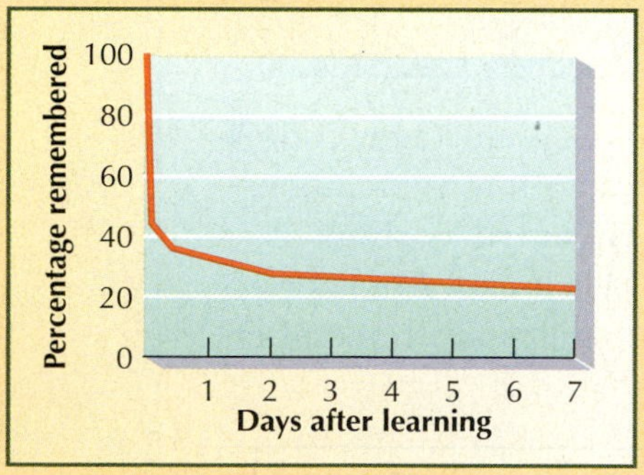

2. Early research by Ebbinghaus showed that we tend to forget newly learned information quickly but that we relearn the information more readily the second time.

3. Six factors that contribute to forgetting are the misinformation effect, the serial position effect (primacy and recency effects), source amnesia, the sleeper effect, spacing of practice (distributed versus massed practice), and culture.

4 Memory Distortions

1. People shape, rearrange, and distort memories in order to create logic, consistency, and efficiency. Despite all their problems and biases, our memories are normally accurate and usually serve us well.

2. When memory errors involve the criminal justice system, they can have serious legal and social consequences. Problems with eyewitness recollections are well established. Judges often allow expert testimony on the unreliability of eyewitnesses.

3. Memory **repression** (especially of childhood sexual abuse) is a complex and controversial topic. Critics note that most people who have witnessed or experienced a violent or traumatic event have intense, persistent memories. Critics worry that if a clinician suggests the possibility of abuse, the client's *constructive processes* may lead him or her to create a false memory of being abused. Researchers continue to explore delayed remembering.

4. **Mnemonic devices** help us remember lists and facts. Three popular mnemonic techniques are method of loci, the peg-word mnemonic, and acronyms.

KEY TERMS

- **memory** p. 174
- **information-processing model** p. 174
- **encoding** p. 175
- **storage** p. 175
- **retrieval** p. 175
- **traditional three-stage memory model** p. 175
- **sensory memory** p. 175
- **short-term memory (STM)** p. 177
- **chunking** p. 177
- **maintenance rehearsal** p. 177
- **working memory** p. 179
- **long-term memory (LTM)** p. 179
- **explicit/declarative memory** p. 179
- **implicit/nondeclarative memory** p. 180
- **elaborative rehearsal** p. 180
- **retrieval cue** p. 180
- **encoding specificity principle** p. 182
- **long-term potentiation (LTP)** p. 182
- **retrograde amnesia** p. 184
- **anterograde amnesia** p. 184
- **Alzheimer's disease (AD)** p. 184
- **retroactive interference** p. 187
- **proactive interference** p. 187
- **tip-of-the-tongue phenomenon** p. 187
- **misinformation effect** p. 188
- **serial position effect** p. 188
- **source amnesia** p. 188
- **sleeper effect** p. 189
- **distributed practice** p. 189
- **massed practice** p. 189
- **repression** p. 192
- **mnemonic devices** p. 193

CRITICAL AND CREATIVE THINKING QUESTIONS

1. If you were forced to lose one type of memory—sensory, short-term, or long-term—which would you select? Why?
2. Why might students do better on a test if they take it in the same seat and classroom where they originally studied the material?
3. What might be the evolutionary benefit of heightened (but not excessive) arousal enhancing memory?
4. Why might advertisers of shoddy services or products benefit from "channel surfing" if the television viewer is skipping from news programs to cable talk shows to infomercials?
5. As an eyewitness to a crime, how could you use information in this chapter to improve your memory for specific details? If you were a juror, what would you say to the other jurors about the reliability of eyewitness testimony?

(See the following page for *What is happening in this picture?*)

What is happening in this picture?

- What phenomenon accounts for why many people feel that they can remember the September 11, 2001, terrorist attacks (or the explosion of the space shuttle *Challenger*, or the assassination of President John F. Kennedy) with perfect clarity?
- What biological process accounts for this kind of intense memory?
- Are memories like this impervious to distortion?

SELF-TEST

(Check your answers in Appendix B.)

1. In a computer model of memory, (a) ______________ would happen at the keyboard, (b) ______________ on the monitor, and (c) ______________ on the hard drive.

2. Label the three-stage memory model in the correct sequence on the figure below:

   a. ______________ b. ______________ c. ______________

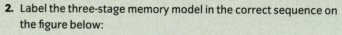

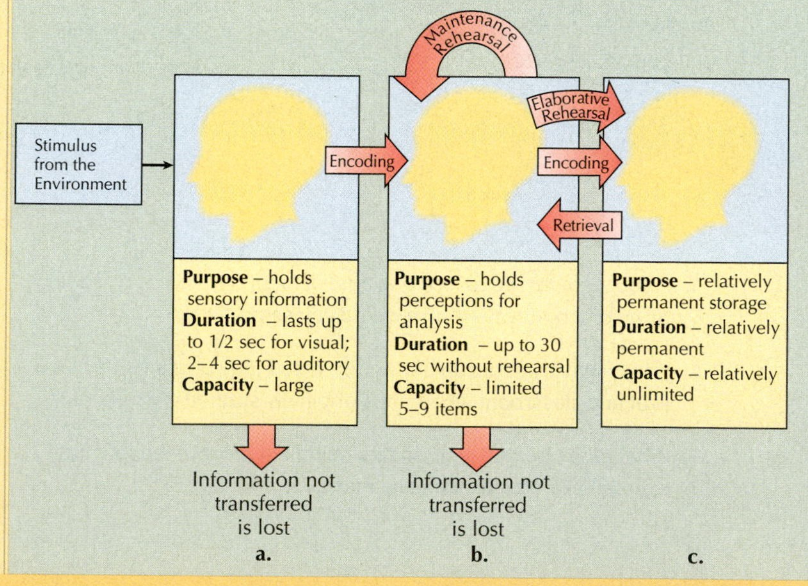

3. The following descriptions are characteristic of ________________: information lasts only a few seconds or less, and a relatively large (but not unlimited) storage capacity.
   a. perceptual processes
   b. short-term storage
   c. working memory
   d. sensory memory

4. ________________ is the process of grouping separate pieces of information into a single unit.
   a. Chunking
   b Cheating
   c. Collecting
   d. Dual-coding

5. Label the two major systems of long-term memory (LTM) on the figure below.

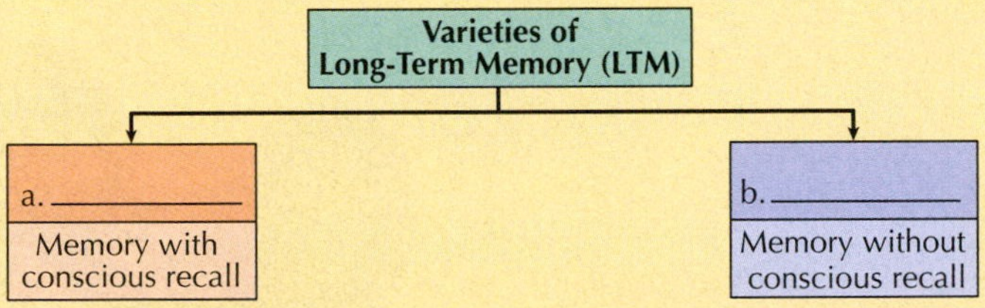

6. In answering this question, the correct multiple-choice option may serve as a ________________ for recalling accurate information from your long-term memory.
   a. specificity code
   b. priming pump
   c. retrieval cue
   d. flashbulb stimulus

7. The encoding specificity principle says that information retrieval is improved when ________________.
   a. both maintenance and elaborative rehearsal are used
   b. reverberating circuits consolidate information
   c. conditions of recovery are similar to encoding conditions
   d. long-term potentiation is accessed

8. The long-lasting increase in neural excitability believed to be a biological mechanism for learning and memory is called ________________.
   a. maintenance rehearsal
   b. adrenaline activation
   c. long-term potentiation
   d. the reverberating response

9. A progressive mental deterioration characterized by severe memory loss that occurs most commonly in the elderly is called ________________.
   a. retrieval loss
   b. prefrontal cortex deterioration
   c. Alzheimer's disease
   d. age-related amnesia

10. Label the two types of amnesia on the figure below:

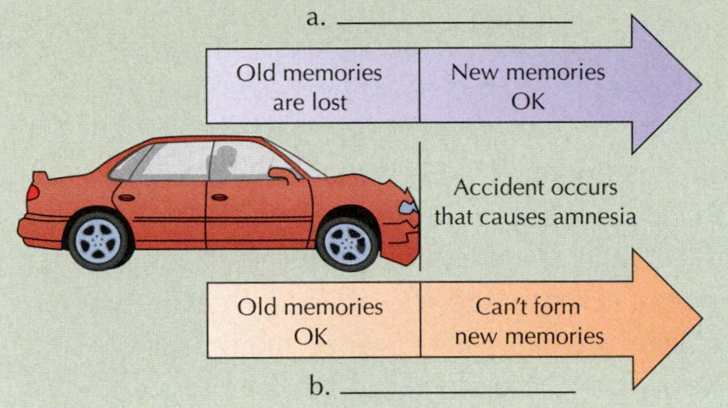

11. List the 5 major theories of forgetting:
   a. ________________
   b. ________________
   c. ________________
   d. ________________
   e. ________________

12. Distributed practice is a learning technique in which ________________.
   a. subjects are distributed across equal study sessions
   b. learning periods alternate with nonlearning rest periods
   c. learning decays faster than it can be distributed
   d. several students study together, distributing various subjects according to their individual strengths

13. Researchers have demonstrated that it is ________________ to create false memories.
   a. relatively easy
   b. moderately difficult
   c. rarely possible
   d. never possible

14. ________________ memories are related to anxiety-provoking thoughts or events that are prevented from reaching consciousness.
   a. Suppressed
   b. Flashback
   c. Motivated
   d. Repressed

15. ________________ devices improve memory by encoding items in a special way.
   a. Eidetic imagery
   b. Mnemonic
   c. Reverberating circuit
   d. ECS

Thinking, Language, and Intelligence 8

On July 9, 2005, professional skateboarder Danny Way rocketed down a 120-foot ramp at almost 50 miles per hour and leapt a 61-foot gap across the Great Wall of China. He did it not once but five times. Do you think that this feat reflects merely athletic skill and daring, or does it also speak to Way's intelligence?

When we think of intelligence, many of us think of Nobel Prize winners, great inventors, or chess champions. But success as a professional skateboarder also requires intelligence—perhaps of a different kind than people generally associate with being "smart." Skateboarding enthusiasts admire the creativity and technical innovation that Way brings to the sport. He has broken many skateboarding records and has devised stunts—including a jump from a helicopter—that few others would try.

The three topics of this chapter—thinking, language, and intelligence—are often studied together under the broader umbrella of cognition, the mental activities involved in acquiring, storing, retrieving, and using knowledge. In a real sense, we discuss cognition throughout this text because psychology is "the scientific study of behavior and *mental processes*" (Chapter 1). For example, the chapters on sensation and perception (Chapter 4), consciousness (Chapter 5), learning (Chapter 6), and memory (Chapter 7) all focus on cognition. In this chapter, we emphasize thinking, language, and intelligence. As you will discover, each is a complex phenomenon that is greatly affected by numerous factors.

NATIONAL GEOGRAPHIC

CHAPTER OUTLINE

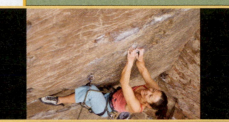

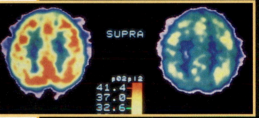

Thinking

LEARNING OBJECTIVES

Explain the role of the prefrontal cortex in thinking.

Describe how mental images, concepts, and prototypes enable thinking.

Identify the three steps involved in problem solving.

Summarize how biases and heuristics can hamper problem solving.

Explain Sternberg and Lubart's investment theory of creativity.

Thinking, language, and intelligence are closely related facets of **cognition**. Every time you take information and mentally act on it, you are thinking. Our thought processes are distributed throughout our brains in networks of neurons. However, they are also localized. For example, during problem solving or decision making, our brains are active in the **prefrontal cortex**. This region associates complex ideas; makes plans; forms, initiates, and allocates attention; and supports multitasking. The prefrontal cortex also links to other areas of the brain, such as the limbic system (Chapter 2), to synthesize information from several senses (Carlson, 2008; Heyder, Suchan, & Daum, 2004; Sachetti, Sacco, & Strata, 2007).

cognition Mental activities involved in acquiring, storing, retrieving, and using knowledge.

COGNITIVE BUILDING BLOCKS

Imagine yourself lying relaxed in the warm, gritty sand on an ocean beach. Do you see palms swaying in the wind? Can you smell the salty sea and taste the dried salt on your lips? Can you hear children playing in the surf? What you've just created is a **mental image** (FIGURE 8.1), a mental representation of a previously stored sensory experience, which includes visual, auditory, olfactory, tactile, motor, and gustatory imagery (McKellar, 1972). We all have a mental space where we visualize and manipulate our sensory images (Hamm, Johnson, & Corballis, 2004).

In addition to mental images, our thinking also involves forming **concepts**, or mental representations of a group or category (Smith, 1995). Concepts can be concrete (*car, concert*) or abstract (*intelligence, pornography*).

Mental imagery FIGURE 8.1

Some of our most creative and inspired moments come when we're forming and manipulating mental images. This mountain climber is probably visualizing her next move, and her ability to do so is critical to her success.

Some birds are "birdier" than others FIGURE 8.2

A Most people have a *prototype* bird that captures the essence of "birdness" and allows us to quickly classify flying animals correctly.

B When we encounter an example that doesn't quite fit our prototype, we need time to review our artificial concept. Because the penguin doesn't fly, it's harder to classify than a robin.

They are essential to thinking and communication because they simplify and organize information. Normally, when you see a new object or encounter a new situation, you relate it to your existing conceptual structure and categorize it according to where it fits. For example, if you see a metal box with four wheels driving on the highway, you know it is a car, even if you've never seen that particular model before.

We create some of our concepts from logical rules or definitions. Some concepts, such as *triangle*, are called "artificial" (or "formal") because the rules for inclusion are sharply defined.

Artificial concepts are often found in science and academe, but in real life, we seldom use precise, artificial definitions. When we see birds in the sky, we don't think *warm-blooded animals that fly, have wings, and lay eggs*—an artificial concept. Instead, we use natural concepts, or **prototypes**, which are based on a typical representative of that concept (Rosch, 1973) (FIGURE 8.2).

Some of our concepts also develop when we create **hierarchies**, or group specific concepts as subcategories within broader concepts (FIGURE 8.3). This process makes mastering new material faster and easier. For example, when you learn that all animals have mitochondria in their cells, you don't need to relearn that fact every time you learn a new animal species.

When we first learn something, we rely primarily on basic-level concepts (Rosch, 1978). Thus, children tend to learn *bird* before they learn the higher-order concept *animal* or the lower-order concept *parakeet*. Even as adults, when shown a picture of a parakeet, we classify it as a *bird* first.

A concept hierarchy FIGURE 8.3

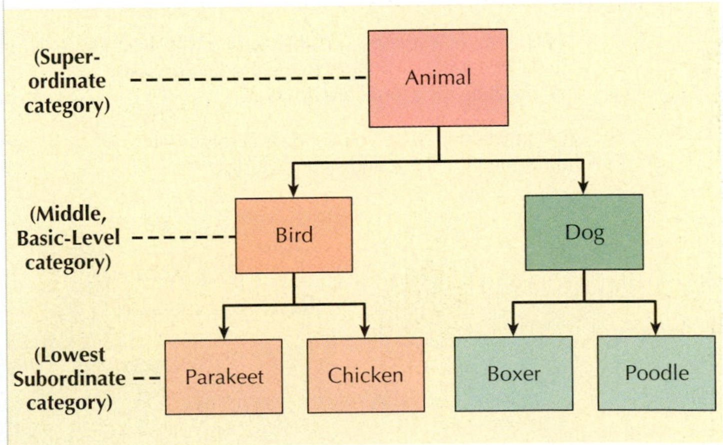

Process Diagram

Three steps to the goal FIGURE 8.4

1 PREPARATION

Imagine that you are determined to find a new home close to work. There are at least three separate components to successful preparation.

1. *Identifying given facts.* Decide what are your most basic, nonnegotiable limits and desires.

Must be able to walk to work.
I prefer a house to a large apartment building.

Must allow cats.

2. *Separating relevant from irrelevant facts.* What are your negotiable items? What do you consider irrelevant and easily compromised?

A fireplace would be a plus.

3. *Defining the ultimate goal.*

2 PRODUCTION

During the *production step*, the problem solver produces possible solutions, called *hypotheses*. Two major approaches to generating hypotheses are *algorithms* and *heuristics*.

Algorithm: Answers every ad.

Houses For Rent

An **algorithm** is a logical, step-by-step procedure (well suited for computers) that will always produce the solution. (For example, an algorithm for solving the problem 2×4 is $2 + 2 + 2 + 2$.) For complex problems, algorithms may take a long time.

A **heuristic** is a simple rule or shortcut that does not guarantee a solution. Heuristics include working backward from the solution (a known condition) and *creating subgoals*, stepping-stones to the original goal.

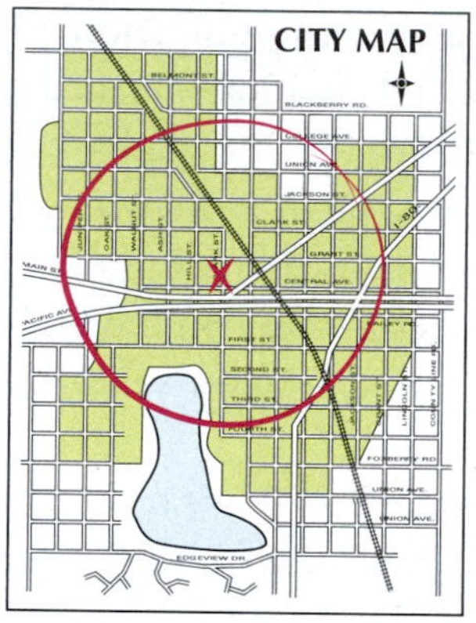

CITY MAP

Heuristic: Works backward—start by drawing a 1-mile radius around work to narrow search.

algorithm A set of steps that, if followed correctly, will eventually solve the problem.

heuristic A simple rule used in problem solving and decision making that does not guarantee a solution but offers a likely shortcut to it.

3 EVALUATION

1. Did your hypotheses (possible solutions) solve the problem? If not, then you must return to the production stage and produce more possible solutions.

2. Take action. Once you've identified the right home, sign the lease and start packing.

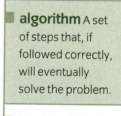

π - Move

SOLVING PROBLEMS

Several years ago in Los Angeles, a 12-foot-high tractor-trailer got stuck under a bridge that was 6 inches too low. After hours of towing, tugging, and pushing, the police and transportation workers were stumped. Then a young boy happened by and asked, "Why don't you let some air out of the tires?" It was a simple, creative suggestion—and it worked.

Our lives are filled with problems, some simple, some difficult. In all cases, **problem solving** requires moving from a given state (the problem) to a goal state (the solution), a process that usually involves three steps (Bourne, Dominowski, & Loftus, 1979) (FIGURE 8.4): *preparation, production,* and *evaluation.*

"Repurposing" as art FIGURE 8.6

Some people look at a pile of garbage and see only trash. Others see possibility. This photo shows how an artist recycled old hubcaps. Have you ever "repurposed" a discarded object such as an old wheelbarrow for use as a garden planter? If so, you have overcome *functional fixedness*.

BARRIERS TO PROBLEM SOLVING

We all encounter barriers to solving problems. We stick to problem-solving strategies (**mental sets**) that worked in the past, rather than trying new, possibly more effective ones (FIGURE 8.5). Or we fail to let our inventive instincts run free, thinking of objects as functioning only in their prescribed, customary way—a phenomenon called **functional fixedness** (FIGURE 8.6).

Other barriers to effective problem solving stem from our tendency to ignore important information. Have you ever caught yourself agreeing with friends who support your political opinions, and discounting conflicting opinions? This inclination to seek confirmation for our preexisting beliefs and to overlook contradictory evidence is known as the **confirmation bias** (Jonas et al., 2008; Kerschreiter et al., 2008; Nickerson, 1998; Reich, 2004).

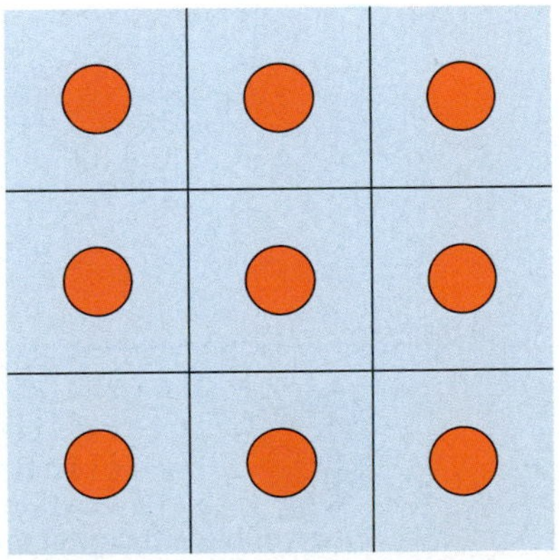

The nine-dot problem FIGURE 8.5

Draw no more than four lines that run through all nine dots on this page without lifting your pencil from the paper (see page 220 for the solution).

British researcher Peter Wason (1968) first demonstrated the confirmation bias. He asked participants to generate a list of numbers that conformed to the same rule that applied to this set of numbers:

2 4 6

Hypothesizing that the rule was "numbers increasing by two," most participants generated sets such as (4, 6, 8) or (1, 3, 5). Each time, Wason assured them that their sets of numbers conformed to the rule but that the rule "numbers increasing by two" was incorrect. The problem was that the participants were searching only for information that confirmed their hypothesis. Proposing a series such as (1, 3, 4) would have led them to reject their initial hypothesis and discover the correct rule: "numbers in increasing order of magnitude."

Cognitive psychologists Amos Tversky and Daniel Kahneman found that heuristics, as handy as they can be, can lead us to ignore relevant information (Kahneman, 2003; Tversky & Kahneman, 1974, 1993). When we use the **availability heuristic**, we judge the likelihood of an event based on how easily recalled (available) other instances of the event are (Buontempo & Brockner, 2008; Caruso, 2008; Oppenheimer, 2004). Shortly after the September 11, 2001, terrorist attacks, one study found that the average American believed that he or she had a 20.5 percent chance of being hurt in a terrorist attack within a year (Lerner et al., 2003). Can you see how intense media coverage of the attacks created this erroneously high perception of risk?

Tversky and Kahneman also demonstrated that the **representativeness heuristic** sometimes hinders problem solving. Using this heuristic, we estimate the probability of something based on how well the circumstances match (represent) our prototype (Fisk, Burg, & Holden, 2006; Greene & Ellis, 2008). For example, if John is six feet, five inches tall, we may guess that he is an NBA basketball player rather than, say, a bank president. But in this case, the representative heuristic ignores *base rate information*—the probability of a characteristic occurring in the general population. In the United States, bank presidents outnumber NBA players by about 50 to 1. So despite his height, John is much more likely to be a bank president.

CREATIVITY: FINDING UNIQUE SOLUTIONS

What makes a person creative? Conceptions of creativity are subject to cultural relevance and to whether a solution or performance is considered useful at the time. In general, three characteristics are associated with **creativity**: originality, fluency, and flexibility. Thomas Edison's invention of the light bulb offers a prime example of each of these characteristics (TABLE 8.1).

> **creativity** The ability to produce valued outcomes in a novel way.

Most tests of creativity focus on **divergent thinking**, a type of thinking where many possibilities are developed from a single starting point (Baer, 1994). For example, in the Unusual Uses Test, people are asked to think of as many uses as possible for an object, such as a brick. In the Anagrams Test, people are asked to reorder the letters in a word to make as many new words as possible.

A classic example of divergent thinking is the decision of Xiang Yu, a Chinese general in the third century

Applying Psychology

Are You Creative?

Everyone exhibits a certain amount of creativity in some aspects of life. Even when doing ordinary tasks, like planning an afternoon of errands, you are being somewhat creative. Similarly, if you've ever tightened a screw with a penny or used a telephone book on a chair as a booster seat for a child, you've found creative solutions to problems.

Would you like to test your own creativity?

- Find 10 coins and arrange them in the configuration shown here. By moving only two coins, form two rows that each contain 6 coins (see page 220 for the solution).
- In five minutes, see how many words you can make using the letters in the word *hippopotamus.*
- In five minutes, list all the things you can do with a paper clip.

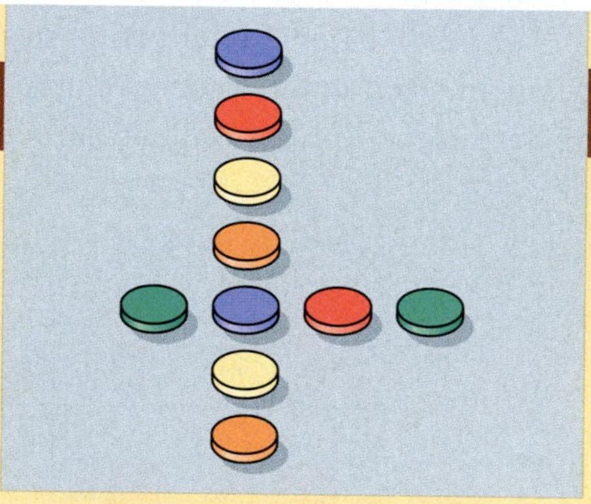

CRITICAL THINKING **Here are two interesting questions:**

1. How did you do? If you had trouble with any of these tasks, can you use the list of *resources of creative people* on page 205 to identify the resources you lack? Is there anything you could do to increase your share of these resources?
2. Creativity is usually associated with art, poetry, and the like. Can you think of other areas in which creativity is highly valued?

Three elements of creative thinking TABLE 8.1

	Explanations	Thomas Edison Examples
Originality	Seeing unique or different solutions to a problem	After noting that electricity passing through a conductor produces a glowing red or white heat, Edison imagined using this light for practical uses.
Fluency	Generating a large number of possible solutions	Edison tried literally hundreds of different materials to find one that would heat to the point of glowing white heat without burning up.
Flexibility	Shifting with ease from one type of problem-solving strategy to another	When he couldn't find a long-lasting material, Edison tried heating it in a vacuum—thereby creating the first light bulb.

B.C., to crush his troops' cooking pots and burn their ships. One would think that no general in his right mind might make such a decision, but Xiang Yu explained that his purpose was to focus the troops on moving forward, as they had no hope of retreating. His divergent thinking was rewarded with victory on the battlefield.

One prominent theory of creativity is Robert J. Sternberg and Todd Lubart's **investment theory** (1992, 1996). According to this theory, creative people tend to "buy low" in the realm of ideas, championing ideas that others dismiss (much like a bold entrepreneur might invest in low-priced, unpopular stocks, believing that their value will rise). Once their creative ideas are highly valued, they "sell high" and move on to another unpopular but promising idea. Investment theory also suggests that creativity requires the coming together of six interrelated resources (Kaufman, 2002; Sternberg & Lubart, 1996). These resources are summarized in TABLE 8.2. One way to improve your personal creativity is to study this list and then strengthen yourself in those areas you think need improvement.

Resources of creative people TABLE 8.2

Intellectual Ability	Enough intelligence to see problems in a new light
Knowledge	Sufficient basic knowledge of the problem to effectively evaluate possible solutions
Thinking Style	Novel ideas and ability to distinguish between the worthy and worthless
Personality	Willingness to grow and change, take risks, and work to overcome obstacles
Motivation	Sufficient motivation to accomplish the task and more internal than external motivation
Environment	An environment that supports creativity

Which resources best explain Robin Williams's phenomenal success?

CONCEPT CHECK STOP

What is the difference between artificial and natural concepts?

Why are heuristics sometimes more appropriate for problem solving than algorithms? When might the reverse be true?

What is an example of the availability heuristic?

NATIONAL GEOGRAPHIC

Language

LEARNING OBJECTIVES

Identify the building blocks of language.

Describe the prominent theories of how language and thought interact.

Describe the major stages of language development.

Review the evidence that nonhuman animals are able to learn and use language.

Language enables us to mentally manipulate symbols, thereby expanding our thinking, and to communicate our thoughts, ideas, and feelings. To produce language, we first build words using **phonemes** and **morphemes**. Then we string words into sentences using rules of **grammar** (syntax and semantics) (FIGURE 8.7).

language A form of communication using sounds and symbols combined according to specified rules.

phoneme [FO-neem] The smallest basic unit of speech or sound. The English language has about 40 phonemes.

LANGUAGE AND THOUGHT: A COMPLEX INTERACTION

Does the fact that you speak English instead of Spanish—or Chinese instead of Swahili—determine how you reason, think, and perceive the world? Linguist Benjamin Whorf (1956) believed so. As evidence for his **linguistic relativity hypothesis**,

morpheme [MOR-feem] The smallest meaningful unit of language, formed from a combination of phonemes.

grammar Rules that specify how phonemes, morphemes, words, and phrases should be combined to express thoughts. These rules include syntax and semantics.

Building blocks of language FIGURE 8.7

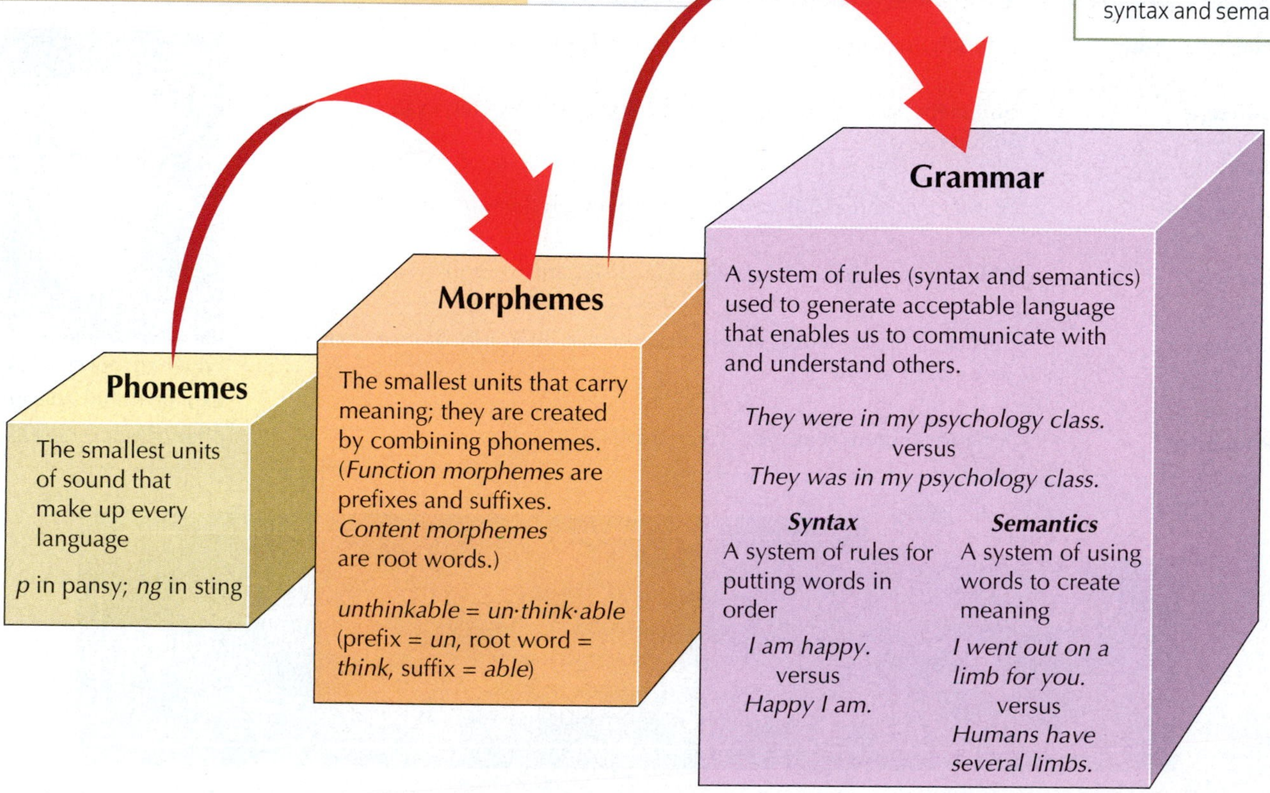

Phonemes
The smallest units of sound that make up every language
p in pansy; ng in sting
Morphemes
The smallest units that carry meaning; they are created by combining phonemes. (Function morphemes are prefixes and suffixes. Content morphemes are root words.)
unthinkable = un·think·able (prefix = un, root word = think, suffix = able)
Grammar
A system of rules (syntax and semantics) used to generate acceptable language that enables us to communicate with and understand others.
They were in my psychology class.
versus
They was in my psychology class.
Syntax
A system of rules for putting words in order
I am happy.
versus
Happy I am.
Semantics
A system of using words to create meaning
I went out on a limb for you.
versus
Humans have several limbs.

Early nonverbal communication FIGURE 8.8

Even before they begin to communicate verbally, infants use nonverbal signals to "speak" to others. Infants as young as 2½ months express basic emotions such as interest, joy, anger, and sadness. Children who are born blind and deaf exhibit the same emotional expressions as sighted and hearing children, supporting the contention that these expressions are universal and innate. What emotions are each of these infants expressing?

Whorf offered a now classic example: Because Inuits (previously known as Eskimos) supposedly have many words for snow (*apikak* for "first snow falling," *pukak* for "snow for drinking water," and so on), they can perceive and think about snow differently from English speakers, who have only one word—*snow*.

Though intriguing, Whorf's hypothesis has not fared well. He apparently exaggerated the number of Inuit words for snow (Pullum, 1991) and ignored the fact that English speakers have a number of terms to describe various forms of snow, such as *slush, sleet, hard pack,* and *powder*. Other research has directly contradicted Whorf's theory. For example, Eleanor Rosch (1973) found that although people of the Dani tribe in New Guinea possess only two color names—one indicating cool, dark colors, and the other describing warm, bright colors—they discriminate among multiple hues as well as English speakers do.

Whorf apparently was mistaken in his belief that language determines thought. But there is no doubt that language *influences* thought (Hoff, 2009). People who speak both Chinese and English report that the language they're currently using affects their sense of self (Matsumoto & Juang, 2008). When using Chinese, they tend to conform to Chinese cultural norms; when speaking English, they tend to adopt Western norms.

Our words also influence the thinking of those who hear them. That's why companies avoid *firing* employees; instead, employees are *outplaced* or *nonrenewed*. Similarly, the military uses terms like *preemptive strike* to cover the fact that they attacked first and *tactical redeployment* to refer to a retreat. And research has shown that consumers who receive a "rebate" are less likely to spend the money than those who receive a "bonus" (Epley, 2008).

LANGUAGE DEVELOPMENT: FROM CRYING TO TALKING

From birth, a child communicates through facial expressions, eye contact, and body gestures (FIGURE 8.8). Babies only hours old begin to "teach" their caregivers when and how they want to be held, fed, and played with.

Eventually, children also communicate verbally, progressing through several distinct stages of language acquisition. These stages are summarized in STUDY ORGANIZER 8.1. By age 5, most children have mastered basic grammar and typically use about 2,000 words (a level of mastery considered adequate for getting by in any given culture). Past this point, vocabulary and grammar gradually improve throughout life.

Psychological Science

Baby Signs

Many babies and toddlers—beginning as young as 9 months of age—learn to communicate by using a modified form of sign language, sometimes called "Baby Sign." Symbolic gestures for basic ideas such as "more," "milk," or "love" enhance parents' and caregivers' interactions with children who cannot yet talk. Many parents find that signing with their infant or toddler gives them a fascinating "window" into the baby's mind—and eliminates a lot of frustration for both them and baby! Some researchers believe that teaching babies to sign helps foster better language comprehension and can speed up the process of learning to talk.

CRITICAL THINKING

Here are two interesting questions:
1. **Would you teach your own child "baby signs"? Why or why not?**
2. **How could signing be combined with a child's early attempts to talk to enhance language development?**

Some theorists believe that language capability is innate, primarily a matter of maturation. Noam Chomsky (1968, 1980) suggests that children are "prewired" with a neurological ability known as a **language acquisition device (LAD)** that enables a child to analyze language and to extract the basic rules of grammar. This mechanism needs only minimal exposure to adult speech to unlock its potential. As evidence for this nativist position, Chomsky observes that children everywhere progress through the same stages of language development at about the same ages. He also notes that babbling is the same in all languages and that deaf babies babble just like hearing babies.

"Nurturists" argue that the nativist position doesn't fully explain individual differences in language development. They hold that children learn language through a complex system of rewards, punishments, and imitation. For example, parents smile and encourage any vocalizations from a very young infant. Later, they respond even more enthusiastically when the infant babbles "mama" or "dada." In this way, parents unknowingly use *shaping* (Chapter 6) to help babies learn language (FIGURE 8.9).

Nature or nurture? FIGURE 8.9

Both sides of the nature-versus-nurture debate have staunch supporters. However, most psychologists believe that language acquisition is a combination of both biology (nature) and environment (nurture) (Hoff, 2009; Plomin, De Fries, & Fulker, 2007). Can you see how both might contribute to this child's pretend phone conversation?

Study Organizer 8.1 Language Acquisition

Developmental Stage	Age	Language Features	Example
Prelinguistic stage 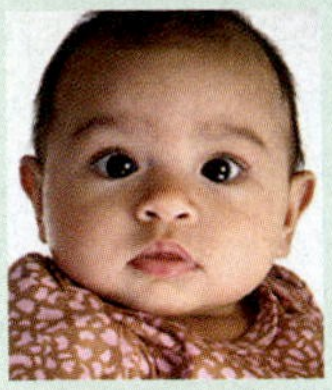	Birth to ~12 months	Crying (reflexive in newborns; soon, crying becomes more purposeful)	hunger cry anger cry pain cry
	2 to 3 months	***Cooing*** (vowel-like sounds)	"ooooh" "aaaah"
	4 to 6 months	***Babbling*** (consonants added)	"bahbahbah" "dahdahdah"
Linguistic stage 	~12 months	Babbling begins to resemble language of the child's home. Child seems to understand that sounds relate to meaning.	
		At first, speech is limited to one-word utterances.	"Mama" "juice" "Daddy" "up"
		Expressive ability more than doubles once child begins to join words into short phrases.	"Daddy, milk" "no night-night!"
	~ 2 years	Child sometimes ***overextends*** (using words to include objects that do not fit the word's meaning).	all men = "Daddy" all furry animals = doggy
	~2 years to ~5 years	Child links several words to create short but intelligible sentences. This speech is called ***telegraphic speech*** because (like telegrams) it omits nonessential connecting words.	"Me want cookie" "Grandma go bye-bye?"
		Vocabulary increases at a phenomenal rate.	
		Child acquires a wide variety of rules for grammar.	adding *-ed* for past tense adding *s* to form plurals
		Child sometimes ***overgeneralizes*** (applying the basic rules of grammar even to cases that are exceptions to the rule).	"I goed to the zoo" "Two mans"

CAN HUMANS TALK TO NONHUMAN ANIMALS?

Without question, nonhuman animals communicate, sending warnings, signaling sexual interest, sharing locations of food sources, and so on. But can nonhuman animals master the complexity of human language? Since the 1930s, many language studies have attempted to answer this question by probing the language abilities of chimpanzees and gorillas (e.g., Barner et al., 2008; Fields et al., 2007; Savage-Rumbaugh, 1990; Segerdahl, Fields, & Savage-Rumbaugh, 2006).

One of the most successful early studies was conducted by Beatrice and Allen Gardner (1969), who recognized chimpanzees' manual dexterity and ability to imitate gestures. The Gardners used American Sign Language (ASL) with a chimp named Washoe. By the time Washoe was 4 years old, she had learned 132 signs and was able to combine them into simple sentences such as "Hurry, gimme toothbrush" and "Please tickle more."

Signing FIGURE 8.10

According to her teacher, Penny Patterson, Koko has used ASL to converse with others, talk to herself, joke, express preferences, and even lie (Linden, 1993; Patterson, 2002).

The famous gorilla Koko also uses ASL to communicate; she reportedly uses more than 1,000 words (**FIGURE 8.10**).

In another well-known study, a chimp named Lana learned to use symbols on a computer to get things she wanted, such as food, a drink, a tickle from her trainers, and having her curtains opened (Rumbaugh et al., 1974) (**FIGURE 8.11**).

Dolphins are also the subject of interesting language research. Communication with dolphins is done by means of hand signals or audible commands transmitted through an underwater speaker system. In one typical study, trainers gave dolphins commands made up of two- to five-word sentences, such as "Big ball—square—return," which meant that they should go get the big ball, put it in the floating square, and return to the trainer (Herman, Richards, & Woltz, 1984). By varying the syntax (for example, the order of the words) and specific content of the commands, the researchers showed that dolphins are sensitive to these aspects of language.

Psychologists disagree about how to interpret these findings on apes and dolphins. Most psychologists believe that nonhuman animals communicate but that their ideas are severely limited. Critics claim that apes and dolphins are unable to convey subtle meanings, use language creatively, or communicate at an abstract level (Jackendoff, 2003; Siegala & Varley, 2008).

Other critics claim that these animals do not truly understand language but are simply operantly conditioned (Chapter 6) to imitate symbols to receive rewards (Savage-Rumbaugh, 1990; Terrace, 1979). Finally, other critics suggest that data regarding animal language has not always been well documented (Lieberman, 1998; Willingham, 2001; Wynne, 2007).

Proponents of animal language respond that apes can use language creatively and have even coined some words of their own. For example, Koko signed "finger bracelet" to describe a ring and "eye hat" to describe a mask (Patterson & Linden, 1981). Proponents also argue that, as demonstrated by the dolphin studies, animals can be taught to understand basic rules of sentence structure.

Still, the gap between human and nonhuman animals' language is considerable. Current evidence suggests that, at best, nonhuman animal language is less complex, less creative, and has fewer rules than any language used by humans.

NATIONAL GEOGRAPHIC

Computer-aided communication FIGURE 8.11

Apes lack the necessary anatomical structures to vocalize the way humans do. For this reason, language research with chimps and gorillas has focused on teaching the animals to use sign language or to "speak" by pointing to symbols on a keyboard. Do you think this amounts to using language the same way humans do?

CONCEPT CHECK STOP

What are phonemes, morphemes, and grammar?

Why do some psychologists believe that language is an innate ability, whereas others believe that it is learned through imitation and reinforcement?

What are some differences between human and nonhuman animal language?

Intelligence

LEARNING OBJECTIVES

Review the history of theorizing about single versus multiple intelligences.

Identify the similarities and differences between two commonly used IQ tests.

Explain why standardization, reliability, and validity are all essential to scientifically acceptable psychological tests.

Many people equate intelligence with "book smarts." For others, what is intelligent depends on the characteristics and skills that are valued in a particular social group or culture (Matsumoto & Juang, 2008; Sternberg, 2008, 2009). For example, the Mandarin word that corresponds most closely to the word "intelligence" is a character meaning "good brain and talented" (Matsumoto, 2000). The word is associated with traits like imitation, effort, and social responsibility (Keats, 1982).

Even among Western psychologists there is debate over the definition of intelligence. In this discussion, we rely on a formal definition developed by psychologist David Wechsler (pronounced "WEX-ler") (1944, 1977). Wechsler defined **intelligence** as the global capacity to think rationally, act purposefully, and deal effectively with the environment.

intelligence The global capacity to think rationally, act purposefully, and deal effectively with the environment.

DO WE HAVE ONE OR MANY INTELLIGENCES?

One of the central debates in research on intelligence concerns whether intelligence is a single ability or a collection of many specific abilities.

In the 1920s, British psychologist Charles Spearman first observed that high scores on separate tests of mental abilities tend to correlate with each other. Spearman (1923) thus proposed that intelligence is a single factor, which he termed **general intelligence (*g*)**. He believed that *g* underlies all intellectual behavior, including reasoning, solving problems, and performing well in all areas of cognition. Spearman's work laid the foundations for today's standardized intelligence tests (Goldstein, 2008; Johnson et al., 2004).

About a decade later, L. L. Thurstone (1938) proposed seven primary mental abilities: verbal comprehension, word fluency, numerical fluency, spatial visualization, associative memory, perceptual speed, and reasoning. J. P. Guilford (1967) later expanded this number, proposing that as many as 120 factors were involved in the structure of intelligence.

Around the same time, Raymond Cattell (1963, 1971) reanalyzed Thurstone's data and argued against the idea of multiple intelligences. He believed that two subtypes of *g* exist:

- **Fluid intelligence (*gf*)** refers to innate, inherited reasoning abilities, memory, and speed of information processing. Fluid intelligence is relatively independent of education and experience, and like other biological capacities it declines with age (Bugg et al., 2006; Daniels et al., 2006; Rozencwajg et al., 2005).

- **Crystallized intelligence (*gc*)** refers to the store of knowledge and skills gained through experience and education (Goldstein, 2008). Crystallized intelligence tends to increase over the life span.

Today there is considerable support for the concept of *g* as a measure of academic smarts. However, many contemporary cognitive theorists believe that intelligence is not a single general factor but a collection of many separate specific abilities.

One of these cognitive theorists, Howard Gardner, believes that people have many kinds of intelligences. The fact that brain-damaged patients often lose some intellectual abilities while retaining others suggests to Gardner that different intelligences are located in discrete areas throughout the brain. According to Gardner's (1983, 1999, 2008) **theory of multiple intelligences**, people have different profiles of intelligence because they are stronger in some areas than others (**Table 8.3**). They also use their intelligences differently to learn new material, perform tasks, and solve problems.

Robert Sternberg's **triarchic theory of successful intelligence** also involves multiple abilities. As discussed in **Table 8.4**, Sternberg theorized that three separate, learned aspects of intelligence exist: (1) analytic intelligence, (2) creative intelligence, and (3) practical intelligence (Sternberg, 1985, 2007, 2008, 2009).

Sternberg (1985, 1999) emphasizes the process underlying thinking rather than just the product. He also stresses the importance of applying mental abilities to real-world situations rather than testing mental abilities in isolation (e.g., Sternberg, 2005; Sternberg & Hedlund, 2002). Sternberg (1998) introduced the term *successful intelligence* to describe the ability to adapt to, shape, and select environments in order to accomplish personal and societal goals.

Gardner's multiple intelligences and possible careers Table 8.3

Linguistic: language, such as speaking, reading a book, writing a story	**Spatial:** mental maps, such as figuring out how to pack multiple presents in a box or how to draw a floor plan	**Bodily/kinesthetic:** body movement, such as dancing, gymnastics, or figure skating	**Intrapersonal:** understanding oneself, such as setting achievable goals or recognizing self-defeating emotions	**Logical/mathematical:** problem solving or scientific analysis, such as following a logical proof or solving a mathematical problem	**Musical:** musical skills, such as singing or playing a musical instrument	**Interpersonal:** social skills, such as managing diverse groups of people	**Naturalistic:** being attuned to nature, such as noticing seasonal patterns or using environmentally safe products	*(Possible)* **Spiritual/existential:** attunement to meaning of life and death and other conditions of life
Careers: novelist, journalist, teacher	**Careers:** engineer, architect, pilot	**Careers:** athlete, dancer, ski instructor	**Careers:** increased success in almost all careers	**Careers:** mathematician, scientist, engineer	**Careers:** singer, musician, composer	**Careers:** salesperson manager, therapist, teacher	**Careers:** biologist, naturalist	**Careers:** philosopher, theologian

Source: Adapted from Gardner, 1983, 1999.

Who's intelligent? What kinds of intelligence do talk show host Oprah Winfrey, daredevil skateboarder Danny Way, and chess champion Garry Kasparov demonstrate?

Sternberg's triarchic theory of successful intelligence TABLE 8.4

	Analytical Intelligence	*Creative Intelligence*	*Practical Intelligence*
Sample skills	Good at analysis, evaluation, judgment, and comparison skills	Good at invention, coping with novelty, and imagination skills	Good at application, implementation, execution, and utilization skills
Methods of assessment	These skills are assessed by intelligence or scholastic aptitude tests. Questions ask about meanings of words based on context and how to solve number-series problems.	These skills are assessed in many ways, including open-ended tasks, writing a short story, drawing a piece of art, or solving a scientific problem requiring insight.	Although these skills are more difficult to assess, they can be measured by asking for solutions to practical and personal problems.

MEASURING INTELLIGENCE

Many college admissions offices, scholarship committees, and employers use scores from intelligence tests as a significant part of their selection criteria. How well do these tests predict student and employee success? Different IQ tests approach the measurement of intelligence from different perspectives. However, most are designed to predict grades in school. Let's look at the most commonly used IQ tests.

The **Stanford-Binet Intelligence Scale** is loosely based on the first IQ tests developed in France around the turn of the last century by Alfred Binet. In the United States, Lewis Terman (1916) developed the Stanford-Binet (at Stanford University) to test the intellectual ability of U.S.-born children ages 3 to 16. The test is revised periodically—most recently in 1985. The test is administered individually and consists of such tasks as copying geometric designs, identifying similarities, and repeating number sequences.

In the original version of the Stanford-Binet, results were expressed in terms of a mental age. For example, if a 7-year-old's score equaled that of an average 8-year-old, the child was considered to have a mental age of eight. To determine the child's **intelligence quotient (IQ)**, mental age was divided by the child's chronological age (actual age in years) and multiplied by 100.

intelligence quotient (IQ) A subject's mental age divided by his or her chronological age and multiplied by 100.

Today, most intelligence test scores are expressed as a comparison of a single person's score to a national sample of similar-aged people (FIGURE 8.12). Even though the actual

The distribution of scores on the Stanford-Binet Intelligence Test
FIGURE 8.12

Sixty-eight percent of children score within one standard deviation (16 points) above or below the national average, which is 100 points. About 16 percent score above 116, and about 16 percent score below 84.

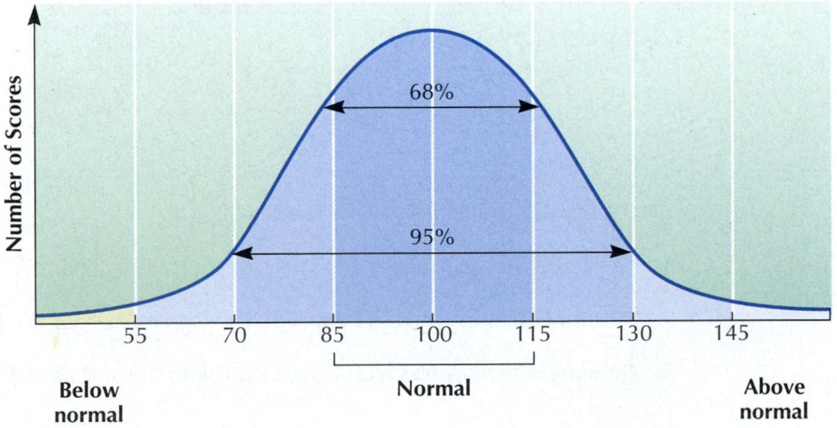

Visualizing

A modern intelligence test FIGURE 8.13

The most widely used intelligence test, the Wechsler Adult Intelligence Scale (WAIS), was developed by David Wechsler in the early 1900s. He later created a similar test for school-age children, the Wechsler Intelligence Scale for Children, now in its fourth edition (WISC-III), and one for preschool children, the Wechsler Preschool and Primary Scale of Intelligence (WPPSI).

Like the Stanford-Binet, Wechsler's tests yield an overall intelligence score, but they have separate scores for verbal intelligence (such as vocabulary, comprehension, and knowledge of general information) and performance (such as arranging pictures to tell a story or arranging blocks to match a given pattern).

Verbal Subtests

Information	How many senators are elected from each state?
Similarities	How are computers and books alike?
Arithmetic	If one baseball card costs three cents, how much will five baseball cards cost?
Vocabulary	Define *lamp*.
Comprehension	What should you do if you accidentally break a friend's toy?

Picture Completion
What is missing from this ambulance?

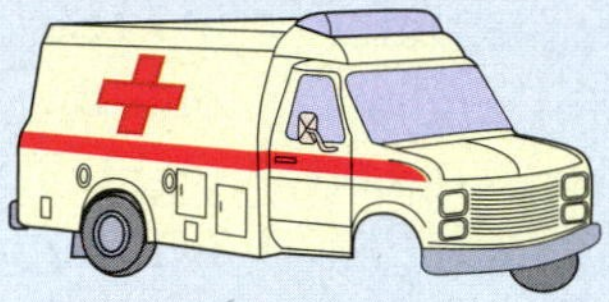

Coding
Write the appropriate number above each symbol.

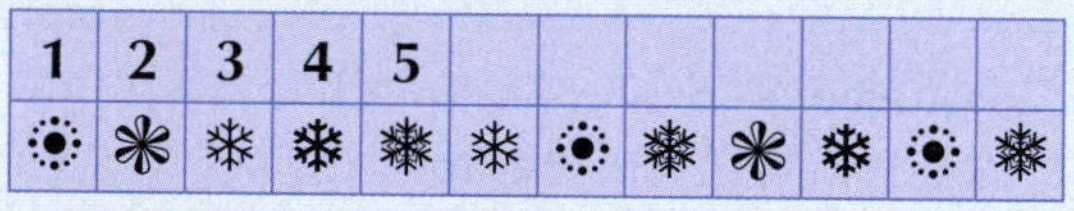

Picture Arrangement
Arrange these pictures in chronological order.

Block Design
Copy this design with blocks.

Object Assembly
Assemble this small jigsaw puzzle.

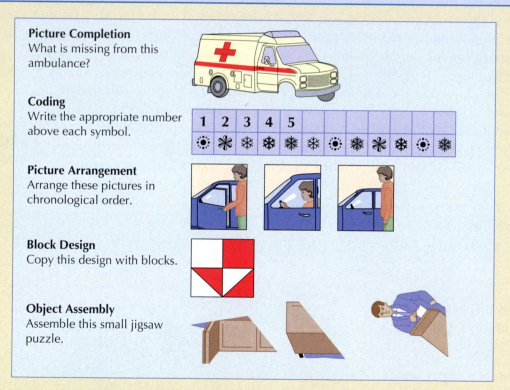

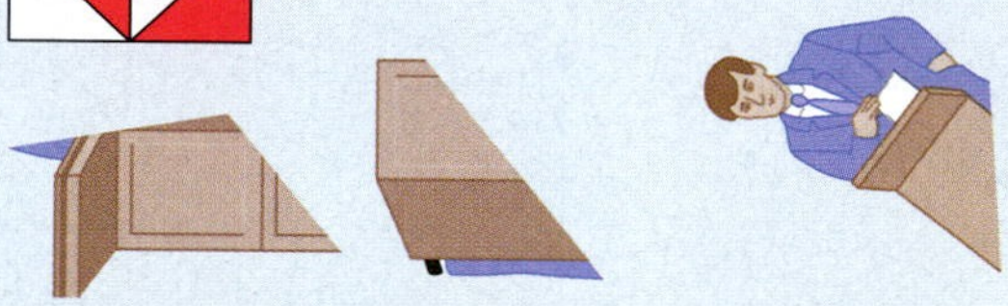

Wechsler's approach has three advantages:

1. The WAIS was specifically designed for adults, not children.
2. Different abilities can be evaluated either separately or together.
3. Non-English speakers can still be tested, with the verbal portion omitted.

NATIONAL GEOGRAPHIC

IQ is no longer calculated using the original formula comparing mental and chronological ages, the term *IQ* remains as a shorthand expression for intelligence test scores.

What makes a good test? How are the tests developed by Binet and Wechsler (FIGURE 8.13) any better than those published in popular magazines and presented on television programs? To be scientifically acceptable, all psychological tests must fulfill three basic requirements:

- **Standardization**. Intelligence tests (as well as personality, aptitude, and most other tests) must be standardized in two senses (Hogan, 2006). First, every test must have norms, or average scores, developed by giving the test to a representative sample of people (a diverse group of people who resemble those for whom the test is intended). Second, testing procedures must be standardized. All test takers must be given the same instructions, questions, and time limits, and all test administrators must follow the same objective score standards.

standardization Establishment of the norms and uniform procedures for giving and scoring a test.

- **Reliability**. To be trustworthy, a test must be consistent, or reliable. Reliability is usually determined by retesting subjects to see whether their test scores change significantly (Hogan, 2006, p. 142). Retesting can be done via the **test-retest method**, in which participants' scores on two separate administrations of the same test are compared, or via the **split-half method**, which involves splitting a test into two equivalent parts (e.g., odd and even questions) and determining the degree of similarity between the two halves.

reliability A measure of the consistency and stability of test scores when a test is readministered.

- **Validity**. Validity is the ability of a test to measure what it is designed to measure. The most important type of validity is **criterion-related validity**, or the accuracy with which test scores can be used to predict another variable of interest (known as the criterion). Criterion-related validity is expressed as the *correlation* (Chapter 1) between the test score and the criterion. If two variables are highly correlated, then one variable can be used to predict the other. Thus, if a test is valid, its scores will be useful in predicting people's behavior in some other specified situation. One example of this is using intelligence test scores to predict grades in college.

validity The ability of a test to measure what it was designed to measure.

Can you see why a test that is standardized and reliable but not valid is worthless? For example, a test for skin sensitivity may be easy to standardize (the instructions specify exactly how to apply the test agent), and it may be reliable (similar results are obtained on each retest). But it certainly would not be valid for predicting college grades.

CONCEPT CHECK STOP

What is the difference between Gardner's and Sternberg's theories of intelligence?

Why is criterion validity important?

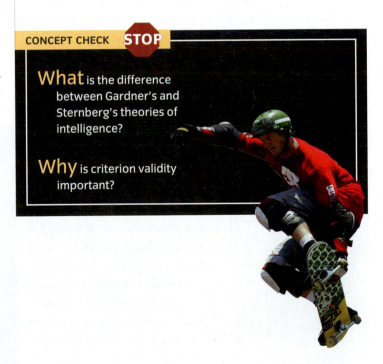

The Intelligence Controversy

LEARNING OBJECTIVES

Explain why extremes in intelligence provide support for the validity of IQ testing.

Review research on how brain functioning is related to intelligence.

Describe how genetics and environment interact to shape intelligence.

Summarize the debate over ethnicity and intelligence.

Psychologists have long debated two important questions related to intelligence: What causes some people to be more intelligent than others and what factors—environmental or hereditary—influence an individual's intelligence? A related question is whether IQ tests are culturally biased. These questions, and the controversies surrounding them, are discussed in this section.

EXTREMES IN INTELLIGENCE: MENTAL RETARDATION AND GIFTEDNESS

One of the best methods for judging the validity of a test is to compare people who score at the extremes. Despite the uncertainties discussed in the previous section, intelligence tests provide one of the major criteria for assessing mental ability at the extremes—specifically, for diagnosing **mental retardation** and **giftedness**.

The clinical label *mentally retarded* is applied when someone is significantly below average in intellectual functioning and has significant deficits in adaptive functioning (such as communicating, living independently, social or occupational functioning, or maintaining safety and health) (American Psychiatric Association, 2000).

Fewer than 3 percent of people are classified as having mental retardation. Of this group, 85 percent have only mild retardation and many become self-supporting, integrated members of society. Furthermore, people can score low on some measures of intelligence and still be average or even gifted in others. The most dramatic examples are people with **savant syndrome** (FIGURE 8.14).

savant syndrome A condition in which a person who has mental retardation exhibits exceptional skill or brilliance in some limited field.

Some forms of retardation stem from genetic abnormalities, such as Down syndrome, fragile-X syndrome, and phenylketonuria (PKU). Other causes are environmental, including prenatal exposure to alcohol and other drugs, extreme deprivation or neglect in early life, and brain damage from accidents. However, in many cases, there is no known cause of retardation.

At the other end of the intelligence spectrum are people with especially high IQs (typically defined as being in the top 1 or 2 percent).

An unusual form of intelligence FIGURE 8.14

Although people with *savant syndrome* score very low on IQ tests (usually between 40 and 70), they demonstrate exceptional skills or brilliance in specific areas, such as rapid calculation, art, memory, or musical ability (Bor et al., 2007; Iavarone, 2007; Miller, 2005; Pring et al., 2008). Brittany Maier has autism and severe visual impairment; she's also a gifted composer and pianist who performs publicly and has recorded a CD. Her musical repertoire includes more than 15,000 songs.

In 1921, Lewis Terman identified 1,500 gifted children—affectionately nicknamed the "Termites"—with IQs of 140 or higher and tracked their progress through adulthood. The number who became highly successful professionals was many times the number a random group would have provided (Leslie, 2000; Terman, 1954). Those who were most successful tended to have extraordinary motivation, and they also had someone at home or school who was especially encouraging (Goleman, 1980). There were some notable failures, and the "Termites" became alcoholics, got divorced, and committed suicide at close to the national rate (Leslie, 2000). A high IQ is no guarantee of success in every endeavor; it only offers more intellectual opportunities.

THE BRAIN'S INFLUENCE ON INTELLIGENCE

A basic tenet of neuroscience is that all mental activity (including intelligence) results from neural activity in the brain. Most recent research on the biology of intelligence has focused on brain functioning. For example, neuroscientists have found that people who score highest on intelligence tests also respond more quickly on tasks involving perceptual judgments (Bowling & Mackenzie, 1996; Deary & Stough, 1996, 1997; Posthuma et al., 2001).

Other research using positron emission tomography (PET) to measure brain activity (Chapter 1) suggests that intelligent brains work smarter, or more efficiently, than less-intelligent brains (Jung & Haier, 2007; Neubauer et al., 2004; Posthuma et al., 2001) (FIGURE 8.15).

Does size matter? It makes logical sense that bigger brains would be smarter—after all, humans have larger brains than less intelligent species, such as dogs. (Some animals, such as whales and dolphins, have larger brains than humans, but our brains are larger relative to our body size.) In fact, brain-imaging studies have found a significant correlation between brain size (adjusted for body size) and intelligence (Christensen et al., 2008; Deary et al., 2007; Ivanovic et al., 2004; Stelmack, Knott, & Beauchamp, 2003). On the other hand, Albert Einstein's brain was no larger than normal (Witelson, Kigar, & Harvey, 1999). In fact, some of Einstein's brain areas were actually smaller than average, but the area responsible for processing mathematical and spatial information was 15 percent larger than average.

Do intelligent brains work more efficiently?
FIGURE 8.15

When given problem-solving tasks, people of low intelligence (PET scans on the left) show more activity (red and yellow indicate more brain activity) in relevant brain areas than people of higher intelligence (PET scans on right).

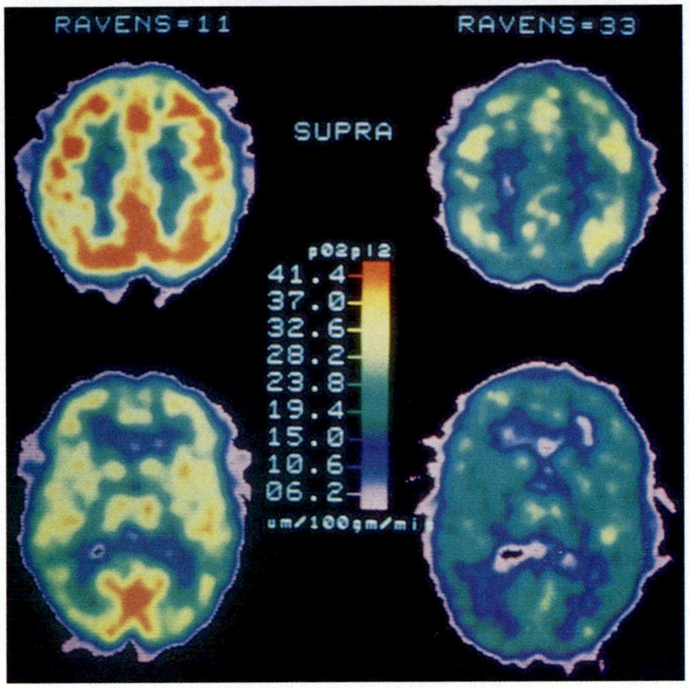

GENETIC AND ENVIRONMENTAL INFLUENCES ON INTELLIGENCE

Similarities in intelligence between family members are due to a combination of hereditary (shared genetic material) and environmental factors (similar living arrangements and experiences). Researchers who are interested in the role of heredity in intelligence often focus on identical (monozygotic) twins because they share 100 percent of their genetic material, as described in *What a Psychologist Sees* on the next page.

The long-running Minnesota Study of Twins, an investigation of identical twins raised in different homes and reunited only as adults (Bouchard, 1994, 1999; Bouchard et al., 1998; Johnson et al., 2007), found that genetic factors appear to play a surprisingly large role in the IQ scores of identical twins.

What a Psychologist Sees

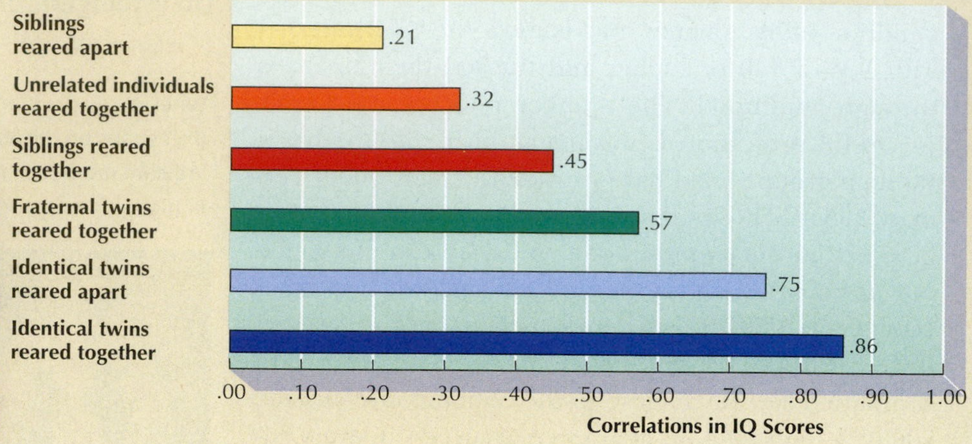

Family Studies of Intelligence

Why are some people intellectually gifted while others struggle? As twin studies demonstrate, genetics plays an important role. In the figure here, note the higher correlations between identical twins' IQ scores compared to the correlations between all other pairs (Bouchard & McGue, 1981; Bouchard et al., 1998; McGue et al., 1993). Although heredity equips each of us with innate intellectual capabilities, the environment significantly influences whether a person will reach his or her full intellectual potential (Dickens & Flynn, 2001; Sangwan, 2001). For example, early malnutrition can retard a child's brain development, which in turn affects the child's curiosity, responsiveness to the environment, and motivation for learning—all of which can lower the child's IQ. The reverse is also true: An enriching early environment can set the stage for intellectual growth.

However, such results are not conclusive. Adoption agencies tend to look for similar criteria in their choice of adoptive parents. Therefore, the homes of these "reared apart" twins were actually quite similar. In addition, these twins also shared the same nine-month prenatal environment, which also might have influenced their brain development and intelligence (White, Andreasen, & Nopoulos, 2002).

ETHNICITY AND INTELLIGENCE: ARE IQ TESTS CULTURALLY BIASED?

How would you answer the following questions?

1. A symphony is to a composer as a book is to a(n) ______. (a) musician; (b) editor; (c) novel; (d) author
2. If you throw a pair of dice and they land with a 7 on top, what is on the bottom? (a) snake eyes; (b) box cars; (c) little Joes; (d) 11

Can you see how the content and answers to these questions might reflect cultural bias? Which of these two questions do you think is most likely to appear on standard IQ tests?

One of the most controversial issues in psychology involves group differences in intelligence test scores and what they really mean. In 1969, Arthur Jensen began a heated debate when he argued that genetic factors are "strongly implicated" as the cause of ethnic differences in intelligence. Richard J. Herrnstein and Charles Murray's book *The Bell Curve: Intelligence and Class Structure in American Life* reignited this debate in 1994; in it, the authors claimed that African Americans score below average in IQ because of their "genetic heritage."

Herrnstein and Murray's book provoked considerable discussion and research. Although there is no clear answer in the debate, psychologists have made several important points:

- IQ tests may be culturally biased, making them an inaccurate measure of true capability (Manly et al., 2004; Naglieri & Ronning, 2000).

stereotype threat Negative stereotypes about minority groups cause some members to doubt their abilities.

- **Stereotype threat** can significantly reduce the test scores of people in stereotyped groups (Bates, 2007; Keller & Bless, 2008; Steele, 2003).
- African Americans and most other minorities are especially likely to live in poverty, hampering intellectual development (Solan & Mozlin, 2001).

Psychological Science

Stereotype Threat: Potential Pitfall for Minorities?

Obama's election as President vs. stereotype threat Preliminary research has found a so-called "Obama effect," which may offset problems related to the stereotype threat. (Dillon, 2009).

In the first study of stereotype threat, Claude Steele and Joshua Aronson (1995) recruited African American and white Stanford University students (with similar ability levels) to complete a "performance exam" that supposedly measured intellectual abilities. The exam's questions were similar to those on the Graduate Record Exam (GRE). Results showed that African American students performed far below white students. In contrast, when the researchers told students it was a "laboratory task," there was no difference between African American and white scores.

Subsequent work showed that stereotype threat occurs because members of stereotyped groups are anxious that they will fulfill their group's negative stereotype. This anxiety in turn hinders their performance on tests. Some people cope with stereotype threat by **disidentifying**, telling themselves they don't care about the test scores (Major et al., 1998). Unfortunately, this attitude lessens motivation, decreasing performance.

Stereotype threat affects many social groups, including African Americans, women, Native Americans, Latinos, low-income people, elderly people, and white male athletes (e.g., Bates, 2007; Ford et al., 2004; Keller & Bless, 2008; Klein et al., 2007; Steele, 2003; Steele, James, & Barnett, 2002). This research helps explain some group differences in intelligence and achievement tests. As such, it underscores why relying solely on such tests to make critical decisions affecting individual lives—for example, in hiring, college admissions, or clinical application—is unwarranted and possibly even unethical.

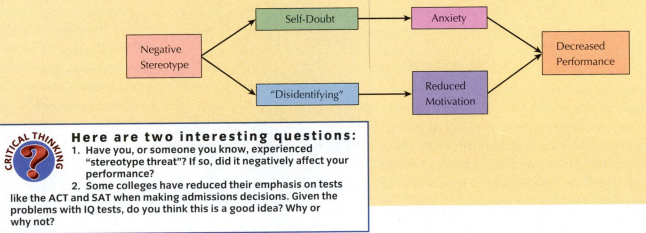

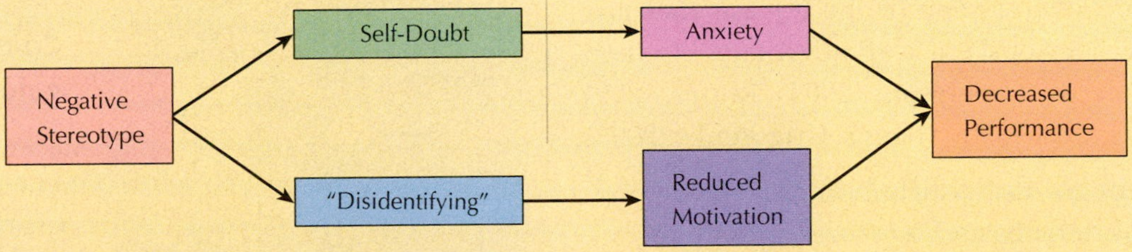

CRITICAL THINKING

Here are two interesting questions:

1. Have you, or someone you know, experienced "stereotype threat"? If so, did it negatively affect your performance?
2. Some colleges have reduced their emphasis on tests like the ACT and SAT when making admissions decisions. Given the problems with IQ tests, do you think this is a good idea? Why or why not?

- Race and ethnicity, like intelligence itself, are almost impossible to define. For example, depending on the definition that you use, there are between 3 and 300 races, and no race is pure in a biological sense (Navarro, 2008; Sternberg & Grigorenko, 2008; Yee et al., 1993).
- Intelligence (as measured by IQ tests) is not a fixed trait. Around the world, IQ scores have increased over the last half century, and this well-established phenomenon is known as the **Flynn effect** in honor of New Zealand researcher James Flynn. Because these increases have occurred in a relatively short period, the cause or causes cannot be due to genetics or heredity. Other possible factors include improved nutrition, better public education, more proficient test-taking skills, and rising levels of education for a greater percentage of the world's population (Flynn, 1987, 2006, 2007; Huang & Hauser, 1998; Mingroni, 2004; Resing & Nijland, 2002). Further evidence for the lack of stability in IQ scores comes from even more recent international research. In the last decade, this rise in scores has reversed itself, and several countries are now reporting a *decline* in IQ scores (Lynn & Harvey, 2008; Teasdale & Owen, 2008). Possible causes for this so-called *negative Flynn effect* are poorly understood. But whether IQ scores rise or fall, the important point is that *intelligence is not a fixed trait.*
- Differences in IQ scores reflect motivational and language factors. In some ethnic groups, a child who excels in school is ridiculed for trying to be different from his or her classmates. Moreover, if children's own language and dialect do not match their education system or the IQ tests they take, they are obviously at a disadvantage (Cathers-Shiffman & Thompson, 2007; Rutter, 2007; Sternberg, 2007; Sternberg & Grigorenko, 2008).
- Different groups' distribution of IQ scores overlap considerably, and IQ scores and intelligence have their greatest relevance in terms of individuals, not groups (Garcia & Stafford, 2000; Myerson et al., 1998; Reifman, 2000). Many individual African Americans receive higher IQ scores than many individual white Americans.
- Traditional IQ tests do not measure many of our multiple intelligences (Gardner, 2002; Manly et al., 2004; Naglieri & Ronning, 2000; Rutter, 2007; Sternberg, 2007, 2009; Sternberg & Grigorenko, 2008).

The ongoing debate over the nature of intelligence and its measurement highlights the complexities of studying *cognition.* In this chapter, we've explored three cognitive processes: thinking, language, and cognition. As you've seen, all three processes are greatly affected by numerous interacting factors.

CONCEPT CHECK **STOP**

Why might more-intelligent people show less activity in cognitive-processing areas than less-intelligent people?

How have twin studies improved researchers' understanding of intelligence?

What is stereotype threat, and why does it occur?

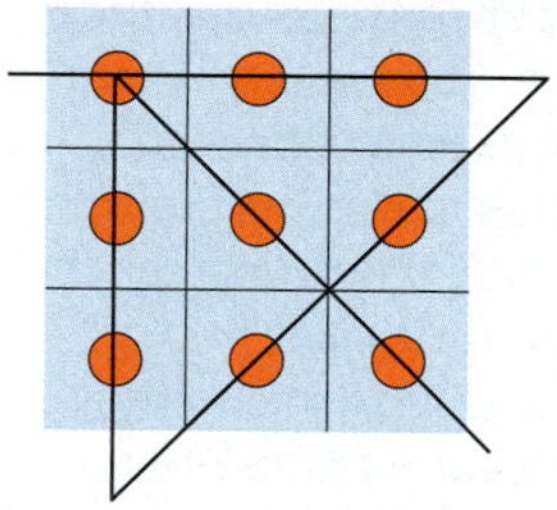

Solution to the nine-dot problem

People find this puzzle difficult because they see the arrangement of dots as a square—a mental set that limits possible solutions.

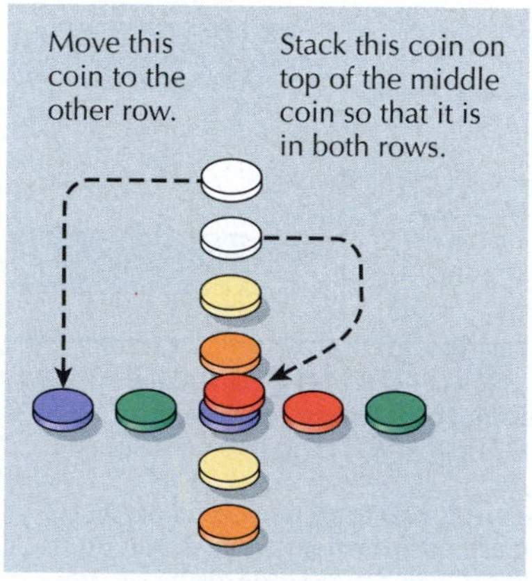
Move this coin to the other row.
Stack this coin on top of the middle coin so that it is in both rows.

Coin problem solution

Stack one coin on top of the middle coin so that it shares both the row and the column.

SUMMARY

1 Thinking

1. Thinking is a central aspect of **cognition**. Thought processes are distributed throughout the brain in neural networks. Mental images, concepts (both artificial and natural), and hierarchies aid our thought processes.
2. Problem solving usually involves three steps: preparation, production, and evaluation. **Algorithms** and **heuristics** help us produce solutions.
3. Barriers to problem solving include mental sets, functional fixedness, confirmation bias, availability heuristic, and representativeness heuristic.
4. **Creativity** is the ability to produce valued outcomes in a novel way. Tests of creativity usually focus on divergent thinking. One prominent theory of creativity is investment theory.

2 Language

1. **Language** supports thinking and enables us to communicate. To produce language, we use **phonemes**, **morphemes**, and **grammar** (syntax and semantics).
2. According to Whorf's linguistic relativity hypothesis, language determines thought. Generally, Whorf's hypothesis is not supported. However, language does strongly influence thought.
3. Children communicate nonverbally from birth. Their verbal communication proceeds in stages: prelinguistic (crying, cooing, babbling) and linguistic (single utterances, telegraphic speech, and acquisition of rules of grammar).
4. According to Chomsky, humans are "prewired" with a language acquisition device that enables language development. "Nurturists" hold that children learn language through rewards, punishments, and imitation. Most psychologists hold an intermediate view.
5. Research with apes and dolphins suggests that these animals can learn and use basic rules of language. However, nonhuman animal language is less complex, less creative, and not as rule-laden as human language.

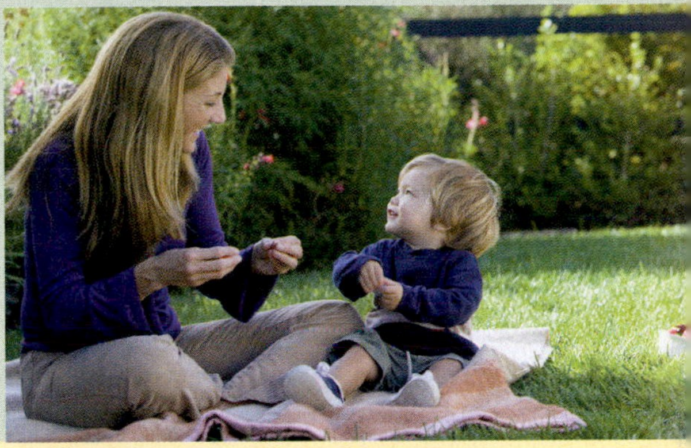

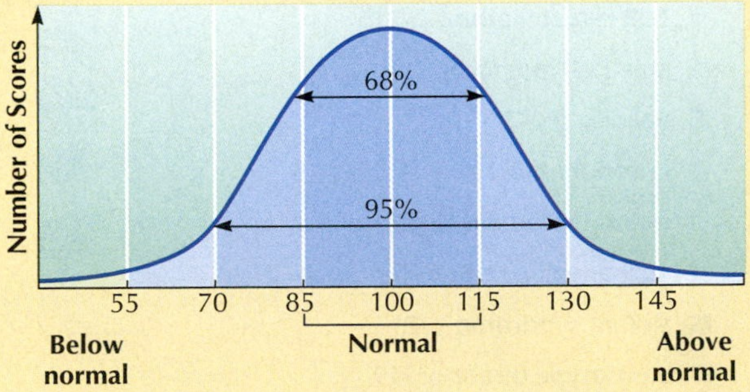

3 Intelligence

1. There is considerable debate over the meaning of **intelligence**. Here it is defined as the global capacity to think rationally, act purposefully, and deal effectively with the environment.
2. Spearman proposed that intelligence is a single factor, which he termed general intelligence (*g*). Thurstone and Guilford argued that intelligence included distinct abilities. Cattell proposed two subtypes of *g*: fluid intelligence and crystallized intelligence. Many contemporary cognitive theorists, including Gardner and Sternberg, believe that intelligence is a collection of many separate specific abilities.
3. Early intelligence tests involved computing a person's mental age to arrive at an **intelligence quotient (IQ)**. Today, two of the most widely used intelligence tests are the Stanford-Binet Intelligence Scale and the Wechsler Adult Intelligence Scale (WAIS).
4. To be scientifically acceptable, all psychological tests must fulfill three basic requirements: **standardization**, **reliability**, and **validity**.

4 The Intelligence Controversy

1. Intelligence tests provide one of the major criteria for assessing mental retardation and giftedness. Mental retardation exists on a continuum. In **savant syndrome**, a person who has mental retardation is exceptional in some limited field. Studies of people who are intellectually gifted found that they had more intellectual opportunities and tended to excel professionally. However, a high IQ is no guarantee of success in every endeavor.

2. Most recent research on the biology of intelligence has focused on brain functioning, not size. Research indicates that intelligent people's brains respond especially quickly and efficiently.

3. Both hereditary and environmental factors contribute to intelligence. Researchers interested in the role of heredity in intelligence often focus on identical twins. Twin studies have found that genetics plays an important role in intelligence. However, the environment significantly influences whether a person will reach his or her full intellectual potential.

4. Claims that genetic factors underlie ethnic differences in intelligence have caused heated debate and have received intense scrutiny. Among other arguments, some psychologists argue that IQ tests may be culturally biased; that **stereotype threat** can significantly reduce test scores of people in stereotyped groups; that socioeconomic factors heavily influence intellectual development; and that traditional IQ tests do not measure many of our multiple intelligences.

KEY TERMS

- **cognition** p. 200
- **prefrontal cortex** p. 200
- **mental image** p. 200
- **concepts** p. 200
- **prototypes** p. 201
- **hierarchies** p. 201
- **algorithm** p. 202
- **heuristic** p. 202
- **problem solving** p. 203
- **mental sets** p. 203
- **functional fixedness** p. 203
- **confirmation bias** p. 203
- **availability heuristic** p. 204
- **representativeness heuristic** p. 204
- **creativity** p. 204
- **divergent thinking** p. 204
- **investment theory** p. 205
- **language** p. 206
- **phoneme** p. 206
- **morpheme** p. 206
- **grammar** p. 206
- **linguistic relativity hypothesis** p. 206
- **language acquisition device (LAD)** p. 208
- **intelligence** p. 211
- **general intelligence (*g*)** p. 211
- **fluid intelligence (*gf*)** p. 211
- **crystallized intelligence (*gc*)** p. 211
- **theory of multiple intelligences** p. 211
- **triarchic theory of successful intelligence** p. 212
- **Stanford-Binet intelligence scale** p. 213
- **intelligence quotient (IQ)** p. 213
- **standardization** p. 215
- **reliability** p. 215
- **test-retest method** p. 215
- **split-half method** p. 215
- **validity** p. 215
- **criterion-related validity** p. 215
- **mental retardation** p. 216
- **giftedness** p. 216
- **savant syndrome** p. 216
- **stereotype threat** p. 219
- **disidentifying** p. 219
- **Flynn effect** p. 220

CRITICAL AND CREATIVE THINKING QUESTIONS

1. During problem solving, do you use primarily algorithms or heuristics? What are the advantages of each?
2. Would you like to be more creative? Can you do anything to acquire more of the resources of creative people?
3. Do you believe that we are born with an innate "language acquisition device" or that language development is a function of our environments?
4. Do you think apes and dolphins have true language? Why or why not?
5. Physicians, teachers, musicians, politicians, and people in many other occupations may become more successful with age and can continue working well into old age. Which kind of general intelligence might explain this phenomenon?
6. If Gardner's and Sternberg's theories of multiple intelligences are correct, what are the implications for intelligence testing and for education?

What is happening in this picture ?

Jerry Levy and Mark Newman, twins separated at birth, first met as adults at a firefighter's convention.

- **What factors might explain why they both became firefighters?**
- **Does the brothers' choosing the same uncommon profession seem like a case of "telepathy"? How might the *confirmation bias* contribute to this perception?**

SELF-TEST

(Check your answers in Appendix B.)

1. The mental activities involved in acquiring, storing, retrieving, and using knowledge are collectively known as ________________.

a. perception
b. cognition
c. consciousness
d. awareness

2. Mental representations of previously stored sensory experiences are called ________________.

a. illusions
b. psychoses
c. mental images
d. mental propositions

3. Label the three stages of problem-solving on the figure below:

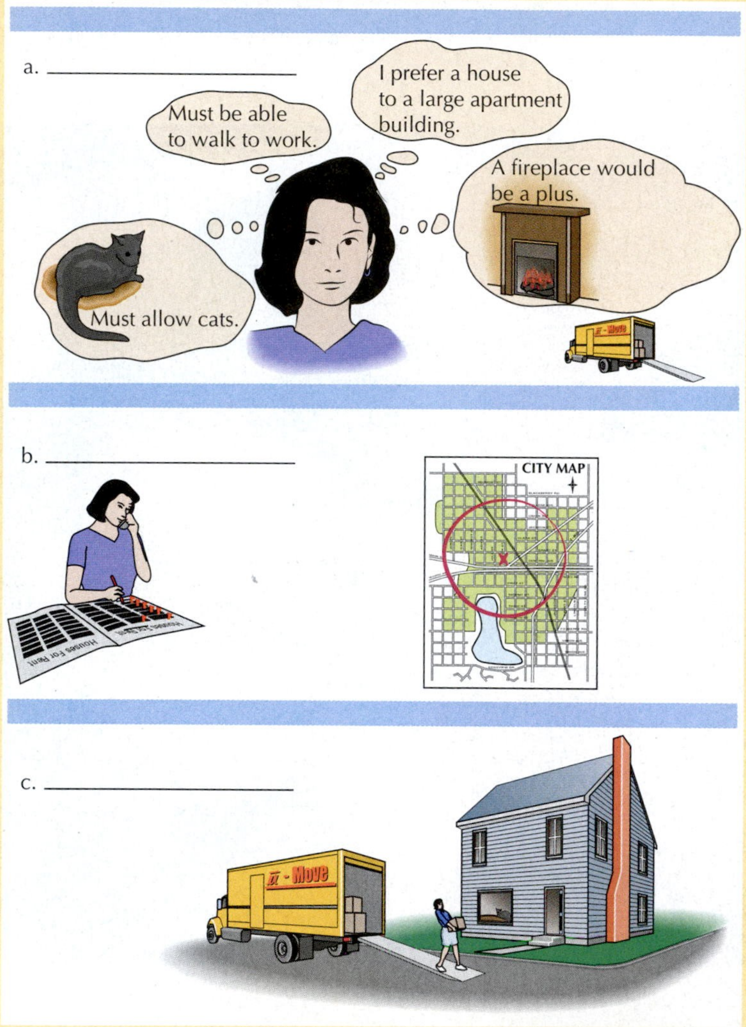

4. This is a logical step-by-step procedure that, if followed, will always produce the solution.

a. algorithm
b. heuristic
c. problem-solving set
d. brainstorming

5. ________________ is the ability to produce valuable outcomes in a novel way.

a. Problem-solving
b. Incubation
c. Functional flexibility
d. Creativity

6. ________________ is the set of rules that specify how phonemes, morphemes, words, and phrases should be combined to express meaningful thoughts.

a. Syntax
b. Pragmatics
c. Semantics
d. Grammar

7. Label the three building blocks of language on the figure below:

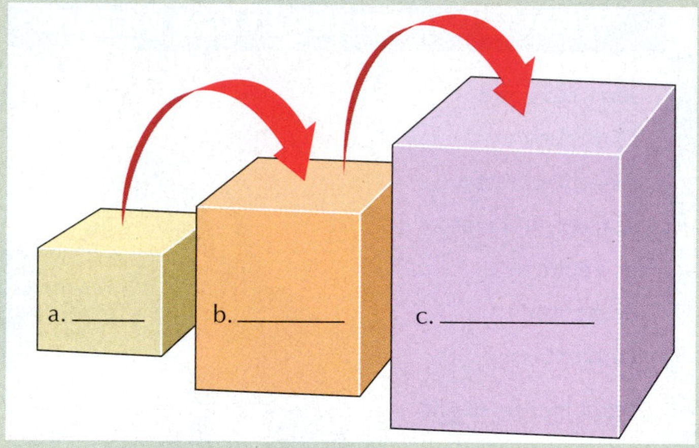

8. According to Chomsky, the innate mechanism that enables a child to analyze language is known as ________________.

a. telegraphic understanding device (TUD)
b. language acquisition device (LAD)
c. language and grammar translator (LGT)
d. overgeneralized neural net (ONN)

9. The definition of *intelligence* stated in your textbook stresses the global capacity to ________________.

a. perform in school and on the job
b. read, write, and make computations
c. perform verbally and physically
d. think rationally, act purposefully, and deal effectively with the environment

10. The IQ test sample in the figure below is from the ________________, the most widely used intelligence test.

a. Wechsler Intelligence Scale for Children
b. Wechsler Adult Intelligence Scale
c. Stanford-Binet Intelligence Scale
d. Binet-Terman Intelligence Scale

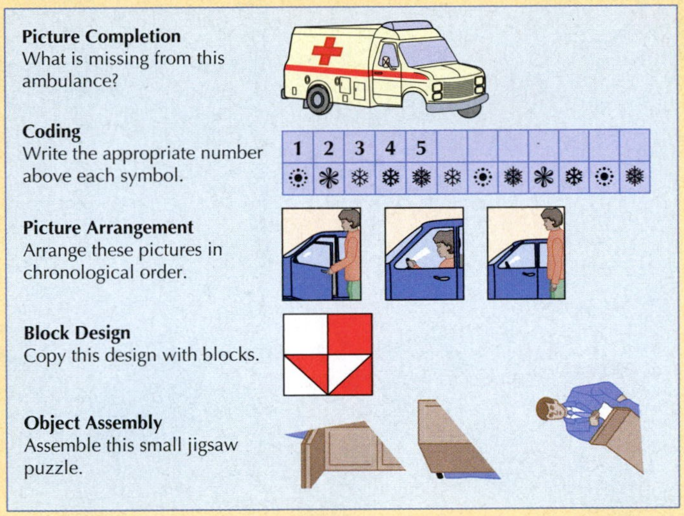

Picture Completion
What is missing from this ambulance?
Coding
Write the appropriate number above each symbol.
1 2 3 4 5
Picture Arrangement
Arrange these pictures in chronological order.
Block Design
Copy this design with blocks.
Object Assembly
Assemble this small jigsaw puzzle.

11. The development of standard procedures for administering and scoring a test is called ________________.

a. norming
b. standardization
c. procedural protocol
d. normalization

12. People with mental retardation who demonstrate exceptional ability in specific areas are called ________________.

a. savants
b. idiot geniuses
c. mildly retarded
d. connoisseurs

13. Speed of response is ________________ correlated with IQ scores.

a. negatively
b. positively
c. highly
d. not

14. Ethnic group differences in IQ scores can be due to ________________.

a. genetic influences
b. socioeconomic differences
c. cultural biases in IQ tests
d. all of these options

15. Stereotype threat affects the IQ scores of which of the following groups?

a. women
b. white male athletes
c. the elderly
d. all of these options

Life Span Development I

9

If you have ever visited a daycare center, you may have been struck by the sight of exuberant children laughing, shrieking, and jostling for attention from caregivers. Yet amid this barely controlled chaos, babies learn to crawl, toddlers learn not to bite one another, and preschoolers learn their ABCs. Day by day, every child grows a little stronger and a little more independent.

Over a lifetime, every person undergoes many physical changes. These changes are most striking in early childhood because they happen so rapidly and are so visible. But everyone is in a state of constant change and development. The typical human will be many different people in his or her lifetime—infant, child, teenager, adult, and senior citizen.

Would you like to know more about yourself at each of these ages? In the next two chapters, we will explore research in developmental psychology. In this chapter, we will look at changes in physical and cognitive development from conception to death. In Chapter 10, we will examine important aspects of social, moral, and personality development across the life span. To emphasize that development is an ongoing, lifelong process, throughout these chapters we will trace physical, cognitive, social, moral, and personality development—one at a time—from conception to death. This topical approach will allow us to see how development affects an individual over the entire life span.

CHAPTER OUTLINE

Studying Development

LEARNING OBJECTIVES

Summarize the three most important debates or questions in developmental psychology.

Define maturation and critical periods.

Contrast the cross-sectional research method with the longitudinal research method.

Explain the limitations of cross-sectional and longitudinal research.

We begin our study of human development by focusing on some key theoretical issues and debates. We then discuss two basic research methods and their advantages and disadvantages.

THEORETICAL ISSUES: ONGOING DEBATES

The three most important debates or questions in **developmental psychology** are about nature versus nurture, continuity versus stages, and stability versus change.

> **developmental psychology** The study of age-related changes in behavior and mental processes from conception to death.

Nature or nurture

The issue of "nature versus nurture" has been with us since the beginning of psychology (Chapter 1). Even the ancient Greeks had the same debate—Plato argued that humans are born with innate knowledge and abilities, while Aristotle held that learning occurs through the five senses. Early philosophers proposed that at birth our minds are a *tabula rasa* (blank slate) and that the environment determines what messages are written on that slate.

According to the nature position, human behavior and development are governed by automatic, genetically predetermined signals in a process known as **maturation**. Just as a flower unfolds in accord with its genetic blueprint, we humans crawl before we walk and walk before we run. Furthermore, there is an optimal period shortly after birth, one of several **critical periods** during our lifetime, when an organism is especially sensitive to certain experiences that shape the capacity for future development. On the other side of the debate, those who hold an extreme nurturist position argue that development occurs by learning through personal experience and observation of others.

> **maturation** Development governed by automatic, genetically predetermined signals.

> **critical period** A period of special sensitivity to specific types of learning that shapes the capacity for future development.

Continuity or stages

Continuity proponents believe that development is continuous, with new abilities, skills, and knowledge being gradually added at a relatively uniform pace. Therefore, the continuity model suggests that adult thinking and intelligence differ quantitatively from a child's. That is, adults simply have more math and verbal skills than children do. Stage theorists, on the other hand, believe that development occurs at different rates, alternating between periods of little change and periods of abrupt, rapid change. In this chapter, we discuss one stage theory, Piaget's theory of cognitive development. In Chapter 10, we discuss Erikson's psychosocial theory of personality development and Kohlberg's theory of moral development.

Stability or change

Have you generally maintained your personal characteristics as you matured from infant to adult (stability)? Or does your current personality bear little resemblance to the personality you displayed during infancy (change)? Psychologists who emphasize stability in development hold that measurements of personality taken during childhood are

Psychological Science

Deprivation and Development

What happens if a child is deprived of appropriate stimulation during critical periods of development? Consider the story of Genie, the so-called wild child. From the time she was 20 months old until authorities rescued her at age 13, Genie was locked alone in a tiny, windowless room. By day, she sat naked and tied to a chair with nothing to do and no one to talk to. At night, she was put in a kind of straitjacket and "caged" in a covered crib. Genie's abusive father forbade anyone to speak to her for those 13 years. If Genie made noise, her father beat her while he barked and growled like a dog.

Genie's tale is a heartbreaking account of the lasting scars from a disastrous childhood. In the years after her rescue, Genie spent thousands of hours receiving special training, and by age 19 she could use public transportation and was adapting well to her foster home and special classes at school (Rymer, 1993). Genie was far from normal, however. Her intelligence scores were still close to the cutoff for mental retardation. And although linguists and psychologists worked with her for many years, she never progressed beyond sentences like "Genie go" (Curtiss, 1977; Rymer, 1993). To make matters worse, she was also subjected to a series of foster home placements, one of which was abusive. At last report, Genie was living in a home for mentally retarded adults (Rymer, 1993).

Here are two interesting questions:

1. Does Genie's case prove that there is a critical period for language development? Why or why not?
2. Which side of the "nature versus nurture" debate does Genie's case best support? Briefly explain your answer.

important predictors of adult personality. Of course, psychologists who emphasize change disagree.

Which of these positions is more correct? Most psychologists do not take a hard line either way. Rather, they prefer an **interactionist perspective**. For example, in the age-old nature versus nurture debate, psychologists generally agree that development emerges both from each individual's unique genetic predisposition and from individual experiences in the environment (Hartwell, 2008; Hudziak, 2008; Rutter, 2007). More recently, the interactionist position has evolved into the *biopsychosocial model* mentioned throughout this text. In this model, biological factors (genetics, brain functions, biochemistry, and evolution), psychological influences (learning, thinking, emotion, personality, and motivation), and social forces (family, school, culture, ethnicity, social class, and politics) all affect and are affected by one another.

Like the nature versus nurture debate, the debates about continuity versus stages and stability versus change are not a matter of either-or. Physical development and motor skills, for example, are believed to be primarily continuous in nature, whereas cognitive skills usually develop in discrete stages. Similarly, some traits are stable, whereas others vary greatly across the life span.

RESEARCH METHODS: TWO BASIC APPROACHES

cross-sectional method Research design that measures individuals of various ages at one point in time and gives information about age differences.

longitudinal method Research design that measures a single individual or a group of same-aged individuals over an extended period and gives information about age changes.

To study development, psychologists use either a cross-sectional or longitudinal method. The **cross-sectional method** examines individuals of various ages (e.g., 20, 40, 60, and 80) at one point in time and gives information about age differences. The **longitudinal method** follows a single individual or a group of same-aged individuals over an extended period and gives information about age changes (FIGURE 9.1).

Imagine that you are a developmental psychologist interested in studying intelligence in adults. Which method would you choose—cross-sectional or longitudinal? Before you decide, note the different research results shown in FIGURE 9.2.

Why do the two methods show such different results? Researchers suggest that the different results may reflect a central problem with cross-sectional studies. These studies often confuse genuine age differences with **cohort effects**, differences that result from specific histories of the age group studied (Elder, 1998). As Figure 9.2 shows, the 81-year-olds measured by the cross-sectional method have dramatically lower scores than the 25-year-olds. But is this due to aging or to broad environmental differences, such as less formal education or poorer nutrition? Because the different age groups, called *cohorts*, grew up in different historical periods, their results may not apply to people growing up at other times. With the cross-sectional method, age effects and cohort effects are hopelessly tangled.

Cross-sectional versus longitudinal research FIGURE 9.1

Note that cross-sectional research uses different participants and is interested in age-related differences, whereas longitudinal research studies the same participants over time to find age-related changes.

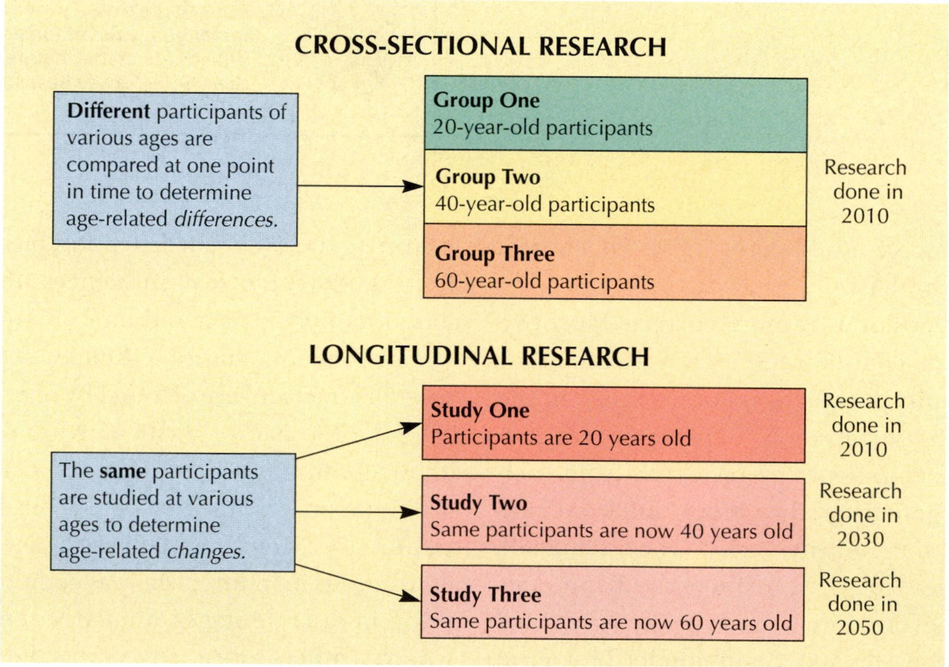
CROSS-SECTIONAL RESEARCH
Different participants of various ages are compared at one point in time to determine age-related differences.
Group One
20-year-old participants
Group Two
40-year-old participants
Group Three
60-year-old participants
Research done in 2010
LONGITUDINAL RESEARCH
The same participants are studied at various ages to determine age-related changes.
Study One
Participants are 20 years old
Research done in 2010
Study Two
Same participants are now 40 years old
Research done in 2030
Study Three
Same participants are now 60 years old
Research done in 2050

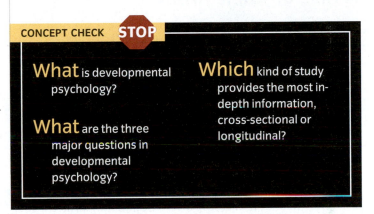

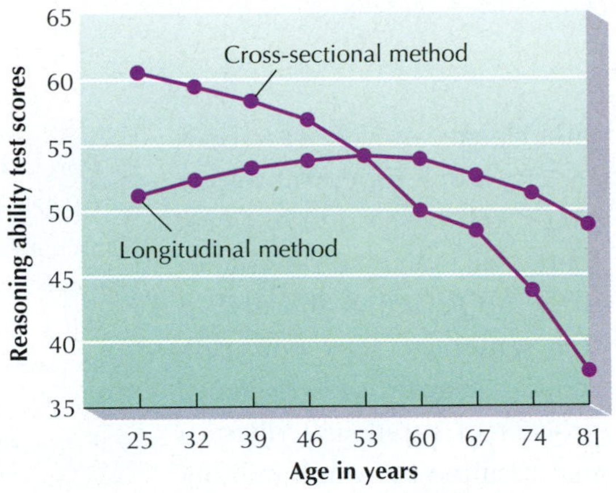

Which results are "true"? FIGURE 9.2

Cross-sectional studies have shown that reasoning and intelligence reach their peak in early adulthood and then gradually decline. In contrast, longitudinal studies have found that a marked decline does not begin until about age 60. (Adapted from Schaie, 1994, with permission.)

Longitudinal studies also have their limits. They are expensive in terms of time and money, and their results are restricted in generalizability. Because participants often drop out or move away during the extended test period, the experimenter may end up with a self-selected sample that differs from the general population in important ways. Each method of research has its own strengths and weaknesses (TABLE 9.1). Keep these differences in mind when you read the findings of developmental research.

Human development, like most areas of psychology, cannot be studied outside its sociocultural context. In fact, researchers have found that culture may be the most important determinant of development. If a child grows up in an individualistic/independent culture (such as North America or most of Western Europe), we can predict that this child will probably be competitive and question authority as an adult. If this same child were reared in a collectivist/interdependent culture (common in Africa, Asia, and Latin America), she or he would most likely grow up to be cooperative and respectful of others (Delgado-Gaitan, 1994; Berry et al., 2002).

CONCEPT CHECK STOP

What is developmental psychology?

What are the three major questions in developmental psychology?

Which kind of study provides the most in-depth information, cross-sectional or longitudinal?

Advantages and disadvantages of cross-sectional and longitudinal research designs TABLE 9.1

	Cross-sectional	*Longitudinal*
Advantages	Gives information about age differences Quick Less expensive Typically larger sample	Gives information about age changes Increased reliability More in-depth information per participant
Disadvantages	Cohort effects are difficult to separate Restricted generalizability (measures behaviors at only one point in time)	More expensive Time consuming Restricted generalizability (typically smaller sample and dropouts over time)

Physical Development

LEARNING OBJECTIVES

Describe the three phases of prenatal physical development.

Identify some important teratogens and their effects on prenatal development.

Summarize physical development during early childhood.

Describe the physical changes that occur during adolescence and adulthood.

In this section, we will explore the fascinating processes of physical development from conception through childhood, adolescence, and adulthood.

PRENATAL AND EARLY CHILDHOOD: A TIME OF RAPID CHANGE

The early years of development are characterized by rapid and unparalleled change. In fact, if you continued to develop at the same rapid rate that marked your first two years of life, you would weigh several tons and be over 12 feet tall as an adult! Thankfully, physical development slows, yet it is important to note that change continues until the very moment of death. Let's look at some of the major physical changes that occur throughout the life span.

Your prenatal development began at conception, when your mother's egg, or ovum, united with your father's sperm cell (FIGURE 9.3). At that time, you were a single cell barely 1/175 of an inch in diameter—smaller than the period at the end of this sentence. This new cell, called a **zygote**, then began a process of rapid cell division that resulted in a multimillion-celled infant (you) some nine months later.

The vast changes that occur during the nine months of a full-term pregnancy are usually divided into three stages: the **germinal period**, the **embryonic period**, and the **fetal period** (FIGURE 9.4). Prenatal growth, as well as growth during the first few years after birth, is **proximodistal** (near to far), with the head and upper body developing before the lower body.

During pregnancy, the **placenta** (the vascular organ that unites the fetus to the mother's uterus) serves as the link for food and excretion of wastes. It also screens out some, but not all, harmful substances. Environmental hazards

germinal period The first stage of prenatal development, which begins with conception and ends with implantation in the uterus (the first two weeks).

embryonic period The second stage of prenatal development, which begins after uterine implantation and lasts through the eighth week.

fetal period The third, and final, stage of prenatal development (eight weeks to birth), which is characterized by rapid weight gain in the fetus and the fine detailing of bodily organs and systems.

Conception FIGURE 9.3

A Note the large number of sperm surrounding the ovum (egg).

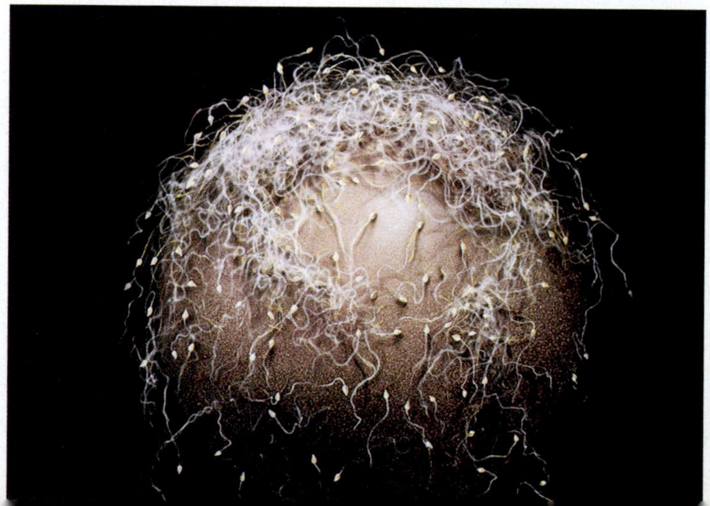

B Although a joint effort is required to break through the outer coating, only one sperm will actually fertilize the egg.

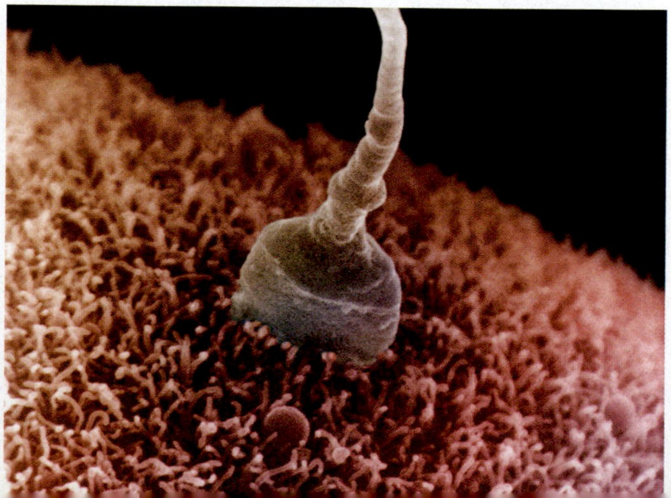

Prenatal development FIGURE 9.4

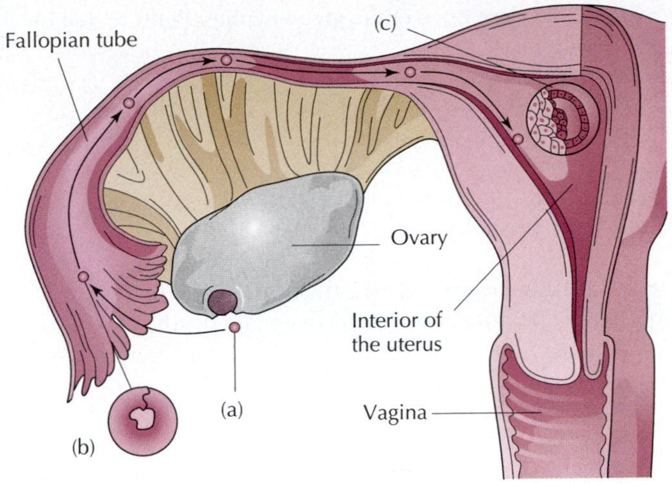

1 Germinal period: From ovulation to implantation. After discharge from either the left or right ovary (a), the ovum travels to the opening of the fallopian tube.

If fertilization occurs (b), it normally takes place in the first third of the fallopian tube. The fertilized ovum is referred to as a zygote.

When the zygote reaches the uterus, it implants itself in the wall of the uterus (c) and begins to grow tendril-like structures that intertwine with the rich supply of blood vessels located there. After implantation, the organism is known as an embryo.

Process Diagram

2 Embryonic period. This stage lasts from implantation to eight weeks. At eight weeks, the major organ systems have become well differentiated. Note that at this stage, the head grows at a faster rate than other parts of the body.

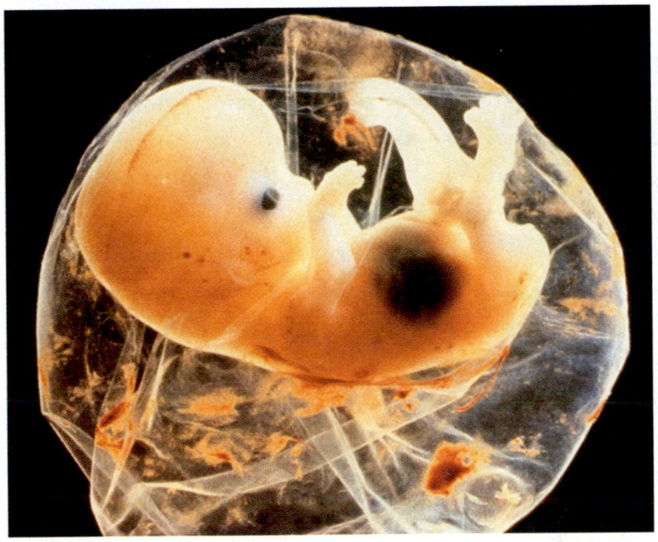

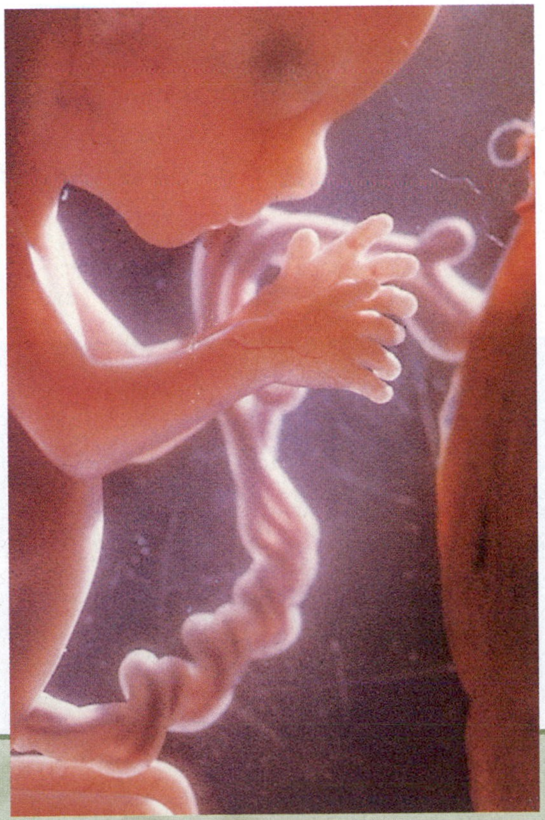

3 Fetal period. This is the period from the end of the second month to birth. At four months, all the actual body parts and organs are established. The fetal stage is primarily a time for increased growth and "fine detailing."

VIEW THIS IN ACTION in your WileyPLUS course

Sample environmental conditions that endanger the child TABLE 9.2

Maternal factors	Possible effects on embryo, fetus, newborn, or young child
Malnutrition	Low birth weight, malformations, less developed brain, greater vulnerability to disease
Stress exposure	Low birth weight, hyperactivity, irritability, feeding difficulties
Exposure to x-rays	Malformations, cancer
Legal and illegal drugs	Inhibition of bone growth, hearing loss, low birth weight, fetal alcohol syndrome, mental retardation, attention deficits in childhood, death
Diseases: German measles (rubella), herpes, AIDS, toxoplasmosis	Blindness, deafness, mental retardation, heart and other malformations, brain infection, spontaneous abortion, premature birth, low birth weight, death

Source: Abadinsky, 2008; Hyde & DeLamater, 2008; Howell, Coles, & Kable, 2008; Leventhal, 2008.

such as x-rays or toxic waste, drugs, and diseases such as rubella (German measles) can cross the placental barrier. These influences generally have their most devastating effects during the first three months of pregnancy, making this a critical period in development.

The pregnant mother plays a primary role in prenatal development because her health directly influences the child she is carrying (TABLE 9.2). Almost everything she ingests can cross the placental barrier (a better term might be *placental sieve*). However, the father also plays a role (other than just fertilization). Environmentally, the father's smoking may pollute the air the mother breathes, and genetically, he may transmit heritable diseases. In addition, research suggests that alcohol, opiates, cocaine, various gases, lead, pesticides, and industrial chemicals can damage sperm (Baker & Nieuwenhuijsen, 2008; Bandstra et al., 2002; Ferretti et al., 2006).

Perhaps the most important—and generally avoidable—danger to the fetus comes from drugs, both legal and illegal. Nicotine and alcohol are major **teratogens**, environmental agents that cause damage during prenatal development. Mothers who smoke tobacco during pregnancy have significantly higher rates of premature births, low-birth-weight infants, and fetal deaths (American Cancer Society, 2004; Bull, 2003; Oliver, 2002). Their children also show increased behavioral abnormalities and cognitive problems (Abadinsky, 2008; Fryer, Crocker, & Mattson, 2008; Howell, Coles, & Kable, 2008). Alcohol also readily crosses the placenta, affects fetal development, and can result in a neurotoxic syndrome called **fetal alcohol syndrome (FAS)** (FIGURE 9.5). About one in a hundred babies in the United States is born with FAS or other birth defects resulting from the mother's alcohol use during

Fetal alcohol syndrome FIGURE 9.5

Compare the healthy newborn infant brain (left) to the brain of a newborn (right) whose mother drank while pregnant. Prenatal exposure to alcohol can cause facial abnormalities and stunted growth. But the most disabling features of FAS are neurobehavioral problems, ranging from hyperactivity and learning disabilities to mental retardation, depression, and psychoses (Pellegrino & Pellegrino, 2008; Sowell et al., 2008; Wass, 2008).

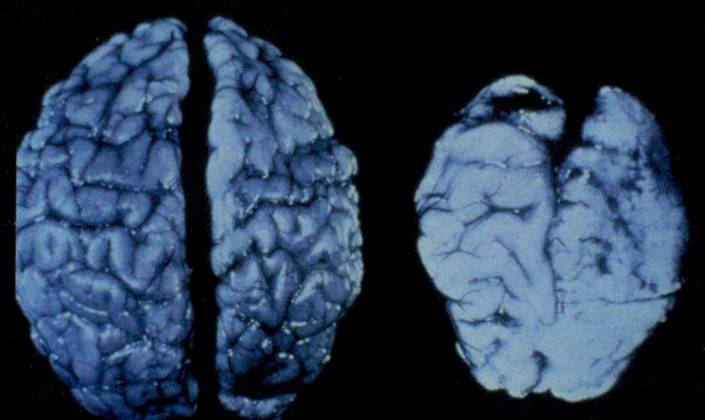

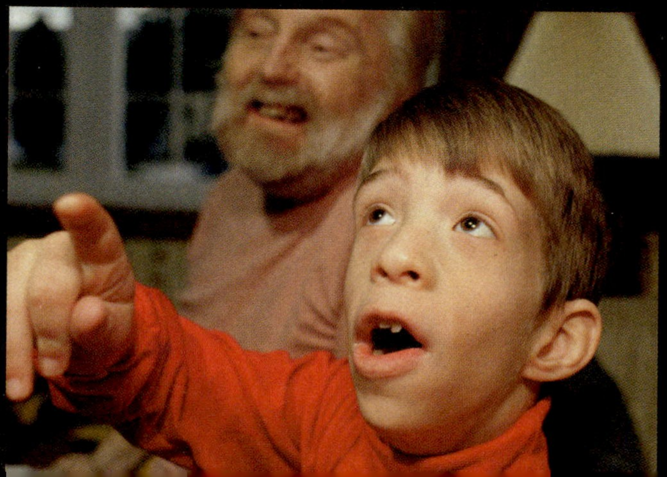

Body proportions FIGURE 9.6

As noted earlier in the chapter, a large part of human development results from the orderly sequence of genetically designed biological processes that we call *maturation*. Notice how our body proportions change as we grow older. At birth, an infant's head is one-fourth its total body's size, whereas in adulthood, the head is one-eighth.

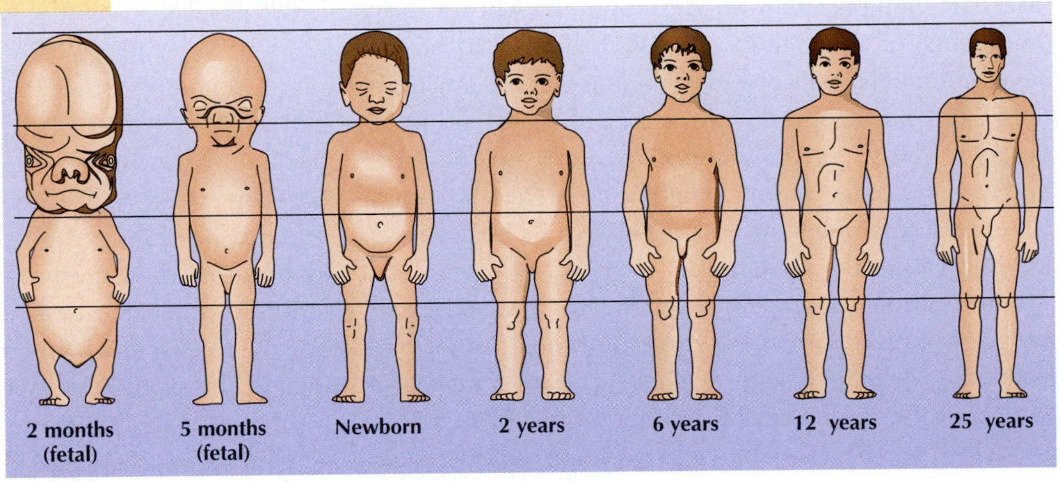

pregnancy (National Organization on Fetal Alcohol Syndrome, 2008).

Like the prenatal period, early childhood is also a time of rapid physical development. Although Shakespeare described infants as capable of only "mewling and puking in the nurse's arms," they are actually capable of much more. Let's explore three key areas of change in early childhood: brain, motor, and sensory/perceptual development.

Brain development The brain and other parts of the nervous system grow faster than any other part of the body during both prenatal development and the first two years of life. At birth, a healthy newborn's brain is one-fourth its full adult size, and it will grow to about 75 percent of its adult weight and size by the age of two. At age six, the child's brain is nine-tenths its full adult weight (FIGURE 9.6).

Rapid brain growth during infancy and early childhood slows down in later childhood. Further brain development and learning occur primarily because neurons grow in size and because the number of axons and dendrites, as well as the extent of their connections, increases (DiPietro, 2000) (FIGURE 9.7).

Brain growth in the first two years FIGURE 9.7

As children learn and develop, synaptic connections between active neurons strengthen, and dendritic connections become more elaborate. *Synaptic pruning* (reduction of unused synapses) helps support this process. *Myelination,* the accumulation of fatty tissue coating the axons of nerve cells, continues until early adulthood. Myelin increases the speed of neural impulses, and the speed of information processing shows a corresponding increase (Chapter 2). In addition, synaptic connections in the frontal lobes and other parts of the brain continue growing and changing throughout the entire life span (Chapters 2 and 6).

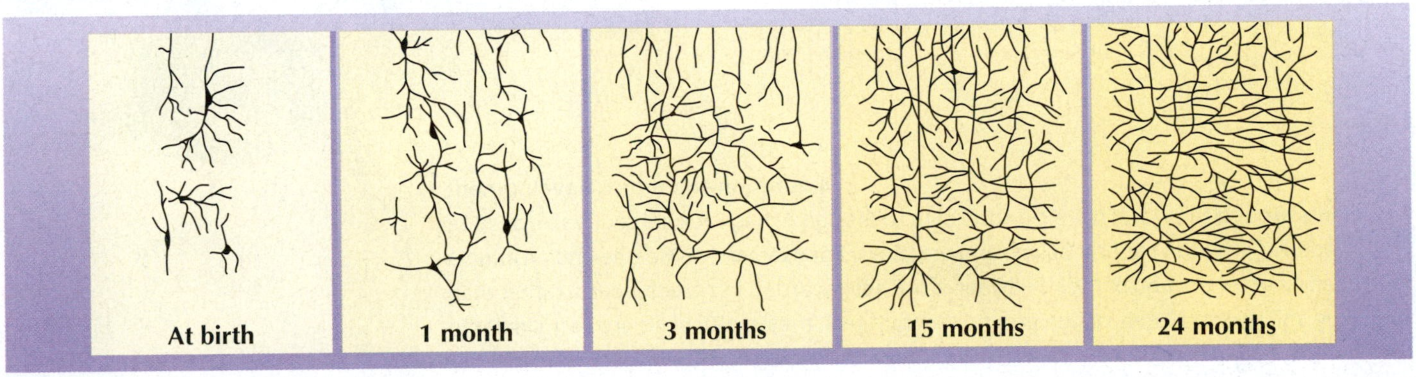

Motor development Compared to the hidden, internal changes in brain development, the orderly emergence of active movement skills, known as motor development, is easily observed and measured. The newborn's first motor abilities are limited to *reflexes*, or involuntary responses to stimulation. For example, the rooting reflex occurs when something touches a baby's cheek: the infant will automatically turn its head, open its mouth, and root for a nipple.

In addition to simple reflexes, the infant soon begins to show voluntary control over the movement of various body parts (see box below). Thus, a helpless newborn who cannot even lift her head is soon transformed into an active toddler capable of crawling, walking, and climbing. Keep in mind that motor development is largely due to natural maturation, but it can also be affected by environmental influences like disease and neglect, as well as by cultural differences (FIGURE 9.8).

Sensory and perceptual development At birth, a newborn can smell most odors and distinguish between sweet, salty, and bitter tastes. Breast-fed newborns also recognize and show preference for the odor and taste of their mother's milk over another mother's (DiPietro, 2000; Rattaz et al., 2005). In addition, the newborn's sense of touch and pain is highly developed, as evidenced by reactions to heel pricks for blood testing and to circumcision (Williamson, 1997).

The newborn's sense of vision, however, is poorly developed. At birth, an infant is estimated to have vision be-

Motor development and culture FIGURE 9.8

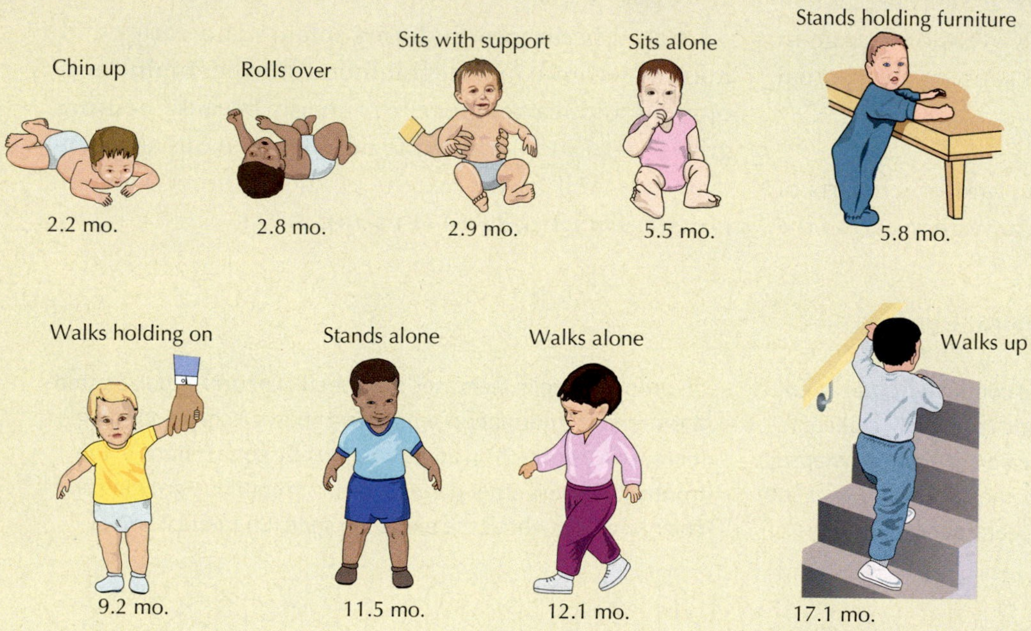

A Milestones in motor development
In the typical progression of motor abilities, "chin up" occurs at 2.2 months. However, no two children are exactly alike; all follow their own individual timetables for physical development. (Adapted from Frankenburg et al., 1992, with permission.)

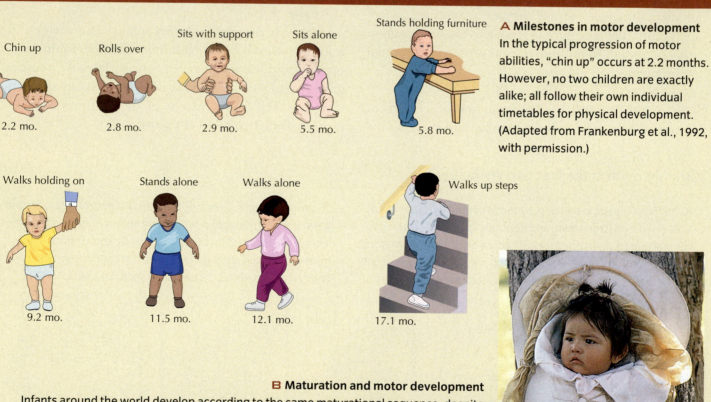

B Maturation and motor development
Infants around the world develop according to the same maturational sequence, despite wide variations in cultural beliefs and practices. For example, some Hopi Indian infants spend a great portion of their first year of life being carried in a cradleboard, rather than crawling and walking freely on the ground. Yet by age 1, their motor skills are very similar to those of infants who have not been restrained in this fashion (Dennis & Dennis, 1940).

tween 20/200 and 20/600 (Haith & Benson, 1998). Imagine what the infant's visual life is like: the level of detail you see at 200 or 600 feet (if you have 20/20 vision) is what they see at 20 feet. Within the first few months, vision quickly improves, and by six months it is 20/100 or better. At two years, visual acuity reaches a near-adult level of 20/20 (Courage & Adams, 1990).

One of the most interesting findings in infant sensory and perceptual research concerns hearing. Not only can the newborn hear quite well at birth (Matlin & Foley, 1997) but also, during the last few months in the womb, the fetus can apparently hear sounds outside the mother's body (Vaughan, 1996). This raises the interesting possibility of fetal learning, and some have advocated special stimulation for the fetus as a way of increasing intelligence, creativity, and general alertness (e.g., Van de Carr & Lehrer, 1997). For more on infant perception, read *What a Psychologist Sees.*

Studies on possible fetal learning have found that newborn infants easily recognize their own mother's voice over that of a stranger (Kisilevsky et al., 2003). They also show a preference for children's stories (such as *The Cat in the Hat* or *The King, the Mice, and the Cheese*) that were read to them while they were still in the womb (DeCasper & Fifer, 1980; Karmiloff & Karmiloff-Smith, 2002). On the other hand, some experts caution that too much or the wrong kind of stimulation before birth can be stressful for both the mother and fetus. They suggest that the fetus gets what it needs without any special stimulation.

What a Psychologist Sees

How an Infant Perceives the World

Because infants cannot talk or follow directions, researchers have had to create ingenious experiments to measure their perceptual abilities and preferences. One of the earliest experimenters, Robert Fantz (1956, 1963), designed a "looking chamber" to measure how long infants stared at stimuli. Research using this apparatus indicates that infants prefer complex rather than simple patterns and pictures of faces rather than nonfaces.

Researchers also use newborns' heart rates and innate abilities, such as the sucking reflex, to study learning and perceptual development (Bendersky & Sullivan, 2007). To study the sense of smell, researchers measure changes in the newborns' heart rates when odors are presented. Presumably, if they can smell one odor but not another, their heart rates will change in the presence of the first but not the second. Brain scans, such as fMRI, MRI, and CTs, also help scientists study changes in the infant's brain. From research such as this, we now know that the senses develop very early in life.

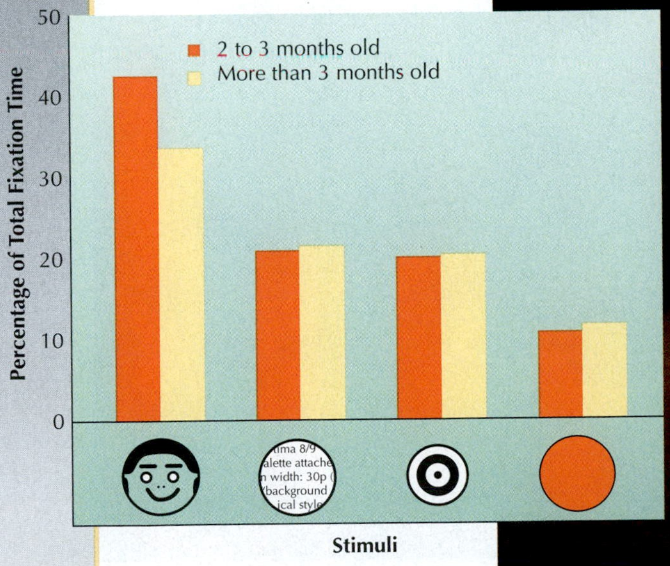

ADOLESCENCE AND ADULTHOOD: A TIME OF BOTH DRAMATIC AND GRADUAL CHANGE

Think back for a moment to your teen years. Were you concerned about the physical changes you were going through? Did you worry about how you differed from your classmates? Changes in height and weight, breast development and menstruation for girls, and a deepening voice and beard growth for boys are important milestones for adolescents. **Puberty**, the period of adolescence when a person becomes capable of reproduction, is a major physical milestone for everyone. It is a clear biological signal of the end of childhood.

Although commonly associated with puberty, **adolescence** is the loosely defined psychological period of development between childhood and adulthood. In the United States, it roughly corresponds to the teenage years. The concept of adolescence and its meaning varies greatly across cultures (FIGURE 9.9).

The clearest and most dramatic physical sign of puberty is the **growth spurt**, which is characterized by rapid increases in height, weight, and skeletal growth (FIGURE 9.10), and by significant changes in reproductive structures and sexual characteristics. Maturation and hormone secretion cause rapid development of the ovaries, uterus, and vagina and the onset of menstruation (**menarche**) in the adolescent female. In the adoles-

Ready for responsibility? FIGURE 9.9

Adolescence is not a universal concept. Unlike in the United States and other Western nations, some nonindustrialized countries have no need for a slow transition from childhood to adulthood; children simply assume adult responsibilities as soon as possible. Can you see how this cultural difference is reflected in the two photos below?

A Teenage girls spend the day on the boardwalk at the Maryland shore.

B A girl in Thailand hangs dyed silk skeins to dry outside a workshop.

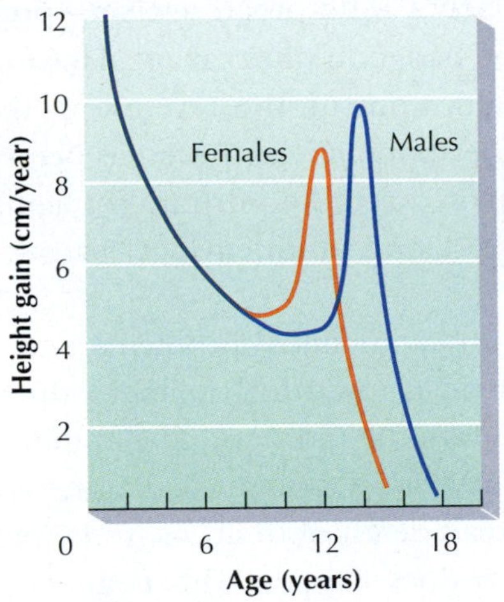

Adolescent growth spurt FIGURE 9.10

Note the gender differences in height gain during puberty. Most girls are about two years ahead of boys in their growth spurt and are therefore taller than most boys between the ages of 10 and 14.

cent male, the testes, scrotum, and penis develop, and he experiences his first ejaculation (**spermarche**). The testes and ovaries in turn produce hormones that lead to the development of secondary sex characteristics, such as the growth of pubic hair, deepening of the voice and growth of facial hair in men, and the growth of breasts in women (FIGURE 9.11).

Age-related physical changes that occur after the obvious pubertal changes are less dramatic. Other than some modest increases in height and muscular development during the late teens and early twenties, most individuals experience only minor physical changes until middle age.

For women, **menopause**, the cessation of the menstrual cycle, which occurs somewhere between the ages of 45 and 55, is the second most important life milestone in physical development. The decreased production of estrogen (the dominant female hormone) produces

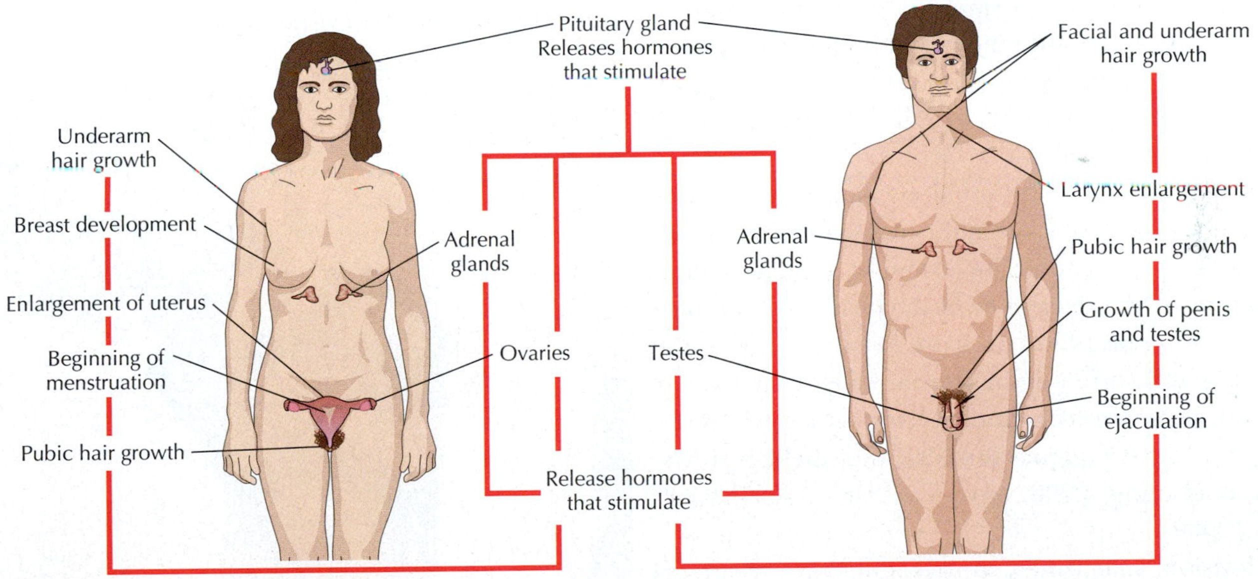

Secondary sex characteristics FIGURE 9.11

Complex physical changes in puberty primarily result from hormones secreted from the ovaries and testes, the pituitary gland in the brain, and adrenal glands near the kidneys.

certain physical changes. However, the popular belief that menopause (or "the change of life") causes serious psychological mood swings, loss of sexual interest, and depression is not supported by current research (Matlin, 2008; Tom, 2008). In fact, most studies find that postmenopausal women report relief, increased libido, and other positive reactions to the end of their menstrual cycles (Chrisler, 2008; Leon et al., 2007). When psychological problems do arise, they may reflect the social devaluation of aging women, not the physiological process of menopause. Given that in Western society women are highly valued for their youth and beauty, a biological process such as aging can be difficult for some women. Women in cultures that derogate aging tend to experience more anxiety and depression during menopause (Mingo, Herman, & Jasperse, 2000; Sampselle et al., 2002; Winterich, 2003).

For men, youthfulness is less important and the physical changes of middle age are less obvious. Beginning in middle adulthood, men experience a gradual decline in the production of sperm and testosterone (the dominant male hormone), although they may remain capable of reproduction into their eighties or nineties. Physical changes such as unexpected weight gain, decline in sexual responsiveness, loss of muscle strength, and graying or loss of hair may lead some men (and women as well) to feel depressed and to question their life progress. They often see these alterations as a biological signal of aging and mortality. Such physical and psychological changes in some men are reportedly known as the **male climacteric**.

After middle age, most physical changes in development are gradual and occur in the heart and arteries and in the sensory receptors. For example, cardiac output (the volume of blood pumped by the heart each minute) decreases, whereas blood pressure increases due to the thickening and stiffening of arterial walls. Visual acuity and depth perception decline, hearing acuity lessens (especially for high-frequency sounds), and smell sensitivity decreases (Chung, 2006; Snyder & Alain, 2007; Whitbourne, 2009).

Television, magazines, movies, and advertisements generally portray aging as a time of balding and graying hair, sagging parts, poor vision, hearing loss, and, of course, no sex life. Such negative portrayals contribute to our society's widespread **ageism**—prejudice or discrimination based on physical age. However, as advertising companies pursue the revenue of the huge aging baby boomer population, there has been a recent shift toward a more accurate portrayal of aging as a time of vigor, interest, and productivity (FIGURES 9.12 and 9.13).

What about memory problems and inherited genetic tendencies toward Alzheimer's disease and other serious diseases of old age? The public and most researchers have long thought that aging is accompanied by widespread death of neurons in the brain. Although this decline does happen with degenerative disorders like Alzheimer's disease, this is no longer believed to be a part of normal aging (Chapter 2). It is also important to remember that age-related memory problems are not on a continuum with Alzheimer's disease (Wilson et al., 2000). That is, normal forgetfulness does not mean that serious dementia is around the corner.

Aging does seem to take its toll on the speed of information processing (Chapter 7). Decreased speed of

Use it or lose it? FIGURE 9.12

Contrary to the unfortunate (and untrue) stereotype that, "You can't teach an old dog new tricks," our cognitive abilities generally grow and improve throughout our lifespan.

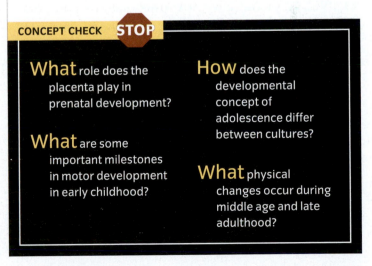

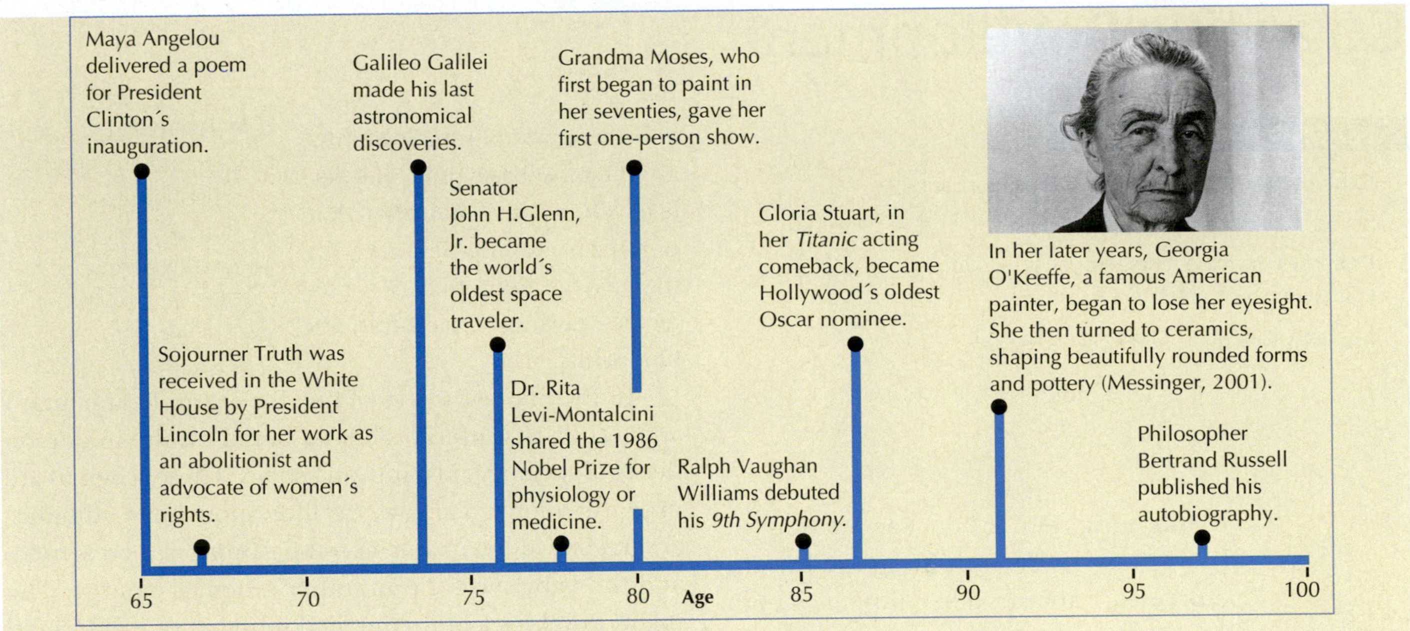
Maya Angelou delivered a poem for President Clinton´s inauguration.
Galileo Galilei made his last astronomical discoveries.
Grandma Moses, who first began to paint in her seventies, gave her first one-person show.
Senator John H.Glenn, Jr. became the world´s oldest space traveler.
Gloria Stuart, in her *Titanic* acting comeback, became Hollywood´s oldest Oscar nominee.
In her later years, Georgia O'Keeffe, a famous American painter, began to lose her eyesight. She then turned to ceramics, shaping beautifully rounded forms and pottery (Messinger, 2001).
Sojourner Truth was received in the White House by President Lincoln for her work as an abolitionist and advocate of women´s rights.
Dr. Rita Levi-Montalcini shared the 1986 Nobel Prize for physiology or medicine.
Ralph Vaughan Williams debuted his *9th Symphony*.
Philosopher Bertrand Russell published his autobiography.
65
70
75
80
Age
85
90
95
100

Achievement in later years FIGURE 9.13

processing may reflect problems with encoding (putting information into long-term storage) and retrieval (getting information out of storage). If memory is like a filing system, older people may have more filing cabinets, and it may take them longer to initially file and later retrieve information. Although mental speed declines with age, general information processing and memory ability is largely unaffected by the aging process (Lachman, 2004; Whitbourne, 2009).

What causes us to age and die? If we set aside contributions from **secondary aging** (changes resulting from disease, disuse, or neglect), we are left to consider **primary aging** (gradual, inevitable age-related changes in physical and mental processes). There are two main theories explaining primary aging and death: *programmed theory* and *damage theory*.

According to the **programmed theory**, aging is genetically controlled. Once the ovum is fertilized, the program for aging and death is set and begins to run. Researcher Leonard Hayflick (1977, 1996) found that human cells seem to have a built-in life span. He observed that after doubling about 50 times, laboratory-cultured cells ceased to divide—they reached the Hayflick limit. The other explanation of primary aging is **damage theory**, which proposes that an accumulation of damage to cells and organs over the years ultimately causes death.

Whether aging is genetically controlled or caused by accumulated damage over the years, scientists generally agree that humans appear to have a maximum life span of about 110 to 120 years. Although we can try to control secondary aging in an attempt to reach that maximum, so far we have no means to postpone primary aging.

CONCEPT CHECK **STOP**

What role does the placenta play in prenatal development?

What are some important milestones in motor development in early childhood?

How does the developmental concept of adolescence differ between cultures?

What physical changes occur during middle age and late adulthood?

Cognitive Development

LEARNING OBJECTIVES

Explain the role of schemas, assimilation, and accommodation in cognitive development.

Describe the major characteristics of Piaget's four stages of cognitive development.

Identify two critiques of Piaget's theory.

The following fan letter was written to Shari Lewis (1963), a children's television performer, about her puppet Lamb Chop:

Dear Shari:
All my friends say Lamb Chop isn't really a little girl that talks. She is just a puppet you made out of a sock. I don't care even if it's true. I like the way Lamb Chop talks. If I send you one of my socks will you teach it how to talk and send it back?

Randi

Randi's understanding of fantasy and reality is certainly different from an adult's. Just as a child's body and physical abilities change, his or her way of knowing and perceiving the world also grows and changes. This seems intuitively obvious, but early psychologists—with one exception—focused on physical, emotional, language, and personality development. The one major exception was Jean Piaget (pronounced pee–ah– ZHAY).

Piaget demonstrated that a child's intellect is fundamentally different from an adult's. He showed that an infant begins at a cognitively "primitive" level and that intellectual growth progresses in distinct stages, motivated by an innate need to know. Piaget's theory, developed in the 1920s and 1930s, has proven so comprehensive and insightful that it remains the major force in the cognitive area of developmental psychology today.

To appreciate Piaget's contributions, we need to consider three major concepts: schemas, assimilation, and accommodation. **Schemas** are the most basic units of intellect. They act as patterns that organize our interactions with the environment, like an architect's drawings or a builder's blueprints.

schemas Cognitive structures or patterns consisting of a number of organized ideas that grow and differentiate with experience.

In the first few weeks of life, for example, the infant apparently has several schemas based on the innate reflexes of sucking, grasping, and so on. These schemas are primarily motor and may be little more than stimulus-and-response mechanisms—the nipple is presented, and the baby sucks. Soon, other schemas emerge. The infant develops a more detailed schema for eating solid food, a different schema for the concepts of "mother" and "father," and so on. Schemas, our tools for learning about the world, expand and change throughout our lives. For example, music lovers who were previously accustomed to LP records and cassette tapes have had to develop schemas for playing CDs and MP3s.

Assimilation and **accommodation** are the two major processes by which schemas grow and change over time. Assimilation is the process of absorbing new information into existing schemas. For instance, infants use their sucking schema not only in sucking nipples but also in sucking blankets and fingers.

assimilation In Piaget's theory, the process of absorbing new information into existing schemas.

accommodation In Piaget's theory, the process of adjusting old schemas or developing new ones to better fit with new

Accommodation occurs when new information or stimuli cannot be assimilated and new schemas are developed or old schemas are changed to better fit with the new information. An infant's first attempt to eat solid food with a spoon is a good example of accommodation. When the spoon first enters her mouth, the child attempts to assimilate it by using the previously successful sucking schema—shaping the lips and tongue around the spoon as if they were around a nipple. After repeated trials, she accommodates by adjusting her lips and tongue in a way that moves the food off the spoon and into her mouth.

STAGES OF COGNITIVE DEVELOPMENT: BIRTH TO ADOLESCENCE

According to Piaget, all children go through approximately the same four stages of cognitive development, regardless of the culture in which they live (FIGURE 9.14). Stages cannot be skipped because skills acquired at earlier stages are essential to mastery at later stages. Let's take a closer look at these four stages: sensorimotor, preoperational, concrete operational, and formal operational.

Piaget's four stages of cognitive development FIGURE 9.14

Process Diagram

1 *Sensorimotor stage (birth to age 2)*

Abilities:

Uses senses and motor skills to explore and develop cognitively.

Limits:

Beginning of stage lacks *object permanence* (understanding that things continue to exist even when not seen, heard, or felt).

2 *Preoperational stage (ages 2 to 7)*

Abilities:

Has significant language and thinks symbolically.

Limits:

- Cannot perform "operations."
- *Egocentric* thinking (inability to consider another's point of view).
- *Animistic* thinking (believing all things are living).

3 *Concrete operational stage (ages 7 to 11)*

Abilities:

- Can perform "operations" on concrete objects.
- Understands conservation (realizing that changes in shape or appearance can be reversed).

Limits:

Cannot think abstractly and hypothetically.

4 *Formal operational stage (age 11 and over)*

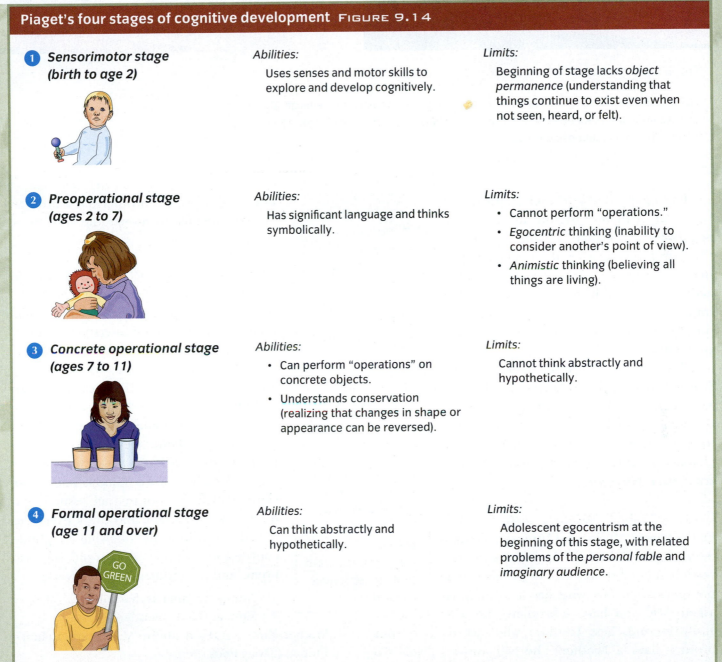

Abilities:

Can think abstractly and hypothetically.

Limits:

Adolescent egocentrism at the beginning of this stage, with related problems of the *personal fable* and *imaginary audience*.

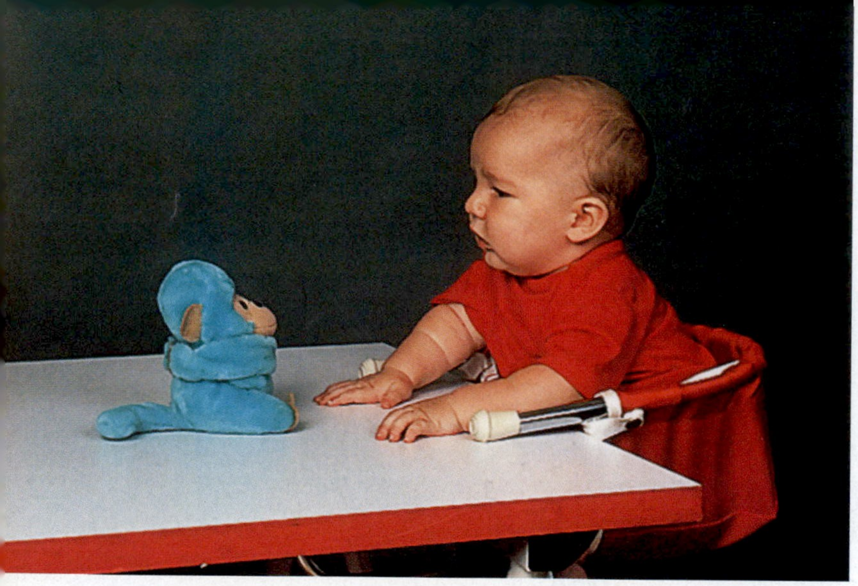

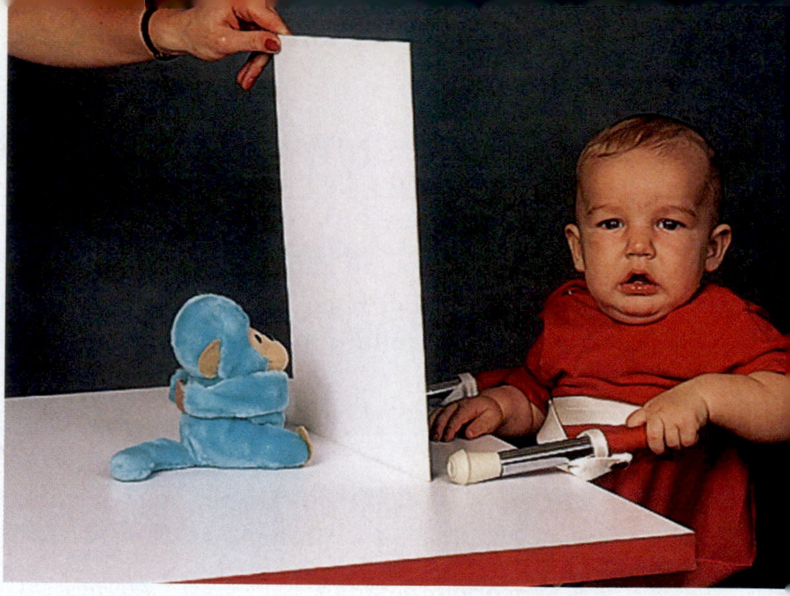

Object permanence FIGURE 9.15

At birth and for the next three or four months, children lack object permanence. They seem to have no schemas for objects they cannot see, hear, or touch—out of sight is truly out of mind. Why do you suppose this happens?

During the **sensorimotor stage**, which lasts from birth until "significant" language acquisition (about age 2), children explore the world and develop their schemas primarily through their senses and motor activities—hence the term *sensorimotor*. One important concept that infants acquire during the sensorimotor stage is **object permanence** (FIGURE 9.15).

During the **preoperational stage** (roughly ages 2 to 7), language advances significantly, and the child begins to think symbolically—using symbols, such as words, to represent concepts. Three other qualities characterize this stage: *lack of operations*, *egocentrism*, and *animism*.

sensorimotor stage Piaget's first stage (birth to approximately age 2), in which schemas are developed through sensory and motor activities.

preoperational stage Piaget's second stage (roughly ages 2 to 7 years), which is characterized by the ability to employ significant language and to think symbolically, though the child lacks operations (reversible mental processes), and thinking is egocentric and animistic.

Concepts are not yet operational

Piaget labeled this period "preoperational" because the child lacks **operations**, or reversible mental processes. For instance, if a preoperational boy who has a brother is asked, "Do you have a brother?" he will easily respond, "Yes." However, when asked, "Does your brother have a brother?" he will answer, "No!" To understand that his brother has a brother, he must be able to reverse the concept of "having a brother."

Thinking is egocentric

Children at this stage have difficulty understanding that there are points of view other than their own. **Egocentrism** refers to the preoperational child's limited ability to distinguish between his or her own perspective and someone else's. (It does not mean "selfishness" in the ordinary sense of the word.) The preschooler who moves in front of you to get a better view of the TV or repeatedly asks questions while you are talking on the telephone is demonstrating egocentrism. Children of this age assume that others see, hear, feel, and think exactly as they do. Consider the following telephone conversation between a 3-year-old, who is at home, and her mother, who is at work:

Mother: Emma, is that you?
Emma: (Nods silently.)
Mother: Emma, is Daddy there? May I speak to him?
Emma: (Twice nods silently.)

Egocentric preoperational children fail to understand that the phone caller cannot see their nodding head. Charming as this is, preoperational children's egocentrism also sometimes leads them to believe that their "bad thoughts" caused their sibling or parent to get sick or that their misbehavior caused their parents' marital problems. Because they think the world centers on them, they often cannot separate reality from what goes on inside their own heads.

Thinking is animistic Children in the preoperational stage believe that objects such as the sun, trees, clouds, and bars of soap have motives, feelings, and intentions (for example, "dark clouds are angry" and "soap sinks because it is tired"). **Animism** refers to the belief that all things are living (or animated). Our earlier example of Randi's letter asking puppeteer Shari Lewis to teach her sock to talk like Lamb Chop is also an example of animistic thinking.

Can preoperational children be taught how to use operations and to avoid egocentric and animistic thinking? Although some researchers have reported success in accelerating the preoperational stage, Piaget did not believe in pushing children ahead of their own developmental schedules. He believed that children should grow at their own pace, with minimal adult interference (Elkind, 1981, 2000). In fact, Piaget thought that Americans were particularly guilty of pushing their children, calling American childhood the "Great American Kid Race."

Following the preoperational stage, at approximately age 7, children enter the **concrete operational stage**. During this stage, many important thinking skills emerge. Unlike the preoperational stage, concrete operational children perform operations on concrete objects. Because they understand the concept of reversibility, they recognize that certain physical attributes (such as volume) remain unchanged when the outward appearance of an object is altered, a process known as **conservation** (FIGURE 9.16).

concrete operational stage Piaget's third stage (roughly ages 7 to 11), in which the child can perform mental operations on concrete objects and understand reversibility and conservation, though abstract thinking is not yet present.

Test for conservation FIGURE 9.16

A In the classic conservation of liquids test, the child is first shown two identical glasses with liquid at the same level.

B The liquid is then poured from one of the short, wide glasses into the tall, thin one.

C When asked whether the two glasses now have the same amount, or if one has more than the other, a preoperational child will reply that the tall, thin glass has more. This demonstrates a failure to conserve volume.

Applying Psychology

Putting Piaget to the Test

If you have access to children in the preoperational or concrete operational stages, try some of the following experiments, which researchers use to test Piaget's various forms of conservation. The equipment is easily obtained, and you will find their responses fascinating. Keep in mind that this should be done as a game. The child should not feel that he or she is failing a test or making a mistake.

Type of Conservation Task (Average age at which concept is grasped)	**Your task as experimenter . . .**	**Child is asked . . .**
Length (ages 6–7)	**Step 1** Center two sticks of equal length. Child agrees that they are of equal length. 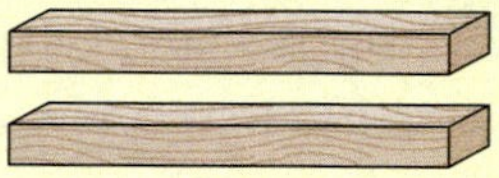 **Step 2** Move one stick. 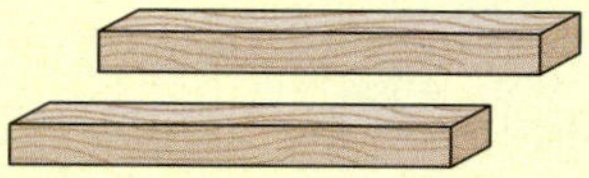	**Step 3** *"Which stick is longer?"* Preoperational child will say that one of the sticks is longer. Child in concrete stage will say that they are both the same length.
Substance amount (ages 6–7)	**Step 1** Center two identical clay balls. Child acknowledges that the two have equal amounts of clay. 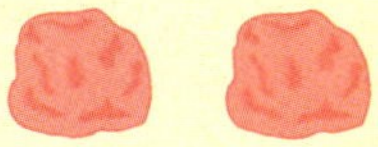 **Step 2** Flatten one of the balls. 	**Step 3** *"Do the two pieces have the same amount of clay?"* Preoperational child will say that the flat piece has more clay. Child in concrete stage will say that the two pieces have the same amount of clay.
Area (ages 8–10)	**Step 1** Center two identical sheets of cardboard with wooden blocks placed on them in identical positions. Child acknowledges that the same amount of space is left open on each piece of cardboard. 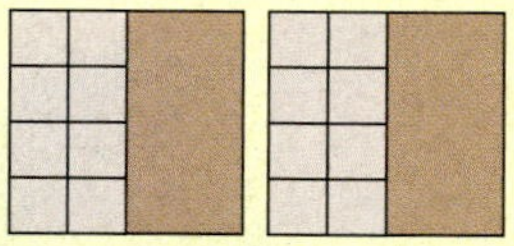 **Step 2** Scatter the blocks on one piece of the cardboard. 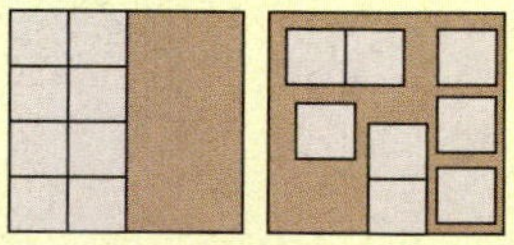	**Step 3** *"Do the two pieces of cardboard have the same amount of open space?"* Preoperational child will say that the cardboard with scattered blocks has less open space. Child in concrete stage will say that both pieces have the same amount of open space.

CRITICAL THINKING

Here are two interesting questions:

1. **Based on their responses, are the children you tested in the preoperational or concrete stage?**
2. **If you repeat the same tests with each child, do their answers change? Why or why not?**

formal operational stage Piaget's fourth stage (around age 11 and beyond), which is characterized by abstract and hypothetical thinking.

The final stage in Piaget's theory is the **formal operational stage**, which typically begins around age 11. In this stage, children begin to apply their operations to abstract concepts in addition to concrete objects. They also become capable of hypothetical thinking ("What if?"), which allows systematic formulation and testing of concepts.

For example, before filling out applications for part-time jobs, adolescents may think about possible conflicts with school and friends, the number of hours they want to work, and the kind of work for which they are qualified. Formal operational thinking also allows the adolescent to construct a well-reasoned argument based on hypothetical concepts and logical processes. Consider the following argument:

1. If you hit a glass with a feather, the glass will break.
2. You hit the glass with a feather.

What is the logical conclusion? The correct answer, "The glass will break," is contrary to fact and direct experience. Therefore, the child in the concrete operational stage would have difficulty with this task, whereas the formal operational thinker understands that this problem is about abstractions that need not correspond to the real world.

Along with the benefits of this cognitive style come several problems. Adolescents in the early stages of the formal operational period demonstrate a type of egocentrism different from that of the preoperational child. Although adolescents recognize that others have unique thoughts and perspectives, they often fail to differentiate between what they are thinking and what others are thinking. This adolescent egocentrism has two characteristics that may affect social interactions as well as problem solving:

- *Personal fable.* Because of their unique form of egocentrism, adolescents may conclude that they alone are having insights or difficulties and that no one else understands or sympathizes with them. David Elkind (1976, 2001, 2007) described this as the formation of a **personal fable**, an intense investment in an adolescent's own thoughts and feelings and a belief that these thoughts are unique. For example, one young woman remembered being very upset in middle school when her mother tried to comfort her over the loss of an important relationship. "I felt like she couldn't possibly know how it felt—no one could. I couldn't believe that anyone had ever suffered like this or that things would ever get better."

  Several forms of risk taking, such as engaging in sexual intercourse without contraception, driving dangerously, and experimenting with drugs, seem to arise from the personal fable (Alberts, Elkind, & Ginsburg, 2007; Flavell, Miller, & Miller, 2002; Greene et al., 2000). Adolescents have a sense of uniqueness, invulnerability, and immortality. They recognize the dangers of these activities but think the rules don't apply to them (FIGURE 9.17).

The personal fable FIGURE 9.17

Thanks to advances in brain imaging, scientists now know that the prefrontal cortex of the adolescent's brain is one of the later areas to develop (Giedd, 2008; Steinberg, 2008). Given that the prefrontal cortex is responsible for higher processes, such as planning ahead and controlling emotions, does this help explain Piaget's concept of the *personal fable* and its consequent higher risk taking?

- *Imaginary audience.* In early adolescence, people tend to believe that they are the center of others' thoughts and attentions, instead of considering that everyone is equally wrapped up in his or her own concerns and plans. In other words, adolescents feel that all eyes are focused on their behaviors. Elkind referred to this as the **imaginary audience**.

If the imaginary audience results from an inability to differentiate the self from others, the personal fable is a product of differentiating too much. Thankfully, these two forms of adolescent egocentrism tend to decrease during later stages of the formal operational period.

ASSESSING PIAGET'S THEORY: CRITICISMS AND CONTRIBUTIONS

As influential as Piaget's account of cognitive development has been, it has received significant criticisms. Let's look briefly at two major areas of concern: underestimated abilities and underestimated genetic and cultural influences.

Research shows that Piaget may have underestimated young children's cognitive development (FIGURES 9.18 and 9.19). For example, researchers report that very young infants have a basic concept of how objects move, are aware that objects continue to exist even when screened from view, and can recognize speech sounds (Baillargeon, 2000, 2008; Charles, 2007).

Nonegocentric responses also appear in the earliest days of life. For example, newborn babies tend to cry in response to the cry of another baby (Diego & Jones, 2007; Dondi, Simon, & Caltran, 1999). Also, preschoolers will adapt their speech by using shorter, simpler expressions when talking to 2-year-olds than to adults.

Piaget's model, like other stage theories, has also been criticized for not sufficiently taking into account genetic and cultural differences (Cole & Gajdamaschko, 2007; Matusov & Hayes, 2000; Maynard & Greenfield,

Visualizing

Infant imitation FIGURE 9.18

In a series of well-known studies, Andrew Meltzoff and M. Keith Moore (1977, 1985, 1994) found that newborns could imitate such facial movements as tongue protrusion, mouth opening, and lip pursing. At 9 months, infants will imitate facial actions a full day after seeing them (Heimann & Meltzoff, 1996). Can you see how this early infant facial expression raises questions about Piaget's estimates of early infant abilities?

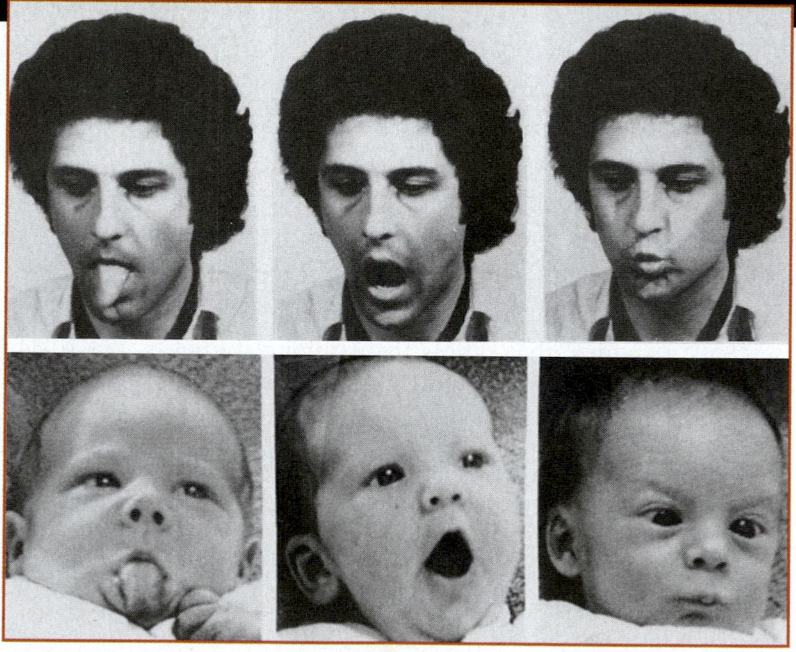

When an adult models a facial expression, even very young infants will respond with a similar expression. Is this true imitation or a simple stimulus-response reflex?

Are preoperational children always egocentric?

FIGURE 9.19

Some toddlers and preschoolers clearly demonstrate empathy for other people. How does this ability to take others' perspective contradict Piaget's beliefs about egocentrism in very young children?

2003). During Piaget's time, the genetic influences on cognitive abilities were poorly understood, but there has been a rapid explosion of information in this field in recent years. In addition, formal education and specific cultural experiences can also significantly affect cognitive development. Consider the following example from a researcher attempting to test the formal operational skills of a farmer in Liberia (Scribner, 1977):

> *Researcher:* All Kpelle men are rice farmers. Mr. Smith is not a rice farmer. Is he a Kpelle man?
>
> *Kpelle farmer:* I don't know the man. I have not laid eyes on the man myself.

Instead of reasoning in the "logical" way of Piaget's formal operational stage, the Kpelle farmer reasoned according to his specific cultural and educational training, which apparently emphasized personal knowledge. Not knowing Mr. Smith, the Kpelle farmer did not feel qualified to comment on him. Thus, Piaget's theory may have underestimated the effect of culture on a person's cognitive functioning.

Despite criticisms, Piaget's contributions to psychology are enormous. As one scholar put it, "assessing the impact of Piaget on developmental psychology is like assessing the impact of Shakespeare on English literature or Aristotle on philosophy—impossible" (Beilin, 1992, p. 191).

CONCEPT CHECK STOP

When does accommodation occur?

How do egocentrism and animism limit children's thinking during the preoperational stage?

Why might the personal fable explain risky behavior among adolescents?

What criticisms has Piaget's theory received?

SUMMARY

1 Studying Development

1. **Developmental psychology** is the study of age-related changes in behavior and mental processes from conception to death. Development is an ongoing, lifelong process.
2. The three most important debates or questions in human development are about nature versus nurture (including studies of **maturation** and **critical periods**), continuity versus stages, and stability versus change. For each question, most psychologists prefer an interactionist perspective.
3. Developmental psychologists use two research methods: the **cross-sectional method** and the **longitudinal method**. Although both have valuable attributes, each also has disadvantages. Cross-sectional studies can confuse true developmental effects with cohort effects. On the other hand, longitudinal studies are expensive and time-consuming, and their results are restricted in generalizability.

2 Physical Development

1. The early years of development are characterized by rapid change. Prenatal development begins at conception and is divided into three stages: the **germinal period**, the **embryonic period**, and the **fetal period**. During pregnancy, the placenta serves as the link for food and the excretion of wastes, and it screens out some harmful substances. However, some environmental hazards (teratogens), such as alcohol and nicotine, can cross the placental barrier and endanger prenatal development.

2. Early childhood is also a time of rapid physical development, including brain, motor, and sensory/perceptual development.

3. During adolescence, both boys and girls undergo dramatic changes in appearance and physical capacity. Puberty is the period of adolescence when a person becomes capable of reproduction. The clearest and most dramatic physical sign of puberty is the growth spurt, characterized by rapid increases in height, weight, and skeletal growth and by significant changes in reproductive structures and sexual characteristics.

4. During adulthood, most individuals experience only minor physical changes until middle age. Around age 45–55, women experience menopause, the cessation of the menstrual cycle. At the same time, men experience a gradual decline in the production of sperm and testosterone, as well as other physical changes, known as the male climacteric.

5. After middle age, most physical changes in development are gradual and occur in the heart and arteries and in the sensory receptors. Most researchers, as well as the public in general, have long thought that aging is accompanied by widespread death of neurons in the brain. Although this decline does happen with degenerative disorders like Alzheimer's disease, this is no longer believed to be a part of normal aging. Although mental speed declines with age, general information processing and much of memory ability are largely unaffected by the aging process. There are two main theories explaining primary aging and death—programmed theory and damage theory.

3 Cognitive Development

1. In the 1920s and 1930s, Piaget conducted groundbreaking work on children's cognitive development. Piaget's theory remains the major force in the cognitive area of developmental psychology today.

2. Three major concepts are central to Piaget's theory: **schemas**, **assimilation**, and **accommodation**. According to Piaget, all children progress through four stages of cognitive development: the **sensorimotor stage**, the **preoperational stage**, the **concrete operational stage**, and the **formal operational stage**. As they progress through these stages, children acquire progressively more sophisticated ways of thinking.

3. Piaget's account of cognitive development has been enormously influential, but it has also received significant criticisms. In particular, research shows that Piaget may have underestimated infants' and young children's cognitive abilities, and he may have underestimated genetic and cultural influences on cognitive development.

KEY TERMS

- developmental psychology p. 228
- maturation p. 228
- critical period p. 228
- interactionist perspective p. 229
- cross-sectional method p. 230
- longitudinal method p. 230
- cohort effects p. 230
- zygote p. 232
- germinal period p. 232
- embryonic period p. 232
- fetal period p. 232
- proximodistal p. 232
- placenta p. 232
- teratogens p. 234
- fetal alcohol syndrome (FAS) p. 234
- puberty p. 238
- adolescence p. 238
- growth spurt p. 238
- menarche p. 238
- spermarche p. 239
- menopause p. 239
- male climacteric p. 240
- ageism p. 240
- secondary aging p. 241
- primary aging p. 241
- programmed theory p. 241
- damage theory p. 241
- schemas p. 242
- assimilation p. 242
- accommodation p. 242
- sensorimotor stage p. 244
- object permanence p. 244
- preoperational stage p. 244
- operations p. 244
- egocentrism p. 244
- animism p. 245
- concrete operational stage p. 245
- conservation p. 245
- formal operational stage p. 247
- personal fable p. 247
- imaginary audience p. 248

CRITICAL AND CREATIVE THINKING QUESTIONS

1. Based on what you have learned about the advantages and disadvantages of cross-sectional and longitudinal methods, can you think of a circumstance when each might be preferable over the other?
2. If a mother knowingly ingests a quantity of alcohol that causes her child to develop fetal alcohol syndrome (FAS), is she guilty of child abuse? Why or why not?
3. From an evolutionary perspective, why might babies prefer looking at more complex patterns and at faces, rather than simple patterns?
4. Based on what you have learned about development during late adulthood, do you think that this period is necessarily a time of physical and mental decline?
5. Piaget's theory states that all children progress through all of the discrete stages of cognitive development in order and without skipping any. Do you agree with this theory? Do you know any children who seem to contradict this theory?

What is happening in this picture?

- What stage of cognitive development does this child's behavior typify?
- What schemas might the child build by "exploring" her food in this way?

SELF-TEST

(Check your answers in Appendix B.)

1. The study of age-related changes in behavior and mental processes from conception to death is called ________________.
a. thanatology
b. neo-gerontology
c. developmental psychology
d. longitudinal psychology

2. This is development governed by automatic, genetically predetermined signals.
a. growth
b. natural progression
c. maturation
d. tabula rasa

3. Label the two basic types of research studies in the figure below:

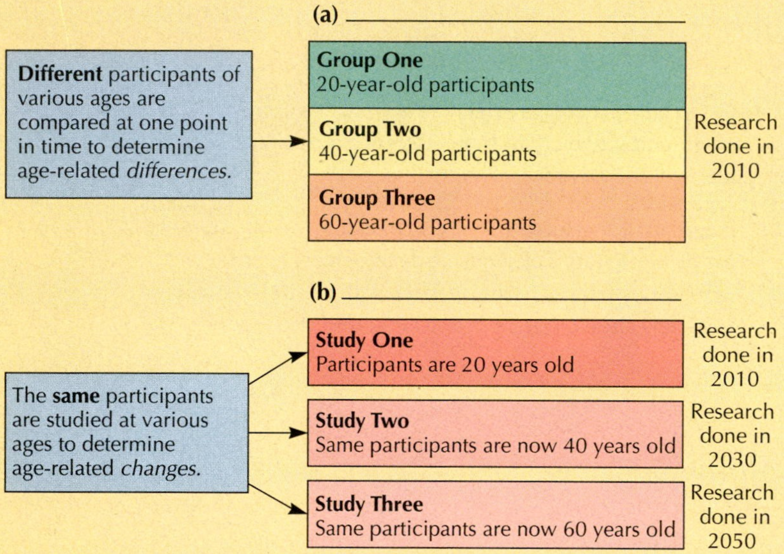

4. This is the first stage of prenatal development, which begins with conception and ends with implantation in the uterus (the first two weeks).
a. embryosis
b. zygote stage
c. critical period
d. germinal period

5. At birth, an infant's head is ________________ its body's size, whereas in adulthood, the head is ________________ its body's size.
a. 1/3; 1/4
b. 1/3; 1/10
c. 1/4; 1/10
d. 1/4; 1/8

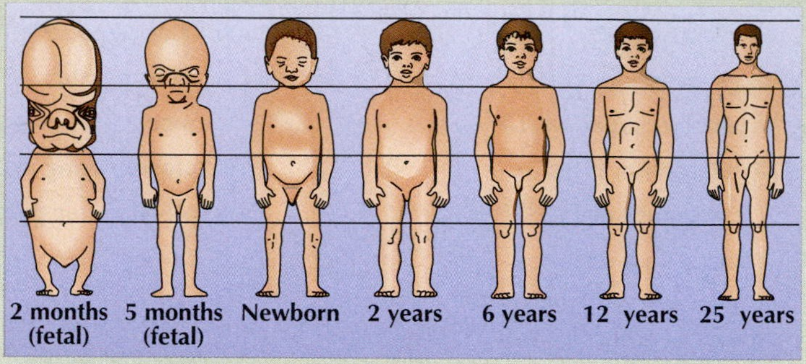

6. Which of the following is **NOT** true regarding infant sensory and perceptual development?
a. Vision is almost 20/20 at birth.
b. A newborn's sense of pain is highly developed at birth.
c. An infant can recognize, and prefers, its own mother's breast milk by smell.
d. An infant can recognize, and prefers, its own mother's breast milk by taste.

7. The clearest and most physical sign of puberty is the ________________, characterized by rapid increases in height, weight, and skeletal growth.
a. menses
b. spermarche
c. growth spurt
d. age of fertility

8. Some employers are reluctant to hire 50- to 60-year-old workers because of a generalized belief that they are sickly and will take too much time off. This is an example of ________________.
a. discrimination
b. prejudice
c. ageism
d. all of these options

9. ________________ was one of the first scientists to demonstrate that a child's intellect is fundamentally different from an adult's.
   a. Baumrind
   b. Beck
   c. Piaget
   d. Elkind

10. ________________ occurs when existing schemas are used to interpret new information, whereas ________________ involves changes and adaptations of the schemas.
   a. Adaptation; accommodation
   b. Adaptation; reversibility
   c. Egocentrism; postschematization
   d. Assimilation; accommodation

11. Label the four stages of Piaget's cognitive development model on the table below.

Piaget's four stages of cognitive development

(a) ______________________________

(b) ______________________________

(c) ______________________________

(d) ______________________________

12. The photos below are an example of ________________.
   a. sensory permanence
   b. perceptual constancy
   c. perceptual permanence
   d. object permanence

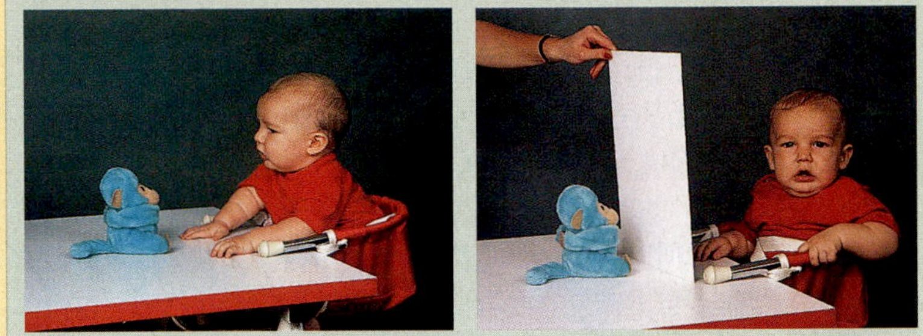

13. The ability to think abstractly or hypothetically occurs in Piaget's ________________ stage.
   a. egocentric
   b. postoperational
   c. formal operational
   d. concrete operational

14. Extreme forms of self-consciousness and concerns for physical appearance are common in adolescents, who tend to exhibit the early formal operational characteristic called ________________.
   a. the personal fable
   b. adolescent egocentrism
   c. the imaginary audience
   d. the egocentric fable

15. Critics have suggested that Piaget underestimated ________________.
   a. the cognitive abilities of infants and young children
   b. genetic influences on cognition
   c. the effect of cultural experiences on cognition
   d. all of these options

Life Span Development II

10

Imagine that you are standing on a bridge over a railroad track when you see that a runaway train is about to kill five people. Coincidentally, you are standing next to a switching mechanism, and you realize that by simply throwing a switch, you can divert the train onto a spur, allowing the five to survive. But here is the catch: diverting the train will condemn *one* person, who is standing on the spur, to death. What would you do? Would you allow one person to die in order to save five others? Would your answer be different if that person were your mother, father, or some other much-loved person? What might lead different people to make different decisions?

It's unlikely that you will ever have to make such a gruesome choice. Yet everyone encounters moral dilemmas from time to time. (Should you remind the cable company that they forgot to disconnect the cable after you discontinued the service? Should you give spare change to a panhandler in your neighborhood?) Likewise, everyone has personal relationships and particular character traits that color our decisions. How we approach moral dilemmas—as well as many other events and circumstances throughout our lives—reflects several key facets of our personal growth: our social, moral, and personality development. In this chapter, we'll look at these aspects of development. We'll also examine how sex, gender, and culture influence our development. Finally, we'll explore several key developmental challenges during adulthood.

NATIONAL GEOGRAPHIC

CHAPTER OUTLINE

Social, Moral, and Personality Development

LEARNING OBJECTIVES

Describe how attachment influences our social development.

Summarize the central characteristics of Kohlberg's theory of moral development.

Identify Erikson's eight stages of psychosocial development.

In addition to physical and cognitive development (Chapter 9), developmental psychologists study social, moral, and personality development by looking at how social forces and individual differences affect development over the life span. The poet John Donne wrote, "No man is an island, entire of itself." In this section, we focus on three major facets of development that shed light on how we affect each other: attachment, Kohlberg's stages of moral development, and Erikson's psychosocial stages.

Imprinting FIGURE 10.1

Lorenz's studies on imprinting demonstrated that baby geese attach to, and then follow, the first large moving object they see during a certain critical period in their development.

SOCIAL DEVELOPMENT: THE IMPORTANCE OF ATTACHMENT

An infant arrives in the world with a multitude of behaviors that encourage a strong bond of **attachment** with primary caregivers.

attachment A strong affectional bond with special others that endures over time.

In studying attachment behavior, researchers are often divided along the lines of the nature-versus-nurture debate. Those who advocate the nativist, or innate, position cite John Bowlby's work (1969, 1989, 2000). He proposed that newborn infants are biologically equipped with verbal and nonverbal behaviors (such as crying, clinging, and smiling) and with "following" behaviors (such as crawling and walking after the caregiver) that elicit instinctive nurturing responses from the caregiver. Konrad Lorenz's (1937) early studies of **imprinting** further support the biological argument for attachment (FIGURE 10.1). Attachment is discussed in more detail in *What a Psychologist Sees.*

What if a child does not form an attachment? Research shows that infants raised in impersonal surroundings (such as in institutions that do not provide the stimulation and love of a regular caregiver) or under abusive conditions suffer from a number of problems. They seldom cry, coo, or babble; they become rigid when picked up; and they have few language skills. They also tend to form shallow or anxious relationships. Some appear forlorn, withdrawn, and uninterested in their caretakers, whereas others seem insatiable in their need for affection (Zeanah, 2000). Finally, they also tend to show intellectual, physical, and perceptual retardation; increased susceptibility to infection; and neurotic "rocking" and isolation behaviors. In some cases, they die from lack of attachment (Bowlby, 1973, 1982, 2000; Combrink-Graham & McKenna, 2006; Nelson, Zeanah, & Fox, 2007; Spitz & Wolf, 1946; Zeanah, 2000).

Although most children are never exposed to such extreme institutional conditions, developmental psychologist Mary Ainsworth and her colleagues (1967, 1978) have found significant differences in the typical levels of attachment between infants and their mothers.

Attachment: The Power of Touch

In a classic experiment involving infant rhesus monkeys, Harry Harlow and Robert Zimmerman (1959) investigated the variables that might affect attachment. They created two types of wire-framed surrogate (substitute) "mother" monkeys: one covered by soft terry cloth and one left uncovered. (A) The infant monkeys were fed by either the cloth or the wire mother, but they otherwise had access to both mothers. The researchers found that monkeys "reared" by a cloth mother clung frequently to the soft material of their surrogate mother and developed greater emotional security and curiosity than did monkeys assigned to the wire mother.

In later research (Harlow & Harlow, 1966), monkey babies were exposed to rejection. Some of the "mothers" contained metal spikes that would suddenly protrude from the cloth covering and push the babies away; others had air jets that would sometimes blow the babies away. Nevertheless, the infant monkeys waited until the rejection was over and then clung to the cloth mothers as tightly as before. From these and related findings, Harlow concluded that **contact comfort**, the pleasurable tactile sensations provided by a soft and cuddly "parent," is a powerful contributor to attachment.

Maternal contact comfort

Several studies suggest that contact comfort between human infants and mothers is similarly important. For example, touching and massaging premature infants produce significant physical and emotional benefits (Field, 1998, 2007; Feldman, 2007; Hernandez-Reif, Diego, & Field, 2007). Mothers around the world tend to kiss, nuzzle, nurse, comfort, clean, and respond to their children with lots of physical contact. (B) Although almost all research on attachment and contact comfort has focused on mothers and infants, recent research shows that the same results also apply to fathers and other caregivers (C) (Diener et al., 2008; Grossmann et al., 2002; Lindberg, Axelsson, & Öhrling, 2008; Martinelli, 2006).

Fathers matter too.

Using a method called the **strange situation procedure**, in which a researcher observes infants in the presence or absence of their mother and a stranger, Ainsworth found that children could be divided into three groups: securely attached, avoidant, and anxious/ambivalent (FIGURE 10.2).

Ainsworth found that infants with a secure attachment style had caregivers who were sensitive and responsive to their signals of distress, happiness, and fatigue (Ainsworth et al., 1967, 1978; Gini et al., 2007; Higley, 2008; Völker, 2007). On the other hand, avoidant infants had caregivers who were aloof and distant, and anxious/ambivalent infants had inconsistent caregivers who alternated between strong affection and indifference. Follow-up studies found that, over time, securely attached children were the most sociable, emotionally aware, enthusiastic, cooperative, persistent, curious, and competent (Bar-Haim et al., 2007; Brown & Whiteside, 2008; Johnson, Dweck, & Chen, 2007). Interestingly, scientists also have found that an infant's attachment pattern may have lasting effects on their adult romantic relationships (see the *Psychological Science* box).

PARENTING STYLES: THEIR EFFECT ON DEVELOPMENT

How much of our personality comes from the way our parents treat us as we're growing up? Researchers since the 1920s have studied the effects of different methods of childrearing on children's behavior, development, and mental health. Studies done by Diana Baumrind (1980, 1991, 1995) found that parenting styles could be reliably divided into three broad patterns, *permissive, authoritarian,* and *authoritative,* which could be identified by their degree of *control/demandingness* (C) and *warmth/responsiveness* (W) (STUDY ORGANIZER 10.1).

Degrees of attachment FIGURE 10.2

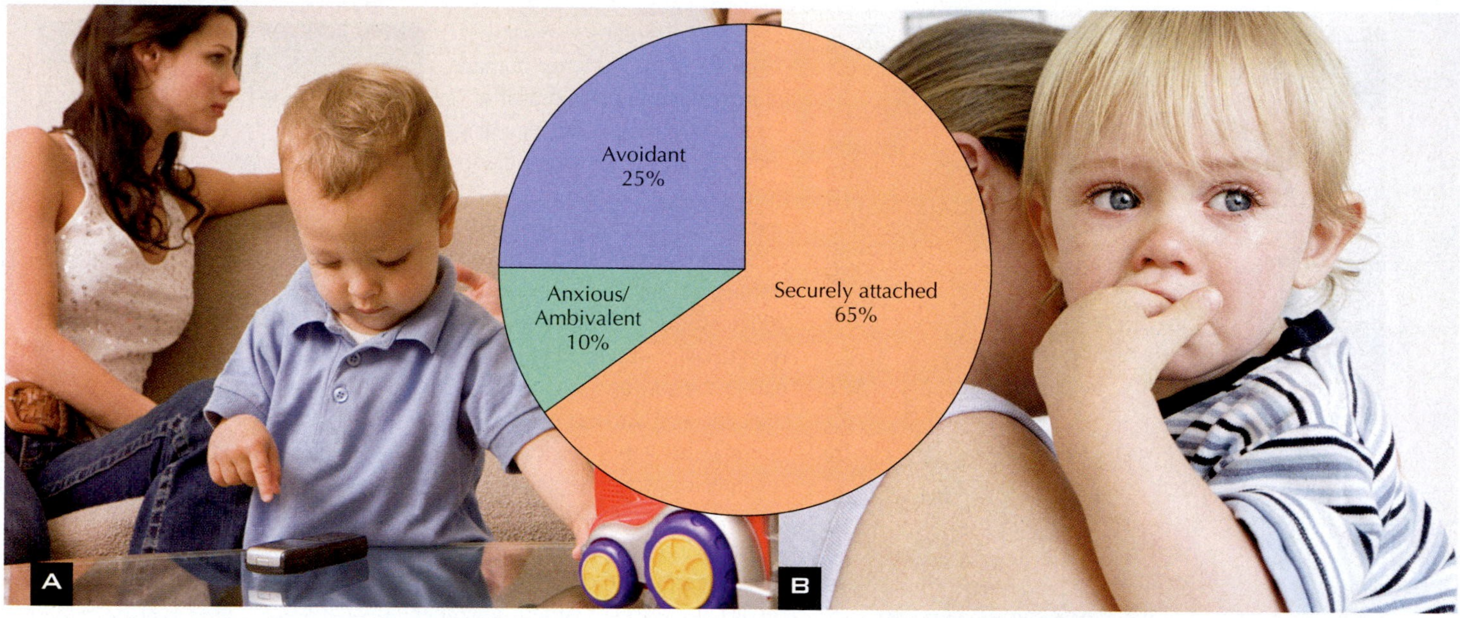

Ainsworth's research identified three different degrees of attachment.

- *Securely attached (65 percent).* When exposed to the stranger, the infant seeks closeness and contact with the mother, (A) uses the mother as a safe base from which to explore, (B) shows moderate distress on separation from the mother, and is happy when the mother returns.
- *Avoidant (25 percent).* The infant does not seek closeness or contact with the mother, treats the mother much like a stranger, and rarely cries when the mother leaves the room.
- *Anxious/Ambivalent (10 percent).* The infant becomes very upset when the mother leaves the room. When she returns, the infant seeks close contact and then squirms angrily to get away.

Study Organizer 10.1 Parenting styles

Parenting Style	*Description*	*Example*	*Effect on Children*
Permissive-neglectful (permissive-indifferent) (low C, low W)	Parents make few demands, with little structure or monitoring, and show little interest or emotional support; may be actively rejecting.	"I don't care about you—or what you do."	Children tend to have poor social skills and little self-control (being overly-demanding and disobedient).
Permissive-indulgent (low C, high W)	Parents set few limits or demands, but are highly involved and emotionally connected.	"I care about you—and you're free to do what you like!"	Children often fail to learn respect for others and tend to be impulsive, immature, and out of control.
Authoritarian (high C, low W)	Parents are rigid and punitive, while also being low on warmth and responsiveness.	"I don't care what you want. Just do it my way, or else!"	Children tend to be easily upset, moody, aggressive, and often fail to learn good communication skills.
Authoritative (high C, high W)	Parents generally set and enforce firm limits, while also being highly involved, tender, and emotionally supportive.	"I really care about you, but there are rules and you need to be responsible."	Children become self-reliant, self-controlled, high achieving, and emotionally well-adjusted; also seem more content, goal oriented, friendly, and socially competent.

Sources: Coplan, Arbeau, & Armer, 2008; Driscoll, Russell, & Crockett, 2008; Martin & Fabes, 2009; McKinney, Donnelly, & Renk, 2008; Shields, 2008.

Psychological Science

Attachment Across the Life Span

If you've been around young children, you've probably noticed how often they share toys and discoveries with a parent and how they seem much happier when a parent is nearby. You may also have thought how cute and sweet it is when infants and parents coo and share baby talk with each other. But have you noticed that these very same behaviors often occur between adults in romantic relationships?

Intrigued by these parallels, several researchers have studied the relationship between an infant's attachment to a parent figure and an adult's love for a romantic partner (Bachman & Zakahi, 2000; Diamond, 2004; Myers & Vetere, 2002). In one study, Cindy Hazan and Phillip Shaver (1987, 1994) discovered that adults who had an avoidant pattern in infancy find it hard to trust others and to self-disclose, and they rarely report finding "true love" (Clulow, 2007; Duncan, 2007; Lele, 2008). In short, they block intimacy by being emotionally aloof and distant.

Anxious/ambivalent infants tend to be obsessed with their romantic partners as adults, fearing that their intense love will not be reciprocated. As a result, they tend to smother intimacy by being possessive and emotionally demanding.

In contrast, individuals who are securely attached as infants easily become close to others, expect intimate relationships to endure, and perceive others as generally trustworthy. As you may expect, the securely attached lover has intimacy patterns that foster long-term relationships and is the most desired partner by the majority of adults, regardless of their own attachment styles (Lele, 2008; Mikulincer & Goodman, 2006; Vorria et al., 2007).

Early attachment experiences may predict romantic love style. How do you think this young woman will relate to her new husband?

As you consider these correlations between infant attachment and adult romantic love styles, remember that it is always risky to infer causation from correlation. Accordingly, the relationship between romantic love style and early infant attachment is subject to several alternative explanations. Also, be aware that early attachment experiences may predict the future, but they do not determine it. Throughout life, we can learn new social skills and different attitudes toward relationships.

CRITICAL THINKING

Here are two interesting questions:

1. **What other factors might account for a romantic love style that does not reflect early infant attachment?**
2. **How do you think an anxious/ambivalent attachment style might affect an individual's relationships later in life?**

MORAL DEVELOPMENT: KOHLBERG'S STAGES

Developing a sense of right and wrong, or morality, is a part of psychological development. Consider the following situation in terms of what you would do.

> *In Europe, a woman was near death from a special kind of cancer. There was one drug that doctors thought might save her. It was a form of radium that a druggist in the same town had recently discovered. The drug was expensive to make, but the druggist was charging 10 times what the drug cost him to make. He paid $200 for the radium and charged $2,000 for a small dose of the drug. The sick woman's husband, Heinz, went to everyone he knew to borrow the money, but he could gather together only about $1,000, half of what it cost. He told the druggist that his wife was dying and asked him to sell it cheaper or let him pay later. But the druggist said, "No, I discovered the drug, and I'm going to make money from it." So Heinz got desperate and broke into the man's store to steal the drug for his wife. (Kohlberg, 1964, pp. 18–19)*

Was Heinz right to steal the drug? What do you consider moral behavior? Is morality "in the eye of the beholder," or are there universal truths and principles? Whatever your answer, your ability to think, reason, and respond to Heinz's dilemma demonstrates another type of development that is very important to psychology—morality.

One of the most influential researchers in moral development was Lawrence Kohlberg (1927–1987). He presented what he called "moral stories" like the Heinz dilemma to people of all ages, and on the basis of his findings, he developed a model of moral development (1964, 1984).

What is the right answer to Heinz's dilemma? Kohlberg was interested not in whether participants judged Heinz to be right or wrong but in the reasons they gave for their decisions. On the basis of participants' responses, Kohlberg proposed three broad levels in the evolution of moral reasoning, each composed of two distinct stages. Individuals at each stage and level may or may not support Heinz's stealing of the drug, but their reasoning changes from level to level.

Kohlberg believed that, like Piaget's stages of cognitive development (Chapter 9), his stages of moral development are universal and invariant. That is, they supposedly exist in all cultures, and everyone goes through each of the stages in a predictable fashion. The age trends that are noticed tend to be rather broad.

Preconventional level

(Stages 1 and 2—birth to adolescence). At the **preconventional level**, moral judgment is self-centered. What is right is what one can get away with or what is personally satisfying. Moral understanding is based on rewards, punishments, and the exchange of favors. This level is called "preconventional" because children have not yet accepted society's (conventional) rule-making processes.

> **preconventional level** Kohlberg's first level of moral development, in which morality is based on rewards, punishment, and the exchange of favors.

- *Stage 1 (punishment-obedience orientation).* Children at this stage focus on self-interest—obedience to authority and avoidance of punishment. Because they also have difficulty considering another's point of view, they ignore people's intentions in their moral judgments. Thus, a 5-year-old will often say that accidentally breaking 15 cups is "badder" and should receive more punishment than intentionally breaking 1 cup.

- *Stage 2 (instrumental-exchange orientation).* During this stage, children become aware of others' perspectives, but their morality is based on reciprocity—an equal exchange of favors. "I'll share my lunch with you because if I ever forget mine you'll share yours with me." The guiding philosophy is "You scratch my back and I'll scratch yours."

Conventional level (Stages 3 and 4—adolescence and young adulthood). At the **conventional level**, moral reasoning advances from being self-centered to other-centered. The individual personally accepts conventional societal rules because they help ensure social order and judges morality in terms of compliance with these rules and values.

conventional level Kohlberg's second level of moral development, in which moral judgments are based on compliance with the rules and values of society.

- *Stage 3 ("good child" orientation).* At Stage 3, the primary moral concern is with being nice and gaining approval. People are also judged by their intentions and motives ("His heart was in the right place").
- *Stage 4 (law-and-order orientation).* During this stage, the individual takes into account a larger perspective—societal laws. Stage 4 individuals understand that if everyone violated laws, even with good intentions, there would be chaos. Thus, doing one's duty and respecting law and order are highly valued. According to Kohlberg, Stage 4 is the highest level attained by most adolescents and adults.

Postconventional level (Stages 5 and 6—adulthood). At the **postconventional level**, individuals develop personal standards for right and wrong. They also define morality in terms of abstract principles and values that apply to all situations and societies. A 20-year-old who judges the "discovery" and settlement of North America by Europeans as immoral because it involved the theft of land from native peoples is thinking in postconventional terms.

postconventional level Kohlberg's highest level of moral development, in which individuals develop personal standards for right and wrong, and they define morality in terms of abstract principles and values that apply to all situations and societies.

- *Stage 5 (social-contract orientation).* Individuals at Stage 5 appreciate the underlying purposes served by laws. When laws are consistent with interests of the majority, they are obeyed because of the "social contract." However, laws can be morally disobeyed if they fail to express the will of the majority or fail to maximize social welfare (FIGURE 10.3).

Postconventional moral reasoning
FIGURE 10.3

Would you travel hundreds of miles to participate in a political demonstration? Why or why not? Would you be willing to be arrested for violating the law to express your moral convictions?

Kohlberg's stages of moral development FIGURE 10.4

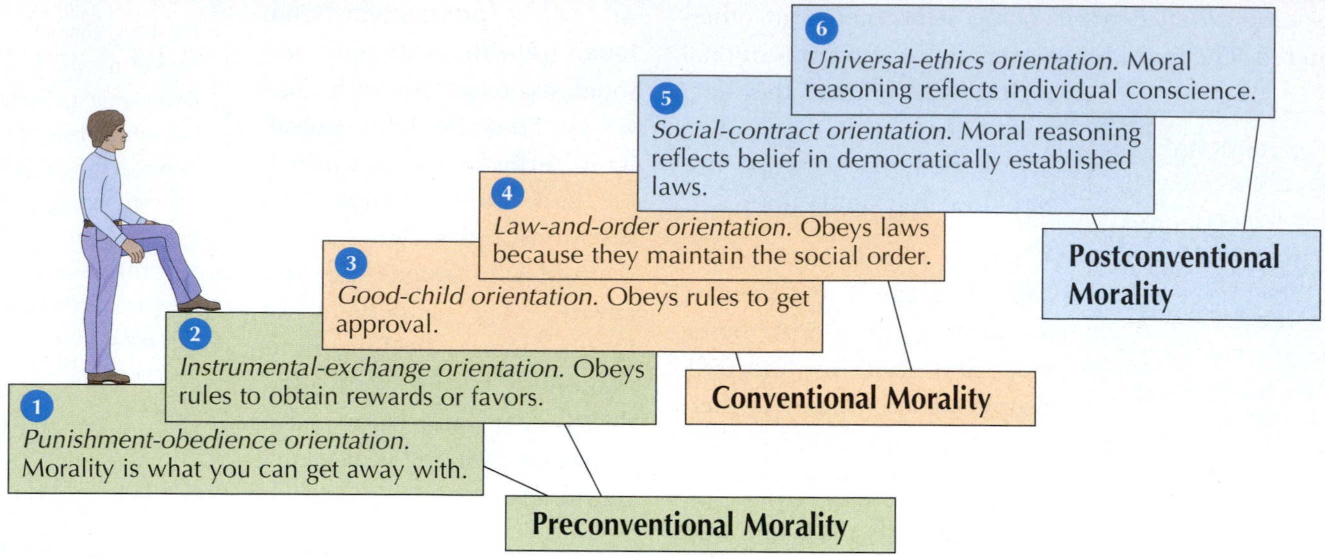

Preconventional level (Stages 1 and 2—birth to adolescence) Moral judgment is *self-centered*. What is right is what one can get away with, or what is personally satisfying. Moral understanding is based on rewards, punishments, and the exchange of favors.

1  Focus is on self-interest—obedience to authority and avoidance of punishment. Because children at this stage have difficulty considering another's point of view, they also ignore people's intentions.

2  Children become aware of others' perspectives, but their morality is based on reciprocity—an equal exchange of favors.

Conventional level (Stages 3 and 4—adolescence and young adulthood) Moral reasoning is *other-centered*. Conventional societal rules are accepted because they help ensure the social order.

3  Primary moral concern is being nice and gaining approval, and judges others by their intentions—"His heart was in the right place."

4  Morality based on a larger perspective—societal laws. Understanding that if everyone violated laws, even with good intentions, there would be chaos.

Postconventional level (Stages 5 and 6—adulthood) Moral judgments based on *personal standards for right and wrong*. Morality also defined in terms of abstract principles and values that apply to all situations and societies.

5  Appreciation for the underlying purposes served by laws. Societal laws are obeyed because of the "social contract," but they can be morally disobeyed if they fail to express the will of the majority or fail to maximize social welfare.

6  "Right" is determined by universal ethical principles (e.g., nonviolence, human dignity, freedom) that *all* religions or moral authorities might view as compelling or fair. These principles apply whether or not they conform to existing laws.

Sources: Adapted from Kohlberg. L. "Stage and Sequence: The Cognitive Developmental Approach to Socialization," in D. A. Goslin, *The Handbook of Socialization Theory and Research.* Chicago: Rand McNally, 1969, p. 376 (Table 6.2).

- *Stage 6 (universal ethics orientation).* At this stage, "right" is determined by universal ethical principles that all religions or moral authorities might view as compelling or fair, such as nonviolence, human dignity, freedom, and equality, whether or not they conform to existing laws. Thus, Mohandas Gandhi, Martin Luther King, Jr., and Nelson Mandela intentionally broke laws that violated universal principles, such as human dignity. Few individuals actually achieve Stage 6 (about 1 or 2 percent of those tested worldwide), and Kohlberg found it difficult to separate Stages 5 and 6. So, in time, he combined the stages (Kohlberg, 1981) (FIGURE 10.4).

Assessing Kohlberg's theory Kohlberg has been credited with enormous insights and contributions, but his theories have also been the focus of three major areas of criticism:

- *Moral reasoning versus behavior.* Are people who achieve higher stages on Kohlberg's scale really more moral than others, or do they just "talk a good game"? Some studies show a positive correlation between higher stages of reasoning and higher levels of moral behavior (Borba, 2001; Rest et al., 1999), but others have found that situational factors are better predictors of moral behavior (FIGURE 10.5) (Bandura, 1986, 1991, 2008; Kaplan, 2006; Satcher, 2007; Slováckova & Slovácek, 2007). For example, research participants are more likely to steal when they are told the money comes from a large company rather than from individuals (Greenberg, 2002). And both men and women will tell more sexual lies during casual relationships than during close relationships (Williams, 2001).
- *Possible gender bias.* Researcher Carol Gilligan criticized Kohlberg's model because on his scale women tend to be classified at a lower level of moral reasoning than men. Gilligan suggested that this difference occurred because Kohlberg's theory emphasizes values more often held by men, such as rationality and independence, while de-emphasizing common female values, such as concern for others and belonging (Gilligan, 1977, 1990, 1993; Kracher & Marble, 2008). Most follow-up studies of Gilligan's theory, however, have found few, if any, gender differences in level or type of moral reasoning (Hoffman, 2000; Hyde, 2007; Pratt, Skoe, & Arnold, 2004; Smith, 2007).
- *Cultural differences.* Cross-cultural studies confirm that children from a variety of cultures generally follow Kohlberg's model and progress sequentially from his first level, the preconventional, to his second, the conventional (Rest et al., 1999; Snarey, 1985, 1995). At the same time, studies find differences among cultures. For example, cross-cultural comparisons of responses to Heinz's moral dilemma show that Europeans and Americans tend to consider whether they like or identify with the victim in questions of morality. In contrast, Hindu Indians consider social responsibility and personal concerns to be separate issues (Miller & Bersoff, 1998). Researchers suggest that the difference reflects the Indians' broader sense of social responsibility.

"I swear I wasn't looking at smut—I was just stealing music."

Morality gap? FIGURE 10.5

What makes people who normally behave ethically willing to steal intellectual property such as music or software?

Process Diagram

Erikson's eight stages of psychosocial development FIGURE 10.6

Stage 1
Trust versus mistrust (birth–age 1)

Infants learn to *trust* or *mistrust* their caregivers and the world based on whether or not their needs—such as food, affection, safety—are met.

Stage 2
Autonomy versus shame and doubt (ages 1–3)

Toddlers start to assert their sense of independence (*autonomy*). If caregivers encourage this self-sufficiency, the toddler will learn to be independent versus feelings of *shame* and *doubt*.

Stage 3
Initiative versus guilt (ages 3–6)

Preschoolers learn to *initiate* activities and develop self-confidence and a sense of social responsibility. If not, they feel irresponsible, anxious, and *guilty*.

Stage 4
Industry versus inferiority (ages 6–12)

Elementary school-aged children who succeed in learning new, productive life skills, develop a sense of pride and competence (industry). Those who fail to develop these skills feel inadequate and unproductive (*inferior*).

PERSONALITY DEVELOPMENT: ERIKSON'S PSYCHOSOCIAL THEORY

Like Piaget and Kohlberg, Erik Erikson (1902–1994) developed a stage theory of development. He identified eight **psychosocial stages** of social development (FIGURE 10.6), each marked by a "psychosocial" crisis or conflict related to a specific developmental task. The name given to each stage reflects a specific crisis encountered at that stage and the two possible outcomes. For example, the crisis or task of most young adults is intimacy versus isolation. This age group's developmental task is developing deep, meaningful relationships with others. Those who don't meet this developmental challenge risk social isolation. Erikson believed that the more successfully we overcome each psychosocial crisis, the better chance we have to develop in a healthy manner (Erikson, 1950).

Many psychologists agree with Erikson's general idea that psychosocial crises contribute to personality development

psychosocial stages The eight developmental stages, each involving a crisis that must be successfully resolved, that individuals pass through in Erikson's theory of psychosocial development.

Stage 5
Identity versus role confusion (ages 12–20)

Adolescents develop a coherent and stable self-definition (identity) by exploring many roles and deciding who or what they want to be in terms of career, attitudes, etc. Failure to resolve this **identity crisis** may lead to apathy, withdrawal and/or *role confusion*.

Stage 6
Intimacy versus isolation (young adulthood)

Young adults form lasting, meaningful relationships, which help them develop a sense of connectednesss and *intimacy* with others. If not, they become psychologically *isolated*.

Stage 7
Generativity versus stagnation (middle adulthood)

The challenge for middle-aged adults is to be nurturant of the younger generation. Failing to meet this challenge leads to self-indulgence and a sense of *stagnation*.

Stage 8
Ego integrity versus despair (late adulthood)

During this stage, older adults reflect on their past. If this reflection reveals a life well-spent, the person experiences self-acceptance and satisfaction (*ego integrity*). If not, he or she experiences regret and deep dissatisfaction (*despair*).

(Berzoff, 2008; Markstrom & Marshall, 2008; Torges, Stewart, & Duncan, 2008). However, some critics argue that his theory oversimplifies development. Others observe that the eight stages do not apply equally to all groups. For example, in some cultures, *autonomy* is highly preferable to *shame and doubt,* but in others, the preferred resolution might be *dependence* or *merging relations* (Matsumoto & Juang, 2008).

Despite their limits, Erikson's stages have greatly contributed to the study of North American and European psychosocial development. By suggesting that development continues past adolescence, Erikson's theory has encouraged further research.

CONCEPT CHECK **STOP**

How do the securely attached, avoidant, and anxious/ambivalent attachment styles differ?

What criticisms have been leveled at Kohlberg's theory of moral development?

What are some cross-cultural limitations to Erikson's psychosocial theory?

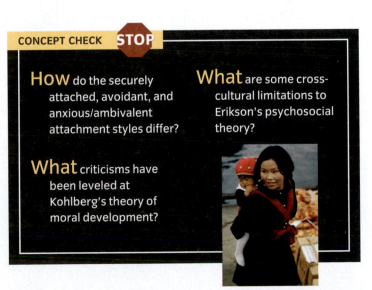

How Sex, Gender, and Culture Affect Development

LEARNING OBJECTIVES

Identify biological sex differences in physical development.

Describe how gender differences are related to cognitive, personality, and social development.

Explain how individualistic versus collectivistic cultures shape personality development.

Imagine for a moment what your life would be like if you were a member of the other sex. Would you think differently? Would you be more or less sociable and outgoing? Would your career plans or friendship patterns change? Most people believe that whether we are male or female has a strong impact on many facets of development. But why is that? Why is it that the first question most people ask after a baby is born is, "Is it a girl or a boy?"

SEX AND GENDER INFLUENCES ON DEVELOPMENT

In this section, we will explore how our development is affected by **sex** (a biological characteristic determined at the moment of conception) and by **gender** and **gender roles** (cultural meanings that accompany biological sex).

sex Biological maleness and femaleness, including chromosomal sex. Also, activities related to sexual behaviors, such as masturbation and intercourse.

gender Psychological and sociocultural meanings added to biological maleness or femaleness.

gender roles Societal expectations for normal and appropriate male and female behavior.

Sex differences

Physical anatomy is the most obvious biological sex difference between men and women (FIGURE 10.7). Men and women also differ in secondary sex characteristics, such as facial hair and breast growth; signs of reproductive capability, such as menstruation and ejaculation of sperm; and physical responses to middle age or the end of reproduction, such as *menopause* and the *male climateric.* There are also several functional and structural differences in the brains of men and women. These result partly from the influence of prenatal sex hormones on the developing fetal brain.

Gender differences

In addition to biological sex differences, scientists have found numerous gender differences, which affect our cognitive, social, and personality development.

For example, females tend to score higher on tests of verbal skills, whereas males score higher on math and visuospatial tests (Castelli, Corazzini, & Geminiani, 2008; Reynolds et al., 2008; van der Sluis et al., 2008). Keep in mind that these differences are statistically small and represent few meaningful differences.

Some researchers suggest that these differences in cognitive ability may reflect biological factors, including structural differences in the cerebral hemispheres, hormones, or the degree of hemispheric specialization.

One argument against this biological model, however, is that male-female differences in verbal ability and math scores have declined in recent years (Brown & Josephs, 1999; Halpern, 1997, 2000; Lizarraga & Ganuza, 2003). However, the gap has not been narrowed with regard to men in the highest IQ range. Men in this group still score higher on SAT math scores than women do.

Like cognitive ability, aggressive behavior also differs slightly between the genders. For example, young boys are more likely to engage in mock fighting and rough-and-tumble play, and as adolescents and adults, they are somewhat more likely to commit aggressive crimes (Campbell & Muncer, 2008; Giancola & Parrott, 2008; Ostrov & Keating, 2004). But gender differences are clearer for physical aggression (like hitting) than for other forms of aggression. Early research

Visualizing

Major physical differences between the sexes FIGURE 10.7

Body Size and Shape

The average man is 35 pounds heavier, has less body fat, and is 5 inches taller than the average woman. Men tend to have broader shoulders, slimmer hips and slightly longer legs in proportion to their height.

Brain

The corpus callosum, the bridge joining the two halves of the brain, is larger in women. This size difference is interpreted by some to mean that women can more easily integrate information from the two halves of the brain and more easily perform more than one task simultaneously.

An area of the hypothalamus causes men to have a relatively constant level of sex hormones; whereas women have cyclic sex hormone production and menstrual cycles. Differences in the cerebral hemispheres may help explain reported sex differences in verbal and spatial skills.

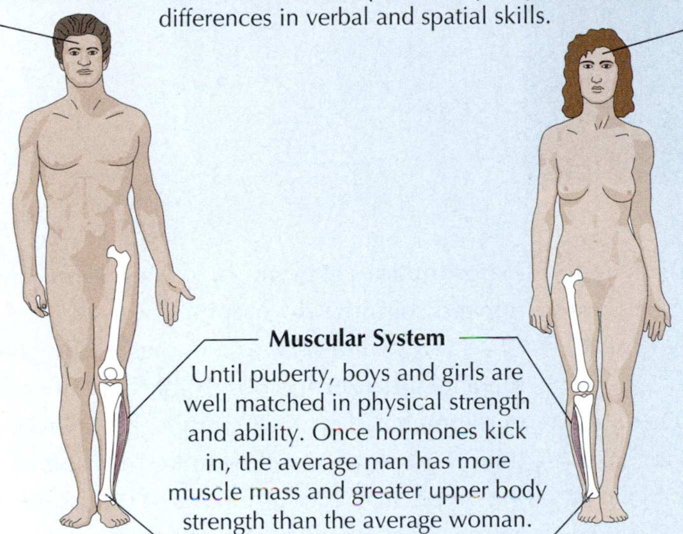

Muscular System

Until puberty, boys and girls are well matched in physical strength and ability. Once hormones kick in, the average man has more muscle mass and greater upper body strength than the average woman.

Skeletal System

Men produce testosterone throughout their life span, whereas estrogen production virtually stops when a women goes through menopause. Because estrogen helps rejuvenate bones, women are more likely to have brittle bones. Women also are more prone to knee damage because a woman's wider hips may place a greater strain on the ligaments joining the thigh to the knee.

Source: Miracle, Tina S., Miracle, Andrew W., and Baumeister, R. F., *Study Guide: Human Sexuality: Meeting Your Basic Needs*, 2nd edition. © 2006. Reprinted by permission of Pearson Education, Inc., Upper Saddle River, NJ.

suggested that females were more likely to engage in more indirect and relational forms of aggression, such as spreading rumors and ignoring or excluding someone (Bjorkqvist, 1994; Ostrov & Keating, 2004). But more recent studies have not found such clear differences (Marsee, Weems, & Taylor, 2008; Shahim, 2008).

Some researchers believe that biological factors cause gender differences in aggression—a nativist position. Several studies have linked the male hormone testosterone to aggressive behavior (Hermans, Ramsey, & van Honk, 2008; Popma et al., 2007; Trainor, Bird, & Marler, 2004). Other studies have found that aggressive

Modeling gendered behavior FIGURE 10.8

Some computer and video games include a great deal of violence and are often criticized for modeling and encouraging violence. In a new game called *Bully*, shown here, the main character ultimately takes on bullies rather than become one. The game includes plenty of fighting, but it also claims to show that actions have consequences. Do you think such an approach might encourage or discourage violent behavior?

men have disturbances in their levels of serotonin, a neurotransmitter that is inversely related to aggression (Berman, Tracy, & Coccaro, 1997; Holtzworth-Munroe, 2000; Nelson & Chiavegatto, 2001). In addition, studies on identical twins have found that genetic factors account for about 50 percent of aggressive behavior (Bartels et al., 2007; Cadoret, Leve, & Devor, 1997; Segal & Bouchard, 2000).

Other researchers take a more nurturist position. They suggest that gender differences in aggressiveness result from environmental experiences with social dominance and pressures that encourage "sex appropriate" behaviors and skills (Rowe et al., 2004) (FIGURE 10.8). TABLE 10.1 summarizes the gender differences between men and women.

Gender-role development

By age 2, children are well aware of gender roles. They recognize that boys "should" be strong, independent, aggressive, dominant, and achieving, whereas girls "should" be soft, dependent, passive, emotional, and "naturally" interested in children (Kimmel, 2000; Renzetti, Curran, & Kennedy-Bergen, 2006). The existence of similar gender roles in many cultures suggests that evolution and biology may play a role. However, most research emphasizes two major theories of gender-role development: social learning and cognitive developmental.

Social learning theorists emphasize the power of the immediate situation and observable behaviors on gender-role development. They suggest that girls learn how to be "feminine" and boys learn how to be "masculine" in two major ways: (1) They receive rewards or punishments for specific gender-role behaviors and (2) they watch and imitate the behavior of others, particularly the same-sex parent (Bandura, 1989, 2000; Fredricks & Eccles, 2005; Kulik, 2005). A boy who puts on his father's tie or baseball cap wins big, indulgent smiles from his parents. But what would happen if he put on his mother's nightgown or lipstick? Parents, teachers, and friends generally reward or punish behaviors according to traditional gender-role expectations. Thus, a child "socially learns" what it means to be male or female.

According to the *cognitive developmental theory*, social learning is part of gender-role development, but it's much more than a passive process of receiving rewards or punishments and modeling others. Instead, cognitive developmentalists argue that children actively observe, interpret, and judge the world around them (Bem, 1981, 1993; Cherney, 2005; Giles & Heyman, 2005). As children process information about the world, they also create internal rules governing correct behaviors for boys versus girls. On the

Research-supported sex and gender differences Table 10.1

Behavior	More often shown by men	More often shown by women
Sexual	• Begin masturbating sooner in life cycle and have higher overall occurrence rates. • Start sexual life earlier and have first orgasm through masturbation. • Are more likely to recognize their own sexual arousal. • Experience more orgasm consistency in their sexual relations.	• Begin masturbating later in life cycle and have lower overall occurrence rates. • Start sexual life later and have first orgasm from partner stimulation. • Are less likely to recognize their own sexual arousal. • Experience less orgasm consistency in their sexual relations.
Touching	• Are touched, kissed, and cuddled less by parents. • Exchange less physical contact with other men and respond more negatively to being touched. • Are more likely to initiate both casual and intimate touch with sexual partner.	• Are touched, kissed, and cuddled more by parents. • Exchange more physical contact with other women and respond more positively to being touched. • Are less likely to initiate either casual or intimate touch with sexual partner.
Friendship	• Have larger number of friends and express friendship by shared activities.	• Have smaller number of friends and express friendship by shared communication about self.
Personality	• Are more self-confident of future success. • Attribute success to internal factors and failures to external factors. • Achievement is task oriented; motives are mastery and competition. • Are more self-validating. • Have higher self-esteem.	• Are less self-confident of future success. • Attribute success to external factors and failures to internal factors. • Achievement is socially directed with emphasis on self-improvement; have higher work motives. • Are more dependent on others for validation. • Have lower self-esteem.

Sources: Crooks & Baur, 2008; Hyde & DeLamater, 2008; King, 2009; Masters & Johnson, 1961, 1966, 1970; Matlin, 2008.

basis of these rules, they form **gender schemas** (mental images) of how they should act (FIGURE 10.9).

Androgyny

One way to overcome rigid or destructive gender-role stereotypes is to express both the "masculine" and "feminine" traits found in each individual—for example, being assertive and aggressive when necessary but also gentle and nurturing. Combining characteristics in this way is known as **androgyny** [an-DRAW-juh-nee].

Researchers have found that masculine and androgynous individuals generally have higher self-esteem and creativity, are more socially competent and motivated to achieve, and exhibit better overall mental health than those with traditional feminine traits (Choi, 2004; Hittner & Daniels, 2002; Venkatesh et al., 2004).

Developing gender schemas FIGURE 10.9

How would social learning theory and cognitive developmental theory explain why children tend to choose stereotypically "appropriate" toys?

What makes androgyny and masculinity adaptive? Research shows that traditional masculine characteristics (analytical thinking, independence) are more highly valued than traditional feminine traits (affectivity, cheerfulness) (FIGURE 10.10). For example, when college students in 14 countries were asked to describe their "current self" and their "ideal self," the ideal self-descriptions for both men and women contained more masculine than feminine qualities (Williams & Best, 1990).

Recent studies show that gender roles are becoming less rigidly defined (Kimmel, 2000; Loo & Thorpe, 1998). Asian American and Mexican American groups show some of the largest changes toward androgyny, and African Americans remain among the most androgynous of all ethnic groups (Denmark, Rabinovitz, & Sechzer, 2005; Duval, 2006; Renzetti et al., 2006).

Androgyny FIGURE 10.10

For both children and adults, it is generally more difficult for males to express so-called female traits like nurturance and sensitivity than it is for women to adopt such traditionally male traits as assertiveness and independence (Kimmel, 2000; Leaper, 2000; Wood et al., 1997). Can you see how being more androgynous may help many couples better meet the demands of modern life?

NATIONAL GEOGRAPHIC

CULTURAL INFLUENCES ON DEVELOPMENT

individualistic cultures Cultures in which the needs and goals of the individual are emphasized over the needs and goals of the group.

collectivistic cultures Cultures in which the needs and goals of the group are emphasized over the needs and goals of the individual.

Our development is rooted in the concept of **self**—how we define and understand ourselves. Yet the very concept of self reflects our culture. In **individualistic cultures**, the needs and goals of the individual are emphasized over the needs and goals of the group. When asked to complete the statement "I am . . . ," people from individualistic cultures tend to respond with personality traits ("I am shy"; "I am outgoing") or their occupation ("I am a teacher"; "I am a student").

In **collectivistic cultures**, however, the opposite is true. The person is defined and understood primarily by looking at his or her place in the social unit (Laungani, 2007; Matsumoto & Juang, 2008; McCrae, 2004). Relatedness, connectedness, and interdependence are valued, as opposed to separateness, independence, and individualism. When asked to complete the statement "I am . . . ," people from collectivistic cultures tend to mention their families or nationality ("I am a daughter"; "I am Chinese").

If you are North American or Western European, you are more likely to be individualistic than collectivistic (TABLE 10.2). And you may find the concept of a self defined in terms of others almost contradictory. A core selfhood probably seems intuitively obvious to you. Recognizing that over 70 percent of the world's popula-

A worldwide ranking of cultures Table 10.2

Individualistic cultures	*Intermediate cultures*	*Collectivistic cultures*
United States	Israel	Hong Kong
Australia	Spain	Chile
Great Britain	India	Singapore
Canada	Argentina	Thailand
Netherlands	Japan	West Africa region
New Zealand	Iran	El Salvador
Italy	Jamaica	Taiwan
Belgium	Arab region	South Korea
Denmark	Brazil	Peru
France	Turkey	Costa Rica
Sweden	Uruguay	Indonesia
Ireland	Greece	Pakistan
Norway	Philippines	Colombia
Switzerland	Mexico	Venezuela

How might these two groups differ in their physical, socioemotional, cognitive, and personality development?

tion lives in collectivistic cultures, however, may improve your cultural sensitivity and prevent misunderstandings (Singelis et al., 1995). For example, North Americans generally define *sincerity* as behaving in accordance with one's inner feelings, whereas Japanese see it as behavior that conforms to a person's role expectations (carrying out one's duties) (Yamada, 1997). Can you see how Japanese behavior might appear insincere to a North American and vice versa?

CONCEPT CHECK **STOP**

What is the difference between sex and gender?

How do people develop gender roles?

How is the concept of "self" different in individualistic versus collectivistic cultures?

Developmental Challenges Through Adulthood

LEARNING OBJECTIVES

Describe the factors that ensure realistic expectations for marriage and long-term committed relationships.

Explain the factors that affect life satisfaction during the adult working years and retirement.

Describe the three basic concepts about death and dying that people learn to understand through the course of development.

In this section, we will explore three of the most important developmental tasks that people face during adulthood: developing a loving, committed relationship with another person; finding rewarding work and a satisfying retirement; and coping with death and dying.

COMMITTED RELATIONSHIPS: OVERCOMING UNREALISTIC EXPECTATIONS

One of the most important tasks faced during adulthood is that of establishing some kind of continuing, loving sexual relationship with another person. Yet navigating such partnerships is often very challenging. For example, nearly half of all marriages in the United States end in divorce, with serious implications for both adult and child development (Amato, 2006; Gagné et al., 2007; Huurre, Junkkari, & Aro, 2006; Li, 2008; Osler et al., 2008; Siegel, 2007). For the adults, both spouses generally experience emotional as well as practical difficulties and are at high risk for depression and physical health problems. In many cases, these problems are present even before the marital disruption.

Realistic expectations are a key ingredient in successful relationships (Gottmann & Levenson, 2002; Waller & McLanahan, 2005). Yet many people harbor unrealistic expectations about marriage and the roles of husband and wife, opening the door to marital problems (FIGURE 10.11).

Did the Brady Bunch set unrealistic expectations?
FIGURE 10.11

Where do unrealistic marital expectations originate? Women are more likely than men to try to model their family lives after what they have seen on TV programs and to expect their partners to act like the men they have seen on TV (Morrison & Westman, 2001). Men's expectations are more likely to be driven by certain myths about marriage, such as "Men are from Mars and women are from Venus" or "Affairs are the main cause of divorce."

Applying Psychology

Are Your Relationship Expectations Realistic?

To evaluate your own expectations, answer the following questions about traits and factors common to happy marriages and committed long-term relationships (Amato, 2007; Gottman & Levenson, 2002; Gottman & Notarius, 2000; Marks et al., 2008; Rauer, 2007):

1. Established "love maps"

Yes ___ *No* ___ *Do you believe that emotional closeness "naturally" develops when two people have the right chemistry?*

In successful relationships, both partners are willing to share their feelings and life goals. This sharing leads to detailed "love maps" of each other's inner emotional life and the creation of shared meaning in the relationship.

2. Shared power and mutual support

Yes ___ *No* ___ *Have you unconsciously accepted the imbalance of power promoted by many TV sitcoms, or are you willing to fully share power and to respect your partner's point of view, even if you disagree?*

The common portrayal of husbands as "head of household" and wives as the "little women" who secretly wield the true power may help create unrealistic expectations for marriage.

3. Conflict management

Yes ___ *No* ___ *Do you expect to "change" your partner or to be able to resolve all your problems?*

Successful couples work hard (through negotiation and accommodation) to solve their solvable conflicts, to accept their unsolvable ones, and to know the difference.

4. Similarity

Yes ___ *No* ___ *Do you believe that "opposites attract?"*

Although we all know couples who are very different but are still happy, similarity (in values, beliefs, religion, and so on) is one of the best predictors of long-lasting relationships (Chapter 15).

5. Supportive social environment

Yes ___ *No* ___ *Do you believe that "love conquers all"?*

Unfortunately, several environmental factors can overpower or slowly erode even the strongest love. These include age (younger couples have higher divorce rates), money and employment (divorce is higher among the poor and unemployed), parents' marriages (divorce is higher for children of divorced parents), length of courtship (longer is better), and premarital pregnancy (no pregnancy is better, and waiting a while after marriage is even better).

6. Positive emphasis

Yes ___ *No* ___ *Do you believe that an intimate relationship is a place where you can indulge your bad moods and openly criticize one another?*

Think again. Positive emotions, positive mood, and positive behavior toward one's partner are vitally important to a lasting, happy relationship.

WORK AND RETIREMENT: HOW THEY AFFECT US

Throughout most of our adult lives, work defines us in fundamental ways. It affects our health, our friendships, where we live, and even our leisure activities. Too often, however, career choices are made based on dreams of high income. In a 1995 survey conducted by the Higher Education Research Institute, nearly 74 percent of college freshmen said that being "very well-off financially" was "very important" or "essential." Seventy-one percent felt the same way about raising a family. These young people understandably hope to combine both family and work roles and "live the good life," but many will find themselves in unsatisfying jobs, having to work long hours just to keep up with the rate of inflation.

Choosing an occupation is one of the most important decisions in a person's life, and the task is becoming ever more complex. The *Dictionary of Occupational Titles,* a government publication, currently lists more than 200,000 job categories. According to psychologist John Holland's **personality-job fit theory,** a match (or "good fit") between our individual personalities and our career choices is a major factor in determining job success and satisfaction. Holland's *Self-Directed Search* questionnaire scores each person on six personality types and then matches their individual interests and abilities to the job demands of various occupations (Holland, 1985, 1994; see also Borchers, 2007; Donohue, 2006; Gottfredson & Duffy, 2008; Kieffer, Schinka, & Curtiss, 2004) (TABLE 10.3).

Work and career are a big part of adult life and self-identity, but the large majority of men and women in the United States choose to retire sometime in their sixties. What helps people successfully navigate this important life change? According to the **activity theory of aging**, successful aging

activity theory of aging Successful aging is fostered by a full and active commitment to life.

Are you in the right job? Table 10.3

Personality characteristics	*Holland personality type*	*Matching/congruent occupation*
Shy, genuine, persistent, stable, conforming, practical	1. *Realistic*: Prefers physical activities that require skill, strength, and coordination	Mechanic, drill press operator, assembly-line worker, farmer
Analytical, original, curious, independent	2. *Investigative*: Prefers activities that involve thinking, organizing, and understanding	Biologist, economist, mathematician, news reporter
Sociable, friendly, cooperative, understanding	3. *Social*: Prefers activities that involve helping and developing others	Social worker, counselor, teacher, clinical psychologist
Conforming, efficient, practical, unimaginative, inflexible	4. *Conventional*: Prefers rule-regulated, orderly, and unambiguous activities	Accountant, bank teller, file clerk, corporate manager
Imaginative, disorderly, idealistic, emotional, impractical	5. *Artistic*: Prefers ambiguous and unsystematic activities that allow creative expression	Painter, musician, writer, interior decorator
Self-confident, ambitious, energetic, domineering	6. *Enterprising*: Prefers verbal activities with opportunities to influence others and attain power	Lawyer, real estate agent, public relations specialist, small business manager

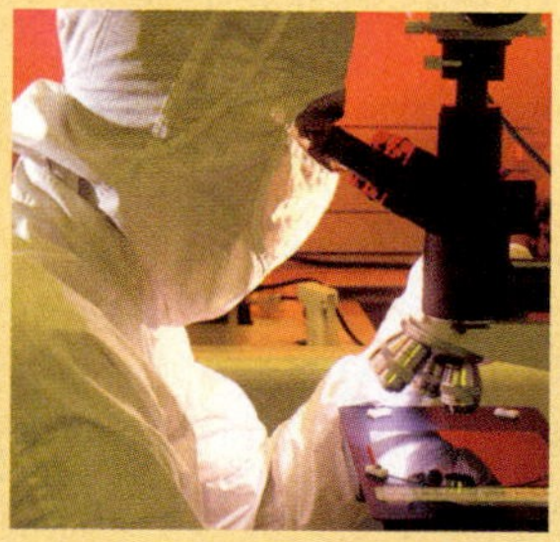

Reproduced by special permission of the publisher, Psychological Assessment Resources, Inc., 16204 North Florida Avenue, Lutz, Florida 33549, from the Self-Directed Search Form R by John L. Holland, Ph.D, copyright 1985. Further reproduction is prohibited without permission from PAR, Inc.

Satisfaction after retirement FIGURE 10.12

Active involvement is a key ingredient to a fulfilling old age.

appears to be most strongly related to good health, control over one's life, social support, and participation in community services and social activities (Warr, Butcher, & Robertson, 2004; Yeh & Lo, 2004) (**FIGURE 10.12**).

The activity theory of aging has largely displaced the older notion that successful aging entails a natural and graceful withdrawal from life (Achenbaum & Bengtson, 1994; Cummings & Henry, 1961; Heckhausen, 2005; Lemus, 2008; Menec, 2003; Neugarten, Havighurst, & Tobin, 1968; Riebe et al., 2005; Sanchez, 2006). This **disengagement theory** has been seriously questioned and largely abandoned; we mention it because of its historical relevance and because of its connection to an influential modern theory,

> ■ **disengagement theory** Successful aging is characterized by mutual withdrawal between the elderly and society.

socioemotional selectivity theory. This perspective helps explain the predictable decline in social contact that almost everyone experiences as they move into their older years (Carstensen, 2006; Charles & Carstensen, 2007) (**FIGURE 10.13**). According to this theory, we don't naturally withdraw from society in our later years—we just become more selective with our time. We deliberately decrease our total number of social contacts in favor of familiar people who provide emotionally meaningful interactions.

> ■ **socioemotional selectivity theory** A natural decline in social contact as older adults become more selective with their time.

As we've seen, there are losses and stress associated with the aging process—although much less than most people think. Perhaps the greatest challenge for the elderly, at least in the United States, is the ageism they encounter. In societies that value older people as wise elders or keepers of valued traditions, the stress of aging is much less than in societies that view them as mentally slow and socially useless. In cultures like that in the United States in which youth, speed, and progress are strongly emphasized, a loss or decline in any of these qualities is deeply feared and denied (Powell, 1998).

Socioemotional selectivity FIGURE 10.13

During infancy, emotional connection is essential to our survival. During childhood, adolescence, and early adulthood, information gathering is critical and the need for emotional connection declines. During late adulthood, emotional satisfaction is again more important—we tend to invest our time in those who can be counted on in times of need.

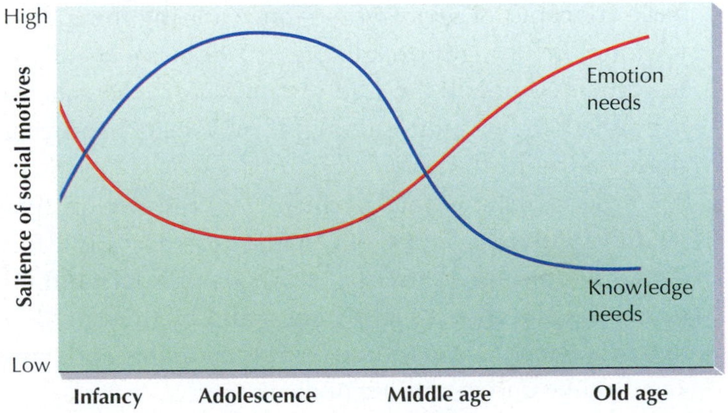

Psychological Science

Myths of Development

A number of popular beliefs about age-related crises are not supported by research. The popular idea of a *midlife crisis* began largely as a result of Gail Sheehy's national best-seller *Passages* (1976). Sheehy drew on the theories of Daniel Levinson (1977, 1996) and psychiatrist Roger Gould (1975), as well as her own interviews. She popularized the idea that almost everyone experiences a "predictable" crisis" at about age 35 for women and 40 for men. Middle age often *is* a time of reexamining one's values and lifetime goals. However, Sheehy's book led many people to automatically expect a midlife crisis with drastic changes in personality and behavior. Research suggests that a severe reaction or crisis may actually be quite rare and not typical of what most people experience during middle age (Horton, 2002; Lachman, 2004).

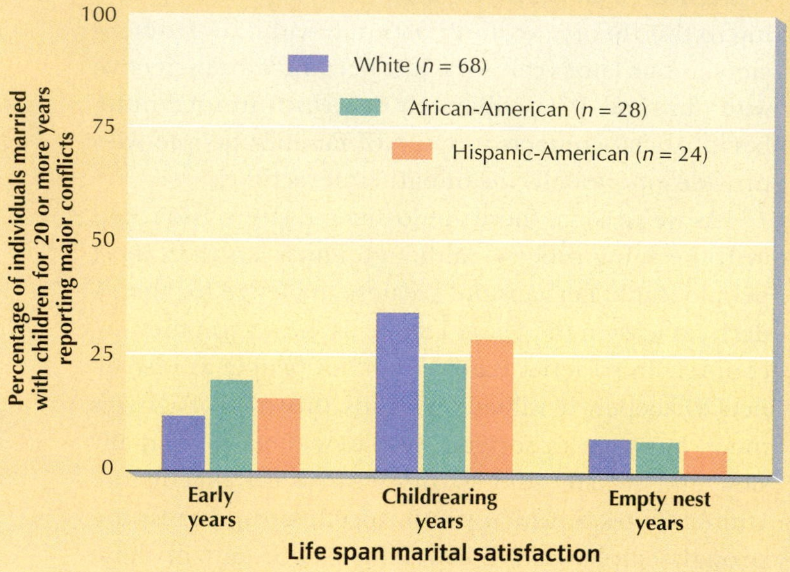

Many people also believe that when the last child leaves home, most parents experience an *empty nest syndrome*—a painful separation and time of depression for the mother, the father, or both parents. Again, research suggests that the empty nest syndrome may be an exaggeration of the pain experienced by a few individuals and an effort to downplay positive reactions (White & Rogers, 1997; Whyte, 1992). For example, one major benefit of the empty nest is an increase in marital satisfaction. Moreover, parent-child relationships do continue once the child leaves home. As one mother said, "The empty nest is surrounded by telephone wires" (Troll, Miller, & Atchley, 1979).

CRITICAL THINKING

Here are two interesting questions:

1. Have you heard people saying that someone you know is "having a midlife crisis"? If so, do you think the "crisis" is genuine?
2. If your last child "left the nest," how do you think you would react?

DEATH AND DYING: OUR FINAL DEVELOPMENTAL CRISIS

One unavoidable part of life is death. In this section, we will look at developmental changes in how people understand and prepare for their own deaths and for the deaths of loved others.

As adults, we understand death in terms of three basic concepts: (1) *permanence*—once a living thing dies, it cannot be brought back to life; (2) *universality*—all living things eventually die; and (3) *nonfunctionality*—all living functions, including thought, movement, and vital signs, end at death.

Research shows that permanence, the notion that death cannot be reversed, is the first and most easily understood concept (FIGURE 10.14). Understanding of universality comes slightly later, and by the age of 7, most children have mastered nonfunctionality and have an adultlike understanding of death.

Although parents may fear that discussing death with children and adolescents will make them unduly anxious, those who are offered open, honest discussions of death have an easier time accepting it (Corr, Nabe, & Corr, 2009; Kastenbaum, 2007).

The same is true for adults—in fact, avoiding thoughts and discussion of death and associating aging with death contribute to ageism (Atchley, 1997). Moreover, the better we understand death and the more wisely we approach it, the more fully we can live. Since the late 1990s, right-to-die and death-with-dignity advocates have brought death out in the open, and mental health professionals have suggested that understanding the psychological processes of death and dying may play a significant role in good adjustment. Cultures around the world interpret and respond to death in widely different ways: "Funerals are the occasion for avoiding people or holding parties, for fighting or having sexual orgies, for weeping or laughter, in a thousand combinations" (Metcalf & Huntington, 1991) (FIGURE 10.15).

Confronting our own death is the last major crisis we face in life. What is it like? Is there a "best" way to prepare to die? Is there such a thing as a "good" death? After spending hundreds of hours at the bedsides of the terminally ill, Elisabeth Kübler-Ross developed her stage theory of the psychological processes surrounding death (1983, 1997, 1999). She proposed that most people go through five sequential stages when facing death:

- *Denial* of the terminal condition ("This can't be true; it's a mistake!")
- *Anger* ("Why me? It isn't fair!")
- *Bargaining* ("God, if you let me live, I'll dedicate my life to you!")
- *Depression* ("I'm losing everyone and everything I hold dear.")
- *Acceptance* ("I know that death is inevitable and my time is near.")

Critics of the stage theory of dying stress that the five-stage sequence has not been scientifically validated and that each person's death is a unique experience (Kastenbaum, 2007). Others worry that popularizing such a stage theory will cause more avoidance and stereotyping of the dying ("He's just in the anger stage right now.").

How do children understand death?
FIGURE 10.14

Preschoolers seem to accept the fact that the dead person cannot get up again, perhaps because of their experiences with dead butterflies and beetles found while playing outside (Furman, 1990). Later, they begin to understand all that death entails and that they, too, will someday die.

Culture influences our response to death
FIGURE 10.15

In October 2006, a dairy truck driver took over a one-room Amish schoolhouse in Pennsylvania, killed and gravely injured several young girls, then shot himself. Instead of responding in rage, his Amish neighbors attended his funeral. Amish leaders urged forgiveness for the killer and called for a fund to aid his wife and three children. Rather than creating an on-site memorial, the schoolhouse was razed, to be replaced by pasture. What do you think of this response? The fact that many Americans were offended, shocked, or simply surprised by the Amish reaction illustrates how strongly culture affects our emotion, beliefs, and values.

Kübler-Ross (1983, 1997, 1999) agrees that not all people go through the same stages in the same way and regrets that anyone would use her theory as a model for a "good" death.

In spite of the potential abuses, Kübler-Ross's theory has provided valuable insights and spurred research into a long-neglected topic. **Thanatology**, the study of death and dying, has become a major topic in human development. Thanks in part to thanatology research, the dying are being helped to die with dignity by the **hospice** movement, which has created special facilities and trained staff and volunteers to provide loving support for the terminally ill and their families (McGrath, 2002; Parker-Oliver, 2002).

CONCEPT CHECK STOP

Why do people develop unrealistic expectations about marriage?

What is the socioemotional selectivity theory?

What are the stages of Kübler-Ross's theory of the psychological processes surrounding death?

SUMMARY

1 Social, Moral, and Personality Development

1. Nativists believe that **attachment** is innate, whereas nurturists believe that it is learned. Studies of imprinting support the biological argument for attachment. Harlow and Zimmerman's research with monkeys raised by cloth or wire "mothers" found that contact comfort might be the most important factor in attachment. The majority of infants are securely attached to their caregivers, but some exhibit either an avoidant or an anxious/ambivalent attachment style. These styles appear to carry over into adult relationships.

2. Kohlberg proposed three levels in the evolution of moral reasoning: the **preconventional level** (Stages 1 and 2), the **conventional level** (Stages 3 and 4), and the **postconventional level** (Stages 5 and 6). According to Kohlberg, Stage 4 is the highest level attained by most adolescents and adults.

3. Erikson identified eight **psychosocial stages** of development, each marked by a crisis (such as the adolescent identity crisis) related to a specific developmental task. Although some critics argue that Erikson's theory oversimplifies development and does not hold true across all cultures, many psychologists agree with Erikson's general idea that psychosocial crises do contribute to personality development.

2 How Sex, Gender, and Culture Affect Development

1. In addition to biological **sex** differences, scientists have found numerous **gender** differences relevant to cognitive, social, and personality development (for example, in cognitive ability and aggressive behavior) and have proposed both biological and environmental explanations for these differences. Most research emphasizes two major theories of **gender-role** development: social learning and cognitive developmental. One way to overcome gender-role stereotypes is to combine "masculine" and "feminine" traits (androgyny).

2. Our culture shapes how we define and understand ourselves. In **individualistic cultures**, the needs and goals of the individual are emphasized; in **collectivistic cultures**, the person is defined and understood primarily by looking at his or her place in the social unit.

3 Developmental Challenges Through Adulthood

1. Having unrealistic expectations about marriage can open the door to marital problems. Developing realistic expectations is critical to a happy marriage.

2. Choosing an occupation is an important and complex decision. According to Holland's personality-job fit theory, a match between personality and career choice is a major factor in determining job success and satisfaction. The **activity theory of aging** proposes that successful aging depends on having control over one's life, having social support, and participating in community services and social activities. The older **disengagement theory** has been largely abandoned. According to the **socioemotional selectivity theory**, people don't naturally withdraw from society in late adulthood but deliberately decrease the total number of their social contacts in favor of familiar people who provide emotionally meaningful interactions.

3. Adults understand death in terms of three basic concepts: permanence, universality, and nonfunctionality. Children most easily understand the concept of permanence; understanding of universality and nonfunctionality come later. For children, adolescents, and adults, understanding the psychological processes of death and dying are important to good adjustment. Kübler-Ross proposed that most people go through five sequential stages when facing death: denial, anger, bargaining, depression, and acceptance. But she has emphasized that her theory should not be interpreted as a model for a "good" death.

KEY TERMS

CRITICAL AND CREATIVE THINKING QUESTIONS

1. Were Harlow and his colleagues acting ethically when they separated young rhesus monkeys from their mothers and raised them with either a wire or cloth "mother"? Why or why not?
2. According to Erikson's psychosocial theory, what developmental crisis must you now resolve in order to successfully move on to the next stage? Are there any earlier crises that you feel you may not have successfully resolved?
3. Can you think of instances where you have adopted traditional gender roles? What about times where you have expressed more androgyny?
4. According to Holland's personality-job fit theory, what occupations might be a good match for your personality?
5. Would you like to know ahead of time that you were dying? Or would you prefer to die suddenly with no warning? Briefly explain your choice.

What is happening in this picture?

Native Americans generally revere and respect the elder members of their tribe.

- **How might viewing old age as an honor and a blessing—as opposed to a dreaded process—affect the experience of aging?**

SELF-TEST

(Check your answers in Appendix B.)

1. ________ is a strong affectional bond with special others that endures over time.
   a. Bonding
   b. Attachment
   c. Love
   d. Intimacy

2. In Ainsworth's studies on infant attachment, the ________ infants sought their mothers' comfort while also squirming to get away when the mother returned to the room.
   a. anxious/ambivalent
   b. caregiver-based
   c. insecurely attached
   d. dependent

3. Kohlberg believed his stages of moral development to be ________.
   a. universal and invariant
   b. culturally bound, but invariable within a culture
   c. universal, but variable within each culture
   d. culturally bound and variable

4. Once an individual has accepted and complied with the rules and values of society, that person has advanced to the ________ level of moral development.
   a. instrumental-exchange
   b. social-contract
   c. conventional
   d. postconventional

5. According to Erikson, humans progress through eight stages of psychosocial development. Label the **CORRECT** sequence on the figure below for the "successful" completion of the first four stages.

Stage 1 ________

Stage 2 ________

Stage 3 ________

Stage 4 ________

6. Label the **CORRECT** sequence on the figure below for the "successful" completion of the second four stages of Erikson's eight stages of psychosocial development.

Stage 5 ________

Stage 6 ________

Stage 7 ________

Stage 8 ________

7. Your ________ is related to societal expectations for normal and appropriate male or female behavior.
   a. gender identity
   b. gender role
   c. sexual identity
   d. sexual orientation

8. On the basis of internal rules governing correct behaviors for boys versus girls, children form ________ of how they should act.
   a. sexual identities
   b. gender identities
   c. gender schemas
   d. any of these options

9. Cultures in which the needs and goals of the group are emphasized over the needs and goals of the individual are ________________.
   a. collectivistic cultures
   b. individualistic cultures
   c. worldwide cultures
   d. all of the above

10. ________________ has consistently been found to be a key factor in successful relationships.
   a. Realistic expectations
   b. The ability to fight for equity
   c. Satisfying sexual activity
   d. Emotional stability

11. The ________________ theory of aging suggests that successful adjustment is fostered by a full and active commitment to life.
   a. activity
   b. commitment
   c. engagement
   d. life-affirming

12. The ________________ theory of aging suggests that adjustment to retirement is fostered by mutual withdrawal between the elderly and society.
   a. disengagement
   b. withdrawal
   c. disengagement
   d. death-preparation

13. Which of the following concepts is considered permanent, universal, and nonfunctional?
   a. marriage
   b. taxes
   c. your grade on this test
   d. death

14. Which of the following is **TRUE** about a child's understanding of death?
   a. Preschoolers understand that death is permanent.
   b. Preschoolers may not understand that death is universal.
   c. Children understand that death is nonfunctional by the age of seven.
   d. All of these options are true.

15. Which of the following is the correct sequence for Kübler-Ross's stage theory of dying?
   a. denial; anger; bargaining; depression; acceptance
   b. anger; denial; depression; bargaining; acceptance
   c. denial; bargaining; anger; depression; acceptance
   d. bargaining; denial; depression; anger; acceptance

Motivation and Emotion

11

In 1974, Bangladeshi economics professor Muhammad Yunus lent $27 to a group of women weavers so they could buy their own materials, rather than borrowing from a "middleman" who charged exorbitant interest. Within a year, everyone had paid him back. Yunus went on to found a bank to extend small loans to poor people. Since 1983, his Grameen Bank has lent billions of dollars to some 17 million people worldwide. In 2006, Yunus and the Grameen Bank were awarded the Nobel Peace Prize for their pioneering work.

In accepting the prize, Yunus said that he planned to use his share of the $1.4 million award to establish a company to sell food to the poor for a nominal price and to build an eye hospital for the poor. What motivates such generosity? When you read about philanthropists like Yunus, are you awestruck, joyful, happy, sad? Why? Research on motivation and emotion attempts to answer such questions.

Motivation and emotion are inseparable. Consider this example. If you saw your loved one in the arms of another, you might experience a variety of emotions (jealousy, fear, sadness, anger), and your motives would determine how you would respond. Your desire for revenge might lead you to look for another partner, but your need for love and belonging might motivate you to explain your partner's behavior and protect your relationship. As you will see in this chapter, considerable overlap exists not only between motivation and emotion but also among motivation, emotion, and most areas of psychology.

CHAPTER OUTLINE

Theories and Concepts of Motivation

LEARNING OBJECTIVES

Summarize three biologically-based theories of motivation.

Explain how incentives, attributions, and expectations affect motivation.

Describe Maslow's hierarchy of needs.

We begin our study of motivation by examining several theories of motivation that fall into three general categories—biological, psychosocial, and biopsychosocial. While studying these theories, try to identify which theory best explains your personal behaviors, such as going to college or choosing a major.

BIOLOGICAL THEORIES: LOOKING FOR INTERNAL "WHYS" OF BEHAVIOR

Many theories of **motivation** focus on inborn biological processes that control behavior. Among these biologically oriented theories are instinct, *drive-reduction*, and *arousal* theories.

In the earliest days of psychology, researchers like William McDougall (1908) proposed that humans had numerous "instincts," such as repulsion, curiosity, and self-assertiveness. Other researchers later added their favorite "instincts," and by the 1920s, the list of recognized instincts had become impossibly long. One researcher found listings for over 10,000 human instincts (Bernard, 1924).

motivation A set of factors that activate, direct, and maintain behavior, usually toward a goal.

instincts Behavioral patterns that are unlearned, always expressed in the same way, and universal in a species.

In addition, the label *instinct* led to unscientific, circular explanations—"men are aggressive because they act aggressively" or "women are maternal because they act maternally." However, in recent years, a branch of biology called sociobiology has revived the case for **instincts** when strictly defined (FIGURE 11.1).

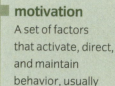

Instincts FIGURE 11.1

▲ **A** Instinctual behaviors are obvious in many animals. Birds build nests, bears hibernate, and salmon swim upstream to spawn.

B Sociobiologists such as Edward O. Wilson (1975, 1978) believe ▶ that humans also have instincts, like competition or aggression, that are genetically transmitted from one generation to the next.

Drive-reduction theory FIGURE 11.2

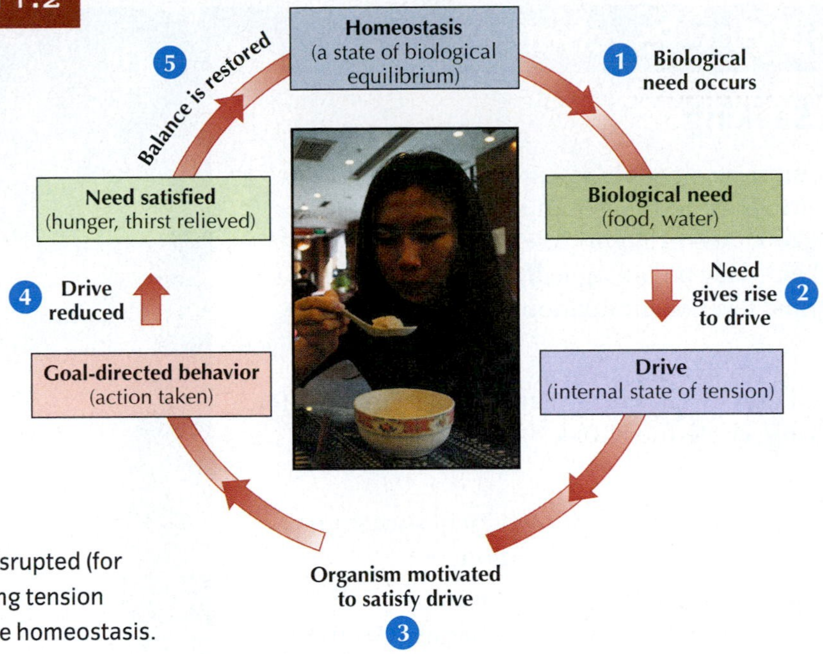

When our normal state of equilibrium is disrupted (for example, when we need food), the resulting tension creates a drive that motivates us to restore homeostasis.

Process Diagram

In the 1930s, the concepts of drive and drive reduction began to replace the theory of instincts. According to **drive-reduction theory** (Hull, 1952), when biological needs (such as food, water, and oxygen) are unmet, a state of tension (known as a *drive*) is created, and the organism is motivated to reduce it. Drive-reduction theory is based largely on the biological concept of **homeostasis**, a term that literally means "standing still" (FIGURE 11.2).

> **homeostasis**
> A body's tendency to maintain a relatively stable state, such as a constant internal temperature, blood sugar, oxygen level, or water balance.

In addition to our obvious biological needs, humans and other animals are innately curious and require a certain amount of novelty and complexity from the environment. This need for sensory stimulation begins shortly after birth and continues throughout the life span. Infants prefer complex versus simple visual stimuli, and adults pay more attention, and for a longer period of time, to complex and changing stimuli. Similarly, research shows that monkeys will work hard at tasks such as opening latches simply for the pleasure of satisfying their curiosity (Butler, 1954; Harlow, Harlow, & Meyer, 1950).

However, the need for arousal is not limitless. According to **arousal theory**, organisms are motivated to achieve and maintain an optimal level of arousal that maximizes their performance. Either too much or too little arousal diminishes performance (FIGURE 11.3).

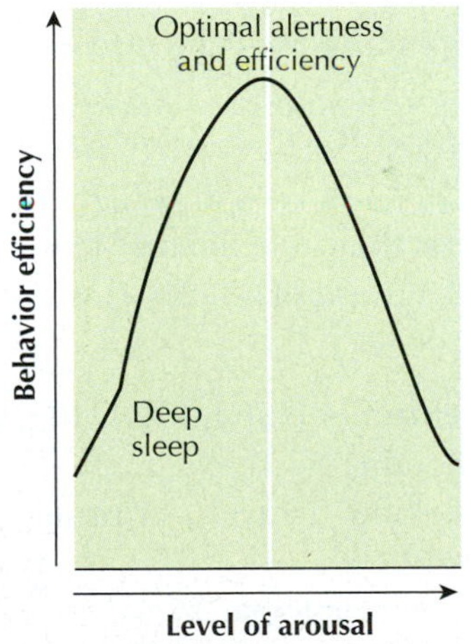

Optimal level of arousal FIGURE 11.3

Have you ever experienced a situation in which you were too aroused or "wired" to function well? Our need for stimulation (the arousal motive) suggests that behavior efficiency increases as we move from deep sleep to increased alertness. However, once we pass the maximum level of arousal, our performance declines.

Applying Psychology

Sensation Seeking

What motivates people to kayak down dangerous waterfalls or go bungee jumping over deep canyons? According to research, these "high-sensation seekers" may be biologically "prewired" to need a higher than usual level of stimulation (Zuckerman, 1979, 1994, 2004).

To sample the kinds of questions that are asked on tests for sensation seeking, circle the choice (**A** or **B**) that BEST describes you:

1. **A** I would like a job that requires a lot of traveling.
   B I would prefer a job in one location.
2. **A** I get bored seeing the same old faces.
   B I like the comfortable familiarity of everyday friends.
3. **A** The most important goal of life is to live it to the fullest and experience as much as possible.
   B The most important goal of life is to find peace and happiness.
4. **A** I would like to try parachute jumping.
   B I would never want to try jumping out of a plane, with or without a parachute.
5. **A** I prefer people who are emotionally expressive even if they are a bit unstable.
   B I prefer people who are calm and even-tempered.

Source: Zuckerman, M. (1978, February). The search for high sensation, *Psychology Today,* pp. 38–46. Copyright © 1978 by the American Psychological Association. Reprinted by permission.

Research suggests that four distinct factors characterize sensation seeking (Legrand et al., 2007; Wallerstein, 2008; Zuckerman, 2004, 2008):

1. Thrill and adventure seeking (skydiving, driving fast, or trying to beat a train)
2. Experience seeking (travel, unusual friends, drug experimentation)
3. Disinhibition ("letting loose")
4. Susceptibility to boredom (lower tolerance for repetition and sameness)

Being very high or very low in sensation seeking might cause problems in relationships with individuals who score toward the other extreme. This is true not just between partners or spouses but also between parent and child and therapist and patient. There might also be job difficulties for high-sensation seekers in routine clerical or assembly line jobs or for low-sensation seekers in highly challenging and variable occupations.

CRITICAL THINKING

Here are two interesting questions:

1. If your answers to the brief quiz above indicate that you are a high-sensation seeker, what do you do to satisfy that urge, and what can you do to make sure it doesn't get out of control?
2. If you are low in sensation seeking, has this trait interfered with some aspect of your life? If so, what could you do to improve your functioning in this area?

PSYCHOSOCIAL THEORIES: INCENTIVES AND COGNITIONS

Instinct and drive-reduction theories explain some motivations but not all. For example, why do we continue to eat after our biological need is completely satisfied? Or why does someone work overtime when his or her salary is sufficient to meet all basic biological needs? These questions are better answered by psychosocial theories that emphasize incentives and cognition.

Incentive theory holds that external stimuli motivate people to act to obtain desirable goals or to avoid undesirable events. People initially eat because their hunger "pushes" them, but they continue to eat because the sight of apple pie or ice cream "pulls" them.

According to **cognitive theories**, motivation is directly affected by **attributions**, or how we interpret or think about our own and others' actions. If you receive a high grade in your psychology course, for example, you can interpret that grade in several ways. You earned it because you really studied. You "lucked out." Or the textbook was exceptionally interesting and helpful (our preference!). People who attribute their successes to personal ability and effort tend to work harder toward their goals than people who attribute their successes to luck (Houtz et al., 2007; Hsieh, 2005; Meltzer, 2004; Weiner, 1972, 1982).

Expectancies, or what we believe will happen, are also important to motivation (Haugen, Ommundsen, & Lund, 2004; Schunk, 2008) (FIGURE 11.4). If you anticipate that you will receive a promotion at work, you're more likely to work overtime for no pay than if you expect no promotion.

Expectancies as psychosocial motivation
FIGURE 11.4

What expectations might these students have that would motivate them to master a new language?

BIOPSYCHOSOCIAL THEORIES: INTERACTIONISM ONCE AGAIN

Research in psychology generally emphasizes either biological or psychosocial factors (nature or nurture). But in the final analysis, biopsychosocial factors almost always provide the best explanation. Theories of motivation are no exception. One researcher who recognized this was Abraham Maslow (1954, 1970, 1999). Maslow believed that we all have numerous needs that compete for fulfillment but that some needs are more important than others. For example, your need for food and shelter is generally more important than college grades.

Maslow's hierarchy of needs FIGURE 11.5

Maslow's theory of motivation suggests we all share a compelling need to "move up"—to grow, improve ourselves, and ultimately become "self-actualized."

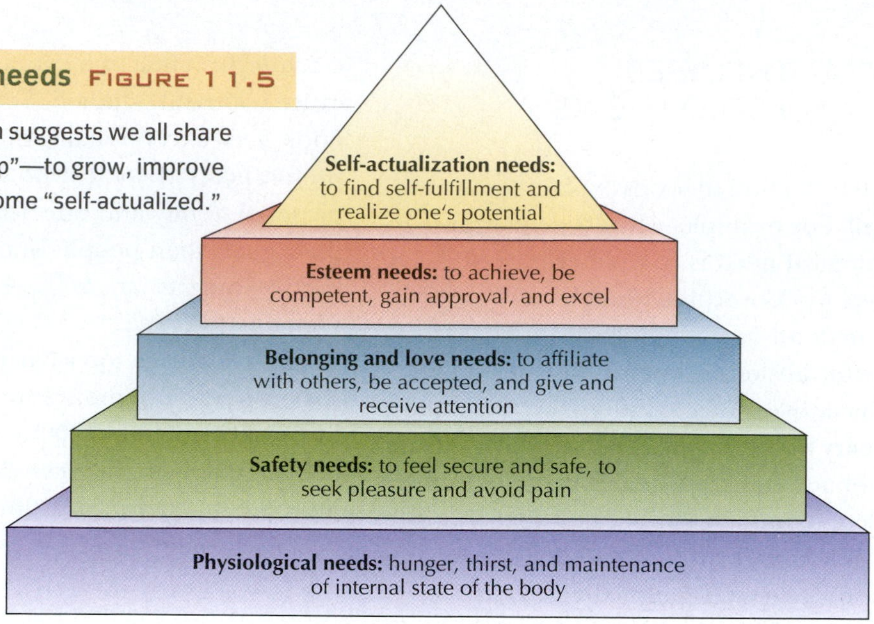

Self-actualization needs: to find self-fulfillment and realize one's potential
Esteem needs: to achieve, be competent, gain approval, and excel
Belonging and love needs: to affiliate with others, be accepted, and give and receive attention
Safety needs: to feel secure and safe, to seek pleasure and avoid pain
Physiological needs: hunger, thirst, and maintenance of internal state of the body

hierarchy of needs Maslow's theory of motivation that some motives (such as physiological and safety needs) must be met before going on to higher needs (such as belonging and self-actualization).

Maslow's **hierarchy of needs** prioritizes needs, with survival needs at the bottom (needs that must be met before others) and social, spiritual needs at the top (FIGURE 11.5).

Maslow's hierarchy of needs seems intuitively correct—a starving person would first look for food, then love and friendship, and then self-esteem. This prioritizing and the concept of **self-actualization** are important contributions to the study of motivation (Frick, 2000; Harper, Harper, & Stills, 2003). But critics argue that parts of Maslow's theory are poorly researched and biased toward Western individualism. Critics also note that people sometimes seek to satisfy higher-level needs even when their lower-level needs have not been met (Cullen & Gotell, 2002; Hanley & Abell, 2002; Neher, 1991) (FIGURE 11.6).

Bypassing basic needs FIGURE 11.6

What higher-level needs do these people seem to be trying to fulfill? Does it appear that their lower-level needs have been met?

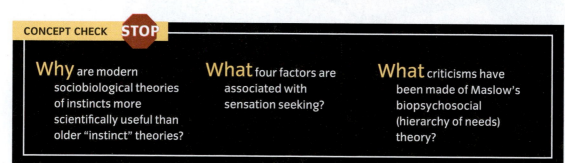

CONCEPT CHECK STOP

Why are modern sociobiological theories of instincts more scientifically useful than older "instinct" theories?

What four factors are associated with sensation seeking?

What criticisms have been made of Maslow's biopsychosocial (hierarchy of needs) theory?

Motivation and Behavior

LEARNING OBJECTIVES

Describe how internal (biological) and external (psychosocial) factors direct hunger and eating as well as how they are involved in serious eating disorders.

Explain why some people are more highly achievement motivated than others.

Summarize what happens to the human body during sexual activity.

Explain why providing extrinsic rewards undermines intrinsic motivation and when it does not.

Why do people put themselves in dangerous situations? Why do salmon swim upstream to spawn? Behavior results from many motives. For example, we discuss the need for sleep in Chapter 5, and we look at aggression, altruism, and interpersonal attraction in Chapter 15. Here, we will focus on three basic motives: hunger and eating, achievement, and sexuality. Then we'll turn to a discussion of how different kinds of motivation affect our intrinsic interests and performance.

HUNGER AND EATING: MULTIPLE FACTORS

What motivates hunger? Is it your growling stomach? Or is it the sight of a juicy hamburger or the smell of a freshly baked cinnamon roll?

The stomach Early hunger researchers believed that the stomach controls hunger, contracting to send hunger signals when it is empty. Today, we know that it's more complicated. As dieters who drink lots of water to keep their stomachs feeling full have been disappointed to discover, sensory input from an empty stomach is not essential for feeling hungry. In fact, humans and nonhuman animals without stomachs continue to experience hunger.

However, there is a connection between the stomach and feeling hungry. Receptors in the stomach and intestines detect levels of nutrients, and specialized pressure receptors in the stomach walls signal feelings of emptiness or **satiety** (fullness or satiation). The stomach and other parts of the gastrointestinal tract also release chemical signals that play a role in hunger (Donini, Savina, & Cannella, 2003; Näslund & Hellström, 2007; Nogueiras & Tschöp, 2005).

Biochemistry Like the stomach, the brain and other parts of the body produce numerous neurotransmitters, hormones, enzymes, and other chemicals that affect hunger and satiety (e.g., Arumugam et al., 2008; Cummings, 2006; Wardlaw & Hampl, 2007). Research in this area is complex because of the large number of known (and unknown) bodily chemicals and the interactions among them. It's unlikely that any one chemical controls our hunger and eating. Other internal factors, such as **thermogenesis** (the heat generated in response to food ingestion), also play a role (Subramanian & Vollmer, 2002).

The brain In addition to its chemical signals, particular brain structures also influence hunger and eating. Let's look at the hypothalamus, which helps regulate eating, drinking, and body temperature.

Early research suggested that one area of the hypothalamus, the lateral hypothalamus (LH), stimulates eating, while another area, the ventromedial hypothalamus (VMH), creates feelings of satiation, signaling the animal to stop eating. When the VMH area was destroyed in rats, researchers found that the rats overate to the point of extreme obesity (FIGURE 11.7). In contrast, when the LH area was destroyed, the animals starved to death if they were not force-fed.

How the brain affects eating FIGURE 11.7

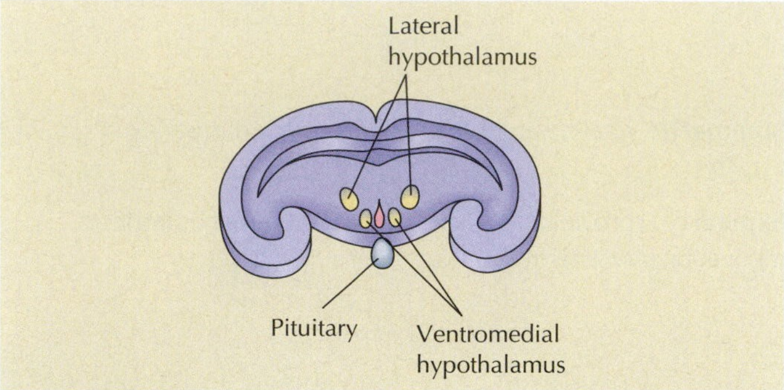

A This diagram shows a section of a rat's brain, including the ventromedial hypothalamus (VMH) and the lateral hypothalamus (LH).

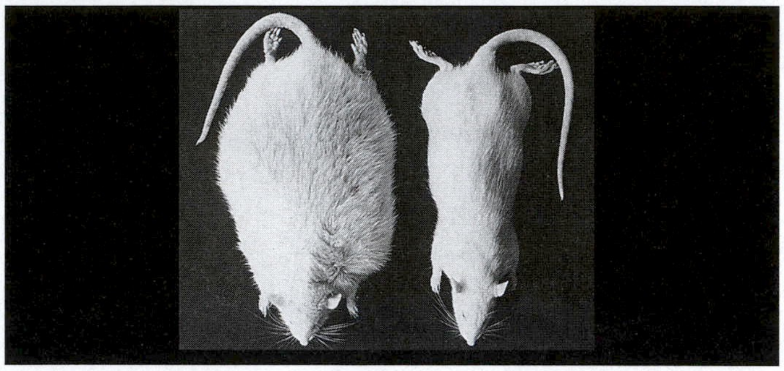

B The rat on the right is of normal weight. In contrast, the ventromedial area of the hypothalamus of the rat on the left was destroyed, which led to the tripling of its body weight.

Later research, however, showed that the LH and VMH areas are not simple on-off switches for eating. For example, lesions (damage) to the VMH make animals picky eaters—they reject food that doesn't taste good. The lesions also increase insulin secretion, which may cause overeating (Challem et al., 2000). Today, researchers know that the hypothalamus plays an important role in hunger and eating, but it is not the brain's "eating center." In fact, hunger and eating, like virtually all behavior, are influenced by numerous neural circuits that run throughout the brain (Berthoud & Morrison, 2008).

Psychosocial factors

The internal motivations for hunger we've discussed (stomach, biochemistry, the brain) are powerful. But psychosocial factors—for example, spying a dessert cart or a McDonald's billboard, or even simply noticing that it's almost lunchtime—can be equally important triggers for hunger and eating.

Another important psychosocial influence on when, what, where, and why we eat is cultural conditioning. North Americans, for example, tend to eat dinner at around 6 P.M., whereas people in Spain and South America tend to eat around 10 P.M. When it comes to *what* we eat, have you ever eaten rat, dog, or horse meat? If you are a typical North American, this might sound repulsive to you, yet most Hindus would feel a similar revulsion at the thought of eating meat from cows.

Eating disorders

As you can see, hunger and eating are motivated by numerous biological, psychological, and social factors. These same biopsychosocial forces also play a role in three serious eating disorders: obesity, anorexia nervosa, and bulimia nervosa.

Obesity has reached epidemic proportions in the United States and other developed nations. Well over half of all adults in the United States meet the current criteria for clinical **obesity** (having a body weight 15 percent or more above the ideal for one's height and age). Each year, billions of dollars are spent treating serious and life-threatening medical problems related to obesity, and consumers spend billions more on largely ineffective weight-loss products and services.

A fattening environment FIGURE 11.8

For Americans, controlling weight is a particularly difficult task. We are among the most sedentary people of all nations, and we've become accustomed to "supersized" cheeseburgers, "Big Gulp" drinks, and huge servings of dessert (Carels et al., 2008; Fisher & Kral, 2008; Herman & Polivy, 2008). We've also learned that we should eat three meals a day (whether we're hungry or not); that "tasty" food requires lots of salt, sugar, and fat; and that food is an essential part of all social gatherings. To successfully lose (and maintain) weight, we must make permanent lifestyle changes regarding the amount and types of foods we eat and when we eat them. Can you see how our everday environments, such as in the workplace seen here, might prevent a person from making healthy lifestyle changes?

Given our culture's preference for thinness and our prejudice against fat people, why are so many Americans overweight? The simple answer is overeating and not enough exercise (FIGURE 11.8). However, we all know some people who can eat anything they want and still not add pounds. This may be a result of their ability to burn calories more effectively (thermogenesis), a higher metabolic rate, or other factors. Adoption and twin studies indicate that genes also play a role. Heritability for obesity ranges between 30 and 70 percent (Fernández et al., 2008; Lee et al., 2008; Schmidt, 2004). Unfortunately, identifying the genes for obesity is difficult. Researchers have isolated over 2,000 genes that contribute to normal and abnormal weight (Camarena et al., 2004; Costa, Brennen, & Hochgeschwender, 2002; Devlin, Yanovski, & Wilson, 2000).

anorexia nervosa
An eating disorder characterized by a severe loss of weight resulting from self-imposed starvation and an obsessive fear of obesity.

Interestingly, as obesity has reached epidemic proportions, we've seen a similar rise in two other eating disorders—**anorexia nervosa** and **bulimia nervosa**. Both disorders are serious and chronic conditions that require treatment.

bulimia nervosa
An eating disorder involving the consumption of large quantities of food (bingeing), followed by vomiting, extreme exercise, or laxative use (purging).

More than 50 percent of women in Western industrialized countries show some signs of an eating disorder, and approximately 2 percent meet the clinical criteria for anorexia nervosa or bulimia nervosa (Porzelius et al., 2001). These disorders also are found at all socioeconomic levels. A few men also develop eating disorders, although the incidence is rarer in men (Jacobi et al., 2004; Raevuori et al., 2008).

Anorexia nervosa is characterized by an overwhelming fear of becoming obese, a distorted body image, a need for control, and the use of dangerous weight-loss measures. The resulting extreme malnutrition often leads to emaciation, osteoporosis, bone fractures, interruption of menstruation, and loss of brain tissue. A significant percentage of individuals with anorexia nervosa ultimately die of the disorder (Kaye, 2008; Wentz et al., 2007; Werth et al., 2003).

Distorted body image FIGURE 11.9

In anorexia nervosa, body image is so distorted that even a skeletal, emaciated body is perceived as fat. Many people with anorexia nervosa not only refuse to eat but also take up extreme exercise regimens—hours of cycling or running or constant walking.

The media's role in eating disorders FIGURE 11.10

Can you explain why popular movie and television stars' extreme thinness may unintentionally contribute to eating disorders?

Occasionally, the person suffering from anorexia nervosa succumbs to the desire to eat and gorges on food, then vomits or takes laxatives. However, this type of bingeing and purging is more characteristic of bulimia nervosa. Individuals with this disorder also show impulsivity in other areas, sometimes engaging in excessive shopping, alcohol abuse, or petty shoplifting (Kaye, 2008). The vomiting associated with bulimia nervosa causes dental damage, severe damage to the throat and stomach, cardiac arrhythmias, metabolic deficiencies, and serious digestive disorders.

There are many suspected causes of anorexia nervosa and bulimia nervosa. Some theories focus on physical causes, such as hypothalamic disorders, low levels of various neurotransmitters, and genetic or hormonal disorders. Other theories emphasize psychosocial factors, such as a need for perfection, a perceived loss of control, being teased about body weight, destructive thought patterns, depression, dysfunctional families, distorted body image (see FIGURE 11.9), and sexual abuse (e.g., Behar, 2007; Fairburn et al., 2008; Kaye, 2008; Sachdev et al., 2008).

Cultural perceptions and stereotypes about weight and eating also play important roles in eating disorders (Eddy et al., 2007; Fairburn et al., 2008; Herman & Polivy, 2008). For instance, Asian and African Americans report fewer eating and dieting disorders and greater body satisfaction than do European Americans (Ruffolo et al., 2006; Taylor et al., 2007), and Mexican students report less concern about their own weight and more acceptance of obese people than do other North American students (Crandall & Martinez, 1996).

Although social pressures for thinness certainly contribute to eating disorders (FIGURE 11.10), anorexia nervosa has also been found in nonindustrialized areas like the Caribbean island of Curaçao (Hoek et al., 2005). On that island, being overweight is socially acceptable, and the average woman is considerably heavier than the average woman in North America. However, some women there still have anorexia nervosa. This research suggests that both culture and biology help explain eating disorders.

ACHIEVEMENT: THE NEED FOR SUCCESS

Do you wonder what motivates Olympic athletes to work so hard just for a gold medal? Or what about someone like Thomas Edison, who patented more than 1,000 inventions? What drives some people to high achievement?

The key to understanding what motivates high-achieving individuals lies in what psychologist Henry Murray (1938) identified as a high need for achievement (nAch), or **achievement motivation**.

achievement motivation A desire to excel, especially in competition with others.

Several traits distinguish people who have high achievement motivation (McClelland, 1958, 1987, 1993; Senko, Durik, & Harackiewicz, 2008; Quintanilla, 2007):

- *Preference for moderately difficult tasks.* People high in nAch (need for achievement) avoid tasks that are too easy because they offer little challenge or satisfaction. They also avoid extremely difficult tasks because the probability of success is too low.
- *Competitiveness.* High achievement–oriented people are more attracted to careers and tasks that involve competition and an opportunity to excel.
- *Preference for clear goals with competent feedback.* High achievement–oriented people tend to prefer tasks with a clear outcome and situations in which they can receive feedback on their performance. They also prefer criticism from a harsh but competent evaluator to that which comes from one who is friendlier but less competent.
- *Responsibility.* People with high nAch prefer being personally responsible for a project so that they can feel satisfied when the task is well done.
- *Persistence.* High achievement–oriented people are more likely to persist at a task when it becomes difficult. In one study, 47 percent of high nAch individuals persisted on an "unsolvable task" until time was called, compared with only 2 percent of people with low nAch.
- *More accomplished.* People who have high nAch scores do better than others on exams, earn better grades in school, and excel in their chosen professions.

Achievement orientation appears to be largely learned in early childhood, primarily through interactions with parents (FIGURE 11.11). Highly motivated children tend to have parents who encourage independence and frequently reward successes (Maehr & Urdan, 2000). Our cultural values also affect achievement needs (Lubinski & Benbow, 2000).

Future achiever FIGURE 11.11

A study by Richard de Charms and Gerald Moeller (1962) found a significant correlation between the achievement themes in children's literature and the industrial accomplishments of various countries.

SEXUALITY: THE WORLD'S MOST POWERFUL MOTIVE?

Obviously, there is strong motivation to engage in sexual behavior: it's essential for the survival of our species, and it's also pleasurable. But **sexuality** includes much more than reproduction. For most humans (and some other animals), a sexual relationship fulfills many needs, including the need for connection, intimacy, pleasure, and the release of sexual tension.

William Masters and Virginia Johnson (1966) were the first to conduct laboratory studies on what happens to the human body during sexual activity. They attached recording devices to male and female volunteers and monitored or filmed their physical responses as they

Process Diagram

Masters and Johnson's sexual response cycle FIGURE 11.12

Note that this simplified description does not account for individual variation and should not be used to judge what's "normal."

2 During the **plateau phase**, physiological and sexual arousal continue at heightened levels. In men, the penis becomes more engorged and erect while the testes swell and pull up closer to the body. In the woman, the clitoris pulls up under the clitoral hood and the entrance to the vagina contracts while the uterus rises slightly. This movement of the uterus causes the upper two-thirds of the vagina to balloon, or expand. As arousal reaches its peak, both sexes may experience a feeling that orgasm is imminent and inevitable.

3 The **orgasm phase** involves a highly intense and pleasurable release of tension. In women, muscles around the vagina squeeze the vaginal walls in and out and the uterus pulsates. Muscles at the base of the penis contract in the man, causing ejaculation, the discharge of semen or seminal fluid.

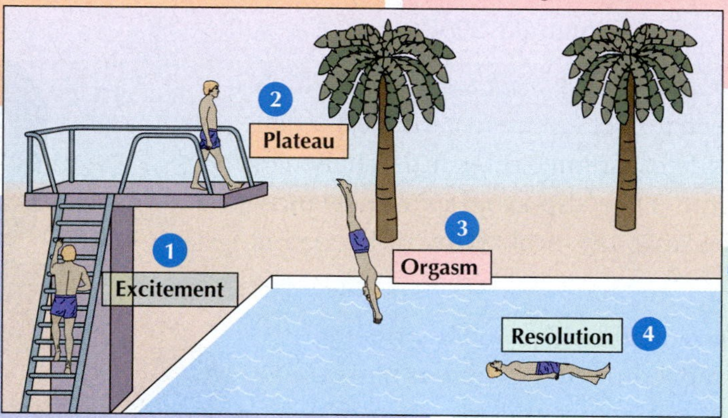

1 The **excitement phase** can last for minutes or hours. Arousal is initiated through touching, fantasy, or erotic stimuli. Heart rate and respiration increase and increased blood flow to the region causes penile or clitoral erection, and vaginal lubrication in women. In both men and women, the nipples may become erect, and both may experience a sex flush (reddening of the upper torso and face).

4 Physiological responses gradually return to normal during the **resolution phase**. After one orgasm, most men enter a **refractory period**, during which further excitement to orgasm is considered impossible. Many women (and some men), however, are capable of multiple orgasms in fairly rapid succession.

A Immediately after orgasm, men generally enter a refractory period, which lasts from several minutes up to a day. ▼

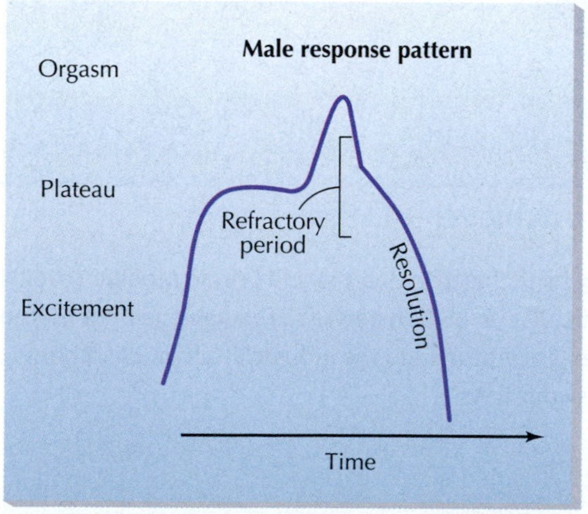

B Female sexual responses generally follow one or more of three basic patterns. ▼

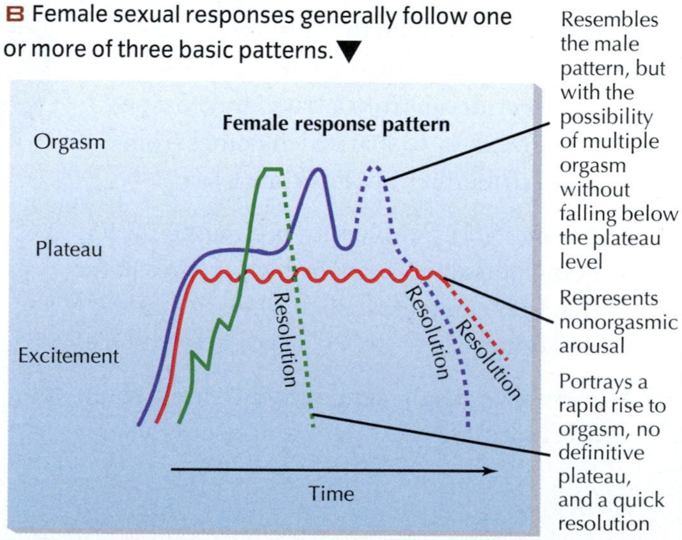

sexual response cycle Masters and Johnson's description of the four-stage bodily response to sexual arousal, which consists of excitement, plateau, orgasm, and resolution.

sexual orientation Primary erotic attraction toward members of the same sex (homosexual, gay or lesbian), both sexes (bisexual), or the other sex (heterosexual).

moved from nonarousal to orgasm and back to nonarousal. They labeled the bodily changes during this series of events a **sexual response cycle** (FIGURE 11.12).

Sexual orientation Of course, an important part of people's sexuality is the question of to whom they are sexually attracted. What leads some people to be homosexual and others to be heterosexual? Unfortunately, the roots of **sexual orientation** are poorly understood. However, most studies suggest that genetics and biology play the dominant role (Bailey, Dunne, & Martin, 2000; Byne, 2007; Ellis et al., 2008; Gooren, 2006; Zucker, 2008).

Studies on identical and fraternal twins and adopted siblings found that if one identical twin was gay, 48 to 65 percent of the time so was the second twin (Hyde, 2005; Kirk et al., 2000). (Note that if the cause were totally genetic, the percentage would be 100.) The rate was 26 to 30 percent for fraternal twins and 6 to 11 percent for brothers and sisters who were adopted into one's family. Estimates of homosexuality in the general population run between 2 and 10 percent.

Research with rats and sheep hints that prenatal hormone levels may also affect fetal brain development and sexual orientation (Bagermihl, 1999). However, the effect of hormones on human fetal development is unknown. Furthermore, no well-controlled study has ever found a difference in adult hormone levels between heterosexuals and gays and lesbians (Hall & Schaeff, 2008; Gooren, 2006; LeVay, 2003).

Research has identified several widespread myths and misconceptions about homosexuality (Bergstrom-Lynch, 2008; Boysen & Vogel, 2007; LeVay, 2003). Although mental health authorities long ago discontinued labeling homosexuality as a mental illness, it continues to be a divisive societal issue (FIGURE 11.13). Gays, lesbians, bisexuals, and transgendered people often confront **sexual prejudice**, and many endure verbal and physical attacks; disrupted family and peer relationships; and high rates of anxiety, depression, and suicide (Espelage et al., 2008; Jellison, McConnell, & Gabriel, 2004; Skinta, 2008; Talley & Bettencourt, 2008; Weber, 2008). Sexual prejudice is a socially reinforced phenomenon, not an individual pathology (as the older term *homophobia* implies).

sexual prejudice Negative attitudes toward an individual because of her or his sexual orientation.

INTRINSIC VERSUS EXTRINSIC MOTIVATION: IS ONE BETTER THAN THE OTHER?

Should parents reward children for getting good grades? Many psychologists are concerned about the widespread practice of giving external, or extrinsic, rewards to motivate behavior (e.g., Deci & Moller, 2005; Markle, 2007;

Gay marriage FIGURE 11.13

Debate, legislation, and judicial action surrounding gay marriage has brought attitudes about homosexuality into full public view. Do you think these activities have served to increase or decrease sexual prejudice? Why?

Prabhu, Sutton, & Sauser, 2008; Reeve, 2005). They're concerned that providing such **extrinsic motivation** will seriously affect the individual's personal, **intrinsic motivation.** Participation in sports and hobbies, like swimming or playing guitar, is usually intrinsically motivated. Going to work is primarily extrinsically motivated.

Research has shown that people who are given extrinsic rewards (money, praise, or other incentives) for an intrinsically satisfying activity, such as watching TV, playing cards, or even engaging in sex, often lose enjoyment and interest and may decrease the time spent on the activity (Hennessey & Amabile, 1998; Kohn, 2000; Moneta & Siu, 2002).

extrinsic motivation Motivation based on obvious external rewards or threats of punishment.

intrinsic motivation Motivation resulting from personal enjoyment of a task or activity.

Not all extrinsic motivation is bad (Banko, 2008; Konheim-Kalkstein & van den Broek, 2008; Moneta & Siu, 2002). Extrinsic rewards are motivating if they are used to inform a person of superior performance or as a special "no strings attached" treat (Deci, 1995). In fact, they may intensify the desire to do well again. Thus, getting a raise or a gold medal can inform us and provide valuable feedback about our performance, which may increase enjoyment. But if rewards are used to control—for example, when parents give children money or privileges for achieving good grades—they inhibit intrinsic motivation (Eisenberger & Armeli, 1997; Eisenberger & Rhoades, 2002; Houlfort, 2006) (FIGURE 11.14).

Motivation is in the eye of the beholder FIGURE 11.14

Do rewards increase motivation, or are they seen as coercion or bribery? It depends. Note how a controlling reward and external pressure both lead to extrinsic motivation. However, an informing award and "no strings" treat produce intrinsic motivation.

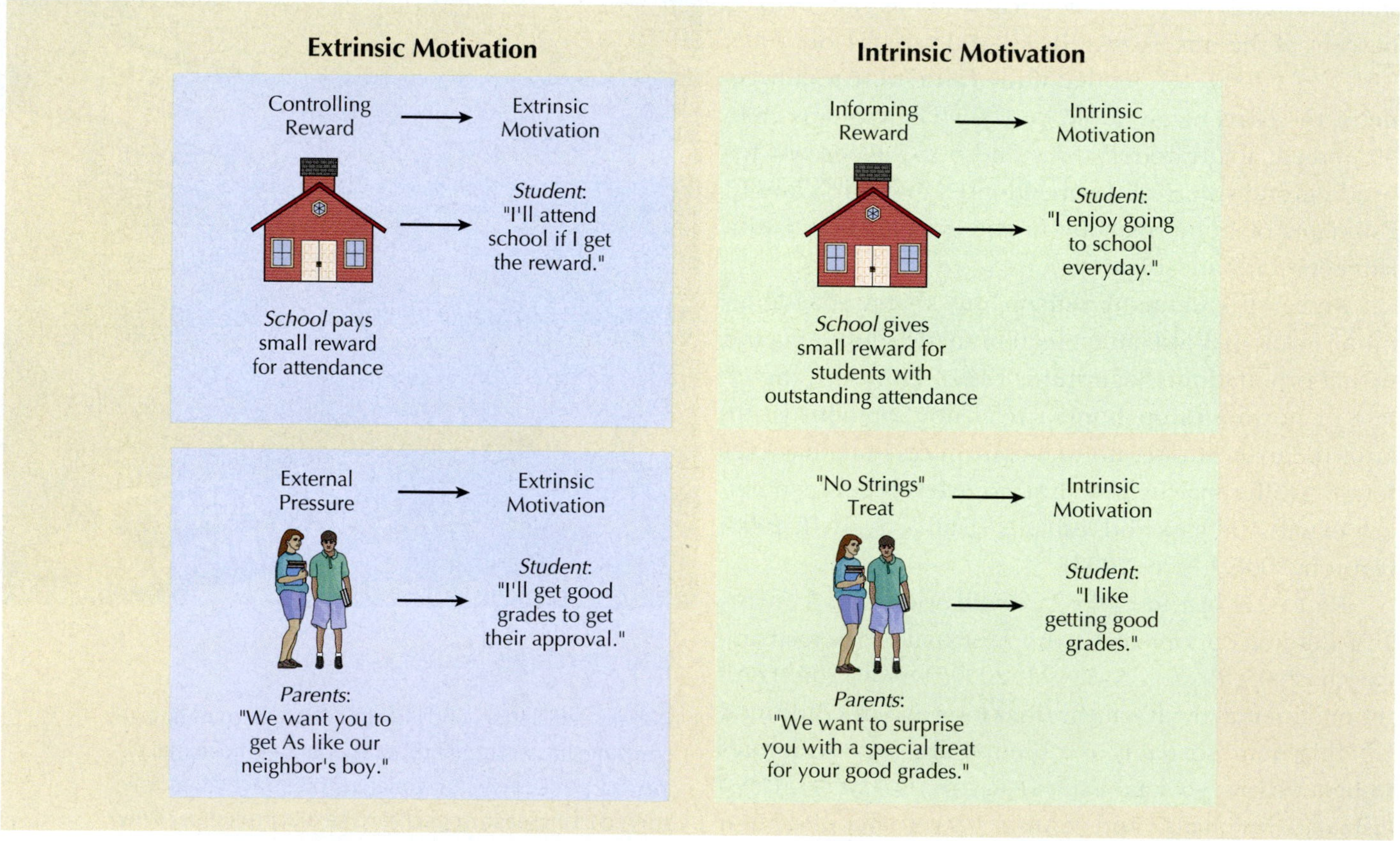

Ψ Psychological Science

Extrinsic versus Intrinsic Rewards

Have you ever noticed that for all the money and glory they receive, professional athletes often don't look like they're enjoying themselves very much? What's the problem? Why don't they appreciate how lucky they are to be able to make a living by playing games?

When we perform a task for no ulterior purpose, we use internal, personal reasons ("I like it"; "It's fun"). But when extrinsic rewards are added, the explanation shifts to external, impersonal reasons ("I did it for the money"; "I did it to please my parents"). This shift generally decreases enjoyment and hampers performance. This is as true for professional athletes as it is for anyone else.

One of the earliest experiments to demonstrate this effect was conducted with preschool children who liked to draw (Lepper, Greene, & Nisbett, 1973). As shown in the figure below, the researchers found that children who were given paper and markers and promised a reward for their drawings were subsequently less interested in drawing than children who were not given a reward or who were given an unexpected reward when they were done. Likewise, for professional athletes, what is a fun diversion for most people can easily become "just a job."

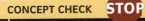

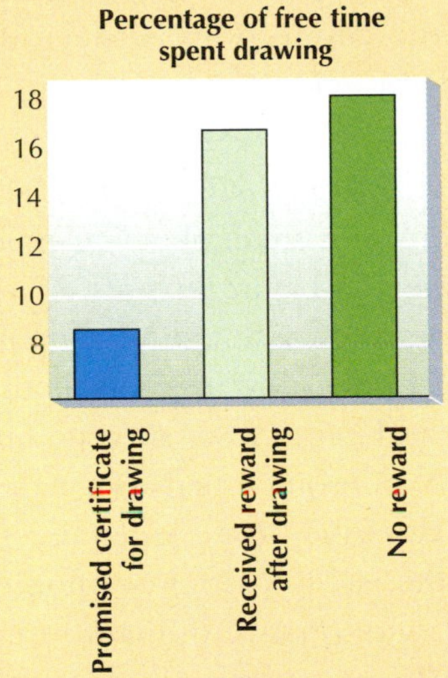

CRITICAL THINKING

Here are two interesting questions:

1. If you play a sport or other extracurricular activity, what do you think motivates you to do so?
2. How do you think employees could be motivated so they don't think of work as "just a job"?

CONCEPT CHECK STOP

What are the biological, psychological, and social factors that motivate hunger and eating?

What traits characterize people with high-achievement motivation?

What is sexual prejudice?

What impact does using extrinsic motivation to reward behavior have?

Theories and Concepts of Emotion

LEARNING OBJECTIVES

Describe the physiological, cognitive, and behavioral components of emotion.

Compare the four major theories of emotion.

Explain cultural similarities and differences in emotion.

Review the problems with relying on polygraph testing as a "lie detector."

THREE COMPONENTS OF EMOTION

emotion A subjective feeling that includes arousal (heart pounding), cognitions (thoughts, values, and expectations), and expressions (frowns, smiles, and running).

Psychologists define and study **emotion** according to three basic components—physiological, cognitive, and behavioral.

The physiological (arousal) component

Internal physical changes occur in our bodies whenever we experience an emotion. Imagine walking alone on a dark street when someone jumps from behind a stack of boxes and starts running toward you. How would you respond? Like most people, you would undoubtedly interpret the situation as threatening and would run. Your predominant emotion, fear, would involve several physiological reactions, such as increased heart rate and blood pressure, perspiration, and goose bumps (piloerection). Such physiological reactions are controlled by certain brain structures and by the autonomic branch of the nervous system (ANS).

Our emotional experiences appear to result from important interactions between several areas of the brain, most particularly the *cerebral cortex* and *limbic system* (Langenecker et al., 2005; LeDoux, 2002; Panksepp, 2005). As we discuss in Chapter 2, the cerebral cortex, the outermost layer of the brain, serves as our body's ultimate control and information-processing center, including our ability to recognize and regulate our emotions.

The limbic system in the brain is also essential to our emotions. For example, several studies have shown that one area of the limbic system, the *amygdala,* plays a key role in emotion—especially fear.

Interestingly, emotional arousal sometimes occurs without conscious awareness. According to psychologist Joseph LeDoux (1996, 2002, 2007), when sensory inputs arrive in the *thalamus* (our brain's sensory switchboard), it sends separate messages up to the cortex (which "thinks" about the stimulus) and the amygdala (which immediately activates the body's alarm system).

As important as the brain is to emotion, it is the *autonomic nervous system* (ANS, Chapter 2) that produces the obvious signs of arousal. These largely automatic responses result from interconnections between the ANS and various glands and muscles (FIGURE 11.15).

The cognitive (thinking) component

Emotional reactions are very individual: what you experience as intensely pleasurable may be boring or aversive to another. To study the cognitive (thought) component of emotions, psychologists typically use self-report techniques, such as paper-and-pencil tests, surveys, and interviews. However, people are sometimes unable or unwilling to accurately describe (or remember) their emotional states. For these reasons, our cognitions about our own and others' emotions are difficult to measure scientifically. This is why many researchers supplement participants' reports on their emotional experiences with methods that assess emotional experience indirectly (for example, by measuring physiological response or behavior).

The behavioral (expressive) component

Emotional expression is a powerful form of communication, and facial expressions may be our most important form of emotional communication. Researchers have developed sensitive techniques to measure subtleties of feeling and to

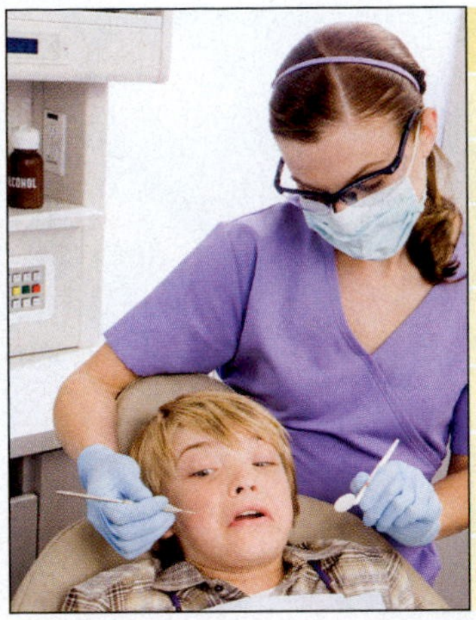

Sympathetic		Parasympathetic
Pupils dilated	**Eyes**	Pupils constricted
Dry	**Mouth**	Salivating
Goose bumps, perspiration	**Skin**	No goose bumps
Respiration increased	**Lungs**	Respiration normal
Increased rate	**Heart**	Decreased rate
Increased epinephrine and norepinephrine	**Adrenal glands**	Decreased epinephrine and norepinephrine
Decreased motility	**Digestion**	Increased motility

Emotion and the autonomic nervous system FIGURE 11.15

During emotional arousal, the sympathetic branch of the autonomic nervous system prepares the body for fight or flight. (The hormones epinephrine and norepinephrine keep the system under sympathetic control until the emergency is over.) The parasympathetic branch returns the body to a more relaxed state (homeostasis).

differentiate honest expressions from fake ones. Perhaps most interesting is the difference between the **social smile** and the **Duchenne smile** (named after French anatomist Duchenne de Boulogne, who first described it in 1862) (FIGURE 11.16). In a false, social smile, our voluntary cheek muscles are pulled back, but our eyes are unsmiling. Smiles of real pleasure, on the other hand, use the muscles not only around the cheeks but also around the eyes.

The Duchenne smile illustrates the importance of nonverbal means of communicating emotion. We all know that people communicate in ways other than speaking or writing. However, few people recognize the full importance of nonverbal signals. Imagine yourself as

Duchenne smile FIGURE 11.16

People who show a Duchenne, or real, smile (A) and laughter elicit more positive responses from strangers and enjoy better interpersonal relationships and personal adjustment than those who use a social smile (B) (Keltner, Kring, & Bonanno, 1999; Prkachin & Silverman, 2002).

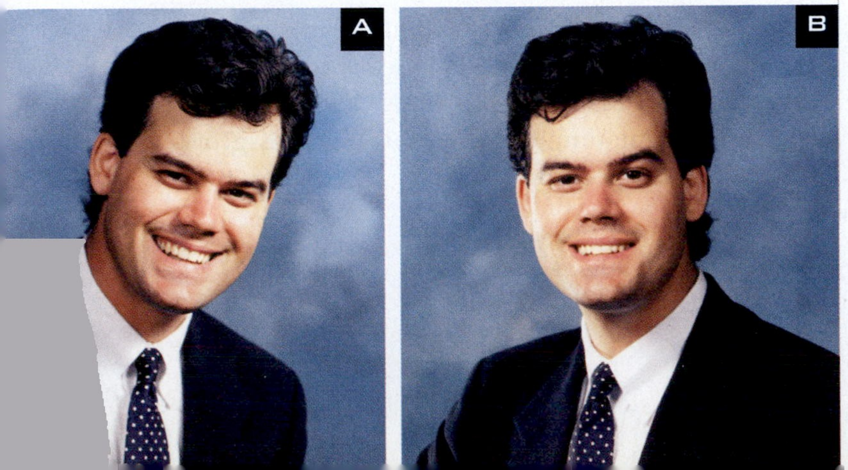

an interviewer. Your first job applicant greets you with a big smile, full eye contact, a firm handshake, and an erect, open posture. The second applicant doesn't smile, looks down, offers a weak handshake, and slouches. Whom do you think you will hire?

Psychologist Albert Mehrabian would say that you're much less likely to hire the second applicant due to his or her "mixed messages." His research suggests that when we're communicating feelings or attitudes, and the verbal and nonverbal dimensions don't match, the receiver trusts the predominant form of communication, which is about 93 percent nonverbal (the way the words are said and facial expression) versus the literal meaning of the words (Mehrabian, 1968, 1971, 2007).

Unfortunately, Mehrabian's research is often overgeneralized, and many people misquote him as saying that "over 90 percent of communication is nonverbal." Clearly, if a police officer says, "Put your hands up," his or her verbal words might carry 100 percent of the meaning. However, when we're confronted with a mismatch between verbal and nonverbal, it is safe to say that we pay far more attention to the nonverbal because we believe it more often tells us what someone is really thinking or feeling. The importance of nonverbal communication, particularly facial expressions, is further illustrated by the popularity of "smileys," and other emotional symbols, in our everyday email and text messages.

Psychological Science

Emotional Intelligence (EI): How Important Are "Emotional Smarts"?

You've heard of IQ, the intelligence quotient, but what do you know about EI—emotional intelligence? Daniel Goleman's research (1995, 2000, 2008) and best-selling books have popularized the concept of **emotional intelligence (EI)**, based on original work by Peter Salovey and John Mayer (1990).

According to the theory, emotional intelligence involves knowing and managing one's emotions, empathizing with others, and maintaining satisfying relationships. In other words, an emotionally intelligent person successfully combines the three components of emotions (cognitive, physiological, and behavioral). Proponents of EI have suggested ways in which the close collaboration between emotion and reason may promote personal well-being and growth (Salovey et al., 2000).

Popular accounts such as Goleman's have suggested that traditional measures of human intelligence ignore a crucial range of abilities that characterize people who excel in real life: self-awareness, impulse control, persistence, zeal and self-motivation, empathy, and social deftness. Goleman also proposes that many societal problems, such as domestic abuse and youth violence, can be attributed to a low EI. Therefore, he argues, EI should be fostered in everyone.

Critics fear that a handy term like EI invites misuse, but their strongest reaction is to Goleman's proposals for teaching EI. For example, Paul McHugh, director of psychiatry at Johns Hopkins University, suggests that Goleman is "presuming that someone has the key to the right emotions to be taught to children. We don't even know the right emotions to be taught to adults" (cited in Gibbs, 1995, p. 68).

CRITICAL THINKING

Here are two interesting questions:

1. Do you think you have high or low emotional intelligence? Why?
2. What value might EI have for a person's functioning in everyday life?

FOUR MAJOR THEORIES OF EMOTION

Researchers generally agree on the three components of emotion (physiological, cognitive, and behavioral), but there is less agreement on how we become emotional. The major competing theories are the James-Lange, Cannon-Bard, facial-feedback, and Schachter's two-factor. As shown in FIGURE 11.17, each of these four theories emphasizes different sequences or aspects of the three elements (cognitions, arousal, and expression).

According to the **James-Lange theory** (originated by psychologist William James and later expanded by physiologist Carl Lange), emotions depend on feedback from our physiological arousal and behavioral expression (FIGURE 11.17A). In other words, as James wrote: "We feel sorry because we cry, angry because we strike, afraid because we tremble" (1890). In short, arousal and expression cause emotion. Without arousal or expression, there is no emotion.

In contrast, the **Cannon-Bard theory** holds that all emotions are physiologically similar and that arousal, cognitions, and expression occur simultaneously. Importantly in this theory, arousal is not a necessary or even major factor in emotion. Walter Cannon (1927) and Philip Bard (1934) proposed that the thalamus sends simultaneous messages to both the ANS and the cerebral cortex (FIGURE 11.17B). Messages to the cortex produce the cognitive experience of emotion (such as fear), whereas messages to the autonomic nervous system produce physiological arousal and behavioral expressions (such as heart palpitations, running, widening eyes, and open mouth).

Cannon supported his position with several experiments in which animals were surgically prevented from experiencing physiological arousal. Yet these animals still showed emotion-like behaviors, such as growling and defensive postures (Cannon, Lewis, & Britton, 1927).

The third major explanation of emotion focuses on the expressive component of emotions. According to the **facial-feedback hypothesis** (FIGURE 11.17C), facial changes not only correlate with and intensify emotions but also cause or initiate the emotions themselves (Adelmann & Zajonc, 1989; Ceschi & Scherer, 2001; Prkachin, 2005; Sigall & Johnson, 2006). Contractions of the various facial muscles send specific messages to the brain, identifying each basic emotion.

Four major theories of emotion FIGURE 11.17

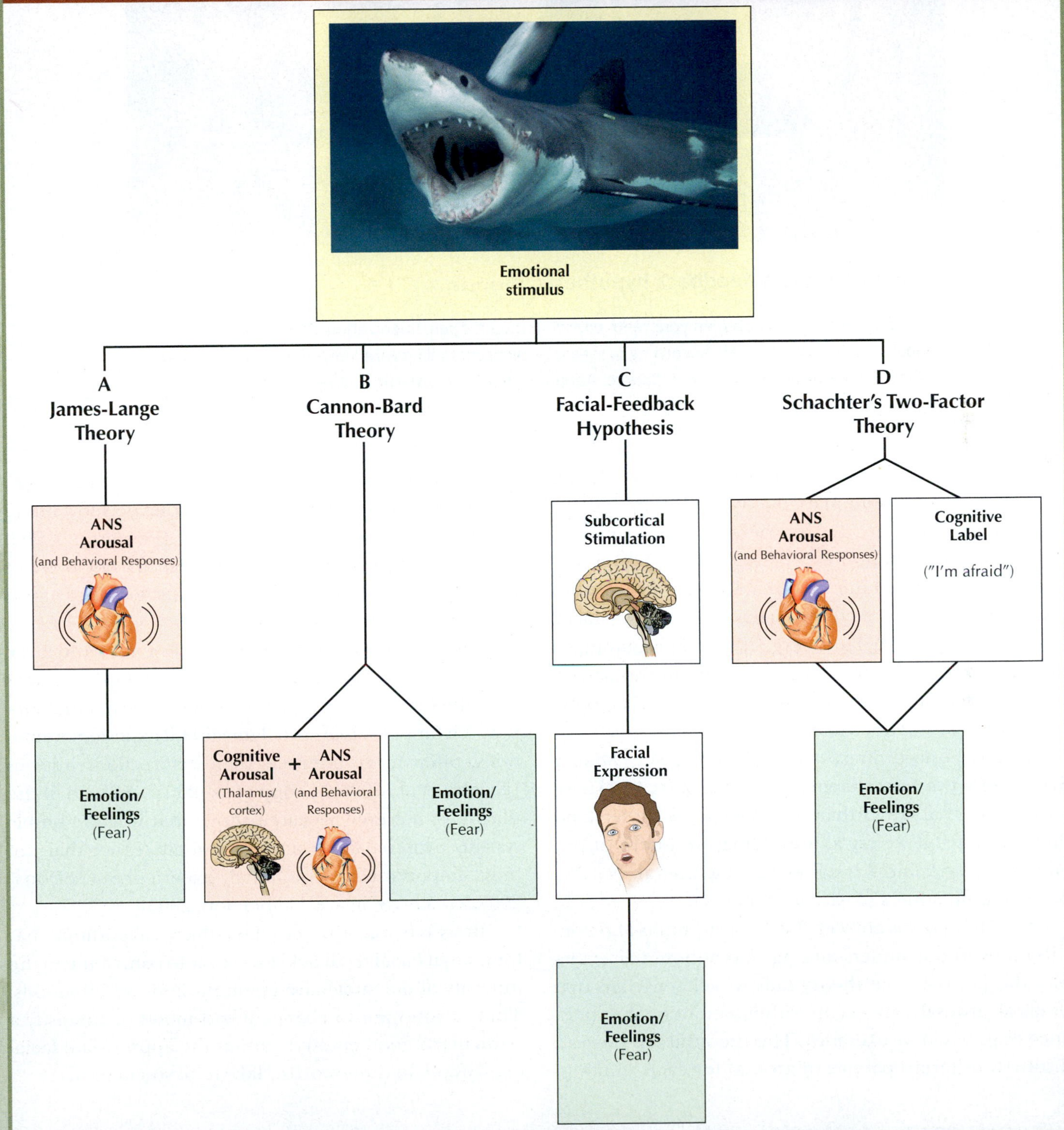

Process Diagram

Testing the facial-feedback hypothesis FIGURE 11.18

Hold a pen or pencil between your teeth with your mouth open. Spend about 30 seconds in this position. How do you feel? According to research, pleasant feelings are more likely when teeth are showing than when they are not. *Source:* Adapted from Starck, Martin, & Stepper, 1988.

The facial-feedback hypothesis also supports Charles Darwin's (1872) evolutionary theory that freely expressing an emotion intensifies it, whereas suppressing outward expression diminishes it (FIGURE 11.18). Interestingly, even watching another's facial expressions causes an automatic, reciprocal change in our own facial muscles (Dimberg & Thunberg, 1998). This automatic matching response can occur even without the participant's attention or conscious awareness (Dimberg, Thunberg, & Elmehed, 2000).

Finally, according to psychologist Stanley **Schachter's two-factor theory** (FIGURE 11.17D), we look to external rather than internal cues to understand our emotions. If we cry at a wedding, we interpret our emotion as happiness, but if we cry at a funeral, we label our emotion sadness (FIGURE 11.19).

Which theory is correct? Each has its limits and contributions to our understanding of emotion. For example, the James-Lange theory fails to acknowledge that physical arousal can occur without emotional experience (e.g., when we exercise). This theory also requires a distinctly different pattern of arousal for each emotion. Otherwise, how do we know whether we are sad, happy, or angry? Although brain-imaging studies do show subtle differences among basic emotions (Levenson, 1992, 2007; Werner et al., 2007), most people are not aware of these slight variations. Thus, there must be other explanations for why we experience emotion.

The Cannon-Bard theory (that the cortex and autonomic nervous system receive simultaneous messages from the thalamus) has received some experimental support. Victims of spinal cord damage still experience emotions, often more intensely than before their injuries (Nicotra et al., 2006; Schopp et al., 2007). Instead of the thalamus, however, research shows that it is the limbic system, hypothalamus, and prefrontal cortex that are most important in emotional experience (LeDoux, 2007; Zillmer, Spiers, & Culberston, 2008).

Research on the facial-feedback hypothesis has found that facial feedback does seem to contribute to the intensity of our subjective emotions and overall moods. Thus, if you want to change a bad mood or intensify a particularly good emotion, adopt the appropriate facial expression. In other words, "fake it 'til you make it."

Visualizing

Schachter's two-factor theory FIGURE 11.19

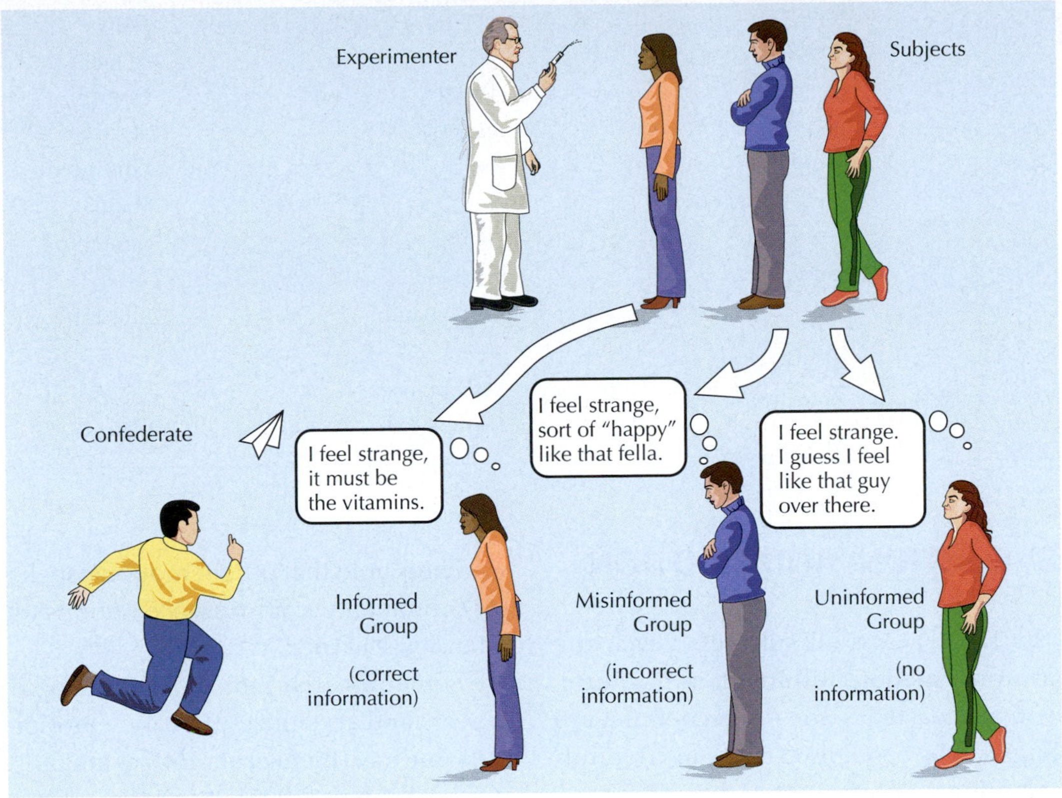

In Schachter and Singer's classic study (1962), participants were given shots of epinephrine and told it was a type of vitamin. One group of participants was correctly informed about the expected effects (hand tremors, excitement, and heart palpitations). A second group was misinformed and told to expect itching, numbness, and headache. A third group was told nothing about the possible effects.

Following the injection, each participant was placed in a room with a confederate (a "stooge" who was part of the experiment but who pretended to be a fellow volunteer) who acted either happy or unhappy.

The results showed that participants who lacked an appropriate cognitive label for their emotional arousal (the misinformed and uninformed groups) tended to look to the situation for an explanation. Thus, those placed with a happy confederate became happy, whereas those with an unhappy confederate became unhappy. Participants in the correctly informed group knew their physiological arousal was the result of the shot, so their emotions were generally unaffected by the confederate.

Finally, Schachter's two-factor theory emphasizes the importance of cognitive processes in emotions, but his findings have been criticized. For example, as mentioned earlier, some neural pathways involved in emotion bypass the cortex and go directly to the limbic system. This and other evidence suggests that emotion is not simply the labeling of arousal (Dimberg, Thunberg, & Elmehed, 2000; LeDoux, 1996b, 2002; Mineka & Oehman, 2002).

The basic human emotions TABLE 11.1

Carroll Izard	Paul Ekman and Wallace Friesen	Robert Plutchik	Silvan Tomkins
Fear	Fear	Fear	Fear
Anger	Anger	Anger	Anger
Disgust	Disgust	Disgust	Disgust
Surprise	Surprise	Surprise	Surprise
Joy	Happiness	Joy	Enjoyment
Shame	—	—	Shame
Contempt	Contempt	—	Contempt
Sadness	Sadness	Sadness	—
Interest	—	Anticipation	Interest
Guilt	—	—	—
—	—	Acceptance	—
—	—	—	Distress

CULTURE, EVOLUTION, AND EMOTION

Are emotions the same across all cultures? Given the seemingly vast array of emotions within our own culture, it may surprise you to learn that some researchers believe that all our feelings can be condensed into 7 to 10 culturally universal emotions (TABLE 11.1). These researchers hold that other emotions, such as love, are simply combinations of primary emotions with variations in intensity (FIGURE 11.20).

Some research indicates that people in all cultures express and recognize the basic emotions in essentially the same way (Biehl et al., 1997; Ekman, 1993, 2004; Matsumoto & Juang, 2008) (FIGURE 11.21).

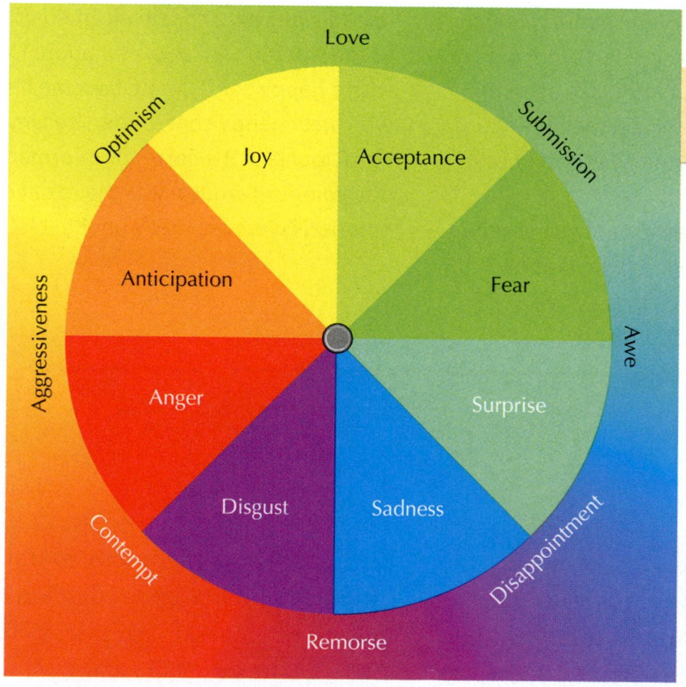

Plutchik's wheel of emotions
FIGURE 11.20

Robert Plutchik (1984, 1994, 2000) suggested that primary emotions (inner circle) combine to form secondary emotions (located outside the circle). Plutchik also found that emotions that lie next to each other are more alike than those that are farther apart. Can you provide examples of this phenomenon from your own life?

Universal emotional expressions FIGURE 11.21

People from very different cultures display remarkably similar facial expressions when experiencing particular emotions. Furthermore, people everywhere can reliably identify at least six basic emotions: happiness, surprise, anger, sadness, fear, and disgust.

From an evolutionary perspective, the idea of universal facial expressions makes adaptive sense because they signal others about our current emotional state (Ekman & Keltner, 1997). Charles Darwin first advanced the evolutionary theory of emotion in 1872. He proposed that expression of emotions evolved in different species as a part of survival and natural selection. For example, fear helps animals avoid danger, whereas expressions of anger and aggression are useful when fighting for mates or resources. Modern evolutionary theory suggests that basic emotions originate in the *limbic system*, which developed earlier than higher brain areas responsible for thought (the cortex).

Studies with infants provide further support for an evolutionary basis for emotions. For example, infants only a few hours old show distinct expressions of emotion that closely match adult facial expressions (Field et al., 1982). And all infants, even those who are born deaf and blind, show similar facial expressions in similar situations (Field et al., 1982; Gelder et al., 2006). In addition, a recent study showed that families may have characteristic facial expressions, shared even by family members who have been blind from birth (Peleg et al., 2006). This collective evidence points to a strong biological, evolutionary basis for emotional expression and decoding. *Psychological Science* on the next page provides more information about the study of happiness around the world.

So how do we explain cultural differences in emotions? Although we all seem to share similar facial expressions for some emotions, each culture has its own *display rules* governing how, when, and where to express emotions (Ekman, 1993, 2004; Fok et al., 2008). For instance, parents pass along their culture's specific display rules when they respond angrily to some emotions in their children, when they are sympathetic to others, and when they simply ignore an expression of emotion. Public physical contact is also governed by display rules. North Americans and Asians are less likely than people in other cultures to touch each other. For example, only the closest family and friends might hug in greeting or farewell. In contrast, Latin Americans and Middle Easterners often embrace and hold hands as a sign of casual friendship (Axtell, 2007).

Psychological Science

Happiness depends on various factors

Social scientists are starting to include relative happiness with hard data on economic status, health, and other factors as they assess quality of life. They rely on surveys of "subjective well-being"—how good people feel about their lives. A world map of one "happiness index" shows many, but not all, wealthy northern countries faring well. Residents of sub-Saharan Africa and the former Soviet Union, meanwhile, report particularly low levels of contentment.

Any attempt to measure happiness will fall short—each life is a series of joys, struggles, and sorrows, and satisfaction can depend as much on outlook as on circumstances. Averages obscure the happy moments in struggling nations, as well as people who suffer from poor health, poverty, or discrimination in countries that rank high. Still, happiness indices can help researchers move beyond simple economics as they track progress—or backsliding—over time.

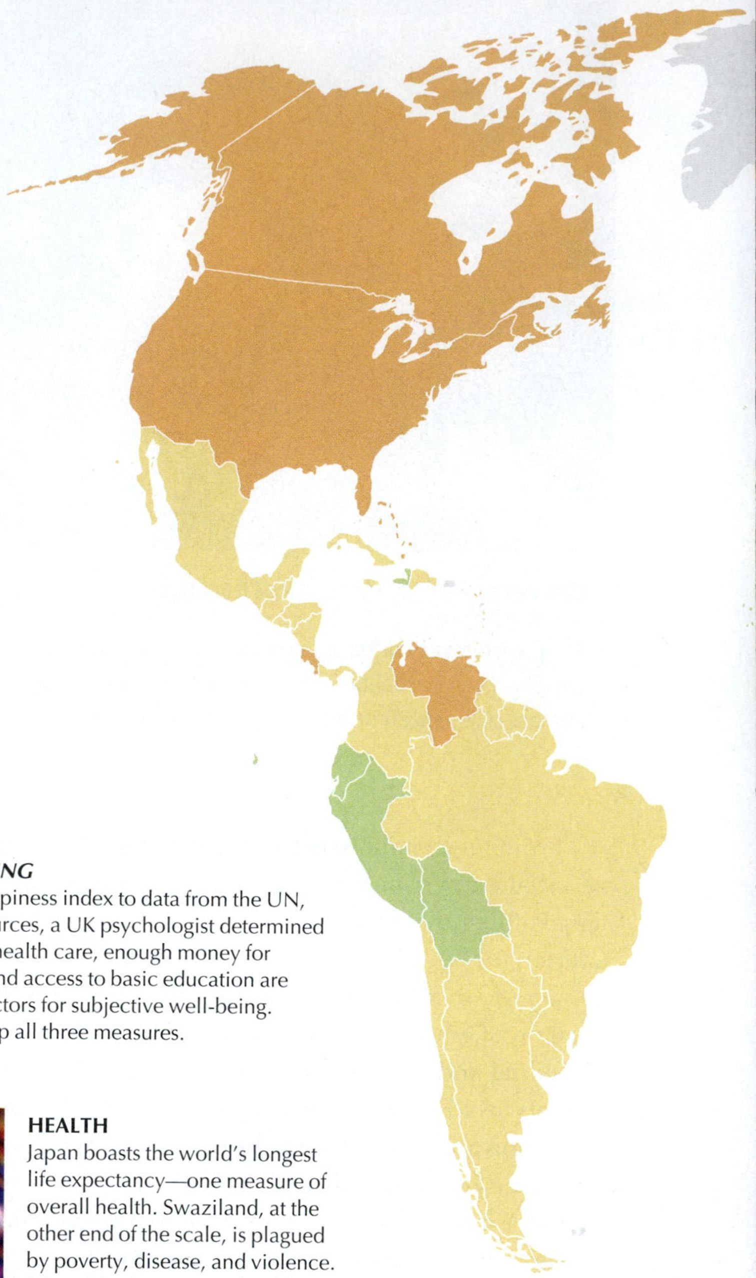

Source: White, A. 2006

MEASURING THE INTANGIBLE
The map is derived from the New Economics Foundation's 2006 "Happy Planet Index," which drew on over 100 surveys of subjective well-being. Its "satisfaction with life scale"—a happiness index—ranks the relative happiness of nations, from a high of 273 (Denmark and Switzerland) to a low of 100 (Burundi).

DEFINING *WELL-BEING*
By comparing the happiness index to data from the UN, the CIA, and other sources, a UK psychologist determined that good health and health care, enough money for fundamental needs, and access to basic education are the most important factors for subjective well-being. European countries top all three measures.

HEALTH
Japan boasts the world's longest life expectancy—one measure of overall health. Swaziland, at the other end of the scale, is plagued by poverty, disease, and violence. Disparities in access to health care divide many countries into haves and have-nots.

NATIONAL GEOGRAPHIC

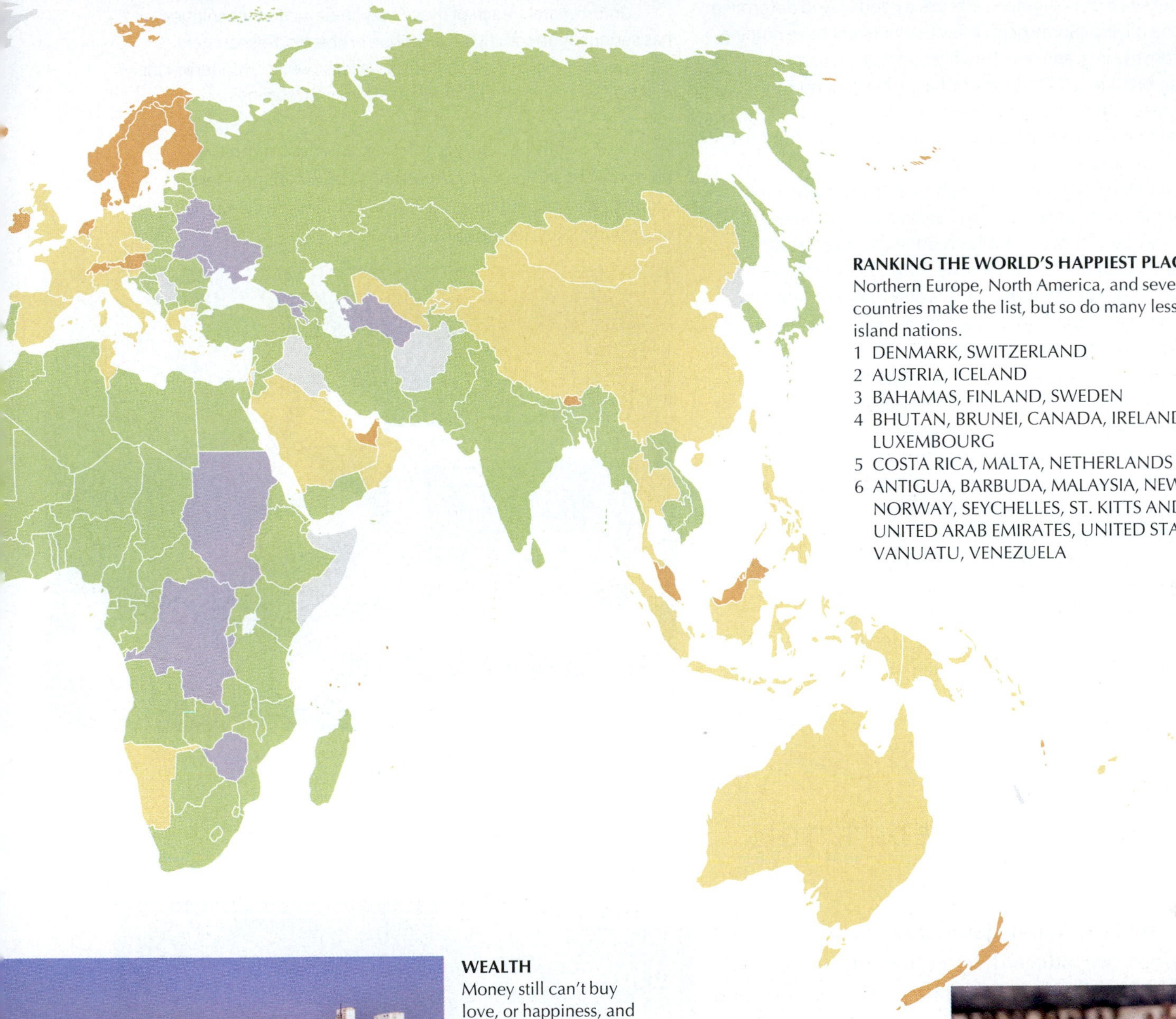

RANKING THE WORLD'S HAPPIEST PLACES

Northern Europe, North America, and several wealthy countries make the list, but so do many less prosperous island nations.

1 DENMARK, SWITZERLAND
2 AUSTRIA, ICELAND
3 BAHAMAS, FINLAND, SWEDEN
4 BHUTAN, BRUNEI, CANADA, IRELAND, LUXEMBOURG
5 COSTA RICA, MALTA, NETHERLANDS
6 ANTIGUA, BARBUDA, MALAYSIA, NEW ZEALAND, NORWAY, SEYCHELLES, ST. KITTS AND NEVIS, UNITED ARAB EMIRATES, UNITED STATES, VANUATU, VENEZUELA

WEALTH

Money still can't buy love, or happiness, and wealthier people aren't always more content. Still, tiny Luxembourg, which takes top rank in per capita gross domestic product (GDP), also rates a 253 on the happiness index. Real poverty means real misery, a fate shared by billions.

EDUCATION

Residents of Australia can expect to spend more time in school—an average of almost 21 years—than citizens of any other country. But only a basic education is needed to see a significant jump in overall happiness. Around the world, hundreds of millions lack even that.

Polygraph Testing

Do you believe that polygraph tests are a good way to determine if someone is lying? Some people say the innocent have nothing to fear from a polygraph test. However, research suggests otherwise (DeClue, 2003; Faigman et al., 1997; Iacono & Lykken, 1997). In fact, although proponents claim that polygraph tests are 90 percent accurate or better, actual tests show error rates ranging between 25 and 75 percent.

Traditional polygraph tests are based on the theory that when people lie, they feel guilty and anxious. Special sensors supposedly detect these testable emotions by measuring sympathetic and parasympathetic nervous system responses (**A** and **B**). The problem is that lying is only loosely related to anxiety and guilt. Some people become nervous even when telling the truth, whereas others remain calm when deliberately lying. A polygraph cannot tell which emotion is being felt (nervousness, excitement, sexual arousal, etc.) or whether a response is due to emotional arousal or something else. One study found that people could affect the outcome of a polygraph by about 50 percent simply by pressing their toes against the floor or biting their tongues (Honts & Kircher, 1994).

For these reasons, most judges and scientists have serious reservations about using polygraphs as lie detectors (DeClue, 2003). Scientific controversy and public concern led the U.S. Congress to pass a bill that severely restricts the use of polygraphs in the courts, in government, and in private industry.

In our post–9/11 world, you can see why tens of millions to hundreds of millions of dollars have been spent on new and improved lie-detection techniques (Kluger & Masters, 2006). Perhaps the most promising new technique is the use of brain scans like the function magnetic imaging (fMRI) (**C**).

Unfortunately, each of these new lie-detection techniques has serious shortcomings and unique problems. Researchers have questioned their reliability and validity, while civil libertarians and judicial scholars question their ethics and legalities.

A During a standard polygraph test, a band around the person's chest measures breathing rate, a cuff monitors blood preasure, and finger electrodes measure sweating, or galvanic skin response (GSR).

THE POLYGRAPH AS A LIE DETECTOR: DOES IT WORK?

We've discussed the four major theories of emotion, and how emotions are affected by culture and evolution. Now, we turn our attention to one of the hottest, and most controversial, topics in emotion research—the **polygraph**. This scientific instrument measures physiological responses, such as respiration, heart rate, and skin conductance (which increases during emotional arousal due to sweat gland activity). Given that we have less control over these physiological responses than over other behaviors, many people believe that when people lie, they feel guilty, fearful, and anxious. These feelings are then supposedly detected by the polygraph machine. *What a Psychologist Sees* describes how and if polygraphs work.

CONCEPT CHECK STOP

How does the autonomic nervous system (ANS) respond to fearful stimuli?

What are the differences between the James-Lange theory and the Cannon-Bard theory of emotion?

Why do some researchers believe that the basic emotions are consistent from one culture to the next?

B Note how the GSR rises sharply in response to the question, "Have you ever taken money from this bank?"

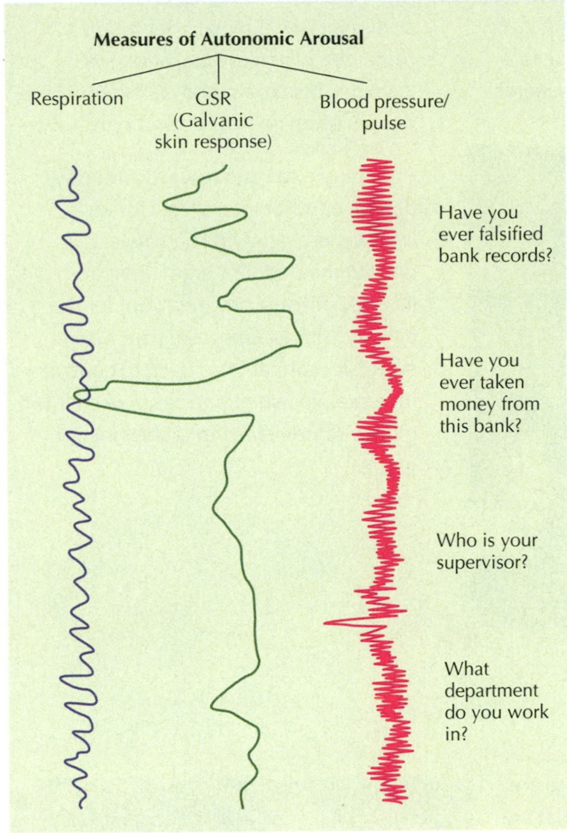

C The three fMRI images appear to show several known and specific areas of the cortex most involved in lying versus truth telling.

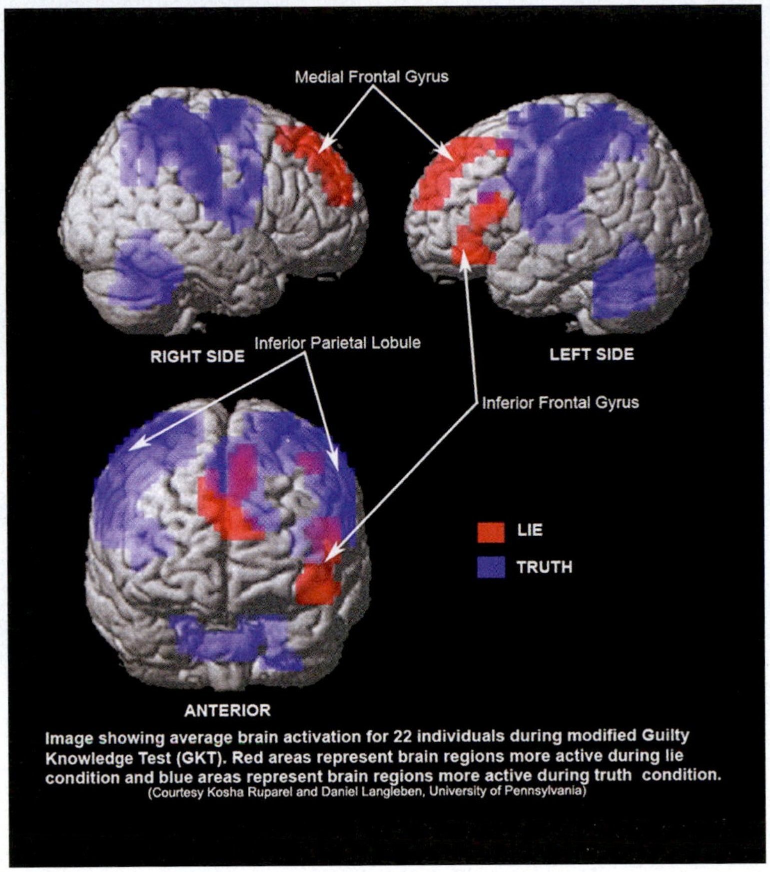

Image showing average brain activation for 22 individuals during modified Guilty Knowledge Test (GKT). Red areas represent brain regions more active during lie condition and blue areas represent brain regions more active during truth condition.
(Courtesy Kosha Ruparel and Daniel Langleben, University of Pennsylvania)

SUMMARY

1 Theories and Concepts of Motivation

1. Because motivation and emotion are closely connected, they are often studied together.
2. The three major biological theories of **motivation** involve **instincts**, drives (produced by the body's need for **homeostasis**), and arousal, or need for novelty and complexity. For people who are high in sensation seeking, the need for stimulation is especially high.
3. Psychosocial theories emphasize the role of incentives, attributions, and expectations in cognition.
4. Maslow's **hierarchy of needs** theory takes a biopsychosocial approach. It prioritizes needs, with survival needs at the bottom and social and spiritual needs at the top. Although the theory has made important contributions, some critics argue that it is poorly researched and biased toward Western individualism.

2 Motivation and Behavior

1. Hunger is one of the strongest motivational drives, and both biological (stomach, biochemistry, the brain) and psychosocial (stimulus cues and cultural conditioning) factors affect hunger and eating. These same factors play a role in obesity, **anorexia nervosa**, and **bulimia nervosa**.

2. The key to understanding what motivates high-achieving individuals lies in a high need for achievement (nAch), or **achievement motivation**, which is learned in early childhood, primarily through interactions with parents.

3. The human motivation for sex is extremely strong. Masters and Johnson first studied and described the **sexual response cycle**, the series of physiological and sexual responses that occurs during sexual activity. Other sex research has focused on the roots of **sexual orientation**, and most studies suggest that genetics and biology play the dominant role. Sexual orientation remains a divisive issue, and gays, lesbians, bisexuals, and transgendered people often confront **sexual prejudice**.

4. Providing **extrinsic rewards** (money, praise, or other incentives) for an intrinsically satisfying activity can undermine people's enjoyment and interest (**intrinsic motivation**) for the activity. This is especially true when extrinsic motivation is used to control—for example, when parents give children money or privileges for achieving good grades.

3 Theories and Concepts of Emotion

1. **Emotion** has physiological (brain and autonomic nervous system), cognitive (thoughts), and behavioral (expressive) components.

2. The major theories of how the components of emotion interact are the James-Lange, Cannon-Bard, facial-feedback, and Schachter's two-factor. Each emphasizes different sequences or aspects of the three elements, and each has its limits and contributions to our understanding of emotion.

3. Some researchers believe that there are 7 to 10 basic, universal emotions that are shared by people of all cultures. Some research indicates that people in all cultures express and recognize the basic emotions in essentially the same way, supporting the evolutionary theory of emotion. Studies with infants provide further support. Although we all seem to share similar facial expressions for some emotions, display rules for emotions (as well as for public physical contact, another aspect of the behavioral element of emotion) vary across cultures.

4. Polygraph tests attempt to detect lying by measuring physiological signs of guilt and anxiety. However, because lying is only loosely related to these emotions, most judges and scientists have serious reservations about using polygraphs as lie detectors.

KEY TERMS

- **motivation** p. 284
- **instincts** p. 284
- **drive-reduction theory** p. 285
- **homeostasis** p. 285
- **arousal theory** p. 285
- **incentive theory** p. 287
- **cognitive theory** p. 287
- **attributions** p. 287
- **expectancies** p. 287
- **hierarchy of needs** p. 288
- **self-actualization** p. 288
- **satiety** p. 289
- **thermogenesis** p. 289
- **obesity** p. 290
- **anorexia nervosa** p. 291
- **bulimia nervosa** p. 291
- **achievement motivation** p. 293
- **sexuality** p. 294
- **excitement phase** p. 294
- **plateau phase** p. 294
- **orgasm phase** p. 294
- **resolution phase** p. 294
- **refractory period** p. 294
- **sexual response cycle** p. 295
- **sexual orientation** p. 295
- **sexual prejudice** p. 295
- **extrinsic motivation** p. 296
- **intrinsic motivation** p. 296
- **emotion** p. 298
- **social smile** p. 299
- **Duchenne smile** p. 299
- **emotional intelligence (EI)** p. 300
- **James-Lange theory** p. 300
- **Cannon-Bard theory** p. 300
- **facial-feedback hypothesis** p. 301
- **Schachter's two-factor theory** p. 302
- **polygraph** p. 308

CRITICAL AND CREATIVE THINKING QUESTIONS

1. Like most Americans, you may have difficulty controlling your weight. Using information from this chapter, can you identify the factors or motives that best explain your experience?
2. How can you restructure elements of your personal, work, or school life to increase intrinsic versus extrinsic motivation?
3. If you were going out on a date with someone or applying for an important job, how might you use the four theories of emotion to increase the chances that things will go well?
4. Have you ever felt depressed after listening to a friend complain about his problems? How might the facial-feedback hypothesis explain this?
5. Why do you think people around the world experience and express the same basic emotions? What evolutionary advantages might help explain these similarities?

What is happening in this picture?

Curiosity is an important aspect of the human experience.

- **Which of the six theories of motivation best explains this behavior?**
- **According to Maslow's hierarchy of needs, what is the likelihood that this person has not fulfilled basic safety needs?**

SELF-TEST

(Check your answers in Appendix B.)

1. *Motivation* is BEST defined as ________.
   a. A set of factors that activate, direct, and maintain behavior, usually toward a goal
   b. the physiological and psychological arousal that occurs when a person really wants to achieve a goal
   c. what makes you do what you do
   d. the conscious and unconscious thoughts that focus a person's behaviors and emotions in the same direction toward a goal

2. The figure below illustrates the ________ theory, in which motivation decreases once homeostasis occurs.
   a. drive-induction
   b. heterogeneity
   c. drive-reduction
   d. biostability

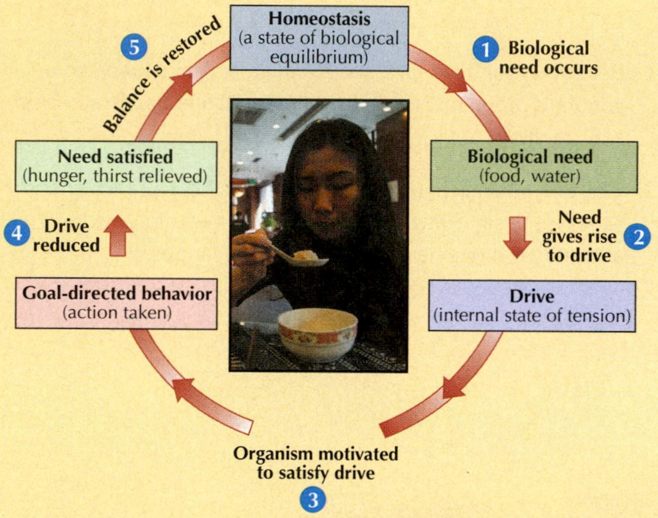

Homeostasis
(a state of biological equilibrium)
1 Biological need occurs
Biological need
(food, water)
Need gives rise to drive 2
Drive
(internal state of tension)
Organism motivated to satisfy drive
3
Goal-directed behavior
(action taken)
4 Drive reduced
Need satisfied
(hunger, thirst relieved)
5
Balance is restored

3. ________ is the body's tendency to maintain a relatively stable state for internal processes.
   a. Homeostasis
   b. Heterogeneity
   c. Drive-induction
   d. Biostability

4. The ________ theory says people are "pulled" by external stimuli to act a certain way.
   a. cognitive
   b. incentive
   c. Maslow's hierarchy of needs
   d. incentive

5. The theory, illustrated in the figure below, that some motives have to be satisfied before a person can advance to fulfilling higher motives is based on ________.
   a. Freud's psychosexual stages of development
   b. Kohlberg's moral stages of development
   c. Erikson's psychosocial stages of development
   d. Maslow's hierarchy of needs

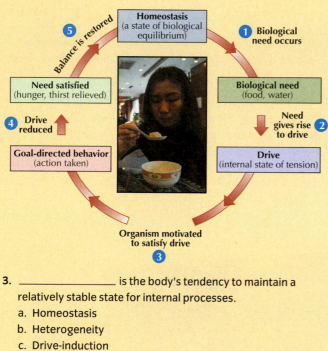

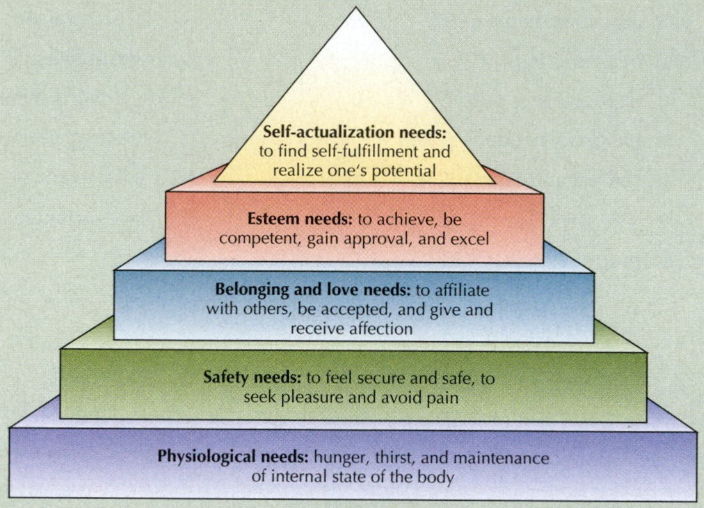

Self-actualization needs: to find self-fulfillment and realize one's potential
Esteem needs: to achieve, be competent, gain approval, and excel
Belonging and love needs: to affiliate with others, be accepted, and give and receive affection
Safety needs: to feel secure and safe, to seek pleasure and avoid pain
Physiological needs: hunger, thirst, and maintenance of internal state of the body

6. The feeling of fullness you get once you have ingested enough food is called ________.
   a. bloating
   b. a bellyful
   c. a bellyache
   d. satiety

7. ________ involves the consumption of large quantities of food followed by self-induced vomiting, the use of laxatives, or extreme exercise.
   a. Anorexia nervosa
   b. The binge-purge syndrome
   c. Bulimia nervosa
   d. Pritikin dieting

8. According to your textbook, the desire to excel, especially in competition with others, is known as ________.
   a. drive
   b. instincts
   c. achievement motivation
   d. all of the above

9. Label the correct sequence of events in Masters and Johnson's sexual response cycle on the figure below.

10. ______ is the term for negative attitudes toward someone based on his or her sexual orientation.
    a. Homophobia
    b. Sexual prejudice
    c. Heterosexism
    d. Sexual phobia

11. Extrinsic motivation is based on ______.
    a. the desire for rewards or threats of punishment
    b. the arousal motive
    c. the achievement motive
    d. all of these options

12. The three components of emotion are ______.
    a. cognitive, physiological, and behavioral
    b. perceiving, thinking, and acting
    c. positive, negative, and neutral
    d. active/passive, positive/negative, and direct/indirect

13. In a ______, as shown in the photo below, the cheek muscles are pulled back and the muscles around the eyes also contract.
    a. Duchenne smile
    b. Madonna smile
    c. Mona Lisa smile
    d. da Vinci smile

14. According to ______, we look to external rather than internal cues to understand emotions.
    a. Cannon-Bard theory
    b. James-Lange theory
    c. the facial feedback hypothesis
    d. Schachter's two-factor theory

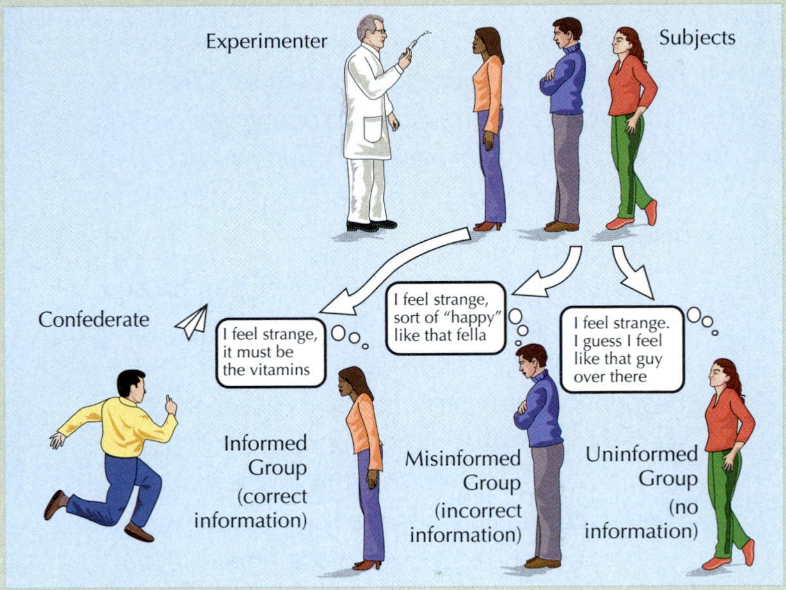

15. Researchers believe all our feelings can be condensed to ______ culturally universal emotions.
    a. 2 to 3
    b. 5 to 6
    c. 7 to 10
    d. 11 to 15

Personality

12

Consider the following personality description. How well does it describe you?

You have a strong need for other people to like and admire you. You tend to be critical of yourself. Although you have some personality weaknesses, you are generally able to compensate for them. At times, you have serious doubts about whether you have made the right decision or done the right thing.

—Adapted from Ulrich, Stachnik, & Stainton, 1963.

Does this sound like you? A high percentage of research participants who read a similar personality description reported that the description was "very accurate"—even after they were informed that it was a phony horoscope (Hyman, 1981). Other research shows that about three-quarters of adults read newspaper horoscopes and that many of them believe that these horoscopes were written especially for them (Halpern, 1998; Wyman & Vyse, 2008).

Why are such spurious personality assessments so popular? In part, it is because they seem to tap into our unique selves. However, the traits they supposedly reveal are characteristics that almost everyone shares. Do you know anyone who doesn't "have a strong need for other people to like and admire [them]"? The traits in horoscopes are also generally flattering, or at least neutral. In this chapter, rather than relying on these unscientific methods, we will focus on research-based methods used by psychologists to assess personality.

Unlike the pseudopsychologies offered in supermarket tabloids, newspaper horoscopes, and Chinese fortune cookies, the descriptions presented by personality researchers are based on empirical studies. In this chapter, we examine the five most prominent theories and findings in personality research and discuss the techniques that psychologists use to assess personality.

NATIONAL GEOGRAPHIC

CHAPTER OUTLINE

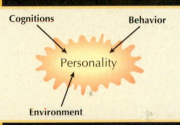

Trait Theories

LEARNING OBJECTIVES

Explain how early trait theorists approached the study of personality.

Identify the "Big Five" personality traits.

Summarize the major critiques of trait theory.

Personality describes you as a person—how you are different from other people and what patterns of behavior are typical of you. You might qualify as an "extrovert," for example, if you are talkative and outgoing most of the time. Or you might be considered "conscientious" if you are responsible and self-disciplined most of the time.

personality Relatively stable and enduring patterns of thoughts, feelings, and actions.

traits Relatively stable and consistent characteristics that can be used to describe someone.

The terms we use to describe other people (and ourselves) are called **traits**. Trait theorists are interested in discovering which traits best describe people and in measuring the degree of variation in traits within individuals and among people.

EARLY TRAIT THEORISTS: ALLPORT, CATTELL, AND EYSENCK

An early study of dictionary terms found almost 4,500 words that described personality traits (Allport & Odbert, 1936). Faced with this enormous list, Gordon Allport (1937) believed that the best way to understand personality was to arrange a person's unique personality traits into a hierarchy, with the most pervasive or important traits at the top.

Later psychologists reduced the list of possible personality traits using a statistical technique called **factor analysis**, in which large arrays of data are grouped into more basic units (factors). Raymond Cattell (1950, 1965, 1990) condensed the list of traits to 30 to 35 basic characteristics. Hans Eysenck (1967, 1982, 1990) reduced the list even further. He described personality as a relationship among three basic types of traits: extroversion-introversion (E), neuroticism (N), and psychoticism (P).

THE FIVE-FACTOR MODEL: FIVE BASIC PERSONALITY TRAITS

Factor analysis was also used to develop the most promising modern trait theory, the **five-factor model (FFM)** (Costa, McCrae, & Martin, 2008; McCrae & Costa, 1990, 1999; McCrae & Sutin, 2007; Wood & Bell, 2008).

five-factor model (FFM) The trait theory that explains personality in terms of the "Big Five" model, which is composed of openness, conscientiousness, extroversion, agreeableness, and neuroticism.

Combining previous research findings and the long list of possible personality traits, researchers discovered that five traits came up repeatedly, even when different tests were used.

These five major dimensions of personality are often dubbed the **Big Five** (FIGURE 12.1). A handy way to remember the five factors is to note that the first letters of each of the five-factor model spell the word *ocean*. The Big Five are:

O *Openness.* People who rate high in this factor are original, imaginative, curious, open to new ideas, artistic, and interested in cultural pursuits. Low scorers tend to be conventional, down-to-earth, narrower in their interests, and not artistic. Interestingly, critical thinkers tend to score higher than others on this factor (Clifford, Boufal, & Kurtz, 2004).

C *Conscientiousness.* This factor ranges from responsible, self-disciplined, organized, and achieving at the high end to irresponsible, careless, impulsive, lazy, and undependable at the other.

E *Extroversion.* This factor contrasts people who are sociable, outgoing, talkative, fun loving, and affectionate at the high end with introverted individuals who tend to be withdrawn, quiet, passive, and reserved at the low end.

A *Agreeableness.* Individuals who score high in this factor are good-natured, warm, gentle, cooperative, trusting, and helpful, whereas low scorers are irritable, argumentative, ruthless, suspicious, uncooperative, and vindictive.

N *Neuroticism* (or emotional stability). People who score high in neuroticism are emotionally unstable and prone to insecurity, anxiety, guilt, worry, and moodiness. People at the other end are emotionally stable, calm, even-tempered, easygoing, and relaxed.

Personality and your career FIGURE 12.1

Are some people better suited for certain jobs than others? According to psychologist John Holland's personality-*job-fit* theory, a match (or "good fit") between our individual personality and our career choice is a major factor in determining job satisfaction (Holland, 1985, 1994). Research shows that a "good fit" between personality and occupation helps increase subjective well-being, job success, and job satisfaction. In other words, people tend to be happier and like their work when they're well matched to their jobs (Borchers, 2007; Donohue, 2006; Gottfredson & Duffy, 2008; Kieffer, Schinka, & Curtiss, 2004).

Applying Psychology

Love and the "Big Five"

Using the figure to the right, plot your personality profile by placing a dot on each line to indicate your degree of openness, conscientiousness, and so on. Do the same for a current, previous, or prospective love partner.

Now look at the two mate preferences lists below. David Buss and his colleagues (1989, 2003) surveyed more than 10,000 men and women from 37 countries and found a surprising level of agreement in the characteristics that men and women value in a mate. Moreover, most of the Big Five personality traits are found at the top of the list. Both men and women prefer dependability (conscientiousness), emotional stability (low neuroticism), pleasing disposition (agreeableness), and sociability (extroversion) to the alternatives. These findings may reflect an evolutionary advantage for people who are open, conscientious, extroverted, agreeable, and free of neuroses.

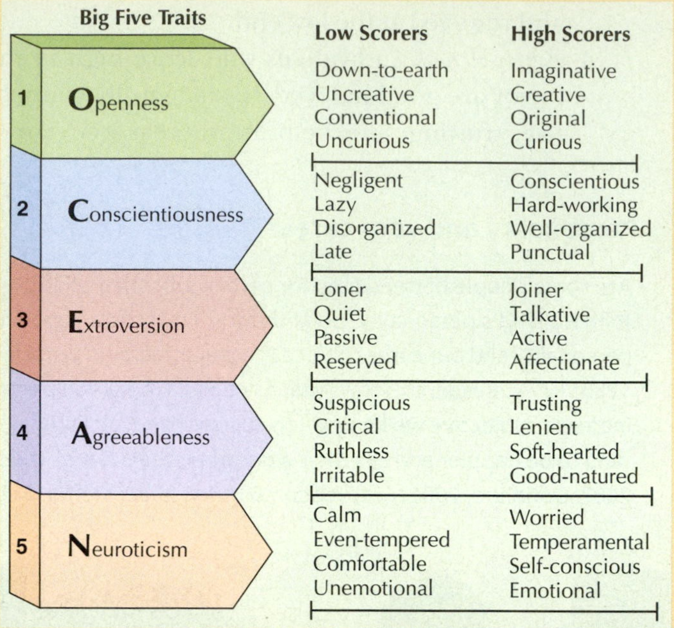

Mate preferences around the world

In the two lists below, note how the top four desired traits are the same for both men and women, as well as how closely their desired traits match those of the five-factor model (FFM).

♂ What Men Want in a Mate

1. Mutual attraction — love
2. Dependable character
3. Emotional stability and maturity
4. Pleasing disposition
5. Good health
6. Education and intelligence
7. Sociability
8. Desire for home and children
9. Refinement, neatness
10. Good looks

♀ What Women Want in a Mate

1. Mutual attraction — love
2. Dependable character
3. Emotional stability and maturity
4. Pleasing disposition
5. Education and intelligence
6. Sociability
7. Good health
8. Desire for home and children
9. Ambition and industriousness
10. Refinement, neatness

Source: Buss et al., "International Preferences in Selecting Mates." *Journal of Cross-Cultural Psychology*, 21, pp. 5–47, 1990. Sage Publications, Inc.

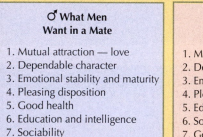

CRITICAL THINKING

Here are two interesting questions:

1. How do your personality traits compare with those of your love partner?
2. If your scores were noticeably different, what might explain the differences?

EVALUATING TRAIT THEORIES

The five-factor model is the first to achieve the major goal of trait theory—to describe and organize personality characteristics using the smallest number of traits. Critics argue, however, that the great variation seen in personalities cannot be accounted for by only five traits and that the Big Five model fails to offer causal explanations for these traits (Friedman & Schustack, 2006; Funder, 2000; Sollod, Monte, & Wilson, 2009).

Critics maintain that, in general, trait theories are good at describing personality, but they have difficulty explaining why people develop these traits or why personality traits differ across cultures. For example, trait theories do not explain why people in cultures that are geographically close tend to have similar personalities or why Europeans and Americans tend to be higher in extroversion and openness to experience and lower in agreeableness than people in Asian and African cultures (Allik & McCrae, 2004).

In addition, some critics have faulted trait theories for their lack of specificity. Although trait theorists have documented a high level of personality stability after age 30 (FIGURE 12.2), they haven't identified which characteristics last a lifetime and which are most likely to change.

Finally, trait theorists have been criticized for ignoring the importance of situational and environmental effects on personality. In one example, psychologists Fred Rogosch and Dante Cicchetti (2004) found that abused and neglected children scored significantly lower in the traits of openness to experience, conscientiousness, and agreeableness and higher in the trait of neuroticism than did children who were not maltreated. Unfortunately, these maladaptive personality traits create significant liabilities that may trouble these children throughout their lifetimes.

Personality change over time FIGURE 12.2

Have you noticed how Madonna's public image and behavior have changed over time? Cross-cultural research has found that neuroticism, extroversion, and openness to experience tend to decline from adolescence to adulthood, whereas agreeableness and conscientiousness increase (McCrae et al., 2004). How would you explain these changes? Do you think they're good or bad?

CONCEPT CHECK STOP

What is the purpose of factor analysis?

What dimensions of personality are central to the five-factor model?

What are some weaknesses of trait theory?

Psychoanalytic/Psychodynamic Theories

LEARNING OBJECTIVES

Identify Freud's most basic and controversial contributions to the study of personality.

Explain how Adler's, Jung's, and Horney's theories differ from Freud's thinking.

Explore the major criticisms of Freud's psychoanalytic theories.

In contrast to trait theories that describe personality as it exists, psychoanalytic (or psychodynamic) theories of personality attempt to explain individual differences by examining how unconscious mental forces interplay with thoughts, feelings, and actions. The founding father of psychoanalytic theory is Sigmund Freud. We will examine Freud's theories in some detail and then briefly discuss three of his most influential followers.

FREUD'S PSYCHOANALYTIC THEORY: THE POWER OF THE UNCONSCIOUS

Who is the best-known figure in all of psychology? Most people immediately name Sigmund Freud, whose theories have been applied not only to psychology but also to anthropology, sociology, religion, medicine, art, and literature. Working from about 1890 until he died in 1939, Freud developed a theory of personality that has been one of the most influential—and most controversial—theories in all of science (Dufresne, 2007; Heller, 2005; Solod, Monte, & Wilson, 2009). Let's examine some of Freud's most basic and debatable concepts.

Freud called the mind the "psyche" and asserted that it contains three **levels of consciousness**, or awareness: the **conscious**, the **preconscious**, and the **unconscious** (FIGURE 12.3).

conscious Freud's term for thoughts or motives that a person is currently aware of or is remembering.

preconscious Freud's term for thoughts or motives that can be easily brought to mind.

unconscious Freud's term for thoughts or motives that lie beyond a person's normal awareness but that can be made available through psychoanalysis.

Freud's three levels of consciousness FIGURE 12.3

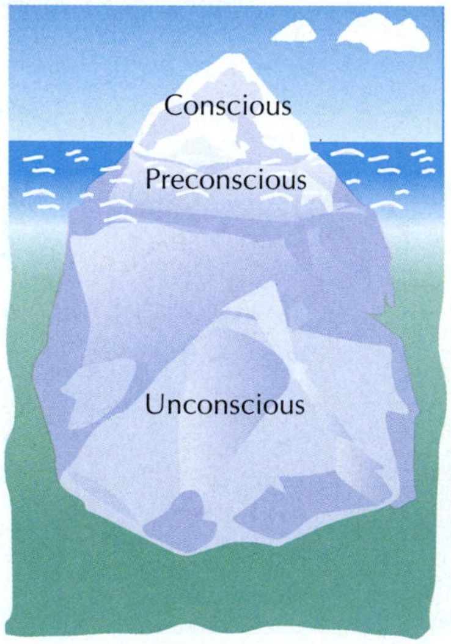

Although Freud never used the analogy himself, his levels of awareness are often compared to an iceberg:

- The tip of the iceberg would be analogous to the *conscious* mind, which is above the water and open to easy inspection.
- The *preconscious* (the area only shallowly submerged) contains information that can be viewed with a little extra effort.
- The large base of the iceberg is somewhat like the *unconscious* mind, completely hidden from personal inspection.

Using the Freudian idea of "levels of consciousness," we can say that at this moment your conscious mind is focusing on this text. However, your preconscious may include feelings of hunger and thoughts of friends you need to contact. Any repressed sexual desires, aggressive impulses, or irrational thoughts and feelings are reportedly stored in your unconscious.

Freud believed that most psychological disorders originate from repressed memories and instincts (sexual and aggressive) that are hidden in the unconscious (FIGURE 12.4). To treat these disorders, Freud developed *psychoanalysis* (Chapter 14).

In addition to proposing that the mind functions at three levels of consciousness, Freud also thought that personality was composed of three mental structures: *id, ego,* and *superego* (FIGURE 12.5).

According to Freud, the **id** is made up of innate, biological instincts and urges. It is immature, impulsive, and irrational. The id is also totally unconscious and serves as the reservoir of mental energy. When its primitive drives build up, the id seeks immediate gratification to relieve the tension—a concept known as the **pleasure principle**.

As a child grows older, the second part of the psyche—the ego—develops. The **ego** is responsible for planning, problem solving, reasoning, and controlling the potentially destructive energy of the id. In Freud's system, the ego corresponds to the *self*—our conscious identity of ourselves as persons.

One of the ego's tasks is to channel and release the id's energy in ways that are compatible with the external world. Thus, the ego is responsible for delaying gratification when necessary. Contrary to the id's pleasure principle, the ego operates on the **reality principle** because it can understand and deal with objects and events in the "real world."

The final part of the psyche to develop is the **superego**, a set of ethical rules for behavior. The superego develops from internalized parental and societal standards. It constantly strives for perfection and is therefore as unrealistic as the id. Some Freudian followers have suggested that the superego operates on the **morality principle** because violating its rules results in feelings of guilt.

defense mechanisms
In Freudian theory, the ego's protective method of reducing anxiety by distorting reality.

When the ego fails to satisfy both the id and the superego, anxiety slips into conscious awareness. Because anxiety is uncomfortable, people avoid it through **defense mechanisms**. For example, an alcoholic who uses his paycheck to buy drinks (a message from the id) may feel very guilty (a response from the superego). He may reduce this conflict by telling himself that he deserves a drink because he works so hard. This is an example of the defense mechanism **rationalization**.

"Good morning, beheaded—uh, I mean beloved."

Freudian slips FIGURE 12.4

Freud believed that a small slip of the tongue (known as a "Freudian slip") can reflect unconscious feelings that we normally keep hidden.

Freud's personality structure FIGURE 12.5

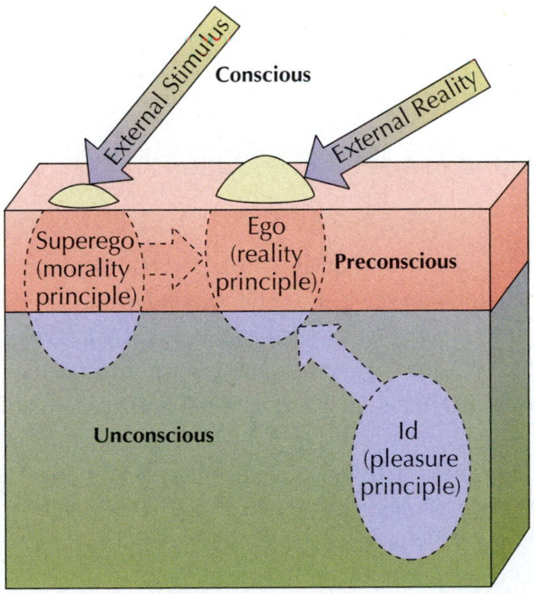

Study Organizer 12.1 Sample psychological defense mechanisms

Defense mechanism	*Description*	*Example*
Repression	Preventing painful or unacceptable thoughts from entering consciousness	Forgetting the details of your parent's painful death
Sublimation	Redirecting unmet desires or unacceptable impulses into acceptable activities	Rechanneling sexual desires into school, work, art, sports, or hobbies that are constructive
Denial	Protecting oneself from an unpleasant reality by refusing to perceive it	Alcoholics refusing to admit their addiction
Rationalization	Substituting socially acceptable reasons for unacceptable ones	Justifying cheating on an exam by saying "everyone else does it"
Intellectualization	Ignoring the emotional aspects of a painful experience by focusing on abstract thoughts, words, or ideas	Emotionless discussion of your divorce while ignoring underlying pain
Projection	Transferring unacceptable thoughts, motives, or impulses to others	Becoming unreasonably jealous of your mate while denying your own attraction to others
Reaction formation	Refusing to acknowledge unacceptable urges, thoughts, or feelings by exaggerating the opposite state	Promoting a petition against adult bookstores even though you are secretly fascinated by pornography
Regression	Responding to a threatening situation in a way appropriate to an earlier age or level of development	Throwing a temper tantrum when a friend doesn't want to do what you'd like
Displacement	Redirecting impulses toward a less threatening person or object	Yelling at a coworker after being criticized by your boss

Although Freud described many kinds of defense mechanisms (STUDY ORGANIZER 12.1), he believed that repression was the most important. **Repression** is the mechanism by which the ego prevents the most unacceptable, anxiety-provoking thoughts from entering consciousness (FIGURE 12.6).

The concept of defense mechanisms has generally withstood the test of time, and it is an accepted part of modern psychology. However, this is not the case for Freud's theory of psychosexual stages of development.

According to Freud, strong biological urges residing within the id push all children through five universal

Is it bad to use defense mechanisms?

FIGURE 12.6

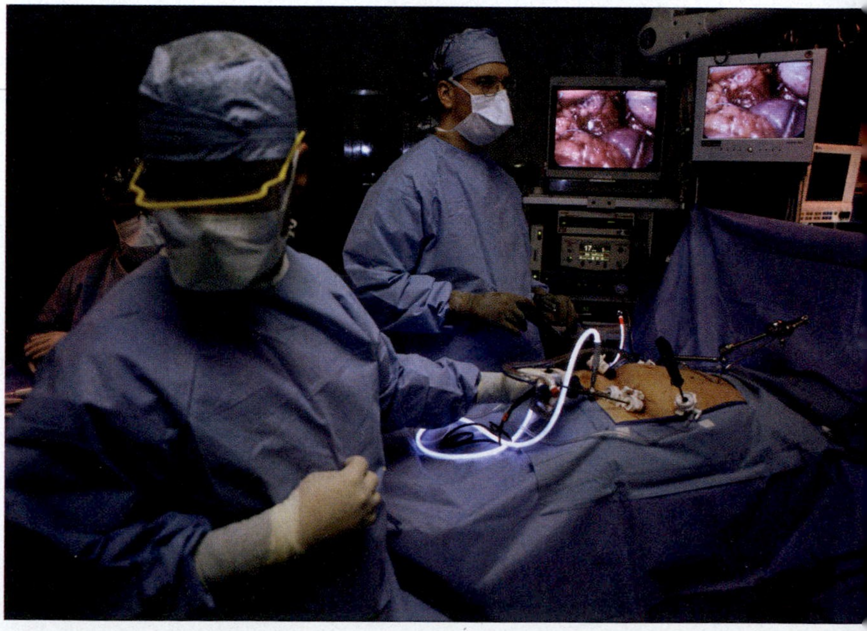

Although defense mechanisms do distort reality, some misrepresentation seems to be necessary for our psychological well-being (Marshall & Brown, 2008; Wenger & Fowers, 2008). During a gruesome surgery, for example, physicians and nurses may **intellectualize** the procedure as an unconscious way of dealing with their personal anxieties. Can you see how focusing on highly objective technical aspects of the situation might help these people avoid becoming emotionally overwhelmed by the potentially tragic circumstances they often encounter?

Freud's five psychosexual stages of development FIGURE 12.7

Process Diagram

Psychosexual Development

	Name of Stage (Approximate age)	Erogenous Zone (Key Conflict or Developmental Task)		Supposed Symptoms of Fixation and/ or Regression
1	**Oral** (0–18 months)	**Mouth** (Weaning from breast or bottle)		Overindulgence reportedly contributes to gullibility ("swallowing" anything), dependence, and passivity. Underindulgence leads to aggressiveness, sadism, and a tendency to exploit others. Freud also believed orally fixated adults may orient their lives around their mouths—overeating, becoming alcoholic, smoking, or talking a great deal.
2	**Anal** (18 months–3 years)	**Anus** (Toilet training)	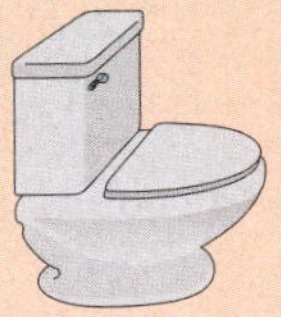	Fixation or regression supposedly leads to highly controlled and compulsively neat (anal-retentive) personality, or messy, disorderly, rebellious, and destructive (anal-expulsive) personality.
3	**Phallic** (3–6 years)	**Genitals** (Overcoming the Oedipus complex by identifying with same-sex parent)		According to Freud, unresolved, sexual longing for the opposite-sex parent can lead to long-term resentment and hostility toward the same-sex parent. Freud also believed that boys develop an **Oedipus complex,** or attraction to their mothers. He thought that young girls develop an attachment to their fathers and harbor hostile feelings towards their mothers, whom they blame for their lack of a penis. According to Freud most girls never overcome **penis envy** or give up their rivalry with their mothers, which leads to enduring moral inferiority.
4	**Latency** (6 years–puberty)	**None** (Interacting with same-sex peers)		The latency stage is a reported period of sexual "dormancy." Children do not have particular psychosexual conflicts that must be resolved during this period.
5	**Genital** (puberty–adult)	**Genitals** (Establishing intimate relationships with the opposite sex)		Unsuccessful outcomes at this stage, supposedly may lead to sexual relationships based only on lustful desires, not on respect and commitment.

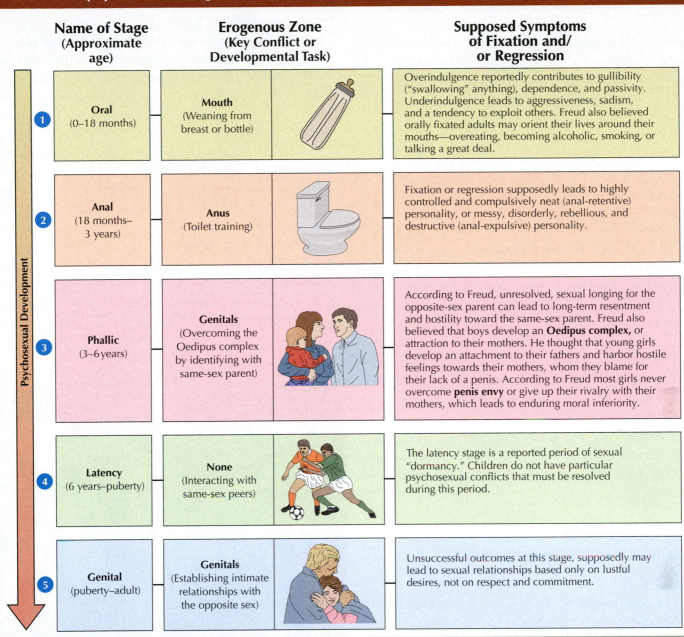

psychosexual stages In Freudian theory, the five developmental periods (oral, anal, phallic, latency, and genital) during which particular kinds of pleasures must be gratified if personality development is to proceed normally.

psychosexual stages (FIGURE 12.7). The term *psychosexual* reflects Freud's belief that children experience sexual feelings from birth (in different forms from those of adolescents or adults).

Freud held that if a child's needs are not met or are overindulged at one particular psychosexual stage, the child may become fixated, and a part of his or her personality will remain stuck at that stage. Furthermore, under stress, individuals may return (or regress) to a stage at which earlier needs were frustrated or overly gratified.

NEO-FREUDIAN/PSYCHODYNAMIC THEORIES: REVISING FREUD'S IDEAS

Some initial followers of Freud later rebelled and proposed theories of their own; they became known as **neo-Freudians**.

Alfred Adler (1870–1937) was the first to leave Freud's inner circle. Instead of seeing behavior as motivated by unconscious forces, he believed that it is purposeful and goal-directed. According to Adler's **individual psychology**, we are motivated by our goals in life—especially our goals of obtaining security and overcoming feelings of inferiority.

Adler believed that almost everyone suffers from an **inferiority complex**, or deep feelings of inadequacy and incompetence that arise from our feelings of helplessness as infants. According to Adler, these early feelings result in a "will-to-power" that can take one of two paths. It can either cause children to strive to develop superiority over others through dominance, aggression, or expressions of envy, or—more positively—it can cause children to develop their full potential and creativity and to gain mastery and control in their lives (Adler, 1964, 1998) (FIGURE 12.8).

Another early Freud follower turned dissenter, Carl Jung (pronounced "yoong"), developed **analytical psychology**. Like Freud, Jung (1875–1961) emphasized unconscious processes, but he believed that the unconscious contains positive and spiritual motives as well as sexual and aggressive forces.

Jung also thought that we have two forms of unconscious mind: the personal unconscious and the collective unconscious. The *personal unconscious* is created from our individual experiences, whereas the *collective unconscious* is identical in each person and is inherited (Jung, 1946, 1959, 1969). The collective unconscious consists of primitive images and patterns of thought, feeling, and behavior that Jung called **archetypes** (FIGURE 12.9).

Because of archetypal patterns in the collective unconscious, we perceive and react in certain predictable ways. One set of archetypes refers to gender roles (Chapter 10). Jung claimed that both males and females have patterns for feminine aspects of personality (*anima*) and masculine aspects of personality (*animus*), which allow us to express both masculine and feminine personality traits and to understand the opposite sex.

An upside to feelings of inferiority?
FIGURE 12.8

Adler suggested that the will-to-power could be positively expressed through social interest—identifying with others and cooperating with them for the social good. Can you explain how these volunteers might be fulfilling their will-to-power interest?

Archetypes in the collective unconscious
FIGURE 12.9

According to Jung, the collective unconscious is the ancestral memory of the human race, which supposedly explains the similarities in religion, art, symbolism, and dream imagery across cultures, such as the repeated symbol of the snake in ancient Egyptian tomb painting and early Australian aboriginal bark painting. Can you think of other explanations?

Like Adler and Jung, psychoanalyst Karen Horney (pronounced "HORN-eye") was an influential follower of Freud's who later came to reject major aspects of Freudian theory. She is remembered mostly for having developed a creative blend of Freudian, Adlerian, and Jungian theory, with added concepts of her own (Horney, 1939, 1945) (FIGURE 12.10).

Horney is also known for her theories of personality development. She believed that adult personality was shaped by the child's relationship to the parents—not by fixation at some stage of psychosexual development, as Freud argued. Horney believed that a child whose needs were not met by nurturing parents would experience extreme feelings of helplessness and insecurity. How people respond to this basic anxiety greatly determines emotional health.

According to Horney, everyone searches for security in one of three ways: We can move toward people (by seeking affection and acceptance from others); we can move away from people (by striving for independence, privacy, and self-reliance); or we can move against people (by trying to gain control and power over others). Emotional health requires a balance among these three styles.

Karen Horney (1885–1952) FIGURE 12.10

Horney argued that most of Freud's ideas about female personality reflected male biases and misunderstanding. She contended, for example, that Freud's concept of penis envy reflected women's feelings of cultural inferiority, not biological inferiority—*power envy*, not penis envy.

Evaluating psychoanalytic theories TABLE 12.1

Criticisms	• ***Difficult to test.*** From a scientific perspective, a major problem with psychoanalytic theory is that most of its concepts—such as the id or unconscious conflicts—cannot be empirically tested (Domhoff, 2004; Esterson, 2002; Friedman & Schustack, 2006). • ***Overemphasizes biology and unconscious forces.*** Modern psychologists believe that Freud did not give sufficient attention to learning and culture in shaping behavior. • ***Inadequate empirical support.*** Freud based his theories almost exclusively on the subjective case histories of his adult patients. Moreover, Freud's patients represented a small and selective sample of humanity: upper-class women in Vienna (Freud's home) who had serious adjustment problems. • ***Sexism.*** Many psychologists (beginning with Karen Horney) reject Freud's theories as derogatory toward women. • ***Lack of cross-cultural support.*** The Freudian concepts that ought to be most easily supported empirically—the biological determinants of personality—are generally not borne out by cross-cultural studies.
Enduring influences	• The emphasis on the unconscious and its influence on behavior. • The conflict among the id, ego, and superego and the resulting defense mechanisms. • Encouraging open talk about sex in Victorian times. • The development of psychoanalysis, an influential form of therapy. • The sheer magnitude of Freud's theory.

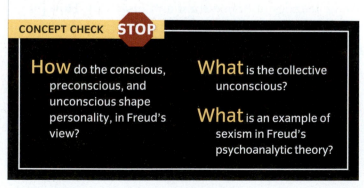

EVALUATING PSYCHOANALYTIC THEORIES: CRITICISMS AND ENDURING INFLUENCE

In this section, we look at major criticisms of Freud's psychoanalytic theories. In addition, we discuss the reasons that Freud has had enormous influence in the field of psychology. According to critics, Freud's theories are problematic for several reasons.

Today there are few Freudian purists. Instead, modern psychodynamic theorists and psychoanalysts use empirical methods and research findings to reformulate and refine traditional Freudian thinking (Knekt et al., 2008; Shaver & Mikulincer, 2005; Tryon, 2008; Westen, 1998).

But wrong as he was on many counts, Freud still ranks as one of the giants of psychology (Heller, 2005; Schülein, 2007; Solod, Monte, & Wilson, 2009). Furthermore, Freud's impact on Western intellectual history cannot be overstated. He attempted to explain dreams, religion, social groupings, family dynamics, neurosis, psychosis, humor, the arts, and literature.

It's easy to criticize Freud if you don't remember that he began his work at the start of the twentieth century and lacked the benefit of modern research findings and technology. We can only imagine how our current theories will look 100 years from now. Right or wrong, Freud has a lasting place among the pioneers in psychology (TABLE 12.1).

CONCEPT CHECK STOP

How do the conscious, preconscious, and unconscious shape personality, in Freud's view?

What is the collective unconscious?

What is an example of sexism in Freud's psychoanalytic theory?

Humanistic Theories

LEARNING OBJECTIVES

Explain the importance of the self in Rogers's theory of personality.

Describe how Maslow's hierarchy of needs affects personality.

Identify three criticisms of humanistic theories.

Humanistic theories of personality emphasize each person's internal feelings, thoughts, and sense of basic worth. In contrast to Freud, humanists believe that people are naturally good (or, at worst, neutral) and that they possess a positive drive toward self-fulfillment.

According to this view, our personality and behavior depend on how we perceive and interpret the world, not on traits, unconscious impulses, or rewards and punishments. Humanistic psychology was developed largely by Carl Rogers and Abraham Maslow.

ROGERS'S THEORY: THE IMPORTANCE OF THE SELF

To psychologist Carl Rogers (1902–1987), the most important component of personality is the *self*—what a person comes to identify as "I" or "me." Today, Rogerians (followers of Rogers) use the term **self-concept** to refer to all the information and beliefs you have regarding your own nature, unique qualities, and typical behaviors.

self-concept Rogers's term for all the information and beliefs that individuals have about their own nature, qualities, and behavior.

Rogers believed that poor mental health and maladjustment developed from a mismatch, or incongruence, between the self-concept and actual life experiences. (See *What a Psychologist Sees.*)

What a Psychologist Sees

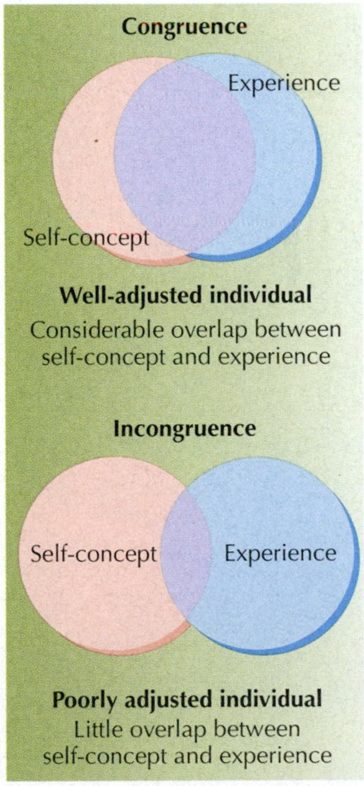

Congruence, Mental Health, and Self-Esteem

According to Carl Rogers, mental health and adjustment are related to the degree of congruence between a person's self-concept and life experiences. He argued that self-esteem—how we feel about ourselves—is particularly dependent on this congruence. Can you see how an artistic child would likely have higher self-esteem if her family valued art highly than if they did not?

Rogers believed that mental health, congruence, and self-esteem are part of our innate, biological capacities. In his view, everyone naturally approaches and values people and experiences that enhance our growth and fulfillment and avoids those that do not. Therefore, Rogers believed that we should trust our feelings to guide us toward mental health and happiness.

Then why do some people have low self-esteem and poor mental health? Rogers believed that these outcomes generally result from early childhood experiences with parents and other adults who make their love and acceptance conditional on behaving in certain ways and expressing only certain feelings.

If a child learns over time that his negative feelings and behaviors (which we all have) are totally unacceptable and unlovable, his self-concept and self-esteem may become distorted. He may always doubt the love and approval of others because they don't know "the real person hiding inside."

unconditional positive regard Rogers's term for positive behavior toward a person with no contingencies attached.

To help children develop to their fullest potential, adults need to create an atmosphere of **unconditional positive regard**—a setting in which children realize that they will be accepted no matter what they say or do.

Some people mistakenly believe that unconditional positive regard means that we should allow people to do whatever they please. But humanists separate the value of the person from his or her behaviors. They accept the person's positive nature while discouraging destructive or hostile behaviors. Humanistic psychologists believe in guiding children to control their behavior so that they can develop a healthy self-concept and healthy relationships with others (FIGURE 12.11).

MASLOW'S THEORY: THE SEARCH FOR SELF-ACTUALIZATION

Like Rogers, Abraham Maslow believed that there is a basic goodness to human nature and a natural tendency toward **self-actualization**. He saw personality as the quest to fulfill basic physiological needs (including safety, belonging and love, and esteem) and then move upward toward the highest level of self-actualization.

According to Maslow, self-actualization is the inborn drive to develop all one's talents and capacities. It involves understanding one's own potential, accepting

Conditional love? FIGURE 12.11

If a child is angry and hits his younger sister, some parents might punish the child or deny his anger, saying, "Nice children don't hit their sisters; they love them!" To gain parental approval, the child has to deny his true feelings of anger, but inside he secretly suspects he is not a "nice boy" because he did hit his sister and (at that moment) did not love her. How might repeated incidents of this type have a lasting effect on someone's self-esteem? What would be a more appropriate response to the child's behavior that acknowledges that it is the behavior that is unacceptable, not the child that is unacceptable?

oneself and others as unique individuals, and taking a problem-centered approach to life situations (Maslow, 1970). Self-actualization is an ongoing process of growth rather than an end product or accomplishment.

Maslow believed that only a few, rare individuals, such as Albert Einstein, Mohandas Gandhi, and Eleanor Roosevelt, become fully self-actualized. However, he saw self-actualization as part of every person's basic hierarchy of needs. (See Chapter 11 for more information on Maslow's theory.)

EVALUATING HUMANISTIC THEORIES: THREE MAJOR CRITICISMS

Humanistic psychology was extremely popular during the 1960s and 1970s. It was seen as a refreshing new perspective on personality after the negative determinism of the psychoanalytic approach and the mechanical nature of learning theories (Chapter 6). Although this early popularity has declined, many humanistic ideas have been incorporated into approaches to counseling and psychotherapy.

At the same time, humanistic theories have also been criticized (e.g., Funder, 2000). Three of the most important criticisms are these:

1. *Naive assumptions.* Critics suggest that the humanists are unrealistic, romantic, and even naive about human nature (FIGURE 12.12).
2. *Poor testability and inadequate evidence.* Like many psychoanalytic terms and concepts, humanistic concepts (such as unconditional positive regard and self-actualization) are difficult to define operationally and to test scientifically.
3. *Narrowness.* Like trait theories, humanistic theories have been criticized for merely describing personality rather than explaining it. For example, where does the motivation for self-actualization come from? To say that it is an "inborn drive" doesn't satisfy those who favor using experimental research and hard data to learn about personality.

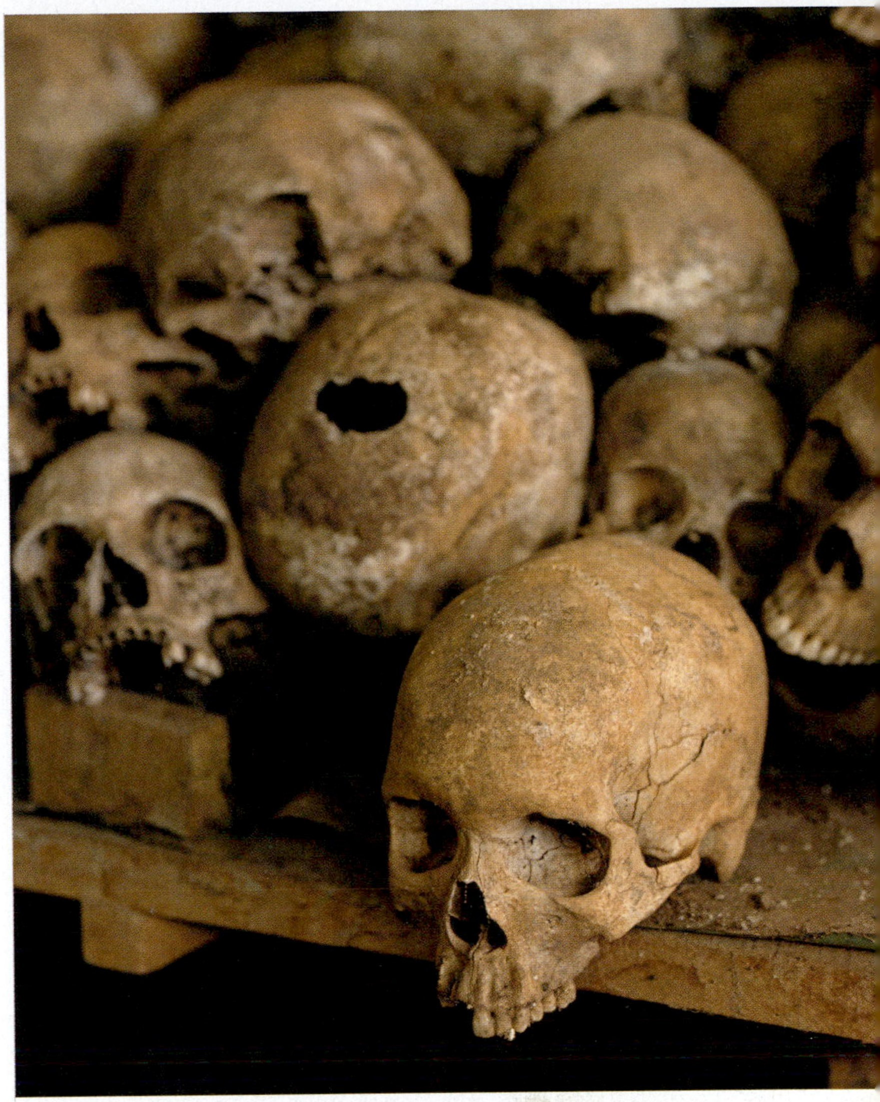

Are all people as inherently good as they say?
FIGURE 12.12

Humankind's continuing history of murders, warfare, and other acts of aggression suggests otherwise.

CONCEPT CHECK STOP

How are self-concept and self-esteem linked in Rogers's theory?

What is self-actualization?

What criticism of humanistic theories is also a weakness of trait theories?

Social-Cognitive Theories

LEARNING OBJECTIVES

Explain Bandura's concepts of self-efficacy and reciprocal determinism and how they affect personality.

Describe the role that Rotter's concept of locus of control plays in personality.

Summarize the attractions and criticisms of the social-cognitive perspective on personality.

According to the social-cognitive perspective, each of us has a unique personality because we have individual histories of interactions with the environment (social) and because we think (cognitive) about the world and interpret what happens to us (Cervone & Shoda, 1999). Two of the most influential social-cognitive theorists are Albert Bandura and Julian Rotter.

BANDURA'S AND ROTTER'S APPROACHES: SOCIAL LEARNING PLUS COGNITIVE PROCESSES

Albert Bandura (also discussed in Chapter 6) has played a major role in reintroducing thought processes into personality theory. Cognition, or thought, is central to his concept of **self-efficacy** (Bandura, 1997, 2000, 2006, 2008).

self-efficacy Bandura's term for the learned belief that one is capable of producing desired results, such as mastering new skills and achieving personal goals.

According to Bandura, if you have a strong sense of self-efficacy, you believe you can generally succeed, regardless of past failures and current obstacles. Your self-efficacy will in turn affect which challenges you choose to accept and the effort you expend in reaching goals. However, Bandura emphasized that self-efficacy is always specific to the situation—it does not necessarily carry across situations. For example, self-defense training significantly improves women's belief that they could escape from or disable a potential assailant or rapist, but it does not lead them to feel more capable in all areas of their lives (Weitlauf et al., 2001). Finally, according to Bandura's concept of **reciprocal determinism**, self-efficacy beliefs will affect how others respond to you, influencing your chances for success (FIGURE 12.13). Thus, a cognition ("I can succeed") will affect behaviors ("I will work hard and ask for a promotion"), which in turn will affect the environment ("My employer recognized my efforts and promoted me").

reciprocal determinism Bandura's belief that cognitions, behaviors, and the environment interact to produce personality.

Julian Rotter's theory is similar to Bandura's in that it suggests that learning experiences create **cognitive expectancies** that guide behavior and influence the environment (Rotter, 1954, 1990). According to Rotter, your behavior or personality is determined by (1) what you expect to happen following a specific action and (2) the reinforcement value attached to specific outcomes.

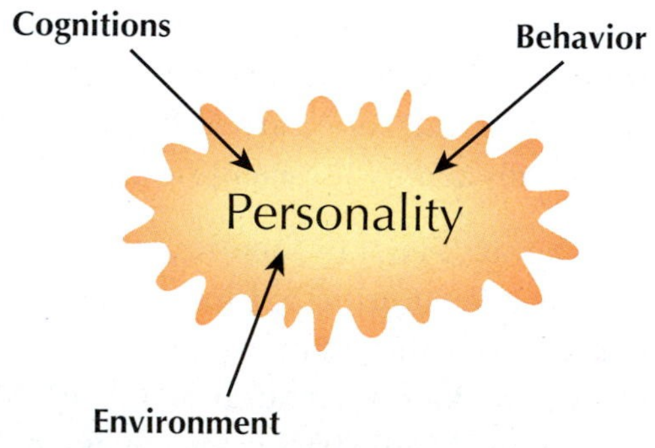

Albert Bandura's theory of reciprocal determinism
FIGURE 12.13

According to Bandura, thoughts (or cognitions), behavior, and the environment all interact to produce personality.

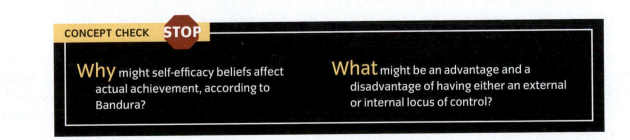

"We're encouraging people to become involved in their own rescue."

Locus of control and achievement

FIGURE 12.14

Research links possession of an internal locus of control with higher achievement and better mental health (Burns, 2008; Jones, 2008; Ruthig et al., 2007). What might this connection imply for human survival?

To understand your personality and behavior, Rotter would use personality tests that measure your internal versus external **locus of control** (Chapter 3). Rotter's tests ask people to respond to statements such as, "People get ahead in this world primarily by luck and connections rather than by hard work and perseverance," and, "When someone doesn't like you, there is little you can do about it." As you may suspect, people with an external locus of control think that environment and external forces have primary control over their lives, whereas people with an internal locus of control think that they can control events in their lives through their own efforts (FIGURE 12.14).

EVALUATING SOCIAL-COGNITIVE THEORY: THE PLUSES AND MINUSES

The social-cognitive perspective holds several attractions. First, it emphasizes how the environment affects and is affected by individuals. Second, it offers testable, objective hypotheses and operationally defined terms, and it relies on empirical data for its basic principles. However, critics note that social-cognitive theory ignores the unconscious and emotional aspects of personality (Mischel, Shoda, & Ayduk, 2008; Westen, 1998).

CONCEPT CHECK STOP

Why might self-efficacy beliefs affect actual achievement, according to Bandura?

What might be an advantage and a disadvantage of having either an external or internal locus of control?

Biological Theories

LEARNING OBJECTIVES

Summarize the roles that brain structures and neurochemistry play in personality.

Describe how researchers study genetic influences on personality.

Describe how the biopsychosocial model integrates different theories of personality.

In this section, we explore how inherited biological factors influence our personalities. We conclude with a discussion of how all theories of personality ultimately interact within the *biopsychosocial model*.

THREE MAJOR CONTRIBUTORS TO PERSONALITY: THE BRAIN, NEUROCHEMISTRY, AND GENETICS

Modern biological research suggests that certain brain areas may contribute to some personality traits. For instance, increased electroencephalographic (EEG) activity in the left frontal lobes of the brain is associated with sociability (or extroversion), whereas greater EEG activity in the right frontal lobes is associated with shyness (introversion) (Tellegen, 1985).

A major limitation of research on brain structures and personality is the difficulty in identifying which structures are uniquely connected with particular personality traits. Neurochemistry seems to offer more precise data on how biology influences personality (Kagan, 1998).

For example, sensation seeking (Chapter 11) has consistently been linked with levels of monoamine oxidase (MAO), an enzyme that regulates levels of neurotransmitters such as dopamine (Íbañez, Blanco, & Sáiz-Ruiz, 2002; Zuckerman, 1994, 2004). Dopamine also seems to be correlated with novelty seeking and extroversion (Dalley et al., 2007; Lang et al., 2007; Levinthal, 2008).

How can neurochemistry have such effects? Studies suggest that high-sensation seekers and extroverts tend to experience less physical arousal than introverts from the same stimulus (Lissek & Powers, 2003). Extroverts' low arousal apparently motivates them to seek out situations that will elevate their arousal. Moreover, it is believed that a higher arousal threshold is genetically transmitted. In other words, personality traits like sensation seeking and extroversion may be inherited.

Finally, psychologists have recently recognized that genetic factors also have an important influence on personality (Congdon & Canli, 2008; Kandler, Riemann, & Kampfe, 2009). This relatively new area, called **behavioral genetics**, attempts to determine the extent to which behavioral differences among people are due to genetics as opposed to environment (Chapter 2).

One way to measure genetic influences is to compare similarities in personality between identical twins and fraternal twins. For example, studies of the heritability of the five-factor model personality traits suggest that genetic factors contribute about 40 to 50 percent of personality (Bouchard, 1997, 2004; Eysenck, 1967, 1990; Jang et al., 2006; McCrae et al., 2004; Plomin, 1990; Weiss, Bates, & Luciano, 2008).

In addition to twin studies, researchers compare the personalities of parents with those of their biological children and their adopted children. Studies of extroversion and neuroticism have found that parents' traits correlate moderately with those of their biological children and hardly at all with those of their adopted children (Bouchard, 1997; McCrae et al., 2000).

At the same time, researchers are careful not to overemphasize genetic influences on personality (Deckers, 2005; Funder, 2001; Sollod, Monte, & Wilson, 2009). Some researchers believe that the importance of the unshared environment (aspects of the environment that differ from one individual to another, even within a family) has been overlooked (Saudino, 1997). Others fear that research on "genetic determinism"—do our genes determine who we are?—could be misused to "prove" that an ethnic or a racial group is inferior, that male dominance is natural, or that social progress is impossible. In short, they worry that an emphasis on genetic determinants of personality and behavior may lead

Psychological Science

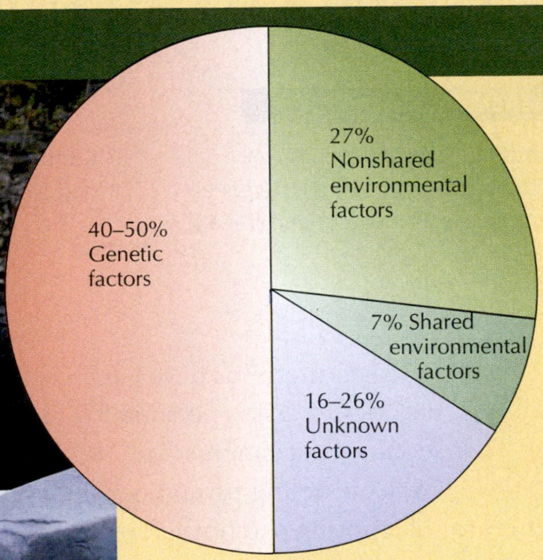

Source: Bouchard, 1997, 2004; Eysenck, 1967, 1990; Jang et al., 2006; McCrae et al., 2004; Plomin, 1990; Weiss, Bates, & Luciano, 2008

Multiple Influences on Personality

What gives a person certain personality characteristics, such as shyness, or conscientiousness, or aggressiveness? As shown in the figure here, research indicates that four major factors overlap to influence personality. These include genetic (inherited) factors; nonshared environmental factors, or how each individual's genetic factors react and adjust to his or her particular environment; shared environmental factors, involving parental patterns and shared family experiences; and error, or unidentified factors or problems with testing.

For example, Hans Eysenck (1990) believed that certain traits (like introversion and extroversion) may reflect inherited patterns of cortical arousal, as well as social learning, cognitive processes, and the environment. Can you see how someone with an introverted personality (and therefore a higher level of cortical arousal) might try to avoid excessive stimulation by seeking friends and jobs with low stimulation levels? Eysenck's work exemplifies how trait, biological, and social-cognitive theories can be combined to provide better insight into personality—the biopsychosocial model.

CRITICAL THINKING

Here are two interesting questions:

1. Can you identify one key inherited influence and one major environmental influence on your own personality?
2. What kinds of jobs would an individual with an extroverted personality be most likely to choose?

people to see themselves as merely "victims" of their genes. Clearly, genetics have produced exciting and controversial results. However, more research is necessary before a cohesive biological theory of personality can be constructed.

THE BIOPSYCHOSOCIAL MODEL: PULLING THE PERSPECTIVES TOGETHER

No one personality theory explains everything we need to know about personality. Each theory offers different insights into how a person develops the distinctive set of characteristics we call "personality." That's why instead of adhering to any one theory, many psychologists believe in the biopsychosocial approach, or the idea that several factors—biological, psychological, and social—overlap in their contributions to personality (Mischel, Shoda, & Ayduk, 2008).

CONCEPT CHECK STOP

What evidence suggests that particular brain areas contribute to some personality traits?

How can traits like sensation seeking and extroversion be related to neurochemistry?

Why is the biopsychosocial model important to research on personality?

Personality Assessment

LEARNING OBJECTIVES

Identify the major methods that psychologists use to assess personality. Then explore the benefits and limitations of each.

Summarize the major features of objective personality tests.

Explain why psychologists use projective tests to assess personality.

Numerous methods have been used over the decades to assess personality. Modern personality assessments are used by clinical and counseling psychologists, psychiatrists, and others for diagnosing psychotherapy patients and for assessing their progress in therapy. Personality assessment is also used for educational and vocational counseling and to aid businesses in making hiring decisions. Personality assessments can be grouped into a few broad categories: interviews, observations, objective tests, and projective tests.

INTERVIEWS AND OBSERVATION

We all use informal "interviews" to get to know other people. When first meeting someone, we usually ask about his or her job, academic interests, family, or hobbies. Psychologists also use interviews. Unstructured interviews are often used for job and college selection and for diagnosing psychological problems. In an unstructured format, interviewers get impressions and pursue hunches or let the interviewee expand on information that promises to disclose personality characteristics. In structured interviews, the interviewer asks specific questions so that the interviewee's responses can be evaluated more objectively (and compared with others' responses).

In addition to interviews, psychologists also assess personality by directly and methodically observing behavior. The psychologist looks for examples of specific behaviors and follows a careful set of evaluation guidelines. For instance, a psychologist might arrange to observe a troubled client's interactions with his or her family. Does the client become agitated by the presence of certain family members and not others? Does he or she become passive and withdrawn when asked a direct question? Through careful observation, the psychologist gains valuable insights into the client's personality, as well as family dynamics (FIGURE 12.15).

OBJECTIVE TESTS

Objective personality tests, or inventories, are the most widely used method of assessing personality, for two reasons. They can be administered to a large number of

objective personality tests Standardized questionnaires that require written responses, usually to multiple-choice or true-false questions.

Behavioral observation FIGURE 12.15

How might careful observation help a psychologist better understand a troubled client's personality and family dynamics?

people relatively quickly, and the tests can be evaluated in a standardized fashion. (An older, and no longer used, method of assessing personality is described in FIGURE 12.16.)

Some objective tests measure one specific personality trait, such as sensation seeking (Chapter 11) or locus of control. However, psychologists in clinical, counseling, and industrial settings often wish to assess a range of personality traits. To do so, they generally use multitrait (or *multiphasic*) inventories.

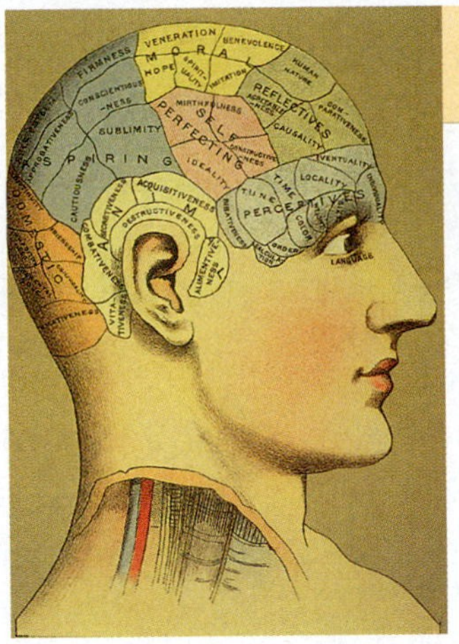

Personality and bumps on the head?
FIGURE 12.16

In the 1800s, if you wanted to have your personality assessed, you would go to a phrenologist, who would determine your personality by measuring the bumps on your skull and comparing the measurements with a chart that associated different areas of the skull with particular traits, such as *sublimity* (ability to squelch natural impulses, especially sexual) and *ideality* (ability to live by high ideals). What traits might be measured if we still believed in phrenology today?

The most widely studied and clinically used multitrait test is the **Minnesota Multiphasic Personality Inventory (MMPI)**—or its revision, the MMPI-2 (Butcher, 2000, 2005; Butcher & Perry, 2008). This test consists of over 500 statements that participants respond to with *True, False*, or *Cannot Say*. The following are examples of the kinds of statements found on the MMPI:

My stomach frequently bothers me.
I have enemies who really wish to harm me.
I sometimes hear things that other people can't hear.
I would like to be a mechanic.
I have never indulged in any unusual sex practices.

Did you notice that some of these questions are about very unusual, abnormal behavior? Although there are many "normal" questions on the full MMPI, the test is designed primarily to help clinical and counseling psychologists diagnose psychological disorders. MMPI test items are grouped into 10 clinical scales, each measuring a different disorder (TABLE 12.2). There are also a number of validity scales designed to reflect the extent to which respondents (1) distort their answers (for example, to fake psychological disturbances or to appear more psychologically healthy than they really are), (2) do not understand the items, and (3) are being uncooperative.

Personality tests like the MMPI are often confused with *career inventories* or vocational interest tests. Career counselors use these latter tests (along with aptitude and achievement tests) to help people identify occupations and careers that match their unique traits, values, and interests.

Subscales of the MMPI-2 TABLE 12.2

Clinical scales	*Typical interpretations of high scores*	*Validity scales*	*Typical interpretations of high scores*
1. Hypochondriasis	Numerous physical complaints	1. L (lie)	Denies common problems, projects a "saintly" or false picture
2. Depression	Seriously depressed and pessimistic	2. F (confusion)	Answers are contradictory
3. Hysteria	Suggestible, immature, self-centered, demanding	3. K (defensiveness)	Minimizes social and emotional complaints
4. Psychopathic deviate	Rebellious, nonconformist	4. ? (cannot say)	Many items left unanswered
5. Masculinity–femininity	Interests like those of other sex		
6. Paranoia	Suspicious and resentful of others		
7. Psychasthenia	Fearful, agitated, brooding		
8. Schizophrenia	Withdrawn, reclusive, bizarre thinking		
9. Hypomania	Distractible, impulsive, dramatic		
10. Social introversion	Shy, introverted, self-effacing		

PROJECTIVE TESTS

Unlike objective tests, **projective tests** use unstructured stimuli that can be perceived in many ways. As the name implies, projective tests supposedly allow each person to project his or her own unconscious conflicts, psychological defenses, motives, and personality traits onto the test materials. Because respondents are unable (or unwilling) to express their true feelings if asked directly, the ambiguous stimuli reportedly provide an indirect "psychological x-ray" of important unconscious processes (Hogan, 2006). Two of the most widely used projective tests are the **Rorschach Inkblot Test** and the **Thematic Apperception Test (TAT)** (FIGURE 12.17).

projective tests Psychological tests that use ambiguous stimuli, such as inkblots or drawings, which allow the test taker to project his or her unconscious thoughts onto the test material.

ARE PERSONALITY MEASUREMENTS ACCURATE?

Let's evaluate the strengths and the challenges of each of the four methods of personality assessment: interviews, observation, objective tests, and projective tests.

Interviews and observations Both interviews and observations can provide valuable insights into personality, but they are time-consuming and expensive. Furthermore, raters of personality tests frequently disagree in their evaluations of the same individuals. Interviews and observations also involve unnatural settings, and, in fact, the very presence of an observer can alter a subject's behavior.

Objective tests Tests like the MMPI-2 provide specific, objective information about a broad range of personality traits in a relatively short period. However, they are also the subject of at least three major criticisms:

1. *Deliberate deception and social desirability bias.* Some items on personality inventories are easy to "see through," so respondents may intentionally, or unintentionally, fake particular personality traits. In addition, some respondents want to look good and will answer questions in ways that they perceive are socially desirable. (The validity scales of the MMPI-2 are designed to help prevent these problems.)

2. *Diagnostic difficulties.* When inventories are used for diagnosis, overlapping items sometimes make it difficult to pinpoint a diagnosis (Graham, 1991). In addition, clients with severe disorders sometimes score within the normal range, and

Visualizing

Projective tests FIGURE 12.17

Responses to projective tests reportedly reflect unconscious parts of the personality that "project" onto the stimuli.

A The Rorschach Inkblot Test was introduced in 1921 by Swiss psychiatrist Hermann Rorschach. With this technique, individuals are shown 10 inkblots like this, one at a time, and are asked to report what figures or objects they see in each of them. ▶

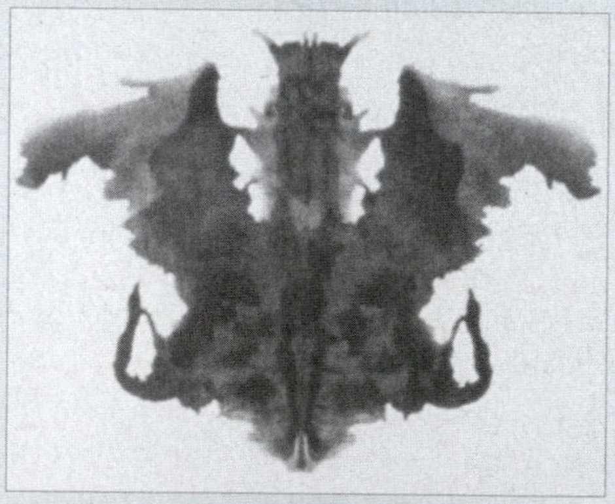

Reproduced with permission. This inkblot is not part of the Rorschach test.

B Created by personality researcher Henry Murray in 1938, the Thematic Apperception Test (TAT) consists of a series of ambiguous black-and-white pictures that are shown to the test taker, who is asked to create a story related to each. Can you think of two different stories that a person might create for the picture of two women here? How might a psychologist interpret each story? ▶

normal clients sometimes score within the elevated range (Gregory, 2007; Weiner, 2008).

3. *Cultural bias and inappropriate use.* Some critics think that the standards for "normalcy" on objective tests fail to recognize the impact of culture. For example, Latinos generally score higher than respondents from North American and Western European cultures on the masculinity-femininity scale of the MMPI-2 (Dana, 2005; Lucio et al., 2001; Lucio-Gomez et al., 1999). However, this tendency reflects traditional gender roles and cultural training more than any individual personality traits.

Projective tests Although projective tests are extremely time-consuming to administer and interpret, their proponents suggest that because they are unstructured, respondents may be more willing to talk honestly about sensitive topics. Critics point out, however, that the reliability and validity (Chapter 8) of projective tests is among the lowest of all tests of personality (Garb et al., 2005; Gacono et al., 2008).

As you can see, each of these methods has its limits. Psychologists typically combine the results from various methods to create a full picture of an individual's personality.

CONCEPT CHECK STOP

How do structured and unstructured interviews differ?

Why are objective personality tests like the MMPI used so widely?

Why might people respond more candidly on projective tests than on some other kinds of personality tests?

SUMMARY

1 Trait Theories

1. Psychologists define **personality** as an individual's relatively stable and enduring patterns of thoughts, feelings, and actions.
2. Allport believed that the best way to understand personality was to arrange a person's unique personality **traits** into a hierarchy. Cattell and Eysenck later reduced the list of possible personality traits using factor analysis.
3. According to the **five-factor model (FFM)**, the five major dimensions of personality are openness, conscientiousness, extroversion, agreeableness, and neuroticism.

2 Psychoanalytic/Psychodynamic Theories

1. Freud, the founder of psychodynamic theory, believed that the mind contained three levels of consciousness: the **conscious**, the **preconscious**, and the **unconscious**. He believed that most psychological disorders originate from unconscious memories and instincts. Freud also asserted that personality was composed of the id, the ego, and the superego. When the ego fails to satisfy both the id and the superego, anxiety slips into conscious awareness, which triggers **defense mechanisms**. Freud's theory of **psychosexual stages** has not withstood the test of time.
2. Neo-Freudians such as Adler, Jung, and Horney were influential followers of Freud who later came to reject major aspects of Freudian theory. Today, few

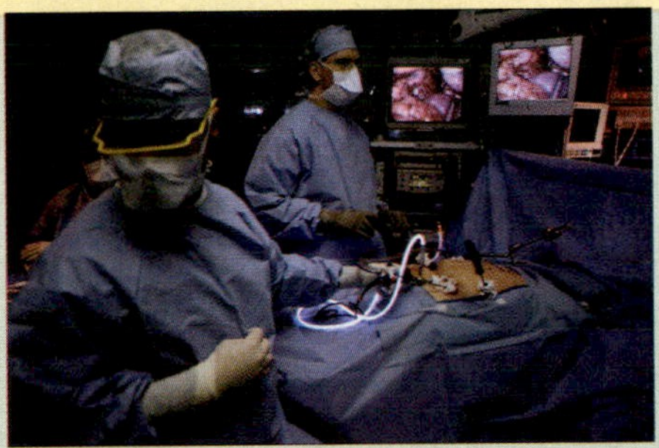

Freudian purists remain, but Freud's impact on psychology and on Western intellectual history cannot be overstated.

3 Humanistic Theories

1. According to Rogers, mental health and self-esteem are related to the degree of congruence between our **self-concept** and life experiences. He argued that poor mental health results when young children do not receive **unconditional positive regard** from caregivers.

2. Maslow saw personality as the quest to fulfill basic physiological needs and to move toward the highest level of self-actualization.

4 Social-Cognitive Theories

1. Cognition is central to Bandura's concept of **self-efficacy**. According to Bandura, self-efficacy affects which challenges we choose to accept and the effort we expend in reaching goals. His concept of **reciprocal determinism** states that self-efficacy beliefs will also affect others' responses to us.

2. Rotter's theory suggests that learning experiences create cognitive expectancies that guide behavior and influence the environment. Rotter believed that having an internal versus external locus of control affects personality and achievement.

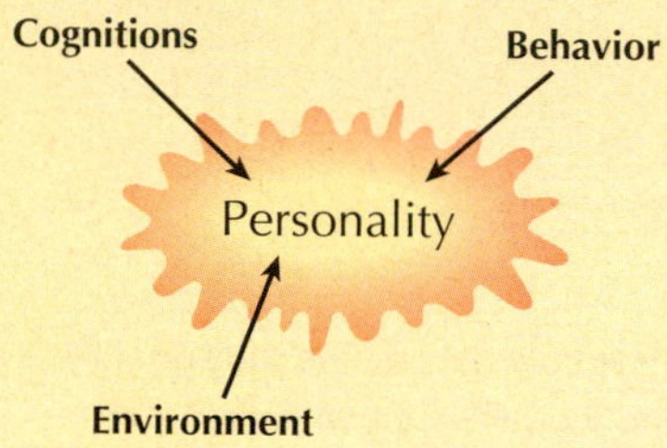

5 Biological Theories

1. There is evidence that certain brain areas may contribute to personality. However, neurochemistry seems to offer more precise data on how biology influences personality. Research in behavioral genetics indicates that genetic factors also strongly influence personality.

2. Instead of adhering to any one theory of personality, many psychologists believe in the biopsychosocial approach—the idea that several factors overlap in their contributions to personality.

6 Personality Assessment

1. In an unstructured interview format, interviewers get impressions and pursue hunches or let the interviewee expand on information that promises to disclose personality characteristics. In structured interviews, the interviewer asks specific questions so that the interviewee's responses can be evaluated more objectively. Psychologists also assess personality by directly observing behavior.

2. **Objective personality tests** are widely used because they can be administered broadly and relatively quickly and because they can be evaluated in a standardized fashion. To assess a range of personality traits, psychologists use multitrait inventories, such as the MMPI.

3. **Projective tests** use unstructured stimuli that can be perceived in many ways. Projective tests supposedly allow each person to project his or her own unconscious conflicts, psychological defenses, motives, and personality traits onto the test materials.

KEY TERMS

- personality p. 316
- traits p. 316
- factor analysis p. 316
- five-factor model (FFM) p. 316
- Big Five p. 316
- levels of consciousness p. 320
- conscious p. 320
- preconscious p. 320
- unconscious p. 320
- id p. 321
- pleasure principle p. 321
- ego p. 321
- reality principle p. 321
- superego p. 321
- morality principle p. 321
- defense mechanisms p. 321
- rationalization p. 321
- repression p. 322
- intellectualization p. 322
- psychosexual stages p. 323
- Oedipus complex p. 323
- penis envy p. 323
- neo-Freudians p. 324
- individual psychology p. 324
- inferiority complex p. 324
- analytical psychology p. 324
- archetypes p. 324
- self-concept p. 327
- unconditional positive regard p. 328
- self-actualization p. 328
- self-efficacy p. 330
- reciprocal determinism p. 330
- cognitive expectancies p. 330
- locus of control p. 331
- behavioral genetics p. 332
- objective personality tests p. 334
- Minnesota Multiphasic Personality Inventory (MMPI) p. 335
- projective tests p. 336
- Rorschach Inkblot Test p. 336
- Thematic Apperception Test (TAT) p. 336

CRITICAL AND CREATIVE THINKING QUESTIONS

1. How do you think you would score on each of the "Big Five" personality dimensions?
2. If scientists have so many problems with Freud, why do you think his theories are still popular with the public? Should psychologists continue to discuss his theories (and include them in textbooks)? Why or why not?
3. In what ways is Adler's individual psychology more optimistic than Freudian theory?
4. How do Bandura's and Rotter's social-cognitive theories differ from biological theories of personality?
5. Which method of personality assessment (interviews, behavioral observation, objective testing, or projective testing) do you think is likely to be most informative? Can you think of circumstances where one kind of assessment might be more effective than the others?

What is happening in this picture?

- How would Freud interpret this woman's aggressive behavior toward her "assailant"?
- How would Bandura's social-cognitive concept of self-efficacy explain the benefits of taking a self-defense course?
- What biological factors might explain why some people choose activities that provide high levels of physiological arousal while others avoid such situations?

SELF-TEST

(Check your answers in Appendix B.)

1. A relatively stable and consistent characteristic that can be used to describe someone is known as a ______________.
   a. character
   b. trait
   c. temperament
   d. personality

2. Label the list of personality traits in the Five-Factor model on the figure below.

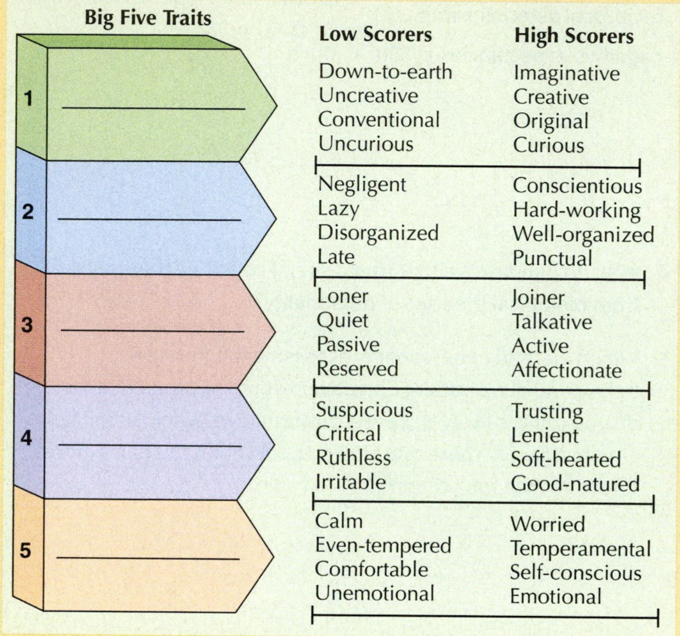

3. Label Freud's three levels of consciousness on the figure below.

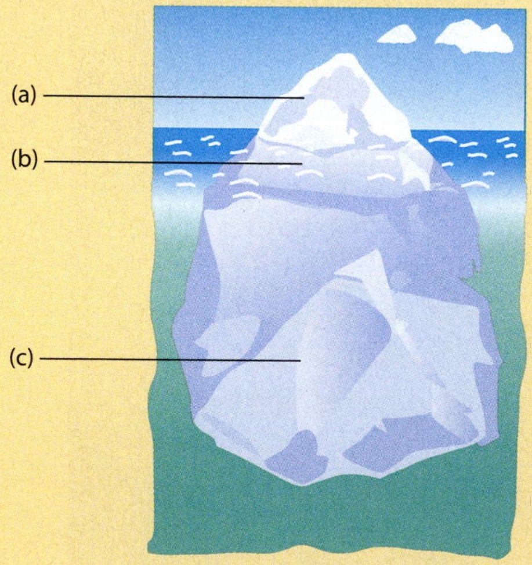

4. According to Freud, the three mental structures that form personality are the ______________.
   a. unconscious, preconscious, and conscious
   b. oral, anal, and phallic
   c. Oedipus, Electra, and sexual/aggressive
   d. id, ego, and superego

5. Used excessively, defense mechanisms can be dangerous because they ______________.
   a. hide true feelings
   b. become ineffective
   c. distort reality
   d. become fixated

6. Label Freud's psychosexual stages of development in correct sequence on the figure below.

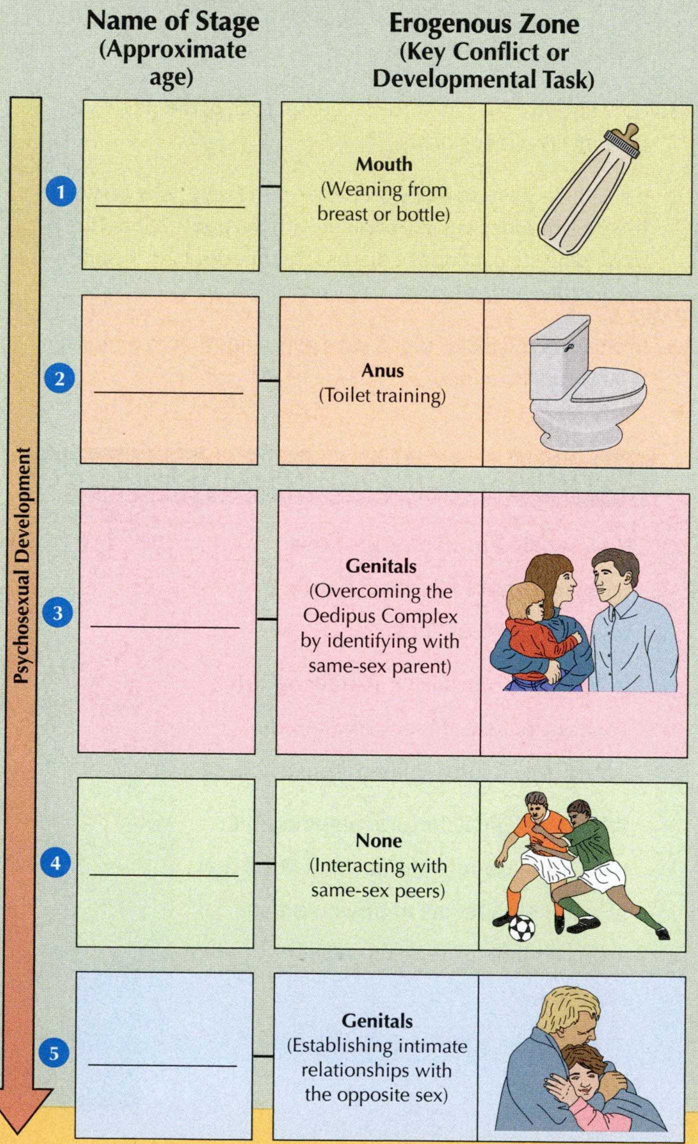

7. Three of the most influential neo-Freudians were ________________.
   a. Plato, Aristotle, and Descartes
   b. Dr. Laura, Dr. Phil, and Dr. Ruth
   c. Pluto, Mickey, and Minnie
   d. Adler, Jung, and Horney

8. Which of the following is NOT a criticism of Freud's psychoanalytic theory?
   a. Most of his concepts prove difficult to test.
   b. Most of his data show a lack of cross-cultural support.
   c. He underemphasized the role of biological determinants.
   d. Some of his views were sexist.

9. *Unconditional positive regard* is a Rogerian term for ________________.
   a. accepting any and all behavior as a positive manifestation of self-actualization
   b. positive behavior toward a person without attaching any contingencies
   c. nonjudgmental listening
   d. phenomenological congruence

10. The personality theorist who believed in the basic goodness of individuals and their natural tendency toward self-actualization was ________________.
    a. Karen Horney
    b. Alfred Adler
    c. Abraham Maslow
    d. Carl Jung

11. According to Bandura, ________________ involves a person's belief about whether he or she can successfully engage in behaviors related to personal goals.
    a. self-actualization
    b. self-esteem
    c. self-efficacy
    d. self-congruence

12. This is the study of the extent to which behavioral differences are due to genetics rather than the environment.
    a. the biobehavioral approach
    b. the genetic-environmental perspective
    c. behavioral genetics
    d. the biopsychosocial model

13. ________________ appear(s) to have the largest influence (40 to 50 percent) on personality.
    a. Nonshared environment
    b. Shared environments
    c. Genetics
    d. Unknown factors

14. The most widely researched and clinically used self-report, personality test is the ________________.
    a. MMPI-2
    b. Rorschach Inkblot Test
    c. TAT
    d. SVII

15. The Rorschach Inkblot Test is an example of a(n) ________________ test.
    a. aptitude
    b. projective personality
    c. the most reliable and valid personality
    d. culturally biased personality

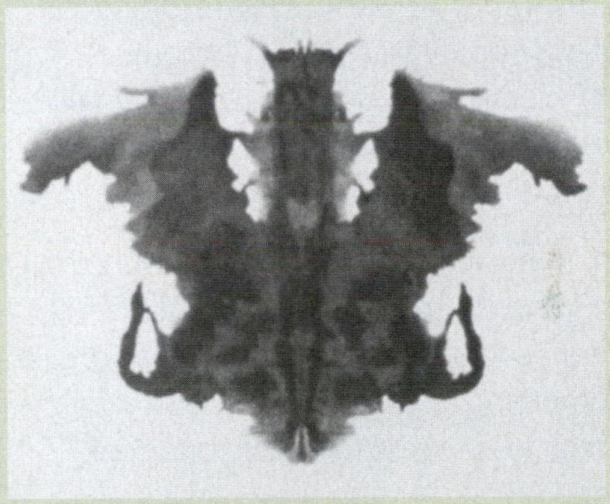

Reproduced with permission. This inkblot is not part of the Rorschach test.

Psychological Disorders

13

Mary's troubles first began in adolescence. She was frequently truant, and her grades declined sharply. During family counseling sessions, it was discovered that Mary had prostituted herself for drug money. . . . She idealized new friends, but when they disappointed her, she angrily cast them aside. . . . Mary's problems, coupled with a preoccupation with inflicting pain on herself (by cutting and burning) and persistent thoughts of suicide, led to her admittance to a psychiatric hospital at age 26 (Kring et al., 2007, pp. 386–387).

Jim is a medical student. Over the last few weeks, he has been noticing that older men appear to be frightened of him when he passes them on the street. Recently, he has become convinced that he is actually the director of the Central Intelligence Agency and that these men are secret agents of a hostile nation. Jim has found confirmatory evidence in the fact that a helicopter flies over his house every day at 8:00 a.m. and at 4:30 p.m. Surely, this surveillance is part of the plot to assassinate him (Bernheim & Lewine, 1979, p. 4).

Both Mary and Jim have severe psychological problems. Was there something in their early backgrounds to explain their later behaviors? Is there something medically wrong with them? What about less severe forms of abnormal behavior?

In this chapter, we discuss how psychological disorders are identified, explained, and classified, and explore six major categories of psychological disorders. We also look at how gender and culture affect mental disorders.

NATIONAL GEOGRAPHIC

CHAPTER OUTLINE

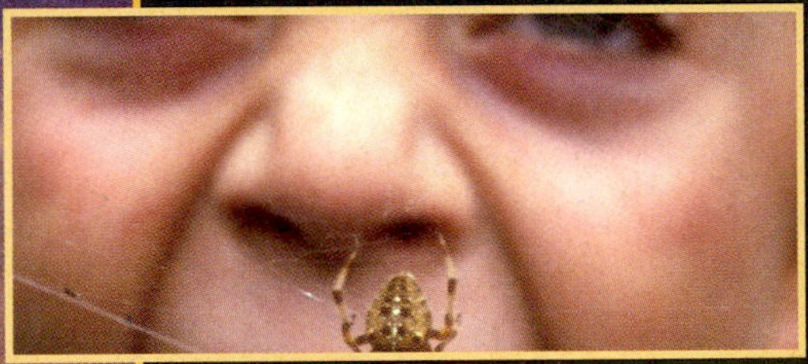

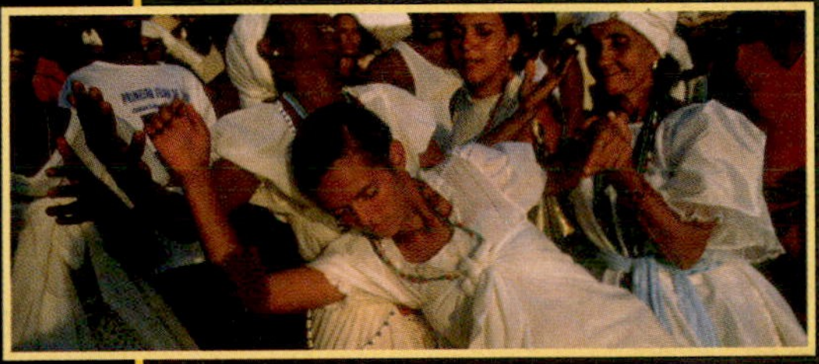

Studying Psychological Disorders

LEARNING OBJECTIVES

List the four criteria for identifying abnormal behavior.

Review how views of abnormal behavior have changed through history.

Explain how the *DSM-IV-TR* is used to classify psychological disorders.

IDENTIFYING ABNORMAL BEHAVIOR: FOUR BASIC STANDARDS

abnormal behavior Patterns of emotion, thought, and action that are considered pathological (diseased or disordered) for one or more of these reasons: statistical infrequency, disability or dysfunction, personal distress, or violation of norms (Davison, Neale, & Kring, 2004).

On the continuum ranging from normal to **abnormal behavior**, people can be unusually healthy or extremely disturbed.

Mental health professionals generally agree on four criteria for abnormal behavior: statistical infrequency, disability or dysfunction, personal distress, and violation of norms (FIGURE 13.1). However, as we consider these criteria, remember that no single criterion is adequate for identifying all forms of abnormal behavior.

EXPLAINING ABNORMALITY: FROM SUPERSTITION TO SCIENCE

What causes abnormal behavior? Historically, evil spirits and witchcraft have been blamed (Goodwin, 2009; Millon, 2004). Stone Age people, for example, believed that abnormal behavior stems from demonic possession; the "therapy" was to bore a hole in the skull so that the evil spirit could escape. During the European Middle Ages, a troubled person was sometimes treated with exorcism, an effort to drive the Devil out through prayer, fasting, noise-making, beating, and drinking terrible-tasting brews. During the fifteenth century, many believed that some individuals chose to consort with the Devil. Many of these supposed witches were tortured, imprisoned for life, or executed.

As the Middle Ages ended, special mental hospitals called *asylums* began to appear in Europe. Initially designed to provide quiet retreats from the world and to protect society (Barlow & Durand, 2009; Millon, 2004), the asylums unfortunately became overcrowded, inhumane prisons.

medical model The perspective that diseases (including mental illness) have physical causes that can be diagnosed, treated, and possibly cured.

psychiatry The branch of medicine that deals with the diagnosis, treatment, and prevention of mental disorders.

Improvement came in 1792 when Philippe Pinel, a French physician in charge of a Parisian asylum, insisted that asylum inmates—whose behavior he believed to be caused by underlying physical illness—be unshackled and removed from their unlighted, unheated cells. Many inmates improved so dramatically that they could be released. Pinel's **medical model** eventually gave rise to the modern specialty of **psychiatry**.

Unfortunately, when we label people "mentally ill," we may create new problems. One of the most outspoken critics of the medical model is psychiatrist Thomas Szasz (1960, 2000, 2004). Szasz believes that the medical model encourages people to believe that they have no responsibility for their actions. He contends that mental illness is a myth used to label individuals who are peculiar or offensive to others (Cresswell, 2008). Furthermore, labels can become self-perpetuating—that is, the person can begin to behave according to the diagnosed disorder.

Despite these potential dangers, the medical model—and the concept of mental illness—remains a founding principle of psychiatry. In contrast, psychology offers a multifaceted approach to explaining abnormal behavior, as described in *What a Psychologist Sees* on page 346.

Visualizing

The four criteria and a continuum of abnormal behavior FIGURE 13.1

Rather than fixed categories, both "abnormal" and "normal" behaviors exist along a continuum (Hansell & Damour, 2008).

(Rare) (Common)

Statistical Infrequency
(e.g., believing others are plotting against you)

Normal Abnormal

- A behavior may be judged abnormal if it occurs infrequently in a given population. *Statistical infrequency* alone does not determine what is normal—for example, no one would classify Albert Einstein's great intelligence or Lance Armstrong's exceptional athletic ability as abnormal.

(Low) (High)

Disability or Dysfunction
(e.g., being unable to go to work due to alcohol abuse)

Normal Abnormal

- People who suffer from psychological disorders may be so *disabled* or *dysfunctional* that they are unable to get along with others, hold a job, eat properly, or clean themselves. Their ability to think clearly and make rational decisions also may be impaired.

(Low) (High)

Personal Distress
(e.g., having thoughts of suicide)

Normal Abnormal

- The *personal distress* criterion focuses on the individual's judgment of his or her level of functioning. Yet many people with psychological disorders deny they have a problem. Also, some serious psychological disorders (such as antisocial personality disorder) cause little or no emotional discomfort. The personal distress criterion by itself is not sufficient for identifying all forms of abnormal behavior.

(Rare) (Common)

Violation of Norms
(e.g., shouting at strangers)

Normal Abnormal

- The fourth approach to identifying abnormal behavior is *violation of norms*, or cultural rules that guide behavior in particular situations. A major problem with this criterion, however, is that cultural diversity can affect what people consider a violation of norms (Lopez & Guarnaccia, 2000).

Seven Psychological Perspectives on Abnormal Behavior

Each of the seven major perspectives in psychology emphasizes different factors believed to contribute to abnormal behavior, but in practice they overlap. Consider the phenomenon of compulsive hoarding. Everyone sometimes makes an impulse purchase, and most people are reluctant to discard some possessions that are of questionable value. But when the acquisition of and inability to discard worthless items becomes extreme, it can interfere with basic aspects of living, such as cleaning, cooking, sleeping on a bed, and moving around one's home. This abnormal behavior is associated with several psychological disorders, but it is most commonly found in people who have obsessive-compulsive disorder, or OCD (an anxiety disorder discussed later in this chapter). Can you imagine how each of the seven major perspectives might explain compulsive hoarding?

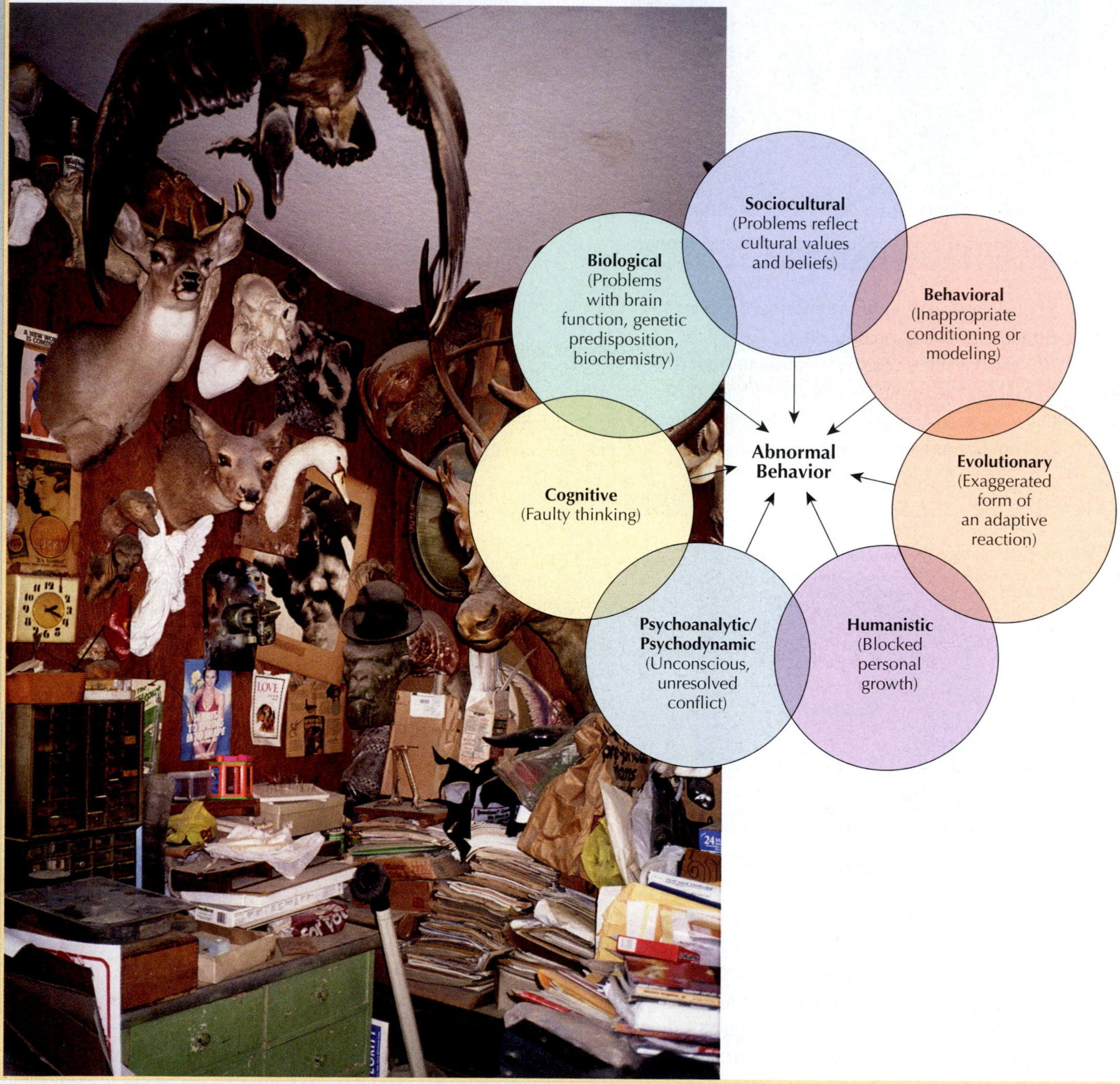

CLASSIFYING ABNORMAL BEHAVIOR: *THE DIAGNOSTIC AND STATISTICAL MANUAL IV-TR*

Without a clear, reliable system for classifying the wide range of psychological disorders, scientific research on them would be almost impossible, and communication among mental health professionals would be seriously impaired. Fortunately, mental health specialists share a uniform classification system, the text revision of the fourth edition of the ***Diagnostic and Statistical Manual of Mental Disorders (DSM-IV-TR)*** (American Psychiatric Association, 2000).

Each revision of the *DSM* has expanded the list of disorders and changed the descriptions and categories to reflect both the latest in scientific research and changes in the way abnormal behaviors are viewed within our social context (First & Tasman, 2004; Smart & Smart, 1997). For example, take the terms **neurosis** and **psychosis**. In previous editions, the term *neurosis* reflected Freud's belief that all neurotic conditions arise from unconscious conflicts (Chapter 12). Now, conditions that were previously grouped under the heading *neurosis* have been formally redistributed as anxiety disorders, somatoform disorders, and dissociative disorders.

Diagnostic and Statistical Manual of Mental Disorders (DSM-IV-TR) The classification system developed by the American Psychiatric Association used to describe abnormal behaviors; the *IV-TR* indicates that it is the text revision (*TR*) of the fourth major edition (*IV*).

Unlike neurosis, the term *psychosis* is still listed in the *DSM-IV-TR* because it is useful for distinguishing the most severe mental disorders, such as schizophrenia and some mood disorders.

Applying Psychology

The Insanity Plea—Guilty of a Crime or Mentally Ill?

On the morning of June 20, 2001, Texas mother Andrea Yates drowned her five children in the bathtub, then calmly called her husband to tell him he should come home. At Yates's trial, both the defense and prosecution agreed that Yates was mentally ill at the time of the murders, yet the jury still found her guilty and sentenced her to life in prison. (An appellate court later overturned this conviction. In 2006, Yates was found not guilty by reason of insanity and committed to a state mental hospital in which she will be held until she is no longer deemed a threat.) How could two juries come to such opposite conclusions? **Insanity** is a complicated legal term. In most states it refers to a person who cannot be held responsible for his or her actions, or is judged incompetent to manage his or her own affairs because of mental illness. Despite high-profile cases like that of Andrea Yates, it's important to keep in mind that the insanity plea is used in less than 1 percent of all cases that reach trial, and when used, it is rarely successful (Kirschner, Litwack, & Galperin, 2004; Slobogin, 2006; West & Lichtenstein, 2006).

CRITICAL THINKING

Here are two interesting questions:

1. If you had been on the jury in Andrea Yates's case, would you have found her guilty? Why or why not?
2. Why do you think the insanity plea is rarely successful?

Understanding the DSM-IV-TR The *DSM-IV-TR* is organized according to five major dimensions called *axes*, which serve as guidelines for making decisions about symptoms (FIGURE 13.2). Axis I describes **state disorders** (the patient's current condition, or "state"), such as anxiety, substance abuse, and depression. Axis II describes **trait disorders** (enduring problems that seem to be an integral part of the self), including long-running personality disorders and mental retardation.

The other three axes are used to record important supplemental information. Medical conditions (Axis III) and psychosocial and environmental stressors (Axis IV) could contribute to moods and emotional problems. Finally, Axis V evaluates a person's overall level of functioning, on a scale from 1 (serious attempt at suicide) to 100 (happy and productive).

The *DSM-IV-TR* contains more than 200 diagnostic categories grouped into 17 subcategories (STUDY ORGANIZER 13.1). And in this chapter, we focus on only the first 6 of the 17 categories. Before we go on, note that the *DSM-IV-TR* classifies disorders, not people. Accordingly, we use terms such as a *person with schizophrenia* rather than describing people as *schizophrenic.*

Evaluating the DSM-IV-TR The *DSM-IV-TR* has been praised for carefully and completely describing symptoms, standardizing diagnoses and treatments, facilitating communication, and serving as a valuable educational tool. Critics, however, suggest that it relies too heavily on the medical model and unfairly labels people (Cooper, 2004; Horwitz, 2007; Mitchell, 2003; Zalaquett et al., 2008). The *DSM-IV-TR* has also been criticized for its possible cultural bias. The manual does provide a culture-specific section and a glossary of culture-bound syndromes, such as amok (Indonesia), genital retraction syndrome (Asia), and windigo psychosis (Native American cultures), which we discuss later in this chapter. However, the classification of most disorders still reflects a Western European and American perspective (Ancis, Chen, & Schultz, 2004; Borra, 2008; Smart & Smart, 1997).

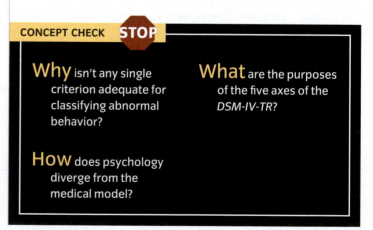

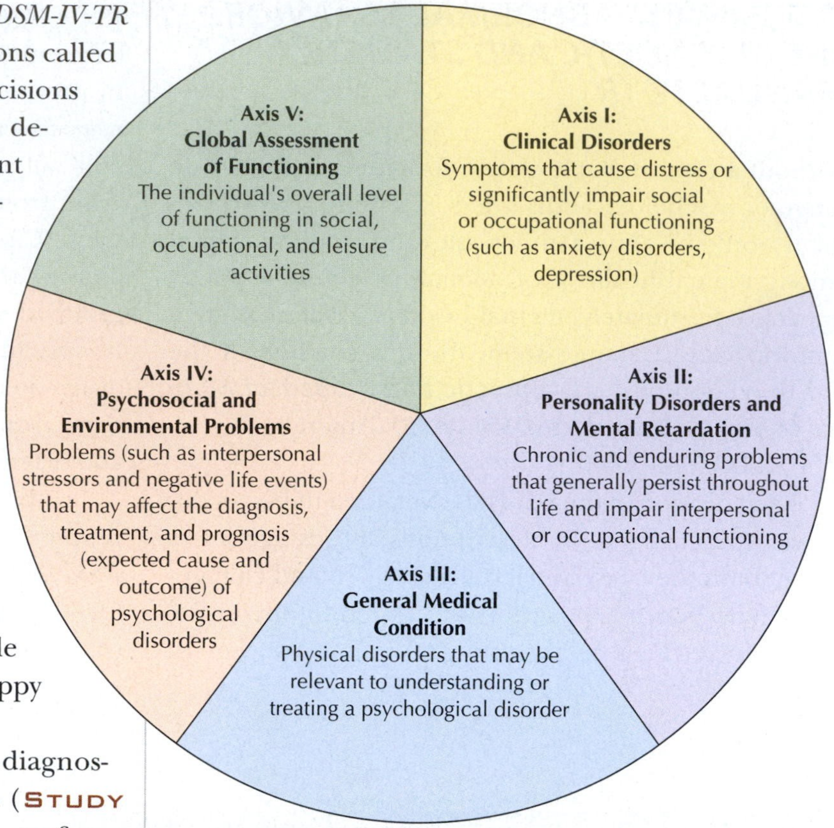

Five axes of the *DSM-IV-TR* FIGURE 13.2

Each axis serves as a broad category that helps organize the wide variety of mental disorders and acts as a guideline for making decisions. (However, the *DSM* does not suggest therapies or treatment.)

CONCEPT CHECK STOP

Why isn't any single criterion adequate for classifying abnormal behavior?

How does psychology diverge from the medical model?

What are the purposes of the five axes of the *DSM-IV-TR*?

Study Organizer 13.1 Main categories of mental disorders in DSM-IV-TR

Category	*Description*	*Examples*
Anxiety Disorders	Problems associated with severe anxiety	phobias, obsessive-compulsive disorder, posttraumatic stress disorder
Mood Disorders	Problems associated with severe disturbances of mood	depression, mania, bipolar disorder
Schizophrenia and Other Psychotic Disorders	A group of disorders characterized by major disturbances in perception, language and thought, emotion, and behavior	schizophrenia, brief psychotic disorder
Dissociative Disorders	Disorders in which the normal integration of consciousness, memory, or identity is suddenly and temporarily altered	amnesia, dissociative identity disorders
Personality Disorders	Problems related to lifelong maladaptive personality traits	antisocial personality disorders, borderline personality disorders
Substance-related Disorders	Problems caused by alcohol, cocaine, tobacco, and other drugs	substance dependence, substance abuse
Somatoform Disorders	Problems related to unusual preoccupation with physical health or physical symptoms with no physical cause	pain disorder, hypochondriasis
Factitious Disorders	Disorders that the individual adopts to satisfy some economic or psychological need	factitious disorder
Sexual and Gender Identity Disorders	Problems related to unsatisfactory sexual activity, finding unusual objects or situations arousing, or gender identity problems	sexual dysfunctions, sexual desire disorders
Eating Disorders	Problems related to food	anorexia nervosa, bulimia
Sleep Disorders	Serious disturbances of sleep	insomnia, sleep terrors, hypersomnia
Impulse Control Disorders (not elsewhere classified)	Disorders involving failure to resist an impulse or temptation to perform an act that is harmful to the person or others	kleptomania, pyromania, pathological gambling
Adjustment Disorders	Problems involving excessive emotional reaction to specific stressors	excessive emotional reaction to divorce, family discord, or unemployment
Disorders usually first diagnosed in infancy, childhood, or early adolescence	Problems that appear before adulthood	mental retardation, language development disorders
Delirium, Dementia, Amnestic, and Other Cognitive Disorders	Problems caused by known damage to the brain	Alzheimer's disease, strokes
Mental disorders due to a general medical condition (not elsewhere classified)	Problems caused by physical deterioration of the brain due to disease, drugs, etc.	personality change due to a general medical condition
Other conditions that may be a focus of clinical attention	Symptoms that may or may not be related to mental disorders but are severe enough to warrant clinical attention	medication-induced movement disorders

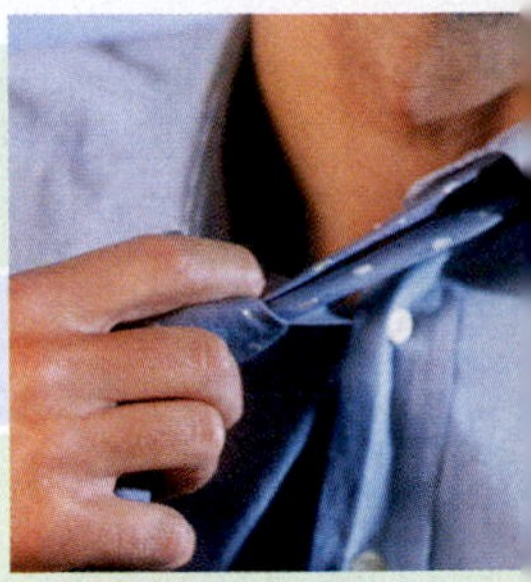

Anxiety disorders

Mood disorders

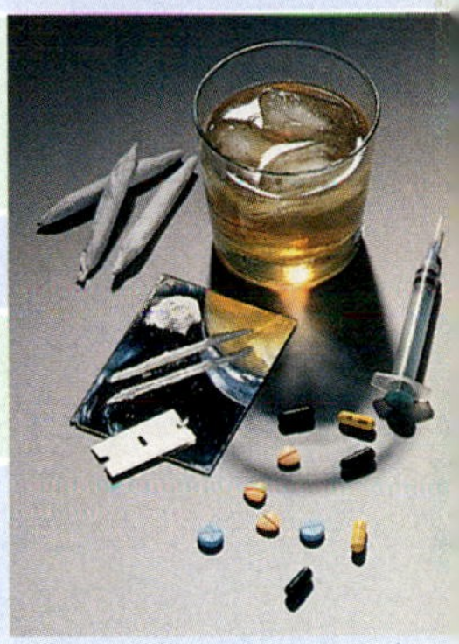

Substance-related disorders

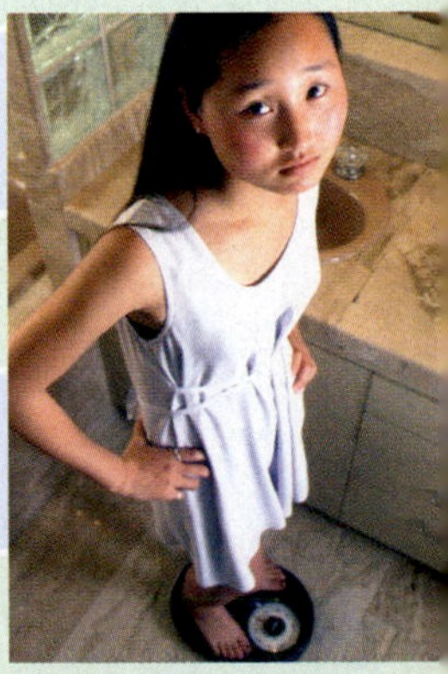

Eating disorders

Anxiety Disorders

LEARNING OBJECTIVES

Describe the symptoms of generalized anxiety disorders, panic disorder, phobias, and obsessive-compulsive disorder.

Summarize how psychological, biological, and sociocultural factors contribute to anxiety disorders.

Anxiety disorders—which are diagnosed twice as often in women as in men—are the most frequently occurring mental disorders in the general population (National Institute of Mental Health, 2008; Swartz & Margolis, 2004). Fortunately, they are also among the easiest disorders to treat and have one of the best chances for recovery (Chapter 14).

anxiety disorder A type of abnormal behavior characterized by unrealistic, irrational fear.

FOUR MAJOR ANXIETY DISORDERS: THE PROBLEM OF FEAR

Symptoms of anxiety, such as rapid breathing, dry mouth, and increased heart rate, plague all of us during stressful moments. But some people experience anxiety that is so intense and chronic it seriously disrupts their lives. They feel threatened, unable to cope, unhappy, and insecure in a world that seems dangerous and hostile. In this section, we consider four anxiety disorders: generalized anxiety disorder, panic disorder, phobias, and obsessive-compulsive disorder (FIGURE 13.3). (Posttraumatic stress disorder, another major anxiety disorder, is discussed in Chapter 3.) Although we discuss these disorders separately, people often have more than one anxiety disorder (Halgin & Whitbourne, 2008).

Generalized anxiety disorder This disorder affects twice as many women as it does men (Brown, O'Leary, & Barlow, 2001). **Generalized anxiety disorder (GAD)** is characterized by chronic, uncontrollable, and

Major anxiety disorders FIGURE 13.3

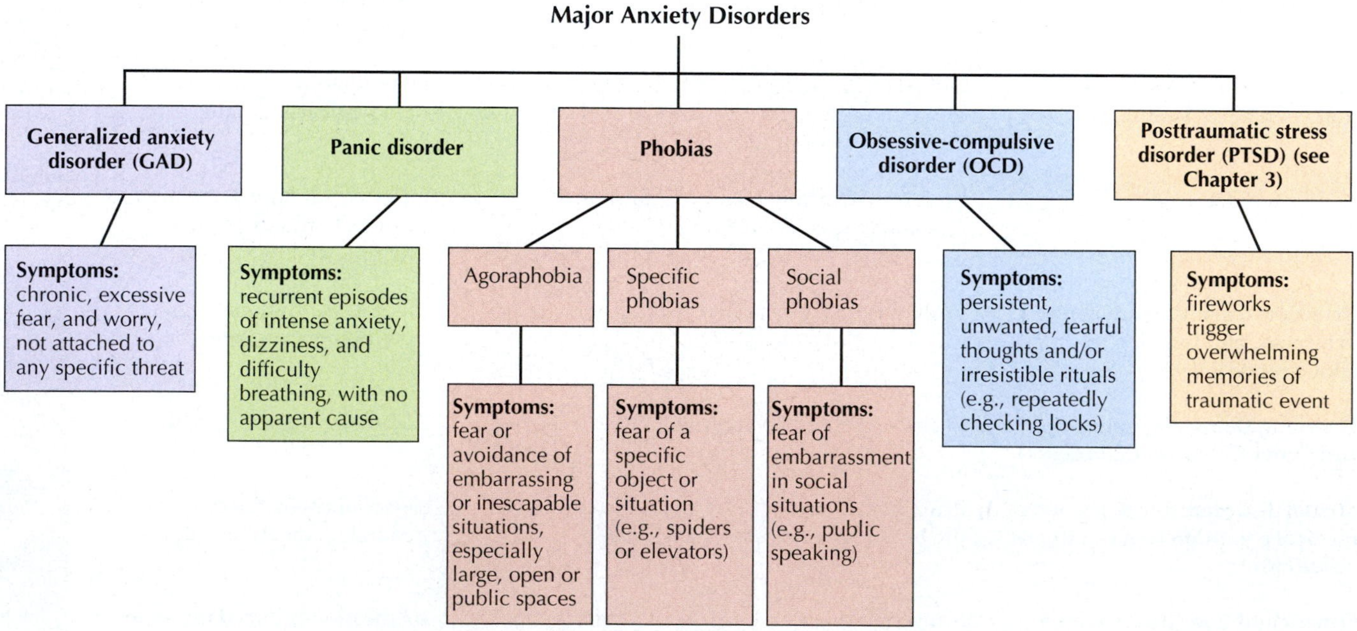

excessive fear and worry that lasts at least six months and that is not focused on any particular object or situation. Because of persistent muscle tension and autonomic fear reactions, people with this disorder may develop headaches, heart palpitations, dizziness, and insomnia, making it even harder to cope with normal daily activities.

Panic disorder Sudden, but brief, attacks of intense apprehension that cause trembling, dizziness, and difficulty breathing are symptoms of **panic disorder**. Panic attacks generally happen after frightening experiences or prolonged stress (and sometimes even after exercise). Panic disorder is diagnosed when several apparently spontaneous panic attacks lead to a persistent concern about future attacks. A common complication of panic disorder is agoraphobia (Cully & Stanley, 2008; Roberge et al., 2008).

Phobias **Phobias** involve a strong, irrational fear and avoidance of objects or situations that are usually considered harmless (fear of elevators or fear of going to the dentist, for example). Although the person recognizes that the fear is irrational, the experience is still one of overwhelming anxiety, and a full-blown panic attack may follow. The *DSM-IV-TR* divides phobic disorders into three broad categories: agoraphobia, specific phobias, and social phobias.

People with *agoraphobia* restrict their normal activities because they fear having a panic attack in crowded, enclosed, or wide-open places where they would be unable to receive help in an emergency. In severe cases, people with agoraphobia may refuse to leave the safety of their homes.

A *specific phobia* is a fear of a specific object or situation, such as needles, rats, spiders, or heights. Claustrophobia (fear of closed spaces) and acrophobia (fear of heights) are the specific phobias most often treated by therapists. People with specific phobias generally recognize that their fears are excessive and unreasonable, but they are unable to control their anxiety and will go to great lengths to avoid the feared stimulus (FIGURE 13.4).

People with *social phobias* are irrationally fearful of embarrassing themselves in social situations. Fear of public speaking and of eating in public are the most common social phobias. The fear of public scrutiny and potential humiliation may become so pervasive that normal life is impossible (Acarturk et al., 2008; Swartz, 2008).

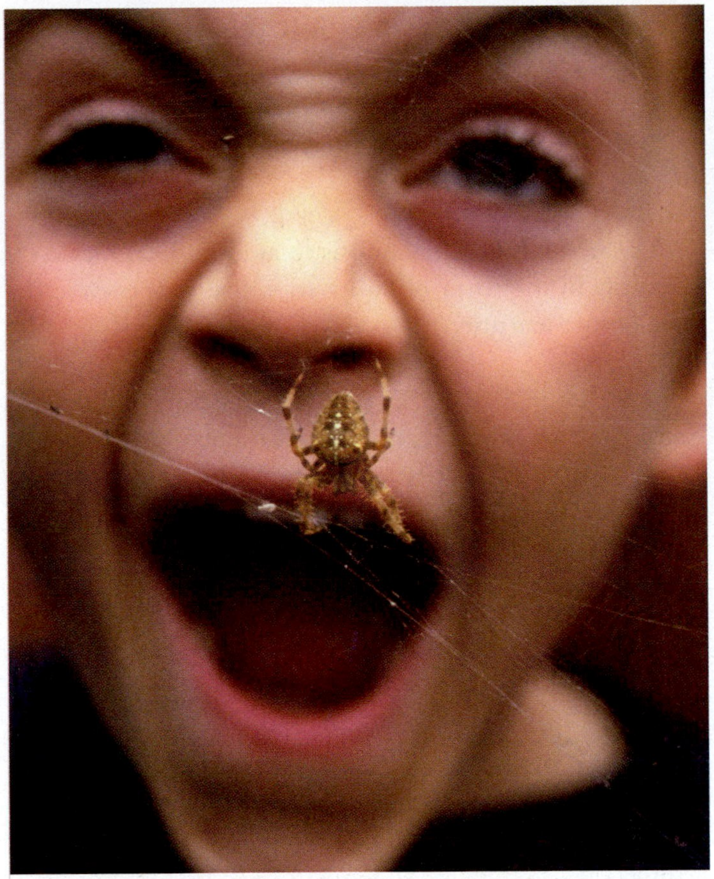

Spider phobia FIGURE 13.4

Can you imagine how it would feel to be so frightened by a spider that you would try to jump out of a speeding car to get away from it? This is how a person suffering from a phobia might feel.

Obsessive-compulsive disorder (OCD) **Obsessive compulsive disorder (OCD)** disorder involves persistent, unwanted, fearful thoughts (obsessions) and/or irresistible urges to perform an act or repeated rituals (compulsions), which help relieve the anxiety created by the obsession. In adults, this disorder is equally common in men and women. However, it is more prevalent among boys when the onset is in childhood (American Psychiatric Association, 2000).

Imagine what it would be like to worry so obsessively about germs that you compulsively wash your hands hundreds of times a day until they are raw and bleeding. Most sufferers of OCD realize that their actions are senseless. But when they try to stop the behavior, they experience mounting anxiety, which is relieved only by giving in to the urges.

EXPLAINING ANXIETY DISORDERS

Why do people develop anxiety disorders? Research has focused on the roles of psychological, biological, and sociocultural processes (the *biopsychosocial model*) (FIGURE 13.5).

Psychological contributions to anxiety disorders are primarily in the form of faulty cognitive processes and maladaptive learning.

Faulty cognitions People with anxiety disorders have habits of thinking, or cognitive habits, that make them prone to fear. They tend to be hypervigilant—they constantly scan their environment for signs of danger and ignore signs of safety. They also tend to magnify ordinary threats and failures and to be hypersensitive to others' opinions of them (FIGURE 13.6).

Maladaptive learning According to learning theorists, anxiety disorders generally result from conditioning and social learning (Chapter 6) (Cully & Stanley, 2008; Mineka & Oehlberg, 2008; Swartz, 2008).

During classical conditioning, for example, a stimulus that is originally neutral (e.g., a harmless spider) becomes paired with a frightening event (a sudden panic attack) so that it becomes a conditioned stimulus that elicits anxiety. The person then begins to avoid spiders in order to reduce anxiety (an operant conditioning process known as negative reinforcement).

Some researchers contend that the fact that most people with phobias cannot remember a specific instance that led to their fear and that frightening experiences do not always trigger phobias suggests that conditioning may not be the only explanation. Social learning theorists propose that some phobias are the result of modeling and imitation.

Phobias may also be learned vicariously (indirectly). In one study, rhesus monkeys viewed videos showing another monkey apparently experiencing extreme fear of a toy snake, a toy rabbit, a toy crocodile, and flowers (Cook & Mineka, 1989). The "viewing" monkeys were later afraid of the snake and crocodile but not of the rabbit or flowers, suggesting that phobias have both learned and biological components.

Biological factors Some researchers believe that phobias reflect an evolutionary predisposition to fear that which was dangerous to our ancestors (Mineka & Oehlberg, 2008; Walker et al., 2008). Some people with panic disorder also seem to be genetically predisposed toward an overreaction of the autonomic nervous system, further supporting the argument for a biological component. In addition, stress and arousal seem to play a role in panic attacks, and drugs such as caffeine or

Anxiety disorders and the biopsychosocial model FIGURE 13.5

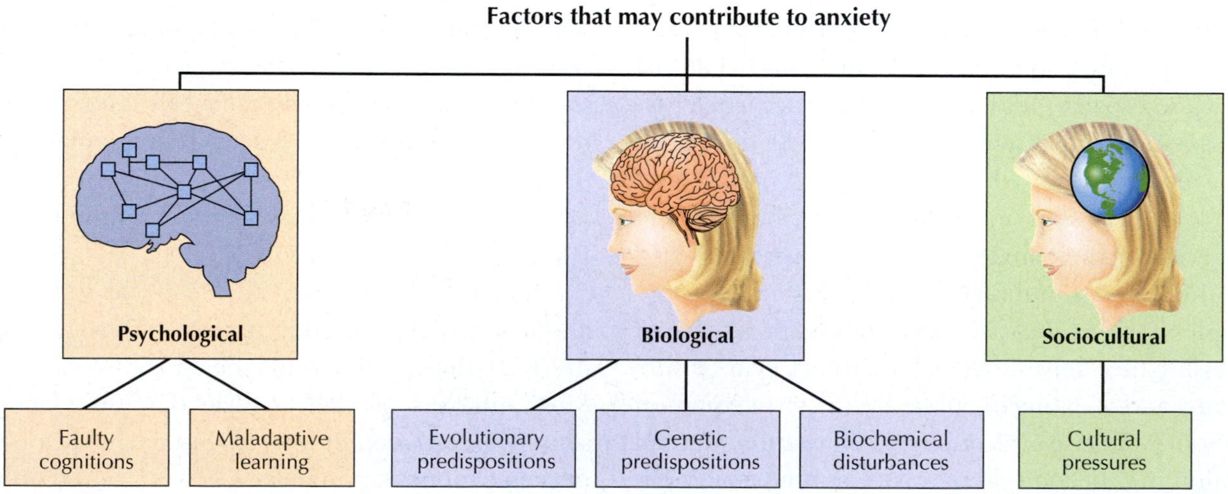

Faulty thinking patterns FIGURE 13.6

People who suffer from social phobias are excessively concerned about others' evaluations, hypersensitive to any criticism, and obsessively worried about potential mistakes. This intense self-preoccupation leads these people to perpetually believe they have failed. What changes in thinking patterns might lessen this anxiety?

nicotine and even hyperventilation can trigger an attack, all suggesting a biochemical disturbance.

Sociocultural factors In addition to psychological and biological components, sociocultural factors can contribute to anxiety. There has been a sharp rise in anxiety disorders in the past 50 years, particularly in Western industrialized countries. Can you see how our increasingly fast-paced lives—along with our increased mobility, decreased job security, and decreased family support—might contribute to anxiety? Unlike the dangers that humans faced in our evolutionary history, today's threats are less identifiable and immediate, which may lead some people to become hypervigilant and predisposed to anxiety disorders (Figure 13.6).

Further support for sociocultural influences on anxiety disorders is that anxiety disorders can have dramatically different forms in other cultures. For example, in a collectivist twist on anxiety, the Japanese have a type of social phobia called *taijin kyofusho* (TKS), which involves morbid dread of doing something to embarrass others. This disorder is quite different from the Western version of social phobia, which centers on a fear of criticism.

CONCEPT CHECK STOP

How do generalized anxiety disorders and phobias differ?

What are examples of the faulty cognitions that characterize anxiety disorders?

Mood Disorders

LEARNING OBJECTIVES

Explain how major depressive disorder and bipolar disorder differ.

Summarize research on the biological and psychological factors that contribute to mood disorders.

UNDERSTANDING MOOD DISORDERS: MAJOR DEPRESSIVE DISORDER AND BIPOLAR DISORDER

As the name implies, **mood disorders** (also known as affective disorders) are characterized by extreme disturbances in emotional states. There are two main types of mood disorders—major depressive disorder and bipolar disorder.

We all feel "blue" sometimes, especially following the loss of a job, end of a relationship, or death of a loved one. People suffering from **major depressive disorder**, however, may experience a lasting and continuously depressed mood without a clear trigger.

Clinically depressed people are so deeply sad and discouraged that they often have trouble sleeping, are likely to lose (or gain) weight, and may feel so fatigued that they cannot go to work or school or even comb their hair and brush their teeth. They may sleep constantly, have problems concentrating, and feel so profoundly sad and guilty that they consider suicide. Depressed individuals have a hard time thinking clearly or recognizing their own problems.

When depression is *unipolar*, the depressive episode eventually ends, and the person returns to a "normal" emotional level. People with **bipolar disorder**, however, rebound to the opposite state, known as *mania* (FIGURE 13.7).

During a manic episode, the person is overly excited, extremely active, and easily distracted. The person exhibits unrealistically high self-esteem, an inflated sense of importance, and poor judgment (FIGURE 13.8). The person is hyperactive and may not sleep for days at a time, yet does not become fatigued. Thinking is speeded up and can change abruptly to new topics, showing "rapid flight of ideas." Speech is also rapid ("pressured speech"), and it is difficult for others to get a word in edgewise.

A manic episode may last a few days or a few months, and it generally ends abruptly. The ensuing depressive episode generally lasts three times as long as the manic episode. The lifetime risk for bipolar disorder is low—somewhere between 0.5 and 1.6 percent—but it can be one of the most debilitating and lethal disorders, with a suicide rate between 10 and 20 percent among sufferers (Carballo et al., 2008; Kinder et al., 2008; Klimes-Dougan et al., 2008).

major depressive disorder Long-lasting depressed mood that interferes with the ability to function, feel pleasure, or maintain interest in life (Swartz & Margolis, 2004).

bipolar disorder Repeated episodes of mania (unreasonable elation and hyperactivity) and depression.

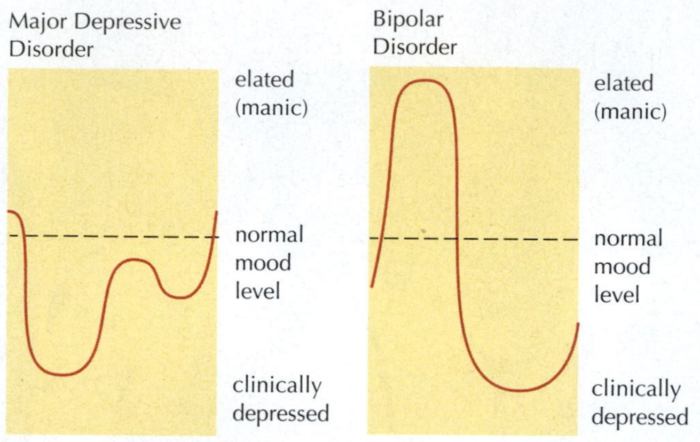

Mood disorders FIGURE 13.7

If major depressive disorders and bipolar disorders were depicted on a graph, they might look something like this.

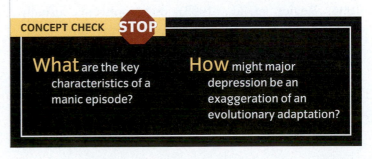

No limits FIGURE 13.8

Poor judgment is common during manic episodes. A person may give away valuable possessions, go on wild spending sprees, or make elaborate plans for becoming rich and famous. How might this behavior contribute to or trigger major depression?

EXPLAINING MOOD DISORDERS: BIOLOGICAL VERSUS PSYCHOSOCIAL FACTORS

Biological factors appear to play a significant role in both major depressive disorder and bipolar disorder. Recent research suggests that structural brain changes may contribute to these mood disorders. Other research points to imbalances of the neurotransmitters serotonin, norepinephrine, and dopamine (Barton et al., 2008; Delgado, 2004; Lyoo et al., 2004; Montgomery, 2008; Wiste et al., 2008). Drugs that alter the activity of these neurotransmitters also decrease the symptoms of depression (and are therefore called antidepressants) (Chuang, 1998).

Some research indicates that mood disorders may be inherited. For example, when one identical twin has a mood disorder, there is about a 50 percent chance that the other twin will also develop the illness (Anisman, Merali, & Stead, 2008; Brent & Melham, 2008; Faraone, 2008; Swartz, 2008). However, it is important to remember that relatives generally have similar environments as well as similar genes.

Finally, the evolutionary perspective suggests that moderate depression may be a normal and healthy adaptive response to a very real loss (such as the death of a loved one). The depression helps us to step back and reassess our goals (Nesse, 2000; Nesse & Jackson, 2006). Clinical, severe depression may just be an extreme version of this generally adaptive response.

Psychosocial theories of depression focus on environmental stressors and disturbances in the person's interpersonal relationships, thought processes, self-concept, and history of learned behaviors (Cheung, Gilbert, & Irons, 2004; Hammen, 2005; Mathews & McLeod, 2005). The psychoanalytic explanation sees depression as anger (stemming from feelings of rejection) turned inward when an important relationship is lost. The humanistic school says that depression results when a person demands perfection of him- or herself or when positive growth is blocked.

According to the **learned helplessness** theory (Seligman, 1975, 1994, 2007), depression occurs when people (and other animals) become resigned to the idea that they are helpless to escape from something painful. Learned helplessness may be particularly likely to trigger depression if the person attributes failure to causes that are internal ("my own weakness"), stable ("this weakness is long-standing and unchanging"), and global ("this weakness is a problem in lots of settings") (Ball, McGuffin, & Farmer, 2008; Gotlieb & Abramson, 1999; Wise & Rosqvist, 2007).

Suicide is a major danger associated with depression. Because of the shame and secrecy associated with suicide, many fail to get or give help.

CONCEPT CHECK STOP

What are the key characteristics of a manic episode?

How might major depression be an exaggeration of an evolutionary adaptation?

Schizophrenia

LEARNING OBJECTIVES

Describe some common symptoms of schizophrenia.

Compare the traditional (four-group) system for classifying different types of schizophrenia with the (two-group) system that has recently emerged.

Summarize the biological and psychosocial factors that contribute to schizophrenia.

Imagine that your daughter has just left for college and that you hear voices inside your head shouting, "You'll never see her again! You have been a bad mother! She'll die." Or what if you saw live animals in your refrigerator? These experiences have plagued "Mrs. T" for decades (Gershon & Rieder, 1993).

Mrs. T suffers from **schizophrenia** (skit-so-FREE-nee-uh). Schizophrenia is often so severe that it is considered a psychosis, meaning that the person is out of touch with reality. People with schizophrenia have serious problems caring for themselves, relating to others, and holding a job. In extreme cases, people with schizophrenia require institutional or custodial care.

schizophrenia A group of psychotic disorders involving major disturbances in perception, language, thought, emotion, and behavior. The individual withdraws from people and reality, often into a fantasy life of delusions and hallucinations.

Schizophrenia is one of the most widespread and devastating mental disorders. Approximately 1 out of every 100 people will develop schizophrenia in his or her lifetime, and approximately half of all people who are admitted to mental hospitals are diagnosed with this disorder (Gottesman, 1991; Kendler et al., 1996; Kessler et al., 1994; Regier et al., 1993). Schizophrenia usually emerges between the late teens and the mid-30s. It seems to be equally prevalent in men and women, but it's generally more severe and strikes earlier in men than in women (Combs et al., 2008; Faraone, 2008; Gottesman, 1991; Mueser & Jeste, 2008; Tsuang, Stone, & Faraone, 2001).

Many people confuse schizophrenia with dissociative identity disorder, which is sometimes referred to as *split* or *multiple personality disorder.* Schizophrenia means "split mind," but when Eugen Bleuler coined the term in 1911, he was referring to the fragmenting of thought processes and emotions, not personalities (Neale, Oltmanns, & Winters, 1983). As we discuss later in this chapter, dissociative identity disorder is the rare condition of having more than one distinct personality.

SYMPTOMS OF SCHIZOPHRENIA: FIVE AREAS OF DISTURBANCE

Schizophrenia is a group of disorders characterized by a disturbance in one or more of the following areas: perception, language, thought, affect (emotions), and behavior.

Perceptual symptoms

The senses of people with schizophrenia may be either enhanced or blunted. The filtering and selection processes that allow most people to concentrate on whatever they choose are impaired, and sensory stimulation is jumbled and distorted.

People with schizophrenia may also experience **hallucinations**—most commonly auditory hallucinations (hearing voices and sounds).

hallucinations Imaginary sensory perceptions that occur without an external stimulus.

On rare occasions, people with schizophrenia will hurt others in response to their distorted perceptions. But a person with schizophrenia is more likely to be self-destructive and suicidal than violent toward others.

Mental illness in public view FIGURE 13.9

Fear of ridicule and discrimination leads many people with psychological disorders to keep their condition secret. But singer-songwriter Trace Moore, who was diagnosed with schizophrenia at age 21, has taken the opposite approach. Even as she combats paranoia and delusional thoughts, Moore continues to nourish her longtime dream of succeeding as a performing artist. *Source:* Roberts, "Idol Dreams," *Schizophrenia Digest* (March 2006), pp. 30–33.

Language and thought disturbances

For people with schizophrenia, words lose their usual meanings and associations, logic is impaired, and thoughts are disorganized and bizarre. When language and thought disturbances are mild, the individual jumps from topic to topic. With more severe disturbances, the person jumbles phrases and words together (into a "word salad") or creates artificial words.

The most common—and frightening—thought disturbance experienced by people with schizophrenia is the lack of contact with reality (psychosis).

delusions Mistaken beliefs based on misrepresentations of reality.

Delusions are also common in people with schizophrenia. We all experience exaggerated thoughts from time to time, such as thinking a friend is trying to avoid us, but the delusions of schizophrenia are much more extreme. For example, Jim (the med student in the chapter opener) was completely convinced that others were trying to assassinate him (a *delusion of persecution*). In *delusions of grandeur*, people believe that they are someone very important, perhaps Jesus Christ or the queen of England. In *delusions of reference*, unrelated events are given special significance, as when a person believes that a radio program is giving him or her a special message. (See FIGURES 13.9 and 13.10.)

A recent trend among individuals suffering from various mental disorders is to join Internet groups where they can share their experiences with others. For example, people with schizophrenia and other psychotic disorders can now share with fellow sufferers their belief that they are being stalked or that their minds are being controlled by technology. Some experts consider Internet groups that offer peer support to be helpful to the mentally ill, but others believe that they may reinforce troubled thinking and impede treatment. Those who use the sites report feeling relieved by the sense that they are not alone in their suffering, but it should be noted that these sites are not moderated by professionals and pose the danger of actually amplifying the symptoms of mentally ill individuals (Kershaw, 2008).

The Jerusalem syndrome FIGURE 13.10

Every year, dozens of tourists to Jerusalem are hospitalized with symptoms of *Jerusalem syndrome*, in which a person becomes obsessed with the significance of Jerusalem and engages in bizarre, deluded behavior. For example, the person might come to believe that he or she is Jesus Christ, transform hotel linens into a long, white robe, and publicly recite Bible verses, a classic example of delusional thought disturbances.

Emotional disturbances Changes in emotion usually occur in people with schizophrenia. In some cases, emotions are exaggerated and fluctuate rapidly. At other times, emotions become blunted. Some people with schizophrenia have *flattened affect*—almost no emotional response of any kind.

Behavioral disturbances Disturbances in behavior may take the form of unusual actions that have special meaning. For example, one patient massaged his head repeatedly to "clear it" of unwanted thoughts. People with schizophrenia may become *cataleptic* and assume a nearly immobile stance for an extended period.

TYPES OF SCHIZOPHRENIA: RECENT METHODS OF CLASSIFICATION

For many years, researchers divided schizophrenia into five subtypes: paranoid, catatonic, disorganized, undifferentiated, and residual (TABLE 13.1). Although these terms are still included in the *DSM-IV-TR*, critics contend that this system does not differentiate in terms of prognosis, cause, or response to treatment and that the undifferentiated type is merely a catchall for cases that are difficult to diagnose (American Psychiatric Association, 2000).

Subtypes of schizophrenia TABLE 13.1

Paranoid	Dominated by delusions (persecution and grandeur) and hallucinations (hearing voices)
Catatonic	Marked by motor disturbances (immobility or wild activity) and echo speech (repeating the speech of others)
Disorganized	Characterized by incoherent speech, flat or exaggerated emotions, and social withdrawal
Undifferentiated	Meets the criteria for schizophrenia but is not any of the above subtypes
Residual	No longer meets the full criteria for schizophrenia but still shows some symptoms

For these reasons, researchers have proposed an alternative classification system:

1. **Positive schizophrenia symptoms** involve additions to or exaggerations of normal thought processes and behaviors, including bizarre delusions and hallucinations.
2. **Negative schizophrenia symptoms** involve the loss or absence of normal thought processes and behaviors, including impaired attention, limited or toneless speech, flattened affect, and social withdrawal.

Positive symptoms are more common when schizophrenia develops rapidly, whereas negative symptoms are more often found in slow-developing schizophrenia. Positive symptoms are associated with better adjustment before the onset and a better prognosis for recovery.

In addition to these two groups, the latest *DSM* suggests adding another dimension to reflect *disorganization of behavior*, including rambling speech, erratic behavior, and inappropriate affect.

EXPLAINING SCHIZOPHRENIA: NATURE AND NURTURE THEORIES

Because schizophrenia comes in many different forms, it probably has multiple biological and psychosocial bases (Walker et al., 2004). Let's look at biological contributions first.

Prenatal viral infections, birth complications, immune responses, maternal malnutrition, and advanced paternal age all may contribute to the development of schizophrenia (Ellman & Cannon, 2008; Meyer et al., 2008; Tandon, Keshavan, & Nasrallah, 2008; Zuckerman & Weiner, 2005). However, most biological theories of schizophrenia focus on genetics, neurotransmitters, and brain abnormalities.

- *Genetics.* Although researchers are beginning to identify specific genes related to schizophrenia, most genetic studies have focused on twins and adoptions (Elkin, Kalidindi, & McGuffin, 2004; Faraone, 2008; Hall et al., 2007). This research indicates that the risk for schizophrenia increases with genetic similarity; that is, people

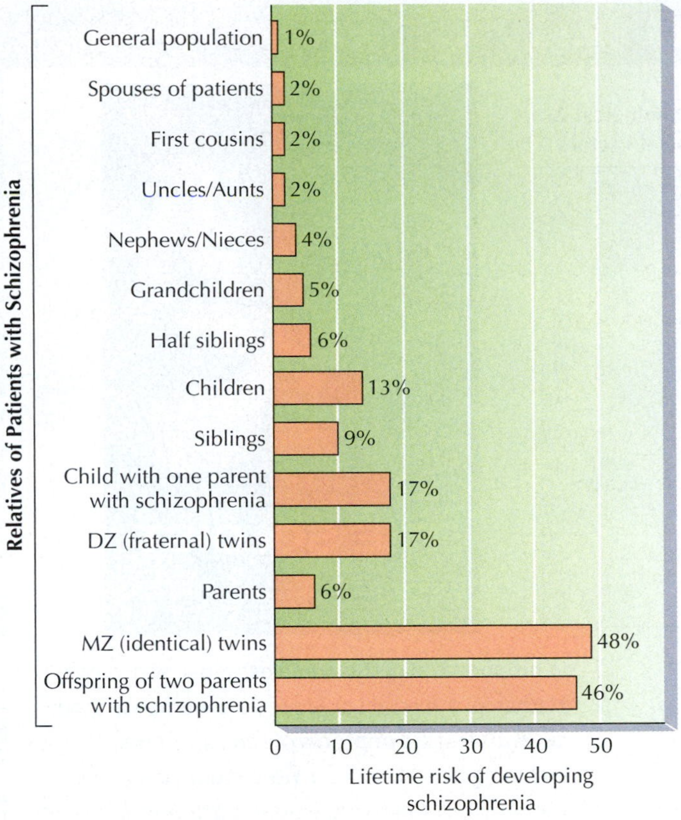

Genetics and schizophrenia FIGURE 13.11

Your lifetime risk of developing schizophrenia depends, in part, on how closely you are genetically related to someone with schizophrenia.

Source: Gottesman, "Schizophrenia Genesis," 1991, W.H. Freeman and Company/ Worth Publishers.

who share more genes with a person who has schizophrenia are more likely to develop the disorder (FIGURE 13.11).

- *Neurotransmitters.* Precisely how genetic inheritance produces schizophrenia is unclear. According to the **dopamine hypothesis**, overactivity of certain dopamine neurons in the brain causes schizophrenia (Ikemoto, 2004; Paquet et al., 2004). This hypothesis is based on two observations. First, administering amphetamines increases the amount of dopamine and can produce (or worsen) some symptoms of schizophrenia, especially in people with a genetic predisposition to the disorder. Second, drugs that reduce dopamine activity in the brain reduce or eliminate some symptoms of schizophrenia.
- *Brain abnormalities.* The third major biological theory for schizophrenia involves abnormalities in brain function and structure. Researchers have found larger cerebral ventricles (fluid-filled spaces in the brain) in some people with schizophrenia (Galderisi et al., 2008; Gaser et al., 2004).

Also, some people with chronic schizophrenia have a lower level of activity in their frontal and temporal lobes—areas that are involved in language, attention, and memory (FIGURE 13.12). Damage in these regions might explain the thought and language disturbances that characterize schizophrenia. This lower level of brain activity, and schizophrenia itself, may also result from an overall loss of gray matter (neurons in the cerebral cortex) (Crespo-Facorro et al., 2007; Gogtay et al., 2004).

Clearly, biological factors play a key role in schizophrenia. But the fact that even in identical twins—who share identical genes—the heritability of schizophrenia is only 48 percent tells us that nongenetic factors must contribute the remaining percentage. Most psychologists believe that there are at least two possible psychosocial contributors.

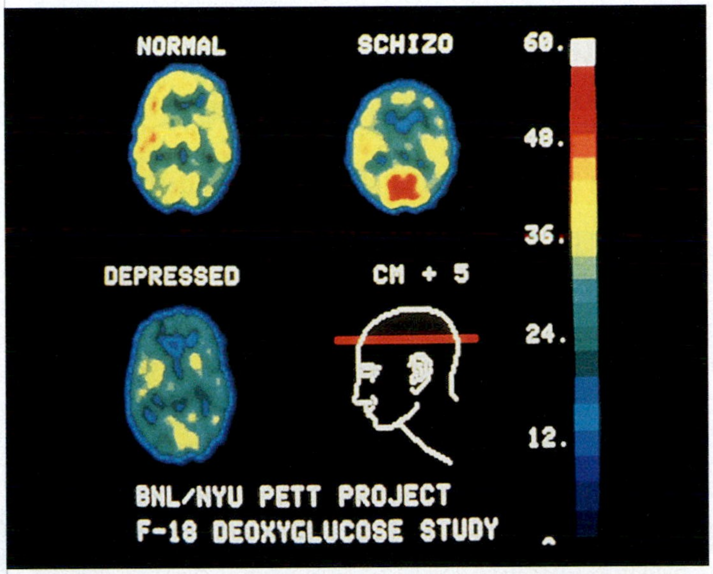

Brain activity in schizophrenia FIGURE 13.12

These positron emission tomography (PET) scans show variations in the brain activity of normal individuals, people with major depressive disorder, and individuals with schizophrenia. Warmer colors (red, yellow) indicate increased activity.

Process Diagram

The biopsychosocial model and schizophrenia FIGURE 13.13

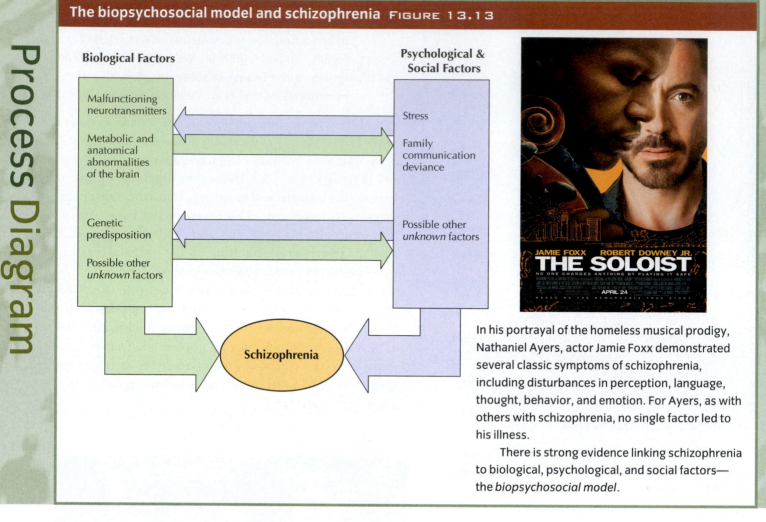

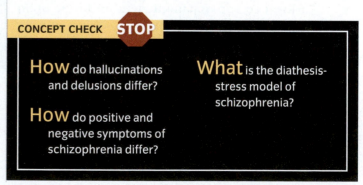

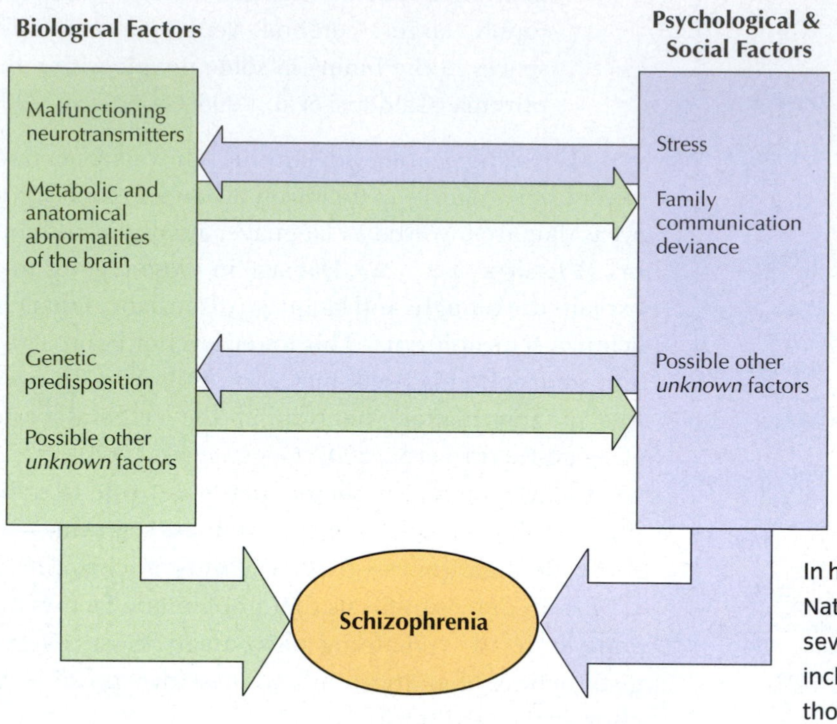

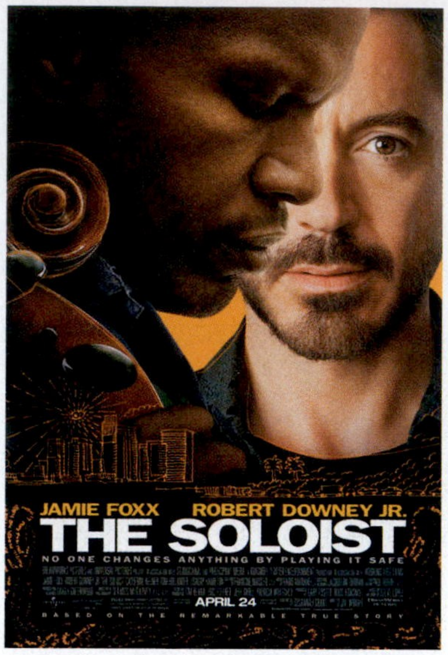

In his portrayal of the homeless musical prodigy, Nathaniel Ayers, actor Jamie Foxx demonstrated several classic symptoms of schizophrenia, including disturbances in perception, language, thought, behavior, and emotion. For Ayers, as with others with schizophrenia, no single factor led to his illness.

There is strong evidence linking schizophrenia to biological, psychological, and social factors—the *biopsychosocial model*.

According to the **diathesis-stress model** of schizophrenia, stress plays an essential role in triggering schizophrenic episodes in people with an inherited predisposition (or diathesis) toward the disease (Jones & Fernyhough, 2007; Reulbach et al., 2007).

Some investigators suggest that communication disorders in family members may also be a predisposing factor for schizophrenia. Such disorders include unintelligible speech, fragmented communication, and parents' frequently sending severely contradictory messages to children. Several studies have also shown greater rates of relapse and worsening of symptoms among hospitalized patients who went home to families that were critical and hostile toward them or overly involved in their lives emotionally (Hooley & Hiller, 2001; McFarlane, 2006).

How should we evaluate the different theories about the causes of schizophrenia? Critics of the dopamine hypothesis and the brain damage theory argue that those theories fit only some cases of schizophrenia. Moreover, with both theories, it is difficult to determine cause and effect. The disturbed-communication theories are also hotly debated, and research is inconclusive. Schizophrenia is probably the result of a combination of known and unknown interacting factors (FIGURE 13.13).

CONCEPT CHECK STOP

How do hallucinations and delusions differ?

How do positive and negative symptoms of schizophrenia differ?

What is the diathesis-stress model of schizophrenia?

Other Disorders

LEARNING OBJECTIVES

Explain how substance abuse and substance dependence differ.

Describe the types of dissociative disorders.

Identify the major characteristics of personality disorders.

Having discussed anxiety disorders, mood disorders, and schizophrenia, we now explore three additional disorders—substance-related, dissociative, and personality.

SUBSTANCE-RELATED DISORDERS

substance-related disorders Abuse of or dependence on a mood- or behavior-altering drug.

The category of **substance-related disorders** is subdivided into two general groups: substance abuse and substance dependence (TABLE 13.2). When alcohol or other drug use interferes with a person's social or occupational functioning, it is called *substance abuse*. Drug use becomes *substance dependence* when it also causes physical reactions, including *tolerance* (requiring more of the drug to get the desired effect) and *withdrawal* (negative physical effects when the drug is removed).

Some people can use alcohol and other drugs and not develop a problem. Unfortunately, researchers have not been able to identify ahead of time those who can use drugs without risk of becoming abusers versus those who are likely to become abusers. Complicating diagnosis and treatment is the fact that substance-related disorders commonly coexist with other mental disorders, including anxiety disorders, mood disorders, schizophrenia, and personality disorders (Cornelius & Clark, 2008; Green et al., 2004; Munro & Edward, 2008; Thomas et al., 2008). This co-occurrence of disorders is called **comorbidity** (FIGURE 13.14).

Comorbidity complicates treatment
FIGURE 13.14

Can you see how comorbidity can cause serious problems? How can the appropriate cause, course, or treatment be identified for someone dealing with a combination of disorders?

DSM-IV-TR substance abuse and substance dependence TABLE 13.2

Criteria for Substance Abuse (alcohol and other drugs)	***Criteria for Substance Dependence (alcohol and other drugs)***
Maladaptive use of a substance shown by one of the following: • Failure to meet obligations • Repeated use in situations where it is physically dangerous • Continued use despite problems caused by the substance • Repeated substance-related legal problems	*Three or more of the following:* • Tolerance • Withdrawal • Substance taken for a longer time or greater amount than intended • Lack of desire or effort to reduce or control use • Social, recreational, or occupational activities given up or reduced • Much time spent in activities to obtain the substance • Use continued despite knowing that psychological or physical problems are worsened by it

What leads to comorbidity? Perhaps the most influential hypothesis is **self-medication**—individuals drink or use drugs to reduce their symptoms (Robinson, 2008). Regardless of the causes, it is critical that patients, family members, and clinicians recognize and deal with comorbidity if treatment is to be effective.

DISSOCIATIVE DISORDERS: WHEN THE PERSONALITY SPLITS APART

dissociative disorders Amnesia, fugue, or multiple personalities resulting from avoidance of painful memories or situations.

The most dramatic psychological disorders are **dissociative disorders**. There are several types of dissociative disorders, but all involve a splitting apart (a *dis-association*) of significant aspects of experience from memory or consciousness. Individuals dissociate from the core of their personality by failing to remember past experiences (*dissociative amnesia*) (FIGURE 13.15), by leaving home and wandering off (*dissociative fugue*), by losing their sense of reality and feeling estranged from the self (*depersonalization disorder*), or by developing completely separate personalities (*dissociative identity disorder*).

Dissociation as an escape FIGURE 13.15

The major force behind all dissociative disorders is the need to escape from anxiety. Imagine witnessing a loved one's death in a horrible car accident. Can you see how your mind might cope by blocking out all memory of the event?

Unlike most other psychological disorders, the primary cause of dissociative disorders appears to be environmental variables, with little or no genetic influence (Waller & Ross, 1997).

The most severe dissociative disorder is **dissociative identity disorder (DID)**—previously known as multiple personality disorder—in which at least two separate and distinct personalities exist within a person at the same time. Each personality has unique memories, behaviors, and social relationships. Transition from one personality to another occurs suddenly and is often triggered by psychological stress. Usually, the original personality has no knowledge or awareness of the alternate personalities, but all of the personalities may be aware of lost periods of time. The disorder is diagnosed more among women than among men. Women with DID also tend to have more identities, averaging 15 or more, compared with men, who average 8 (American Psychiatric Association, 2000).

DID is a controversial diagnosis. Some experts suggest that many cases are faked or result from false memories and an unconscious need to please the therapist (Kihlstrom, 2005; Lawrence, 2008; Pope et al., 2007; Stafford & Lynn, 2002). Other psychologists accept the validity of multiple personality and contend that the condition is underdiagnosed (Dalenberg et al., 2007; Lipsanen et al., 2004; Spiegel & Maldonado, 1999).

PERSONALITY DISORDERS: ANTISOCIAL AND BORDERLINE

What would happen if the characteristics of a personality were so inflexible and maladaptive that they significantly impaired someone's ability to function? This is what happens with **personality disorders**. Several types of personality disorders are included in this category in the *DSM-IV-TR*, but here we will focus on antisocial personality disorder and borderline personality disorder.

personality disorders Inflexible, maladaptive personality traits that cause significant impairment of social and occupational functioning.

Antisocial personality disorder

The term **antisocial personality disorder** is used interchangeably with the terms *sociopath* and *psychopath*. These labels describe behavior so far outside the ethical and legal standards of society that many consider it the most serious of all mental disorders. Unlike people with anxiety, mood disorders, and schizophrenia, those with this diagnosis feel little personal distress (and may not be motivated to change). Yet their maladaptive traits generally bring considerable harm and suffering to others (Hervé et al., 2004; Kirkman, 2002; Nathan et al., 2003). Although serial killers are often seen as classic examples of antisocial personality disorder (FIGURE 13.16), many sociopaths harm people in less dramatic ways—for example, as ruthless businesspeople and crooked politicians.

The four hallmarks of antisocial personality disorder are egocentrism (preoccupation with oneself and insensitivity to the needs of others), lack of conscience, impulsive behavior, and superficial charm (American Psychiatric Association, 2000).

Unlike most adults, individuals with antisocial personality disorder act impulsively, without giving thought to the consequences. They are usually poised when confronted with their destructive behavior and feel contempt for anyone they are able to manipulate. They also change jobs and relationships suddenly, and they often have a history of truancy from school and of being expelled for destructive behavior. People with antisocial personalities can be charming and persuasive, and they have remarkably good insight into the needs and weaknesses of other people.

Twin and adoption studies suggest a possible genetic predisposition to antisocial personality disorder (Bock & Goode, 1996; Jang et al., 2003). Biological contributions are also suggested by studies that have found abnormally low autonomic activity during stress, right hemisphere abnormalities, and reduced gray matter in the frontal lobes (De Oliveira-Souza et al., 2008; Huesmann & Kirwil, 2007; Kendler & Prescott, 2006; Lyons-Ruth et al., 2007; Raine & Yang, 2006).

Evidence also exists for environmental or psychological causes. Antisocial personality disorder is highly correlated with abusive parenting styles and inappropriate modeling (Barnow et al., 2007; De Oliveira-Souza et al., 2008; Grover et al., 2007; Lyons-Ruth et al., 2007). People with antisocial personality disorder often come from homes characterized by emotional deprivation, harsh and inconsistent disciplinary practices, and antisocial parental behavior. Still other studies show a strong interaction between both heredity and environment (Gabbard, 2006; Hudziak, 2008).

The BTK killer FIGURE 13.16

In 2005, Dennis Rader, known as the "BTK killer" for his method of binding, torturing, and killing his victims, pleaded guilty and was sentenced to life in prison for his crimes. Between 1974 and 1991, Rader murdered 10 women in and around Wichita, Kansas. After his killings, he sent taunting letters to police and local media, boasting of the crimes in graphic detail. Despite having a normal outward appearance—he was a husband, father, Cub Scout leader, and church council president—Rader's egocentrism and lack of conscience are primary characteristics of antisocial personality disorder.

Borderline personality disorder

Borderline personality disorder (BPD) is among the most commonly diagnosed personality disorders (Ansell & Grilo, 2007; Bradley, Conklin, & Westen, 2007). The core features of this disorder are impulsivity and instability in mood, relationships, and self-image. Originally, the term implied that the person was on the borderline between neurosis and schizophrenia (Kring et al., 2007). The modern conceptualization no longer has this connotation, but BPD remains one of the most complex and debilitating of all the personality disorders.

Mary's story of chronic, lifelong dysfunction, described in the chapter opener, illustrates the serious problems associated with this disorder. People with borderline personality disorder experience extreme difficulties in relationships. Subject to chronic feelings of depression, emptiness, and intense fear of abandonment, they also engage in destructive, impulsive behaviors, such as sexual promiscuity, drinking, gambling, and eating sprees (Chabrol et al., 2004; Trull et al., 2000). They may attempt suicide and sometimes engage in self-mutilating behavior (Chapman, Leung, & Lynch, 2008; Crowell et al., 2008; Links et al., 2008).

People with BPD tend to see themselves and everyone else in absolute terms—perfect or worthless (Mason & Kreger, 1998). They constantly seek reassurance from others and may quickly erupt in anger at the slightest sign of disapproval. The disorder is also typically marked by a long history of broken friendships, divorces, and lost jobs.

People with borderline personality disorder frequently have a childhood history of neglect; emotional deprivation; and physical, sexual, or emotional abuse (Christopher et al., 2007; Minzenberg, Poole, & Vinogradov, 2008). Borderline personality disorder also tends to run in families, and some data suggest it is a result of impaired functioning of the brain's frontal lobes and limbic system, areas that control impulsive behaviors (Schmahl et al., 2004; Tebartz van Elst et al., 2003).

Although some therapists have had success treating BPD with drug therapy and behavior therapy (Johnson & Murray, 2007; Markovitz, 2004), the general prognosis is not favorable. People with BPD appear to have a deep well of intense loneliness and a chronic fear of abandonment. Sadly, given their troublesome personality traits, friends, lovers, and even family and therapists often do "abandon" them—thus creating a tragic self-fulfilling prophecy.

CONCEPT CHECK STOP

Why does the presence of comorbid disorders complicate the diagnosis and treatment of substance-related disorders?

Which is more important in the development of dissociative disorders, environment or genetics?

How do personality disorders differ from the other psychological disorders discussed in this chapter?

NATIONAL GEOGRAPHIC

How Gender and Culture Affect Abnormal Behavior

LEARNING OBJECTIVES

Identify the biological, psychological, and social factors that might explain gender differences in depression.

Explain why it is difficult to directly compare mental disorders across cultures.

Explain why recognizing the difference between culture-general and culture-bound disorders and symptoms can help prevent ethnocentrism in the diagnosis and treatment of psychological disorders.

Among the Chippewa, Cree, and Montagnais-Naskapi Indians in Canada, there is a disorder called *windigo*—or *wiitiko—psychosis*, which is characterized by delusions and cannibalistic impulses. Believing that they have been possessed by the spirit of a windigo, a cannibal giant with a heart and entrails of ice, victims become severely depressed (Faddiman, 1997). As the malady begins, the individual typically experiences loss of appetite, diarrhea, vomiting, and insomnia, and he or she may see people turning into beavers and other edible animals.

In later stages, the victim becomes obsessed with cannibalistic thoughts and may even attack and kill loved ones in order to devour their flesh (Berreman, 1971).

If you were a therapist, how would you treat this disorder? Does it fit neatly into any of the categories of psychological disorders that you have learned about? We began this chapter discussing the complexities and problems with defining, identifying, and classifying abnormal behavior. Before we close, we need to add two additional confounding factors: gender and culture. In this section, we explore a few of the many ways in which men and women differ in how they experience abnormal behavior. We also look at cultural variations in abnormal behavior.

GENDER AND DEPRESSION: WHY ARE WOMEN MORE DEPRESSED?

In the United States, Canada, and other countries, the rate of severe depression for women is two to three times the rate for men (Barry et al., 2008; Nicholson et al., 2008; Nolen-Hoeksema, Larson, & Grayson, 2000).

Why is there such a disparity between men and women? Research explanations can be grouped under biological influences (hormones, biochemistry, and genetic predisposition), psychological processes (ruminative thought processes), and social factors (greater poverty, work-life conflicts, unhappy marriages, and sexual or physical abuse) (Cooper et al., 2008; Jackson & Williams, 2006; Shear et al., 2007).

According to the *biopsychosocial model*, some women inherit a genetic or hormonal predisposition toward depression. This biological predisposition combines with society's socialization processes to help reinforce behaviors—such as greater emotional expression, passivity, and dependence—that increase the chances for depression (Alloy et al., 1999; Nolen-Hoeksema, Larson, & Grayson, 2000). At the same time, focusing only on classical symptoms of depression (sadness, low energy, and feelings of helplessness) may cause large numbers of depressed men to be overlooked (FIGURE 13.17).

Depression in disguise? FIGURE 13.17

In our society, men are typically socialized to suppress their emotions and to show their distress by acting out (showing aggression), acting impulsively (driving recklessly and committing petty crimes), and engaging in substance abuse. How might such cultural pressures lead us to underestimate male depression?

CULTURE AND SCHIZOPHRENIA: DIFFERENCES AROUND THE WORLD

Peoples of different cultures experience mental disorders in a variety of ways. For example, the reported incidence of schizophrenia varies within different cultures around the world. It is unclear whether these differences result from actual differences in prevalence of the disorder or from differences in definition, diagnosis, or reporting (Hoye et al., 2006).

The symptoms of schizophrenia also vary across cultures (Stompe et al., 2003), as do the particular stresses that may trigger its onset (FIGURE 13.18).

Finally, despite the advanced treatment facilities and methods in industrialized nations, the prognosis for people with schizophrenia is actually better in nonindustrialized societies. This may be because the core symptoms of schizophrenia (poor rapport with others, incoherent speech, etc.) make it more difficult to survive in highly industrialized countries. In addition, in most industrialized nations, families and other support groups are less likely to feel responsible for relatives and friends with schizophrenia (Brislin, 2000; Lefley, 2000).

AVOIDING ETHNOCENTRISM

Most research on psychological disorders originates and is conducted primarily in Western cultures. Such a restricted sampling can limit our understanding of disorders in general and lead to an ethnocentric view of mental disorders.

Fortunately, cross-cultural researchers have devised ways to overcome these difficulties (Matsumoto & Juang, 2008; Triandis, 2007). For example, Robert Nishimoto (1988) has found several **culture-general symptoms** that are useful in diagnosing disorders across cultures (TABLE 13.3).

What is stressful? FIGURE 13.18

A Some stressors are culturally specific, such as feeling possessed by evil forces or being the victim of witchcraft.

B Other stressors are shared by many cultures, such as the unexpected death of a spouse or loss of a job (Al-Issa, 2000; Browne, 2001; Neria et al., 2002; Torrey & Yolken, 1998).

Twelve culture-general symptoms of mental health difficulties TABLE 13.3		
Nervous	Trouble sleeping	Low spirits
Weak all over	Personal worries	Restless
Feel apart, alone	Can't get along	Hot all over
Worry all the time	Can't do anything worthwhile	Nothing turns out right

Source: From *Understanding Culture's Influence on Behavior,* 2nd edition by Brislin. ©2000. Reprinted with permission of Wadsworth, a division of Thomson Learning. www.thomsonrights.com.

In addition, Nishimoto found several **culture-bound symptoms**. For example, the Vietnamese Chinese reported "fullness in head," the Mexican respondents had "problems with [their] memory," and the Anglo-Americans reported "shortness of breath" and "headaches." Apparently, people learn to express their problems in ways that are acceptable to others in the same culture (Brislin, 1997, 2000; Dhikav et al., 2008; Laungani, 2007; Tolin et al., 2007).

This division between culture-general and culture-bound symptoms also helps us understand depression. Certain symptoms of depression (such as intense sadness, poor concentration, and low energy) seem to exist across all cultures (World Health Organization, 2007, 2008). But there is also evidence of some culture-bound symptoms. For example, feelings of guilt are found more often in North America and Europe. And in China, *somatization* (the conversion of depression into bodily complaints) occurs more frequently than it does in other parts of the world (Helms & Cook, 1999).

Just as there are culture-bound and culture-general symptoms, researchers have found that mental disorders are themselves sometimes culturally bound (FIGURE 13.19). The earlier example of windigo psychosis, a disorder limited to a small group of Canadian Indians, illustrates just such a case.

Puerto Rican and other Latin cultures *Ataque de nervios* ("attack of nerves")	**Southeast Asian, Malaysian, Indonesian, Thai** Running amok	**West African** Brain fog	**Ethiopia** Possession by the "Zar"	**South Chinese and Vietnamese** *Koro*	**Western Nations** *Anorexia nervosa* (as other countries become Westernized they're showing some rare cases of anorexia)
Symptoms: trembling, heart palpitations, and seizurelike episodes often associated with the death of a loved one, accidents, or family conflict	**Symptoms:** wild, out-of-control, aggressive behaviors and attempts to injure or kill others	**Symptoms:** "brain tiredness," a mental and physical response to the challenges of schooling	**Symptoms:** involuntary movements, mutism, and incomprehensible language	**Symptoms:** belief that the penis is retracting into the abdomen and that when it is fully retracted, death will result; attempts to prevent the supposed retraction may lead to severe physical damage	**Symptoms:** occurs primarily among young women; preoccupied with thinness, they exercise excessively and refuse to eat; death can result

Culture-bound disorders FIGURE 13.19

Some disorders are fading as remote areas become more Westernized, whereas other disorders (such as anorexia nervosa) are spreading as other countries adopt Western values.

Source: Barlow & Durand, 2009; Dhikav et al., 2008; Gaw, 2001; Kring et al., 2007; Laungani, 2007; Sue & Sue, 2008; Tolin et al., 2007.

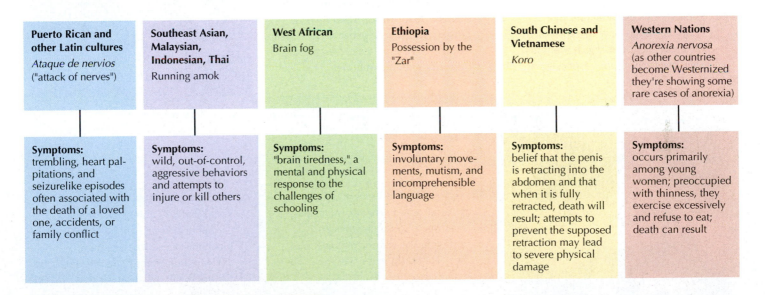

As you can see, culture has a strong effect on mental disorders. Studying the similarities and differences across cultures can lead to better diagnosis and understanding. It also helps mental health professionals who work with culturally diverse populations understand both culturally general and culturally bound symptoms.

CONCEPT CHECK STOP

Why might depression be frequently overlooked in men?

How does schizophrenia differ from one culture to another?

What are some examples of culture-bound disorders?

SUMMARY

1 Studying Psychological Disorders

1. Criteria for **abnormal behavior** include statistical infrequency, disability or dysfunction, personal distress, and violation of norms. None of these criteria alone is adequate for classifying abnormal behavior.
2. Superstitious explanations for abnormal behavior were replaced by the **medical model**, which eventually gave rise to the modern specialty of **psychiatry**. In contrast to the medical model, psychology offers a multifaceted approach to explaining abnormal behavior.
3. The ***Diagnostic and Statistical Manual of Mental Disorders***, fourth edition, text revision (***DSM-IV-TR***) is organized according to five major axes, which serve as guidelines for making decisions about symptoms.

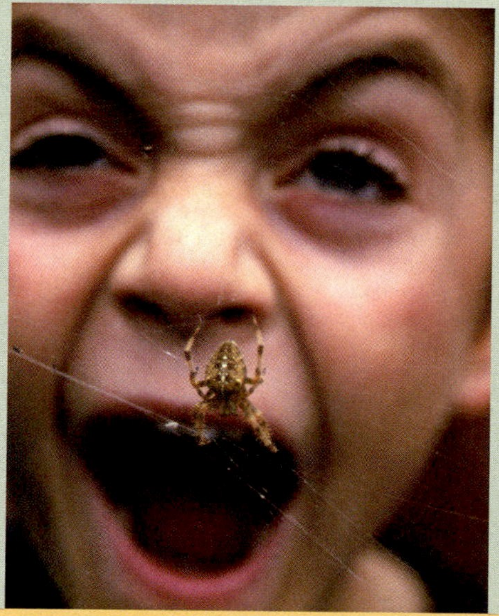

2 Anxiety Disorders

1. Major **anxiety disorders** include generalized anxiety disorder, panic disorder, phobias (including agoraphobia, specific phobias, and social phobias), and obsessive-compulsive disorder (OCD).
2. Psychological (faulty cognitions and maladaptive learning), biological (evolutionary and genetic predispositions, biochemical disturbances), and sociocultural factors (cultural pressures toward hypervigilance) likely all contribute to anxiety.

3 Mood Disorders

1. Mood disorders are characterized by extreme disturbances in emotional states. People suffering from **major depressive disorder** may experience a lasting depressed mood without a clear trigger. In contrast, people with **bipolar disorder** alternate between periods of depression and mania (hyperactivity and poor judgment).

2. Biological factors play a significant role in mood disorders. Psychosocial theories of depression focus on environmental stressors and disturbances in interpersonal relationships, thought processes, self-concept, and learning history (including learned helplessness).

4 Schizophrenia

1. **Schizophrenia** is a group of disorders, each characterized by a disturbance in perception (including **hallucinations**), language, thought (including **delusions**), emotions, and/or behavior.

2. In the past, researchers divided schizophrenia into multiple subtypes. More recently, researchers have proposed focusing instead on positive versus negative symptoms. The latest *DSM-IV-TR* also suggests adding another dimension to reflect disorganization of behavior.

3. Most biological theories of schizophrenia focus on genetics, neurotransmitters, and brain abnormalities. Psychologists believe that there are also at least two possible psychosocial contributors: stress and communication disorders in families.

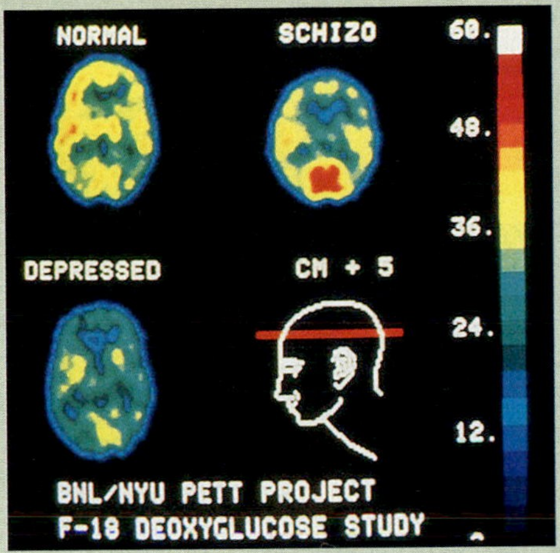

5 Other Disorders

1. **Substance-related disorders** fall into two general groups: substance abuse and substance dependence. Substance-related disorders commonly coexist with other mental disorders (comorbidity), which complicates their diagnosis and treatment.

2. **Dissociative disorders** include dissociative amnesia, dissociative fugue, depersonalization disorder, and dissociative identity disorder (DID). Environmental variables appear to be the primary cause of dissociative disorders.

3. **Personality disorders** occur when inflexible, maladaptive personality traits cause significant impairment of social and occupational functioning. The best-known type of personality disorder is antisocial personality disorder, characterized by egocentrism, lack of conscience, impulsive behavior, and superficial charm. The most common personality disorder is borderline personality disorder (BPD). Its core features are impulsivity and instability in mood, relationships, and self-image. Although some therapists have success with drug therapy and behavior therapy, prognosis is not favorable.

6 How Gender and Culture Affect Abnormal Behavior

1. Men and women differ in how they experience and express abnormal behavior. For example, in North America severe depression is much more common in women than in men. Biological, psychological, and social factors probably combine to explain this phenomenon.

2. Peoples of different cultures experience mental disorders in a variety of ways. For example, the reported incidence of schizophrenia varies within different cultures around the world, as do the disorder's symptoms, triggers, and prognosis.

3. Some symptoms of psychological disorders, as well as some disorders themselves, are culture-general, whereas others are culture-bound.

KEY TERMS

- **abnormal behavior** p. 344
- **medical model** p. 344
- **psychiatry** p. 344
- ***Diagnostic and Statistical Manual of Mental Disorders (DSM-IV-TR)*** p. 347
- **neurosis** p. 347
- **psychosis** p. 347
- **insanity** p. 347
- **state disorders** p. 348
- **trait disorders** p. 348
- **anxiety disorder** p. 350
- **generalized anxiety disorder (GAD)** p. 350
- **panic disorder** p. 351
- **phobias** p. 351
- **obsessive-compulsive disorder (OCD)** p. 351
- **mood disorders** p. 354
- **major depressive disorder** p. 354
- **bipolar disorder** p. 354
- **learned helplessness** p. 355
- **schizophrenia** p. 356
- **hallucinations** p. 356
- **delusions** p. 357
- **positive schizophrenia symptoms** p. 358
- **negative schizophrenia symptoms** p. 358
- **dopamine hypothesis** p. 359
- **diathesis-stress model** p. 360
- **substance-related disorders** p. 361
- **comorbidity** p. 361
- **self-medication** p. 362
- **dissociative disorders** p. 362
- **dissociative identity disorder (DID)** p. 362
- **personality disorders** p. 362
- **antisocial personality disorder** p. 363
- **borderline personality disorder (BPD)** p. 363
- **culture-general symptoms** p. 366
- **culture-bound symptoms** p. 367

CRITICAL AND CREATIVE THINKING QUESTIONS

1. Can you think of cases where each of the four criteria for abnormal behavior might not be suitable for classifying a person's behavior as abnormal?
2. Do you think the insanity plea, as it is currently structured, should be abolished? Why or why not?
3. Why do you suppose that anxiety disorders are among the easiest to treat?
4. Have you ever felt depressed? How would you distinguish between "normal" depression and a major depressive disorder?
5. Culture clearly has strong effects on mental disorders. How does that influence how you think about what is normal or abnormal?
6. Most of the disorders discussed in this chapter have some evidence for a genetic predisposition. What would you tell a friend who has a family member with one of these disorders and fears that he or she might develop the same disorder?

What is happening in this picture ?

- Is this man's behavior abnormal? Which criteria for abnormal behavior do his piercings and tattoos meet? Which do they not?
- Can you think of any behavior *you* exhibit that might be considered abnormal if your own cultural norms were not taken into account?

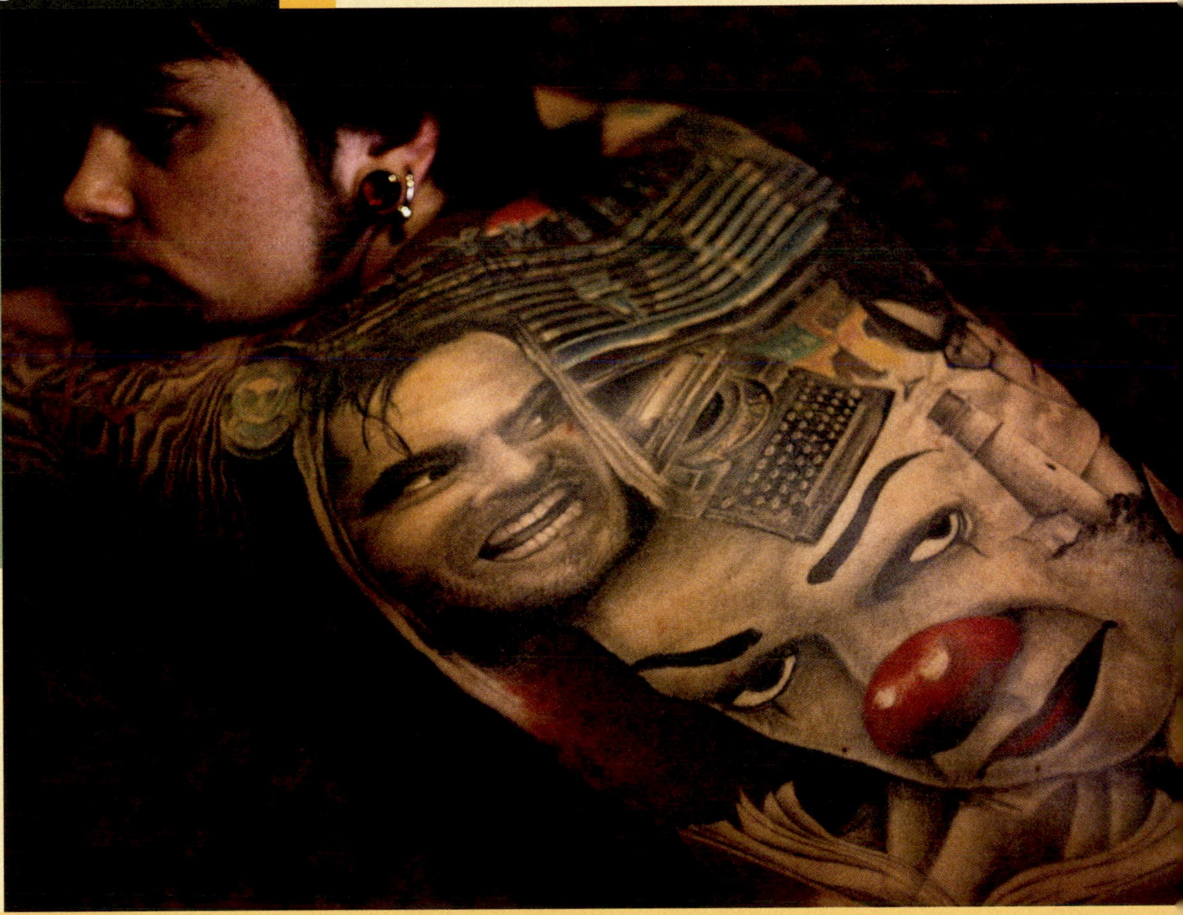

SELF-TEST

(Check answers in Appendix B.)

1. Your textbook defines *abnormal behavior* as ______________.
   a. a statistically infrequent pattern of pathological emotion, thought, or action
   b. patterns of emotion, thought, and action that are considered pathological
   c. a pattern of pathological emotion, thought, or action that causes personal distress or violates social norms
   d. all of these options

2. ______________ is the branch of medicine that deals with the diagnosis, treatment, and prevention of mental disorders.
   a. Psychology
   b. Psychiatry
   c. Psychobiology
   d. Psychodiagnostics

3. Label the five axes of the *Diagnostic and Statistical Manual of Mental Disorders (DSM-IV-TR)* on the figure below.

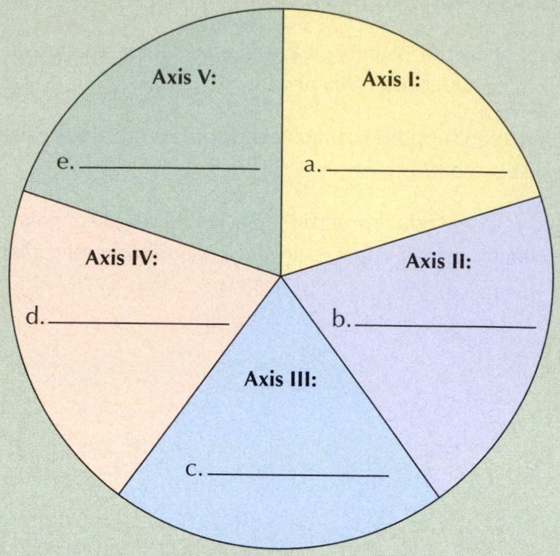

4. Anxiety disorders are ______________.
   a. characterized by unrealistic, irrational fear
   b. the least frequent of the mental disorders
   c. twice as common in men as in women
   d. all of these options

5. Label the five major anxiety disorders on the figure below.

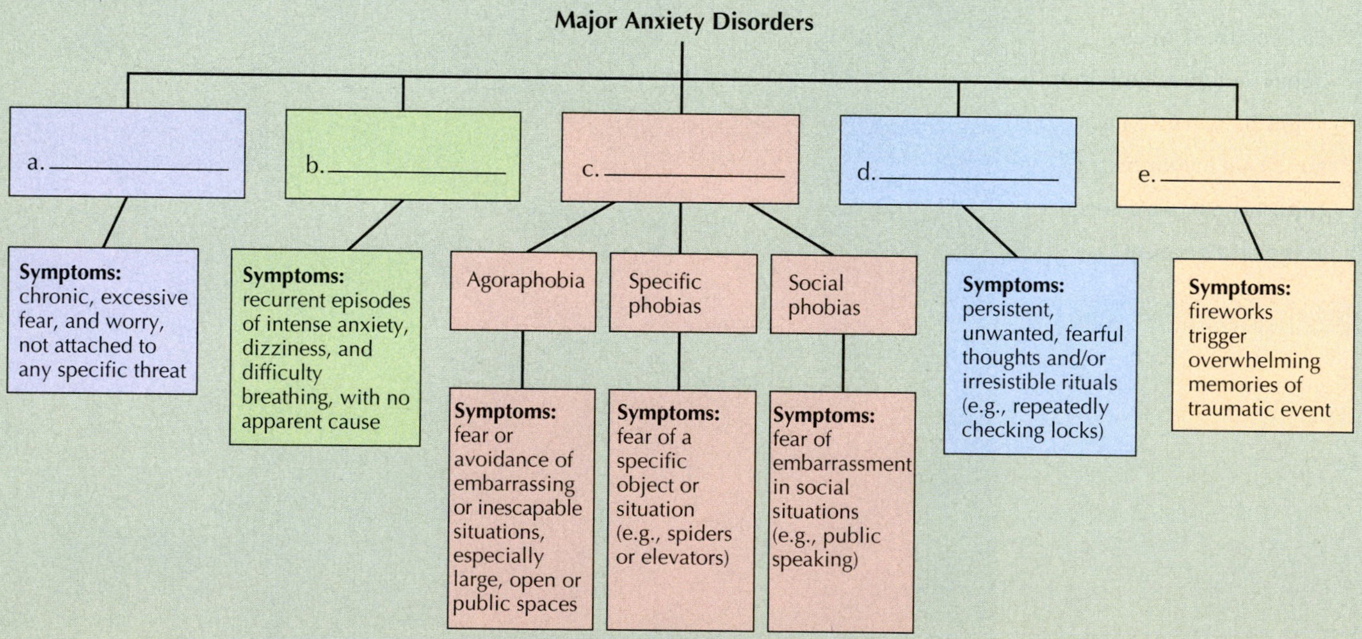

6. The two main types of mood disorders are ______________.
   a. major depressive disorder and bipolar disorder
   b. mania and depression
   c. SAD and MAD
   d. learned helplessness and suicide

7. Someone who experiences repeated episodes of mania or cycles between mania and depression has a ______________.
   a. disruption of circadian rhythms
   b. bipolar disorder
   c. manic-depressive syndrome
   d. cyclothymia disorder

8. According to the theory known as ______________, when faced with a painful situation from which there is no escape, animals and people enter a state of helplessness and resignation.
   a. autonomic resignation
   b. helpless resignation
   c. resigned helplessness
   d. learned helplessness

9. A psychotic disorder that is characterized by major disturbances in perception, language, thought, emotion, and behavior is ______________.
   a. schizophrenia
   b. multiple personality disorder
   c. borderline psychosis
   d. neurotic psychosis

10. Perceptions for which there are no appropriate external stimuli are called ______________, and the most common type among people suffering from schizophrenia is ______________.
    a. hallucinations; auditory
    b. hallucinations; visual
    c. delusions; auditory
    d. delusions; visual

11. Label the five subtypes of schizophrenia on the table below.

Subtypes of schizophrenia

______________	Dominated by delusions (persecution and grandeur) and hallucinations (hearing voices)
______________	Marked by motor disturbances (immobility or wild activity) and echo speech (repeating the speech of others)
______________	Characterized by incoherent speech, flat or exaggerated emotions, and social withdrawal
______________	Meets the criteria for schizophrenia but is not any of the above subtypes
______________	No longer meets the full criteria for schizophrenia but still shows some symptoms

12. Failure to meet obligations may be indicative of alcohol or drug ______________ whereas tolerance and withdrawal may be indicative of alcohol or drug ______________.
    a. use; abuse
    b. abuse; dependence
    c. abuse; abuse
    d. dependence; abuse

13. The disorder that is an attempt to avoid painful memories or situations and is characterized by amnesia, fugue, or multiple personalities is ______________.
    a. dissociative disorder
    b. displacement disorder
    c. disoriented disorder
    d. identity disorder

14. Inflexible, maladaptive personality traits that cause significant impairment of social and occupational functioning is known as ______________.
    a. nearly all mental disorders
    b. the psychotic and dissociative disorders
    c. personality disorders
    d. none of these options

15. Which of the following are examples of culture-general symptoms of mental health difficulties, useful in diagnosing disorders across cultures?
    a. trouble sleeping
    b. can't get along
    c. worry all the time
    d. all of the above

Therapy 14

Since the beginning of the movie age, mentally ill people and their treatment have been the subject of some of Hollywood's most popular and influential films. But consider how people with mental illness are generally portrayed in the movies. They are either cruel, sociopathic criminals (Anthony Hopkins in *Silence of the Lambs*) or helpless, innocent victims (Jack Nicholson in *One Flew Over the Cuckoo's Nest* and Angelina Jolie in *Changeling*). Likewise, popular films about mental illness often feature mad doctors, heartless nurses, and brutal treatment methods. Although these portrayals may boost movie ticket sales, they also perpetuate harmful stereotypes.

In this chapter, we offer a balanced, factual presentation of the latest research on psychotherapy and mental illness. As you will see, modern psychotherapy can be very effective and prevent much needless suffering, not only for people with psychological disorders but also for those seeking help with everyday problems in living.

In addition to psychologists, professionals involved in psychotherapy include psychiatrists, psychiatric nurses, social workers, counselors, and clergy with training in pastoral counseling. There are numerous forms of psychotherapy. According to one expert (Kazdin, 1994), there may be over 400 approaches to treatment. To organize our discussion, we have grouped treatments into three categories: insight therapies, behavior therapies, and biomedical therapies. After exploring these approaches, we conclude with a discussion of issues that are common to all major forms of psychotherapy.

CHAPTER OUTLINE

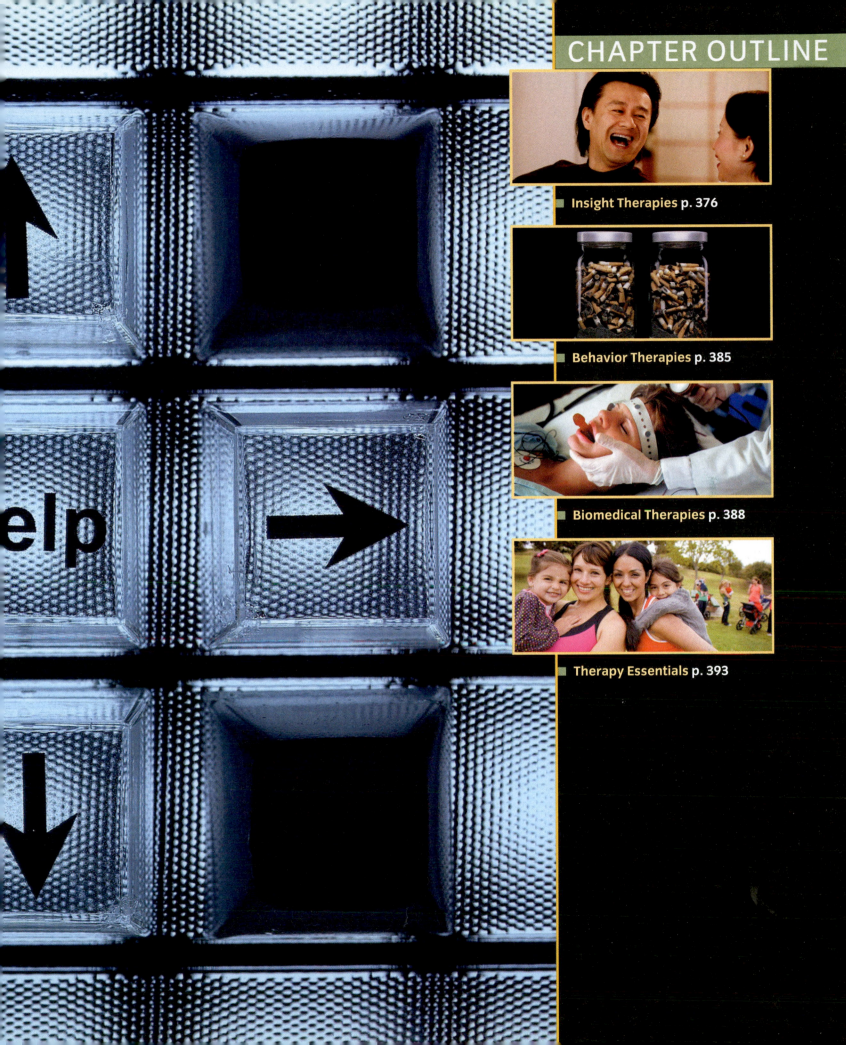

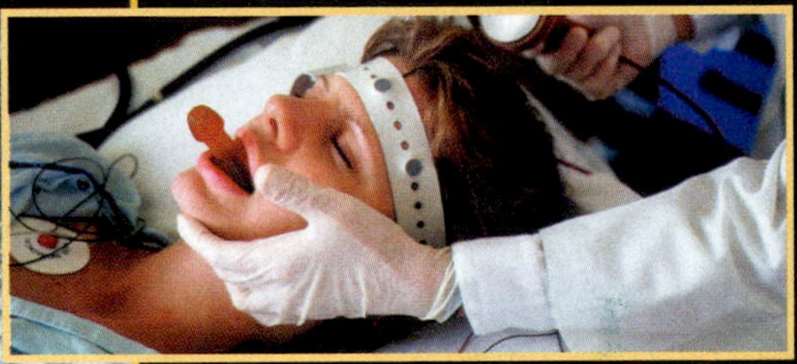

Insight Therapies

LEARNING OBJECTIVES

Describe the core treatment techniques in psychoanalysis and modern psychodynamic treatments.

Explain the principles underlying cognitive therapies.

Summarize the four key qualities of communication in Rogerian therapy.

Identify several benefits of group therapy.

We begin our discussion of professional **psychotherapy** with traditional psychoanalysis and its modern counterpart, psychodynamic therapy. Then we explore cognitive, humanistic, group, and family therapies. Although these therapies differ significantly, they're often grouped together as **insight therapies** because they seek to increase insight into clients' difficulties. The general goal is to help people gain greater control over and improvement in their thoughts, feelings, and behaviors. (See STUDY ORGANIZER 14.1.)

psychotherapy Techniques employed to improve psychological functioning and promote adjustment to life.

PSYCHOANALYSIS/PSYCHODYNAMIC THERAPIES: UNLOCKING THE SECRETS OF THE UNCONSCIOUS

psychoanalysis Freudian therapy designed to bring unconscious conflicts, which usually date back to early childhood experiences, into consciousness. Also, Freud's theoretical school of thought, which emphasizes unconscious processes.

In **psychoanalysis**, a person's *psyche* (or mind) is analyzed. Traditional psychoanalysis is based on Sigmund Freud's central belief that abnormal behavior is caused by unconscious conflicts among the three parts of the psyche—the id, ego, and superego (Chapter 12).

During psychoanalysis, these conflicts are brought to consciousness. The patient comes to understand the reasons for his or her behavior and realizes that the childhood conditions under which the conflicts developed no longer exist. Once this realization (or insight) occurs, the conflicts can be resolved and the patient can develop more adaptive behavior patterns.

How can gaining insight into one's unconscious change behavior? Freud explained that becoming aware of previously hidden conflicts permits a release of tension and anxiety called **catharsis**. He observed that when his patients relived a traumatic incident, the conflict seemed to lose its power to control the person's behavior.

According to Freud, the ego has strong defense mechanisms that block unconscious thoughts from coming to light. Thus, to gain insight into the unconscious, the ego must be "tricked" into relaxing its guard. With that goal, psychoanalysts employ five major methods: free association, dream analysis, analyzing resistance, analyzing transference, and interpretation.

Free association

According to Freud, when you let your mind wander and remove conscious censorship over thoughts—a process called **free association**—interesting and even bizarre connections seem to spring into awareness. Freud believed that the first thing to come to a patient's mind is often an important clue to what the person's unconscious wants to conceal. Having the patient recline on a couch, with only the ceiling to look at, is believed to encourage free association.

Dream analysis

According to Freud, defenses are lowered during sleep, and forbidden desires and unconscious conflicts can be freely expressed. Even while dreaming, however, these feelings and conflicts are recognized as being unacceptable and must be disguised as images that have deeper symbolic meaning. Thus, according to Freudian dream theory, a therapist might interpret a dream of riding a horse or driving a car (the **manifest content**) as a desire for, or concern about, sexual intercourse (the **latent content**).

Study Organizer 14.1 An overview of the three approaches to therapy

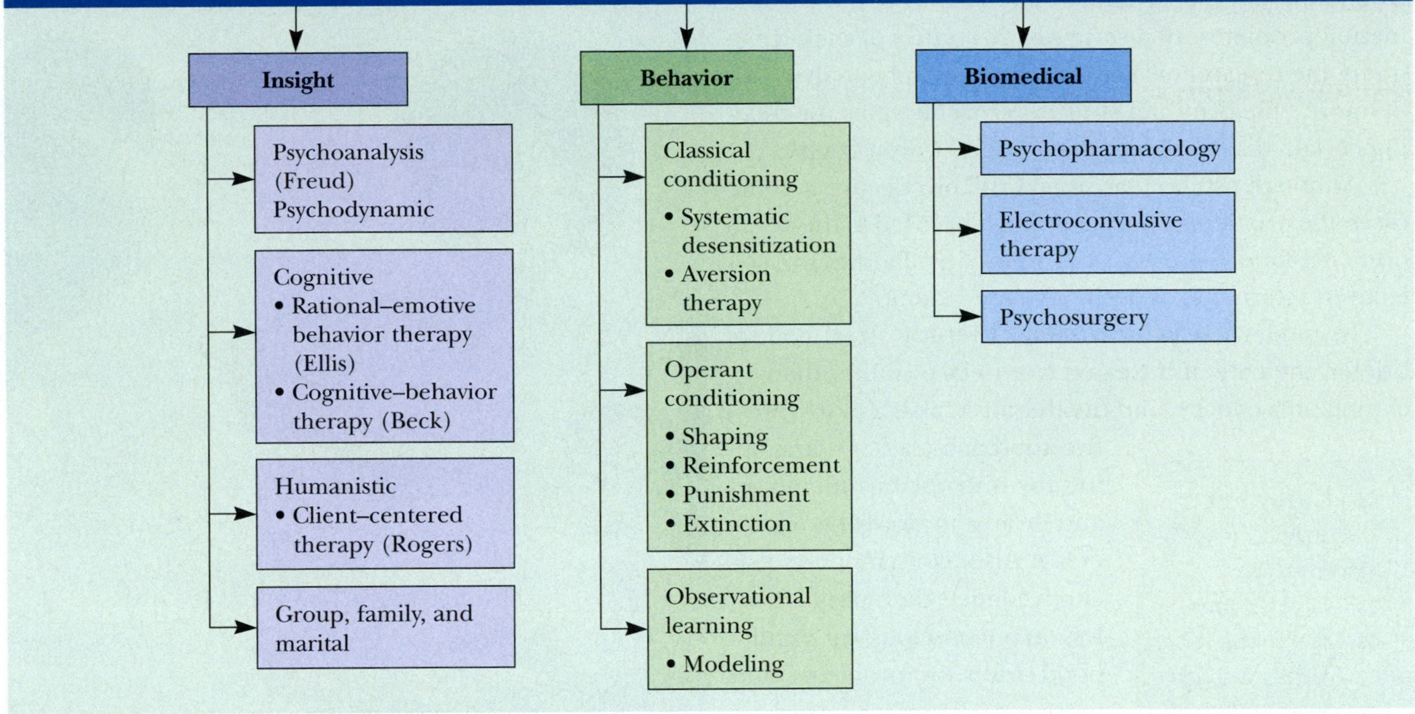

Analyzing resistance

During free association or dream analysis, Freud found that patients often show **resistance**—for example, suddenly "forgetting" what they were saying or completely changing the subject. It is the therapist's job to identify these cases of resistance, help patients face their problems, and help patients learn to deal with them realistically.

Analyzing transference

During psychoanalysis, patients disclose intimate feelings and memories, and the relationship between the therapist and patient may become complex and emotionally charged. As a result, patients often apply (or transfer) some of their unresolved emotions and attitudes from past relationships onto the therapist. The therapist uses this process of **transference** to help the patient "relive" painful past relationships in a safe, therapeutic setting so that he or she can move on to healthier relationships.

Interpretation

The core of all psychoanalytic therapy is **interpretation**. During free association, dream analysis, resistance, and transference, the analyst listens closely and tries to find patterns and hidden conflicts. At the right time, the therapist explains (or interprets) the underlying meanings to the client.

As you can see, most of psychoanalysis rests on the assumption that repressed memories and unconscious conflicts actually exist. But, as noted in Chapters 7 and 12, this assumption is questioned by modern scientists and has become the subject of a heated, ongoing debate. In addition to questioning the validity of repressed memories, critics also point to two other problems with psychoanalysis:

- *Limited applicability.* Critics argue that psychoanalysis seems to suit only a select group of highly motivated, articulate individuals with less severe disorders. Psychoanalysis is also time consuming (often lasting several years with four to five sessions a week) and expensive, and it seldom works well with severe mental disorders, such as schizophrenia, in which verbalization and rationality are significantly disrupted. Finally, critics suggest that spending years chasing unconscious conflicts from the past allows patients to escape from the responsibilities and problems of adult life.
- *Lack of scientific credibility.* Another problem with psychoanalysis is that its "insights" (and therefore its success) cannot be proved or disproved.

Despite these criticisms, there is some evidence that psychoanalysis can be effective in treating some chronic mental problems. In a review of 23 studies of such treatment, the researchers concluded that psychoanalysis can be more effective than shorter-term therapies for treating certain disorders, notably anxiety (Carey, 2008).

Although psychoanalysis may still be effective in some cases, the problems associated with it have led to the development of more streamlined forms of psychotherapy, collectively referred to as psychodynamic therapy.

In modern **psychodynamic therapy**, treatment is briefer, the patient is treated face to face (rather than reclining on a couch), and the therapist takes a more directive approach (rather than waiting for unconscious memories and desires to slowly be uncovered). Also, contemporary psychodynamic therapists focus less on unconscious, early childhood roots of problems and more on conscious processes and current problems. Such refinements have helped make psychoanalysis more available and more effective for an increasing number of people (Knekt et al., 2008; Lehto et al., 2008; Lerner, 2008) (FIGURE 14.1).

psychodynamic therapy A briefer, more directive contemporary form of psychoanalysis, which emphasizes conscious processes and current problems.

Interpersonal therapy (IPT) FIGURE 14.1

Interpersonal therapy (IPT) is an influential, brief form of psychodynamic therapy. As the name implies, interpersonal therapy focuses almost exclusively on the client's current relationships. Its goal is to relieve immediate symptoms and to help the client learn better ways to solve future interpersonal problems. Why do you think many patients might prefer psychodynamic therapy over psychoanalysis?

COGNITIVE THERAPIES: A FOCUS ON FAULTY THOUGHTS AND BELIEFS

Cognitive therapy assumes that faulty thought processes—beliefs that are irrational, overly demanding, or that fail to match reality—create problem behaviors and emotions (Barlow, 2008; Corey, 2009; Davies, 2008; Ellis, 1996, 2003b, 2004; Kellogg & Young, 2008).

cognitive therapy Therapy that focuses on changing faulty thought processes and beliefs to treat problem behaviors.

Like psychoanalysts, cognitive therapists believe that exploring unexamined beliefs can produce insight into the reasons for disturbed behaviors. However, instead of believing that a change in behavior occurs because of insight and catharsis, cognitive therapists believe that insight into negative **self-talk** (the unrealistic things a person tells himself or herself) is most important. Through a process called **cognitive restructuring**, this insight allows clients to challenge their thoughts, change how they interpret events, and modify maladaptive behaviors (FIGURE 14.2).

In addition, whereas psychoanalysts focus primarily on childhood family relationships, cognitive therapists assume that a broad range of events and people—both inside and outside the family—influence beliefs.

One of the best-known cognitive therapists is the late Albert Ellis, who developed an approach known as **rational-emotive therapy (RET)** (1961, 2003a, 2003b, 2004). Ellis

Using cognitive therapy to improve sales FIGURE 14.2

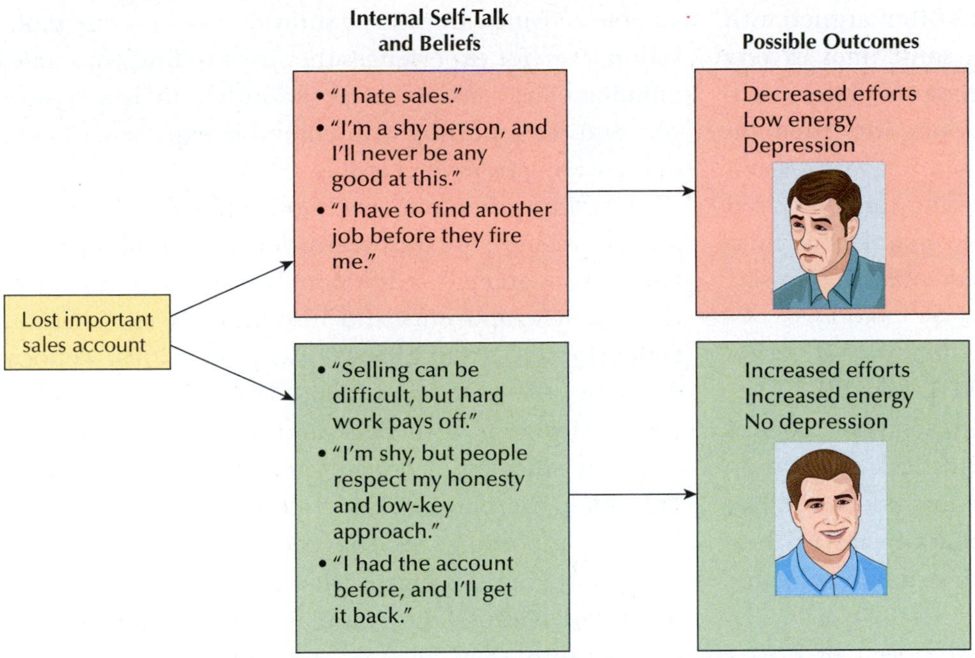

Internal Self-Talk and Beliefs
Possible Outcomes
Lost important sales account
• "I hate sales."
• "I'm a shy person, and I'll never be any good at this."
• "I have to find another job before they fire me."
Decreased efforts
Low energy
Depression
• "Selling can be difficult, but hard work pays off."
• "I'm shy, but people respect my honesty and low-key approach."
• "I had the account before, and I'll get it back."
Increased efforts
Increased energy
No depression

A Note how the negative interpretation and destructive self-talk lead to destructive and self-defeating outcomes.

B Cognitive therapy teaches clients to challenge and change their negative beliefs and negative self-talk into positive ones, which, in turn, leads to more positive outcomes. Can you think of other situations in which such reinterpretation could be helpful?

called RET an A–B–C–D approach, referring to the four steps involved in creating and dealing with maladaptive thinking: an **a**ctivating event, the person's **b**elief system, the emotional and behavioral **c**onsequences that the person experiences, and **d**isputing (or challenging) erroneous beliefs (FIGURE 14.3).

According to Ellis, when people demand certain "musts" ("I must get into graduate school") and "shoulds" ("He should love me") from themselves and others, they create emotional distress and behavioral dysfunction (David, Schnur, & Belloiu, 2002; Ellis, 1997, 2003a, 2003b, 2004).

The development and treatment of irrational misconceptions FIGURE 14.3

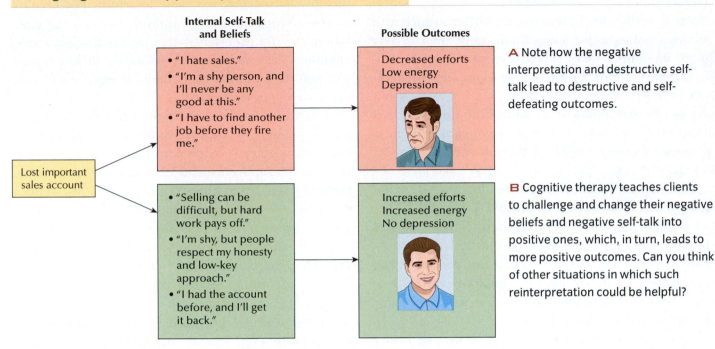

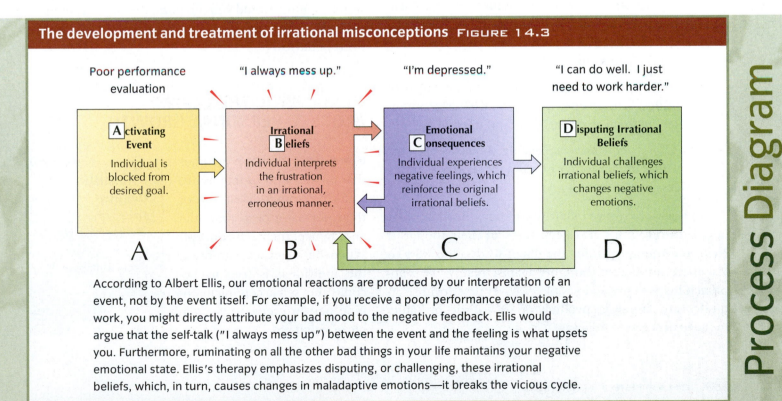

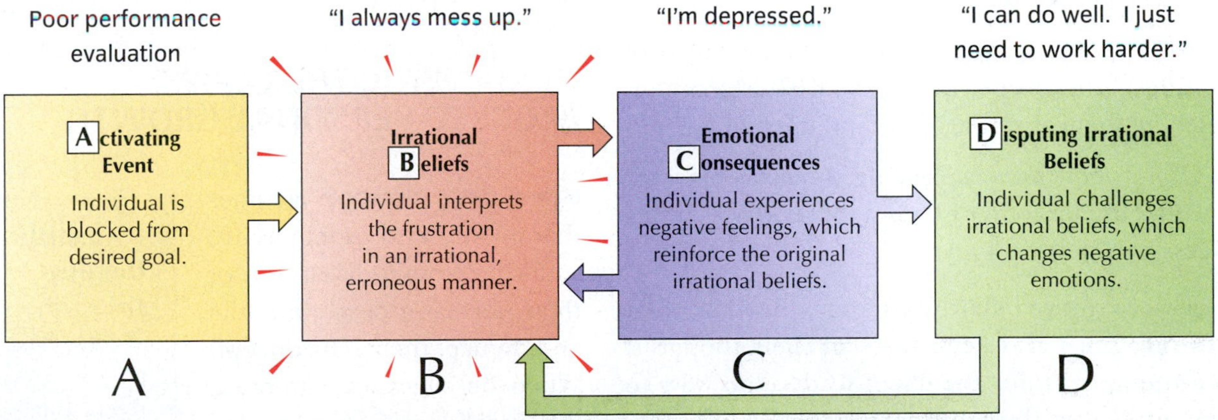

Poor performance evaluation
"I always mess up."
"I'm depressed."
"I can do well. I just need to work harder."
Activating Event
Individual is blocked from desired goal.
Irrational Beliefs
Individual interprets the frustration in an irrational, erroneous manner.
Emotional Consequences
Individual experiences negative feelings, which reinforce the original irrational beliefs.
Disputing Irrational Beliefs
Individual challenges irrational beliefs, which changes negative emotions.
A
B
C
D

According to Albert Ellis, our emotional reactions are produced by our interpretation of an event, not by the event itself. For example, if you receive a poor performance evaluation at work, you might directly attribute your bad mood to the negative feedback. Ellis would argue that the self-talk ("I always mess up") between the event and the feeling is what upsets you. Furthermore, ruminating on all the other bad things in your life maintains your negative emotional state. Ellis's therapy emphasizes disputing, or challenging, these irrational beliefs, which, in turn, causes changes in maladaptive emotions—it breaks the vicious cycle.

Process Diagram

According to Ellis such unrealistic, unproductive self-talk generally goes unexamined unless the client is confronted directly. In therapy, Ellis often argued with clients, cajoling and teasing them, sometimes in very blunt language. Once clients recognize their self-defeating thoughts, Ellis begins working with them on how to behave differently—to test out new beliefs and to learn better coping skills. Reflecting this increased attention to behavioral change, he renamed his therapy **rational-emotive behavior therapy (REBT)** (Crosby, 2003).

Another well-known cognitive therapist is Aaron Beck (1976, 2000; Beck & Grant, 2008). Like Ellis, Beck believes that psychological problems result from illogical thinking and destructive self-talk. But Beck seeks to directly confront and change the behaviors associated with destructive cognitions. Beck's **cognitive-behavior therapy** is designed to reduce *both* self-destructive thoughts *and* self-destructive behaviors.

One of the most successful applications of Beck's theory has been in the treatment of depression. Beck has identified several thinking patterns that he believes are associated with depression-prone people:

- *Selective perception.* Focusing selectively on negative events while ignoring positive events.
- *Overgeneralization.* Overgeneralizing and drawing negative conclusions about one's own self-worth—for example, believing that you are completely worthless because you lost a promotion or failed an exam.
- *Magnification.* Exaggerating the importance of undesirable events or personal shortcomings, seeing them as catastrophic and unchangeable.
- *All-or-nothing thinking.* Seeing things as black or white: totally good or bad, right or wrong, a success or a failure.

In Beck's cognitive-behavior therapy, clients are first taught to recognize and keep track of their thoughts. Next, the therapist trains the client to develop ways to test these automatic thoughts against reality. This approach helps depressed people discover that negative attitudes are largely a product of unrealistic or faulty thought processes (FIGURE 14.4).

At this point, Beck introduces the second phase of therapy—persuading the client to actively pursue pleasurable activities. Depressed individuals often lose motivation, even for experiences they used to find enjoyable. Simultaneously taking an active rather than a passive role and reconnecting with enjoyable experiences help in recovering from depression.

Cognitive therapies are highly effective treatments for depression, anxiety disorders, bulimia nervosa, anger management, addiction, and even some symptoms of schizophrenia and insomnia (Beck & Grant, 2008; Dobson, 2008; Ellis, 2003a, 2003b, 2004; Kellogg & Young, 2008; Neenan, 2008; Palmer & Gyllensten, 2008). However, both Beck and Ellis have been criticized for ignoring or denying the client's unconscious dynamics, overemphasizing rationality, and minimizing the importance of the client's past (Hammack, 2003). Other critics suggest that cognitive therapies are successful because they employ behavior techniques, not because they change the underlying cognitive structure (Bandura, 1969, 1997, 2006, 2008; Laidlaw & Thompson, 2008; Wright & Beck, 1999). Imagine that you sought treatment for depression and learned to construe events more positively and to curb your all-or-nothing thinking. Imagine that your therapist also helped you identify activities and behaviors that would promote greater fulfillment. If you found your depression lessening, would you attribute the improvement to your changing thought patterns or to changes in your overt behavior?

HUMANISTIC THERAPIES: BLOCKED PERSONAL GROWTH

Humanistic therapy assumes that people with problems are suffering from a disruption of their normal growth potential and, hence, their self-concept. When obstacles are removed, the individual is free to become the self-accepting, genuine person everyone is capable of being.

humanistic therapy Therapy that seeks to maximize personal growth through affective restructuring (emotional readjustment).

Tracking faulty thoughts FIGURE 14.4

In cognitive-behavior therapy, clients often record their thoughts in "thought journals" so that, together with the therapist, they can compare their thoughts with reality, detecting and correcting faulty thinking.

Name: Jordan Holley Date: November 11

DYSFUNCTIONAL THOUGHT RECORD

Directions: When you notice your mood getting worse, ask yourself, **"What's going through my mind right now?"** and as soon as possible jot down the thought or mental image in the Automatic Thought Column.

DATE/ TIME	SITUATION	AUTOMATIC THOUGHT(S)	EMOTION(S)	ALTERNATIVE RESPONSE	OUTCOME
	1. What actual event or stream of thoughts, or daydreams, or recollection led to the unpleasant emotion? 2. What (if any) distressing physical sensations did you have?	1. What thought(s) and/or image(s) went through your mind? 2. How much did you believe each one at the time?	1. What emotion(s) (sad, anxious, angry, etc.) did you feel at the time? 2. How intense (0–100%) was the emotion?	1. (optional) What cognitive distortion did you make (e.g., all-or-nothing thinking, mind-reading, catastrophizing)? 2. Use questions at the bottom to compose a response to the automatic thought(s). 3. How much do you believe each response?	1. How much do you now believe each automatic thought? 2. What emotion(s) do you feel now? How intense (0–100%) is the emotion? 3. What will you do? (or did you do?)
Nov. 10, 9 P.M.	My mom called last night. When I saw her number on the caller I.D., I felt my jaw clench and my heart rate go up.	She's going to nag me again to spend the whole Thanksgiving weekend there to make me feel guilty. I was 90% this was it, so I didn't pick up.	I felt angry and frustrated. Intensity = about 80%	Congnitive distortion was mind-reading or maybe all-or-nothing thinking. The evidence that the automatic thoughts were true is from my past talks with my mom. Maybe she was calling for another reason, not to make me feel guilty about Thanksgiving. The worst that could happen is that she'd make me feel guilty again... If so, I could refuse to feel that way. I could also say I didn't like the repeated pressure – it makes me guilty and sad. Maybe she'd understand and we'd break this pattern, or she'd stop nagging me so much. Now, I feel angry whenever she calls and often don't pick up the phone. If I changed my thinking, maybe I'll feel better about her calls, at least sometimes. So, I should work on not expecting the worst when she calls.	1. 50% 2. Now I feel more optimistic, anger is decreased to about 20% 3. Next time I will remind myself of the alternative response before picking up the phone.

Questions to help compose an alternative response:

1. What is the evidence that the automatic thought is true? Not true?
2. Is there an alternative explanation?
3. What's the worst that could happen? If it did happen, how could I cope? What's the best that could happen? What's the most realistic outcome?
4. What's the effect of my believing the automatic thought? What could be the effect of changing my thinking?
5. What should I do about it?
6. If ________ (friend's name) was in this situation and had this thought, what would I tell him/her?

One of the best-known humanistic therapists is Carl Rogers (Rogers, 1961, 1980), who developed an approach that encourages people to actualize their potential and to relate to others in genuine ways. His approach is referred to as **client-centered therapy** (Rogers used the term *client* because he believed the label *patient* implied that one was sick or mentally ill, rather than responsible and competent) (FIGURE 14.5).

Client-centered therapy, like psychoanalysis and cognitive therapies, explores thoughts and feelings as a way to obtain insight into the causes for behaviors. For Rogerian therapists, however, the focus is on providing an accepting atmosphere and encouraging healthy emotional experiences. Clients are responsible for discovering their own maladaptive patterns.

Rogerian therapists create a therapeutic relationship by focusing on four important qualities of communication: empathy, unconditional positive regard, genuineness, and active listening.

Nurturing growth FIGURE 14.5

Imagine how you feel when you are with someone who believes that you are a good person with unlimited potential, a person who believes that your "real self" is unique and valuable. These are the feelings that are nurtured in humanistic therapy.

Applying Psychology

Client-Centered Therapy in Action

This is an excerpt from an actual session. As you can see, humor and informality can be an important part of the therapeutic process.

THERAPIST (TH): What has it been like coming down to the emergency room today?

CLIENT (CL): Unsettling, to say the least. I feel very awkward here, sort of like I'm vulnerable. To be honest, I've had some horrible experiences with doctors. I don't like them.

TH: I see. Well, they scare the hell out of me, too (smiles, indicating the humor in his comment).

CL: (Chuckles) I thought you were a doctor.

TH: I am (pauses, smiles)—that's what's so scary.

CL: (Smiles and laughs)

TH: Tell me a little more about some of your unpleasant experiences with doctors, because I want to make sure I'm not doing anything that is upsetting to you. I don't want that to happen.

CL: Well, that's very nice to hear. My last doctor didn't give a hoot about what I said, and he only spoke in huge words (Shea, 1988, pp. 32–33).

Here are two interesting questions: (CRITICAL THINKING)
1. What techniques of client-centered therapy are being used in this excerpt?
2. If you were the client in this case, would you find this exchange helpful?

Empathy Therapists pay attention to body language and listen for subtle cues to help them understand the emotional experiences of clients. To help clients explore their feelings, the therapist uses open-ended statements such as "You found that upsetting" or "You haven't been able to decide what to do about this," rather than asking questions or offering explanations.

Unconditional positive regard Because humanists believe human nature is positive and each person is unique, clients can be respected and cherished without having to prove themselves worthy of the therapist's esteem. Humanists believe that when people receive unconditional caring from others, they become better able to value themselves in a similar way.

Genuineness Humanists believe that when therapists honestly share their thoughts and feelings with their clients, their clients will in turn develop self-trust and honest self-expression.

Active listening By reflecting, paraphrasing, and clarifying what the client says and means, the clinician communicates that he or she is genuinely interested in what the client is saying (FIGURE 14.6).

Supporters say that there is empirical evidence for the efficacy of client-centered therapy (Hardcastle et al., 2008; Kirschenbaum & Jourdan, 2005; Lein & Wills, 2007; Stiles et al., 2008), but critics argue that outcomes such as self-actualization and self-awareness are difficult to test scientifically. In addition, research on specific therapeutic techniques such as "empathy" and "active listening" has had mixed results (Clark, 2007; Hodges & Biswas-Diener, 2007; Rosenthal, 2007).

Active listening FIGURE 14.6

Noticing a client's brow furrowing and hands clenching while he is discussing his marital problems, a clinician might respond, "It sounds like you're angry with your wife and feeling pretty miserable right now." Can you see how this statement reflects the client's anger, paraphrases his complaint, and gives feedback to clarify the communication?

GROUP, FAMILY, AND MARITAL THERAPIES: HEALING INTERPERSONAL RELATIONSHIPS

In contrast to the therapies described so far, group, marital, and family therapies treat multiple individuals simultaneously. During sessions, therapists often apply psychoanalytic, cognitive, and humanistic techniques.

In **group therapy**, multiple people meet together to work toward therapeutic goals. Typically, a group of 8 to 10 people meet with a therapist on a regular basis to talk about problems in their lives.

group therapy A form of therapy in which a number of people meet together to work toward therapeutic goals.

A variation on group therapy is the **self-help group**. Unlike other group therapy approaches, a professional does not guide these groups. They are simply groups of people who share a common problem (such as alcoholism, single parenthood, or breast cancer) and who meet to give and receive support. Faith-based 12-step programs such as Alcoholics Anonymous, Narcotics Anonymous, and Spenders Anonymous are examples of self-help groups.

Although group members don't get the same level of individual attention found in one-on-one therapies, group and self-help therapies provide their own unique advantages (Corey, 2009; Minuchin, Lee, & Simon, 2007; Qualls, 2008). Compared with one-on-one therapies, they are less expensive and provide a broader base of social support. Group members can learn from each other's mistakes, share insights and coping strategies, and role-play social interactions together.

Therapists often refer their patients to group therapy and self-help groups to supplement individual therapy. Research on self-help groups for alcoholism, obesity, and other disorders suggests that they can be very effective, either alone or in addition to individual psychotherapy (McEvoy, 2007; Oei & Dingle, 2008; Silverman et al., 2008).

Because a family or marriage is a system of interdependent parts, the problem of any one individual unavoidably affects all the others, and therapy can help everyone involved (Minuchin, Lee, & Simon, 2007; Qualls, 2008). The line between marital (or couples) therapy and family therapy is often blurred. Here, our discussion will focus on **family therapy**, in which the primary aim is to change maladaptive family interaction patterns (FIGURE 14.7). All members of the family attend therapy sessions, though at times the therapist may see family members individually or in twos or threes.

family therapy Treatment to change maladaptive interaction patterns within a family.

Family therapy is also useful in treating a number of disorders and clinical problems. As we discussed in Chapter 13, schizophrenic patients are more likely to relapse if their family members express emotions, attitudes, and behaviors that involve criticism, hostility, or emotional overinvolvement (Hooley & Hiller, 2001; Lefley, 2000; Quinn, Barowclough, & Tarrier, 2003). Family therapy can help family members modify their behavior toward the patient. Family therapy can also be the most favorable setting for the treatment of adolescent drug abuse (Minuchin, Lee, & Simon, 2007; Ng et al., 2008; O'Farrell et al., 2008; Sim & Wong, 2008).

Family therapy FIGURE 14.7

Many families initially come into therapy believing that one member is the cause of all their problems. However, family therapists generally find that this "identified patient" is a scapegoat for deeper disturbances. For example, instead of confronting their own problems with intimacy, a couple may focus all their attention and frustration on a delinquent child (Hanna & Brown, 1999). How could changing ways of interacting within the family system promote the health of individual family members and the family as a whole?

CONCEPT CHECK STOP

How does modern psychodynamic therapy differ from traditional psychoanalysis?

What are the four steps of Ellis's RET?

What is the significance of the term *client-centered therapy?*

When might family therapy be more successful than individual psychotherapy?

Behavior Therapies

LEARNING OBJECTIVES

Identify a key difference between these two classical conditioning techniques: systematic desensitization and aversion therapy.

Explore how operant conditioning can be used in therapy.

Summarize how modeling therapy works.

Sometimes having insight into a problem does not automatically solve it. In **behavior therapy**, the focus is on the problem behavior itself, rather than on any underlying causes. Although the person's feelings and interpretations are not disregarded, they are also not emphasized. The therapist diagnoses the problem by listing maladaptive behaviors that occur and adaptive behaviors that are absent. The therapist then attempts to shift the balance of the two, drawing on the learning principles of classical conditioning, operant conditioning, and observational learning (Chapter 6).

behavior therapy A group of techniques based on learning principles that is used to change maladaptive behaviors.

© Sidney Harris

CLASSICAL CONDITIONING TECHNIQUES

Behavior therapists use the principles of classical conditioning to decrease maladaptive behaviors by creating new associations to replace the faulty ones. We will explore two techniques based on these principles: aversion therapy and systematic desensitization.

Aversion therapy uses principles of classical conditioning to create anxiety rather than extinguish it (FIGURE 14.8). People who engage in excessive drinking, for example, build up a number of pleasurable associations. Because these pleasurable associations cannot always be prevented, aversion therapy provides negative associations to compete with the pleasurable ones. Someone who wants to stop drinking, for example, could take a drug called Antabuse, which causes vomiting whenever alcohol enters the system. When the new connection between alcohol and nausea has been classically conditioned, engaging in the once-desirable behavior will cause an immediate negative response.

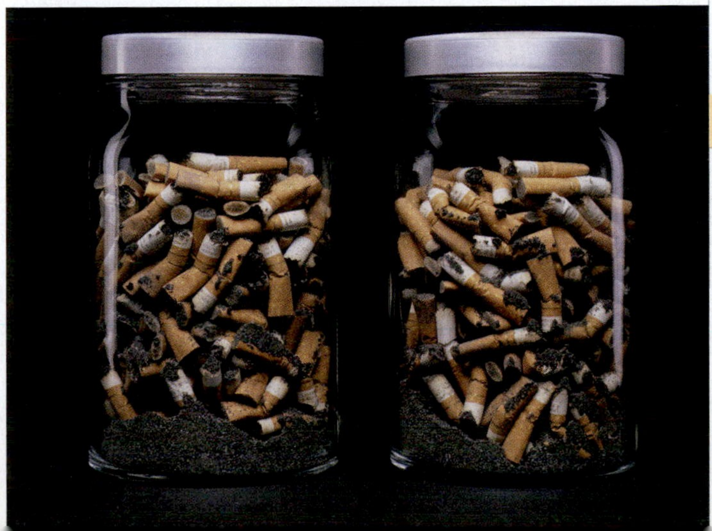

Making a nasty habit nastier FIGURE 14.8

A person who wants to quit smoking could collect a jar full of (smelly) cigarette butts or smoke several cigarettes at once to create an aversion to smoking. Can you see how this aversion therapy uses classical conditioning?

Aversion therapy is controversial. First, some researchers question whether it is ethical to hurt someone (even when the person has given permission). The treatment also has been criticized because it does not provide lasting relief (Seligman, 1994). One reason is that (in the case of the aversion therapy for alcoholism) people understand that the nausea is produced by the Antabuse and do not generalize their learning to the alcohol.

In contrast to aversion therapy, **systematic desensitization** (Wolpe & Plaud, 1997) begins with relaxation training, followed by imagining or directly experiencing various versions of a feared object or situation while remaining deeply relaxed. The goal is to replace an anxiety response with a relaxation response when confronting the feared stimulus (Heriot & Pritchard, 2004). Recently, some therapists have been using advanced technologies to aid in the desensitization process (FIGURE 14.9).

Desensitization is a three-step process (FIGURE 14.10). First, a client is taught how to maintain a state of deep relaxation that is physiologically incompatible with an anxiety response. Next, the therapist and client construct a hierarchy, or ranked listing, of anxiety-arousing images. In the final step, the relaxed client mentally visualizes or physically experiences items in the hierarchy, starting at the bottom and working his or her way to the most anxiety-producing images at the top. If any image or situation begins to create anxiety, the client stops momentarily and returns to a state of complete relaxation. Eventually, the fear response is extinguished.

Virtual reality therapy FIGURE 14.9

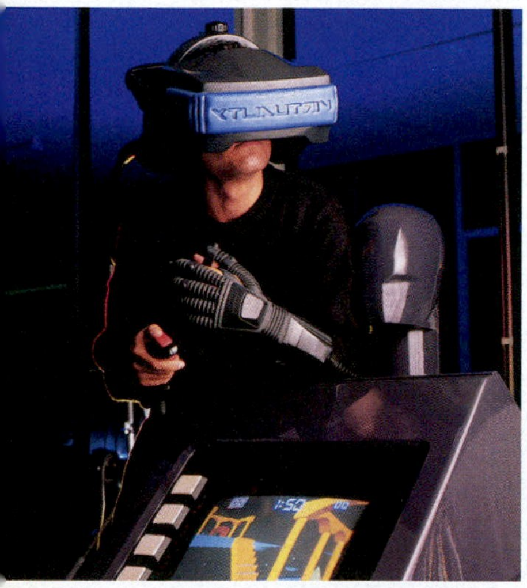

Rather than using mental imaging or actual physical experiences of a fearful situation, modern systematic desensitization can use the latest in computer technology—virtual reality headsets and data gloves. What kind of "virtual" experiences do you think a therapist might provide for a client with a fear of spiders?

OPERANT CONDITIONING TECHNIQUES

One operant conditioning technique for eventually bringing about a desired (or target) behavior is **shaping**—providing rewards for successive approximations of the target behavior (Chapter 6). One of the most successful applications of shaping and reinforcement has been with developing language skills in children with autism. First, the child is rewarded for making any sounds; later, only for forming words and sentences.

Shaping can also help people acquire social skills and greater assertiveness. If you are painfully shy, for example, a clinician might first ask you to role-play simply saying hello to someone you find attractive. Then you might practice behaviors that gradually lead you to suggest a get-together or date. During such role-playing, or behavior rehearsal, the clinician would give you feedback and reinforcement.

Adaptive behaviors can also be taught or increased with techniques that provide immediate reinforcement in the form of tokens (Kazdin, 2008; Tarbox, Ghezzi, & Wilson, 2006). For example, patients in an inpatient treatment facility might at first be given tokens (to be exchanged for primary rewards, such as food, treats, TV time, a private room, or outings) for merely attending group therapy sessions. Later they will be rewarded only for actually participating in the sessions. Eventually, the tokens can be discontinued when the patient receives the reinforcement of being helped by participation in the therapy sessions.

OBSERVATIONAL LEARNING TECHNIQUES

We all learn many things by observing others. Therapists use this principle in **modeling therapy**, in which clients are asked to observe and imitate appropriate models as they perform desired behaviors.

modeling therapy A learning technique in which the subject watches and imitates models who demonstrate desirable behaviors.

Visualizing

Systematic desensitization FIGURE 14.10

Madeleine, a college student, has been offered a lucrative summer job painting commercial buildings. She is deathly afraid of heights, but she needs the money. She goes to see a therapist, who tells her that she can be treated in a relatively brief period of time, through systematic desensitization. To understand how this works, look at the six behavioral steps depicted on the continuum below. Can you see how this technique might help Madeleine gradually overcome her fear of heights?

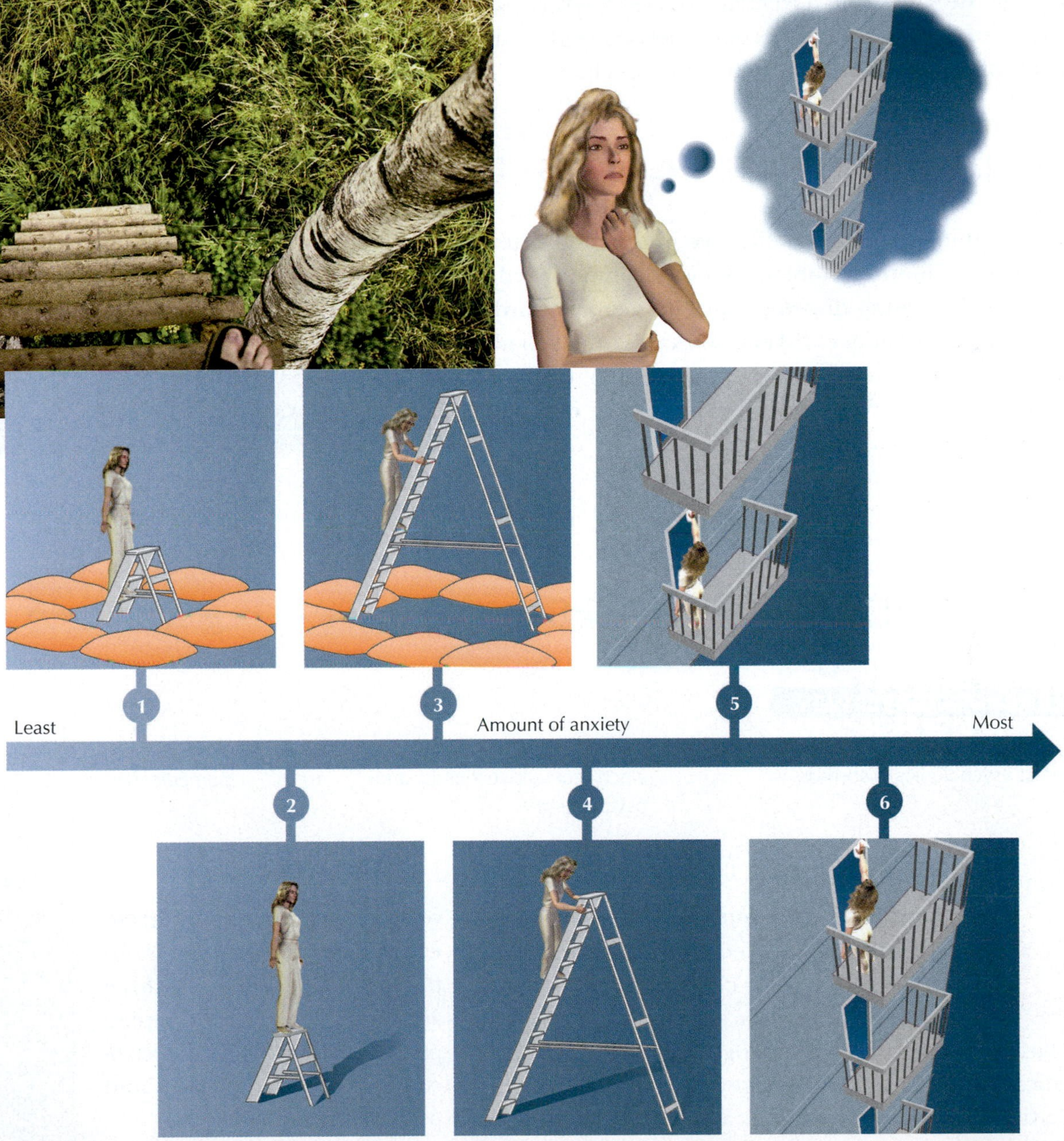

NATIONAL GEOGRAPHIC

For example, Albert Bandura and his colleagues (1969) asked clients with snake phobias to watch other (nonphobic) people handle snakes. After only two hours of exposure, over 92 percent of the phobic observers allowed a snake to crawl over their hands, arms, and necks.

Modeling is also part of social skills training and assertiveness training. Clients learn how to interview for a job by first watching the therapist role-play the part of the interviewee. The therapist models the appropriate language (assertively asking for a job), body posture, and so forth, and then asks the client to imitate the behavior and play the same role. Over the course of several sessions, the client gradually becomes desensitized to the anxiety of interviews and learns interview skills.

EVALUATING BEHAVIOR THERAPIES

Behavior therapy has been effective in treating various problems, including phobias, obsessive-compulsive disorder (OCD), eating disorders, autism, mental retardation, and delinquency (Ekers, Richards, & Gilbody, 2008; Flatt & King, 2009; Miltenberger, 2008). Critics of behavior therapy, however, raise important questions that fall into two major categories:

- *Generalizability.* Critics argue that in the "real world" patients are not consistently reinforced, and their newly acquired behaviors may disappear. To deal with this possibility, behavior therapists work to gradually shape clients toward real-world rewards.
- *Ethics.* Critics contend that it is unethical for one person to control another's behavior. Behaviorists, however, argue that rewards and punishments already control our behaviors and that behavior therapy actually increases a person's freedom by making these controls overt and by teaching people how to change their own behavior.

CONCEPT CHECK STOP

What three types of learning are used in behavior therapy?

How can shaping be used to develop desired behaviors?

What are two criticisms of behavior therapy?

Biomedical Therapies

LEARNING OBJECTIVES

Identify the major types of drugs used to treat psychological disorders.

Explain what happens in electroconvulsive therapy and psychosurgery.

Describe the risks associated with biomedical therapies.

biomedical therapy Using physiological interventions (drugs, electroconvulsive therapy, and psychosurgery) to reduce or alleviate symptoms of psychological disorders.

Biomedical therapies are based on the premise that problem behaviors are caused, at least in part, by chemical imbalances or disturbed nervous system functioning. A physician, rather than a psychologist, must prescribe biomedical therapies, but psychologists do work with patients receiving biomedical therapies and are frequently involved in research programs to evaluate their effectiveness. In this section, we will discuss three aspects of biomedical therapies: psychopharmacology, electroconvulsive therapy (ECT), and psychosurgery.

PSYCHOPHARMACOLOGY: TREATING PSYCHOLOGICAL DISORDERS WITH DRUGS

psychopharmacology The study of drug effects on the mind and behavior.

Since the 1950s, drug companies have developed an amazing variety of chemicals to treat abnormal behaviors. In some cases, discoveries from **psychopharmacology** have helped correct chemical imbalances. In other cases, drugs have been used to relieve or suppress the symptoms of psychological disturbances even when the underlying cause was not thought to be biological. As shown in TABLE 14.1, psychiatric drugs are classified into four major categories: antianxiety, antipsychotic, mood stabilizer, and antidepressant.

- **Antianxiety drugs** (also known as "minor tranquilizers") lower the sympathetic activity of the brain—the crisis mode of operation—so that anxious responses are diminished or prevented and are replaced by feelings of tranquility and calmness (Barlow, 2008; Swartz, 2008).

"Before Prozac, she loathed company."

Common drug treatments for psychological disorders TABLE 14.1

Type of Drug (Chemical Group)	Psychological Disorder	Generic Name	Brand Name
Antianxiety drugs (Benzodiazepines)	Anxiety disorders	alprazolam diazepam lorazepam	Xanax Valium Ativan
Antipsychotic drugs (Phenothiazines Butyrophenones Atypical antipsychotics)	Schizophrenia and bipolar disorders	chlorpromazine fluphenazine thioridazine haloperidol clozapine resperidone quetiapine	Thorazine Prolixin Mellaril Haldol Clozaril Risperdal Seroquel
Mood stabilizer drugs (Antimanic)	Bipolar disorder	lithium carbonate carbamazepine	Eskalith CR Lithobid Tegretol
Antidepressant drugs Tricyclic antidepressants Monoamine oxidase inhibitors (MAOIs) Selective serotonin reuptake inhibitors (SSRIs) Serotonin and norepinephrine reuptake inhibitors (SNRIs) Atypical antidepressants	Depressive disorders	imipramine amitriptyline phenelzine paroxetine fluoxetine venlafaxine duloxetine bupropion mirtazapine	Tofranil Elavil Nardil Paxil Prozac Effexor Cymbalta Wellbutrin Remeron

- **Antipsychotic drugs**, or neuroleptics, are used to treat schizophrenia and other acute psychotic states. They are often referred to as "major tranquilizers," creating the mistaken impression that they invariably have a strong sedating effect. However, the main effect of antipsychotic drugs is to diminish or eliminate psychotic symptoms, including hallucinations, delusions, withdrawal, and apathy. They are not designed to sedate the patient. Traditional antipsychotics work by decreasing activity at the dopamine synapses in the brain. A large majority of patients show marked improvement when treated with antipsychotic drugs.
- **Mood stabilizer drugs**, such as lithium, can help relieve manic episodes and depression for people suffering from bipolar disorder. Because lithium acts relatively slowly—it can be three or four weeks before it takes effect—its primary use is in preventing future episodes and helping to break the manic-depressive cycle.
- **Antidepressant drugs** are used to treat people with depression. There are five types of antidepressant drugs: tricyclics, monoamine oxidase inhibitors (MAOIs), selective serotonin reuptake inhibitors (SSRIs), serotonin and norepinephrine reuptake unhibitors (SNRIs), and atypical antidepressants. Each class of drugs affects neurochemical pathways in the brain in slightly different ways, increasing or decreasing the availability of certain chemicals. SSRIs (such as Paxil and Prozac) are by far the most commonly prescribed antidepressants. The atypical antidepressants are a miscellaneous group of drugs used for patients who fail to respond to the other drugs or for people who experience side effects common to other antidepressants.

ELECTROCONVULSIVE THERAPY AND PSYCHOSURGERY

electroconvulsive therapy (ECT) Biomedical therapy in which electrical current is passed through the brain.

In **electroconvulsive therapy (ECT)**, also known as electroshock therapy (EST), a moderate electrical current is passed through the brain between two electrodes placed on the outside of the head. The current triggers a widespread firing of neurons, or convulsions. The convulsions produce many changes in the central and peripheral nervous systems, including activation of the autonomic nervous system, increased secretion of various hormones and neurotransmitters, and changes in the blood–brain barrier.

During the early years of ECT, some patients received hundreds of treatments (Fink, 1999), but today most receive 12 or fewer treatments. Sometimes the electrical current is applied only to the right hemisphere, which causes less interference with verbal memories and left hemisphere functioning.

Modern ECT is used primarily on patients with severe depression who do not respond to antidepressant drugs or psychotherapy and on suicidal patients because it works faster than antidepressant drugs (Birkenhäger, Renes, & Pluijms, 2004; Goforth & Holsinger, 2007). (ECT is described further in *What a Psychologist Sees.*)

Although clinical studies of ECT conclude that it is effective for very severe depression (Jain et al., 2008; Khalid et al., 2008), its use remains controversial because it triggers massive changes in the brain—and also because we do not understand why it works (Baldwin & Oxlad, 2000; Cloud, 2001; Pearlman, 2002).

The most extreme, and least used, biomedical therapy is **psychosurgery**—brain surgery performed to reduce serious, debilitating psychological problems.

psychosurgery Operative procedures on the brain designed to relieve severe mental symptoms that have not responded to other forms of treatment.

You may have heard of the now-outmoded form of psychosurgery called **lobotomy**. This technique originated in 1936, when Portuguese neurologist Egaz Moniz first treated uncontrollable psychoses by cutting the nerve fibers between the frontal lobes (where association areas for monitoring and planning behavior are found) and the thalamus and hypothalamus.

Although these surgeries did reduce emotional outbursts and aggressiveness, many patients were left with permanent brain damage. In the mid-1950s, when antipsychotic drugs came into use, psychosurgery virtually stopped. Recently, however, psychiatrists have been

Electroconvulsive therapy (ECT)

Although electroconvulsive therapy may seem barbaric, for some severely depressed people it is their only hope for lifting the depression. Unlike portrayals of ECT in movies like *One Flew Over the Cuckoo's Nest* and *The Snake Pit,* patients show few, if any, visible reactions to the treatment, owing to modern muscle-relaxant drugs that dramatically reduce muscle contractions during the seizure. Most ECT patients are also given an anesthetic to block their memories of the treatment, but some patients still find the treatment extremely uncomfortable (Jain et al., 2008; Khalid et al., 2008). However, many others find it lifesaving.

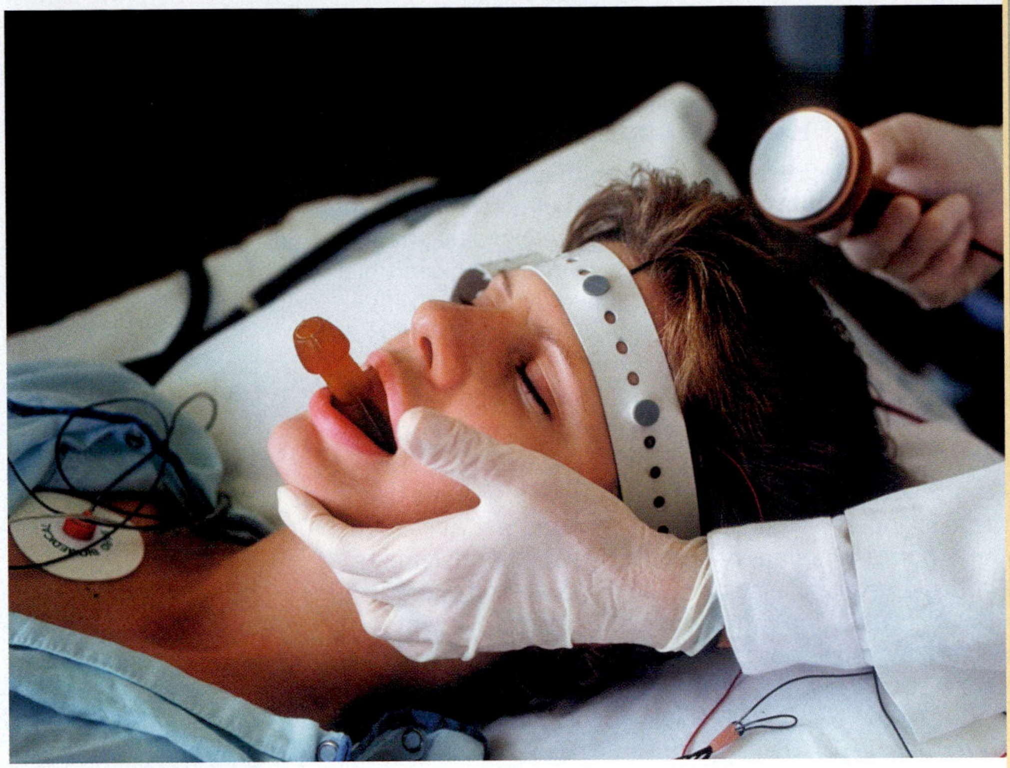

experimenting with a much more limited and precise surgical procedure called *cingulotomy*, in which the cingulum (a small structure in the brain's limbic system known to be involved in emotionality) is partially destroyed. On rare occasions, it is used in the treatment of severely debilitating cases of obsessive-compulsive disorder, severe depression, and chronic, intractable pain (Cohen et al., 2001; Dougherty et al., 2002; Sotres-Bayón & Pellicer, 2000).

EVALUATING BIOMEDICAL THERAPIES

Like all forms of therapy, biomedical therapies have both proponents and critics.

Pitfalls of psychopharmacology Drug therapy poses several potential problems. First, although drugs may relieve symptoms for some people, they seldom provide "cures." In addition, some patients become physically dependent on the drugs. Also, researchers are still learning about the drugs' long-term effects and potential interactions. Furthermore, psychiatric medications can cause a variety of side effects, ranging from mild fatigue to severe impairments in memory and movement.

A final potential problem with drug treatment is that its relative inexpensiveness and generally fast results have led to its overuse in some cases. One report found that antidepressants are prescribed roughly 50 percent of the time a patient walks into a psychiatrist's office (Olfson et al., 1998).

Despite the problems associated with them, psychotherapeutic drugs have led to revolutionary changes in mental health. Before the use of drugs, some patients were destined to spend a lifetime in psychiatric institutions. Today, most patients improve enough to return to their homes and live successful lives if they continue to take their medications to prevent relapse.

Psychological Science

What about Herbal Remedies?

Some recent research suggests that the herbal supplement St. John's wort may be an effective treatment for mild to moderate depression, with fewer side effects than traditional medications (Butterweck, 2003; Lecrubier et al., 2002; Mayers et al., 2003). However, other studies have found the drug to be ineffective for people with major depression (Hypericum Depression Trial Study, 2002). Herbal supplements, like kava, valerian, and gingko biloba, also have been used in the treatment of anxiety, insomnia, and memory problems (e.g., Connor & Davidson, 2002; Parrott et al., 2004). Although many people assume that these products are safe because they are "natural," they can produce a number of potentially serious side effects. For this reason and also because the U.S. Food and Drug Administration does not regulate herbal supplements, researchers advise a wait-and-see approach (Crone & Gabriel, 2002; Swartz, 2008).

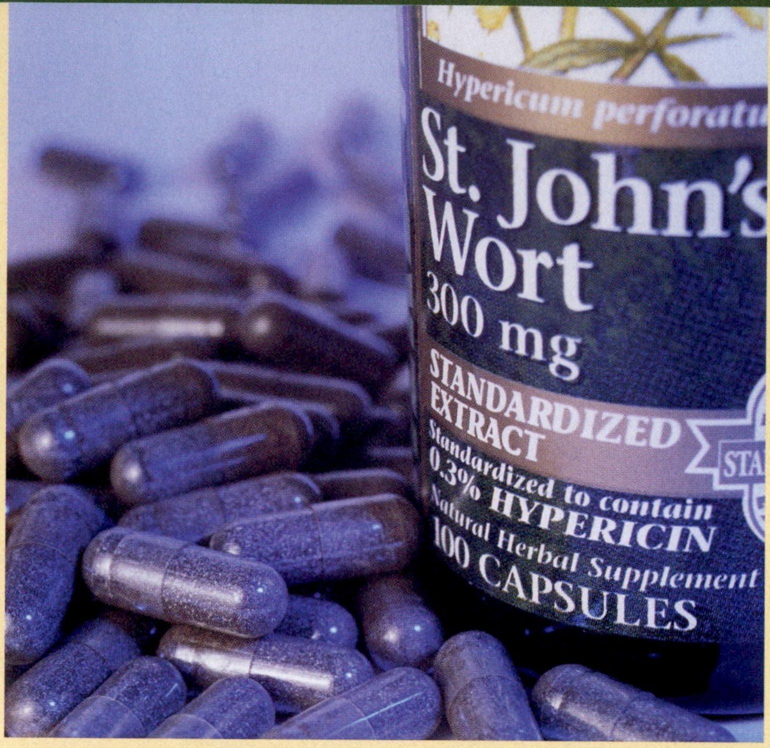

CRITICAL THINKING

Here are two interesting questions:

1. Do you agree with the FDA's policy of not regulating herbal remedies?
2. Do you think these products should carry labels warning that they can produce potentially serious side effects?

Challenges to ECT and psychosurgery As we mentioned, ECT is a controversial treatment for several reasons. However, problems with ECT may become obsolete, thanks to **repetitive transcranial magnetic stimulation (rTMS)**, which delivers a brief (but powerful) electrical current through a coil of wire placed on the head. Unlike ECT, which passes a strong electrical current directly through the brain, the rTMS coil creates a strong magnetic field that is applied to certain areas in the brain. When used to treat depression, the coil is usually placed over the prefrontal cortex, a region linked to deeper parts of the brain that regulate mood. Currently, the benefits of rTMS over ECT remain uncertain (Bloch et al., 2008; Knapp et al., 2008; Wasserman, Epstein, & Ziemann, 2008).

Because all forms of psychosurgery have potentially serious or fatal side effects and complications, some critics suggest that they should be banned altogether. Furthermore, the consequences are irreversible. For these reasons, psychosurgery is considered experimental and remains a highly controversial treatment.

CONCEPT CHECK STOP

When are drug therapies used?

How does modern ECT differ from the therapy's early use?

What is cingulotomy?

Therapy Essentials

LEARNING OBJECTIVES

Summarize the goals that are common to all major forms of psychotherapy.

Describe some key cross-cultural similarities and differences in therapy.

Explain why therapists need to be sensitive to gender issues that pertain to mental illness.

Explore some alternatives to long-term institutionalization for people with severe psychological disorders.

Earlier, we mentioned that there may be more than 400 forms of therapy. How would you choose one for yourself or someone you know? In this section, we help you to synthesize the material in this chapter and to put what you have learned about each of the major forms of therapy into a broader context.

THERAPY GOALS AND EFFECTIVENESS

All major forms of therapy are designed to help the client in five specific areas (FIGURE 14.11).

Although most therapists work with clients in several of these areas, the emphasis varies according to the therapist's training (psychodynamic, cognitive, humanistic, behaviorist, or biomedical). Clinicians who regularly borrow freely from various theories are said to take an **eclectic approach**.

Does therapy work? After years of controlled research and *meta-analysis* (a method of statistically combining and analyzing data from many studies), we have fairly clear evidence that it does. Forty to 90 percent of people who receive treatment are better off than people who do not. Furthermore, short-term treatments can be as effective as long-term treatments (Castonguay & Hill, 2007; Cleaves & Latner, 2008; Knekt et al., 2008; Loewenthal & Winter, 2006; Stiles et al., 2008; Wachtel, 2008).

Some therapies are more effective than others for specific problems. For example, phobias seem to respond best to systematic desensitization, and OCD can

The five most common goals of therapy FIGURE 14.11

Most therapies focus on one or more of these five goals. Can you identify which would be of most interest to a psychoanalyst, a cognitive therapist, a behaviorist, and a psychiatrist?

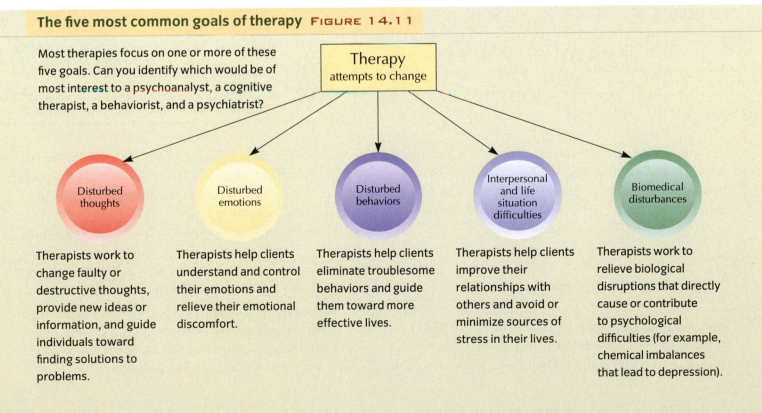

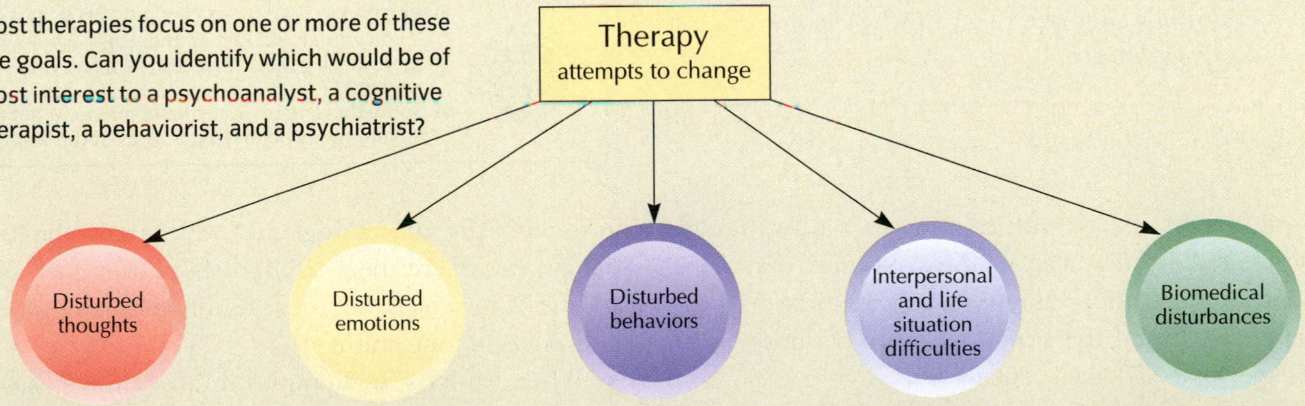

Therapists work to change faulty or destructive thoughts, provide new ideas or information, and guide individuals toward finding solutions to problems.

Therapists help clients understand and control their emotions and relieve their emotional discomfort.

Therapists help clients eliminate troublesome behaviors and guide them toward more effective lives.

Therapists help clients improve their relationships with others and avoid or minimize sources of stress in their lives.

Therapists work to relieve biological disruptions that directly cause or contribute to psychological difficulties (for example, chemical imbalances that lead to depression).

Applying Psychology

Finding a Therapist

When stresses overwhelm our natural coping mechanisms, professional therapy can provide invaluable relief. But how do we find a good therapist for our specific needs? If you have the time (and the money) to explore options, search for a therapist who is best suited to your specific goals. However, if you are in a crisis—you have suicidal thoughts, or are the victim of abuse—get help fast. Most communities have telephone hotlines that provide counseling on a 24-hour basis. (In addition, most colleges and universities have counseling centers that provide immediate, short-term therapy to students free of charge.)

If you are encouraging someone else to get therapy, you might offer to help locate a therapist and to go with him or her for the first visit. If he or she refuses help and the problem affects you, it is often a good idea to seek therapy yourself. You will gain insights and skills that will help you deal with the situation more effectively. Finally, if you're dealing with someone who is clearly suffering from depression, keep in mind a few simple rules:

1. ***Don't trivialize the disease.*** Depression, like cancer or heart disease, is a critical, life-threatening illness. Asking someone "What do you have to be depressed about?" or encouraging him to "pull up his socks" is akin to asking the cancer patient why she has cancer or why she doesn't just smile and exercise more.

2. ***Don't be a cheerleader or a Mr. or Ms. Fix-it.*** You can't pep-talk someone out of deep depression, and offering cheap advice or solutions is the best way to ensure that you'll be the last person he or she will turn to for help.
3. ***Don't equate normal, everyday "down times" with clinical depression.*** Virtually everyone has experienced down moods and times of loss and deep sadness. Unless you have shared true, clinical depression, comments like, "I know just how you feel," only makes it clear that you don't understand what clinical depression is all about.

CRITICAL THINKING

Here are two interesting questions:
1. If you were looking for a therapist, would you want the therapist's gender to be the same as yours? Why or why not?
2. Some therapists treat clients online. What might be the advantages and disadvantages of this approach?

be significantly relieved with cognitive-behavior therapy accompanied by medication. Most studies that have compared medication alone versus medication plus therapy have found the combination to be more effective (e.g., Doyle & Pollack, 2004).

CULTURAL ISSUES IN THERAPY

The therapies described in this chapter are based on Western European and North American culture. Does this mean they are unique? Or do our psychotherapists do some of the same things that, say, a native healer or shaman does? Are there similarities in therapies across cultures? Conversely, are there fundamental therapeutic differences among cultures?

When we look at therapies in all cultures, we find that they have certain key features in common (Laungani, 2007; Lee, 2002; Sue & Sue, 2008). Richard Brislin (2000) has summarized some of these features:

- *Naming the problem.* People often feel better just by knowing that others experience the same problem and that the therapist has had experience with their particular problem.

- *Qualities of the therapist.* Clients must feel that the therapist is caring, competent, approachable, and concerned with finding solutions to their problem.
- *Establishing credibility.* Word-of-mouth testimonials and status symbols (such as diplomas on the wall) establish the therapist's credibility. Among native healers, in lieu of diplomas, credibility may be established by having served as an apprentice to a revered healer.
- *Placing the problem in a familiar framework.* If the client believes that evil spirits cause psychological disorders, the therapist will direct treatment toward eliminating these spirits. Similarly, if the client believes in the importance of early childhood experiences and the unconscious mind, psychoanalysis will be the likely treatment of choice.
- *Applying techniques to bring relief.* In all cultures, therapy involves action. Either the client or the therapist must do something. Moreover, what therapists do must fit the client's expectations—whether it is performing a ceremony to expel demons or talking with the client about his or her thoughts and feelings.
- *A special time and place.* The fact that therapy occurs outside the client's everyday experiences seems to be an important feature of all therapies.

Although there are basic similarities in therapies across cultures, there are also important differences. In the traditional Western European and North American model, the emphasis is on the "self" and on independence and control over one's life—qualities that are highly valued in individualistic cultures. In collectivist cultures, however, the focus of therapy is on interdependence and accepting the realities of one's life (Sue & Sue, 2008) (FIGURE 14.12).

Not only does culture affect the types of therapy that are developed. It also influences the perceptions of the therapist. What one culture considers abnormal behavior may be quite common—and even healthy—in others. For this reason, recognizing cultural differences is very important for building trust between therapists and clients and for effecting behavioral change (Laungani, 2007; Sue & Sue, 2008; Tseng, 2004).

WOMEN AND THERAPY

Within our individualistic Western culture, men and women present different needs and problems to therapists. For example, compared with men, women are more comfortable and familiar with their emotions, have fewer negative attitudes toward therapy, and are more likely to seek psychological help (Komiya, Good, & Sherrod, 2000). Research has identified five major concerns related to women and psychotherapy (Halbreich & Kahn, 2007; Hall, 2007; Hyde, 2007; Matlin, 2008; Russo & Tartaro, 2008).

1. *Rates of diagnosis and treatment of mental disorders.* Women are diagnosed and treated for mental illness at a much higher rate than men. Is this because women are "sicker" than men as a group, or are they just more willing to admit their problems? Or perhaps the categories for illness are biased against women. More research is needed to answer this question.

Emphasizing interdependence FIGURE 14.12

In Japanese Naikan therapy, patients sit quietly from 5:30 A.M. to 9:00 P.M. for seven days and are visited by an interviewer every 90 minutes. During this time, they reflect on their relationships with others in order to discover personal guilt for having been ungrateful and troublesome and developing gratitude toward those who have helped them (Nakamura, 2006; Ozawa-de Silva, 2007; Ryback, Ikerni, & Miki, 2001). How do these methods and goals differ from those of the therapies described in this chapter?

Meeting women's unique needs FIGURE 14.13

Therapists must be sensitive to possible connections between clients' problems and their gender. Rather than prescribing drugs to relieve depression in women, for example, it may be more appropriate for therapists to explore ways to relieve the stresses of multiple roles or poverty. Can you see how helping a single mother identify parenting resources, such as play groups, parent support groups, and high-quality child care, might be just as effective at relieving depression as prescribing drugs?

2. *Stresses of poverty.* Women are disproportionately likely to be poor. Poverty contributes to stress, which is directly related to many psychological disorders.
3. *Stresses of multiple roles.* Women today are mothers, wives, homemakers, wage earners, students, and so on. The conflicting demands of their multiple roles often create special stresses (FIGURE 14.13).
4. *Stresses of aging.* Aging brings special concerns for women. Elderly women, primarily those with age-related dementia, account for over 70 percent of the chronically mentally ill who live in nursing homes in the United States.
5. *Violence against women.* Rape, incest, and sexual harassment—which are much more likely to happen to women than to men—may lead to depression, insomnia, posttraumatic stress disorders, eating disorders, and other problems.

INSTITUTIONALIZATION

Despite Hollywood film portrayals, forced institutionalization of the mentally ill poses serious ethical problems and is generally reserved for only the most serious and life-threatening situations or when there is no reasonable, less restrictive alternative.

In emergencies, mental health professionals can authorize temporary commitment for 24 to 72 hours—time enough for laboratory tests to be performed to rule out medical illnesses that could be causing the symptoms. The patient can also receive psychological testing, medication, and short-term therapy during this time.

Today, many states have a policy of **deinstitutionalization**—discharging patients from mental hospitals as soon as possible and discouraging admissions. As an alternative to institutionalizing people in state hospitals, most clinicians suggest expanding and improving community care (FIGURE 14.14) (Duckworth & Borus,

Outpatient support FIGURE 14.14

Community mental health (CMH) centers are a prime example of alternative treatment to institutionalization. CMH centers provide outpatient services, such as individual and group therapy and prevention programs. They also coordinate short-term inpatient care and programs for discharged mental patients, such as halfway houses and aftercare services. The downside of CMH centers and their support programs is that they are expensive. Investing in primary prevention programs (such as more intervention programs for people at high risk for mental illness) could substantially reduce these costs.

1999; Lamb, 2000). They recommend that general hospitals be equipped with special psychiatric units where acutely ill patients can receive inpatient care. For less disturbed individuals and chronically ill patients, they recommend walk-in clinics, crisis intervention services, improved residential treatment facilities, and psychosocial and vocational rehabilitation. State hospitals would be reserved for the most unmanageable patients.

CONCEPT CHECK STOP

What is the evidence that therapy works?

How do individualistic and collectivistic cultures differ in their approaches to therapy?

What is deinstitutionalization?

SUMMARY

1 Insight Therapies

1. "Insight therapies" are forms of **psychotherapy** that seek to increase insight into clients' difficulties.
2. In **psychoanalysis**, the therapist seeks to identify the patient's unconscious conflicts and to help the patient resolve them through catharsis. In modern **psychodynamic therapy**, treatment is briefer and the therapist takes a more directive approach (and puts less emphasis on unconscious childhood memories) than in traditional psychoanalysis.
3. **Cognitive therapy** seeks to help clients challenge faulty thought processes and adjust maladaptive behaviors. Ellis's rational-emotive behavior therapy (REBT) and Beck's cognitive-behavior therapy are important examples of cognitive therapy.
4. **Humanistic therapy**, such as Rogers's client-centered therapy, seeks to maximize personal growth, encouraging people to actualize their potential and relate to others in genuine ways.
5. In **group therapy**, multiple people meet together to work toward therapeutic goals. A variation is the self-help group, which is not guided by a professional. Therapists often refer their patients to group therapy and self-help groups to supplement individual therapy.
6. In **family therapy**, the aim is to change maladaptive family interaction patterns. All members of the family attend therapy sessions, though at times the therapist may see family members individually or in twos or threes.

2 Behavior Therapies

1. In **behavior therapy**, the focus is on the problem behavior itself, rather than on any underlying causes. The therapist uses learning principles to change behavior.
2. Classical conditioning techniques include aversion therapy and systematic desensitization.
3. Operant conditioning techniques used to increase adaptive behaviors include shaping and reinforcement.
4. In **modeling therapy**, clients observe and imitate others who are performing the desired behaviors.

3 Biomedical Therapies

1. **Biomedical therapies** are based on the premise that chemical imbalances or disturbed nervous system functioning contribute to problem behaviors.

2. **Psychopharmacology** is the most common form of biomedical therapy. Major classes of drugs used to treat psychological disorders are antianxiety drugs, antipsychotic drugs, mood stabilizer drugs, and antidepressant drugs.

3. In **electroconvulsive therapy (ECT)**, an electrical current is passed through the brain, stimulating convulsions that produce changes in the central and peripheral nervous systems. ECT is used primarily in cases of severe depression that do not respond to other treatments.

4. The most extreme biomedical therapy is **psychosurgery**. Lobotomy, an older form of psychosurgery, is now outmoded. Recently, psychiatrists have been experimenting with a more limited and precise surgical procedure called cingulotomy.

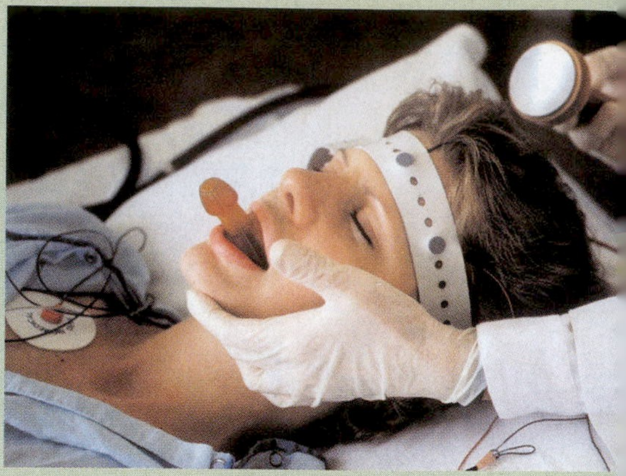

4 Therapy Essentials

1. All major forms of therapy are designed to address disturbed thoughts, disturbed emotions, disturbed behaviors, interpersonal and life situation difficulties, and biomedical disturbances. Research indicates that, overall, therapy does work.

2. Therapies in all cultures have certain key features in common; however, there are also important differences among cultures. Therapists must recognize cultural differences in order to build trust with clients and effect behavioral change.

3. Therapists must also be sensitive to possible gender issues in therapy.

4. Forced institutionalization of the mentally ill is generally reserved for only the most serious and life-threatening situations. Many states have a policy of deinstitutionalization, and most clinicians suggest expanding and improving community care to ensure that people with mental health problems receive appropriate care.

KEY TERMS

- **psychotherapy** p. 376
- **insight therapies** p. 376
- **psychoanalysis** p. 376
- **catharsis** p. 376
- **free association** p. 376
- **manifest content** p. 376
- **latent content** p. 376
- **resistance** p. 377
- **transference** p. 377
- **interpretation** p. 377
- **psychodynamic therapy** p. 378
- **interpersonal therapy (IPT)** p. 378
- **cognitive therapy** p. 378
- **self-talk** p. 378
- **cognitive restructuring** p. 378
- **rational-emotive therapy (RET)** p. 378
- **rational-emotive behavior therapy (REBT)** p. 380
- **cognitive-behavior therapy** p. 380
- **humanistic therapy** p. 380
- **client-centered therapy** p. 382
- **empathy** p. 383
- **unconditional positive regard** p. 383
- **genuineness** p. 383
- **active listening** p. 383
- **group therapy** p. 383
- **self-help group** p. 383
- **family therapy** p. 384
- **behavior therapy** p. 385
- **aversion therapy** p. 385
- **systematic desensitization** p. 386
- **shaping** p. 386
- **modeling therapy** p. 386
- **biomedical therapy** p. 388
- **psychopharmacology** p. 389
- **antianxiety drugs** p. 389
- **antipsychotic drugs** p. 390
- **mood stabilizer drugs** p. 390
- **antidepressant drugs** p. 390
- **electroconvulsive therapy (ECT)** p. 390
- **psychosurgery** p. 390
- **lobotomy** p. 390
- **repetitive transcranial magnetic stimulation (rTMS)** p. 392
- **eclectic approach** p. 393
- **deinstitutionalization** p. 396

CRITICAL AND CREATIVE THINKING QUESTIONS

1. You undoubtedly had certain beliefs and ideas about therapy before reading this chapter. Has studying this chapter changed these beliefs and ideas? Explain.
2. Which form of therapy do you personally find most appealing? Why?
3. What do you consider the most important commonalities among the major forms of therapy described in this chapter? What are the most important differences?
4. Imagine that you were going to use the principles of cognitive-behavioral therapy to change some aspect of your own thinking and behavior. (Maybe you'd like to quit smoking, or be more organized, or overcome your fear of riding in elevators.) How would you identify faulty thinking? What could you do to change your thinking patterns and behavior?

What is happening in this picture?

In the 2008 movie, *The Changeling*, Angelina Jolie portrays a grief-stricken mother (Christine) who loudly and publicly challenges the police and press who try to force her to accept an impostor as her abducted son. The authorities try everything to silence her, including involuntary commitment to the county hospital's "psychopathic ward." While confined, she is forced to take mood-altering drugs and is told that her only chance for release is to admit she was mistaken about the identity of her son.

- **How might this film contribute to the negative stereotypes of psychotherapy? Given that this movie was based on real life events in 1928, what special danger does this "true story" pose?**
- **Can you think of a Hollywood film that offers a positive portrayal of psychotherapy?**

SELF-TEST

(Check your answers in Appendix B.)

1. Psychoanalysis, cognitive, humanistic, group, and family therapies are often grouped together as ________.
   a. insight therapy
   b. behavior therapy
   c. humanistic and operant conditioning
   d. cognitive restructuring

2. The system of psychotherapy developed by Freud that seeks to bring unconscious conflicts into conscious awareness is known as ________.
   a. transference
   b. cognitive restructuring
   c. psychoanalysis
   d. the "hot seat" technique

3. This modern form of therapy emphasizes internal conflicts, motives, and unconscious forces.
   a. self-talk therapy
   b. belief-behavior therapy
   c. psychodynamic therapy
   d. thought analysis

4. The figure below illustrates the process by which the therapist and client work to change destructive ways of thinking called ________.
   a. problem solving
   b. self-talk
   c. cognitive restructuring
   d. rational recovery

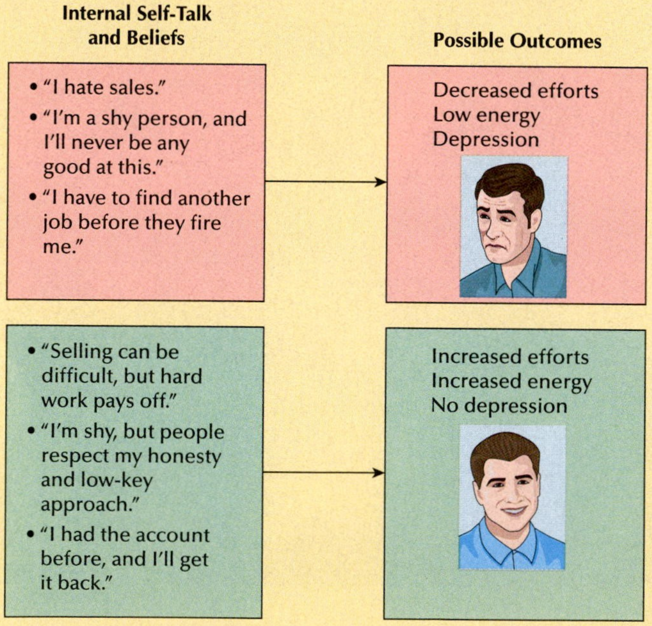

Internal Self-Talk and Beliefs
Possible Outcomes
• "I hate sales."
• "I'm a shy person, and I'll never be any good at this."
• "I have to find another job before they fire me."
Decreased efforts
Low energy
Depression
• "Selling can be difficult, but hard work pays off."
• "I'm shy, but people respect my honesty and low-key approach."
• "I had the account before, and I'll get it back."
Increased efforts
Increased energy
No depression

5. Beck practices ________, which attempts to change not only destructive thoughts and beliefs, but the associated behaviors as well.
   a. psycho-behavior therapy
   b. cognitive-behavior therapy
   c. thinking-acting therapy
   d. belief-behavior therapy

6. ________ therapy seeks to maximize personal growth through affective restructuring.
   a. Cognitive-emotive
   b. Emotive
   c. Humanistic
   d. Actualization

7. In Rogerian therapy, the ________ is responsible for discovering maladaptive patterns.
   a. therapist
   b. analyst
   c. patient
   d. client

8. This type of group does not have a professional leader, and members assist each other in coping with a specific problem.
   a. self-help
   b. encounter
   c. peer
   d. behavior

9. The main focus in behavior therapy is to increase ________ and decrease ________.
   a. positive thoughts and feelings; negative thoughts and feelings
   b. adaptive behaviors; maladaptive behaviors
   c. coping resources; coping deficits
   d. all of these options

10. The three steps in systematic desensitization include all **EXCEPT** ________.
    a. learning to become deeply relaxed
    b. arranging anxiety-arousing stimuli into a hierarchy from least to worst arousing
    c. practicing relaxation to anxiety-arousing stimuli, starting at the top of the hierarchy
    d. all of these options are included

11. This type of therapy involves watching and imitating appropriate models who demonstrate desirable behaviors.
   a. shaping
   b. spectatoring
   c. modeling
   d. systematic desensitization

12. Label the four major categories of psychiatric drugs on the chart below.

Common drug treatments for psychological disorders

Type of drug	*Psychological disorder*
a. ______________	Anxiety disorders
b. ______________	Schizophrenia
c. ______________	Bipolar disorder
d. ______________	Depressive disorders

13. In electroconvulsive therapy (ECT), ______________.
   a. current is never applied to the left hemisphere
   b. convulsions activate the ANS, stimulate hormone and neurotransmitter release, and change the blood–brain barrier
   c. electrical current passes through the brain for up to one minute
   d. most patients today receive hundreds of treatments because it is safer than in the past

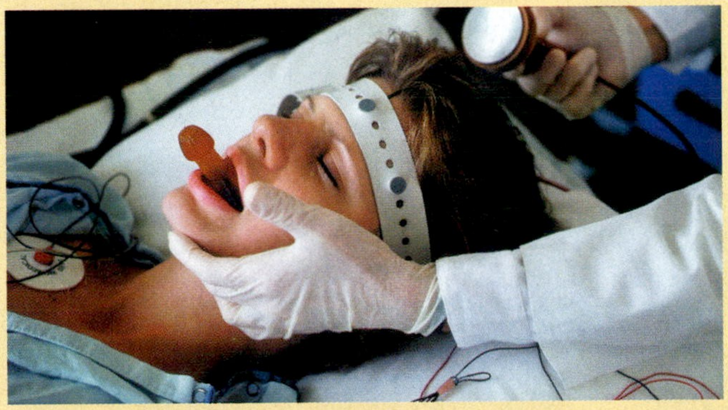

14. Label the five most common goals of therapies on the figure below.

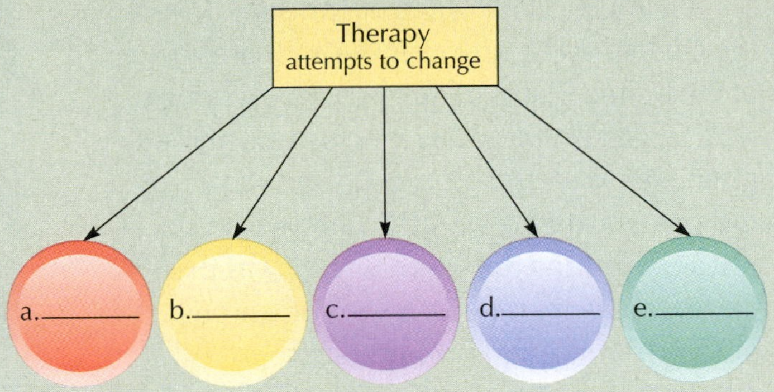

15. Psychiatric professionals may authorize temporary commitment for assessment and treatment of a dangerous or incompetent individual for up to ______________.
   a. 12 to 24 hours
   b. 24 to 72 hours
   c. 3 to 4 days
   d. 3 to 4 weeks

Social Psychology 15

January 20, 2009, was a historic day. Barack H. Obama was sworn in as the 44th president of the United States of America, an event that was personally witnessed by close to 2 million attendees in Washington, D.C., and viewed by millions around the world through coverage on television and the Internet. Why was this particular inauguration such a momentous occasion? And why is it important to psychologists? For many, the election and swearing in of the first black American president signaled a hopeful end to the racial hatred and divisive policies of the past. The outpouring of cheers and congratulations from around the globe gave evidence for a universal desire for leadership and social connection.

Understanding why we admire, like, and love, or dislike and even hate, some people or groups of people is essential to understanding ourselves and the world around us. Social psychology studies both the negative aspects in social behavior, such as prejudice and aggression, and its more positive aspects, such as interpersonal attraction and helping behaviors. Because almost everything we think, feel, or do is social, the subject matter of social psychology is enormous and varied.

We will approach our study by looking at the three major components of social psychology: thoughts, feelings, and actions toward others. We'll conclude with a discussion of how social psychology can help reduce prejudice, discrimination, and destructive forms of behavior.

CHAPTER OUTLINE

Our Thoughts about Others

LEARNING OBJECTIVES

Explain how attributions and attitudes affect the way we perceive and judge others.

Summarize the three components of attitudes.

Describe cultural differences in how people explain behavior.

ATTRIBUTION: EXPLAINING BEHAVIOR

social psychology The study of how other people influence a person's thoughts, feelings, and actions.

attributions Explanations for behaviors or events.

One critical aspect of **social psychology** is the search for reasons and explanations for our own and others' behavior. It's natural to want to understand and explain why people behave as they do and why events occur as they do. Many social psychologists believe that developing logical **attributions** for people's behavior makes us feel safer and more in control (Chiou, 2007; Heider, 1958; Krueger, 2007). To do so, most people begin with the basic question of whether a given action stems mainly from the person's internal disposition or from the external situation.

Mistaken attributions

Making the correct choice between disposition and situation is central to accurately judging why people do what they do. Unfortunately, our attributions are frequently marred by two major errors: the fundamental attribution error and the self-serving bias. Let's explore each of these.

When we consider the environmental influences on people's behavior, we generally make accurate attributions. However, given that people have enduring personality traits (Chapter 12) and a tendency to take cognitive shortcuts (Chapter 8), we more often choose dispositional attributions—that is, we blame the person. For example, if an instructor seems relaxed and talkative in class, you might conclude that she is an extroverted person. You may not realize that in situations that do not demand gregariousness, she tends to be shier. This bias is so common that it is called the **fundamental attribution error (FAE)** (Gebauer, Krempl, & Fleisch, 2008; Kimmel, 2006; Tal-Or & Papirman, 2007).

fundamental attribution error (FAE) Misjudgment of others' behavior as stemming from internal (dispositional) rather than external (situational) causes.

One reason that we tend to jump to internal, personal explanations is that human personalities and behaviors are more salient (or noticeable) than situational factors (FIGURE 15.1). This **saliency bias** helps explain why people sometimes suggest that a raped woman in a short skirt was asking for it or that homeless beggars should get a job—a phenomenon also called "blaming the victim."

When we explain our own behavior, we tend to favor internal attributions for our successes and external attributions for our failures. This **self-serving bias** is motivated by a desire to maintain positive self-esteem and a good public image (Gobbo & Raccanello, 2007; Krusemark, Campbell, & Clementz, 2008; Shepperd, Malone, & Sweeny, 2008). For example, students often take personal credit for doing well on an exam. If they fail the test, however, they tend to blame the instructor, the textbook, or the "tricky" questions.

Culture and attributional biases

Both the fundamental attribution error and the self-serving bias may depend in part on cultural factors (Kudo & Numazaki, 2003; Matsumoto, 2000) (FIGURE 15.2). In highly individualistic cultures, like the United States, people are defined and understood as individual selves—largely responsible for their successes and failures. But in collectivistic cultures, like China, people are primarily defined as members of their social network—responsible for doing as others expect. Accordingly, they tend to be more aware of situational constraints on behavior, making the FAE less likely (Bozkurt & Aydin, 2004; Norenzayan, 2006).

Saliency bias FIGURE 15.1

How would you explain this situation? According to the saliency bias, we're likely to blame homeless individuals for their condition because people are more conspicuous and "salient" than the situational factors that lead to poverty and homelessness.

Culture and attributional biases FIGURE 15.2

Westerners watching a baseball game in Japan who saw the umpire make a bad call would probably make a dispositional attribution ("He's a lousy umpire"), whereas Japanese spectators would tend to make a situational attribution ("He's under pressure") . What accounts for this difference?

The self-serving bias is also much less common in Eastern nations. In Japan, for instance, the ideal person is someone who is aware of his or her shortcomings and continually works to overcome them—not someone who thinks highly of himself or herself (Heine & Renshaw, 2002). In the East (as well as in other collectivistic cultures), self-esteem is not related to doing better than others but to fitting in with the group (Markus & Kitayama, 2003).

ATTITUDES: LEARNED PREDISPOSITIONS TOWARD OTHERS

attitudes Learned predispositions to respond cognitively, affectively, and behaviorally to particular objects in a particular way.

When we observe and respond to the world around us, we are seldom completely neutral. Rather, our responses toward subjects as diverse as pizza, people, AIDS, and abortion reflect our **attitudes**. We learn our attitudes both from direct experience and from watching others.

Social psychologists generally agree that most attitudes have three components: cognitive (thoughts and beliefs), affective (feelings), and behavioral (predisposition to act

Attitude toward Economic "Meltdown"

Cognitive element (believes economic troubles resulted from actions of greedy bankers) → Measured by self-report techniques (surveys and questionnaires)

Affective element (fears further loss of personal savings) → Measured by physiological techniques (heart rate, respiration)

Behavioral element (withdraws savings from local bank) → Measured by self-reported or directly observed behavioral changes

Three components of attitudes FIGURE 15.3

When social psychologists study attitudes, they measure each of the three components: cognitive, affective, and behavioral.

in a certain way toward an object or situation, or attitude object) (FIGURE 15.3).

Although attitudes begin to form in early childhood, they are not permanent (a fact that advertisers and politicians know and exploit). One way to change attitudes is to make direct, persuasive appeals. But an even more efficient strategy is to create **cognitive dissonance** (Cooper & Hogg, 2007; Gringart, Helmes, & Speelman, 2008). Contradictions between our attitudes and behaviors can motivate us to change our attitudes to agree with our behaviors (Festinger, 1957) (FIGURE 15.4).

cognitive dissonance A feeling of discomfort caused by a discrepancy between an attitude and a behavior or between two attitudes.

In one of the best-known tests of cognitive dissonance theory, Leon Festinger & J. Merrill Carlsmith (1959) found that participants who received $1 for lying about a boring experimental task subsequently held more positive attitudes toward the task than those who were paid $20. Why? All participants presumably recognized the discrepancy between their attitude (the task was boring) and their behavior (saying it was interesting). In participants' minds, a $20 payment might justify the lie, resolving cognitive dissonance, but a $1 payment could not. Those paid $1 had only one way to reduce cognitive dissonance: by changing their attitude.

The experience of cognitive dissonance may depend on a distinctly Western way of thinking about and evaluating the self. As we mentioned earlier, people in Eastern cultures tend not to define themselves in terms of their individual accomplishments. For this reason, making a bad decision may not pose the same threat to self-esteem that it would in more individualistic cultures, such as the United States (Choi & Nisbett, 2000; Markus & Kitayama, 1998, 2003).

CONCEPT CHECK STOP

Why do we tend to blame others for their misfortunes but deny responsibility for our own failures?

How can people resolve cognitive dissonance?

What role does culture play in attribution, attitudes, and cognitive dissonance?

Process Diagram

Cognitive dissonance theory FIGURE 15.4

Cognitive Dissonance Theory

People are motivated to maintain consistency in their thoughts, feelings, and actions. → When inconsistencies or conflicts exist between our thoughts, feelings, and actions, they can lead to... → Strong tension and discomfort (cognitive dissonance). → To reduce this cognitive dissonance, we are motivated to change our attitude or behavior.

How would health professionals, who obviously know the dangers of smoking, deal with their cognitive dissonance? They could quit smoking, but, like most people, they probably take the easier route. They change their attitudes about the dangers of smoking by reassuring themselves with examples of people who smoke and live to be 100, or they simply ignore or discount contradictory information.

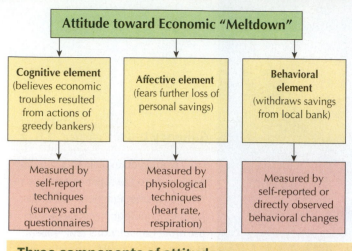

Our Feelings about Others

LEARNING OBJECTIVES

Explain the difference between prejudice and discrimination.

Identify four explanations for why prejudice develops.

Summarize the factors that influence interpersonal attraction.

Explain how loving is different from liking.

Having explored our thoughts about others (attributions and attitudes), we now focus on our feelings about others. First we will examine the negative feelings (and thoughts and actions) associated with prejudice and discrimination. Then we'll explore the generally positive feelings of interpersonal attraction (liking and loving).

PREJUDICE AND DISCRIMINATION

prejudice A learned, generally negative attitude directed toward specific people solely because of their membership in an identified group.

Prejudice, which literally means prejudgment, creates enormous problems for its victims and also limits the perpetrator's ability to accurately judge others and to process information.

Like all attitudes, prejudice is composed of three elements: cognitive (stereotypical beliefs about people in a group), affective (emotions about the group), and behavioral (predisposition to discriminate against members of the group). Although the terms *prejudice* and *discrimination* are often used interchangeably, they are not the same. Prejudice refers to an attitude, whereas **discrimination** refers to action (Fiske, 1998). The two often coincide, but not always (FIGURE 15.5).

How do prejudice and discrimination originate? Why do they persist? Four commonly cited sources of prejudice are learning, mental shortcuts, economic and political competition, and displaced aggression.

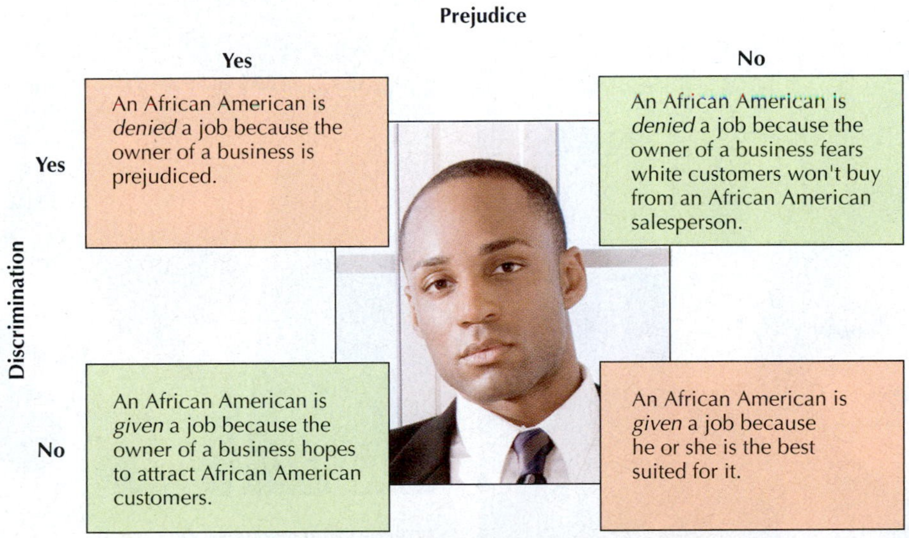
Prejudice
Yes
No
Discrimination
Yes
No
An African American is *denied* a job because the owner of a business is prejudiced.
An African American is *denied* a job because the owner of a business fears white customers won't buy from an African American salesperson.
An African American is *given* a job because the owner of a business hopes to attract African American customers.
An African American is *given* a job because he or she is the best suited for it.

Prejudice and discrimination FIGURE 15.5

Note how prejudice can exist without discrimination, and vice versa.

Learning People learn prejudice the same way they learn all attitudes—through classical and operant conditioning and social learning (Chapter 6). For example, repeated exposure to stereotypical portrayals of minorities and women on television, in movies, and in books and magazines teach children that such images are correct. Hearing parents, friends, and teachers express prejudice also initiates and reinforces prejudice (Anderson & Hamilton, 2005; Bennett et al., 2004; Kassin, Fein, & Markus, 2008; Levitan, 2008; Livingston & Drwecki, 2007; Neto & Furnham, 2005). People also learn prejudice through direct experience. For example, derogating others or receiving attention for expressing racist or sexist remarks can boost one's self-esteem (Fein & Spencer, 1997; Plummer, 2001). Generalizing a single negative experience with a specific member of a group also creates prejudice (Vidmar, 1997).

Mental shortcuts Prejudice may stem from normal attempts to simplify a complex social world (Kulik, 2005; Sternberg, 2007, 2009). Stereotypes allow people to make quick judgments about others, thereby freeing up their mental resources for other activities (FIGURE 15.6A). In fact, stereotypes and prejudice can occur even without a person's conscious awareness or control. This process is known as "automatic" or **implicit bias** (Greenwald et al., 2008; Hofmann et al., 2008; von Hippel, Brener, & von Hippel, 2008).

People use stereotypes to classify others in terms of their membership in a group. Given that people generally classify themselves as part of the preferred group, they also create ingroups and outgroups. An **ingroup** is any category that people see themselves as belonging to; an **outgroup** is any other category (FIGURE 15.6B).

Visualizing

The cost of prejudice FIGURE 15.6

A Harmless stereotypes?
Even seemingly positive forms of prejudice can be destructive. For example, the stereotype that "Asian Americans are good at math" might lead Asian Americans to see few other routes to success. ▼

B Prejudice and war ▲
This group supports Osama Bin Laden and is demonstrating against the United States. How would you explain their loyalty to Bin Laden and prejudice against Americans?

Compared with how they see outgroup members, people tend to see ingroup members as being more attractive, having better personalities, and so on—a phenomenon known as **ingroup favoritism** (Ahmed, 2007; Dunham, 2007; Harth, Kessler, & Leach, 2008). People also tend to recognize greater diversity among members of their ingroup than they do among members of outgroups (Cehajic, Brown, & Castano, 2008; Wegener, Clark, & Petty, 2006). A danger of this **outgroup homogeneity effect** is that when members of minority groups are not recognized as varied and complex individuals, it is easier to treat them in discriminatory ways. During the Vietnam War, for example, labeling Asians "gooks" made it easier for American soldiers to kill large numbers of Vietnamese civilians. As in most wars, facelessness makes it easier to kill large numbers of soldiers, as well as civilians. This type of dehumanization and facelessness also helps to perpetuate our high levels of fear and anxiety associated with terrorism (Haslam et al., 2007; Hodson & Costello, 2007; Zimbardo, 2004, 2007).

Competition for limited resources

Some theorists think that prejudice develops and is maintained because it offers significant economic and political advantages to the dominant group (Esses et al., 2001; Mays, Cochran, & Barnes, 2007; Schaefer, 2008) (FIGURE 15.6C).

Displaced aggression

As we discuss in the next section, frustration often leads people to attack the source of frustration. As history has shown, when the cause of frustration is too powerful and capable of retaliation or is ambiguous, people often redirect their aggression toward an alternative, innocent target, or a scapegoat (FIGURE 15.6D).

Understanding the causes of prejudice is just a first step toward overcoming it. Later in this chapter, we consider several methods that psychologists have developed to reduce prejudice. In the next section, we examine the positive side of our feelings about others.

C Immigration and competition
The immigration question is complex; what roles might competition for limited resources play in anti-immigrant prejudice? Prejudice can lead to atrocities. ▼

D Prejudice and genocide ▲
These skulls are from the recent acts of genocide in Sudan, where thousands of black Africans have died from starvation, disease, and violence.

INTERPERSONAL ATTRACTION

What causes us to feel admiration, liking, friendship, intimacy, lust, or love? All of these social experiences are reflections of **interpersonal attraction**. Psychologists have found three compelling factors in interpersonal attraction: physical attractiveness, proximity, and similarity. Each influences our attraction in different ways.

> **interpersonal attraction** Positive feelings toward another.

Physical attractiveness (size, shape, facial characteristics, and manner of dress) is one of the most important factors in our initial liking or loving of others (Andreoni & Petric, 2008; Buss, 2003, 2005, 2007, 2008; Cunningham, Fink, & Kenix, 2008; Lippa, 2007; Maner et al., 2008). Attractive individuals are seen as more poised, interesting, cooperative, achieving, sociable, independent, intelligent, healthy, and sexually warm than unattractive people (Fink & Penton-Voak, 2002, Swami & Furnham, 2008; Willis, Esqueda, & Schacht, 2008).

Many cultures around the world share similar standards of attractiveness, especially for women—for example, youthful appearance and facial and body symmetry (Fink et al., 2004; Jones, DeBruine, & Little, 2007). From an evolutionary perspective, these findings may reflect the fact that good looks generally indicate good health, sound genes, and high fertility. However, what is considered beautiful also varies from era to era and from culture to culture (FIGURE 15.7).

How do those of us who are not "superstar beautiful" manage to find mates? The good news is that people usually do not hold out for partners who are "ideally attractive." Instead, both men and women tend to select partners whose physical attractiveness approximately matches their own (Regan, 1998; Sprecher & Regan, 2002). Also, people use nonverbal flirting behavior to increase their attractiveness and signal interest to a potential romantic partner (Lott, 2000; Moore, 1998). Although both men and women flirt, in heterosexual couples women generally initiate courtship.

Culture and attraction FIGURE 15.7

Which of these women do you find most attractive? Can you see how your cultural background might train you to prefer one look to the others?

The face in the mirror FIGURE 15.8

We even like ourselves better when we see ourselves in a familiar way. People presented with pictures of themselves and reversed images of themselves strongly prefer the reversed images, which they are used to seeing in the mirror. Close friends prefer the unreversed images (Mita, Dermer, & Knight, 1977).

Proximity also promotes attraction, largely because of **mere exposure**—that is, repeated exposure increases liking (Monin, 2003; Rhodes, Halberstadt, & Brajkovich, 2001) (FIGURE 15.8).

The major cementing factor for both liking and loving relationships is *similarity* in social background and values (Caprara et al., 2007; Morry, 2005; Smithson & Baker, 2008). In other words, "Birds of a feather flock together." However, what about the old saying "Opposites attract"? This adage probably refers to personality traits. That is, attraction to seemingly opposite people is most often based on the recognition that those people offer us something we lack (Dryer & Horowitz, 1997).

We can think of interpersonal attraction as a fundamental building block of how we feel about others. But how do we make sense of love? Many people find the subject to be alternately mysterious, exhilarating, comforting—and even maddening. In this section, we explore three perspectives on love: liking versus loving, romantic love, and companionate love.

Love often develops from initial feelings of friendship and liking. Using a paper-and-pencil test of liking and loving, Zick Rubin (1970) found that liking involves a favorable evaluation reflected in greater feelings of admiration and respect. Love, a more intense experience, involves caring, attachment to the other person, and intimacy (TABLE 15.1). For more about love over the life span, read *What a Psychologist Sees* on page 412.

Couples who scored highest on Rubin's love scale spent more time looking into one another's eyes. Although both partners tended to match each other on their love scores, women liked their dating partners significantly more than they were liked in return (Rubin, 1970).

Sample items from Rubin's liking and loving test TABLE 15.1

Love scale	***Liking scale***
1. I feel that I can confide in _____ about virtually everything.	1. I think that _____ is unusually well adjusted.
2. I would do almost anything for _____.	2. I would highly recommend _____ for a responsible job.
3. If I could never be with _____, I would feel miserable.	3. In my opinion, _____ is an exceptionally mature person.

Source: Rubin, Z. (1970). "Measurement of romantic love," *Journal of Personality and Social Psychology, 16,* 265–273.

What a Psychologist Sees

Love over the Life Span

Even in the most devoted couples, the intense attraction and excitement of **romantic love** generally begin to fade 6 to 30 months after the relationship begins (Hatfield & Rapson, 1996; Livingston, 1999). This is because romantic love is largely based on mystery and fantasy—people often fall in love with what they want another person to be (Fletcher & Simpson, 2000; Levine, 2001). These illusions usually fade with the realities of everyday living.

Companionate love is a strong, lasting attraction based on admiration, respect, trust, deep caring, and commitment. Studies of close friendships show that satisfaction grows with time as we come to recognize the value of companionship and intimacy (Kim & Hatfield, 2004). One tip for maintaining companionate love is to overlook each other's faults. People are more satisfied with relationships when they have a somewhat idealized perception of their partner (Campbell et al., 2001; Fletcher & Simpson, 2000). This makes sense in light of research on cognitive dissonance: Idealizing our mates allows us to believe we have a good deal.

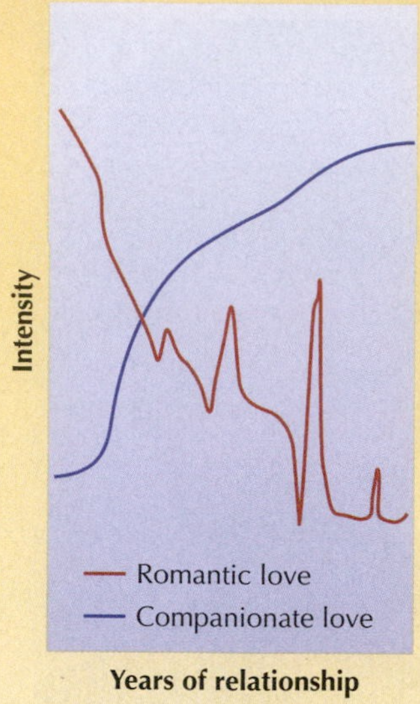

CONCEPT CHECK STOP

How does prejudice differ from discrimination?

What process accounts for someone saying that members of another ethnic group "all look alike"?

What are the major factors that affect our attraction to others?

Our Actions Toward Others

LEARNING OBJECTIVES

Identify the factors that contribute to conformity and obedience.

Explain how groups affect behavior and decision making.

Summarize the biological and psychosocial factors believed to be involved in aggression.

Compare the egoistic model and the empathy-altruism hypothesis.

Kurt Lewin (1890–1947), often considered the "father of social psychology," was among the first to suggest that all behavior results from interactions between the individual and the environment. In this section we consider several examples of such interaction, including social influence, group processes, and aggression.

SOCIAL INFLUENCE: CONFORMITY AND OBEDIENCE

Our society and culture teach us to believe certain things, feel certain ways, and act in accordance with these beliefs and feelings. These influences are so strong and so much a part of who we are that we rarely recognize them. In this section, we discuss two kinds of social influence: conformity and obedience.

Solomon Asch's study of conformity

FIGURE 15.9

Which line (A, B or C) is most like line X? Could anyone convince you otherwise?

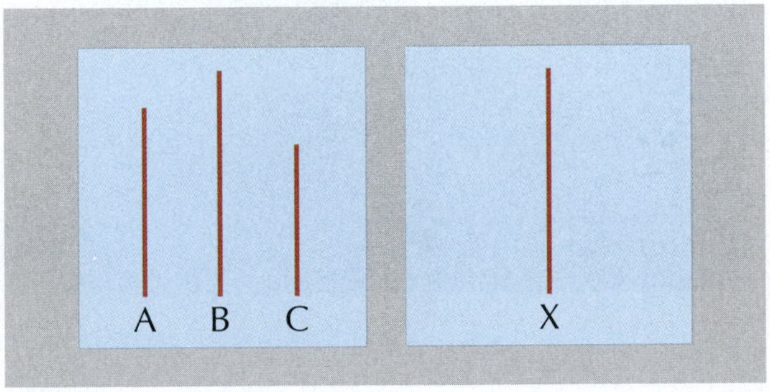

Conformity—Going along with others

Imagine that you have volunteered for a psychology experiment on perception. You are seated around a table with six other students. You are all shown a card with three lines labeled A, B, and C, as in FIGURE 15.9. You are then asked to select the line that is closest in length to a fourth line, X.

At first, everyone agrees on the correct line. On the third trial, however, the first participant selects line A, obviously a wrong answer. When the second, third, fourth, and fifth participants also say line A, you really start to wonder: "What's going on here? Are they blind? Or am I?"

What do you think you would do at this point in the experiment? Would you stick with your convictions and say line B, regardless of what the others have answered? Or would you go along with the group? In the original version of this experiment, conducted by Solomon Asch (1951), six of the seven participants were actually *confederates* of the experimenter (that is, they were working with the experimenter and purposely gave wrong answers). Their incorrect responses were designed to test the participant's degree of **conformity**.

conformity The act of changing behavior as a result of real or imagined group pressure.

More than one-third of Asch's participants conformed—they agreed with the group's obviously incorrect choice. (Participants in a control group experienced no group pressure and almost always chose correctly.) Asch's study has been conducted dozens of times, in at least 17 countries, and always with similar results (Bond & Smith, 1996; Jung, 2006; Takano & Sogon, 2008).

Why would so many people conform? To the onlooker, conformity is often difficult to understand. Even the conformer sometimes has a hard time explaining his or her behavior. Let's look at three factors that drive conformity:

- *Normative social influence.* Often, people conform to group pressure out of a need for approval and acceptance by the group. **Norms** are expected behaviors that are adhered to by members of a group. Most often, norms are quite subtle and implicit. Have you ever asked what others are wearing to a party, or watched your neighbor to be sure you pick up the right fork? Such behavior reflects your desire to conform and the power of normative social influence.

- *Informational social influence.* Have you ever bought a specific product simply because of a friend's recommendation? You conform not to gain their approval (normative social influence) but because you assume they have more information than you do. Given that participants in Asch's experiment observed all the other participants give unanimous decisions on the length of the lines, they also may have conformed because they believed the others had more information.

- *Reference groups.* The third major factor in conformity is the power of **reference groups**—people we most admire, like, and want to resemble. Attractive actors and popular sports stars are paid millions of dollars to endorse products because advertisers know that we want to be as cool as Wesley Snipes or as beautiful as Jennifer Lopez. Of course, we also have more important reference groups in our lives—parents, friends, family members, teachers, religious leaders, and so on.

Psychological Science

Cultural Norms for Personal Space

Culture and socialization have a lot to do with shaping norms for personal space. If someone invades the invisible "personal bubble" around our bodies, we generally feel very uncomfortable. People from Mediterranean, Muslim, and Latin American countries tend to maintain smaller interpersonal distances than do North Americans and Northern Europeans (Axtell, 2007; Steinhart, 1986). Children also tend to stand very close to others until they are socialized to recognize and maintain a greater personal distance. Furthermore, friends stand closer than strangers, women tend to stand closer than men, and violent prisoners prefer approximately three times the personal space of nonviolent prisoners (Axtell, 2007; Gilmour & Walkey, 1981; Lawrence & Andrews, 2004).

CRITICAL THINKING

Here are two interesting questions:

1. How might cultural differences in personal space help explain why Americans traveling abroad are sometimes seen as being "too loud and brassy"?
2. If men and women have different norms for personal space, what effect might this have on their relationships?

The advantages of conformity and obedience FIGURE 15.10

These people willingly obey the firefighters who order them to evacuate a building, and many lives are saved. What would happen to our everyday functioning if people did not go along with the crowd or generally did not obey orders?

Obedience—Following orders

As we've seen, conformity involves going along with the group. A second form of social influence, **obedience**, involves going along with a direct command, usually from someone in a position of authority.

> **obedience** The act of following a direct command, usually from an authority figure.

Conformity and obedience aren't always bad (FIGURE 15.10). In fact, most people conform and obey most of the time because it is in their own best interest (and everyone else's) to do so. Like most North Americans, you stand in line at the movie theatre instead of pushing ahead of others. This allows an orderly purchasing of tickets. Conformity and obedience allow social life to proceed with safety, order, and predictability.

However, on some occasions it is important not to conform or obey. We don't want teenagers (or adults) engaging in risky sex or drug use just to be part of the crowd. And we don't want soldiers (or anyone else) mindlessly following orders just because they were told to do so by an authority figure. Because recognizing and resisting destructive forms of obedience are particularly important to our society, we'll explore this material in greater depth at the end of this chapter.

Imagine that you have responded to a newspaper ad that is seeking volunteers for a study on memory. At the Yale University laboratory, an experimenter explains to you and another participant that he is studying the effects of punishment on learning and memory. You are selected to play the role of the "teacher." The experimenter leads you into a room where he straps the other participant—the "learner"—into a chair. He applies electrode paste to the learner's wrist "to avoid blisters and burns" and attaches an electrode that is connected to a shock generator.

What a Psychologist Sees

What Influences Obedience?

Milgram conducted a series of studies to discover the specific conditions that either increased or decreased obedience to authority.

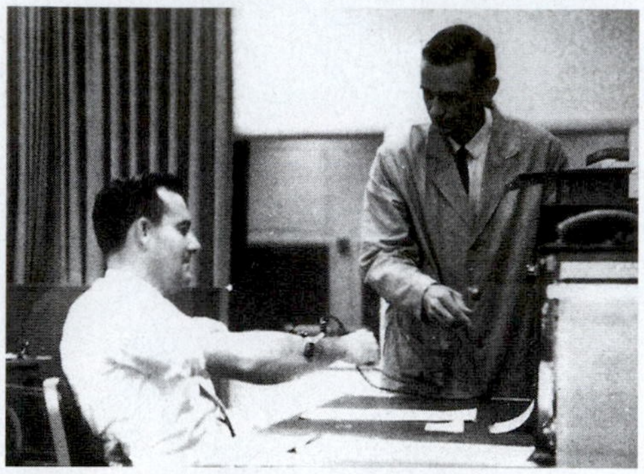

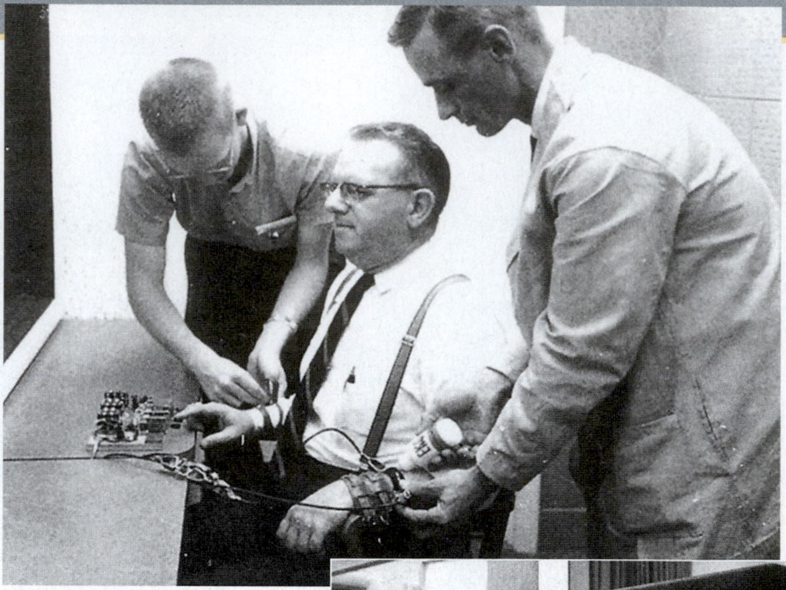

Milgram's Learner
Under orders from an experimenter, would you shock a man with a known heart condition who is screaming and begging to be released? Few people believe they would. But research shows otherwise.

Milgram's Shock Generator

You are shown into an adjacent room and told to sit in front of this same shock generator, which is wired through the wall to the chair of the learner. The shock machine consists of 30 switches representing successively higher levels of shock, from 15 volts to 450 volts. Written labels appear below each group of switches, ranging from "Slight Shock" to "Danger: Severe Shock," all the way to "XXX." The experimenter explains that it is your job to teach the learner a list of word pairs and to punish any errors by administering a shock. With each wrong answer, you are to increase the shock by one level. (The setup for the experiment is illustrated in *What a Psychologist Sees.*)

You begin teaching the word pairs, but the learner's responses are often wrong. Before long, you are inflicting shocks that you can only assume must be extremely painful. After you administer 150 volts, the learner begins to protest: "Get me out of here. . . . I refuse to go on."

You hesitate, and the experimenter tells you to continue. He insists that even if the learner refuses to answer, you must keep increasing the shock levels. But the other person is obviously in pain. What should you do?

Actual participants in this research—the "teachers"—suffered real conflict and distress when confronted with this problem. They sweated, trembled, stuttered, laughed nervously, and repeatedly protested that they did not want to hurt the learner. But still they obeyed.

The psychologist who designed this study, Stanley Milgram, was actually investigating not punishment and learning but obedience to authority: Would participants obey the experimenter's prompts and commands to shock another human being? In Milgram's public survey, fewer than 25 percent thought they would go beyond 150 volts. And no respondents predicted they would go past the 300-volt level. Yet 65 percent of the

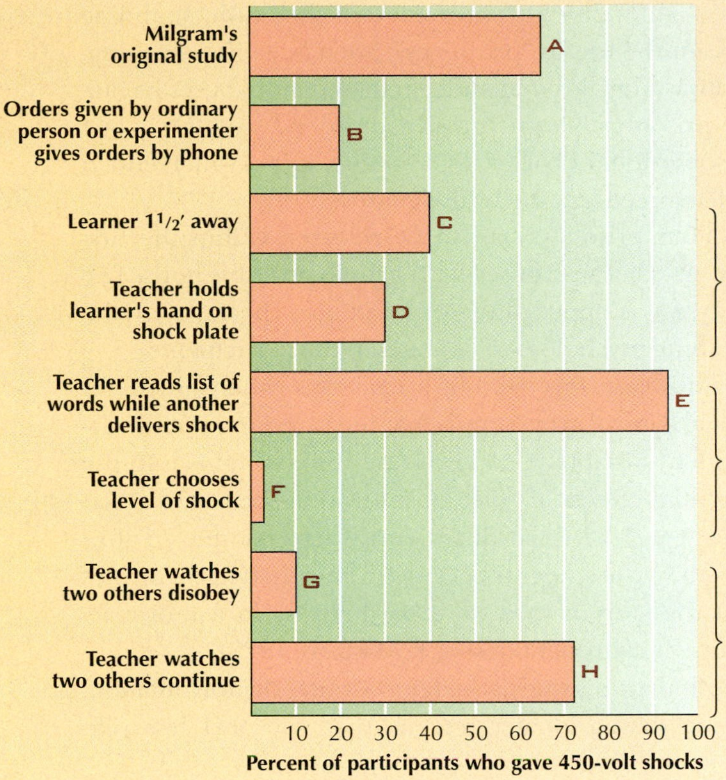

As you can see in the first bar on the graph (**A**), 65 percent of the participants in Milgram's original study gave the learner the full 450-volt level of shocks.

In the second bar (**B**), when orders came from an ordinary person, or from the experimenter by phone, (versus in person) obedience dropped to 20 percent.

Now look at the third and fourth bars (**C** and **D**). Note how the physical closeness of the victim affected obedience. In the original experiment, the learner was seated behind a wall in another room, and 65 percent of the teachers gave the full 450-volt level of shocks. But obedience dropped to 40 percent when the learner was only 1 1/2 feet away, and dropped even further to 30 percent when the teacher was required to hold the learner's hand on the shock plate.

Looking at bars (**E**) and (**F**), how does the teacher's level of responsibility affect obedience? In the original experiment, the teacher was required to actually pull the lever that supposedly delivered shocks to the learner. When the teacher only read the list of words, while another person delivered the shock, obedience increased to 92 percent. In contrast, when the teacher was allowed to choose the level of shock, obedience dropped to 2 percent.

Finally, looking at bars (**G**) and (**H**), note how modeling and imitation affected obedience. When teachers watched two other supposed teachers disobey the experimenter's orders, their own obedience dropped to 10 percent. However, when they watched other teachers follow the experimenter's orders, their obedience increased to over 70 percent (Milgram 1963, 1974).

teacher-participants in this series of studies obeyed completely—going all the way to the end of the scale, even beyond the point when the "learner" (Milgram's confederate, who actually received no shocks at all) stopped responding altogether.

Even Milgram was surprised by his results. Before the study began, he polled a group of psychiatrists, and they predicted that most people would refuse to go beyond 150 volts and that less than 1 percent of those tested would "go all the way." But, as Milgram discovered, most of his participants—men and women, of all ages, and from all walks of life—administered the highest voltage. The study was replicated many times and in many other countries, with similarly high levels of obedience.

In a series of follow-up studies, Milgram found several important factors that influenced obedience: (1) legitimacy and closeness of the authority figure, (2) remoteness of the victim, (3) assignment of responsibility, (4) modeling or imitating others. (Blass, 1991, 2000; Meeus & Raaijmakers, 1989; Snyder, 2003). These are summarized in *What a Psychologist Sees* on page 416.

Although there have been some recent partial replications of Milgram's study (e.g., Burger, 2009), Milgram's original setup could never be undertaken today. Many people are upset both by Milgram's findings and by the treatment of the participants in his research. Deception is a necessary part of some research, but the degree of deception and discomfort of participants in Milgram's research is now viewed as highly unethical. On the other hand, Milgram carefully debriefed every subject after the study and followed up with the participants for several months. Most of his "teachers" reported the experience as being personally informative and valuable.

GROUP PROCESSES

Although we seldom recognize the power of group membership, social psychologists have identified several important ways that groups affect us.

Group membership How do the roles that we play within groups affect our behavior? This question fascinated social psychologist Philip Zimbardo. In his famous study at Stanford University, 24 carefully screened, well-adjusted young college men were paid $15 a day for participating in a two-week simulation of prison life (Haney, Banks, & Zimbardo, 1978; Zimbardo, 1993).

The students were randomly assigned to the role of either prisoner or guard. Prisoners were "arrested," frisked, photographed, fingerprinted, and booked at the police station. They were then blindfolded and driven to the "Stanford Prison." There, they were given ID numbers, deloused, issued prison clothing (tight nylon caps, shapeless gowns, and no underwear), and locked in cells. Participants assigned to be guards were outfitted with official-looking uniforms, billy clubs, and whistles, and they were given complete control.

Not even Zimbardo foresaw how the study would turn out. Although some guards were nicer to the prisoners than others, they all engaged in some abuse of power. The slightest disobedience was punished with degrading tasks or the loss of "privileges" (such as eating, sleeping, and washing). As demands increased and abuses began, the prisoners became passive and depressed. Only one prisoner fought back with a hunger strike, which ended with a forced feeding by the guards.

Four prisoners had to be released within the first four days because of severe psychological reactions. The study was stopped after only six days because of the alarming psychological changes in the participants.

Although this was not a true experiment in that it lacked control groups and clear measurements of the dependent variable (Chapter 1), it offers insights into the potential effects of roles on individual behavior (FIGURE 15.11). According to interviews conducted after the study, the students became so absorbed in their roles that they forgot they were participants in a university study (Zimbardo, Ebbeson, & Maslach, 1977).

Zimbardo's study also demonstrates **deindividuation**. To be deindividuated means that you feel less self-

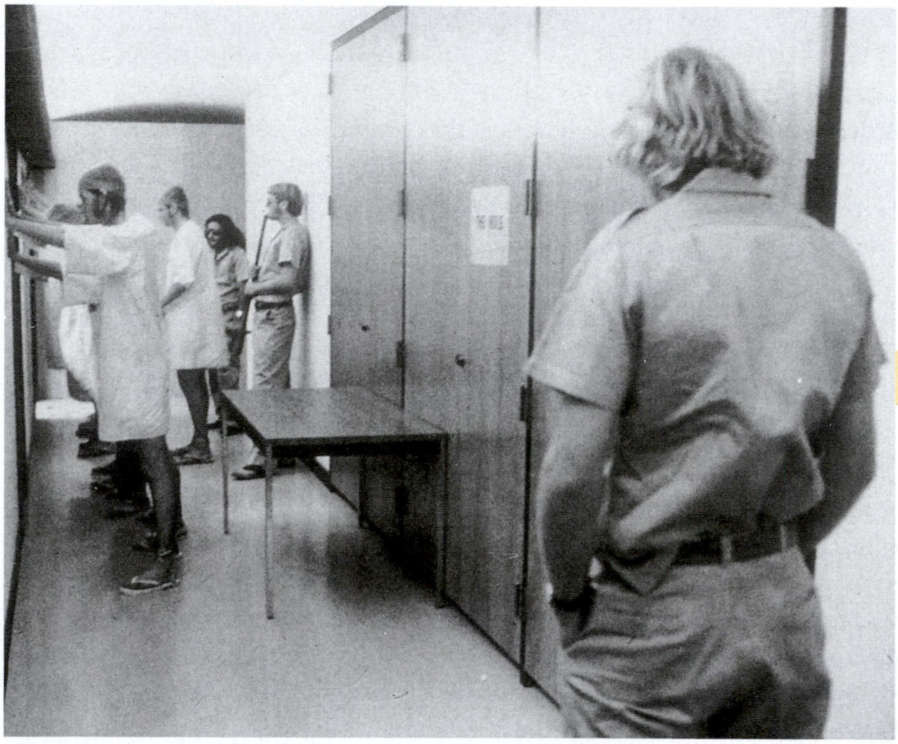

Power corrupts FIGURE 15.11

Zimbardo's prison study showed how the demands of roles and situations could produce dramatic changes in behavior in just a few days. Can you imagine what happens to prisoners during life imprisonment, six-year sentences, or even a few nights in jail?

Lost in the crowd FIGURE 15.12

One of the most compelling explanations for deindividuation is the fact that the presence of others tends to increase arousal and feelings of anonymity. Anonymity is a powerful disinhibitor, helping to explain why vandalism seems to increase on Halloween (when people commonly wear masks) and why most crimes and riots occur at night—under the cover of darkness. Can you imagine your own behavior changing under such conditions?

conscious, less inhibited, and less personally responsible as a member of a group than when you're alone. Although deindividuation can sometimes be positive, it also helps explain angry mobs, rioters, gang rapes, lynchings, and hate crimes (Rodriques, Assmar, & Jablonski, 2005; Zimbardo, 2007). Groups sometimes actively promote deindividuation by requiring its members to wear uniforms, for example, as a way to increase allegiance and conformity (FIGURE 15.12).

Group decision making

We have seen that the groups we belong to and the roles that we play within them influence how we think about ourselves. How do groups affect our decisions? Are two heads truly better than one?

Most people assume that group decisions are more conservative, cautious, and middle-of-the-road than individual decisions. But is this true? Initial investigations indicated that after discussing an issue, groups actually support riskier decisions than decisions they made as individuals before the discussion (Stoner, 1961).

Subsequent research on this **risky-shift** phenomenon, however, shows that some groups support riskier decisions while others support conservative decisions (Liu & Latané, 1998). How can we tell if a given group's decision will be risky or conservative? The final decision (risky or conservative) depends primarily on the dominant preexisting tendencies of the group. That is, as individuals interact and discuss their opinions, their initial positions become more exaggerated (or polarized).

Why are group decisions more exaggerated? This tendency toward **group polarization** stems from increased

group polarization A group's movement toward either riskier or more conservative behavior, depending on the members' initial dominant tendencies.

groupthink Faulty decision making that occurs when a highly cohesive group strives for agreement and avoids inconsistent information.

exposure to persuasive arguments that reinforce the group's original opinion (Liu & Latané, 1998).

A related phenomenon is **groupthink** (FIGURE 15.13). When a group is highly cohesive (a couple, a family, a panel of military advisers, an athletic team), the members' desire for agreement may lead them to ignore important information or points of view held by outsiders or critics (Hergovich & Olbrich, 2003; Vaughn, 1996). During the discussion process, the members:

- come to believe that they are invulnerable,
- tend to develop common rationalizations and stereotypes of the outgroup, and
- exert considerable pressure on anyone who dares to offer a dissenting opinion.

Many presidential errors—from Franklin Roosevelt's failure to anticipate the attack on Pearl Harbor to Ronald Reagan's Iran-Contra scandal—have been blamed on groupthink. Groupthink may have contributed to the losses of the space shuttles *Challenger* and *Columbia,* the terrorist attack of September 11, and the war in Iraq (Barlow 2008; Ehrenreich, 2004; Janis, 1972, 1989; Landy & Kuhnehenn, 2004; Rodriques, Assmar, & Jablonski, 2005).

Process Diagram

How groupthink occurs FIGURE 15.13

Antecedent Conditions

1. A highly cohesive group of decision makers
2. Insulation of the group from outside influences
3. A directive leader
4. Lack of procedures to ensure careful consideration of the pros and cons of alternative actions
5. High stress from external threats with little hope of finding a better solution than that favored by the leader

↓

Strong desire for group consensus—the groupthink tendency

↓

Symptoms of Groupthink

1. Illusion of invulnerability
2. Belief in the morality of the group
3. Collective rationalizations
4. Stereotypes of outgroups
5. Self-censorship of doubts and dissenting opinions
6. Illusion of unanimity
7. Direct pressure on dissenters

↓

Symptoms of Poor Decision Making

1. An incomplete survey of alternative courses of action
2. An incomplete survey of group objectives
3. Failure to examine risks of the preferred choice
4. Failure to reappraise rejected alternatives
5. Poor search for relevant information
6. Selective bias in processing information
7. Failure to develop contingency plans

↓

Low probability of successful outcome

◀ A The process of groupthink begins when group members feel a strong sense of cohesiveness and isolation from the judgments of qualified outsiders. Add a directive leader and little chance for debate, and you have the recipe for a potentially dangerous decision.

▲ B Few people realize that the decision to marry can be a form of groupthink. (Remember that a "group" can have as few as two members.) When planning for marriage, a couple may show symptoms of groupthink such as an illusion of invulnerability ("We're different—we won't ever get divorced"), collective rationalizations ("Two can live more cheaply than one"), shared stereotypes of the outgroup ("Couples with problems just don't know how to communicate"), and pressure on dissenters ("If you don't support our decision to marry, we don't want you at the wedding!").

AGGRESSION

aggression Any behavior intended to harm someone.

Why do people act aggressively? In this section, we explore both biological and psychosocial explanations for **aggression** and how aggression can be controlled or reduced.

Biological explanations

Instincts. Because aggression has such a long history and is found in all cultures, many theorists believe that humans are instinctively aggressive. Evolutionary psychologists and ethologists (scientists who study animal behavior) believe that aggression evolved because it prevents overcrowding and allows the strongest animals to win mates and reproduce (Buss, 2008; Kardong, 2008). Most social psychologists, however, reject the view that instincts drive aggression.

Genes. Twin studies suggest that some individuals are genetically predisposed to have hostile, irritable temperaments and to engage in aggressive acts (Haberstick et al., 2006; Hartwell, 2008; van Lier et al., 2007). Remember, though, that biology interacts with social experience to shape behavior.

The brain and nervous system. Electrical stimulation or the severing of specific parts of an animal's brain has a direct effect on aggression (Delgado, 1960; Delville, Mansour, & Ferris, 1996; Roberts & Nagel, 1996). Research with brain injuries and organic disorders has also identified possible aggression circuits in the brain (Anderson & Silver, 2008; Delgado, 1960; Kotulak, 2008; Siever, 2008).

Substance abuse and other mental disorders. Substance abuse (particularly alcohol) is a major factor in most forms of aggression (Tremblay, Graham, & Wells, 2008; Levinthal, 2008). Homicide rates are also higher among men with schizophrenia and antisocial disorders, particularly if they abuse alcohol (Carballo et al., 2008; Fresán et al., 2007; Garno, Gunawardane, & Goldberg, 2008; Haddock & Shaw, 2008).

Hormones and neurotransmitters. Several studies have linked the male hormone testosterone to aggressive behavior (Hermans, Ramsey, & van Honk, 2008; Juntil, Coats, & Shah, 2008; Popma et al., 2007; Sato et al., 2008). Violent behavior has also been linked with lowered levels of the neurotransmitters serotonin and GABA (gamma-aminobutyric acid) (Kovacic et al., 2008; Siever, 2008).

Psychosocial explanations

Aversive stimuli. Noise, heat, pain, insults, and foul odors all can increase aggression (Anderson, 2001; Anderson, Buckley, & Carnegey, 2008; Monks, Ortega-Ruiz, & Rodríguez-Hidalgo, 2008; Twenge et al., 2001). According to the **frustration-aggression hypothesis**, developed by John Dollard and colleagues (1939), another aversive stimulus—frustration—creates anger, which for some leads to aggression.

Culture and learning. Social learning theory (Chapter 6) suggests that people raised in a culture with aggressive models will learn aggressive responses (Matsumoto & Juang, 2008). Among developed nations, for example, the United States has a high rate of violent crimes, and American children grow up with numerous models for aggression, which they tend to imitate.

Media and video games. The media can contribute to aggression in both children and adults (Anderson, 2004; Bartholow & Anderson, 2002; Kronenberger et al., 2005; Uhlmann & Swanson, 2004) (Figure 15.14). However, the link between media violence and aggression appears to be a two-way street. Laboratory studies, correlational research, and cross-cultural studies have found that exposure to TV violence increases aggressiveness and that aggressive children tend to seek out violent programs (Barlett, Harris, & Bruey, 2008; Carnagey, Anderson, & Bartholow, 2007; Giumetti & Markey, 2007).

Video violence Figure 15.14

Video games such as *Doom*, *Mortal Kombat*, *Resident Evil*, and *Half-Life* all feature realistic sound effects and gory depictions of "lifelike" violence. Research shows that media can contribute to aggression in children and that aggressive children seek out violent television programs.

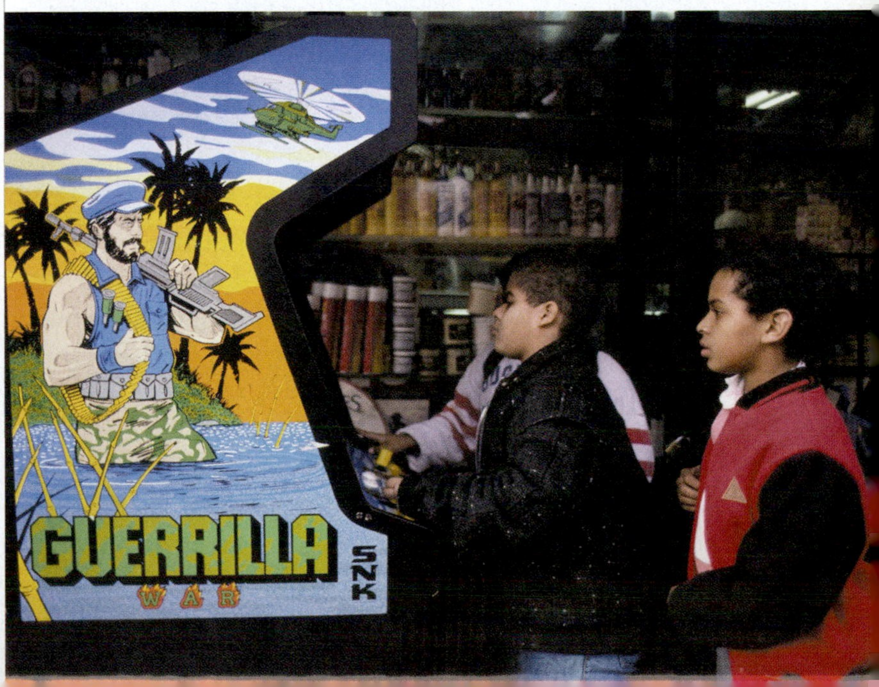

How can people control or eliminate aggression? Some therapists advise people to release their aggressive impulses by engaging in harmless forms of aggression, such as vigorous exercise, punching a pillow, or watching competitive sports. But studies suggest that this type of *catharsis* doesn't really help and may only intensify the feeling (Bushman, 2002; Kuperstok, 2008).

A more effective approach is to introduce incompatible responses. Because certain emotional responses, such as empathy and humor, are incompatible with aggression, purposely making a joke or showing some sympathy for an opposing person's point of view can reduce anger and frustration (Garrick, 2006; Kassin, Fein, & Markus, 2008; Kaukiainen et al., 1999; Oshima, 2000; Weiner, 2006). Improving social and communication skills can also help keep aggression in check.

ALTRUISM: HELPING OTHERS

On January 15, 2009, pilot Chesley B. "Sully" Sullenberger successfully crash-landed a US Airways jet in New York's Hudson River six minutes after departing from LaGuardia airport (FIGURE 15.15). Thanks to the incredible skill and professionalism of Sully Sullenberger and the flight crew, along with incredibly fast responses from local commercial vessels and the New York fire and police forces, all of the passengers and flight crew were rescued safely.

The incredible **altruism** (or **prosocial behavior**) shown by the pilot, flight crew, and rescuers of this airliner has practical consequences and profound implications for our view of human nature. Under what conditions do people respond to, or sometimes ignore, others' pleas for help? Let's look at the three major theories: *evolutionary*, *egoistic*, and *empathy-altruism* (FIGURE 15.16A).

altruism Actions designed to help others with no obvious benefit to the helper.

Evolutionary theory suggests that altruism is an instinctual behavior that has evolved because it favors survival of one's genes (de Waal, 2008; Kardong, 2008; McNamara et al., 2008). By helping your child or other relative, for example, you increase the odds of your genes' survival.

Other research suggests that helping others may actually be self-interest in disguise. According to this **egoistic model**, we help others only because we hope for later reciprocation, because it makes us feel virtuous, or because it helps us avoid feeling guilty (Cialdini, 2009; Williams et al., 1998).

Opposing the evolutionary and egoistic model is the **empathy-altruism hypothesis** (Batson, 1991, 1998, 2006; Batson & Ahmad, 2001). This perspective holds that simply seeing or hearing of another person's suffering can create *empathy*—a subjective grasp of that person's feelings or experiences. And when we feel empathic toward another, we are motivated to help that person for his or her own sake. The ability to empathize may even be innate. Research with infants in the first few hours of life shows that they become distressed and cry at the sound of another infant's cries (Hay, 1994; Hoffman, 1993).

Many theories have been proposed to explain why people help, but few explain why we do not. One of the most comprehensive explanations for helping or not helping comes from the research of Bibb Latané and John Darley (1970) (see FIGURE 15.16B). They found that whether or not someone helps depends on a series of interconnected events and decisions: The potential helper must first notice what is happening, interpret the event as an emergency, accept personal responsibility for helping, decide how to help, and then actually initiate the helping behavior.

A recent example of altruism: The "Miracle on the Hudson" FIGURE 15.15

Imagine yourself as one of these soaked and freezing passengers waiting for help on the wings of the rapidly sinking airliner. Having just survived a horrific fall from the sky, you're now at great risk of being swept away into the Hudson River. Why should anyone come to your aid? What motivated captains of nearby commercial vessels to jeopardize their boats and own safety to rush to help? Why did some passengers risk their own lives to help their fellow travelers?

Visualizing

Helping FIGURE 15.16

A Three Models for Helping

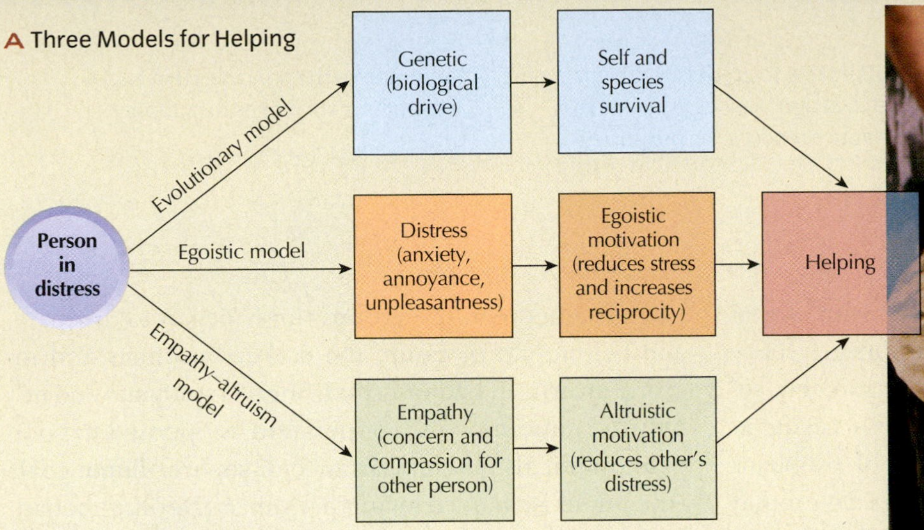

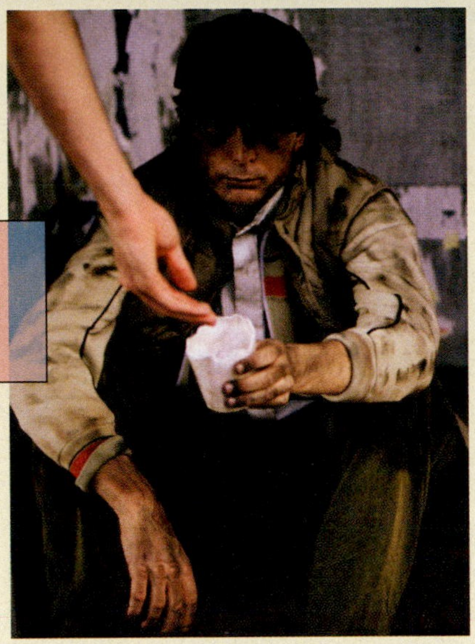

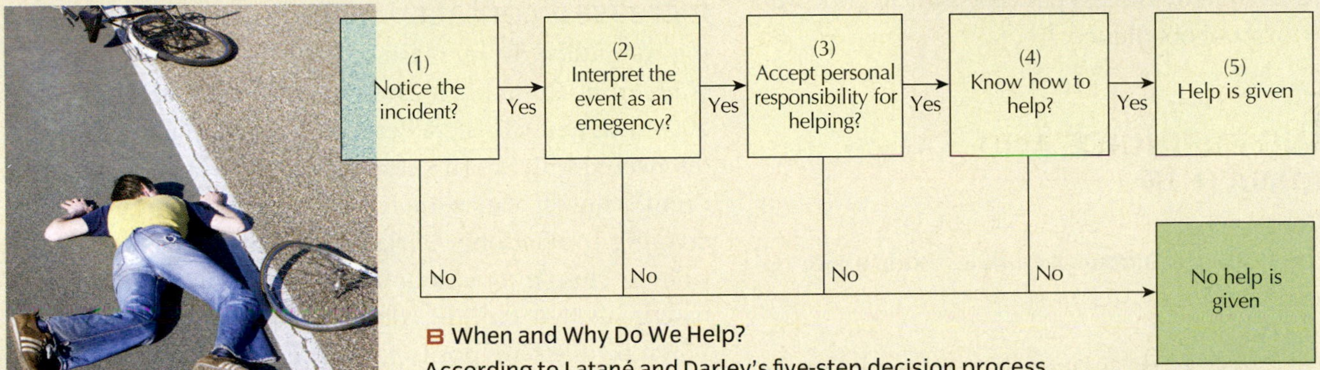

B When and Why Do We Help?
According to Latané and Darley's five-step decision process, if the answer at each step is yes, help is given. If the answer is no at any point, the helping process ends.

NATIONAL GEOGRAPHIC

How does this sequence explain television programs and "caught on tape" situations in which people are robbed or attacked and no one comes to their aid? In follow-up interviews most onlookers report that they failed to intervene because they were certain that someone must have called the police already. Latané and Darley called this the **diffusion of responsibility** phenomenon—the dilution (or diffusion) of personal responsibility for acting by spreading it among all other group members.

CONCEPT CHECK STOP

How do norms, need for information, and reference groups produce conformity?

How does group polarization occur?

What are the most effective ways to control aggression?

What is diffusion of responsibility?

Applying Social Psychology to Social Problems

LEARNING OBJECTIVES

Describe four major approaches to reducing prejudice and discrimination.

Explain how social changes might create cognitive dissonance and eventually promote a reduction in prejudice.

Summarize the principles that explain destructive obedience to authority.

Every day we're confronted with social problems—from noisy neighbors to freeway congestion to terrorism. Unfortunately, social psychology has been more successful in describing, explaining, and predicting social problems than in changing them. However, researchers have found several helpful techniques. In this section, we first explore what scientists have discovered about methods to reduce prejudice, and then we discuss effective ways to cope with destructive forms of obedience.

REDUCING PREJUDICE AND DISCRIMINATION

What can be done to combat prejudice? Four major approaches can be used.

Cooperation and superordinate goals

One of the best ways to combat prejudice is to encourage cooperation rather than competition (Cunningham, 2002; Sassenberg et al., 2007). In a classic study, Muzafer Sherif and his colleagues (1966, 1998) created strong feelings of ingroup loyalty at a summer camp by physically separating two groups of boys into different cabins and assigning different projects to each group.

Once each group developed strong feelings of group identity and allegiance, the researchers set up a series of competitive games and awarded desirable prizes to the winning teams. The groups soon began to pick fights, call each other names, and raid each other's camps.

After using competition to create prejudice between the two groups, the researchers created "minicrises" and tasks that required expertise, labor, and cooperation from both groups, and prizes were awarded to all. The prejudice between the groups slowly began to dissipate, and by the end of camp, the earlier hostilities and ingroup favoritism had vanished. Sherif's study showed not only the importance of cooperation as opposed to competition but also the importance of **superordinate goals** (the minicrises) in reducing prejudice (Der-Karabetian, Stephenson, & Poggi, 1996).

Increased contact

A second approach to reducing prejudice is increasing contact between groups (Cameron, Rutland, & Brown, 2007; Gómez & Huici, 2008; Wagner, Christ, & Pettigrew, 2008). But as you just discovered with Sherif's study of the boys at the summer camp, contact can sometimes increase prejudice. Increasing contact only works under conditions that provide for close interaction, interdependence (superordinate goals that require cooperation), and equal status. When people have positive experiences with one group, they tend to generalize to other groups (Pettigrew, 1998).

Cognitive retraining

One of the most recent strategies for prejudice reduction requires taking another's perspective or undoing associations of negative stereotypical traits (Buswell, 2006; Galinsky & Ku, 2004; Galinsky & Moskowitz, 2000). People can also learn to be unprejudiced if they are taught to selectively pay attention to similarities between groups rather than differences (Phillips & Ziller, 1997).

Cognitive dissonance

One of the most efficient methods of changing an attitude uses the principle of cognitive dissonance (Cook, 2000; D'Alessio & Allen, 2002). Each time we meet someone who does not conform to our

Creating cognitive dissonance FIGURE 15.17

How might social changes, such as school busing, integrated housing, and increased civil rights legislation, initially create cognitive dissonance and then eventually lead to a reduction in prejudice? What other principle of attitude change is illustrated in this photo?

prejudiced views, we experience dissonance—"I thought all gay men were effeminate. This guy is a deep-voiced professional athlete. I'm confused." To resolve the dissonance, we can maintain our stereotypes by saying, "This gay man is an exception to the rule." However, if we continue to come into contact with a variety of gay men, this "exception to the rule" defense eventually breaks down, and attitude change occurs (FIGURE 15.17).

OVERCOMING DESTRUCTIVE OBEDIENCE: WHEN IS IT OKAY TO SAY NO?

Obedience to authority is an important part of our lives. If we routinely refused to obey police officers, firefighters, and other official personnel, our individual safety and social world would collapse. However, there are also many times when obedience may be unnecessary and even destructive—such as obedience to someone who is abusing his or her military power or to a religious cult leader—and should be reduced.

How do we explain (and hopefully reduce) destructive obedience? Let's reexamine some of the major points discussed in this chapter, while also introducing a few new ones.

Socialization Society and culture have a tremendous influence on all our thoughts, feelings, and actions. Obedience is no exception. From very early childhood, we're taught to respect and obey our parents, teachers, and other authority figures, and without this obedience there would be social chaos. Unfortunately, this early (and lifelong) socialization often becomes so deeply ingrained that we no longer recognize it, which helps explain many instances of mindless obedience to immoral requests from people in positions of authority (such as the actions of Nazi prison guards who were "just following orders"). History is replete with instances in which atrocities were committed because people were "just following orders."

Power of the situation Situational factors also have a strong impact on obedience. For example, the roles of police officer or public citizen, teacher or student, and parent or child all have built-in guidelines for appropriate behavior. One person is ultimately "in charge," and the other person is supposed to follow along. Because these roles are so well socialized, we mindlessly play them and find it difficult to recognize the point where they become maladaptive. As we discovered in the Zimbardo prison study, when well-adjusted and well-screened college students were suddenly given the roles of prisoners and guards, their behaviors were dramatically affected.

Groupthink When discussing Milgram's study and other instances of destructive obedience, most people believe that they and their friends would never do such a thing. Can you see how this might be a form of groupthink, a type of faulty thinking that occurs when group members strive for agreement and avoid inconsistent information? When we proclaim that Americans would never follow the orders that some German people did during the holocaust, we're demonstrating several symptoms of groupthink—stereotypes of the outgroup, illusion of unanimity, belief in the morality of the group, and so on. We're also missing one of the most important lessons from Milgram's studies and the cross-cultural follow-ups: As philosopher Hannah Arendt has suggested, the horrifying thing about the Nazis was not that they were so deviant but that they were so "terrifyingly normal."

Foot-in-the-door The gradual nature of many situations involving obedience may also help explain why so many people were willing to give the maximum shocks in Milgram's studies. The initial mild level of shocks may have worked as a **foot-in-the-door technique**, in which a first, small request is used as a setup for later, larger requests. Once Milgram's participants complied with the initial request, they might have felt obligated to continue (Chartrand, Pinckert, & Burger, 1999; Sabini & Silver, 1993). Can you see how this technique helps explain other forms of destructive obedience?

Relaxed moral guard One common intellectual illusion that hinders critical thinking about obedience is the belief that only evil people do evil things, or that evil announces itself. The experimenter in Milgram's study looked and acted like a reasonable person who was simply carrying out a research project. Because he was not seen as personally corrupt and evil, the participants' normal moral guard was down, which can maximize obedience. This relaxed moral guard also may help explain obedience to highly respected military officers and leaders of religious cults.

Disobedient models These forces toward obedience are powerful. It's important to remember that each of us must be personally alert to immoral forms of obedience. On occasion, we also need the courage to stand up and say "No!" One of the most beautiful and historically important examples of this type of bravery occurred in Alabama in 1955. Rosa Parks (FIGURE 15.18) boarded a bus and, as expected in those times, she obediently sat in the back section marked "Negroes." When the bus became crowded, the driver told her to give up her seat for a white man. Surprisingly for those days, Parks refused and was eventually forced off the bus by police and arrested. This single act of disobedience was the catalyst for the small, but growing civil rights movement.

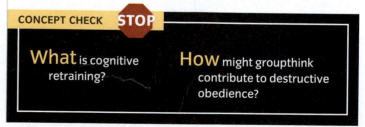

Rosa Parks FIGURE 15.18

Today, Rosa Parks' courageous stand inspires the rest of us to carefully consider when it is appropriate and good to obey authorities versus times when we must resist unethical or dangerous demands.

CONCEPT CHECK STOP

What is cognitive retraining?

How might groupthink contribute to destructive obedience?

SUMMARY

1 Our Thoughts about Others

1. **Social psychology** is the study of how other people influence our thoughts, feelings, and actions. Our **attributions** are frequently marred by the **fundamental attribution error** and the self-serving bias. Both biases may depend in part on cultural factors.

2. **Attitudes** have three components: cognitive, affective, and behavioral. An efficient strategy for changing attitudes is to create **cognitive dissonance**. Like attributional biases, the experience of cognitive dissonance may depend on culture.

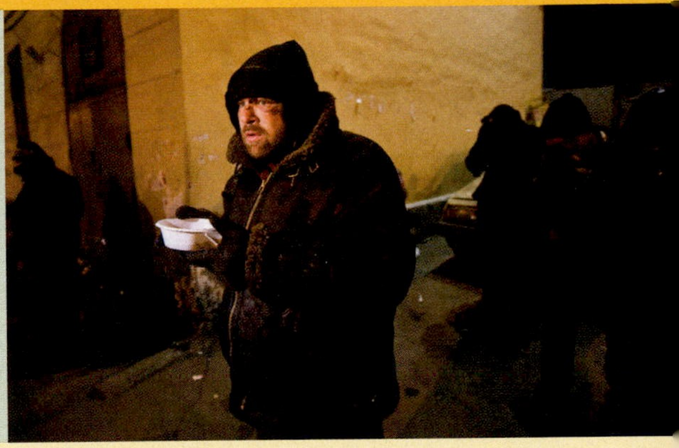

2 Our Feelings about Others

1. Like all attitudes, **prejudice** involves cognitive, affective, and behavioral components. Although the terms *prejudice* and *discrimination* are often used interchangeably, they are not the same. Four commonly cited sources of prejudice are learning, mental shortcuts, economic and political competition, and displaced aggression.

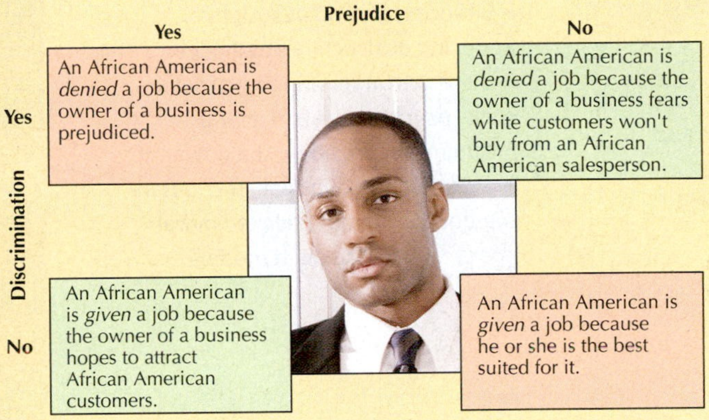
Prejudice
Yes
No
Discrimination
Yes
No
An African American is *denied* a job because the owner of a business is prejudiced.
An African American is *denied* a job because the owner of a business fears white customers won't buy from an African American salesperson.
An African American is *given* a job because the owner of a business hopes to attract African American customers.
An African American is *given* a job because he or she is the best suited for it.

2. Psychologists have found three compelling factors in **interpersonal attraction**: physical attractiveness, proximity, and similarity. Love often develops from initial feelings of friendship and liking. Romantic love is an intense but generally short-lived attraction based on mystery and fantasy, whereas companionate love is a strong, lasting attraction based on admiration, respect, trust, deep caring, and commitment.

3 Our Actions Toward Others

1. Three factors drive **conformity**: normative social influence, informational social influence, and the role of reference groups. Conformity involves going along with the group, while **obedience** involves going along with a direct command, usually from someone in a position of authority. Milgram's research demonstrated the startling power of social situations to create obedience. The degree of deception and discomfort that Milgram's participants were subjected to raises serious ethical questions, and the same study would never be done today.

2. The roles that we play within groups strongly affect our behavior, as Zimbardo's Stanford Prison experiment showed. Zimbardo's study also demonstrated deindividuation. In addition, groups affect our decisions. As individuals interact and discuss their opinions, **group polarization** (a group's movement toward more extreme decisions) and **groupthink** (a group's tendency to strive for agreement and to avoid inconsistent information) tend to occur. Both processes may hinder effective decision making.

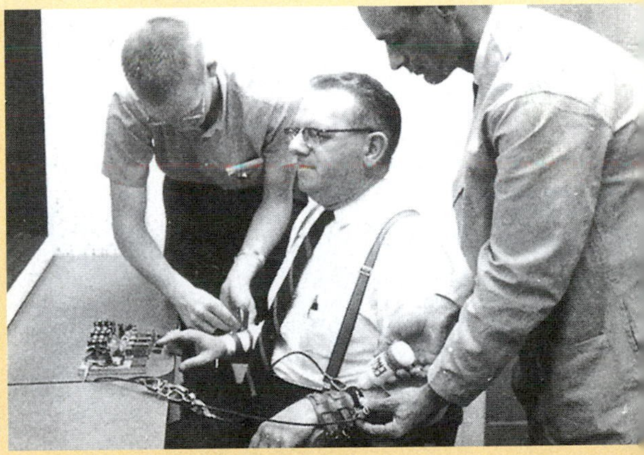

3. Several biological factors may help explain **aggression**, including instincts, genetic predisposition, aggression circuits in the brain and nervous system, mental disorders, and hormones and neurotransmitters.

(Continued on the following page)

Researchers have also proposed several psychosocial explanations for aggression, including aversive stimuli, culture and learning, and media influences. Catharsis does not appear to help release aggressive impulses; more effective ways to control aggression are to introduce incompatible responses and to improve social and communication skills.

4. Evolutionary theory suggests that **altruism** is an evolved, instinctual behavior. Other research suggests that helping may actually be self-interest in disguise (egoistic model). The empathy-altruism hypothesis proposes that although altruism is sometimes based in selfish motivations, it is sometimes truly selfless and motivated by concern for others (empathy). Latané and Darley found that in order for helping to occur, the potential helper must notice what is happening, interpret the event as an emergency, take personal responsibility for helping, decide how to help, and then actually initiate the helping behavior.

4 Applying Social Psychology to Social Problems

1. Four major approaches can be used to combat prejudice: cooperation and superordinate goals, increased contact, cognitive retraining, and cognitive dissonance.

2. There are many times when obedience is unnecessary and destructive. Reducing destructive obedience requires an understanding of several social psychological principles, including socialization, the power of the situation, groupthink, the foot-in-the-door technique, relaxed moral guard, and disobedient models.

KEY TERMS

- **social psychology** p. 404
- **attributions** p. 404
- **fundamental attribution error (FAE)** p. 404
- **saliency bias** p. 404
- **self-serving bias** p. 404
- **attitudes** p. 405
- **cognitive dissonance** p. 406
- **prejudice** p. 407
- **discrimination** p. 407
- **implicit bias** p. 408
- **ingroup** p. 408
- **outgroup** p. 408
- **ingroup favoritism** p. 409
- **outgroup homogeneity effect** p. 409
- **interpersonal attraction** p. 410
- **mere exposure** p. 411
- **romantic love** p. 412
- **companionate love** p. 412
- **conformity** p. 413
- **norms** p. 414
- **reference groups** p. 414
- **obedience** p. 415
- **deindividuation** p. 418
- **risky-shift** p. 419
- **group polarization** p. 419
- **groupthink** p. 420
- **aggression** p. 421
- **frustration-aggression hypothesis** p. 421
- **altruism** p. 422
- **prosocial behavior** p. 422
- **egoistic model** p. 422
- **empathy-altruism hypothesis** p. 422
- **diffusion of responsibility** p. 423
- **superordinate goals** p. 424
- **foot-in-the-door technique** p. 426

CRITICAL AND CREATIVE THINKING QUESTIONS

1. Have you ever changed a strongly held attitude? What caused the change for you?
2. Do you believe that you are free of prejudice? After reading this chapter, which of the many factors that cause prejudice do you think is most important to change?
3. How do Milgram's results—particularly the finding that the remoteness of the victim affected obedience—relate to some aspects of modern warfare?
4. What are some of the similarities between Zimbardo's prison study and the abuses at the Abu Ghraib prison in Iraq?
5. Have you ever done something in a group that you would not have done if you were alone? What happened? How did you feel? What have you learned from this chapter that might help you avoid this behavior in the future?
6. Can you think of situations when the egoistic model of altruism seems most likely correct? What about the empathy-altruism hypothesis?

What is happening in this picture?

From a very early age, Andrew Golden was taught how to fire hunting rifles. At age 11, he and a friend killed four classmates and a teacher in a school shooting.

- **What factors might have contributed to Golden's tragically aggressive act?**
- **Should Golden's parents be held legally or morally responsible for his behavior?**
- **Should very young children be taught how to fire hunting rifles? Why or why not?**

SELF-TEST

(Check your answers in Appendix B.)

1. The study of how other people influence our thoughts, feelings, and actions is called ________.
a. sociology
b. social science
c. social psychology
d. sociobehavioral psychology

2. The two major attribution mistakes people make are ________.
a. the fundamental attribution error and self-serving bias
b. situational attributions and dispositional attributions
c. the actor bias and the observer bias
d. stereotypes and biases

3. Label the three components of attitudes on the figure below.

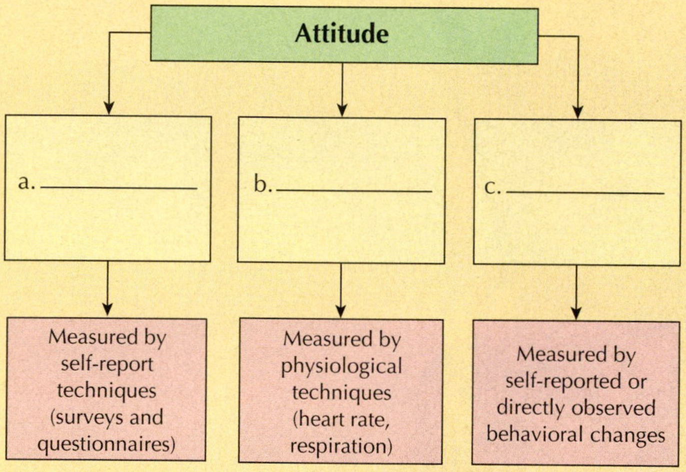

4. This theory says that contradictions between our attitudes and behavior can motivate us to change our attitudes to agree with our behavior.
a. social learning theory
b. cognitive dissonance theory
c. defense mechanisms theory
d. power of inconsistencies theory

5. ________ is a learned, generally negative, attitude toward specific people solely because of their membership in an identified group.
a. Discrimination
b. Stereotyping
c. Cognitive biasing
d. Prejudice

6. The degree of positive feelings you have toward others is called ________.
a. affective relations
b. interpersonal attraction
c. interpersonal attitudes
d. affective connections

7. A strong and lasting attraction characterized by trust, caring, tolerance, and friendship is called ________.
a. companionate love
b. intimate love
c. passionate love
d. all of these options

8. This is the act of changing behavior as a result of real or imagined group pressure.
a. norm compliance
b. obedience
c. conformity
d. mob rule

9. Stanley Milgram was investigating ________ in his classic teacher-learner shock study.
a. the effects of punishment on learning
b. the effects of reinforcement on learning
c. obedience to authority
d. None of these options is correct.

10. During ________________, a person who feels anonymous within a group or crowd experiences an increase in arousal and a decrease in self-consciousness, inhibitions, and personal responsibility.
   a. groupthink
   b. group polarization
   c. authoritarianism
   d. deindividuation

11. Faulty decision making resulting from a highly cohesive group striving for agreement to the point of avoiding inconsistent information is known as ________________.
   a. the risky-shift
   b. group polarization
   c. groupthink
   d. destructive conformity

12. Actions designed to help others with no obvious benefit to the helper are collectively known as ________________.
   a. empathy
   b. sympathy
   c. altruism
   d. egoism

13. Label the three major explanations for helping on the figure below.

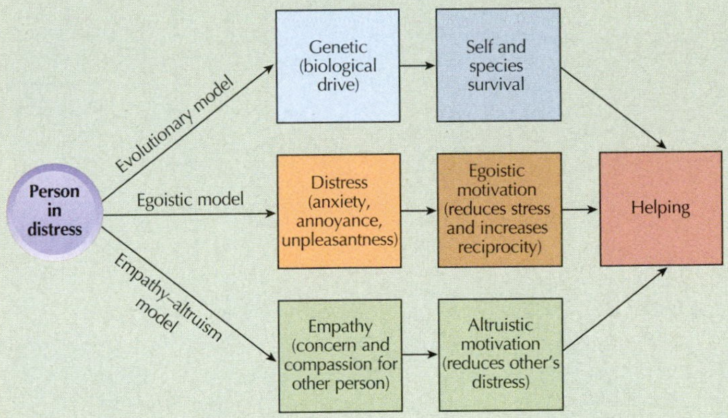

14. Research suggests that one of the best ways to decrease prejudice is to encourage ________________.
   a. cooperation
   b. friendly competition
   c. reciprocity of liking
   d. conformity

15. Which of the following factors may contribute to destructive obedience?
   a. power of the situation
   b. foot-in-the-door
   c. socialization
   d. All of these options are correct.

Appendix A
Statistics and Psychology

We are constantly bombarded by numbers: "On sale for 30 percent off," "70 percent chance of rain," "9 out of 10 doctors recommend . . ." The president uses numbers to try to convince us that the economy is healthy. Advertisers use numbers to convince us of the effectiveness of their products. Psychologists use statistics to support or refute psychological theories and demonstrate that certain behaviors are indeed results of specific causal factors.

When people use numbers in these ways, they are using statistics. **Statistics** is a branch of applied mathematics that uses numbers to describe and analyze information on a subject.

Statistics make it possible for psychologists to quantify the information they obtain in their studies. They can then critically analyze and evaluate this information. Statistical analysis is imperative for researchers to describe, predict, or explain behavior. For instance, Albert Bandura (1973) proposed that watching violence on television causes aggressive behavior in children. In carefully controlled experiments, he gathered numerical information and analyzed it according to specific statistical methods. The statistical analysis helped him substantiate that the aggression of his subjects and the aggressive acts they had seen on television were related, and that the relationship was not mere coincidence.

Drawing by M. Stevens; © 1989 the New Yorker Magazine, Inc.

Although statistics is a branch of applied mathematics, you don't have to be a math whiz to use statistics. Simple arithmetic is all you need to do most of the calculations. For more complex statistics involving more complicated mathematics, computer programs are available for virtually every type of computer. What is more important than learning the mathematical computations, however, is developing an understanding of when and why each type of statistic is used. The purpose of this appendix is to help you understand the significance of the statistics most commonly used.

Gathering and Organizing Data

Psychologists design their studies to facilitate gathering information about the factors they want to study. The information they obtain is known as **data** (**data** is plural; its singular is **datum**). When the data are gathered, they are generally in the form of numbers; if they aren't, they are converted to numbers. After they are gathered, the data must be organized in such a way that statistical analysis is possible. In the following section, we will examine the methods used to gather and organize information.

Variables

When studying a behavior, psychologists normally focus on one particular factor to determine whether it has an effect on the behavior. This factor is known as a **variable,** which is in effect anything that can assume more than one value (see Chapter 1). Height, weight, sex, eye color, and scores on an IQ test or a video game are all factors that can assume more than one value and are therefore variables. Some will vary between people, such as sex (you are either male *or* female but not both at the same time). Some may even vary within one person, such as scores on a video game (the same person might get 10,000 points on one try and only 800 on another). Opposed to a variable, anything that remains the same and does not vary is called a **constant.** If researchers use only females in their research, then sex is a constant, not a variable.

In nonexperimental studies, variables can be factors that are merely observed through naturalistic observation or case studies, or they can be factors about which people are questioned in a test or survey. In experimental studies, the two major types of variables are independent and dependent variables.

Independent variables are those that are manipulated by the experimenter. For example, suppose we were to conduct a study to determine whether the sex of the debater influences the outcome of a debate. In this study, one group of subjects watches a videotape of a debate between a male arguing the "pro" side and a female arguing the "con"; another group watches the same debate, but with the pro and con roles reversed. In such a study, the form of the presentation viewed by each group (whether "pro" is argued by a male or a female) is the independent variable because the experimenter manipulates the form of presentation seen by each group. Another example might be a study to determine whether a particular drug has any effect on a manual dexterity task. To study this question, we would administer the drug to one group and no drug to another. The independent variable would be the amount of drug given (some or none). The independent variable is particularly important when using **inferential statistics,** which we will discuss later.

The **dependent variable** is a factor that results from, or depends on, the independent variable. It is a measure of some outcome or, most commonly, a measure of the subjects' behavior. In the debate example, each subject's choice of the winner of the debate would be the dependent variable. In the drug experiment, the dependent variable would be each subject's score on the manual dexterity task.

Frequency Distributions

After conducting a study and obtaining measures of the variable(s) being studied, psychologists need to organize the data in a meaningful way. Table A.1 presents test scores from a Math Aptitude Test collected from 50 college students. This information is called **raw data** because there is no order to the numbers. They are presented as they were collected and are therefore "raw."

The lack of order in raw data makes them difficult to study. Thus, the first step in understanding the results of an experiment is to impose some order on the raw data. There are several ways to do this. One of the simplest is to create a **frequency distribution,** which shows the number of times a score or event occurs. Although frequency distributions are helpful in several ways, the major advantages are that they allow us to see the data in an organized manner and they make it easier to represent the data on a graph.

The simplest way to make a frequency distribution is to list all the possible test scores, then tally the number of people *(N)* who received those scores. Table A.2 presents a frequency distribution using the raw data from Table

Math Aptitude Test scores for 50 college students TABLE A.1

73	57	63	59	50
72	66	50	67	51
63	59	65	62	65
62	72	64	73	66
61	68	62	68	63
59	61	72	63	52
59	58	57	68	57
64	56	65	59	60
50	62	68	54	63
52	62	70	60	68

Frequency distribution of 50 students on Math Aptitude Test TABLE A.2

Score	*Frequency*
73	2
72	3
71	0
70	1
69	0
68	5
67	1
66	2
65	3
64	2
63	5
62	5
61	2
60	2
59	5
58	1
57	3
56	1
55	0
54	1
53	0
52	2
51	1
50	3
Total	50

Psychology Aptitude Test scores for 50 college students TABLE A.3

1350	750	530	540	750
1120	410	780	1020	430
720	1080	1110	770	610
1130	620	510	1160	630
640	1220	920	650	870
930	660	480	940	670
1070	950	680	450	990
690	1010	800	660	500
860	520	540	880	1090
580	730	570	560	740

A.1. As you can see, the data are now easier to read. From looking at the frequency distribution, you can see that most of the test scores lie in the middle with only a few at the very high or very low end. This was not at all evident from looking at the raw data.

This type of frequency distribution is practical when the number of possible scores is 20 or fewer. However, when there are more than 20 possible scores it can be even harder to make sense out of the frequency distribution than the raw data. This can be seen in Table A.3, which presents the hypothetical Psychology Aptitude Test scores for 50 students. Even though there are only 50 actual scores in this table, the number of possible scores ranges from a high of 1390 to a low of 400. If we included zero frequencies there would be 100 entries in a frequency distribution of this data, making the frequency distribution much more difficult to understand than the raw data. If there are more than 20 possible scores, therefore, a **group** frequency distribution is normally used.

Group frequency distribution of Psychology Aptitude Test scores for 50 college students TABLE A.4

Class Interval	*Frequency*
1300–1390	1
1200–1290	1
1100–1190	4
1000–1090	5
900–990	5
800–890	4
700–790	7
600–690	10
500–590	9
400–490	4
Total	50

FIGURE A.1 **A histogram illustrating the information found in Table A.4.**

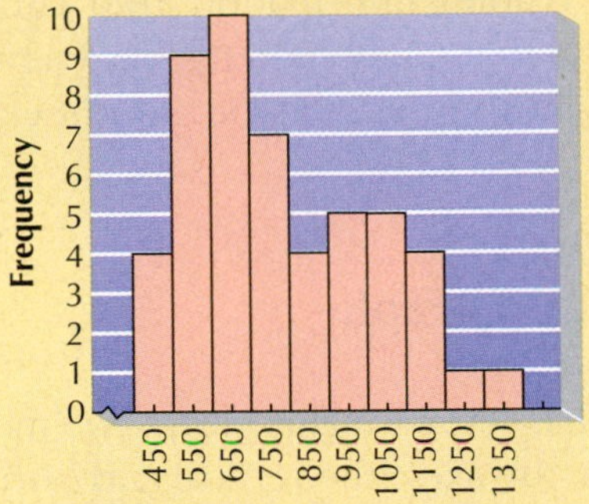

In a **group frequency distribution,** individual scores are represented as members of a group of scores or as a range of scores (see Table A.4). These groups are called **class intervals.** Grouping these scores makes it much easier to make sense out of the distribution, as you can see from the relative ease in understanding Table A.4 as compared to Table A.3. Group frequency distributions are easier to represent on a graph.

When graphing data from frequency distributions, the class intervals are represented along the **abscissa** (the horizontal or *x* axis). The frequency is represented along the **ordinate** (the vertical or *y* axis). Information can be graphed in the form of a bar graph, called a **histogram,** or in the form of a point or line graph, called a **polygon.** Figure A.1 shows a histogram presenting the data from Table A.4. Note that the class intervals are represented along the bottom line of the graph (the *x* axis) and the height of the bars indicates the frequency in each class interval. Now look at Figure A.2. The information presented here is

FIGURE A.2 **A polygon illustrating the information found in Table A.4.**

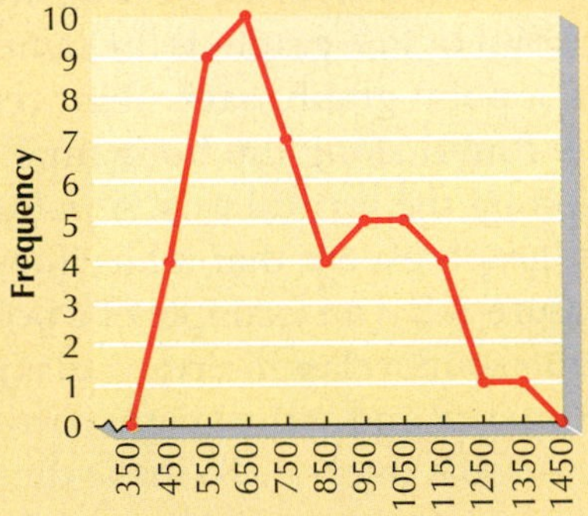

exactly the same as that in Figure A.1 but is represented in the form of a polygon rather than a histogram. Can you see how both graphs illustrate the same information? Even though reading information from a graph is simple, we have found that many students have never learned to read graphs. In the next section we will explain how to read a graph.

How to Read a Graph

Every graph has several major parts. The most important are the labels, the axes (the vertical and horizontal lines), and the points, lines, or bars. Find these parts in Figure A.1.

The first thing you should notice when reading a graph are the labels because they tell what data are portrayed. Usually the data consist of the descriptive statistics, or the numbers used to measure the dependent variables. For example, in Figure A.1 the horizontal axis is labeled "Psychology Aptitude Test Scores," which is the dependent variable measure; the vertical axis is labeled "Frequency," which means the number of occurrences. If a graph is not labeled, as we sometimes see in TV commercials or magazine ads, it is useless and should be ignored. Even when a graph *is* labeled, the labels can be misleading. For example, if graph designers want to distort the information, they can elongate one of the axes. Thus, it is important to pay careful attention to the numbers as well as the words in graph labels.

Next, you should focus your attention on the bars, points, or lines on the graph. In the case of histograms like the one in Figure A.1, each bar represents the class interval. The width of the bar stands for the width of the class interval, whereas the height of the bar stands for the frequency in that interval. Look at the third bar from the left in Figure A.1. This bar represents the interval "600 to 690 SAT Scores," which has a frequency of 10. You can see that this directly corresponds to the same class interval in Table A.4, since graphs and tables are both merely alternate ways of illustrating information.

Reading point or line graphs is the same as reading a histogram. In a point graph, each point represents two numbers, one found along the horizontal axis and the other found along the vertical axis. A polygon is identical to a point graph except that it has lines connecting the points. Figure A.2 is an example of a polygon, where each point represents a class interval and is placed at the center of the interval and at the height corresponding to the frequency of that interval. To make the graph easier to read, the points are connected by straight lines.

Displaying the data in a frequency distribution or in a graph is much more useful than merely presenting raw data and can be especially helpful when researchers are trying to find relations between certain factors. However, as we explained earlier, if psychologists want to make predictions or explanations about behavior, they need to perform mathematical computations on the data.

Uses of the Various Statistics

The statistics psychologists use in a study depend on whether they are trying to describe and predict behavior or explain it. When they use statistics to describe behavior, as in reporting the average score on an aptitude test, they are using **descriptive statistics.** When they use them to explain behavior, as Bandura did in his study of children modeling aggressive behavior seen on TV, they are using **inferential statistics.**

Descriptive Statistics

Descriptive statistics are the numbers used to describe the dependent variable. They can be used to describe characteristics of a **population** (an entire group, such as all people living in the United States) or a **sample** (a part of a group, such as a randomly selected group of 25 students from Cornell University). The major descriptive statistics include measures of central tendency (mean, median, and mode), measures of variation (variance and standard deviation), and correlation.

Measures of Central Tendency

Statistics indicating the center of the distribution are called **measures of central tendency** and include the mean, median, and mode. They are all scores that are typical of the center of the distribution. The **mean** is what most of us think of when we hear the word "average." The **median** is the middle score. The **mode** is the score that occurs most often.

Mean What is your average golf score? What is the average yearly rainfall in your part of the country? What is the average reading test score in your city? When these questions ask for the average, they are really asking for the "mean." The arithmetic mean is the weighted average of all the raw scores, which is computed by totaling all the raw scores and then dividing that total by the number of scores added together. In statistical computation, the mean is represented by an "X" with a

Computation of the mean for 10 IQ scores TABLE A.5

IQ Scores X
143
127
116
98
85
107
106
98
104
116
$\Sigma X = 1{,}100$

$$\text{Mean} = \overline{X} = \frac{\Sigma X}{N} = \frac{1{,}100}{10} = 110$$

bar above it ($\overline{X}$, pronounced "X bar"), each individual raw score by an "X," and the total number of scores by an "N." For example, if we wanted to compute the $\overline{X}$, of the raw statistics test scores in Table A.1, we would sum all the X's (ΣX, with Σ meaning sum) and divide by N (number of scores). In Table A.1, the sum of all the scores is equal to 3,100 and there are 50 scores. Therefore, the mean of these scores is

$$\overline{X} = \frac{3{,}100}{50} = 62$$

Table A.5 illustrates how to calculate the mean for 10 IQ scores.

Median The **median** is the middle score in the distribution once all the scores have been arranged in rank order. If N (the number of scores) is odd, then there actually is a middle score and that middle score is the median. When N is even, there are two middle scores and the median is the mean of those two scores. Table A.6 shows the computation of the median for two different sets of scores, one set with 15 scores and one with 10.

Mode Of all the measures of central tendency, the easiest to compute is the **mode**, which is merely the most frequent score. It is computed by finding the score that occurs most often. Whereas there is always only one mean and only one median for each distribution, there can be more than one mode. Table A.7 shows how to find the mode in a distribution with one mode (unimodal) and in a distribution with two modes (bimodal).

Computation of median for odd and even numbers of IQ scores TABLE A.6

IQ	IQ
139	137
130	135
121	121
116	116
107	108 ← middle score
101	106 ← middle score
98	105
96 ← middle score	101
84	98
83	97
82	N = 10
75	N is even
75	
68	
65	$\text{Median} = \frac{106 + 108}{2} = 107$
N = 15	
N is odd	

There are several advantages to each of these measures of central tendency, but in psychological research the mean is used most often. A book solely covering psychological statistics will provide a more thorough discussion of the relative values of these measures.

Measures of Variation

When describing a distribution, it is not sufficient merely to give the central tendency; it is also necessary to give a **measure of variation**, which is a measure of the spread of the scores. By examining the spread, we can determine

Finding the mode for two different distributions TABLE A.7

IQ	IQ
139	139
138	138
125	125
116 ←	116 ←
116 ←	116 ←
116 ←	116 ←
107	107
100	98 ←
98	98 ←
98	98 ←
Mode = most frequent score	Mode = 116 and 98
Mode = 116	

FIGURE A.3 Three distributions having the same mean but a different variability.

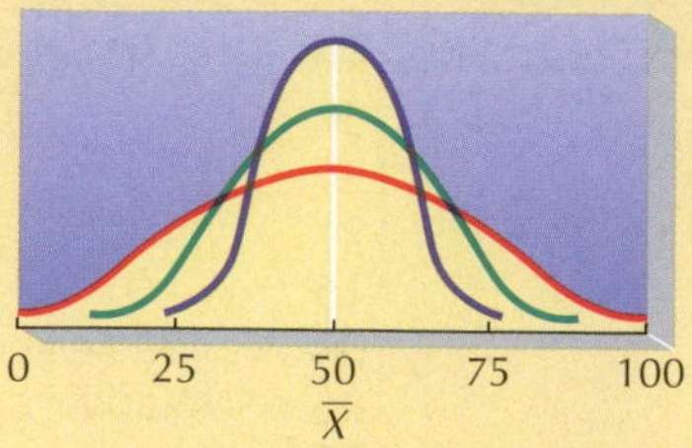

whether the scores are bunched around the middle or tend to extend away from the middle. Figure A.3 shows three different distributions, all with the same mean but with different spreads of scores. You can see from this figure that, in order to describe these different distributions accurately, there must be some measures of the variation in their spread. The most widely used measure of variation is the standard deviation, which is represented by a lowercase s. The standard deviation is a standard measurement of how much the scores in a distribution deviate from the mean. The formula for the standard deviation is

$$s = \sqrt{\frac{\sum (X - \bar{X})^2}{N}}$$

Table A.8 illustrates how to compute the standard deviation.

Computation of the standard deviation for 10 IQ scores TABLE A.8

IQ Scores X	$X - \bar{X}$	$(X - \bar{X})^2$
143	33	1089
127	17	289
116	6	36
98	−12	144
85	−25	625
107	−3	9
106	−4	16
98	−12	144
104	−6	36
116	6	36
$\sum X = 1100$		$\sum (X - \bar{X})^2 = 2424$

Standard Deviation = s

$$= \sqrt{\frac{\sum (X - \bar{X})^2}{N}} = \sqrt{\frac{2424}{10}}$$

$$= \sqrt{242.4} = 15.569$$

Most distributions of psychological data are bell-shaped. That is, most of the scores are grouped around the mean, and the farther the scores are from the mean in either direction, the fewer the scores. Notice the bell shape of the distribution in Figure A.4. Distributions such as this are called **normal** distributions. In normal distributions, as shown in Figure A.4, approximately two-thirds of the scores fall within a range that is one standard deviation below the mean to one standard deviation above the mean. For example, the Wechsler IQ tests (see Chapter 7) have a mean of 100 and a standard deviation of 15. This means that approximately two-thirds of the people taking these tests will have scores between 85 and 115.

Correlation

Suppose for a moment that you are sitting in the student union with a friend. To pass the time, you and your friend decide to play a game in which you try to guess the height of the next male who enters the union. The winner, the one whose guess is closest to the person's actual height, gets a piece of pie paid for by the loser. When it is your turn, what do you guess? If you are like most people, you will probably try to estimate the mean of all the males in the union and use that as your guess. The mean is always your best guess if you have no other information.

Now let's change the game a little and add a friend who stands outside the union and weighs the next male to enter the union. Before the male enters the union, your friend says "125 pounds." Given this new information, will you still guess the mean height? Probably not—you will probably predict *below* the mean. Why? Because you intuitively understand that there is a **correlation**, a relationship, between height and weight, with tall peo-

FIGURE A.4 The normal distribution forms a bell-shaped curve. In a normal distribution, two-thirds of the scores lie between one standard deviation above and one standard deviation below the mean.

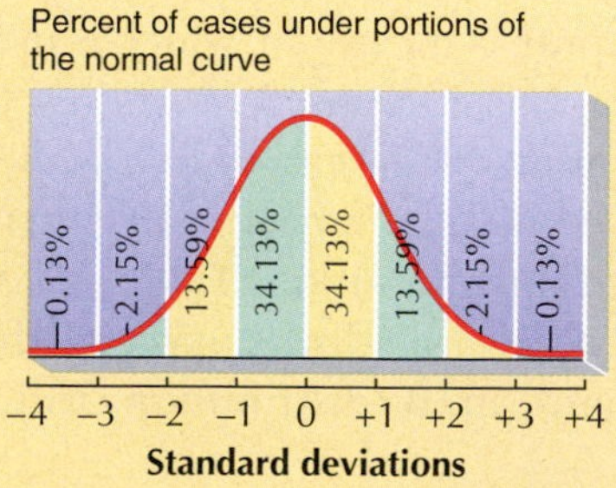

ple usually weighing more than short people. Given that 125 pounds is less than the average weight for males, you will probably guess a less-than-average height. The statistic used to measure this type of relationship between two variables is called a correlation coefficient.

Correlation Coefficient A correlation coefficient measures the relationship between two variables, such as height and weight or IQ and SAT scores. Given any two variables, there are three possible relationships between them: positive, negative, and zero (no relationship). A positive relationship exists when the two variables vary in the same direction (e.g., as height increases, weight normally also increases). A negative relationship occurs when the two variables vary in opposite directions (e.g., as temperatures go up, hot chocolate sales go down). There is no relationship when the two variables vary totally independently of one another (e.g., there is no relationship between peoples' height and the color of their toothbrushes). Figure A.5 illustrates these three types of correlations.

The computation and the formula for a correlation coefficient (correlation coefficient is delineated by the letter "r") are shown in Table A.9. The correlation coefficient (r) always has a value between +1 and −1 (it is never greater than +1 and it is never smaller than −1). When r is close to +1, it signifies a high positive relationship between the two variables (as one variable goes up, the other variable also goes up). When r is close to −1, it signifies a high negative relationship between the two variables (as one variable goes up, the other variable goes down). When r is 0, there is no linear relationship between the two variables being measured.

Correlation coefficients can be quite helpful in making predictions. Bear in mind, however, that predictions are just that: predictions. They will have some error as long as the correlation coefficients on which they are

FIGURE A.5 Three types of correlation. Positive correlation (top): As the number of days of class attendance increases, so does the number of correct exam items. Negative correlation (middle): As the number of days of class attendance increases, the number of incorrect exam items decreases. Zero correlation (bottom): The day of the month on which one is born has no relationship to the number of exam items correct.

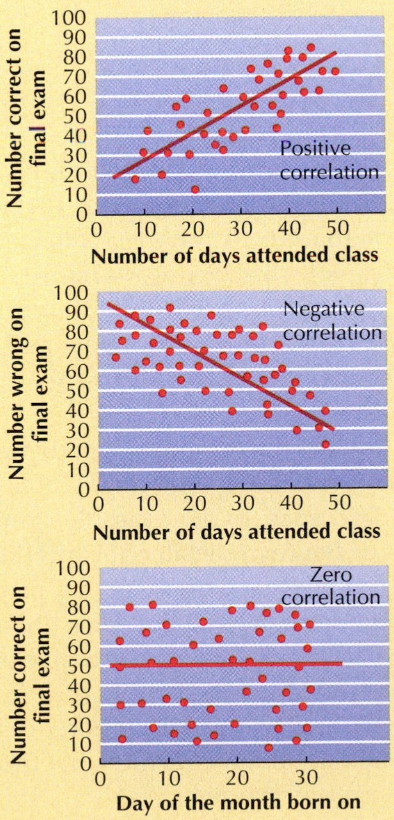

Computation of correlation coefficient between height and weight for 10 males
TABLE A.9

Height (inches) X	X^2	Weight (pounds) Y	Y^2	XY
73	5,329	210	44,100	15,330
64	4,096	133	17,689	8,512
65	4,225	128	16,384	8,320
70	4,900	156	24,336	10,920
74	5,476	189	35,721	13,986
68	4,624	145	21,025	9,860
67	4,489	145	21,025	9,715
72	5,184	166	27,556	11,952
76	5,776	199	37,601	15,124
71	5,041	159	25,281	11,289
700	49,140	1,630	272,718	115,008

$$r = \frac{N \cdot \Sigma XY - \Sigma X \cdot \Sigma Y}{\sqrt{[N \cdot \Sigma X^2 - (\Sigma X)^2]}\sqrt{[N \cdot \Sigma Y^2 - (\Sigma Y)^2]}}$$

$$r = \frac{10 \cdot 115{,}008 - 700 \cdot 1{,}630}{\sqrt{[10 \cdot 49{,}140 - 700^2]}\sqrt{[10 \cdot 272{,}718 - 1{,}630^2]}}$$

$$r = 0.92$$

based are not perfect (+1 or −1). Also, correlations cannot reveal any information regarding causation. Merely because two factors are correlated, it does not mean that one factor causes the other. Consider, for example, ice cream consumption and swimming pool use. These two variables are positively correlated with one another, in that as ice cream consumption increases, so does swimming pool use. But nobody would suggest that eating ice cream causes swimming, or vice versa. Similarly, just because Michael Jordan eats Wheaties and can do a slam dunk it does not mean that you will be able to do one if you eat the same breakfast. The only way to determine the cause of behavior is to conduct an experiment and analyze the results by using inferential statistics.

Inferential Statistics

Knowing the descriptive statistics associated with different distributions, such as the mean and standard deviation, can enable us to make comparisons between various distributions. By making these comparisons, we may be able to observe whether one variable is related to another or whether one variable has a causal effect on another. When we design an experiment specifically to measure causal effects between two or more variables, we use **inferential** statistics to analyze the data collected. Although there are many inferential statistics, the one we will discuss is the t-test, since it is the simplest.

t-Test Suppose we believe that drinking alcohol causes a person's reaction time to slow down. To test this hypothesis, we recruit 20 participants and separate them into two groups. We ask the participants in one group to drink a large glass of orange juice with one ounce of alcohol for every 100 pounds of body weight (e.g., a person weighing 150 pounds would get 1.5 ounces of alcohol). We ask the control group to drink an equivalent amount of orange juice with no alcohol added. Fifteen minutes after the drinks, we have each participant perform a reaction time test that consists of pushing a button as soon as a light is flashed. (The reaction time is the time between the onset of the light and the pressing of the button.) Table A.10 shows the data from this hypothetical experiment. It is clear from the data that there is definitely a difference in the reaction times of the two groups: There is an obvious difference between the means. However, it is possible that this difference is due merely to chance. To determine whether the difference is real or due to chance, we can conduct a t-test. We have run a sample t-test in Table A.10.

Reaction times in milliseconds (msec) for subjects in alcohol and no alcohol conditions and computation of *t* TABLE A.10

RT (msec) Alcohol X_1	*RT (msec) No Alcohol X_2*
200	143
210	137
140	179
160	184
180	156
187	132
196	176
198	148
140	125
159	120
$SX_1 = 1{,}770$	$SX_2 = 1{,}500$
$N_1 = 10$	$N_2 = 10$
$\bar{X}_1 = 177$	$\bar{X}_2 = 150$
$s_1 = 24.25$	$s_2 = 21.86$

$$\Sigma_{\bar{X}_1} = \frac{s}{\sqrt{N_1 - 1}} = 8.08 \qquad \Sigma_{\bar{X}_2} = \frac{s}{\sqrt{N_2 - 1}} = 7.29$$

$$S_{\bar{X}_1 - \bar{X}_2} = \sqrt{S_{\bar{X}_1}^2 + S_{\bar{X}_2}^2} = \sqrt{8.08^2 + 7.29^2} = 10.88$$

$$t = \frac{\bar{X}_1 - \bar{X}_2}{S_{\bar{X}_1 - \bar{X}_2}} = \frac{177 - 150}{10.88} = 2.48$$

$t = 2.48, p < 0.5$

The logic behind a t-test is relatively simple. In our experiment we have two samples. If each of these samples is from the same population (e.g., the population of all people, whether drunk or sober), then any difference between the samples will be due to chance. On the other hand, if the two samples are from different populations (e.g., the population of drunk people and the population of sober people), then the difference is a significant difference and not due to chance.

If there is a significant difference between the two samples, then the independent variable must have caused that difference. In our example, there is a significant difference between the alcohol and the no alcohol groups. We can tell this because p (the probability that this t value will occur by chance) is less than .05. To obtain the p, we need only look up the t value in a statistical table, which is found in any statistics book. In our example, because there is a significant difference between the groups, we can reasonably conclude that the alcohol did cause a slower reaction time.

Appendix B
Answers to Self-Tests

Chapter 1: 1d; 2d; 3c; 4a; 5b; 6, see Fig. 1.6; 7b; 8d; 9c; 10b; 11b; 12. see Fig. 1.14

Chapter 2: 1b; 2d; 3a; 4a; 5, see Fig. 2.5; 6c; 7b; 8c; 9, see Fig. 2.8; 10a; 11d; 12c; 13, see Fig. 2.16; 14, see Fig. 2.19; 15: a. left; b. right; c. right; d. right

Chapter 3: 1c; 2b; 3c; 4c; 5, see Fig. 3.4; 6c; 7c; 8b; 9d; 10d; 11c; 12b; 13, see Fig. 3.10; 14b; 15b

Chapter 4: 1c; 2c; 3, see Fig. 4.4; 4c; 5, see Fig. 4.5; 6c; 7c; 8c; 9: a. photo 1, b. photo 2; 10c; 11b; 12a; 13b; 14c; 15b

Chapter 5: 1b; 2b; 3c; 4d, 5 see Fig. 5.3; 6c; 7c; 8c; 9d; 10a; 11: a-2, b-1, c-3, d-4; 12c; 13d; 14d; 15c

Chapter 6: 1c; 2c; 3d; 4d; 5d; 6c; 7c; 8: a-2, b-1; 9a; 10c; 11d; 12a; 13c; 14a; 15d

Chapter 7: 1: a, encoding, b, retrieval, c, storage; 2: a. sensory memory, b. short-term memory (STM), c. long-term memory (LTM); 3d; 4a; 5: a, explicit and declarative, b, implicit and nondeclarative; 6c; 7c; 8c; 9c; 10: a. retrograde amnesia (top), b. antegrade amnesia (bottom); 11: a. decay theory, b. interference, c. motivated, d. encoding failure, e. retrieval; 12.b; 13a; 14d; 15b

Chapter 8: 1b; 2c; 3: a. preparation, b. identify given facts, c. separate relevant from irrelevant facts; 4a; 5d; 6d; 7: a. phonemes, b. morphemes, c. grammar; 8b; 9d; 10b; 11b; 12a; 13b; 14d; 15d

Chapter 9: 1c; 2c; 3: a. cross-sectional research, b. longitudinal research; 4d; 5d; 6a; 7c; 8c; 9c; 10d; 11: a. sensorimotor, b. preoperational, c. concrete operation, d. formal; 12d; 13c; 14c; 15d

Chapter 10: 1b; 2a; 3a; 4c; 5: a. trust → b. autonomy → c. initiative → d. industry; 6: a. identity → b. intimacy → c. generativity → d. ego integrity; 7b; 8c; 9a; 10a; 11a; 12a; 13d; 14d; 15a

Chapter 11: 1a; 2c; 3a; 4b; 5d; 6d; 7c; 8c; 9: a. excitement → b. plateau → c. orgasm → d. resolution; 10b; 11a; 12a; 13a; 14d; 15c

Chapter 12: 1b; 2: a. openness, b. conscientiousness, c. extroversion, d. agreeableness, e. neuroticism (OCEAN); 3: a. conscious, b. preconscious, c. unconscious; 4d; 5c; 6: a. oral → b. anal → c. phallic → d. latency → e. genital; 7d; 8c; 9b; 10c; 11c; 12c; 13c; 14a; 15b

Chapter 13: 1d; 2b; 3: Axis I. clinical disorders, Axis II. personality disorders and mental retardation, Axis III. general medical conditions, Axis IV. psychosocial and environmental problems, Axis V. global assessment of functioning; 4a; 5: a. generalized anxiety disorder, b. panic disorder, c. phobias, d. obsessive-compulsive disorder, e. posttraumatic stress disorder; 6a; 7b; 8d; 9a; 10a; 11: a. paranoid, b. catatonic, c. disorganized, d. undifferentiated, e. residual; 12b; 13a; 14c; 15d

Chapter 14: 1.a; 2c; 3c; 4c; 5b; 6c; 7d; 8a; 9b; 10c; 11c; 12: a. antianxiety drugs, b. antipsychotic drugs, c. mood stabilizer drugs, antidepressant drugs; 13b; 14: a. disturbed thoughts, b. disturbed emotions, c. disturbed behaviors, d. interpersonal and life situation difficulties, e. biomedical disturbances; 15b

Chapter 15: 1c; 2a; 3: a. cognitive element, b. affective element, c. behavioral element; 4b; 5d; 6b; 7a; 8c; 9c; 10d; 11c; 12c; 13: a. egoistic model, b. empathy-altruism; 14a; 15d

GLOSSARY

abnormal behavior Patterns of emotion, thought, and action that are considered pathological (diseased or disordered) for one or more of these reasons: statistical infrequency, disability or dysfunction, personal distress, or violation of norms (Davison, Neale, & Kring, 2004).

accommodation In Piaget's theory, the process of adjusting old schemas or developing new ones to better fit with new information.

achievement motivation A desire to excel, especially in competition with others.

activity theory of aging Successful aging is fostered by a full and active commitment to life.

aggression Any behavior intended to harm someone.

algorithm A set of steps that, if followed correctly, will eventually solve the problem.

altered state of consciousness (ASC) Mental states found during sleep, dreaming, psychoactive drug use, hypnosis, and so on.

altruism Actions designed to help others with no obvious benefit to the helper.

Alzheimer's disease (AD) Progressive mental deterioration characterized by severe memory loss.

anorexia nervosa An eating disorder characterized by a severe loss of weight resulting from self-imposed starvation and an obsessive fear of obesity.

anxiety disorder A type of abnormal behavior characterized by unrealistic, irrational fear.

applied research Research designed to solve practical problems.

assimilation In Piaget's theory, the process of absorbing new information into existing schemas.

attachment A strong affectional bond with special others that endures over time.

attitudes Learned predispositions to respond cognitively, affectively, and behaviorally to particular objects in a particular way.

attributions Explanations for behaviors or events.

autonomic nervous system (ANS) Subdivision of the peripheral nervous system (PNS) that controls involuntary functions. It includes the *sympathetic* nervous system and the *parasympathetic* nervous system.

basic research Research conducted to advance scientific knowledge rather than for practical application.

behavior therapy A group of techniques based on learning principles that is used to change maladaptive behaviors.

behavioral genetics The study of the relative effects of heredity and environment on behavior and mental processes.

binge drinking Occurs when a man consumes five or more drinks, or a woman consumes four or more drinks in about two hours.

biological preparedness Built-in (innate) readiness to form associations between certain stimuli and responses.

biological research Scientific studies of the brain and other parts of the nervous system.

biomedical therapy Using physiological interventions (drugs, electroconvulsive therapy, and psychosurgery) to reduce or alleviate symptoms of psychological disorders.

biopsychosocial model A unifying theme of modern psychology that considers biological, psychological, and social processes.

bipolar disorder Repeated episodes of mania (unreasonable elation and hyperactivity) and depression.

bulimia nervosa An eating disorder involving the consumption of large quantities of food (bingeing), followed by vomiting, extreme exercise, or laxative use (purging).

central nervous system (CNS) The brain and spinal cord.

cerebral cortex Thin surface layer on the cerebral hemispheres that regulates most complex behavior, including receiving sensations, motor control, and higher mental processes.

chronic pain Continuous or recurrent pain lasting six months or longer.

chunking The act of grouping separate pieces of information into a single unit (or chunk).

circadian [ser-KAY-dee-an] rhythms Biological changes that occur on a 24-hour cycle (in Latin, *circa* means about, and *dies* means day).

classical conditioning Learning that occurs when a neutral stimulus (NS) becomes paired (associated) with an unconditioned stimulus (UCS) to elicit a conditioned response (CR).

coding Process that converts a particular sensory input into a specific sensation.

cognition Mental activities involved in acquiring, storing, retrieving, and using knowledge.

cognitive dissonance A feeling of discomfort caused by a discrepancy between an attitude and a behavior or between two attitudes.

cognitive therapy Therapy that focuses on changing faulty thought processes and beliefs to treat problem behaviors.

cognitive-social theory A perspective that emphasizes the roles of thinking and social learning in behavior.

collectivistic cultures Cultures in which the needs and goals of the group are emphasized over the needs and goals of the individual.

concrete operational stage Piaget's third stage (roughly ages 7 to 11), in which the child can perform mental operations on concrete objects and understand reversibility and conservation, though abstract thinking is not yet present.

conditioning The process of learning associations between environmental stimuli and behavioral responses.

conformity The act of changing behavior as a result of real or imagined group pressure.

conscious Freud's term for thoughts or motives that a person is currently aware of or is remembering.

consciousness An organism's awareness of its own self and surroundings.

conventional level Kohlberg's second level of moral development, in which moral judgments are based on compliance with the rules and values of society.

correlational research Scientific study in which the researcher observes or measures (without directly manipulating) two or more variables to find relationships between them.

creativity The ability to produce valued outcomes in a novel way.

critical period A period of special sensitivity to specific types of learning that shapes the capacity for future development.

critical thinking The process of objectively evaluating, comparing, analyzing, and synthesizing information.

cross-sectional method Research design that measures individuals of various ages at one point in time and gives infomation about age differences.

debriefing Informing participants after a study about the purpose of the study, the nature of the anticipated results, and any deception used.

defense mechanisms In Freudian theory, the ego's protective method of reducing anxiety by distorting reality.

delusions Mistaken beliefs based on misrepresentations of reality.

dependent variable (DV) Variable that is measured; it is affected by (or dependent on) the independent variable.

descriptive research Research methods used to observe and record behavior (without producing causal explanations).

developmental psychology The study of age-related changes in behavior and mental processes from conception to death.

Diagnostic and Statistical Manual of Mental Disorders (DSM-IV-TR) The classification system developed by the American Psychiatric Association used to describe abnormal behaviors; the *IV-TR* indicates that it is the text revision (*TR*) of the fourth major edition (*IV*).

disengagement theory Successful aging is characterized by mutual withdrawal between the elderly and society.

dissociative disorders Amnesia, fugue, or multiple personalities resulting from avoidance of painful memories or situations.

dyssomnias Problems in the amount, timing, and quality of sleep, including insomnia, sleep apnea, and narcolepsy.

elaborative rehearsal The process of linking new information to previously stored material.

electroconvulsive therapy (ECT) Biomedical therapy in which electrical current is passed through the brain.

embryonic period The second stage of prenatal development, which begins after uterine implantation and lasts through the eighth week.

emotion A subjective feeling that includes arousal (heart pounding), cognitions (thoughts, values, and expectations), and expressions (frowns, smiles, and running).

emotion-focused forms of coping Managing one's emotional reactions to a stressor.

empirical evidence Information acquired by direct observation and measurement using systematic scientific methods.

encoding Processing information into the memory system.

encoding specificity principle Retrieval of information is improved when the conditions of recovery are similar to the conditions that existed when the information was encoded.

evolutionary/circadian theory As a part of circadian rhythms, sleep evolved to conserve energy and to serve as protection from predators.

evolutionary psychology A branch of psychology that studies how natural selection and adaptation help explain behavior and mental processes.

experiment A carefully controlled scientific procedure that determines whether variables manipulated by the experimenter have a causal effect on other variables.

explicit/declarative memory The subsystem within long-term memory that consciously stores facts, information, and personal life experiences.

external locus of control Belief that chance or outside forces beyond one's control determine one's fate.

extrinsic motivation Motivation based on obvious external rewards or threats of punishment.

family therapy Treatment to change maladaptive interaction patterns within a family.

feature detectors Specialized brain cells that respond only to certain sensory information.

fetal period The third, and final, stage of prenatal development (eight weeks to birth), which is characterized by rapid weight gain in the fetus and the fine detailing of bodily organs and systems.

five-factor model (FFM) The trait theory that explains personality in terms of the "Big Five" model, which is composed of openness, conscientiousness, extroversion, agreeableness, and neuroticism.

formal operational stage Piaget's fourth stage (around age 11 and beyond), which is characterized by abstract and hypothetical thinking.

fundamental attribution error (FAE) Misjudgment of others' behavior as stemming from internal (dispositional) rather than external (situational) causes.

gate-control theory Theory that pain sensations are processed and altered by mechanisms within the spinal cord.

gender Psychological and sociocultural meanings added to biological maleness or femaleness.

gender roles Societal expectations for normal and appropriate male and female behavior.

general adaptation syndrome (GAS) Selye's three-stage (alarm, resistance, exhaustion) reaction to chronic stress.

germinal period The first stage of prenatal development, which begins with conception and ends with implantation in the uterus (the first two weeks).

grammar Rules that specify how phonemes, morphemes, words, and phrases should be combined to express thoughts. These rules include syntax and semantics.

group polarization A group's movement toward either riskier or more conservative behavior, depending on the members' initial dominant tendencies.

group therapy A form of therapy in which a number of people meet together to work toward therapeutic goals.

groupthink Faulty decision making that occurs when a highly cohesive group strives for agreement and avoids inconsistent information.

habituation Tendency of the brain to ignore environmental factors that remain constant.

hallucinations Imaginary sensory perceptions that occur without an external stimulus.

health psychology The study of how biological, psychological, and social factors affect health and illness.

heuristic A simple rule used in problem solving and decision making that does not guarantee a solution but offers a likely shortcut to it.

hierarchy of needs Maslow's theory of motivation that some motives (such as physiological and safety needs) must be met before going on to higher needs (such as belonging and self-actualization).

homeostasis A body's tendency to maintain a relatively stable state, such as a constant internal temperature, blood sugar, oxygen level, or water balance.

hormones Chemicals manufactured by endocrine glands and circulated in the bloodstream to produce bodily changes or maintain normal bodily function.

humanistic therapy Therapy that seeks to maximize personal growth through affective restructuring (emotional readjustment).

hypnosis A trancelike state of heightened suggestibility, deep relaxation, and intense focus.

implicit/nondeclarative memory The subsystem within long-term memory that consists of unconscious procedural skills, simple classically conditioned responses and priming.

independent variable (IV) Variable that is manipulated to determine its causal effect on the dependent variable.

individualistic cultures Cultures in which the needs and goals of the individual are emphasized over the needs and goals of the group.

informed consent A participant's agreement to take part in a study after being told what to expect.

insight Sudden understanding of a problem that implies the solution.

instinctive drift The tendency of some conditioned responses to shift (or drift) back toward innate response pattern.

instincts Behavioral patterns that are unlearned, always expressed in the same way, and universal in a species.

intelligence The global capacity to think rationally, act purposefully, and deal effectively with the environment.

intelligence quotient (IQ) A subject's mental age divided by his or her chronological age and multiplied by 100.

internal locus of control Belief that one controls one's own fate.

interpersonal attraction Positive feelings toward another.

intrinsic motivation Motivation resulting from personal enjoyment of a task or activity.

language A form of communication using sounds and symbols combined according to specified rules.

latent learning Hidden learning that exists without behavioral signs.

law of effect Thorndike's rule that the probability of an action being repeated is strengthened when followed by a pleasant or satisfying consequence.

learning A relatively permanent change in behavior or mental processes because of practice or experience.

localization of function Specialization of various parts of the brain for particular functions.

longitudinal method Research design that measures a single individual or a group of same-aged individuals over an extended period and gives information about age changes.

long-term memory (LTM) This third memory stage stores information for long periods. Its capacity is limitless; its duration is relatively permanent.

long-term potentiation (LTP) Long-lasting increase in neural excitability believed to be a biological mechanism for learning and memory.

maintenance rehearsal Repeating information to maintain it in short-term memory (STM).

major depressive disorder Long-lasting depressed mood that interferes with the ability to function, feel pleasure, or maintain interest in life.

maturation Development governed by automatic, genetically predetermined signals.

medical model The perspective that diseases (including mental illness) have physical causes that can be diagnosed, treated, and possibly cured.

meditation A group of techniques designed to refocus attention, block out all distractions, and produce an alternate state of consciousness.

memory An internal record or representation of some prior event or experience.

misinformation effect Distortion of a memory by misleading post-event information.

modeling therapy A learning technique in which the subject watches and imitates models who demonstrate desirable behaviors.

morpheme [MOR-feem] The smallest meaningful unit of language, formed from a combination of phonemes.

motivation A set of factors that activate, direct, and maintain behavior, usually toward a goal.

nature-nurture controversy Ongoing dispute over the relative contributors of nature (heredity) and nurture (environment) to the development of behavior and mental processes.

neurogenesis The division and differentiation of nonneuronal cells to produce neurons.

neuron A nerve cell that receives and conducts electrical impulses from the brain.

neuroplasticity The brain's ability to reorganize and change its structure and function through the life span.

neuroscience An interdisciplinary field studying how biological processes relate to behavioral and mental processes.

neurotransmitters Chemicals that neurons release, which affect other neurons.

obedience The act of following a direct command, usually from an authority figure.

objective personality tests Standardized questionnaires that require written responses, usually to multiple-choice or true-false questions.

observational learning Learning new behavior or information by watching others (also known as social learning or modeling).

operant conditioning Learning in which voluntary responses are controlled by their consequences (also known as instrumental or Skinnerian conditioning).

opponent-process theory Theory that color perception is based on three systems of color receptors, each of which responds in an on-off fashion to opposite-color stimuli: blue-yellow, red-green, and black-white.

parasomnias Abnormal disturbances occurring during sleep, including nightmares, night terrors, sleepwalking, and sleep talking.

peripheral nervous system (PNS) All nerves and neurons connecting the CNS to the rest of the body.

personality disorders Inflexible, maladaptive personality traits that cause significant impairment of social and occupational functioning.

personality Relatively stable and enduring patterns of thoughts, feelings, and actions.

phoneme [FO-neem] The smallest basic unit of speech or sound. The English language has about 40 phonemes.

postconventional level Kohlberg's highest level of moral development, in which individuals develop personal standards for right and wrong, and they define morality in terms of abstract principles and values that apply to all situations and societies.

posttraumatic stress disorder (PTSD) An anxiety disorder following exposure to a life-threatening or other extreme event that evoked great horror or helplessness. It is characterized by flashbacks, nightmares, and impaired functioning.

preconscious Freud's term for thoughts or motives that can be easily brought to mind.

preconventional level Kohlberg's first level of moral development, in which morality is based on rewards, punishment, and the exchange of favors.

prejudice A learned, generally negative attitude directed toward specific people solely because of their membership in an identified group.

preoperational stage Piaget's second stage (roughly ages 2 to 7 years), which is characterized by the ability to employ significant language and to think symbolically, though the child lacks operations (reversible mental processes), and thinking is egocentric and animistic.

problem-focused forms of coping Dealing directly with a stressor to decrease or eliminate it.

projective tests Psychological tests that use ambiguous stimuli, such as inkblots or drawings, that allow the test taker to project his or her unconscious thoughts onto the test material.

psychiatry The branch of medicine that deals with the diagnosis, treatment, and prevention of mental disorders.

psychoactive drugs Chemicals that change conscious awareness, mood, or perception.

psychoanalysis Freudian therapy designed to bring unconscious conflicts, which usually date back to early childhood experiences, into consciousness. Also, Freud's theoretical school of thought, which emphasizes unconscious processes.

psychodynamic therapy A briefer, more directive contemporary form of psychoanalysis which emphasizes conscious processes and current problems.

psychology The scientific study of behavior and mental processes.

psychoneuroimmunology [sye-koh-NEW-roh-IM-you-NOLL-oh-gee] The interdisciplinary field that studies the effects of psychological factors on the immune system.

psychopharmacology The study of drug effects on the mind and behavior.

psychosexual stages In Freudian theory, the five developmental periods (oral, anal, phallic, latency, and genital) during which particular kinds of pleasures must be gratified if personality development is to proceed normally.

psychosocial stages The eight developmental stages, each involving a crisis that must be successfully resolved, that individuals pass through in Erikson's theory of psychosocial development.

psychosurgery Operative procedures on the brain designed to relieve severe mental symptoms that have not responded to other forms of treatment.

psychotherapy Techniques employed to improve psychological functioning and promote adjustment to life.

punishment A consequence that weakens a response and makes it less likely to recur.

reciprocal determinism Bandura's belief that cognitions, behaviors, and the environment interact to produce personality.

reinforcement A consequence that strengthens a response and makes it more likely to recur.

reliability A measure of the consistency and stability of test scores when a test is readministered.

repair/restoration theory Sleep serves a recuperative function, allowing organisms to repair or replenish key factors.

retrieval Recovering information from memory storage.

retrieval cue A clue or prompt that helps stimulate recall and retrieval of a stored piece of information from long-term memory.

savant syndrome A condition in which a person who has mental retardation exhibits exceptional skill or brilliance in some limited field.

schemas Cognitive structures or patterns consisting of a number of organized ideas that grow and differentiate with experience.

schizophrenia A group of psychotic disorders involving major disturbances in perception, language, thought, emotion, and behavior. The individual withdraws from people and reality, often into a fantasy life of delusions and hallucinations.

selective attention Filtering out and attending only to important sensory messages.

self-concept Rogers's term for all the information and beliefs that individuals have about their own nature, qualities, and behavior.

self-efficacy Bandura's term for the learned belief that one is capable of producing desired results, such as mastering new skills and achieving personal goals.

sensorimotor stage Piaget's first stage (birth to approximately age 2), in which schemas are developed through sensory and motor activities.

sensory adaptation Repeated or constant stimulation decreases the number of sensory messages sent to the brain, which causes decreased sensation.

sensory memory This first memory stage holds sensory information. It has a relatively large capacity, but duration is only a few seconds.

sensory reduction Filtering and analyzing incoming sensations before sending a neural message to the cortex.

sex Biological maleness and femaleness, including chromosomal sex. Also, activities related to sexual behaviors, such as masturbation and intercourse.

sexual orientation Primary erotic attraction toward members of the same sex (homosexual, gay or lesbian), both sexes (bisexual), or the other sex (heterosexual).

sexual prejudice Negative attitudes toward an individual because of her or his sexual orientation.

sexual response cycle Masters and Johnson's description of the four-stage bodily response to sexual arousal, which consists of excitement, plateau, orgasm, and resolution.

shaping Reinforcement by a series of successively improved steps leading to desired response.

short-term memory (STM) This second memory stage temporarily stores sensory information and decides whether to send it on to long-term memory (LTM). Its capacity is limited to five to nine items, and its duration is about 30 seconds.

social psychology The study of how other people influence a person's thoughts, feelings, and actions.

socioemotional selectivity theory A natural decline in social contact as older adults become more selective with their time.

somatic nervous system (SNS) Subdivision of the peripheral nervous system (PNS). The SNS connects the sensory receptors and controls the skeletal muscles.

standardization Establishment of the norms and uniform procedures for giving and scoring a test.

stem cells Precursor (immature) cells that give birth to new specialized cells; a stem cell holds all the information it needs to make bone, blood, brain—any part of a human body—and can also copy itself to maintain a stock of stem cells.

stereotype threat Negative stereotypes about minority groups cause some members to doubt their abilities.

storage Retaining information over time.

stress The body's nonspecific response to any demand made on it; physical and mental arousal to situations or events that we perceive as threatening or challenging.

substance-related disorders Abuse of or dependence on a mood- or behavior-altering drug.

traits Relatively stable and consistent characteristics that can be used to describe someone.

transduction Process by which a physical stimulus is converted into neural impulses.

trichromatic theory Theory that color perception results from mixing three distinct color systems—red, green, and blue.

Type A personality Behavior characteristics that include intense ambition, competition, exaggerated time urgency, and a cynical, hostile outlook.

Type B personality Behavior characteristics consistent with a calm, patient, relaxed attitude.

unconditional positive regard Rogers's term for positive behavior toward a person with no contingencies attached.

unconscious Freud's term for thoughts or motives that lie beyond a person's normal awareness but that can be made available through psychoanalysis.

validity The ability of a test to measure what it was designed to measure.

REFERENCES

Aarts, H. (2007). Unconscious authorship ascription: The effects of success and effect-specific information priming on experienced authorship. *Journal of Experimental Social Psychology, 43,* 119–126.

Abadinsky, H. (2008). *Drug use and abuse: A comprehensive introduction* (6th ed.). Belmont, CA: Cengage.

Abbott, A. (2004). Striking back. *Nature, 429* (6990), 338–339.

Aboa-Éboulé, C. (2008). Job strain and recurrent coronary heart disease events—Reply. *Journal of the American Medical Association, 299,* 520–521.

Aboa-Éboulé, C., Brisson, C., Maunsell, E., Benoît, M., Bourbonnais, R., Vézina, M., Milot, A., Théroux, P., & Dagenais, G. R. (2008). Job strain and risk of acute recurrent coronary heart disease events. *Journal of the American Medical Association, 298,* 1652–1660.

About James Randi. (2002). *Detail biography.* [On-line]. Available: http://www.randi.org/jr/bio.html.

Acarturk, C., de Graaf, R., van Straten, A., ten Have, M., & Cuijpers, P. (2008). Social phobia and number of social fears, and their association with comorbidity, health-related quality of life and help seeking: A population-based study. *Social Psychiatry and Psychiatric Epidemiology, 43,* 273–279.

Achenbaum, W. A., & Bengtson, V. L. (1994). Re-engaging the disengagement theory of aging: On the history and assessment of theory development in gerontology. *Gerontologist, 34,* 756–763.

Adelmann, P. K., & Zajonc, R. B. (1989). Facial efference and the experience of emotion. *Annual Review of Psychology, 40,* 249–280.

Adler, A. (1964). The individual psychology of Alfred Adler. In H. L. Ansbacher & R. R. Ansbacher (Eds.), *The individual psychology of Alfred Adler.* New York: Harper & Row.

Adler, A. (1998). *Understanding human nature.* Center City: MN: Hazelden Information Education.

Aftanas, L. I., & Golosheikin, S. A. (2003). Changes in cortical activity in altered states of consciousness: The study of meditation by high-resolution EEG. *Human Physiology, 29,* 143–151.

Ahlsén, E. (2008). Embodiment in communication–Aphasia, apraxia, and the possible role of mirroring and imitation. *Clinical Linguistics & Phonetics, 22,* 311–315.

Ahmed, A. M. (2007). Group identity, social distance, and intergroup bias. *Journal of Economic Psychology, 28,* 324–337.

Ainsworth, M. D. S. (1967). *Infancy in Uganda: Infant care and the growth of love.* Baltimore: Johns Hopkins University Press.

Ainsworth, M. D. S., Blehar, M., Waters, E., & Wall, S. (1978). *Patterns of attachment: Observations in the strange situation and at home.* Hillsdale, NJ: Erlbaum.

Akers, R. M., & Denbow, D. (2008). *Anatomy and physiology of domestic animals.* Hoboken, NJ: Wiley.

Al'absi, M., Hugdahl, K., & Lovallo, W. R. (2002). Adrenocortical stress responses and altered working memory performance. *Psychophysiology, 39*(1), 95–99.

Al-Issa, I. (2000). Culture and mental illness in Algeria. In I. Al-Issa (Ed.), Al-Junun: *Mental illness in the Islamic world* (pp. 101–119). Madison, CT: International Universities Press.

Alberts, A., Elkind, D., & Ginsberg, S. (2007). The personal fable and risk-taking in early adolescence. *Journal of Youth and Adolescence, 36,* 71–76.

Allan, K., & Gabbert, F. (2008). I still think it was a banana: Memorable "lies" and forgettable "truths." *Acta Psychologica, 127,* 299–308.

Allik, J., & McCrae, R. R. (2004). Toward a geography of personality traits: Patterns of profiles across 36 cultures. *Journal of Cross-Cultural Psychology, 35*(1), 13–28.

Alloy, L. B., Abramson, L. Y., Whitehouse, W. G., Hogan, M. E., Tashman, N. A., Steinberg, D. L., Rose, D. T., & Donovan, P. (1999). Depressogenic cognitive styles: Predictive validity, information processing and personality characteristics, and developmental origins. *Behavior Research and Therapy, 37,* 503–531.

Allport, G. W. (1937). *Personality: A psychological interpretation.* New York: Holt, Rinehart and Winston.

Allport, G. W., & Odbert, H. S. (1936). Trait-names: A psycho-lexical study. *Psychological Monographs: General and Applied, 47,* 1–21.

Amato, P. R. (2006). Marital discord, divorce, and children's well-being: Results from a 20 year longitudinal study of two generations. In A. Clarke-Stewart & J. Dunn (Eds.), *Families count: Effects on child and adolescent development* (pp. 179–202). *The Jacobs Foundation series on adolescence.* New York: Cambridge University Press.

Amato, P. R. (2007). Transformative processes in marriage: Some thoughts from a sociologist. *Journal of Marriage and Family, 69,* 305–309.

American Cancer Society (2004). *Women and smoking.* [On-line]. Available: http://cancer.org/docroot/PED/content/PED_10_2X_Women_and_Smoking. asp?sitearea?PED.

American Counseling Association. (2006). *Crisis fact sheet: 10 ways to recognize Post-Traumatic Stress Disorder.* [On-line]. Available: http://www.counseling.org/PressRoom/PressReleases.aspx?AGuid=69c6fad2-c05e-4be3-af4e-3a1771414044.

American Heart Association. (2008). *Heart disease and stroke statistics—2008 update.* [Online]. Available: http://www.americanheart.org/presenter.jhtml?identifier=3054076

American Medical Association. (2008). *Alcohol and other drug abuse.* [On-line]. Available: http://www.ama-assn.org/ama/pub/category/3337.html.

American Psychiatric Association. (2000). *Diagnostic and statistical manual of mental disorders* (4th ed. TR). Washington, DC: American Psychiatric Press.

American Psychiatric Association. (2002). *APA Let's talk facts about posttraumatic stress disorder.* [On-line]. Available: http://www.psych.org/disasterpsych/fs/ptsd.cfm.

American Psychological Associaiton. (2002). *Ethical Principles of Psychologists and Code of Conduct.* (2002). [On-line]. Available: http://www.apa.org/ethics/code2002.html.

American Psychological Association. (1984). *Behavioral research with animals.* Washington, DC: Author.

Amir, N., Cobb, M., & Morrison, A. S. (2008). Threat processing in obsessive-compulsive disorder: Evidence from a modified negative priming task. *Behaviour Research and Therapy, 46,* 728–736.

Amundson, J. K., & Nuttgens, S. A. (2008). Strategic eclecticism in hypnotherapy: Effectiveness research considerations. *American Journal of Clinical Hypnosis, 50,* 233–245.

Ancis, J. R., Chen, Y., & Schultz, D. (2004). Diagnostic challenges and the so-called culture bound syndromes. In J. R. Ancis (Ed.), *Culturally responsive interventions: Innovative approaches to working with diverse populations* (pp. 197–209). New York: Brunner-Routledge.

Anderson, C. A. (2001). Heat and violence. *Current Directions in Psychological Science, 10*(1), 33–38.

Anderson, C. A. (2004). An update on the effects of playing violent video games. *Journal of Adolescence, 27*(1), 113–122.

Anderson, C. A., & Bushman, B. J. (2001). Effects of violent video games on aggressive behavior, aggressive cognition, aggressive affect, physiological arousal, and prosocial behavior: A meta-analytic review of the scientific literature. *Psychological Science, 12*(5), 353–359.

Anderson, C. A., & Dill, K. E. (2000). Video games and aggressive thoughts, feelings, and behavior in the laboratory and in life. *Journal of Personality and Social Psychology, 78*(4), 772–790.

Anderson, C. A., & Hamilton, M. (2005). Gender role stereotyping of parents in children's picture books: The invisible father. *Sex Roles, 52,* 145–151.

Anderson, C. A., Buckley, K. E., & Carnegey, N. L. (2008). Creating your own hostile environment: A laboratory examination of trait aggressiveness and the violence escalation cycle. *Personality and Social Psychology Bulletin, 34,* 462–473.

Anderson, L. E., & Silver, J. M. (2008). Neurological and medical disorders. In R. I. Simon & K. Tardiff (Eds.), *Textbook of violence assessment and management* (pp. 185–209). Arlington, VA: American Psychiatric Publishing.

Anderson, M. C., Ochsner, K. N., Kuhl, B., Cooper, J., Robertson, E., Gabrieli, S. W., Glover, G. H., & Gabrieli, J. D. E. (2004). Neural systems underlying the suppression of unwanted memories. *Science, 303*(5655), 232–235.

Andrade, T. G. C. S., & Graeff, F. G. (2001). Effect of electrolytic and neurotoxic lesions of the median raphe nucleus on anxiety and stress. *Pharmacology, Biochemistry & Behavior, 70*(1), 1–14.

Andrasik, F. (2006). Psychophysiological disorders: Headache as a case in point. In F. Andrasik (Ed.), *Comprehensive handbook of personality and psychopathology: Vol. 2: Adult Psychopathology* (pp. 409–422). Hoboken, NJ: Wiley.

Andreassi, J. K., & Thompson, C. A. (2007). Dispositional and situational sources of control: Relative impact on work-family conflict and positive spillover. *Journal of Managerial Psychology, 22*(8), 722–740.

Andreoni, J., & Petrie, R. (2008). Beauty, gender and stereotypes: Evidence from laboratory experiments. *Journal of Economic Psychology, 29,* 73–93.

Anisman, H., Merali, Z., & Stead, J. D. H. (2008). Experiential and genetic contributions to depressive- and anxiety-like disorders: Clinical and experimental studies. *Neuroscience & Biobehavioral Reviews, 32,* 1185–1206.

Ansell, E. B., & Grilo, C. M. (2007). Personality disorders. In M. Hersen, S. M. Turner, & D. C. Beidel (Eds.), *Adult psychopathology and diagnosis* (5th ed.) (pp. 633–678). Hoboken, NJ: Wiley.

Aou, S., (2006). Role of medial hypothalamus on peptic ulcer and depression. In C. Kubo & T. Kuboki (Eds), *Psychosomatic medicine: Proceedings of the 18th World Congress on Psychosomatic Medicine.* New York: Elsevier Science.

Appel, M., & Richter, T. (2007). Persuasive effects of fictional narratives increase over time. *Media Psychology, 10,* 113–134.

Appiah, K. A. (2008). *Experiments in ethics.* Cambridge, MA: Harvard University Press.

Arbib, M. A., & Mundhenk, T. N. (2005). Schizophrenia and the mirror system: An essay. *Neuropsychologia, 43,* 268–280.

Arbib, M. A., & Mundhenk, T. N. (2005). Schizophrenia and the mirror system: An essay. *Neuropsychologia, 43,* 268–280.1

Arriagada, O., Constandil, L., Hernández, A., Barra, R., Soto-Moyano, R., & Laurido, C. (2007). Brief communication: Effects of interleukin-1B on spinal cord nociceptive transmission in intact and propentofylline-treated rats. *International Journal of Neuroscience, 117,* 617–625.

Arumugam, V., Lee, J-S., Nowak, J. K., Pohle, R. J., Nyrop, J. E., Leddy, J. J., & Pelkman, C. L. (2008). A high-glycemic meal pattern elicited increased subjective appetite sensations in overweight and obese women. *Appetite, 50,* 215–222.

Asch, S. E. (1951). Effects of group pressure upon the modification and distortion of judgment. In H. Guetzkow (Ed.), *Groups, leadership, and men.* Pittsburgh: Carnegie Press.

Asghar, A. U. R., Chiu, Y-C., Hallam, G., Liu, S., Mole, H., Wright, H., & Young, A. W. (2008). An amygdala response to fearful faces with covered eyes. *Neuropsychologia, 46,* 2364–2370.

Astin, J. A. (2004). Mind-body therapies for the management of pain. *Clinical Journal of Pain, 20*(1), 27–32.

Atchley, R. C. (1997). *Social forces and aging* (8th ed.). Belmont, CA: Wadsworth.

Atkinson, R. C., & Shiffrin, R. M. (1968). Human memory: A proposed system and its control processes. In K. W. Spence & J. T. Spence (Eds.), *The psychology of learning and motivation* (Vol. 2). New York: Academic Press.

Aubert, A., & Dantzer, R. (2005). The taste of sickness: Lipopolysaccharide-induced finickiness in rats. *Physiology & Behavior, 84*(3), 437–444.

Axtell, R. E. (1998). *Gestures: The do's and taboos of body language around the world,* revised and expanded ed. New York: Wiley.

Axtell, R. E. (2007). *Essential do's and taboos: The complete guide to international business and leisure travel.* Hoboken, NJ: Wiley.

Ayers, S., Baum, A., McManus, C., Newman, S., Wallston, K., Weinman, J., & West, R. (Eds.). (2007). *Cambridge handbook of psychology, health, and medicine.* New York: Cambridge University Press.

Azizian, A., & Polich, J. (2007). Evidence for attentional gradient in the serial position memory curve from event-related potentials. *Journal of Cognitive Neuroscience, 19,* 2071–2081.

Bachman, G., & Zakahi, W. R. (2000). Adult attachment and strategic relational communication: Love schemas and affinity-seeking. *Communication Reports, 13*(1), 11–19.

Baddeley, A. D. (1992). Working memory. *Science, 255,* 556–559.

Baddeley, A., & Jarrold, C. (2007). Working memory and Down syndrome. *Journal of Intellectual Disability Research, 51,* 925–931.

Baer, J. (1994). Divergent thinking is not a general trait: A multi-domain training experiment. *Creativity Research Journal, 7,* 35–36.

Bagermihl, B. (1999). *Biological exuberance: Animal homosexuality and natural diversity.* New York: St Martins Press.

Bailey, J. M., Dunne, M. P., & Martin, N. G. (2000). Genetic and environmental influences on sexual orientation and its correlates in an Australian twin sample. *Journal of Personality and Social Psychology, 78*(3), 524–536.

Bailey, K. R., & Mair, R. G. (2005). Lesions of specific and nonspecific thalamic nuclei affect prefrontal cortex-dependent spects of spatial working memory. *Behavioral Neuroscience, 119*(2), 410–419.

Baillargeon, R. (2000). Reply to Bogartz, Shinskey, and Schilling; Schilling; and Cashon and Cohen. *Infancy, 1,* 447–462.

Baillargeon, R. (2008). Innate ideas revisited: For a principle of persistence in infants' physical reasoning. *Perspectives on Psychological Science, 3,* 2–13.

Baker, D., & Nieuwenhuijsen, M. J. (Eds.) (2008). *Environmental epidemiology: Study methods and application.* New York: Oxford University Press.

Baldwin, S., & Oxlad, M. (2000). *Electroshock and minors: A fifty year review.* New York: Greenwood Publishing Group.

Ball, H. A., McGuffin, P., & Farmer, A. E. (2008). Attributional style and depression. *British Journal of Psychiatry,* 192, 275–278.

Bandstra, E. S., Morrow, C. E., Vogel, A. L., Fifer, R. C., Ofir, A. Y., Dausa, A. T., Xue, L., & Anthony, J. C. (2002). Longitudinal influence of prenatal cocaine exposure on child language functioning. *Neurotoxicology & Teratology, 24*(3), 297–308.

Bandura, A. (1969). *Principles of behavior modification.* New York: Holt, Rinehart and Winston.

Bandura, A. (1986). *Social foundations of thought and action: A social cognitive theory.* Englewood Cliffs, NJ: Prentice Hall.

Bandura, A. (1989). Social cognitive theory. In R. Vasta (Ed.), *Annals of child development* (Vol. 6). Greenwich, CT: JAI Press.

Bandura, A. (1991). Social cognitive theory of moral thought and action. In W. M. Kurtines & J. L. Gewirtz (Eds.), *Handbook of moral behavior and development: Vol. 1. Theory.* Hillsdale, NJ: Erlbaum.

Bandura, A. (1997). *Self-efficacy: The exercise of control.* New York: Freeman.

Bandura, A. (2000). Exercise of human agency through collective efficacy. *Current Directions in Psychological Science, 9*(3), 75–83.

Bandura, A. (2006). Going global with social cognitive theory: From prospect to paydirt. In D. E. Berger & K. Pezdek (Eds.), *The rise of applied psychology: New frontiers and rewarding careers.* Mahwah, NJ: Erlbaum.

Bandura, A. (2008). Reconstrual of "free will" from the agentic perspective of social cognitive theory. In J. Baer, J. C. Kaufman, & R. F. Baumeister (Eds.), *Are we free? Psychology and free will.* New York: Oxford University Press.

Bandura, A., Ross, D., & Ross, S. (1961). Transmission of aggression through imitation of aggressive models. *Journal of Abnormal & Social Psychology, 63,* 575–582.

Bandura, A., & Walters, R. H. (1963). *Social learning and personality development.* New York: Holt, Rinehart and Winston.

Banko, K. M. (2008). Increasing intrinsic motivation using rewards: The role of the social context. *Dissertation Abstracts International: Section B: The Sciences and Engineering, 68*(10-B), 7005.

Banks, M. S., & Salapatek, P. (1983). Infant visual perception. In M. M. Haith & J. J. Campos (Eds.), *Handbook of child psychology.* New York: Wiley.

Bar-Haim, Y., Dan, O., Eshel, Y., & Sagi-Schwartz, A. (2007). Predicting childrens' anxiety from early attachment relationships. *Journal of Anxiety Disorders, 21,* 1061–1068.

Bard, C. (1934). On emotional expression after decortication with some remarks on certain theoretical views. *Psychological Review, 41,* 309–329.

Bargai, N., Ben-Shakar, G., & Shalev, A. Y. (2007). Posttraumatic stress disorder and depression in battered women: The mediating role of learned helplessness. *Journal of Family Violence, 22*(5), 267–275.

Barlett, C. P., Harris, R. J., & Bruey, C. (2008). The effect of the amount of blood in a violent video game on aggression, hostility, and arousal. *Journal of Experimental Social Psychology, 44,* 539–546.

Barlow, D. H. (Ed.). (2008). *Clinical handbook of psychological disorders: A step-by-step treatment manual (4th ed.).* New York: Guilford Press.

Barlow, D. H., & Durand, V. M. (2009). *Abnormal psychology: An integrative approach* (5th ed.). Belmont, CA: Cengage.

Barner, D, Wood, J., Hauser, M., & Carey, S. (2008). Evidence for a non-linguistic distinction between singular and plural sets in rhesus monkeys. *Cognition, 107,* 603–622.

Barnow, S., Ulrich, I., Grabe, H-J., Freyberger, H. J., & Spitzer, C. (2007). The influence of parental drinking behaviour and antisocial personality disorder on adolescent behavioural problems: Results of the Greifswalder family study. *Alcohol and Alcoholism, 42,* 623–628.

Barry, L. C., Allore, H. G., Guo, Z., Bruce, M. L., & Gill, T. M. (2008). Higher burden of depression among older women: The effect of onset, persistence, and mortality over time. *Archives of General Psychiatry, 65,* 172–178.

Bartels, M., van Beijsterveldt, C. E. M., Derks, E. M., Stroet, T. M., Polderman, T. J. C., Hudziak, J. J., & Boomsma, D. I. (2007). Young Netherlands Twin Register (Y-NTR): A longitudinal multiple informant study of problem behavior. *Twin Research and Human Genetics, 10,* 3–11.

Bartholow, B. D., & Anderson, C. A. (2002). Effects of violent video games on aggressive behavior: Potential sex differences. *Journal of Experimental Social Psychology, 38*(3), 283–290.

Barton, D. A., Esler, M. D., Dawood, T., Lambert, E. A., Haikerwal, D., et al. (2008). Elevated brain serotonin turnover in patients with depression: Effect of genotype and therapy. *Archives of General Psychiatry, 65,* 38-46.

Bates, A. L. (2007). How did you get in? Attributions of preferential selection in college admissions. *Dissertation Abstracts International: Section B: The Sciences and Engineering, 68*(4-B), 2694.

Batson, C. D. (1991). *The altruism question: Toward a social-psychological answer.* Hillsdale, NJ: Erlbaum.

Batson, C. D. (1998). Altruism and prosocial behavior. In D.T. Gilbert, S.T. Fiske, & G. Lindzey (Eds.). *The handbook of social psychology, Vol. 2* (4th ed.) (pp. 282–316). Boston, MA: McGraw-Hill.

Batson, C. D. (2006). "Not all self-interest after all": Economics of empathy-induced altruism. In D. De Cremer, M. Zeelenberg, & J. K. Murnighan (Eds.), *Social psychology and economics* (pp. 281–299). Mahwah, NJ: Erlbaum.

Batson, C. D., & Ahmad, N. (2001). Empathy-induced altruism in a prisoner's dilemma II: What if the target of empathy has defected? *European Journal of Social Psychology, 31*(1), 25–36.

Baumrind, D. (1980). New directions in socialization research. *American Psychologist, 35,* 639–652.

Baumrind, D. (1991). Effective parenting during the early adolescent transition. In P. A. Cowan & E. M. Hetherington (Eds.), *Family transition* (pp. 111–163). Hillsdale, NJ: Erlbaum.

Baumrind, D. (1995). *Child maltreatment and optimal caregiving in social contexts.* New York: Garland.

Bearer, C. F., Stoler, J. M., Cook, J. D., & Carpenter, S. J. (2004–2005). Biomarkers of alcohol use in pregnancy. *Alcohol Research & Health, 28*(1), 38–43.

Beck, A. T. (1976). *Cognitive therapy and the emotional disorders.* New York: International Universities Press.

Beck, A. T. (2000). *Prisoners of hate.* New York: Harper Perennial.

Beck, A. T., & Grant, P. M. (2008). Negative self-defeating attitudes: Factors that influence everyday impairment in individuals with schizophrenia. *American Journal of Psychiatry, 165,* 772.

Becker, S. I. (2008). The mechanism of priming: Episodic retrieval or priming of popout? *Acta Psychologica, 127,* 324–339.

Begg, I. M., Needham, D. R., & Bookbinder, M. (1993). Do backward messages unconsciously affect listeners? No. *Canadian Journal of Experimental Psychology, 47,* 1–14.

Behar, R. (2007). Gender related aspects of eating disorders: A psychosocial view. In J. S. Rubin (Ed.), *Eating disorders and weight loss research* (pp. 39–65). Hauppauge, NY: Nova Science Publishers.

Beilin, H. (1992). Piaget's enduring contribution to developmental psychology. *Developmental Psychology, 28,* 191–204.

Bem, S. L. (1981). Gender schema theory: A cognitive account of sex typing. *Psychological Review, 88,* 354–364.

Bem, S. L. (1993). *The lenses of gender: Transforming the debate on sexual inequality.* New Haven, CT: Yale University Press.

Ben-Eliyahu, S., Page, G. G., & Schleifer, S. J. (2007). Stress, NK cells, and cancer: Still a promissory note. *Brain, Behavior, & Immunity, 21,* 881–887.

Benazzi, F. (2004). Testing early-onset chronic atypical depression subtype. *Neuropsychopharmacology, 29*(2), 440–441.

Bendersky, M., & Sullivan, M. W. (2007). Basic methods in infant research. In A. Slater & M. Lewis (Eds.), *Introduction to infant development* (2nd ed.). New York: Oxford University Press.

Benloucif, S., Orbeta, L., Ortiz, R., Janssen, I., Finkel, S., Bleiberg, J., & Zee, P. C. (2004). Morning or evening activity improves neuropsychological performance and subjective sleep quality in older adults. *Sleep, 27*(8), 1542–1551.

Bennett, M., Barrett, M., Karakozov, R., Kipiani, G., Lyons, E., Pavlenko, V., & Riazanova, T. (2004). Young children's evaluations of the ingroup and of outgroups: A multi-national study. *Social Development, 13*(1), 124–141.

Benton, T. R., McDonnell, S., Ross, D. F., Thomas III, W. N., & Bradshaw, E. (2007). Has eyewitness research penetrated the American legal system? A synthesis of case history, juror knowledge, and expert testimony. In R. C. L. Lindsay, D. F. Ross, J. D. Read, & M. P. Toglia (Eds.), *The handbook of eyewitness psychology, Vol II: Memory for people* (pp. 453–500). Mahwah, NJ: Erlbaum.

Berchtold, N. C. (2008). Exercise, stress mechanisms, and cognition. In W. W. Spirduso, L. W. Poon, & W. Chodzko-Zajko (Eds), *Exercise and its mediating effects on cognition* (pp. 47–67). *Aging, exercise, and cognition series.* Champaign, IL: Human Kinetics.

Berger, M., Speckmann, E.J., Pape, H. C., & Gorji, A. (2008). Spreading depression enhances human neocortical excitability in vitro. *Cephalalgia, 28*, 558–562.

Bergstrom-Lynch, C. A. (2008). Becoming parents, remaining childfree: How same-sex couples are creating families and confronting social inequalities. *Dissertation Abstracts International Section A: Humanities and Social Sciences, 68*(8-A), 3608.

Berman, M. E., Tracy, J. I., & Coccaro, E. F. (1997). The serotonin hypothesis of aggression revisited. *Clinical Psychology Review, 17*(6), 651–665.

Bernard, L. L. (1924). *Instinct.* New York: Holt.

Bernheim, K. F., & Lewine, R. R. J. (1979). *Schizophrenia: Symptoms, causes, and treatments.* New York: Norton.

Berreman, G. (1971). *Anthropology today.* Del Mar, CA: CRM Books.

Berry, J. W., Poortinga, Y. H., Segall, M. H., & Dasen, P. R. (2002). *Cross-cultural psychology: Research and applications* (2nd ed.). New York: Cambridge University Press.

Berthoud, H-R., & Morrison, C. (2008). The brain, appetite, and obesity. *Annual Review of Psychology, 59*, 55–92.

Berzoff, J. (2008). Psychosocial ego development: The theory of Erik Erikson. In J. Berzoff, L. M. Flanagan, & P. Hertz (Eds.), *Inside out and outside in: Psychodynamic clinical theory and psychopathology in contemporary multicultural contexts* (2nd ed.) (pp. 99–120). Lanham, MD: Jason Aronson.

Best, J. B. (1999). *Cognitive psychology* (5th ed.). Belmont, CA: Wadsworth.

Bhattacharya, S. K., & Muruganandam, A. V. (2003). Adaptogenic activity of Withania somnifera: An experimental study using a rat model of chronic stress. *Pharmacology, Biochemistry & Behavior, 75*(3), 547–555.

Bianchi, M., Franchi, S., Ferrario, P., Sotgiu, M. L., & Sacerdote, P. (2008). Effects of the bisphosphonate ibandronate on hyperalgesia, substance P, and cytokine levels in a rat model of persistent inflammatory pain. *European Journal of Pain, 12*, 284–292.

Bieber, C., Müller, K. G., Blumenstiel, K., Hochlehnert, A., Wilke, S., Hartmann, M., & Eich, W. (2008). A shared decision-making communication training program for physicians treating fibromyalgia patients: Effects of a randomized controlled trial. *Journal of Psychosomatic Research, 64*, 13–20.

Biehl, M., Matsumoto, D., Ekman, P., Hearn, V., Heider, K., Kudoh, T., & Ton, V. (1997). Matsumoto and Ekman's Japanese and Caucasian facial expressions of emotion (JACFEE): Reliability data and cross-national differences. *Journal of Nonverbal Behavior, 21*, 3–21.

Billiard, M. (2007). Sleep disorders. In L. Candelise, R. Hughes, A. Liberati, B. M. J. Uitdehaag, & C. Warlow (Eds.), Evidence-based neurology: Management of neurological disorders (pp. 70–78). *Evidence-based medicine.* Malden, MA: Blackwell Publishing.

Birkenhäger, T. K., Renes, J., & Pluijms, E. M. (2004). One-year follow-up after successful ECT: A naturalistic study in depressed inpatients. *Journal of Clinical Psychiatry, 65*(1), 87–91.

Bjorkqvist, K. (1994). Sex differences in physical, verbal, and indirect aggression: A review of recent research. *Sex Roles, 30*, 177–188.

Blakemore, C., & Cooper, G. F. (1970). Development of the brain depends on the visual environment. *Nature, 228*, 477–478.

Blanco, C., Schneier, F. R., Schmidt, A., Blanco-Jerez, C. R., Marshall, R. D., Sanchez-Lacay, A., & Liebowitz, M. R. (2003). Pharmacological treatment of social anxiety disorder: A meta-analysis. *Depression & Anxiety, 18*(1), 29–40.

Blass, T. (1991). Understanding behavior in the Milgram obedience experiment: The role of personality, situations, and their interactions. *Journal of Personality and Social Psychology, 60*(3), 398–413.

Blass, T. (2000). Stanley Milgram. In A. E. Kazdin (Ed.), *Encyclopedia of psychology* (Vol. 5) (pp. 248–250). Washington, DC: American Psychological Association.

Bloch, Y., Grisaru, N., Hard, E. V., Beitler, G., Faivel, N., Ratzoni, G., Stein, D., & Levkovitz, Y. (2008). Repetitive transcranial magnetic stimulation in the treatment of depression in adolescents: An open-label study. *Journal of ECT, 24*, 153–159.

Blum, K., Braverman, E. R., Holder, J. M., Lubar, J. F., Monastra, V. J., Miller, D., Lubar, J. O., Chen, T. J. H., & Comings, D. E. (2000). Reward deficiency syndrome: A biogenetic model for the diagnosis and treatment of impulsive, addictive, and compulsive behaviors. *Journal of Psychoactive Drugs, 32*(Suppl), 1–68.

Bob, P. (2008). Pain, dissociation and subliminal self-representations. *Consciousness and Cognition: An International Journal, 17*, 355–369.

Boccato, G., Capozza, D., Falvo, R., & Durante, F. (2008). Capture of the eyes by relevant and irrelevant onsets. *Social Cognition, 26*, 224–234.

Bock, G. R., & Goode, J. A. Eds. (1996). *Genetics of criminal and antisocial behavior.* Chichester, England: Wiley.

Bohm-Starke, N., Brodda-Jansen, G., Linder, J., & Danielson, I. (2007). The result of treatment on vestibular and general pain thresholds in women with provoked vestibulodynia. *Clinical Journal of Pain, 23*(7), 598–604.

Bohus, M., Haaf, B., Simms, T., Limberger, M. F., Schmahl, C., Unckel, C., Lieb, K., & Linehan, M. M. (2004). Effectiveness of inpatient dialectical behavioral therapy for borderline personality disorder: A controlled trial. *Behaviour Research & Therapy, 42*(5), 487–499.

Bond, F. W., & Bunce, D. (2000). Mediators of change in emotion-focused and problemfocused worksite stress management intervention. *Journal of Occupational Health Psychology, 5*, 153–163.

Bond, M. H., & Smith, P. B. (1996). Cross-cultural social and organizational psychology. *Annual Review of Psychology, 47*, 205–235.

Bor, D., Billington, J., & Baron-Cohen, S. (2007). Savant memory for digits in a case of synaesthesia and Asberger Syndrome is related to hyperactivity in the lateral prefrontal cortex. *Neurocase, 13*, 311–319.

Borba, M. (2001). *Building moral intelligence: The seven essential virtues that teach kids to do the right thing.* San Francisco: Jossey-Bass.

Borbely, A. A. (1982). Circadian and sleep dependent processes in sleep regulation. In J. Aschoff, S. Daan, & G. A. Groos (Eds.), *Verte-*

brate circadian rhythms (pp. 237–242). Berlin: Springer/Verlag.

Borchers, B. J. (2007). Workplace environment fit, commitment, and job satisfaction in a nonprofit association. *Dissertation Abstracts International: Section B: The Sciences and Engineering, 67*(7-B), 4139.

Borra, R. (2008). Working with the cultural formulation in therapy. *European Psychiatry, 23,* S43–S48.

Borrego, J., Ibanez, E. S., Spendlove, S. J., & Pemberton, J. R. (2007). Treatment acceptability among Mexican American parents. *Behavior Therapy, 38*(3), 218–227.

Bouchard, T. J., Jr. (1994). Genes, environment, and personality. *Science, 264,* 1700–1701.

Bouchard, T. J., Jr. (1997). The genetics of personality. In K. Blum & E. P. Noble (Eds.), *Handbook of psychiatric genetics.* Boca Raton, FL: CRC Press.

Bouchard, T. J., Jr. (1999). The search for intelligence. *Science, 284,* 922–923.

Bouchard, T. J., Jr. (2004). Genetic influence on human psychological traits: A survey. *Current Directions in Psychological Science, 13*(4), 148–151.

Bouchard, T. J., Jr. & McGue, M. (1981). Familial studies of intelligence: A review. *Science, 212*(4498), 1055–1059.

Bouchard, T. J., Jr., McGue, M., Hur, Y., & Horn, J. M. (1998). A genetic and environmental analysis of the California Psychological Inventory using adult twins reared apart and together. *European Journal of Personality, 12,* 307–320.

Boucher, L., & Dienes, Z. (2003). Two ways of learning associations. *Cognitive Science, 27*(6), 807–842.

Bourne, L. E., Dominowski, R. L., & Loftus, E. F. (1979). *Cognitive processes.* Englewood Cliffs, NJ: Prentice Hall.

Bouton, M. E. (1994). Context, ambiguity, and classical conditioning. *Current Directions in Psychological Science, 2,* 49–53.

Bowers, K. S., & Woody, E. Z. (1996). Hypnotic amnesia and the paradox of intentional forgetting. *Journal of Abnormal Psychology, 105,* 381–390.

Bowlby, J. (1969). *Attachment and loss: Vol. 1. Attachment.* New York: Basic Books.

Bowlby, J. (1973). *Attachment and loss: Vol. 2. Separation and anxiety.* New York: Basic Books.

Bowlby, J. (1982). Attachment and loss: Retrospect and prospect. *American Journal of Orthopsychiatry, 52,* 664–678.

Bowlby, J. (1989). *Secure attachment.* New York: Basic Books.

Bowlby, J. (2000). *Attachment.* New York: Basic Books.

Bowling, A. C., & Mackenzie, B. D. (1996). The relationship between speed of information processing and cognitive ability. *Personality & Individual Differences, 20*(6), 775–800.

Boyle, S. H., Williams, R. B., Mark, D. B., Brummett, B. H., Siegler, I. C., Helms, M. J., & Brady, S. S. (2007). Young adults' mediause and attitudes toward interpersonal andinstitutional forms of aggression. *Aggressive Behavior, 33*(6), 519–525.

Boysen, G. A., & Vogel, D. L. (2007). Biased assimilation and attitude polarization in response to learning about biologicalexplanations of homosexuality. *Sex Roles, 56,* 755–762.

Bozkurt, A. S., & Aydin, O. (2004). Temel yükleme hatasinin degisik yas ve iki alt kültürde incelenmesi. [A developmental investigation of fundamental attribution error in two subcultures.] *Türk Psikoloji Dergisi, 19,* 91–104.

Bradley, R., Conklin, C. Z., & Westen, D. (2007). Borderline personality disorder. In W. O'Donohue, K. A. Fowler, S. O. Lilienfeld (Eds.), *Personality disorders: Toward the DSM-V* (pp. 167–201). Thousand Oaks, CA: Sage.

Bragdon, A. D., & Gamon, D. (1999). *Building left-brain power: Left-brain conditioning exercises and tips to strengthen language, math, and uniquely human skills.* Thousand Oaks, CA: Brainwaves Books.

Brandon, T. H., Collins, B. N., Juliano, L. M., & Lazev, A. B. (2000). Preventing relapse among former smokers: A comparison of minimal interventions through telephone and mail. *Journal of Consulting and Clinical Psychology, 68*(1), 103–113.

Breland, K., & Breland, M. (1961). The misbehavior of organisms. *American Psychologist, 16,* 681–684.

Bremner, J. D., Vythilingam, M., Vermetten, E., Anderson, G., Newcomer, J. W. & Charney, D. S. (2004). Effects of glucocorticoids on declarative memory function in major depression. *Biological Psychiatry, 55*(8), 811–815.

Brent, D. A., Melhem, N. (2008). Familial transmission of suicidal behavior. *Psychiatric Clinics of North America, 31,* 157–177.

Brewer, J. B., Zhao, Z., Desmond, J. E., Glover, G. H., & Gabrieli, J. D. (1998). Making memories: Brain activity that predicts how well visual experience will be remembered. *Science, 281,* 1185–1187.

Brim, O. (1999). *The MacArthur Foundation study of midlife development.* Vero Beach, FL: MacArthur Foundation.

Brislin, R. W. (1997). *Understanding culture's influence on behavior* (2nd ed.). San Diego: Harcourt Brace.

Brislin, R. W. (2000). *Understanding culture's influence on behavior.* Ft. Worth, TX: Harcourt.

Brkich, M., Jeffs, D., & Carless, S. A. (2002). A global self-report measure of person-job fit. *European Journal of Psychological Assessment, 18*(1), 43–51.

Brody, A., Olmstead, R. E., London, E. D., Farahi, J., Meyer, J. H., Grossman, P., Lee, G. S., Huang, J., Hahn, E. L., & Mandelkern, M. A. (2004). Smoking-induced ventral striatum dopamine release. *American Journal of Psychiatry, 161*(7), 1211–1218.

Brown, A., & Whiteside, S. P. (2008). Relations among perceived parental rearing behaviors, attachment style, and worry in anxious children. *Journal of Anxiety Disorders, 22,* 263–272.

Brown, E., Deffenbacher, K., & Sturgill, K. (1977). Memory for faces and the circumstances of encounter. *Journal of Applied Psychology, 62,* 311–318.

Brown, P. K., & Wald, G. (1964). Visual pigments in single rods and cones of the human retina. Direct measurements reveal mechanisms of human night and color vision. *Science, 144,* 45–52.

Brown, R. P., & Josephs, R. A. (1999). A burden of proof: Stereotype relevance and gender differences in math performance. *Journal of Personality and Social Psychology, 76*(2), 246–257.

Brown, R., & Kulik, J. (1977). Flashbulb memories. *Cognition, 5,* 73–99.

Brown, T. A., O'Leary, T. A., & Barlow, D. H. (2001). Generalized anxiety disorder. In D. H. Barlow (Ed.), *Clinical handbook of psychological disorders: A step-by-step treatment manual* (3rd ed.) (pp. 154–208). New York: Guilford Press.

Browne, K. O. (2001). Cultural formulation of psychiatric diagnoses. *Culture, Medicine & Psychiatry, 25*(4), 411–425.

Bugg, J. M., Zook, N. A., DeLosh, E. L., Davalos, D. B., & Davis, H. P. (2006). Age differences in fluid intelligence: Contributions of general slowing and frontal decline. *Brain and Cognition, 62,* 9–16.

Bull, L. (2003). What can be done to prevent smoking in pregnancy? A literature review. *Early Child Development & Care, 173*(6), 661–667.

Bunde, J., & Suls, J. (2006). A quantitative analysis of the relationship between the cook-medley hostility scale and traditional coronary artery disease risk factors. *Health Psychology, 25,* 493–500.

Buontempo, G., & Brockner, J. (2008). Emotional intelligence and the ease of recall judgment bias: The mediating effect of private self-focused attention. *Journal of Applied Social Psychology, 38,* 159–172.

Burger, J. M. (2009). Replicating Milgram: Would people still obey today? *American Psychologist, 64(1),* 1–11.

Burns, S. M. (2008). Unique and interactive predictors of mental health quality of life among men living with prostate cancer. *Dissertation Abstracts International: Section B: The Sciences and Engineering, 68*(10-B), 6953.

Bushman, B. J. (2002). Does venting anger feed or extinguish the flame? Catharsis, rumination, distraction, anger and aggressive responding. *Personality & Social Psychology Bulletin, 28(6),* 724–731.

Buss, D. M. (1989). Sex differences in human mate preferences: Evolutionary hypotheses tested in 37 cultures. *Behavioral and Brain Sciences, 12,* 1–49.

Buss, D. M. (2003). *The evolution of desire: Strategies of human mating.* New York: Basic Books.

Buss, D. M. (2005). *The handbook of evolutionary psychology.* Hoboken, NJ: Wiley.

Buss, D. M. (2007). The evolution of human mating. *Acta Psychologica Sinica. 39,* 502–512.

Buss, D. M. (2008). *Evolutionary psychology: The new science of the mind* (3rd ed.). Boston: Allyn & Bacon.

Buss, D. M. and 40 colleagues. (1990). International preferences in selecting mates: A study of 37 cultures. *Journal of Cross-Cultural Psychology, 21,* 5–47.

Buswell, B. N. (2006). The role of empathy, responsibility, and motivations to respond without prejudice in reducing prejudice. *Dissertation Abstracts International: Section B: The Sciences and Engineering, 66,* 6968.

Butcher, J. N. (2000). Revising psychological tests: Lessons learned from the revision of the MMPI. *Psychological Assessment, 12*(3), 263–271.

Butcher, J. N. (2005). *A beginner's guide to the MMPI-2 (2nd ed.).* Washington, DC: American Psychological Association.

Butcher, J. N., & Perry, J. N. (2008). *Personality assessment in treatment planning: Use of the MMPI-2 and BTPI.* New York: Oxford University Press.

Butler, R. A. (1954, February). Curiosity in monkeys. *Scientific American, 190,* 70–75.

Butterweck, V. (2003). Mechanism of action of St John's Wort in depression: What is known? *CNS Drugs, 17*(8), 539–562.

Byne, W. (2007). Biology and sexual minority status. In I. H. Meyer & M. E. Northridge (Eds.), *The health of sexual minorities: Public health perspectives on lesbian, gay, bisexual, and transgender populations* (pp. 65–90). New York: Springer Science + Business Media.

Byne, W., Dracheva, S., Chin, B., Schmeidler, J. M., Davis, K. L., & Haroutunian, V. (2008). Schizophrenia and sex associated differences in the neuronal and oligodendrocyte-specific genes in individual thalamic nuclei. *Schizophrenia Research, 98,* 118–128.

Cadoret, R. J., Leve, L. D., & Devor, E. (1997). Genetics of aggressive and violent behavior. *Psychiatric Clinics of North America, 20,* 301–322.

Cai, W-H., Blundell,J., Han, J., Greene, R. W., & Powell, C. M. (2006). Postreactivation glucocorticoids impair recall of established fear memory. *Journal of Neuroscience, 26*(37), 9560–9566.

Camarena, B., Santiago, H., Aguilar, A., Ruvinskis, E., González-Barranco, J., & Nicolini, H. (2004). Family-based association study between the monoamine oxidase A gene and obesity: Implications for psychopharmacogenetic studies. *Neuropsychobiology, 49*(3), 126–129.

Cameron, L., Rutland, A. & Brown, R. (2007). Promoting children's positive intergroup attitudes towards stigmatized groups: Extended contact and multiple classification skills training. *International Journal of Behavioral Development, 31,* 454–466.

Campbell, A., & Muncer, S. (2008). Intent to harm or injure? Gender and the expression of anger. *Aggressive Behavior, 34,* 282–293.

Campbell, L., Simpson, J. A., Kashy, D. A., & Fletcher, G. J. O. (2001). Ideal standards, the self, and flexibility of ideals in close relationships. *Personality & Social Psychology Bulletin, 27*(4), 447–462.

Cannon, W. B. (1927). The James-Lange theory of emotions: A critical examination and an alternative theory. *American Journal of Psychology, 39,* 106–124.

Cannon, W. B., Lewis, J. T., & Britton, S. W. (1927). The dispensability of the sympathetic division of the autonomic nervous system. *Boston Medical Surgery Journal, 197,* 514.

Caprara, G. V., Vecchione, M., Barbaranelli, C., & Fraley, R. C. (2007). When likeness goes with liking: The case of political preference. *Political Psychology, 28,* 609–632.

Carballo, J. J., Harkavy-Friedman, J., Burke, A. K., Sher, L., Baca-Garcia, E., Sullivan, G. M., Gruneman, M. F., Parsey, R. V., Mann, J. J., & Oquendo, M. A. (2008). Family history of suicidal behavior and early traumatic experiences: Addictive effect on suicidality and course of bipolar illness? *Journal of Affective Disorders, 109,* 57–63.

Carels, R. A., Konrad, K., Young, K. M., Darby, L. A., Coit, C., Clayton, A. M., & Oemig, C. K. (2008). Taking control of your personal eating and exercise environment: A weight maintenance program. *Eating Behaviors, 9,* 228–237.

Carey, B. (2008, October 1). Psychoanalytic therapy wins backing. *New York Times.* [On-line]. Available: http://www.nytimes.com/2008/10/01/health/01psych.html.

Carey, B. (2008, December 4). H.M. an unforgettable amnesiac, dies at 82. *New York Times.* [On-line]. Available: http:/www.nytimes.com/2008/12/05/us/05hm.html.

Carlson, L. E., Speca, M., Faris, P., & Patel, K. D. (2007). One year pre-post intervention follow-up of psychological, immune, endocrine and blood pressure outcomes of mindfulness- based stress reduction (MBSR) in breast and prostate cancer outpatient. *Brain, Behavior, and Immunity, 21,* 1038–1049.

Carlson, N. R. (2008). *Foundations of physiological psychology: International edition.* Boston: Allyn & Bacon.

Carnagey, N. L., Anderson, C. A., & Bartholow, B. D. (2007). Media violence and social neuroscience: New questions and new opportunities. *Current Directions in Psychological Science, 16,* 178–182.

Carstensen, L. L. (2006). The influence of a sense of time on human development. *Science, 312,* 1913–1915.

Caruso, E. M. (2008). Use of experienced retrieval ease in self and social judgments. *Journal of Experimental Social Psychology, 44,* 148–155.

Carvalho, C., Mazzoni, G., Kirsch, I., Meo, M., & Santandrea, M. (2008). The effect of posthypnotic suggestion, hypnotic suggestibility, and goal intentions on adherence to medical instructions. *International Journal of Clinical and Experimental Hypnosis, 56,* 143–155.

Castelli, L., Corazzini, L. L., & Geminiani, G. C. (2008). Spatial navigation in large-scale virtual environments: Gender differences in survey tasks. *Computers in Human Behavior, 24,* 1643–1667.

Castillo, R. J. (2003). Trance, functional psychosis, and culture. *Psychiatry: Interpersonal & Biological Processes, 66*(1), 9–21.

Castonguay, L. G., & Hill, C. (Eds.) (2007). *Insight in psychotherapy.* Washington, DC: American Psychological Association.

Cathers-Schiffman, T. A., & Thompson, M. S. (2007). Assessment of English-and Spanish-speaking students with the WISC-III and Leiter-R. *Journal of Psychoeducational Assessment, 25,* 41–52.

Cattell, R. B. (1950). *Personality: A systematic, theoretical, and factual study.* New York: McGraw-Hill.

Cattell, R. B. (1963). Theory of fluid and crystallized intelligence: A critical experiment. *Journal of Educational Psychology, 54*, 1–22.

Cattell, R. B. (1965). *The scientific analysis of personality.* Baltimore: Penguin.

Cattell, R. B. (1971). *Abilities: Their structure, growth, and action.* Boston: Houghton Mifflin.

Cattell, R. B. (1990). Advances in Cattellian personality theory. In L. A. Pervin (Ed.), *Handbook of personality: Theory and research.* New York: Guilford Press.

Cehajic, S., Brown, R., & Castano, E. (2008). Forgive and forget? Antecedents and consequences of intergroup forgiveness in Bosnia and Herzegovina. *Political Psychology, 29*, 351–367.

Cervone, D., & Shoda, Y. (1999). Beyond traits in the study of personality coherence. *Current Directons in Psychological Science, 8*(a), 27–32.

Cesaro, P, & Ollat, H. (1997). Pain and its treatments. *European Neurology, 38*, 209–215.

Ceschi, G., & Scherer, K. R. (2001). Contrôler l'expression faciale et changer l'émotion: Une approche développementale. The role of facial expression in emotion: A developmental perspective. *Enfance, 53*(3), 257–269.

Chabrol, H., Montovany, A., Ducongé, E., Kallmeyer, A., Mullet, E., & Leichsenring, F. (2004). Factor structure of the borderline personality inventory in adolescents. *European Journal of Psychological Assessment, 20*(1), 59–65.

Challem, J., Berkson, B., Smith, M. D., & Berkson, B. (2000). *Syndrome X: The complete program to prevent and reverse insulin resistance.* New York: Wiley.

Champtiaux, N., Kalivas, P. W., & Bardo, M. T. (2006). Contribution of dihydro-betaerythroidine sensitive nicotinic acetylcholine receptors in the ventral tegmental area to cocaine-induced behavioral sensitization in rats. *Behavioural Brain Research, 168*, 120–126.

Chandrashekar, J., Hoon, M. A., Ryba, N. J. P., & Zuker, C. S. (2006). The receptors and cells for mammalian taste. *Nature, 444*, 288–294.

Chang, G., Orav, J., McNamara, T. K., Tong, MY., & Antin, J. H. (2005). Psychosocial function after hematopoietic stem cell transplantation. *Psychosomatics: Journal of Consultation Liaison Psychiatry, 46*(1), 34–40.

Chapman, A. L., Leung, D. W., & Lynch, T. R. (2008). Impulsivity and emotion dysregulation in borderline personality disorder. *Journal of Personality Disorders, 22*, 148–164.

Charles, E. P. (2007). Object permanence, an ecological approach. *Dissertation Abstracts International: Section B: The Sciences and Engineering, 67*(8-B), 4737.

Charles, S. T., & Carstensen, L. L. (2007). Emotion regulation and aging. In J. J. Gross (Ed.), *Handbook of emotion regulation.* New York: Guilford Press.

Chartrand, T., Pinckert, S., & Burger, J. M. (1999). When manipulation backfires: The effects of time delay and requester on the foot-in-the-door technique. *Journal of Applied Social Psychology, 29*, 211–221.

Cherney, I. D. (2005). Children's and adults' recall of sex-stereotyped toy pictures: Effects of presentation and memory task. *Infant & Child Development, 14*(1), 11–27.

Cheung, M. S., Gilbert, P., & Irons, C. (2004). An exploration of shame, social rank, and rumination in relation to depression. *Personality & Individual Differences, 36*(5), 1143–1153.

Chiou, W-B. (2007). Customers' attributional judgments towards complaint handling in airline service: A confirmatory study based on attribution theory. *Psychological Reports, 100*, 1141–1150.

Chiricozzi, F. R., Clausi, S., Molinari, M., & Leggio, M. G. (2008). Phonological short-term store impairment after cerebellar lesion: A single case study. *Neuropsychologia, 46*, 1940–1953.

Choi, I., & Nisbett, R. E. (2000). Cultural psychology of surprise: Holistic theories and recognition of contradiction. *Journal of Personality and Social Psychology, 79*(6), 890–905.

Choi, N. (2004). Sex role group differences in specific, academic, and general selfefficacy. *Journal of Psychology: Interdisciplinary & Applied, 138*(2), 149–159.

Chomsky, N. (1968). *Language and mind.* New York: Harcourt, Brace, World.

Chomsky, N. (1980). *Rules and representations.* New York: Columbia University Press.

Chrisler, J. C. (2008). The menstrual cycle in a biopsychosocial context. In F. Denmark & M. A. Paludi (Eds.), *Psychology of women: A handbook of issues and theories* (2nd ed.) (pp. 400–439). *Women's psychology.* Westport, CT: Praeger/Greenwood.

Christensen, H., Anstey, K. J., Leach, L. S., & Mackinnon, A. J. (2008). Intelligence, education, and the brain reserve hypothesis. In F. I. M. Craik & T. A. Salthouse (Eds.), *The handbook of aging and cognition* (3rd ed.) (pp. 133–188). New York: Psychology Press.

Christopher, K., Lutz-Zois, C. J., & Reinhardt, A. R. (2007). Female sexual-offenders: Personality pathology as a mediator of the relationship between childhood sexual abuse history and sexual abuse. *Child Abuse & Neglect, 31*, 871–883.

Chung, J. C. C. (2006). Measuring sensory processing patterns of older Chinese people: Psychometric validation of the adult sensory profile. *Aging & Mental Health, 10*, 648–655.

Cialdini, R. B. (2009). *Influence: Science and practice* (5th ed.). Boston: Allyn & Bacon.

Clark, A. J. (2007). *Empathy in counseling and psychotherapy: Perspectives and practices.* Mahwah, NJ: Erlbaum.

Clark, K. B., & Clark, M. P. (1939). The development of consciousness of self and theemergence of racial identification in Negropreschool children. *Journal of Social Psychology, 10*, 591–599.

Cleaves, D. H., & Latner, J. D. (2008). Evidence-based therapies for children and adolescents with eating disorders. In R. G. Steele, D. T. Elkin, & M. C. Roberts (Eds.), *Handbook of evidence-based therapies for children and adolescents: Bridging science and practice (pp. 335–353). Issues in clinical child psychology.* New York: Springer Science + Business Media.

Cleeremans, A., & Sarrazin, J. C. (2007). Time, action, and consciousness. *Human Movement Science, 26*, 180–202.

Clifford, J. S., Boufal, M. M., & Kurtz, J. E. (2004). Personality traits and critical thinking: Skills in college students empirical tests of a two-factor theory. *Assessment, 11*(2), 169–176.

Clinton, S. M., & Meador-Woodruff, J H. (2004). Thalamic dysfunction in schizophrenia: Neurochemical, neuropathological, and in vivo imaging abnormalities. *Schizophrenia Research, 69*(2–3), 237–253.

Cloud, J. (2001, February 26). New sparks over electroshock. *Time*, 60–62.

Clulow, C. (2007). John Bowlby and couple psychotherapy. *Attachment & Human Development, 9*, 343–353.

Cohen, D., Mason, K., & Farley, T. A. (2004). Beer consumption and premature mortality in Louisiana: An ecologic analysis. *Journal of Studies on Alcohol, 65*(3), 398–403.

Cohen, R. A., Paul, R., Zawacki, T. M., Moser, D. J., Sweet, L., & Wilkinson, H. (2001). Emotional and personality changes following cingulotomy. *Emotion, 1*(1), 38–50.

Cohen, S., Hamrick, N., Rodriguez, M. S., Feldman, P. J., Rabin, B. S., & Manuck, S. B. (2002). Reactivity and vulnerability to stress associated risk for upper respiratory illness. *Psychosomatic Medicine, 64*(2), 302–310.

Cohen, S., & Lemay, E. P. (2007). Why would social networks be linked to affect and health practices? *Health Psychology, 26*, 410–417.

Cole, M., & Gajdamaschko, N. (2007). Vygotsky and culture. In H. Daniels, J. Wertsch, & M. Cole (Eds.), *The Cambridge companion to Vygotsky.* New York: Cambridge University Press.

Combrink-Graham, L., & McKenna, S. B. (2006). Families with children with disrupted attachments. In L. Combrinck-Graham (Ed.), *Children in family contexts: Perspectives on treatment* (pp. 242–264). New York: Guilford Press.

Combs, D. R., Basso, M. R., Wanner, J. L., & Ledet, S. N. (2008). Schizophrenia. In M. Hersen & J. Rosqvist (Eds.), *Handbook of psychological assessment, case conceptualization, and treatment, Vol 1: Adults* (pp. 352–402). Hoboken, NJ: Wiley.

Congdon, E., & Canli, T. (2008). A neurogenetic approach to impulsivity. *Journal of Personality, 76,* 1447–1484.

Connolly, S. (2000). *LSD (just the facts).* Baltimore: Heinemann Library.

Connor, K. M., & Davidson, J. R. T. (2002). A placebo-controlled study of Kava kava in generalized anxiety disorder. *International Clinical Psychopharmacology, 17*(4), 185–188.

Constantino, M. J., Manber, R., Ong, J., Kuo, T. F., Huang, J. S., & Arnow, B. A. (2007). Patient expectations and therapeutic alliance as predictors of outcome in group cognitive-behavioral therapy for insomnia. *Behavioral Sleep Medicine, 5,* 210–228.

Cook, M., & Mineka, S. (1989). Observational conditioning of fear to fear-relevant versus fear-irrelevant stimuli in rhesus monkeys. *Journal of Abnormal Psychology, 98,* 448–459.

Cook, P. F. (2000). Effects of counselors' etiology attributions on college students' procrastination. *Journal of Counseling Psychology, 47*(3), 352–361.

Coombs, R. H. (2004). *Handbook of addictive disorders.* Hoboken, NJ: Wiley.

Cooper, C., Bebbington, P. E., Meltzer, H., Bhugra, D., Brugha, T., Jenkins, R., Farrell, M., & King, M. (2008). Depression and common mental disoders in lone parents: Results of the 2000 National Psychiatric Morbidity Survey. *Psychological Medicine, 38,* 335–342.

Cooper, J., & Hogg, M. A. (2007). Feeling the anguish of others: A theory of vicarious dissonance. In M. P. Zanna (Ed.), *Advances in experimental social psychology.* San Diego, CA: Elsevier.

Cooper, R. (2004). What is wrong with the DSM? *History of Psychiatry, 15*(1), 5–25.

Cooper, W. E. Jr., Pérez-Mellado, V., Vitt, L. J., & Budzinsky, B. (2002). Behavioral responses to plant toxins in two omnivorous lizard species. *Physiology & Behavior, 76*(2), 297–303.

Coplan, R. J., Arbeau, K. A., & Armer, M. (2008). Don't fret, be supportive! Maternal characteristics linking child shyness to psychosocial and school adjustment in kindergarten. *Journal of Abnormal Child Psychology, 36,* 359–371.

Corey, G. (2009). *Theory and practice of counseling and psychotherapy* (8th ed.). Belmont, CA: Cengage.

Corkin, S. (2002). What's new with the amnesic patient H.M.? *Nature Reviews Neuroscience, 3,* 153–160.

Cornelius, J. R., & Clark, D. B. (2008). Depressive disorders and adolescent substance use disorders. In Y. Kaminer & O. G. bukstein (Eds.), *Adolescent substance abuse: Psychiatric comorbidity and high-risk behaviors* (pp. 221–242). New York: Routledge/Taylor & Francis Group.

Corr, C. A., Nabe, C. M., & Corr, D. M. (2009). *Death and dying: Life and living* (6th ed.). Belmont, CA: Wadsworth.

Corydon, H. D. (2007). What is neurofeedback? *Journal of Neurotherapy, 10,* 25–36.

Costa, J. L., Brennen, M. B., & Hochgeschwender, U. (2002). The human genetics of eating disorders: Lessons from the leptin/melanocortin system. *Child & Adolescent Psychiatric Clinics of North America, 11*(2), 387– 397.

Costa Jr., P.T., McCrae, R. R., & Martin, T. A. (2008). Incipient adult personality: The NEO-PI-3 in middle-school-aged children. *British Journal of Developmental Psychology, 26,* 71–89.

Courage, M. L., & Adams, R. J. (1990). Visual acuity assessment from birth to three years using the acuity card procedures: Cross-sectional and longitudinal samples. *Optometry and Vision Science, 67,* 713–718.

Coyne, S. M., Archer, J., & Eslea, M. (2004). Cruel intentions on television and in real life: Can viewing indirect aggression increase viewers' subsequent indirect aggression? *Journal of Experimental Child Psychology, 88*(3), 234–253.

Craig, A. D., & Bushnell, M. C. (1994). The thermal grill illusion: Unmasking the burn of cold pain. *Science, 265,* 252–255.

Crandall, C. S., & Martinez, R. (1996). Culture, ideology, and antifat attitudes. *Personality and Social Psychology Bulletin, 22,* 1165–1176.

Crawford, C. S. (2008). Ghost in the machine: A genealogy of phantom-prosthetic relations (amputation, dismemberment, prosthetic). *Dissertation Abstracts International Section A: Humanities and Social Sciences, 68*(7-A), 3173.

Crespo-Facorro, B., Barbadillo, L., Pelayo-Terán, J., Rodríguez-Sánchez, J. M., & Teran, J. M. (2007). Neuropsychological functioning and brain structure in schizophrenia. *International Review of Psychiatry, 19,* 325–336.

Cresswell, M. (2008). Szasz and his interlocutors: Reconsidering Thomas Szasz's "Myth of Mental Illness" thesis. *Journal for the Theory of Social Behaviour, 38,* 23–44.

Crews, F. (1997). The verdict on Freud. *Psychological Science, 7*(2), 63–68.

Crews, F. T., Collins, M. A., Dlugos, C., Littleton, J., Wilkins, L., Neafsey, E. J., Pentney, R., Snell, L. D., Tabakoff, B., Zou, J., & Noronha, A. (2004). Alcohol-induced neurodegeneration: When, where and why? *Alcoholism: Clinical & Experimental Research, 28*(2), 350–364.

Crone, C. C., & Gabriel, G. (2002). Herbal and nonherbal supplements in medical-psychiatric patient populations. *Psychiatric Clinics of North America, 25*(1), 211–230.

Crooks, R. & Bauer, K. (2008). *Our sexuality* (10th ed.). Belmont, CA: Cengage.

Crosby, B. (2003). Case studies in rational emotive behavior therapy with children and adolescents. *Journal of Cognitive Psychotherapy, 17*(3), 289–291.

Crowell, S. E., Beauchaine, T. P., & Lenzenweger, M. F. (2008). The development of borderline personality disorder and self-injurious behavior. In T. P. Beauchaine & S. P. Hinshaw (Eds.), *Child and adolescent psychopathology* (pp. 510–539). Hoboken, NJ: Wiley.

Cubelli, R., & Della Sala, S. (2008). Flashbulb memories: Special but not iconic. *Cortex, 44,* 908–909.

Cullen, D., & Gotell, L. (2002). From orgasms to organizations: Maslow, women's sexuality and the gendered foundations of the needs hierarchy. *Gender, Work & Organization, 9*(5), 537–555.

Cully, J. A., & Stanley, M. A. (2008). Assessment and treatment of anxiety in later life. In K. Laidlaw & B. Knight (Eds.), *Handbook of emotional disorders in later life: Assessment and treatment.* New York: Oxford University Press.

Cummings, D. E. (2006). Ghrelin and the short- and long-term regulation of appetite and body weight. *Physiology & Behavior, 89,* 71–84.

Cummings, E., & Henry, W. E. (1961). *Growing old: The process of disengagement.* New York: Basic Books.

Cunningham, G. B. (2002). Diversity and recategorization: Examining the effects of cooperation on bias and work outcomes. *Dissertation Abstracts International Section A: Humanities and Social Sciences, 63,* 1288.

Cunningham, G. B., Fink, J. S., & Kenix, L. J. (2008). Choosing an endorser for a women's sporting event: The interaction of attractiveness and expertise. *Sex Roles, 58,* 371–378.

Curtiss, S. (1977). *Genie: A psycholinguistic study of a modern-day "wild child."* New York: Academic Press.

D'Alessio, D., & Allen, M. (2002). Selective exposure and dissonance after decisions. *Psychological Reports, 91*(2), 527–532.

Dackis, C. A., & O'Brien, C. P. (2001). Cocaine dependence: A disease of the brain's reward centers. *Journal of Substance Abuse Treatment, 21*(3), 111–117.

Dalenberg, C., Loewenstein, R., Spiegel, D., Brewin, C., Lanius, R., Frankel, S., Gold, S., Van der Kolk, B., Simeon, D., Vermetten, E., Butler, L., Koopman, C., Courtois, C., Dell, P., Nijenhuis, E., Chu, J., Sar, V., Palesh, O., Cuevas, C., & Paulson, K. (2007). Scientific study of the dissociative disorders. *Psychotherapy and Psychosomatics, 76*, 400–401.

Dalley, J. W., Fryer, T. D., Brichard, L., Robinson, E. S. J, Theobald, D. E. H., Lääne, K., Peña, Y., Murphy, E. R., Shah, Y., Probst, K., Abakumova, I., Aigbirhio, F. I., Richards, H. K., Hong, Y., Baron, J-C., Everitt, B. J., & Robbins, T. W. (2007). Nucleus accumbens D2/3 receptors predict trait impulsivity and cocaine reinforcement. *Science, 315*, 1267–1270.

Damasio, A. R. (1999). *The feeling of what happens: Body and emotion in the making of consciousness.* New York: Harcourt Brace.

Dana, R. H. (2005). *Multicultural assessment: Principles, applications, and examples.* Mahwah, NJ: Erlbaum.

Daniels, K., Toth, J., & Jacoby, J. (2006). The aging of executive functions. In E. Bialystok & F. I. M. Craik (Eds.) *Lifespan cognition: Mechanisms of change.* New York: Oxford University Press.

Dantzer, R., O'Connor, J. C., Freund, G. G., Johnson, R. W., & Kelley, K. W. (2008). From inflammation to sickness and depression: When the immune system subjugates the brain. *Nature Reviews Neuroscience, 9*, 46–57.

Dapretto, M., Davies, M. S., Pfeifer, J. H., Scott, A. A., Sigman, M., Bookheimer, S. Y., & Iacoboni, M. (2006). Understanding emotions in others: Mirror neuron dysfunction in children with autism spectrum disorders. *Nature Neuroscience, 9*, 28–30.

Darmani, N. A., & Crim, J. L. (2005) Delta9-tetrahydrocannabinol prevents emesis more potently than enhanced locomotor activity produced by chemically diverse dopamine D2/D3 receptor agonists in the least shrew (Cryptotis parva). *Pharmacology Biochemistry and Behavior, 80*, 35–44.

Darwin, C. (1859). *On the origin of species.* London: Murray.

Darwin, C. (1872). *The expression of the emotions in man and animals.* London: Murray.

Dasí, C., Soler, M. J., Cervera, T., & Ruiz, J. C. (2008). Influence of articulation rate on two memory tasks in young and older adults. *Perceptual and Motor Skills, 106*, 579–589.

David, D., Schnur, J. E., & Belloiu, A. (2002). Another search for the "hot" cognitions: Appraisal, irrational beliefs, attributions, and their relation to emotion. *Journal of Rational-Emotive & Cognitive Behavior Therapy, 20*(2), 93–132.

Davidson, P. S. R., Anaki, D., Ciaramelli, E., Cohn, M., Kim, A. S. N., Murphy, K. J., Troyer, A. K., Moscovitch, M., & Levine, B. (2008). Does lateral parietal cortex support episodic memory? Evidence from focal lesion patients. *Neuropsychologia, 46*, 1743–1755.

Davies, I. (1998). A study of colour grouping in three languages: A test of the linguistic relativity hypothesis. *British Journal of Psychology, 89*, 433–452.

Davies, J. M. (1996). Dissociation, repression and reality testing in the countertransference: the controversey over memory and false memory in the psychoanalytic treatment of adult survivors of childhood sexual abuse. *Psychoanalytic Dialogues, 6*, 189–218.

Davies, M. F. (2008). Irrational beliefs and unconditional self-acceptance. III. The relative importance of different types of irrational belief. *Journal of Rational-Emotive & Cognitive Behavior Therapy, 26*, 102–118.

Dawson, K. A. (2004). Temporal organization of the brain: Neurocognitive mechanisms and clinical implications. *Brain & Cognition, 54*(1), 75–94.

De Charms, R., & Moeller, G. H. (1962). Values expressed in American children's readers: 1800–1950. *Journal of Abnormal and Social Psychology, 64*(2), 136–142.

De Coteau, T. J., Hope, D. A., & Anderson, J. (2003). Anxiety, stress, and health in northern plains Native Americans. *Behavior Therapy, 34*(3), 365–380.

de Oliveira-Souza, R., Moll, J., Ignácio, F. A., & Hare, R. D. (2008). Psychopathy in a civil psychiatric outpatient sample. *Criminal Justice and Behavior, 35*, 427–437.

de Pinho, R. S. N., da Silva-Júnior, F. P., Bastos, J. P. C., Maia, W. S., de Mello, M. T., de Bruin, V. M. S., & de Bruin, P. F. C. (2006). Hypersomnolence and accidents in truck drivers: A cross-sectional study. *Chronobiology International, 23*, 963–971.

de Waal, F. B. M. (2008). Putting the altruism back into altruism: The evolution of empathy. *Annual Review of Psychology, 59*, 279–300.

Deary, I. J., & Stough, C. (1996). Intelligence and inspection time: Achievements, prospects, and problems. *American Psychologist, 51*, 599–608.

Deary, I. J., & Stough, C. (1997). Looking down on human intelligence. *American Psychologist, 52*, 1148–1149.

Deary, I. J., Ferguson, K. J., Bastin, M. E., Barrow, G. W. S., Reid, L. M., Seckl, J. R., Wardlaw, J. M., & MacLullich, A. M. J. (2007). Skull size and intelligence, and King Robert Bruce's IQ. *Intelligence, 35*, 519–528.

DeCasper, A. J., & Fifer, W. D. (1980). Of human bonding: Newborns prefer their mother's voices. *Science, 208*, 1174–1176.

Deci, E. L. (1995). *Why we do what we do: The dynamics of personal autonomy.* New York: Putnam's Sons.

Deci, E. L., & Moller, A. C. (2005). The concept of competence: A starting place for understanding intrinsic motivation and self-determined extrinsic motivation. In A. J. Elliot & C. S. Dweck (Eds.), *Handbook of competence and motivation* (pp. 579–597). New York: Guilford.

Deckers, L. (2005). *Motivation: Biological, psychological, and environmental* (2nd ed.). Boston: Allyn & Bacon/Longman.

DeClue, G. (2003). The polygraph and lie detection. *Journal of Psychiatry & Law, 31*(3), 361–368.

Delgado, J. M. R. (1960). Emotional behavior in animals and humans. *Psychiatric Research Report, 12*, 259–271.

Delgado, P. L. (2004). How antidepressants help depression: Mechanisms of action and clinical response. *Journal of Clinical Psychiatry, 65*, 25–30.

Delgado-Gaitan, C. (1994). Socializing young children in Mexican-American families: An intergenerational perspective. In P. M. Greenfield & R. R. Cocking (Eds.), *Cross-cultural roots of minority child development* (pp. 55–86). Hillsdale, NJ: Erlbaum.

Deller, T., Haas, C. A., Freiman, T. M., Phinney, A., Jucker, M., & Frotscher, M. (2006). Lesion-induced axonal sprouting in the central nervous system. *Advances in Experimental Medicine and Biology, 557*, 101–121.

Delville, Y., Mansour, K. M., & Ferris. C. F. (1996). Testosterone facilitates aggression by modulating vasopressin receptors in the hypothalamus. *Physiology and Behavior, 60*, 25–29. *Engendering psychology: Women and gender revisited* (2nd ed.). Boston: Allyn and Bacon.

Dembe, A. E., Erickson, J. B., Delbos, R. G., & Banks, S. M. (2006). Nonstandard shift schedules and the risk of job-related injuries. *Scandinavian Journal of Work, Environment, & Health, 32*, 232–240.

Dement, W. C., & Vaughan, C. (1999). *The promise of sleep.* New York: Delacorte Press. den Boer, J. A. (2000). Social anxiety disorder/social phobia: Epidemiology, diagnosis, neurobiology, and treatment. *Comprehensive Psychiatry, 41*(6), 405–415.

Denmark, F. L., Rabinovitz, V. C., & Sechzer, J. A. (2005). *Engendering psychology: Women and gender revisited* (2nd ed.). Boston: Allyn and Bacon.

Dennis, W., & Dennis, M. G. (1940). Cradles and cradling customs of the Pueblo Indians. *American Anthropologist, 42*, 107–115.

Der-Karabetian, A., Stephenson, K., & Poggi, T. (1996). Environmental risk perception, activism and world-mindedness among samples of British and U. S. college students. *Perceptual and Motor Skills, 83*(2), 451–462.

DeValois, R. L. (1965). Behavioral and electrophysiological studies of primate vision. In W. D. Neff (Ed). *Contributions to sensory physiology* (Vol. 1). New York: Academic Press.

Devlin, M. J., Yanovski, S. Z., & Wilson, G. T. (2000). Obesity: What mental health professionals need to know. *American Journal of Psychiatry, 157*(6), 854–866.

Dhikav, V., Aggarwal, N., Gupta, S., Jadhavi, R., & Singh, K. (2008). Depression in Dhat syndrome. *Journal of Sexual Medicine, 5*, 841–844.

Diamond, A., & Amso, D. (2008). Contributions of neuroscience to our understanding of cognitive development. *Current Directions in Psychological Science, 17*, 136–141.

Diamond, L. M. (2004). Emerging perspectives on distinctions between romantic love and sexual desire. *Current Directions in Psychological Science, 13*(3), 116–119.

Diaz-Berciano, C., de Vicente, F., & Fontecha, E. (2008). *Modulating effects in learned helplessness of dyadic dominance-submission relations. Aggressive Behavior, 34*(3), 273–281.

Dickens, W. T., & Flynn, J. R. (2001). Heritability estimates versus large environmental effects: The IQ paradox resolved. *Psychological Review, 108*(2), 346–369.

Dickinson, D. J., O'Connell, D. Q., & Dunn, J. S. (1996). Distributed study, cognitive study strategies and aptitude on student learning. *Psychology: A Journal of Human Behavior, 33*(3), 31–39.

Diego, M. A., & Jones, N. A. (2007). Neonatal antecedents for empathy. In T. Farrow & P. Woodruff (Eds.), *Empathy in mental illness* (pp. 145–167). New York: Cambridge University Press.

Diener, E. (2008). Myths in the science of happiness, and directions for future research. In M. Eid & R. J. Larsen (Eds.), *The science of subjective well-being.* New York: Guilford Press.

Diener, M. L., Mengelsdorf, S. C., McHale, J. L., & Frosch, C. A. (2002). Infants' behavioral strategies for emotion regulation with fathers and mothers: Associations with emotional expressions and attachment quality. *Infancy, 3*(2), 153–174.

Diener, M. L., Isabella, R. A., Behunin, M. G., & Wong, M. S. (2008). Attachment to mothers and fathers during middle childhood: Associations with child gender, grade, and competence. *Social Development, 7*, 84–101.

Dijksterhuis, A., Aarts, H., & Smith, P. K. (2005). The power of the subliminal: On subliminal persuasion and other potential applications. In R. R. Hassin, J. S. Uleman, & J. A. Bargh (Eds.), The new unconscious (pp. 77–106). *Oxford series in social cognition and social neuroscience.* New York: Oxford University Press.

Dill, K. E., & Thill, K. P. (2007). Video game characters and the socialization of gender roles: Young people's perceptions mirror sexist media depictions. *Sex Roles, 57*, 851–864.

Dillon, S. (2009). Study sees an Obama effect as lifting black test-takers. New York Time. [Online]. Available: http://www.nytimes.com/2009/01/23/education/23gap.html

Dimberg, U., & Thunberg, M. (1998). Rapid facial reactions to emotion facial expressions. *Scandinavian Journal of Psychology, 39*(1), 39–46.

Dimberg, U., Thunberg, M., & Elmehed, K. (2000). Unconscious facial reactions to emotional facial expressions. *Psychological Science, 11*(1), 86–89.

DiPietro, J. A. (2000). Baby and the brain: Advances in child development. *Annual Review of Public Health, 21*, 455–71.

Dobson, K. S. (2008). Cognitive therapy for depression. In M. A. Whisman (Ed.), *Adapting cognitive therapy for depression: Managing complexity and comorbidity* (pp. 3–35). New York: Guilford.

Dollard, J., Doob, L., Miller, N., Mowrer, O. H., & Sears, R. R. (1939). *Frustration and aggression.* New Haven, CT: Yale University Press.

Domhoff, G. W. (2005). A reply to Hobson. (2005). *Dreaming, 15*(1), 30–32.

Domhoff, G. W. (2007). Realistic simulation and bizarreness in dream content: Past findings and suggestions for future research. In D. Barrett & P. McNamara (Eds.), *The new science of dreaming Volume 2. Content, recall, and personality correlates* (pp. 1–27). Praeger perspectives. Westport, CT: Praeger.

Domjan, M. (2005). Pavlovian conditioning: A functional perspective. *Annual Review of Psychology, 56*, 179–206.

Dondi, M., Simion, F., & Caltran, G. (1999). Can newborns discriminate between their own cry and the cry of another newborn infant? *Developmental Psychology, 35*, 418–426.

Donini, L. M., Savina, C., & Cannella, C. (2003). Eating habits and appetite control in the elderly: The anorexia of aging. *International Psychogeriatrics, 15*(1), 73–87.

Donohue, R. (2006). Person-environment congruence in relation to career change and career persistence. *Journal of Vocational Behavior, 68*, 504–515.

Dougal, S., Phelps, E. A., & Davachi, L. (2007). The role of medial temporal lobe in item recognition and source recollection of emotional stimuli. *Cognitive, Affective & Behavioral Neuroscience, 7*, 233–242.

Dougherty, D. D., Baer, L., Cosgrove, G. R., Cassem, E. H., Price, B. H., Nierenberg, A. A., Jenike, M. A., & Rauch, S. L. (2002). Prospective long-term follow-up of 44 patients who received cingulotomy for treatment-refractory obsessive-compulsive disorder. *American Journal of Psychiatry, 159*(2), 269–275.

Doyle, A., & Pollack, M. H. (2004). Long-term management of panic disorder. *Journal of Clinical Psychiatry. 65*(Suppl5), 24–28.

Driscoll, A. K., Russell, S. T., & Crockett, L. J. (2008). Parenting styles and youth wellbeing across immigrant generations. *Journal of Family Issues, 29*, 185–209.

Dryer, D. C., & Horowitz, L. M. (1997). When do opposites attract? Interpersonal complementarity versus similarity. *Journal of Personality and Social Psychology, 72*, 592–603.

Duckworth, K., & Borus, J. F. (1999). Population-based psychiatry in the public sector and managed care. In A. M. Nicholi (Ed.), *The Harvard guide to psychiatry.* Cambridge, MA: Harvard University Press.

Dufresne, T. (2007). *Against Freud: Critics talk back.* Palo Alto, CA: Stanford University Press.

Duncan, T. B. (2007). Adult attachment and value orientation in marriage. *Dissertation Abstracts International: Section B: The Sciences and Engineering, 68*, 3447.

Dunham, Y. C. (2007). Assessing the automaticity of intergroup bias. *Dissertation Abstracts International: Section B: The Sciences and Engineering, 68*(6-B), 4153.

Duval, D. C. (2006). The relationship between African centeredness and psychological androgyny among African American women in middle adulthood. *Dissertation Abstracts International: Section B: The Sciences and Engineering. 67*(2-B), 1146.

Eaker, E. D., Sullivan, L. M., Kelly-Hayes, M., D'Agostino, R. B., & Benajmin, E. J. (2007). Marital status, marital strain, and risk of coronary heart disease or total mortality: The Framingham offspring study. *Psychosomatic Medicine, 69*, 509–513.

Eddy, K. T., Hennessey, M., & Thompson-Brenner, H. (2007). Eating pathology in East African women: The role of media exposure and globalization. *Journal of Nervous and Mental Disease, 195,* 196–202.

Edwards, B. (1999). *The new drawing on the right side of the brain.* Baltimore: J P Tarcher.

Edwards, S. (2007). *50 plus one ways to improve your study habits,* Chicago, IL: New Eductaion Encouragement Press.

Ehrenreich, B. (2004, July 15). *All together now.* [On-line]. Available: http://select.nytimes.com/gst/abstract.html?res=F00E16FA3C5E0C768DDDAE0894DC404482.

Eisenberger, R., & Armeli, S. (1997). Can salient reward increase creative performance without reducing intrinsic creative interest? *Journal of Personality and Social Psychology, 72,* 652–663.

Eisenberger, R., & Rhoades, L. (2002). Incremental effects of reward on creativity. *Journal of Personality and Social Psychology, 81*(4), 728–741.

Ekers, D., Richards, D., & Gilbody, S. (2008). A meta-analysis of randomized trials of behavioural treatment of depression. *Psychological Medicine, 38,* 611–623.

Ekman, P. (1993). Facial expression and emotion. *American Psychologist, 48,* 384–392.

Ekman, P. (2004). *Emotions revealed: Recognizing faces and feelings to improve communication and emotional life.* Thousand Oaks, CA: Owl Books.

Ekman, P., & Keltner, D. (1997). Universal facial expressions of emotion: An old controversy and new findings. In U. C. Segerstrale & P. Molnar (Eds.), *Nonverbal communication: Where nature meets culture.* Mahwah, NJ: Erlbaum.

Elder, G. (1998). The life course as developmental theory. *Current Directions in Psychological Science, 69,* 1–12.

Elkin, A., Kalidindi, S., & McGuffin, P. (2004). Have schizophrenia genes been found? *Current Opinion in Psychiatry, 17(2),* 107–113.

Elkind, D. (1967). Egocentrism in adolescence. *Child Development, 38,* 1025–1034.

Elkind, D. (1981). *The hurried child.* Reading, MA: Addison-Wesley.

Elkind, D. (2000). A quixotic approach to issues in early childhood education. *Human Development, 43*(4–5), 279–283.

Elkind, D. (2007). *The hurried child: Growing up too fast too soon* (25th anniversary ed.). Cambridge, ME: Da Capo Press.

Ellis, A. (1961). *A guide to rational living.* Englewood Cliffs, NJ: Prentice-Hall.

Ellis, A. (1996). *Better, deeper, and more enduring brief therapy.* New York: Institute for Rational Emotive Therapy.

Ellis, A. (1997). Using rational emotive behavior therapy techniques to cope with disability. *Professional Psychology: Research and Practice, 28,* 17–22.

Ellis, A. (2003a). Early theories and practices of rational emotive behavior therapy and how they have been augmented and revised during the last three decades. *Journal of Rational-Emotive & Cognitive Behavior Therapy, 21*(3–4), 219–243.

Ellis, A. (2003b). Similarities and differences between rational emotive behavior therapy and cognitive therapy. *Journal of Cognitive Psychotherapy, 17*(3), 225–240.

Ellis, A. (2004). Why rational emotive behavior therapy is the most comprehensive and effective form of behavior therapy. *Journal of Rational Emotive & Cognitive Behavior Therapy, 22*(2), 85–92.

Ellis, L., Ficek, C., Burke, D., & Das, S. (2008). Eye color, hair color, blood type, and the rhesus factor: Exploring possible genetic links to sexual orientation. *Archives of Sexual Behavior, 37,* 145–149.

Ellman, L. M., & Cannon, T. D. (2008). Environmental pre-and perinatal influences in etiology. In K. T. Mueser & D. V. Jeste (Eds.), *Clinical handbook of schizophrenia* (pp. 65–73). New York: Guilford.

Epley, N. (2008) Rebate psychology. *New York Times.* [On-line]. Available: http://select.nytimes.com/mem/tnt.html?tntget=2008/01/31/opinion/31epley.html.

Erdelyi, M. H., & Applebaum, A. G. (1973). Cognitive masking: The disruptive effect of an emotional stimulus upon the perception of contiguous neutral items. *Bulletin of the Psychonomic Society, 1,* 59–61.

Erikson, E. (1950). *Childhood and society.* New York: Norton.

Erlacher, D., & Schredl, M. (2004). Dreams reflecting waking sport activities: A comparison of sport and psychology students. *International Journal of Sport Psychology, 35*(4), 301–308.

Ertekin-Taner, N. (2007). Genetics of Alzheimer's disease: A centennial review. *Neurologic Clinics, 25,* 611–667.

Espelage, D. L., Aragon, S. R., Birkett, M., & Koenig, B. W. (2008). Homophobic teasing, psychological outcomes, and sexual orientation among high school students: What influence do parents and schools have? *School Psychology Review, 37,* 202–216.

Esses, V. M., Dovidio, J. F., Jackson, L. M., & Armstrong, T. L. (2001). The immigration dilemma: The role of perceived group competition, ethnic prejudice, and national identity. *Journal of Social Issues, 57*(3), 389–412.

Esterson, A. (2002). The myth of Freud's ostracism by the medical community in 1896–1905: Jeffrey Masson's assault on truth. *History of Psychology, 5*(2), 115–134.

Evans, S., Ferrando, S., Findler, M., Stow-ell, C., Smart, C., & Haglin, D. (2008). Mindfulness-based cognitive therapy for generalized anxiety disorder. *Journal of Anxiety Disorders, 22,* 716–721.

Eysenck, H. J. (1967). *The biological basis of personality.* Springfield, IL: Charles C Thomas.

Eysenck, H. J. (1982). *Personality, genetics, and behavior: Selected papers.* New York: Prager.

Eysenck, H. J. (1990). Biological dimensions of personality. In L. A. Pervin (Ed.), *Handbook of personality: Theory and research.* New York: Guilford Press.

Faddiman, A. (1997). *The spirit catches you and you fall down.* New York: Straus & Giroux.

Faigman, D. L., Kaye, D., Saks, M. J., & Sanders, J. (1997). *Modern scientific evidence: The law and science of expert testimony.* St. Paul, MN: West.

Fairburn, C. G., Cooper, Z., Shafran, R., & Wilson, G. T. (2008). Eating disorders: A transdiagnostic protocol. In D. H. Barlow (Ed.), *Clinical handbook of psychological disorders: A step-by-step treatment manual* (4th ed.) (pp. 578–614). New York: Guilford Press.

Fang, X., & Corso, P. S. (2007). Child maltreatment, youth violence, and intimate partner violence: Developmental relationships. *American Journal of Preventive Medicine, 33,* 281–290.

Fantz, R. L. (1956). A method for studying early visual development. *Perceptual and Motor Skills, 6,* 13–15.

Fantz, R. L. (1963). Pattern vision in newborn infants. *Science, 140,* 296–297.

Faraone, S. V. (2008). Statistical and molecular genetic approaches to developmental psychopathology: The pathway forward. In J. J. Hudziak (Ed.), *Developmental psychopathology and wellness: Genetic and environmental influences* (pp. 245–265). Arlington, VA: American Psychiatric Publishing.

Fassler, O., Lynn, S. J., & Knox, J. (2008). Is hypnotic suggestibility a stable trait? *Consciousness and Cognition: An International Journal, 17,* 240–253.

Fehr, C., Yakushev, I., Hohmann, N., Buchholz, H-G., Landvogt, C., Deckers, H., Eberhardt, A., Kläger, M., Smolka, M. N., Scheurich, A., Dielentheis, T., Schmidt, L. G., Rösch, F., Bartehstein, P., Gründer, G., & Schreckenberger, M. (2008). Association of low striatal dopamine D-sub-2 receptor availability with nicotine dependence similarto that seen with other drugs of abuse. *American Journal of Psychiatry, 165,* 507–514.

Fein, S. & Spencer, S. J. (1997). Prejudice as self-image maintenance: Affirming the self through derogating others. *Journal of Personality and Social Psychology, 73*(1), 31–44.

Feldman, R. (2007). Maternal-infant contact and child development: Insights from the kangaroo intervention. In L. L'Abate (Ed.), *Low-cost approaches to promote physical and mental health: Theory, research, and practice* (pp. 323–351). New York: Springer.

Fernández, J. R., Casazza, K., Divers, J., & López-Alarcón, M. (2008). Disruptions inenergy balance: Does nature overcome nurture? *Physiology & Behavior, 94*, 105–112.

Ferrari, P. F., Rozzi, S., & Fogassi, L. (2005). Mirror neurons responding to observation of actions made with tools in monkey ventral premotor cortex. *Journal of Cognitive Neuroscience, 17*, 212–226.

Ferretti, P., Copp, A., Tickle, C., & Moore, G. (Eds.) (2006). *Embryos, genes, and birth defects* (2nd ed.). Hoboken, NJ: Wiley.

Festinger, L. A. (1957). *A theory of cognitive dissonance.* Palo Alto, CA: Stanford University Press.

Festinger, L. A., & Carlsmith, L. M. (1959). Cognitive consequences of forced compliance. *Journal of Abnormal and Social Psychology, 58*, 203–210.

Field, A. P. (2006). Is conditioning a useful framework for understanding the development and treatment of phobias? *Clinical Psychology Review, 26*, 857–875.

Field, K. M., Woodson, R., Greenberg, R., & Cohen, D. (1982). Discrimination and imitation of facial expressions by neonates. *Science, 218*, 179–181.

Field, T. M. (1998). Massage Therapy effects. *American Psychologist*, 53, 1270–1281.

Field, T., Diego, M., & Hernandez-Reif, M. (2007). Massage therapy research. *Developmental Review, 27*, 75–89.

Fields, R. (2007). *Drugs in perspective* (6th ed.). New York: McGraw-Hill.

Fields, W. M., Segerdahl, P., & Savage-Rumbaugh, S. (2007). The material practices of ape language research. In J. Valsiner & A. Rosa (Eds.), *The Cambridge handbook of sociocultural psychology* (pp. 164–186). New York: Cambridge University Press.

Fink, B., & Penton-Voak, I. (2002). Evolutionary psychology of facial attractiveness. *Current Directions in Psychological Science*, 11(5), 154–158.

Fink, B., Manning, J. T., Neave, N., & Grammer, K. (2004). Second to fourth digit ratio and facial asymmetry. *Evolution and Human Behavior, 25*(2), 125–132.

Fink, M. (1999). *Electroshock: Restoring the mind.* London: Oxford University Press.

First, M., & Tasman, A. (2004). *DSM-IV-TR mental disorders: Diagnosis, etiology, and treatment.* Hoboken, NJ: Wiley.

Fisher, J. O. & Kral, T. V. E. (2008). Super-size me: Portion size effects on young children's eating. *Physiology & Behavior, 94*, 39–47.

Fish, J. E., Bury, A. S., & Holden, R. (2006). Reasoning about complex probabilistic concepts in childhood. *Scandinavian Journal of Psychology, 47*, 497–504.

Fiske, S. T. (1998). Stereotyping, prejudice, and discrimination. In D. T. Gilbert, S. T. Fiske, and G. Lindzey (Eds.), *The handbook of social psychology* (4th ed., Vol. 2, pp. 357–411). New York: Oxford University Press.

Flatt, N., & King, N. (2009). Building the case for brief psychointerventions in the treatment of specific phobia in children and adolescents. *Behavior Change*, 25, 191–200.

Flavell, J. H., Miller, P. H., & Miller, S. A. (2002). *Cognitive development* (4th ed.). Upper Saddle River, NJ: Prentice-Hall. Fletcher, G. J. O., & Simpson, J. A. (2000). Ideal standardsin close relationships: Their structure and functions. *Current Directions in Psychological Science, 9*(3), 102–105.

Fleischmann, B. K., & Welz, A. (2008). Cardiovascular regeneration and stem cell therapy. *Journal of the American Medical Association, 299*(6), 700–701.

Fletcher, G. J. O., & Simpson, J. A. (2000). Ideal standards in close relationships: Their structure and functions. *Current Directions in Psychological Science, 9*, 102–105.

Flynn, J. R. (1987). Massive IQ gains in 14 nations: What IQ tests really measure. *Psychological Bulletin, 101*, 171–191.

Flynn, J. R. (2006). The history of the American mind in the 20th century: A scenario to explain gains over time and a case for the irrelevance of g. In P. C. Kyllonen, R. D. Roberts, & L. Stankov (Eds.), *Extending intelligence.* Mahwah, NJ: Erlbaum.

Flynn, J. R. (2007). *What is intelligence?: Beyond the Flynn Effect.* New York: Cambridge University Press.

Fogarty, A., Rawstorne, P., Prestage, G., Crawford, J., Grierson, J., & Kippax, S. (2007). Marijuana as therapy for people living with HIV/AIDS: Social and health aspects. *AIDS Care, 19*, 295–301.

Fogassi, L., Ferrari, P. F., Gesierich, B., Rozzi, S., Chersi, F., & Rizzolatti, G. (2005). Parietal lobe: From action understanding to intention understanding. *Science, 308*, 662–667.

Fok, H. K., Hui, C. M., Bond, M. H., Matsumoto, D., & Yoo, S. H. (2008). Integrating personality, context, relationship, and emotion type into a model of display rules. *Journal of Research in Personality, 42*, 133–150.

Folk, C. L., & Remington, R. W. (1998). Selectivity in distraction by irrelevant featural singletons: Evidence for two forms of attentional capture. *Journal of Experimental Psychology: Human Perception and Performance, 24*, 1–12.

Ford, T. E., Ferguson, M. A., Brooks, J. L., & Hagadone, K. M. (2004). Coping sense of humor reduces effects of stereotype threat on women's math performance. *Personality & Social Psychology Bulletin, 30*(5), 643–653.

Frankenburg, W., Dodds, J., Archer, P., Shapiro, H., & Bresnick, B. (1992). The Denver II: A major revision and restandardization of the Denver Developmental Screening Test. *Pediatrics, 89*, 91–97.

Fredricks, J. A., & Eccles, J. S. (2005). Family socialization, gender, and sport motivation and involvement. *Journal of Sport & Exercise Psychology, 27*(1), 3–31.

Fresán, A., Apiquian, R., García-Anaya, M., de la Fuente-Sandoval, C., Nicolini, H., & Graff-Guerrero, A. (2007). The P50 auditory evoked potential in violent and non-violent patients with schizophrenia. *Schizophrenia Research, 97*, 128–136.

Frick, W. B. (2000). Remembering Maslow: Reflections on a 1968 interview. *Journal of Humanistic Psychology, 40*(2), 128–147.

Friedman, H., & Schustack, M. (2006). *Personality: Classic theories and modern research* (3rd ed.). Boston: Allyn & Bacon/Longman.

Fryer, S. L., Crocker, N.A., & Mattson, S. N. (2008). Exposure to teratogenic agents as a risk factor for psychopathology. In T. P. Beauchaine & S. P. Hinshaw (Eds.), *Child and adolescent psychopathology* (pp. 180–207). Hoboken, NJ: Wiley.

Funder, D. C. (2000). Personality. *Annual Review of Psychology, 52*, 197–221.

Funder, D. C. (2001). The really, really fundamental attribution error. *Psychological Inquiry, 12*(1), 21–23.

Furman, E. (1990, November). Plant a potato, learn about life (and death). *Young Children, 46*(1), 15–20.

Gabbard, G. O. (2006). Mente, cervello e disturbi di personalita. / Mind, brain, and personality disorders. *Psicoterapia e scienze umane, 40*, 9–26.

Gabry, K. E., Chrousos, G. P., Rice, K. C., Mostafa, R. M., Sternberg, E., Negrao, A. B., Webster, E. L., McCann, S. M., & Gold, P. W. (2002). Marked suppression of gastric ulcerogenesis and intestinal responses to stress by a novel class of drugs. *Molecular Psychiatry*, 7(5), 474–483.

Gacono, C. B., Evans, F. B., & Viglione, D. J. (2008). Essential issues in the forensic use of the Rorschach. In C. B. Gacono (Ed.), F. B. Evans (Ed.), N. Kaser-Boyd (Col.), & L. A. Gacono (Col.), *The handbook of forensic Rorschach assessment (pp. 3–20). The LEA series in personality and clinical psychology.* New York: Routledge/Taylor & Francis Group. `

Gaetz, M., Weinberg, H., Rzempoluck, E., & Jantzen, K. J. (1998). Neural network classifications and correlational analysis of EEG and MEG activity accompanying spontaneous reversals of the Necker Cube. *Cognitive Brain Research, 6,* 335–346.

Gagné, M-H., Drapeau, S., Melançon, C., Saint-Jacques, M-C., & Lépine, R. (2007). Links between parental psychological violence, other family disturbances, and children's adjustment. *Family Process, 46,* 523–542.

Galderisi, S., Quarantelli, M., Volpe, U., Mucci, A., Cassano, G. B., Invernizzi, G., Rossi, A., Vita, A., Pini, S., Cassano, P., Daneluzzo, E., De Peri, L., Stratta, P., Brunetti, A., & Maj, M. (2008). Patterns of structural MRI abnormalities in deficit and nondeficit schizophrenia. *Schizophrenia Bulletin, 34,* 393–401.

Galinsky, A. D., & Moskowitz, G. B. (2000). Perspective-taking: Decreasing stereotype expression, stereotype accessibility, and in-group favoritism. *Journal of Personality and Social Psychology, 78*(4), 708–724.

Galinsky, A. D., & Ku, G. (2004). The effects of perspective-taking on prejudice: The moderating role of self-evaluation. *Personality & Social Psychology Bulletin, 30*(5), 594–604.

Galloway, J. L. (1999, March 8). Into the heart of darkness. *U. S. News and World Report,* pp. 25–32.

Garb, H. N., Wood, J. M., Lilienfeld, S. O., & Nezworski, M. T. (2005). Roots of the Rorschach controversy. *Clinical Psychology Review, 25,* 97–118.

Garcia, G. M., & Stafford, M. E. (2000). Prediction of reading by Ga and Gc specific cognitive abilities for low-SES White and Hispanic English-speaking children. *Psychology in the Schools, 37*(3), 227–235.

Garcia, J. (2003). Psychology is not an enclave. In R. Sternberg (Ed.), *Psychologists defying the crowd: Stories of those who battled the establishment and won* (pp. 67–77). Washington, DC: American Psychological Association.

Garcia, J., & Koelling, R. S. (1966). Relation of cue to consequence in avoidance learning. *Psychonomic Science, 4,* 123–124.

Gardner, H. (1983). *Frames of mind.* New York:Basic Books.

Gardner, H. (1999, February). Who owns intelligence? *Atlantic Monthly,* pp. 67–76.

Gardner, H. (2008). Who owns intelligence? In M. H. Immordino-Yang (Ed.), *Jossey-Bass Education Team. The Jossey-Bass reader on the brain and learning* (pp. 120–132). San Francisco: Jossey-Bass.

Garno, J. L., Gunawardane, N., & Goldberg, J. F. (2008). Predictors of trait aggression in bipolar disorder. *Bipolar Disorders, 10,* 285–292.

Garrick, J. (2006). The humor of trauma survivors: Its application in a therapeutic milieu. *Journal of Aggression, Maltreatment & Trauma, 12,* 169–182.

Garry, M., & Gerrie, M. P. (2005). When photographs create false memories. *Current Directions in Psychological Science, 14,* 321–324.

Gaser, C., Nenadic, I., Buchsbaum, B. R., Hazeltt, E. A., & Buchsbaum, M. S. (2004). Ventricular enlargement in schizophrenia related to volume reduction of the thalamus, striatum, and superior temporal cortex. *American Journal of Psychiatry, 161*(1), 154–156.

Gasser, S., & Raulet, D. H. (2006). Activation and self-tolerance of natural killer cells. *Immunology Review, 214,* 130–142.

Gaw, A. C. (2001). *Concise guide to cross-cultural psychiatry. Concise guides.* Washington, DC: American Psychiatric Association.

Gay, P. (2000). *Freud for historians.* Boston: Replica Books.

Gebauer, H., Krempl, R., & Fleisch, E. (2008). Exploring the effect of cognitive biases on customer support services. *Creativity and Innovation Management, 17,* 58–70.

Gelder, B. D., Meeren, H. K., Righart, R. Stock, J. V., van de Riet, W. A., & Tamietto, M. (2006). Beyond the face: Exploring rapid influences of context on face processing. *Progress in Brain Research, 155,* 37–48.

Gerber, A. J., Posner, J., Gorman, D., Colibazzi, T., Yu, S., Wang, Z., Kangarlu, A., Zhu, H., Russell, J., & Peterson, B. S. (2008). An affective circumplex model of neural systems subserving valence, arousal, and cognitive overlay during the appraisal of emotional faces. *Neuropsychologia, 46,* 2129–2139.

Gershon, E. S., & Rieder, R. O. (1993). Major disorders of mind and brain. *Mind and brain: Readings from Scientific American Magazine* (pp. 91–100). New York: Freeman.

Giacobbi, P. Jr., Foore, B., & Weinberg, R. S. (2004). Broken clubs and expletives: The sources of stress and coping responses of skilled and moderately skilled golfers. *Journal of Applied Sport Psychology, 16*(2), 166–182.

Giancola, P. R., & Parrott, D. J. (2008). Further evidence for the validity of the Taylor aggression paradigm. *Aggressive Behavior, 34,* 214–229.

Gibbs, N. (1995, October 2). The EQ factor. *Time,* pp. 60–68.

Gibson, E. J., & Walk, R. D. (1960). The visual cliff. *Scientific American, 202*(2), 67–71.

Giedd, J. N. (2008). The teen brain: Insights from neuroimaging. *Journal of Adolescent Health, 42,* 335–343.

Giles, J. W., & Heyman, G. D. (2005). Young children's beliefs about the relationship between gender and aggressive behavior. *Child Development, 76*(1), 107–121.

Gilligan, C. (1977). In a different voice: Women's conception of morality. *Harvard Educational Review, 47*(4), 481–517.

Gilligan, C. (1990). Teaching Shakespeare's sister. In C. Gilligan, N. Lyons, & T. Hanmer (Eds.), *Mapping the moral domain* (pp. 73–86). Cambridge, MA: Harvard University Press.

Gilligan, C. (1993). Adolescent development reconsidered. In A. Garrod (Ed.), *Approaches to moral development: New research and emerging themes.* New York: Teachers College Press.

Gilmour, D. R., & Walkey, F. H. (1981). Identifying violent offenders using a video measure of interpersonal distance. *Journal of Consulting and Clinical Psychology, 49,* 287–291.

Gini, M., Oppenheim, D., & Sagi-Schwartz, A. (2007). Negotiation styles in mother-child narrative co-construction in middle childhood: Associations with early attachment. *International Journal of Behavioral Development, 31,* 149–160.

Girardi, P., Monaco, E., Prestigiacomo, C., Talamo, A., Ruberto, A., & Tatarelli, R. (2007). Personality and psychopathological profiles in individuals exposed to mobbing. *Violence & Victims, 22,* 172–188.

Giumetti, G. W., & Markey, P. M. (2007). Violent video games and anger as predictors of aggression. *Journal of Research in Personality, 41,* 1234–1243.

Gluck, M. A. (2008). Behavioral and neural correlates of error correction in classical conditioning and human category learning. In M. A. Gluck, J. R. Anderson, & S. M. Kosslyn (Eds.), *Memory and mind: A festschrift for Gordon H. Bower* (pp. 281–305). Mahwah, NJ: Lawrence Erlbaum.

Gobbo, C., & Raccanello, D. (2007). How children narrate happy and sad events: Does affective state count? *Applied Cognitive Psychology, 21,* 1173–1190.

Godden, D. R., & Baddeley, A. D. (1975). Context-dependent memory in two natural environments: On land and underwater. *British Journal of Psychology, 66,* 325–331.

Goforth, H. W., & Holsinger, T. (2007). Response to effect of the first ECT in a series by C. Kellner, MD. *Journal of ECT, 23,* 209.

Gogtay, N., Sporn, A., Clasen, L. S., Nugent, T. F. III, Greenstein, D., Nicolson, R., Giedd, J. N., Lenane, M., Gochman, P., Evans, A., & Rapoport, J. L. (2004). Comparison of progressive cortical gray matter loss in childhood-onset schizophrenia with that in childhood-onset atypical psychoses. *Archives of General Psychiatry, 61*(1), 17–22.

Golden, W. L. (2006). Hypnotherapy for anxiety, phobias and psychophysiological disorders. In R. A. Chapman (Ed.), *The clinical use of hypnosis in cognitive behavior therapy: A practitioner's casebook* (pp. 101–137). New York: Springer.

Goldstein, E. G. (2008). *Cognitive psychology: Connecting mind, research, and everyday experience* (2nd ed.). Belmont, CA: Cengage.

Goleman, D. (1980, February). 1,528 little geniuses and how they grew. *Psychology Today,* pp. 28–53.

Goleman, D. (1995). *Emotional intelligence: Why it can matter more than IQ.* New York: Bantam.

Goleman, D. (2000). *Working with emotional intelligence.* New York: Bantam Doubleday.

Goleman, D. (2008). Leading resonant teams. In F. Hesselbein & A. Shrader (Eds.), *Leader to leader 2: Enduring insights on leadership from the Leader to Leader Institute's award-winning journal* (pp. 186–195). Leader to Leader Institute. San Francisco, CA: Jossey-Bass.

Gómez, Á., & Huici, C. (2008). Vicarious intergroup contact and role of authorities in prejudice reduction. *The Spanish Journal of Psychology, 11,* 103–114.

Goodman, G. S., Ghetti, S., Quas, J. A., Edelstein, R. S., Alexander, K. W., Redlich, A. D., Cordon, I. M., & Jones, D. P. H. (2003). A prospective study of memory for child sexual abuse: New findings relevant to the repressedmemory controversy. *Psychological Science, 14*(2), 113–118.

Goodwin, C. J. (2005). *A history of modern psychology* (2nd ed.). Hoboken, NJ: Wiley.

Goodwin, C. J. (2009). *A history of modern psychology* (3rd ed.). Hoboken, NJ: Wiley.

Gooren, L. (2006). The biology of human psychosexual differentiation. *Hormones and Behavior, 50,* 589–601.

Gotlieb, I. H., & Abramson, L. Y. (1999). Attributional theories of emotion. In T. Dalgleish & M. Power (Eds.), *Handbook of cognition and emotion.* New York: Wiley.

Gottesman, I. I. (1991). *Schizophrenia genesis: The origins of madness.* New York: Freeman.

Gottfredson, G. D., & Duffy, R. D. (2008). Using a theory of vocational personalities and work environments to explore subjective well-being. *Journal of Career Assessment, 16,* 44–59.

Gottman, J. M., & Levenson, R. W. (2002). A two-factor model for predicting when a couple will divorce: Exploratory analyses using 14-year longitudinal data. *Family Process, 41*(1), 83–96.

Gottman, J. M., & Notarius, C. I. (2000). Decade review: Observing marital interaction. *Journal of Marriage & the Family, 62*(4), 927–947.

Gould, R. L. (1975, August). Adult life stages: Growth toward self-tolerance. *Psychology Today,* pp. 74–78.

Gracely, R. H., Farrell, M. J., & Grant, M. A. (2002). Temperature and pain perception. In H. Pashler & S. Yantis (Eds.), *Steven's handbook of experimental psychology: Vol. 1. Sensation and perception* (3rd ed.). Hoboken, NJ: Wiley.

Graham, J. R. (1991). Comments on Duckworth's review of the Minnesota Multiphasic Personality Inventory-2. *Journal of Counseling and Development, 69,* 570–571.

Green, A. I., Tohen, M. F., Hamer, R. M., Strakowski, S. M., Lieberman, J. A., Glick, I., Clark, W. S., & HGDH Research Group. (2004). First episode schizophrenia-related psychosis and substance use disorders: Acute response to olanzapine and haloperidol. *Schizophrenia Research, 66*(2–3), 125–135.

Greenberg, D. L. (2004). President Bush's false "flashbulb" memory of 9/11/01. *Applied Cognitive Psychology, 18*(3), 363–370.

Greenberg, J. (2002). Who stole the money, and when? Individual and situational determinants of employee theft. *Organizational Behavior and Human Decision Processes, 89*(1), 985–1003.

Greene, E., & Ellis, L. (2008). *Decision making in criminal justice.* In D. Carson, R. Milne, F. Pakes, K. Shalev, & A. Shawyer (Eds.), *Applying psychology to criminal justice* (pp. 183–200). New York: Wiley.

Greene, K., Kremar, M., Walters, L. H., Rubin, D. L., Hale, J., & Hale, L. (2000). Targeting adolescent risk-taking behaviors: The contributions of egocentrism and sensation-seeking. *Journal of Adolescence,* 23, 439–461.

Greenwald, A.G., Poehlman, A., Uhlmann, E., Banaji, M. R. (2008). Understanding and interpreting the Implicit Association Test III: Meta-analysis of predictive validity. *Journal of Personality and Social Psychology* (in press).

Gregory, R. J. (2007). *Psychological testing: History, principles, and applications (5th ed.).* Needham Heights, MA: Allyn and Bacon.

Gresack, J. E., Kerr, K. M., & Frick, K. M. (2007). Life-long environmental enrichment differentially affects the mnemonic response to estrogen in young, middle-aged, and aged female mice. *Neurobiology of Learning & Memory, 88,* 393–408.

Gringart, E., Helmes, E., & Speelman, C. (2008). Harnessing cognitive dissonance to promote positive attitudes toward older workers in Australia. *Journal of Applied Social Psychology, 38,* 751–778.

Grondin, S., Ouellet, B., & Roussel, M. (2004). Benefits and limits of explicit counting for discriminating temporal intervals. *Canadian Journal of Experimental Psychology, 58*(1), 1–12.

Grossmann, K., Grossmann, K. E., Fremmer-Bombik, E., Kindler, H., Scheuerer-Englisch, H., & Zimmermann, P. (2002). The uniqueness of the child-father attachment relationship: Fathers' sensitive and challenging play as a pivotal variable in a 16-year longitudinal study. *Social Development, 11*(3), 307–331.

Grover, K. E., Carpenter, L. L., Price, L. H., Gagne, G. G., Mello, A. F., Mello, M. F., & Tyra, A. R. (2007). The relationship between childhood abuse and adult personality disorder symptoms. *Journal of Personality Disorders, 21,* 442–447.

Guidelines for Ethical Conduct in the Care and Use of Animals. (2008). [On-line]. Available: http://www.apa.org/science/anguide. html. Guidelines for the treatment of animals in behavioural research and teaching. (2005). *Animal Behaviour, 69*(1), i–vi.

Guilford, J. P. (1967). *The nature of human intelligence.* New York: McGraw-Hill.

Gunzerath, L., Faden, V., Zakhari, S., & Warren, K. (2004). National Institute on Alcohol Abuse and Alcoholism report on moderate drinking. *Alcoholism: Clinical & Experimental Research, 28*(6), 829–847.

Gustavson, C. R., & Garcia, J. (1974, August). Pulling a gag on the wily coyote. *Psychology Today,* pp. 68–72.

Haberstick, B. C., Schmitz, S. Young, S. E., & Hewitt, J. K. (2006). Genes and developmental stability of aggressive behavior problems at home and school in a community sample of twins aged 7–12. *Behavior Genetics, 36,* 809–819.

Haddock, G., & Shaw, J. J. (2008). Understanding and working with aggression, violence, and psychosis. In K. T. Mueser & D. V. Jeste (Eds.), *Clinical handbook of schizophrenia* (pp. 398–410). New York: Guilford Press.

Haith, M. M., & Benson, J. B. (1998). Infant cognition. In W. Damon & R. M. Lerner (Eds.), *Handbook of child psychology (Vol. 1).* New York: Wiley.

Halbreich, U., & Kahn, L. S. (2007). Atypical depression, somatic depression and anxious depression in women: Are they gender-preferred phenotypes? *Journal of Affective Disorders, 102,* 245–258.

Haley, A. P. (2005). Effects of orally administered glucose on hippocampal metabolites and cognition in Alzheimer's disease. *Dissertation Abstracts International: Section B: The Sciences and Engineering, 66*(3-B), 1719.

Halgin, R. P., & Whitbourne, S. K. (2008). *Abnormal psychology: Clinical perspectives on psychological disorders.* New York: McGraw-Hill.

Hall, K. (2007). Sexual dysfunction and childhood sexual abuse: Gender differences and treatment implications. In S. R. Leiblum (Ed.), *Principles and practice of sex therapy* (4th ed.) (pp. 350–378). New York: Guilford.

Hall, M-H., Rijsdijk, F., Picchioni, M., Schulze, K., Ettinger, U., Toulopoulou, T., Bramon, E., Murray, R. M., & Sham, P. (2007). Substantial shared genetic influences on schizophrenia and event-related potentials. *American Journal of Psychiatry, 164,* 804–812.

Hall, P. A., Schaeff, C. M. (2008). Sexual orientation and fluctuating asymmetry in men and women. *Archives of Sexual Behavior, 37,* 158–165.

Haller, S., Radue, E. W., Erb, M., Grodd, W., & Kircher, T. (2005). Overt sentence production in event-related fMRI. *Neuropsychologia, 43*(5), 807–814.

Halpern, D. F. (1997). Sex differences in intelligence: Implications for education. *American Psychologist, 52,* 1091–1102.

Halpern, D. F. (1998). Teaching critical thinking for transfer across domains. *American Psychologist, 53,* 449–455.

Halpern, D. F. (2000). *Sex differences in cognitive abilities.* Hillsdale, NJ: Erlbaum.

Hamilton, J. P., & Gotlib, I. H. (2008). Neural substrates of increased memory sensitivity for negative stimui in major depression. *Biological Psychiatry, 63,* 1155–1162.

Hamm, J. P., Johnson, B. W., & Corballis, M. C. (2004). One good turn deserves another: An event-related brain potential study of rotated mirror-normal letter discriminations. *Neuropsychologia, 42*(6), 810–820.

Hammack, P. L. (2003). The question of cognitive therapy in a postmodern world. *Ethical Human Sciences & Services, 5*(3), 209–224.

Hammen, C. (2005). Stress and depression. *Annual Review of Clinical Psychology, 1,* 293–319.

Hammond, D. C. (2005). Neurofeedback with anxiety and affective disorders. *Child & Adolescent Psychiatric Clinics of North America, 14*(1), 105–123.

Hammond, D. C. (2007). What is neurofeedback? Journal of Neurotherapy, 10, 25–36.

Hampton, T. (2006). Stem cells probed as diabetes treatment. *Journal of American Medical Association, 296,* 2785–2786.

Hampton, T. (2007). Stem cells ease Parkinson symptoms in monkeys. *Journal of American Medical Association, 298,* 165.

Haney, C., Banks, C., & Zimbardo, P. (1978). Interpersonal dynamics in a simulated prison. *International Journal of Criminology and Penology, 1,* 69–97.

Hanley, S. J., & Abell, S. C. (2002). Maslow and relatedness: Creating an interpersonal model of self-actualization. *Journal of Humanistic Psychology, 42*(4), 37–56.

Hanna, S. M., & Brown, J. H. (1999). *The practice of family therapy: Key elements across models* (2nd ed.). Belmont, CA: Brooks/Cole.

Hansell, J. H., & Damour, L. K. (2008). *Abnormal psychology* (2nd ed.). Hoboken, NJ: Wiley.

Hardcastle, S., Taylor, A., Bailey, M., & Castle, R. (2008). A randomized controlled trial on the effectiveness of a primary health care based counseling intervention on physical activity, diet and CHD risk. *Patient Education and Counseling, 70,* 31–39.

Harlow, H. F., Harlow, M. K., & Meyer, D. R. (1950). Learning motivated by a manipulation drive. *Journal of Experimental Psychology, 40,* 228–234.

Harlow, H. F., & Zimmerman, R. R. (1959). Affectional responses in the infant monkey. *Science, 130,* 421–432.

Harlow, H. F., & Harlow, M. K. (1966). Learning to love. *American Scientist, 54,* 244–272.

Harper, F. D., Harper, J. A., & Stills, A. B. (2003). Counseling children in crisis based on Maslow's hierarchy of basic needs. *International Journal for the Advancement of Counselling,* 25(1), 10–25.

Harrison, E. (2005). *How meditation heals: Scientific evidence and practical applications* (2nd ed.). Berkeley, CA: Ulysses Press.

Hart Jr., J., & Kraut, M. A. (2007). Neural hybrid model of semantic object memory (version 1.1). In J. Hart Jr. & M. A. Kraut (Eds.), *Neural basis of semantic memory* (pp. 331–359). New York: Cambridge University Press.

Hartenbaum, N., Collop, N., Rosen, I. M., Phillips, B., George, C. F. P., Rowley, J. A., Freedman, N., Weaver, T. E., Gurubhagavatula, I., Strohl, K., Leaman, H. M., Moffitt, G. L., & Rosekind, M. R. (2006). Sleep apnea and commercial motor vehicle operators: Statement from the Joint Task Force of the American College of Chest Physicians, American College of Occupational and Environmental Medicine, and the National Sleep Foundation. *Journal of Occupational & Environmental Medicine, 48,* S4-S37.

Harth, N. S., Kessler, T., & Leach, C. W. (2008). Advantaged group's emotional reactions to intergroup inequality: The dynamics of pride, guilt, and sympathy. *Personality and Social Psychology Bulletin, 34,* 115–129.

Hartwell, L. (2008). *Genetics* (3ed ed.). New York: McGraw-Hill.

Haslam, N., Loughnan, S., Reynolds, C., & Wilson, S. (2007). Dehumanization: A new perspective. *Social and Personality Psychology Compass, 1,* 409–422.

Hatfield, E., & Rapson, R. L. (1996). *Love and Sex: Cross-cultural perspectives.* Needham Heights, MA: Allyn & Bacon.

Hatsukami, D. K. (2008). Nicotine addiction: Past, present and future. *Drug and Alcohol Dependence, 92,* 312–316.

Haugen, R., Ommundsen, Y., & Lund, T. (2004). The concept of expectancy: A central factor in various personality dispositions. *Educational Psychology, 24*(1), 43–55.

Hay, D. F. (1994). Prosocial development. *Journal of Child Psychology and Psychiatry, 35,* 29–71.

Hayflick, L. (1977). The cellular basis for biological aging. In C. E. Finch & L. Hayflick (Eds.), *Handbook of the biology of aging*(pp. 159–186). New York: Van Nostrand Reinhold.

Hayflick, L. (1996). *How and why we age.* New York: Ballantine Books.

Hays, W. S. T. (2003). Human pheromones: Have they been demonstrated? *Behavioral Ecology & Sociobiology, 54*(2), 89–97.

Hazan, C., & Shaver, P. (1987). Romantic love conceptualized as an attachment process. *Journal of Personality and Social Psychology, 52,* 511–524.

Hazan, C., & Shaver, P. R. (1994). Attachment as an organizational framework for research on close relationships. *Psychological Inquiry, 5,* 1–22.

Healy, A. F., Shea, K. M., Kole, J. A., & Cunningham, T. F. (2008). Position distinctiveness, item familiarity, and presentation frequency affect reconstruction of order in immediate episodic memory. *Journal of Memory and Language, 58,* 746–764.

Heckhausen, J. (2005). Competence and motivation in adulthood and old age: Making the most of changing capacities and resources. In A. J. Elliot & C. S. Dweck (Eds.), *Handbook of competence and motivation* (pp. 240–256). New York: Guilford.

Hedges, D., & Burchfield, C. (2006). *Mind, brain, and drug: An introduction to psychopharmacology.* Boston: Allyn & Bacon/Longman.

Heffelfinger, A. K., & Newcomer, J. W. (2001). Glucocorticoid effects on memory function over the human life span. *Development & Psychopathology, 13*(3), 491–513.

Heider, F. (1958). *The psychology of interpersonal relations.* New York: Wiley.

Heimann, M., & Meltzoff, A. N. (1996). Deferred imitation in 9-and 14-month-old infants. *British Journal of Developmental Psychology, 14,* 55–64.

Heine, S. J., & Renshaw, K. (2002). Inter-judge agreement, self-enhancement, and liking: Cross-cultural divergences. *Personality & Social Psychology Bulletin, 28*(5), 578–587.

Heller, S. (2005). *Freud A to Z.* Hoboken, NJ: Wiley.

Helms, J. E., & Cook, D. A. (1999). *Using race and culture in counseling and psychotherapy: Theory and process.* Boston: Allyn & Bacon.

Hennessey, B. A., & Amabile, T. M. (1998). Reward, intrinsic motivation, and creativity. *American Psychologist, 53,* 674–675.

Hergovich, A., & Olbrich, A. (2003). The impact of the Northern Ireland conflict on social identity, groupthink and integrative complexity in Great Britain. *Review of Psychology, 10*(2), 95–106.

Heriot, S. A., & Pritchard, M. (2004). Test of time: Reciprocal inhibition as the main basis of psychotherapeutic effects' by Joseph Wolpe (1954) [Book review]. *Clinical Child Psychology and Psychiatry, 9*(2), 297–307.

Herman, C. P., & Polivy, J. (2008). External cues in the control of food intake in humans: The sensory-normative distinction. *Physiology & Behavior, 94,* 722–728.

Herman, L. M., Richards, D. G., & Woltz, J. P. (1984). Comprehension of sentences by bottlenosed dolphins. *Cognition, 16,* 129–139.

Hermans, E. J., Ramsey, N. F., & van Honk, J. (2008). Exogenous testosterone enhances responsiveness to social threat in the neural circuitry of social aggression in humans. *Biological Psychiatry, 63*(3), 263–270.

Hernandez-Reif, M., Diego, M., & Field, T. (2007). Preterm infants show reduced stress behaviors and activity after 5 days of massage therapy. *Infant Behavior & Development, 30,* 557–561.

Herrnstein, R. J., & Murray, C. (1994). *The bell curve: Intelligence and class structure in American life.* New York: Free Press.

Hervé, H., Mitchell, D., Cooper, B. S., Spidel, A., & Hare, R. D. (2004). Psychopathy and unlawful confinement: An examination of perpetrator and event characteristics. *Canadian Journal of Behavioural Science, 36*(2), 137–145.

Hewitt, J. K. (1997). The genetics of obesity: What have genetic studies told us about the environment? *Behavior Genetics, 27,* 353–358.

Heyder, K. Suchan, B., & Daum, I. (2004). Cortico-subcortical contributions to executive control. *Acta Psychologica, 115*(2–3), 271–289.

Higley, E. R. (2008). Nighttime interactions and mother-infant attachment at one year. *Dissertation Abstracts International: Section B: The Sciences and Engineering,. 68,* 2008, 5575.

Hilgard, E. R. (1978). Hypnosis and consciousness. *Human Nature, 1,* 42–51.

Hilgard, E. R. (1992). Divided consciousness and dissociation. *Consciousness and Cognition, 1,* 16–31.

Hill, E. L. (2004). Evaluating the theory of executive dysfunction in autism. *Developmental Review, 24*(2), 189–233.

Hinton, E. C., Parkinson, J. A., Holland, A. J., Arana, F. S., Roberts, A. C., & Owen, A. M. (2004). Neural contributions to the motivational control of appetite in humans. *European Journal of Neuroscience, 20*(5), 1411–1418.

Hittner, J. B., & Daniels, J.R. (2002). Gender-role orientation, creative accomplishments and cognitive styles. *Journal of Creative Behavior, 36*(1), 62–75.

Hobson, J. A. (1988). *The dreaming brain.* New York: Basic Books.

Hobson, J. A. (1999). *Dreaming as delirium: How the brain goes out of its mind.* Cambridge, MA: MIT Press.

Hobson, J. A. (2002). *Dreaming: An introduction to the science of sleep.* New York: Oxford University Press.

Hobson, J. A. (2005). In bed with Mark Solms? What a nightmare! A reply to Domhoff. *Dreaming, 15*(1), 21–29.

Hobson, J. A., & McCarley, R. W. (1977). The brain as a dream state generator: An activation-synthesis hypothesis of the dream process. *American Journal of Psychiatry, 134,* 1335–1348.

Hobson, J. A., & Silvestri, L. (1999). Parasomnias. *The Harvard Mental Health Letter, 15*(8), 3–5.

Hodges, S. D., & Biswas-Diener, R. (2007). Balancing the empathy expense account: Strategies for regulating empathic response. In T. Farrow & P. Woodruff (Eds.), *Empathy in mental illness* (pp. 389–407). New York: Cambridge University Press.

Hodson, G., & Costello, K. (2007). Interpersonal disgust, ideological orientations, and dehumanization as predictors of intergroup attitudes. *Psychological Science, 18,* 691–698.

Hoek, H. W., van Harten, P. N., Hermans, K. M. E., Katzman, M. A., Matroos, G. E., & Susser, E. S. (2005). The incidence of anorexia nervosa on Curacao. *American Journal of Psychiatry, 162,* 748–752.

Hofer, A., Siedentopf, C. M., Ischebeck, A., Rettenbacher, M. A., Verius, M., Golaszewski, S. M., Felber, S., & Fleischhacker, W. W. (2007). Neural substrates for episodic encoding and recognition of unfamiliar faces. *Brain and Cognition, 63,* 174–181.

Hoff, E. (2009). *Language development* (4th ed.).Belmont, CA: Wadsworth.

Hoffman, M. L. (1993). Empathy, social cognition, and moral education. In A. Gar-rod (Ed.), *Approaches to moral development: New research and emerging themes.* New York: Teachers College Press.

Hoffman, M. L. (2000). *Empathy and moral development: Implications for caring and justice.* New York: Cambridge University Press.

Hofmann, W., Gschwendner, T., Castelli, L., & Schmitt, M. (2008). Implicit and explicit attitudes and interracial interaction: The moderating role of situationally available control resources. *Group Processes & Intergroup Relations, 11,* 69–87.

Hogan, T. P. (2006). *Psychological testing: A practical Introduction (2nd ed.).* Hoboken, NJ: Wiley.

Holland, J. L. (1985). *Making vocational choices: A theory of vocational personalites and work environments* (2nd ed). Englewood Cliffs, NJ: Prentice Hall.

Holland, J. L. (1994). *Self-directed search form R.* Lutz, Fl: Psychological Assessment Resources.

Holmes, T. H., & Rahe, R. H. (1967). The social readjustment rating scale. *Journal of Psychosomatic Research, 11,* 213–218.

Holtzworth-Munroe, A. (2000). A typology of men who are violent toward their female partners: Making sense of the heterogeneity in husband violence. *Current Directions in Psychological Science, 9*(4), 140–143.

Honts, C. R., & Kircher, J. C. (1994). Mental and physical countermeasures reduce the accuracy of polygraph tests. *Journal of Applied Psychology, 79*(2), 252–259.

Hooley, J. M., & Hiller, J. B. (2001). Family relationships and major mental disorder: Risk factors and preventive strategies. In B. R. Sarason & S. Duck (Eds.), *Personal relationships: implications for clinical and community psychology.* New York: Wiley.

Horney, K. (1939). *New ways in psychoanalysis.* New York: International Universities Press.

Horney, K. (1945). *Our inner conflicts: A constructive theory of neurosis.* New York: Norton.

Horton, S. L. (2002). Conceptualizing transition: The role of metaphor in describing the experience of change at midlife. *Journal of Adult Development, 9*(4), 277–290.

Horwitz, A. V. (2007). Transforming normality into pathology: The DSM and the outcomes of stressful social arrangements. *Journal of Health and Social Behavior, 48,* 211–222.

Houlfort, N. (2006). The impact of performance-contingent rewards on perceived autonomy and intrinsic motivation. *Dissertation*

Abstracts International Section A: Humanities and Social Sciences, 67(2-A), 460.

Houtz, J. C., Matos, H., Park, M-K. S., Scheinholtz, J., & Selby, E. (2007). Problem-solving style and motivational attributions. *Psychological Reports, 101*, 823–830.

Hovland, C. I. (1937). The generalization of conditioned responses: II. The sensory generalization of conditioned responses with varying intensities of tone. *Journal of Genetic Psychology, 51*, 279–291.

Howell, K. K., Coles, C. D., & Kable, J. A. (2008). The medical and developmental consequences of prenatal drug exposure. In J. Brick (Ed.), *Handbook of the medical consequences of alcohol and drug abuse* (2nd ed.) (pp. 219–249). *The Haworth Press series in neuropharmacology.* New York: Haworth Press/Taylor and Francis Group.

Howes, M. B. (2007). *Human memory: Structures and images.* Thousand Oaks, CA: Sage Publications.

Hoye, A., Rezvy, G., Hansen, V., & Olstad, R. (2006). The effect of gender in diagnosing early schizophrenia: An experimental case simulation study. *Social Psychiatry and Psychiatric Epidemiology, 41*, 549–555.

Hsieh, P-H. (2005). How college students explain their grades in a foreign language course: The interrelationship of attributions, self-efficacy, language learning beliefs, and achievement. *Dissertation Abstracts International Section A: Humanities and Social Sciences, 65*(10-A), 3691.

Huang, M., & Hauser, R. M. (1998). Trends in Black–White test-score differentials: II. The WORDSUM Vocabulary Test. In U. Neisser (Ed.), *The rising curve: Long-term gains in IQ and related measures* (pp. 303–334). Washington, DC: American Psychological Association.

Hubel, D. H., & Wiesel, T. N. (1965). Receptive fields and the functional architecture in two nonstriate visual areas (18 and 19) of the cat. *Journal of Neurophysiology, 28*, 229–289.

Hubel, D. H., & Wiesel, T. N. (1979). Brain mechanisms of vision. *Scientific American, 241*, 150–162.

Hudziak, J. J. (Ed). (2008). *Developmental psychopathology and wellness: Genetic and environmental influences.* Arlington, VA: American Psychiatric Publishing.

Huesmann, L. R., & Kirwil, L. (2007). Why observing violence increases the risk of violent behavior by the observer. In D. J. Flannery, A. T. Vazsonyi, & I. D. Waldman (Eds.), *The Cambridge handbook of violent behavior and aggression* (pp. 545–570), New York: Cambridge University Press.

Huffman, C. J., Matthews, T. D., & Gagne, P. E. (2001). The role of part-set cuing in the recall of chess positions: Influence of chunking in memory. *North American Journal of Psychology, 3*(3), 535–542.

Hull, C. (1952). *A behavior system.* New Haven, CT: Yale University Press.

Hurley, S. (2008). The shared circuits model (SCM): How control, mirroring, and simulation can enable imitation, and deliberation, and mindreading. *Behavioral and Brain Sciences, 31*, 1–22.

Hutchinson-Phillips, S., Gow, K., & Jamieson, G. A. (2007). Hypnotizability, eating behaviors, attitudes, and concerns: A literature survey. *International Journal of Clinical and Experimental Hypnosis, 55*, 84–113.

Huurre, T., Junkkari, H., & Aro, H. (2006). Long-term psychosocial effects of parental divorce: A follow-up study from adolescence to adulthood. *European Archives of Psychiatry and Clinical Neuroscience, 256*, 256–263.

Hyde, J. S. (2005). The genetics of sexual orientation. In J. S. Hyde (Ed.), *Biological substrates of human sexuality.* Washington, DC: American Psychological Association.

Hyde, J. S. (2007). *Half the human experience: The psychology of women* (7th ed.). Boston: Houghton Mifflin.

Hyde, J. S., & DeLamater, J. D. (2008). *Understanding human sexuality* (10th ed.). New York: McGraw-Hill.

Hyman, R. (1981). Cold reading: How to convince strangers that you know all about them. In K. Fraizer (Ed.), *Paranormal borderlands of science* (pp. 232–244). Buffalo, NY: Prometheus.

Hyman, R. (1996). The evidence for psychic functioning: Claims vs. reality. *Skeptical Inquirer, 20*, 24–26.

Hypericum Depression Trial Study Group. (2002). Effect of Hypericum performatum (St John's wort) in major depressive disorder: A randomized controlled trial. *JAMA: Journal of the American Medical Association, 287*(14), 1807–1814.

Iacono, W. G., & Lykken, D. T. (1997). The validity of the lie detector: Two surveys of scientific opinion. *Journal of Applied Psychology, 82*(3), 426–433.

Iavarone, A., Patruno, M., Galeone, F., Chieffi, S., & Carlomagno, S. (2007). Brief report: Error pattern in an autistic savant calendar calculator. *Journal of Autism and Developmental Disorders, 37*, 775–779.

Ibáñez, A., Blanco, C., & Sáiz-Ruiz, J. (2002). Neurobiology and genetics of pathological gambling. *Psychiatric Annals, 32*(3), 181–185.

Ikemoto, K. (2004). Significance of human striatal D-neurons: Implications in neuropsychiatric functions. *Neuropsychopharmacology, 29*(4), 429–434.

Irwin, M., Mascovich, A., Gillin, J. C., Willoughby, R., Pike, J., & Smith, T. L. (1994). Partial sleep deprivation reduced natural killer cell activity in humans. *Psychosomatic Medicine, 56*(6), 493–498.

Ivanovic, D. M., Leiva, B. P., Pérez, H. T., Olivares, M. G., Díaz, N. S., Urrutia, M. S. C., Almagià, A. F., Toro, T. D., Miller, P. T., Bosch, E. O., & Larraín, C. G. (2004). Head size and intelligence, learning, nutritional status and brain development: Head, IQ, learning, nutrition and brain. *Neuropsychologia, 42*(8), 1118–1131.

Jackendoff, R. (2003). Foundations of language, brain, meaning, grammar, evolution. *Applied Cognitive Psychology, 17*(1), 121–122.

Jackson, E. D. (2008). Cortisol effects on emotional memory: Independent of stress effects. *Dissertation Abstracts International: Section B: The Sciences and Engineering, 68* (11-B), 7666.

Jackson, P. B., & Williams, D. R. (2006). Culture, race/ethnicity, and depression. In C. L. M. Keyes & S. H. Goodman (Eds), *Women and depression: A handbook for the social, behavioral, and biomedical sciences* (pp. 328–359). New York: Cambridge University Press.

Jacob, P. (2008). What do mirror neurons contribute to human social cognition? *Mind & Language, 23*, 90–223.

Jacobi, C., Hayward, C., de Zwaan, M., Kraemer, H. C., & Agras, W. S. (2004). Coming to terms with risk factors for eating disorders: Application of risk terminology and suggestions for a general taxonomy. *Psychological Bulletin, 130*(1), 19–65.

Jain, G., Kumar, V., Chakrabarti, S., & Grover, S. (2008). The use of electroconvulsive therapy in the elderly: A study from the psychiatric unit of a North Indian teaching hospital. *Journal of ECT, 24*, 122–127.

James, F. O., Cermakian, N., & Boivin, D. B. (2007). Circadian rhythms of melatonin, cortisol, and clock gene expression during simulated night shift work. *Sleep: Journal of Sleep and Sleep Disorders Research, 30*(11), 1427–1436.

James, W. (1890). *The principles of psychology* (Vol. 2). New York: Holt.

Jamieson, G. A., & Hasegawa, H. (2007). *New paradigms of hypnosis research.* In G. A. Jamieson (Ed.), *Hypnosis and conscious states: The cognitive neuroscience perspective* (pp. 133–144). New York: Oxford University Press.

Jang, K. L., Livesley, W. J., Anso, J., Yamagata, S., Suzuki, A., Angleitner, A., Ostendort, F., Riemann, R., & Spinath, F. (2006). Behavioral genetics of the higher-order factors of the Big Five. *Personality and Individual Differences, 41*, 261–272.

Jang, K. L., Stein, M. B., Taylor, S., Asmundson, G. J. G., & Livesley, W. J. (2003). Exposure to traumatic events and experiences: Aetiological relationships with personality function. *Psychiatry Research, 120*(1), 61–69.

Janis, I. L. (1972). *Victims of groupthink: A psychological study of foreign-policy decisions and fiascoes.* Boston: Houghton Mifflin.

Janis, I. L. (1989). *Crucial decisions: Leadership in policymaking and crisis management.* New York: Free Press.

Jellison, W. A., McConnell, A. R., & Gabriel, S. (2004). Implicit and explicit measures of sexual orientation attitudes: Ingroup preferences and related behaviors and beliefs among gay and straight men. *Personality & Social Psychology Bulletin, 30*(5), 629–642.

Jensen, M. P., Hakimian, S., Sherlin, L. H., & Fregni, F. (2008). New insights into neuromodulatory approaches for the treatment of pain. *The Journal of Pain, 9*, 193–199.

Johnson, D. M., Delahanty, D. L., & Pinna, K. (2008). The cortisol awakening response as a function of PTSD severity and abuse chronicity in sheltered battered women. *Journal of Anxiety Disorders, 22*, 793–800.

Johnson, N. J. (2008). Leadership styles and passive-aggressive behavior in organizations. *Dissertation Abstracts International: Section B: The Sciences and Engineering, 68*(7-B), 4828.

Johnson, S. C., Dweck, C. S., & Chen, F. S. (2007). Evidence for infants' internal working methods of attachment. *Psychological Science, 18*, 501–502.

Johnson, W., Bouchard Jr., T. J., McGue, M., Segal, N. L., Tellegen, A., Keyes, M., & Gottesman, I. I. (2007). Genetic and environmental influences on the Verbal-Perceptual-Image Rotation (VPR) model of the structure of mental abilities in the Minnesota study of twins reared apart. *Intelligence, 35*, 542–562.

Johnson, W., Bouchard, T. J. Jr., Krueger, R. F., McGue, M., & Gottesman, I. I. (2004). Just one g: Consistent results from three test batteries. *Intelligence, 32*(1), 95–107.

Johnson, W., McGue, M., Krueger, R. F., & Bouchard, T. J. Jr. (2004). Marriage and personality: A genetic analysis. *Journal of Personality and Social Psychology. 86*(2), 285–294.

Johnson, W. B., & Murray, K. (2007). *Crazy love: Dealing with your partner's problem personality.* Atascadero, CA: Impact Publishers.

Jolliffe, C. D., & Nicholas, M. K. (2004). Verbally reinforcing pain reports: An experimental test of the operant conditioning of chronic pain. *Pain, 107*, 167–175.

Jonas, E., Traut-Mattausch, E., Frey, D., & Greenberg, J. (2008). The path or the goal? Decision vs. information focus in biased information seeking after preliminary decisions. *Journal of Experimental Social Psychology, 44*, 1180–1186.

Jones, B. C., DeBruine, L. M., & Little, A. C. (2007). The role of symmetry in attraction to average faces. *Perception & Psychophysics, 69*, 1273–1277.

Jones, D. G., Anderson, E. R., & Galvin, K. A. (2003). Spinal cord regeneration: Moving tentatively towards new perspectives. *NeuroRehabilitation, 18*(4), 339–351.

Jones, E. (2008). Predicting performance in first-semester college basic writers: Revisiting the role of self-beliefs. *Contemporary Educational Psychology, 33*, 209–238.

Jones, S. R., & Fernyhough, C. (2007). A new look at the neural diathesis-stress model of schizophrenia: The primacy of social-evaluative and uncontrollable situations. *Schizophrenia Bulletin, 33*, 1171–1177.

Jonides, J., Lewis, R. L., Nee, D. E., Lustig, C. A., Berman, M. G., & Moore, K. S. (2008). The mind and brain of short-term memory. *Annual Review of Psychology, 59*, 193–224.

Jung, C. (1969). The concept of collective unconscious. In *Collected Works* (Vol. 9, Part 1). Princeton, NJ: Princeton University Press (Original work published 1936).

Jung, H. (2006). Assessing the influence of cultural values on consumer susceptibility to social pressure for conformity: Self-image enhancing motivations vs. information searching motivation. In L. R. Kahle & C-H. Kim (Eds.), *Creating images and the psychology of marketing communication (pp. 309–329). Advertising and Consumer Psychology.* Mahwah, NJ: Erlbaum.

Jung, R. E., & Haier, R. J. (2007). The Parieto-Frontal Integration Theory (P-FIT) of intelligence: Converging neuroimaging evidence. *Behavioral and Brain Sciences, 30*, 135–154.

Juntii, S. A., Coats, J. K., & Shah, N. M. (2008). A genetic approach to dissect sexually dimorphic behaviors. *Hormones and Behavior, 53*, 627–637.

Kagan, J. (1998). Biology and the child. In W. Damon & R. M. Lerner (Eds.), *Handbook of child psychology (Vol. 1).* New York: Wiley.

Kahneman, D. (2003). Experiences of collaborative research. *American Psychologist, 58*(9), 723–730.

Kandler, C., Riemann, R., & Kämpfe, N. (2009). Genetic and environmental mediation between measures of personality and family environment in twins reared together. *Behavioral Genetics,* 39, 24–35.

Kaplan, L. E. (2006). Moral reasoning of MSW social workers and the influence of education. *Journal of Social Work Education, 42*, 507–522.

Kapner, D. A. (2004). *Infofacts resources: Alcohol and other drugs on campus.* [On-line]. Available: http://www.edc.org/hec//hec/pubs/factsheets/scope.html.

Kardong, K. (2008). *Introduction to biological evolution* (2nd ed.). New York: McGraw-Hill.

Kareev, Y. (2000). Seven (indeed, plus or minus two) and the detection of correlations. *Psychological Review, 107*(2):397–402.

Karmiloff, K, & Karmiloff-Smith, A. (2002). *Pathways to language: From fetus to adolescent.* Cambridge, MA: Harvard University Press.

Karon, B. P., & Widener, A. J. (1998). Repressed memories: The real story. Professional Psychology: *Research & Practice, 29*, 482–487.

Karremans, J. C., Stroebe, W., & Claus, J. (2006). Beyond Vicary's fantasies: The impact of subliminal priming and brand choice. *Journal of Experimental Social Psychology, 42*, 792–798.

Kassin, S., Fein, S., & Markus, H. R. (2008). *Social psychology* (7th ed.). Belmont, CA: Cengage

Kastenbaum, R. J. (2007). *Death, society, and human experience* (9th ed.). Upper Saddle River, NJ: Prentice Hall.

Kaufman, J. C. (2002). Dissecting the golden goose: Components of studying creative writers. *Creativity Research Journal, 14*(1), 27–40.

Kaukiainen, A., Björkqvist, K., Lagerspetz, K., Österman, K., Salmivalli, C., Rothberg, S., et al. (1999). The relationships between social intelligence, empathy, and three types of aggression. *Aggressive Behavior, 25*(2), 81–89.

Kaye, W. (2008). Neurobiology of anorexia and bulimia nervosa. *Physiology & Behavior, 94*, 121–135.

Kaysen, D., Pantalone, D. W., Chawla, N., Lindgren, K. P., Clum, G. A., Lee, C., & Resick, P. A. (2008). Posttraumatic stress disorder, alcohol use, and physical health concerns. *Journal of Behavioral Medicine, 31*, 115–125.

Kazdin, A. E. (1994). Methodology, design, and evaluation in psychotherapy research. In A.E. Bergin & S.L. Garfield (Eds.), *Handbook of psycho and behavior change* (4th ed.). New York: Wiley.

Kazdin, A. E. (2008). *Behavior modification in applied settings.* Long Grove, IL: Waveland Press.

Keats, D. M. (1982). Cultural bases of concepts of intelligence: A Chinese versus Australian comparison. In P. Sukontasarp, N. Yongsiri, P. Intasuwan, N. Jotiban, & C. Suvannathat (Eds.), *Proceedings of the Second Asian Work-*

shop on Child and Adolescent Development (pp. 67–75). Bangkok: Burapasilpa Press.

Keefe, F.J., Abernethy, A. P., & Campbell, L. C. (2005). Psychological approaches to understanding and treating disease-related pain. *Annual Review of Psychology, 56,* 601–630.

Keller, J., & Bless, H. (2008). *The interplay of stereotype threat and regulatory focus.* In Y. Kashima, K. Fiedler, & P. Freytag (Eds.), *Stereotype dynamics: Language-based approaches to the formation, maintenance, and transformation of stereotypes* (pp. 367–389). Mahwah, NJ: Erlbaum.

Kellogg, S. H., & Young, J. E. (2008). Cognitive therapy. In J. L. Lebow (Ed.), *Twenty-first century psychotherapies: Contemporary approaches to theory and practice* (pp. 43–79). Hoboken, NJ: Wiley.

Kelly, G., Brown, S., Todd, J., & Kremer, P. (2008). Challenging behaviour profiles of people with acquired brain injury living in community settings. *Brain Injury, 22,* 457–470.

Keltner, D., Kring, A. M., & Bonanno, G. A. (1999). Fleeting signs of the course of life: Facial expression and personal adjustment. *Current Directions in Psychological Science, 8*(1), 18–22.

Kemeny, M. E. (2007). Psychoneuroimmunology. In H. S. Friedman & R. C. Silver (Eds.), *Foundations of health psychology.* New York: Oxford University Press.

Kendler, K. S., & Prescott, C. A. (2006). *Genes, environment, and psychopathology: Understanding the causes of psychiatric and substance use disorders.* New York: Guilford Press.

Kendler, K. S., Gallagher, T. J., Abelson, J. M., & Kessler, R. C. (1996). Lifetime prevalence, demographic risk factors, and diagnostic validity of nonaffective psychosis as assessed in a U. S. community sample. *Archives of General Psychiatry, 53,* 1022–1031.

Kenealy, P. M. (1997). Mood-state-dependent retrieval: The effects of induced mood on memory reconsidered. *Quarterly Journal of Experimental Psychology: Human Experimental Psychology, 50A,* 290–317.

Kerschreiter, R., Schulz-Hardt, S., Mojzisch, A., & Frey, D. (2008). Biased information search in homogeneous groups: Confidence as a moderator for the effect of anticipated task requirements. *Personality & Social Psychology Bulletin, 34,* 679–691.

Kershaw, S. (2008). Sharing their demons on the Web. *New York Times.* [On-line]. Available: http://www.nytimes.com/2008/11/13/fashion/13psych.html

Kessler, R. C., McGonagle, K. A., Zhao, S., Nelson, C. B., Hughes, M., Eshleman, S., Wittchen, H., & Kendler, K. S. (1994). Lifetime and 12-month prevalence of DSM-IIIR psychiatric disorders in the United States. *Archives of General Psychiatry, 51,* 8–19.

Khalid, N., Atkins, M., Tredget, J., Giles, M., Champney-Smith, K., & Kirov, G. (2008). The effectiveness of electroconvulsive therapy in treatment-resistant depression. *Journal of ECT, 24,* 141–145.

Kieffer, K. M., Schinka, J. A., & Curtiss, G. (2004). Person-environment congruence and personality domains in the prediction of job performance and work quality. *Journal of Counseling. Psychology, 51*(2), 168–177.

Kihlstrom, J. F. (2004). An unbalanced balancing act: Blocked, recovered, and false memories in the laboratory and clinic. *Clinical Psychology: Science & Practice, 11*(1), 34–41.

Kihlstrom, J. F. (2005). Dissociative disorders. *Annual Review of Clinical Psychology, 1,* 227–253.

Killen, M., & Hart, D. (1999). *Morality in everyday life: Developmental perspectives.* New York: Cambridge University Press.

Kim, J., & Hatfield, E. (2004). Love types and subjective well-being: A cross cultural study. *Social Behavior & Personality, 32*(2), 173–182.

Kim, S. U. (2004). Human neural stem cells genetically modified for brain repair in neurological disorders. *Neuropathology, 24*(3), 159–171.

Kim, Y. H. (2008). Rebounding from learned helplessness: A measure of academic resilience using anagrams. *Dissertation Abstracts International: Section B: The Sciences and Engineering. 68*(10-B), 6947.

Kimmel, M. S. (2000). *The gendered society.* London: Oxford University Press.

Kimmel, P. R. (2006). Culture and conflict. In M. Deutsch, P. T. Coleman, & E. C. Marcus (Eds.), *The handbook of conflict resolution: Theory and practice* (2nd ed.) (pp. 625–648). Hoboken, NJ: Wiley.

Kinder, L. S., Bradley, K. A., Katon, W. J., Ludman, E., McDonnell, M. B., & Bryson, C. L. (2008). Depression, posttraumatic stress disorder, and mortality. *Psychosomatic Medicine, 70,* 20–26.

King, B. M. (2009). *Human sexuality today* (6th ed.). Boston: Allyn & Bacon.

Kirk, K. M., Bailey, J. M., Dunne, M. P., & Martin, N. G. (2000). Measurement models for sexual orientation in a community twin sample. *Behavior Genetics, 30*(4), 345–356.

Kirkman, C. A. (2002). Non-incarcerated psychopaths: Why we need to know more about the psychopaths who live amongst us. *Journal of Psychiatric & Mental Health Nursing, 9*(2), 155–160.

Kirsch, I., & Braffman, W. (2001). Imaginative suggestibility and hypnotizability. *Current Directions in Psychological Science, 10*(2), 57–61.

Kirsch, I., Mazzoni, G., & Montgomery, G. H. (2006). Remembrance of hypnosis past. *American Journal of Clinical Hypnosis, 49,* 171–178.

Kirschenbaum, H. & Jourdan, A. (2005). The current status of Carl Rogers and the person-centered approach. *Psychotherapy: Theory, Research Practice, Training, 42*(1), 37–51.

Kirschner, S. M., Litwack, T. R., & Galperin, G. J. (2004). The defense of extreme emotional disturbance: A qualitative analysis

Kisilevsky, B. S., Hains, S. M. J., Lee, K., Xie, X., Huang, H., Ye, H., Zhang, K., & Wang, Z. (2003). Effects of experience on fetal voice recognition. *Psychological Science, 14*(3), 220–224.

Klatzky, R. L. (1984). *Memory and awareness.* New York: Freeman.

Kleider, H. M., Pezdek, K., Goldinger, S. D., & Kirk, A. (2008). Schema-driven source misattribution errors: Remembering the expected from a witnessed event. *Applied Cognitive Psychology, 22,* 1–20.

Klein, O., Pohl, S., & Ndagijimana, C. (2007). The influence of intergroup comparisons on Africans' intelligence test performance in a job selection context. *Journal of Psychology: Interdisciplinary and Applied, 141,* 453–467.

Klimes-Dougan, B., Lee, C-Y. S., Ronsaville, D., & Martinez, P. (2008). Suicidal risk in young adult offspring of mothers with bipolar or major depressive disorder: A longitudinal family risk study. *Journal of Clinical Psychology, 64,* 531–540.

Kluger, J., & Masters, C. (2006, August 28). How to spot a liar. *Time,* 46–48.

Knapp, M., Romeo, R., Mogg, A., Eranti, S., Pluck, G., Purvis, R., Brown, R. G., Howard, R., Philpot, M., Rothwell, J., Edwards, D., & McLoughlin, D. M. (2008). Cost-effectiveness of transcranial magnetic stimulation vs. electroconvulsive therapy for severe depression: A multi-centre randomized controlled trial. *Journal of Affective Disorders, 109,* 273–285.

Knekt, P., Lindfors, O., Laaksonen, M. A., Raitasalo, R., Haaramo, P., Järvikoski, A., & The Helsinki Psychotherapy Study Group, Helsinki, Finlan. (2008). Effectiveness of short-term and long-term psychotherapy on work ability and functional capacity—A randomized clinical trial on depressive and anxiety disorders. *Journal of Affective Disorders, 107,* 95–106.

Knoblauch, K., Vital-Durand, F., & Barbur, J. L. (2000). Variation of chromatic sensitivity across the life span. *Vision Research, 41*(1), 23–36.

Kobasa, S. (1979). Stressful life events, personality, and health: An inquiry into hardiness. *Journal of Personality and Social Psychology, 37,* 1–11.

Köfalvi, A. (Ed). (2008). *Cannabinoids and the brain.* New York: Springer Science & Business Media.

Kohlberg, L. (1964). Development of moral character and moral behavior. In L. W. Hoffman & M. L. Hoffman (Eds.), *Review of child development research (Vol. 1).* New York: Sage.

Kohlberg, L. (1981). *The meaning and measurement of moral development.* Worcester, MA: Clark University Press.

Kohlberg, L. (1984). *The psychology of moral development: Essays on moral development (Vol. 2).* San Francisco: Harper & Row.

Köhler, W. (1925). *The mentality of apes.* New York: Harcourt, Brace.

Kohn, A. (2000). *Punished by rewards: The trouble with gold stars, incentive plans, A's, and other bribes.* New York: Houghton Mifflin.

Komiya, N., Good, G. E., & Sherrod, N. B. (2000). Emotional openness as a predictor of college students' attitudes toward seeking psychological help. *Journal of Counseling Psychology, 47*(1), 138–143.

Konheim-Kalkstein, Y. L., & van den Broek, P. (2008). The effect of incentives on cognitive processing of text. *Discourse Processes, 45,* 180–194.

Koop, C. E., Richmond, J., & Steinfeld, J. (2004). America's choice: Reducing tobacco addiction and disease. *American Journal of Public Health, 94*(2), 174–176.

Kotulak, R. (2008). The effect of violence and stress in kids' brains. In M. H. Immordino-Yang (Ed.), *Jossey-Bass Education Team. The Jossey-Bass reader on the brain and learning* (pp. 216–225). San Francisco, CA: Jossey-Bass.

Kovacic, Z., Henigsberg, N., Pivac, N., Nedic, G., & Borovecki, A. (2008). Platelet serotonin concentration and suicidal behavior in combat related posttraumatic stress disorder. *Progress in Neuro-Psychopharmacology & Biological Psychiatry, 32,* 544–551.

Kraaij, V., Arensman, E., & Spinhoven, P. (2002). Negative life events and depression in elderly persons: A meta-analysis. *Journals of Gerontology: Series B: Psychological Sciences & Social Sciences, 57B*(1), 87–94.

Kracher, B., & Marble, R. P. (2008). The significance of gender in predicting the cognitive moral development of business practitioners using the Socioemotional Reflection Objective Measure. *Journal of Business Ethics, 78,* 503–526.

Kramer, A. F., Hahn, S., Irwin, D. E., & Theeuwes, J. (2000). Age differences in the control of looking behavior. *Psychological Science, 11*(3), 210–217.

Krantz, D. S., & McCeney, M. K. (2002). Effects of psychological and social factors on organic disease: A critical assessment of research on coronary heart disease. *Annual Review of Psychology, (1),* 341–369.

Kring, A. M., Davison, G. C., Neale, J. M., & Johnson, S. L. (2007). *Abnormal psychology (10th ed.).* Hoboken, NJ: Wiley.

Krishna, G. (1999). *The dawn of a new science.* Los Angeles: Institute for Consciousness Research.

Kronenberger, W. G., Mathews, V. P., Dunn, D. W., Wang, Y., Wood, E. A., Larsen, J. J., Rembusch, M. E., Lowe, M. J., Giauque, A. L., & Lurito, J. T. (2005). Media violence exposure in aggressive and control adolescents: Differences in self-and parent-reported exposure to violence on television and in video games. *Aggressive Behavior, 31*(3), 201–216.

Krueger, J. I. (2007). From social projection to social behaviour. *European Review of Social Psychology, 18,* 1–35.

Krusemark, E. A., Campbell, W. K., & Clementz, B. A. (2008). Attributions, deception, and event related potentials: An investigation of the self-serving bias. *Psychophysiology, 45,* 511–515.

Ksir, C. J., Hart, C. I., & Ray, O. S. (2008). *Drugs, society, and human behavior.* New York: McGraw-Hill.

Kubiak, T., Vögele, C., Siering, M., Schiel, R., & Weber, H. (2008). Daily hassles and emotional eating in obese adolescents under restricted dietary conditions—The role of ruminative thinking. *Appetite, 51,* 206–209.

Kübler-Ross, E. (1983). *On children and death.* New York: Macmillan.

Kübler-Ross, E. (1997). *Death: The final stage of growth.* New York: Simon & Schuster.

Kübler-Ross, E. (1999). *On death and dying.* New York: Simon & Schuster.

Kudo, E., & Numazaki, M. (2003). Explicit and direct self-serving bias in Japan. Reexamination of self-serving bias for success and failure. *Journal of Cross-Cultural Psychology, 34*(5), 511–521.

Kuhn, C., Swartzwelder, S., & Wilson, W. (2003). *Buzzed: The straight facts about the most used and abused drugs from alcohol to ecstasy* (2nd ed.). New York: Norton.

Kulik, L. (2005). Intrafamiliar congruence in gender role attitudes and ethnic stereotypes: The Israeli case. *Journal of Comparative Family Studies, 36*(2), 289–303.

Kumkale, G. T., & Albarracín, D. (2004). The sleeper effect in persuasion: A metaanalytic review. *Psychological Bulletin, 130*(1), 143–172.

Kuperstok, N. (2008). Effects of exposure to differentiated aggressive films, equated for levels of interest and excitation, and the vicarious hostility catharsis hypothesis. *Dissertation Abstracts International: Section B: The Sciences and Engineering 68*(7-B), 4806.

Lachman, M. E. (2004). Development in midlife. *Annual Review of Psychology, 55,* 305–331.

Lader, M. (2007). Limitations of current medical treatments for depression: Disturbed circadian rhythms as a possible therapeutic target. *European Neuropsychopharacology, 17,* 743–755.

Lader, M., Cardinali, D. P., & Pandi-Perumal, S. R. (Eds.). (2006). *Sleep and sleep disorders: A neuropsychopharmacological approach.* New York: Springer-Verlag.

Lahav. O., & Mioduser, D. (2008). Haptic-feedback support for cognitive mapping of unknown spaces by people who are blind. *International Journal of Human-Computer Studies, 66.* 23–35.

Laidlaw, K., & Thompson, L. W. (2008). Cognitive behavior therapy with depressed older people. In K. Laidlaw & B. Knight (Eds.), *Handbook of emotional disorders in later life: Assessment and treatment.* New York: Oxford University Press.

Lakein, A. (1998). *Give me a moment and I'll change your life: Tools for moment management.* New York: Andrews McMeel Publishing.

Lamb, H. R. (2000). Deinstitutionalization and public policy. In R. W. Menninger & J. C. Nemiah (Eds.), *American psychiatry after World War II.* Washington, DC: American Psychiatric Press.

Landay, J. S., & Kuhnehenn, J. (2004, July 10). Probe blasts CIA on Iraq data. *The Philadelphia Inquirer,* p. AO1.

Landeira-Fernandez, J. (2004). Analysis of the cold-water restraint procedure in gastric ulceration and body temperature. *Physiology & Behavior, 82*(5), 827–833.

Lang, U. E., Bajbouj, M., Sander, T., & Gallinat, J. (2007). Gender-dependent association of the functional catechol-Omethyltransferase Val158Met genotype with sensation seeking personality trait. *Neuropsychopharmacology, 32,* 1950–1955.

Langenecker, S. A., Bieliauskas, L. A., Rapport, L. J., Zubieta, J-K., Wilde, E. A., & Berent, S. (2005). Face emotion perception and executive functioning deficits in depression. *Journal of Clinical & Experimental Neuropsychology, 27*(3), 320–333.

Latane, B., & Darley, J. M. (1970). *The unresponsive bystander: Why doesn't he help?* New York: Appleton-Century-Crofts.

Laungani, P. D. (2007). *Understanding cross-cultural psychology.* Thousand Oaks, CA: Sage.

Lawrence, C., & Andrews, K. (2004). The influence of perceived prison crowding on male inmates' perception of aggressive events. *Aggressive Behavior, 30,* 273–283.

Lawrence, M. (2008). Review of the bifurcation of the self: The history and theory of dissociation and its disorders. *American Journal of Clinical Hypnosis, 50,* 281–282.

Lazar, S. W., Kerr, C. E., Wasserman, R. H., Gray, J. R., Greve, D. N., Treadway, M. T., McGarvey, M., Quinn, B. T., Dusek, J. A., Benson, H., Rauch, S. L., Moore, C. I., Fischl, B. (2005). Meditation experience is associated with increased cortical thickness. *Neuroreport, 16*(17), 1893–1897.

Leaper, C. (2000). Gender, affiliation, assertion, and the interactive context of parent-child play. *Developmental Psychology, 36*(3), 381–393.

Leary, C. E., Kelley, M. L., Morrow, J., & Mikulka, P. J. (2008). Parental use of physical punishment as related to family environment, psychological well-being, and personality in undergraduates. *Journal of Family Violence, 23*(1), 1–7.

Lecrubier, Y., Clerc, G., Didi, R., & Kieser, M. (2002). Efficacy of St. John's wort extract WS 5570 in major depression: A double-blind, placebo-controlled trial. *American Journal of Psychiatry, 159*(8), 1361–1366.

LeDoux, J. (1996a). *The emotional brain: The mysterious underpinnings of emotional life.* New York: Simon & Schuster.

LeDoux, J. E. (1996b). Sensory systems and emotion: A model of affective processing. *Integrative Psychiatry, 4,* 237–243.

LeDoux, J. E. (1998). *The emotional brain.* New York: Simon & Schuster.

LeDoux, J. E. (2002). *Synaptic self: How our brains become who we are.* New York: Viking.

LeDoux, J. E. (2007). Emotional memory. *Scholarpedia, 2,* 180.

Lee, J-H., Kwon, Y-D., Hong, S-H., Jeong, H-J., Kim, H-M., & Um, J-Y. (2008). Interleukin-1 beta gene polymorphism and traditional constitution in obese women. *International Journal of Neuroscience, 118,* 793–805.

Lee, W-Y. (2002). One therapist, four cultures: Working with families in Greater China. *Journal of Family Therapy, 24,* 258-275.

Lefley, H. P. (2000). Cultural perspectives on families, mental illness, and the law. *International Journal of Law and Psychiatry, 23,* 229–243.

Leglise, A. (2008). *Progress in circadian rhythm research.* Hauppauge, NY: Nova Science.

Legrand, F. D., Gomà-i-freixanet, M., Kaltenbach, M. L., & Joly, P. M. (2007). Association between sensation seeking and alcohol consumption in French college students: Some ecological data collected in "open bar" parties. *Personality and Individual Differences, 43,* 1950–1959.

Lehnert, G., & Zimmer, H. D. (2008). Modality and domain specific components in auditory and visual working memory tasks. *Cognitive Processing, 9,* 53–61.

Lehto, S. M., Tolmunen, T., Joensuu, M., Saarinen, P. I., Valkonen-Korhonen, M., Vanninen, R., Ahola, P., Tiihonen, J., Kuikka, J., & Lehtonen, J. (2008). Changes in midbrain serotonin transporter availability in atypically depressed subjects after one year of psychotherapy. Progress in Neuro-*Psychopharmacology & Biological Psychiatry, 32,* 229–237.

Leichtman, M. D. (2006). Cultural and maturational influences on long-term event memory. In L. Balter & C. S. Tamis-LeMonda (Eds.), *Child psychology: A handbook of contemporary issues* (2nd ed.) (pp. 565–589). New York: Psychology Press.

Lein, C., & Wills, C. E. (2007). Using patient-centered interviewing skills to manage complex patient encounters in primary care. *Journal of the American Academy of Nurse Practitioners, 19,* 215–220.

Lele, D. U. (2008). The influence of individual personality and attachment styles on romantic relationships (partner choice and couples' satisfaction). *Dissertation Abstracts International: Section B: The Sciences and Engineering, 68,* 6316.

Lemus, D. R. (2008). Communication during retirement planning: An information-seeking process. *Dissertation Abstracts International Section A: Humanities and Social Sciences, 68*(10-A), 4139.

Leon, P., Chedraui, P., Hidalgo, L., & Ortiz, F. (2007). Perceptions and attitudes toward the menopause among middle-aged women from Guayaquil, Ecuador. *Maturitas, 57,* 233–238.

Leonard, B. E. (2003). *Fundamentals of psychopharmacology* (3rd ed.). Hoboken, NJ: Wiley.

Lepper, M. R., Greene, D., & Nisbett, R. E. (1973). Undermining children's intrinsic interest with extrinsic rewards: A test of the overjustification hypothesis. *Journal of Personality and Social Psychology, 28,* 129–137.

Leri, A., Anversa, P., & Frishman, W. H. (Eds.) (2007). *Cardiovascular regeneration and stem cell therapy.* Hoboken, NJ: Wiley-Blackwell.

Lerner, H. D. (2008). Psychodynamic perspectives. In M. Hersen & A. M. Gross (Eds.), *Handbook of clinical psychology, vol 1: Adults* (pp. 127–160). Hoboken, NJ: Wiley.

Lerner, J. S., Gonzalez, R. M., Small, D. A., & Fischhoff, B. (2003). Effects of fear and anger on perceived risks of terrorism: A national field experiment. *Psychological Science, 14,* 144–150.

Leslie, M. (2000, July/August). The Vexing Legacy of Lewis Terman. *Stanford Magazine.* [On-line]. Available: http://www.stanfordalumni.org/news/magazine/2000/julaug/articles/terman.html.

LeVay, S. (2003). Queer science: The use and abuse of research into homosexuality. *Archives of Sexual Behavior, 32*(2), 187–189.

Levenson, R. W. (1992). Autonomic nervous system differences among emotions. *Psychological Science, 3,* 23–27.

Levenson, R. W. (2007). Emotion elicitation with neurological patients. In J. A. Coan & J. J. B. Allen (Eds.), *Handbook of emotion elicitation and assessment.* (pp. 158–168). *Series in affective science.* New York: Oxford University Press.

Leventhal, H., Weinman, J., Leventhal, E. A., & Phillips, L. A. (2008). Health psychology: The search for pathways between behavior and health. *Annual Review of Psychology, 59,* 477–505.

Levine, J. R. (2001). *Why do fools fall in love: Experiencing the magic, mystery, and meaning of successful relationships.* New York: Jossey-Bass.

Levinson, D. J. (1977). The mid-life transition, *Psychiatry, 40,* 99–112.

Levinson, D. J. (1996). *The seasons of a woman's life.* New York: Knopf.

Levinthal, C. (2008). *Drugs, behavior, and modern society (5th ed.).* Boston: Allyn & Bacon.

Levitan, L. C. (2008). Giving prejudice an attitude adjustment: The implications of attitude strength and social network attitudinal composition for prejudice and prejudice reduction. *Dissertation Abstracts International: Section B: The Sciences and Engineering, 68* (8-B), 5634.

Lewis, S. (1963). *Dear Shari.* New York: Stein & Day.

Li, J-C. A. (2008). Rethinking the case against divorce. *Dissertation Abstracts International Section A: Humanities and Social Sciences, 68,* 4093.

Libon, D. J., Xie, S. X., Moore, P., Farmer, J., Antani, S., McCawley, G., Cross, K., & Grossman, M. (2007). Patterns of neuropsychological impairment in frontotemporal dementia. *Neurology, 68,* 369–375.

Lieberman, P. (1998). *Eve spoke: Human language and human evolution.* New York: Norton.

Lindberg, B., Axelsson, K., & Öhrling, K. (2008).

Adjusting to being a father to an infant born prematurely: Experiences from Swedish fathers. *Scandinavian Journal of Caring Sciences, 22,* 79–85.

Linden, E. (1993). Can animals think? *Time,* pp. 54–61.

Lindwall, M., Rennemark, M., & Berggren, T. (2008). Movement in mind: The relationship of exercise with cognitive status for older adults in the Swedish National Study on Aging and Care (SNAC). *Aging & Mental Health, 12,* 212–220.

Links, P. S., Heslegrave, R., & van Reekum, R. (1998). Prospective follow-up of borderline personality disorder: Prognosis, prediction outcome, and Axis II comorbidity. *Canadian Journal of Psychiatry, 43,* 265–270.

Links, P. S., Eynan, R., Heisel, M. J., & Nisenbaum, R. (2008). Elements of affective instability associated with suicidal behaviour in patients with borderline personality disorder. *The Canadian Journal of Psychiatry 53,* 112–116.

Lippa, R. A. (2007). The preferred traits of mates in a cross-national study of heterosexual and homosexual men and women: An examination of biological and cultural influences. *Archives of Sexual Behavior, 36,* 193–208.

Lipsanen, T., Korkeila, J., Peltola, P., Järvinen, J., Langen, K., & Lauerma, H. (2004). Dissociative disorders among psychiatric patients: Comparison with a nonclinical sample. *European Psychiatry, 19*(1), 53–55.

Lissek, S., & Powers, A. S. (2003). Sensation seeking and startle modulation by physically threatening images. *Biological Psychology, 63*(2), 179–197.

Liu, H., Mantyh, P, & Basbaum, A. I. (1997). NMDA-receptor regulation of substance P release from primary afferent nociceptors. *Nature, 386,* 721–724.

Liu, J. H., & Latané, B. (1998). Extremitization of attitudes: Does thought- and discussion-induced polarization cumulate? *Basic and Applied Social Psychology, 20,* 103–110.

Livingston, R. W., & Drwecki, B. B. (2007). Why are some individuals not racially biased? Susceptibility to affective conditioning predicts nonprejudice toward Blacks. *Psychological Science, 18,* 816–823.

Lizarraga, M. L. S., & Ganuza, J. M. G. (2003). Improvement of mental rotation in girls and boys. *Sex Roles, 49*(5–6), 277–286.

Loewenthal, D., & Winter, D. (Eds.. (2006). *What is psychotherapeutic research?* London, England: Karnac Books.

Loftus, E. (1982). Memory and its distortions. In A. G. Kraut (Ed.), *The G. Stanley Hall Lecture Series Vol. 2,* (pp. 123–154). Washington, DC: American Psychological Association.

Loftus, E., & Ketcham, K. (1994). *The myth of repressed memories: False memories and allegations of sexual abuse.* New York: St. Martin's Press.

Loftus, E. F. (2000). Remembering what never happened. In E. Tulving, et al. (Eds.), *Memory, consciousness, and the brain: The Tallinn Conference,* pp. 106–118. Philadelphia: Psychology Press/Taylor & Francis.

Loftus, E. F. (2001). Imagining the past. *Psychologist, 14*(11), 584–587.

Loftus, E. F. (2007). Memory distortions: Problems solved and unsolved. In M. Garry, & H. Hayne (Eds.), *Do justice and let the skies fall.* Mahwah, NJ: Erlbaum.

Loftus, E. F., & Cahill, L. (2007). Memory distortion from misattribution to rich false memory. In J. S. Nairne (Ed.), *The foundations of remembering: Essays in honor of Henry L. Roediger, III* (pp. 413–425). New York: Psychology Press.

Loo, R., & Thorpe, K. (1998). Attitudes toward women's roles in society. *Sex Roles, 39,* 903–912.

Lopez, J. C. (2002). Brain repair: A spinal scaffold. *Nature Reviews Neuroscience, 3,* 256.

Lopez, S. R., & Guarnaccia, P. J. J. (2000). Cultural psychopathology: Uncovering the social world of mental illness. *Annual Review of Psychology, 51,* 571–598.

Lorenz, K. Z. (1937). The companion in the bird's world. *Auk, 54,* 245–273.

Lores-Arnaiz, S., Bustamante, J., Cz ernizyniec, A., Galeano, P., Gervasoni, M. G., Martinez, A. R., Paglia, N., Cores, V., & Lores-Arnaiz, M. R. (2007). Exposure to enriched environments increases brain nitric oxide synthase and improves cognitive performance in prepubertal but not in young rats. *Behavioural Brain Research, 184,* 117–123.

Lott, D. A. (2000). *The new flirting game.* London: Sage.

Loxton, N. J., Nguyen, D., Casey, L., & Dawe, S. (2008). Reward drive, rash impulsivity and punishment sensitivity in problem gamblers. *Personality & Individual Differences, 45,* 167–173.

Lu, Z. L., Williamson, S. J., & Kaufman, L. (1992). Behavioral lifetime of human auditory sensory memory predicted by physiological measures. *Science, 258,* 1668–1670.

Lubinski, D., & Benbow, C. P. (2000). States of excellence. *American Psychologist, 55*(1), 137–150.

Lucio, E., Ampudia, A., Durán, C., León, I., & Butcher, J. N. (2001). Comparison of the Mexican and American norms of the MMPI-2. *Journal of Clinical Psychology, 57,* 1459–1468.

Lucio-Gómez, E., Ampudia-Rueda, A., Durán Patiño, C., Gallegos-Mejía, L., & León Guzmán, I. (1999). La nueva versión del inventario multifásico de la personalidad de Minnesota para adolescentes mexicanos. [The new version of the Minnesota Multiphasic Personality Inventory for Mexican adolescents]. *Revista Mexicana de Psicología, 16*(2), 217–226.

Luria, A. R. (1968). *The mind of a mnemonist: A little book about a vast memory.* New York: Basic Books.

Lutz, A., Greischar, L. L., Rawlings, N. B., Ricard, M., & Davidson, R. J. (2004). Long-term mediators self-induce high-amplitude gamma synchrony during mental practice. *Proceedings of the National Academy of Sciences, 101*(46) 16369–16373.

Lynn, R., & Harvey, J. (2008). The decline of the world's IQ. *Intelligence, 36,* 112–120.

Lynn, S. J. (2007). Hypnosis reconsidered. *American Journal of Clinical Hypnosis, 49,* 195–197.

Lyons-Ruth, K., Holmes, B. M., Sasvari-Szekely, M., Ronai, Z., Nemoda, Z., & Pauls, D. (2007). Serotonin transporter polymorphism and borderline or antisocial traits among low-income young adults. *Psychiatric Genetics, 17,* 339–343.

Lyoo, I. K., Kim, M. J., Stoll, A. L., Demopulos, C. M., Parow, A. M., Dager, S. R., Friedman, S. D., Dunner, D. L., & Renshaw, P. F. (2004). Frontal lobe gray matter density decreases in Bipolar I Disorder. *Biological Psychiatry, 55*(6), 648–651.

Lytle, M. E., Bilt, J. V., Pandav, R. S., Dodge, H. H., & Ganguli, M. (2004). Exercise level and cognitive decline: The MoVIES project. *Alzheimer Disease & Associated Disorders, 18*(2), 57–64.

Lyubimov, N. N. (1992). *Electrophysiological characteristics of sensory processing and mobilization of hidden brain reserves. 2nd Russian-Swedish Symposium New Research in Neurobiology,* Moscow: Russian Academy of Science Institute of Human Brain.

Maas, J. B. & Wherry, M. L. (1997). *Power sleep.* New York: Villard Books/Random House.

Maccoby, E. E. (2000). Parenting and its effects on children: On reading and misreading behavior genetics. *Annual Review of Psychology, 51,* 1–27.

Macmillan, M. B. (2000). *An odd kind of fame: Stories of Phineas Gage.* Cambridge, MA: MIT Press.

Maddi, S. R. (2004). Hardiness: An operationalization of existential courage. *Journal of Humanistic Psychology, 44*(3), 279–298.

Maddi, S. R., Harvey, R. H., Khoshaba, D. M., Lu, J. L., Persico, M., & Brow, M. (2006). The personality construct of hardiness, III: Relationships with repression, innovativeness, authoritarianism, and performance. *Journal of Personality, 74*, 575–597.

Maehr, M. L., & Urdan, T. C. (2000). *Advances in motivation and achievement: The role of context.* Greewich, CT: JAI Press.

Maisto, S. A., Galizio, M., & Connors, G. J. (2008). *Drug use and abuse* (5th ed.). Belmont: Cengage.

Major, B., Spencer, S., Schmader, T., Wolfe, C., & Crocker, J. (1998). Coping with negative stereotypes about intellectual performance: The role of psychological disengagement. *Personality & Social Psychology Bulletin, 24*(1), 34–50.

Maner, J. K., DeWall, C. N., & Gailliot, M. T. (2008). Selective attention to signs of success: Social dominance and early stage interpersonal perception. *Personality and Social Psychology Bulletin, 34*, 488–501.

Manly, J. J., Byrd, D., Touradji, P., Sanchez, D., & Stern, Y. (2004). Literacy and cognitive change among ethnically diverse elders. *International Journal of Psychology, 39*(1), 47–60.

Manning, J. (2007). The use of meridian-based therapy for anxiety and phobias. *Australian Journal of Clinical Hypnotherapy and Hypnosis, 28*, 45–50.

Markle, A. (2007). Asymmetric disconfirmation in managerial beliefs about employee motivation. *Dissertation Abstracts International Section A: Humanities and Social Sciences, 68*(5-A), 2051.

Markovitz, P. J. (2004). Recent trends in the pharmacotherapy of personality disorders. *Journal of Personality Disorders, 18*(1), 99–101.

Marks, L. D., Hopkins, K. C., Monroe, P. A., Nesteruk, O., & Sasser, D. D. (2008). "Together, we are strong": A qualitative study of happy, enduring African American marriages. *Family Relations, 57*, 172–185.

Markstrom, C. A., & Marshall, S. K. (2007). The psychosocial inventory of ego strengths: Examination of theory and psychometric properties. *Journal of Adolescence, 30*, 63–79.

Markus, H. R., & Kitayama, S. (1998). The cultural psychology of personality. *Journal of Cross-Cultural Psychology, 29*, 63–87.

Markus, H. R., & Kitayama, S. (2003). Culture, self, and the reality of the social. *Psychological Inquiry, 14*(3–4), 277–283.

Marsee, M. A., Weems, C. F., Taylor, L. K. (2008). Exploring the association between aggression and anxiety in youth: A look at aggressive subtypes, gender, and social cognition. *Journal of Child and Family Studies, 17*, 154–168.

Marshall, L., & Born, J. (2007). The contribution of sleep to hippocampus-dependent memory consolidation. *Trends in Cognitive Sciences, 11*, 442–450.

Marshall, M., & Brown, J. D. (2008). On the psychological benefits of self-enhancement. In E. C. Chang (Ed.). *Self-criticism and self-enhancement: Theory, research, and clinical implications* (pp. 19–35). Washington, DC: American Psychological Association.

Martin, C. L., & Fabes, R. (2009). *Discovering child development* (2nd ed.). Belmont, CA: Cengage.

Martin, S. E., Snyder, L. B., Hamilton, M., Fleming- Milici, F., Slater, M. D., Stacy, A., Chen, M., & Grube, J. W. (2002). Alcohol advertising and youth. *Alcoholism: Clinical & Experimental Research, 26*(6), 900–906.

Martineau, J., Cochin, S., Magne, R., & Barthelemy, C. (2008). Impaired cortical activation in autistic children: Is the mirror neuron system involved? *International Journal of Psychophysiology, 68*, 35–40.

Martinelli, E. A. (2006). Paternal role development and acquisition in fathers of pre-term infants: A qualitative study. *Dissertation Abstracts International: Section B: The Sciences and Engineering 66*(9-B), 5125.

Maslow, A. H. (1954). *Motivation and personality.* New York: Harper & Row.

Maslow, A. H. (1970). *Motivation and personality (2nd ed.).* New York: Harper & Row.

Maslow, A. H. (1999). *Toward a psychology of being* (3rd ed.). New York: Wiley.

Mason, P. T., & Kreger, R. (1998). *Stop walking on eggshells: Taking your life back when someone you care about has borderline personality disorder.* New York: New Harbinger Publishers.

Massicotte-Marquez, J., Décary, A., Gagnon, J. F., Vendette, M., Mathieu, A., Postuma, R. B., Carrier, J., & Montplaisir, J. (2008). Executive dysfunction and memory impairment in idiopathic REM sleep behavior disorder. *Neurology, 70*, 1250–1257.

Masters, W. H., & Johnson, V. E. (1961). Orgasm, anatomy of the female. In A. Ellis & A. Abarbonel (Eds.), *Encyclopedia of Sexual Behavior,* Vol. 2. New York: Hawthorn.

Masters, W. H., & Johnson, V. E. (1966). *Human sexual response.* Boston: Little, Brown.

Masters, W. H., & Johnson, V. E. (1970). *Human sexual inadequacy.* Boston: Little, Brown.

Mathews, A., & MacLeod, C. (2005). Cognitive vulnerability to emotional disorders. *Annual Review of Clinical Psychology, 1*, 167–195.

Matlin, M. W. (2008). *The psychology of women* (6th ed.). Belmont, CA: Cengage.

Matlin, M. W., & Foley, H. J. (1997). *Sensation and perception* (4th ed.). Boston: Allyn and Bacon.

Matsumoto, D. (2000). *Culture and psychology: People around the world.* Belmont, CA: Wadsworth.

Matsumoto, D., & Juang, L. (2008). *Culture and psychology* (4th ed.). Belmont, CA: Cengage.

Matusov, E., & Hayes, R. (2000). Sociocultural critique of Piaget and Vygotsky. *New Ideas in Psychology, 18*(2–3), 215–239.

May, A., Hajak, G., Gänflßbauer, S., Steffens, T., Langguth, B., Kleinjung, T., & Eichhammer, P. (2007). Structural brain alterations following 5 days of intervention: Dynamic aspects of neuroplasticity. *Cerebral Cortex, 17*, 205–210.

Mayers, A. G., Baldwin, D. S., Dyson, R., Middleton, R. W., & Mustapha, A. (2003). Use of St John's wort (Hypericum perforatum L) in members of a depression self-help organization: A 12-week open prospective pilot study using the HADS scale. *Primary Care Psychiatry, 9*(1), 15–20.

Maynard, A. E., & Greenfield, P. M. (2003). Implicit cognitive development in cultural tools and children: Lessons from Maya Mexico. *Cognitive Development, 18*(4), 489–510.

Mays, V. M., Cochran, S. D., & Barnes, N. W. (2007). Race, race-based discrimination, and health outcomes among African Americans. *Annual Review of Psychology, 58*, 201–225.

Mazzoni, G., & Memon, A. (2003). Imagination can create false autobiographical memories. *Psychological Science, 14*, 186–188.

Mazzoni, G., & Vannucci, M. (2007). Hindsight bias, the misinformation effect, and false autobiographical memories. *Social Cognition, 25*, 203–220.

McAllister-Williams, R. H., & Rugg, M. D. (2002). Effects of repeated cortisol administration on brain potential correlates of episodic memory retrieval. *Psychopharmacology, 160*(1), 74–83.

McCabe, C., & Rolls, E. T. (2007). Umami: A delicious flavor formed by convergence of taste and olfactory pathways in the human brain. *European Journal of Neuroscience, 25*, 1855–1864.

McClelland, D. C. (1958). Risk-taking in children with high and low need for achievement. In J. W. Atkinson (Ed.), *Motives in fantasy, action, and society.* Princeton, NJ: Van Nostrand.

McClelland, D. C. (1987). Characteristics of successful entrepreneurs. *Journal of Creative Behavior, 3*, 219–233.

McClelland, D. C. (1993). Intelligence is not the best predictor of job performance. *Current Directions in Psychological Science, 2*, 5–6.

McCrae, R. R. (2004). Human nature and culture: A trait perspective. *Journal of Research in Personality, 38*(1), 3–14.

McCrae, R. R., & Costa, P. T. Jr. (1990). *Personality in adulthood.* New York: Guilford Press.
McCrae, R. R., & Costa, P. T. Jr. (1999). A five-factor theory of personality. In L. A. Pervin, & O. P. John (Eds.), *Handbook of personality: Theory and research.* New York: Guilford Press.
McCrae, R. R., Costa, P. T. Jr., Hrebícková, M., Urbánek, T., Martin, T. A., Oryol, V. E., Rukavishnikov, A. A., & Senin, I. G. (2004). Age differences in personality traits across cultures: Self-report and observer perspectives. *European Journal of Personality, 18*(2), 143–157.
McCrae, R. R., Costa, P. T. Jr., Martin, T. A., Oryol, V. E., Rukavishnikov, A. A., Senin, I. G., Hrebícková, M., & Urbánek, T. (2004). Consensual validation of personality traits across cultures. *Journal of Research in Personality, 38*(2), 179–201.
McCrae, R. R., Costa, P. T., Jr., Ostendorf, F., Angleitner, A., Hrebickova, M., Avia, M. D., Sanz, J., Sanchez-Bernardos, M. L., Kusdil, M. E., Woodfield, R., Saunders, P. R., & Smith, P. B. (2000). Nature over nurture: Temperament, personality, and life span development. *Journal of Personality and Social Psychology, 78*(1), 173–186.
McCrae, R. R., & Sutin, A. R. (2007). New frontiers for the Five-Factor Model: A preview of the literature. *Social and Personality Psychology Compass, 1,* 423–440.
McDonald, J. W., Liu, X. Z., Qu, Y., Liu, S., Mickey, S. K., Turetsky, D., Gottlieb, D. I., & Choi, D. W. (1999). Transplanted embryonic stem cells survive, differentiate, and promote recovery in injured rat spinal cord. *Nature & Medicine, 5,* 1410–1412.
McDougall, W. (1908). *Social psychology.* New York: Putnam's Sons.
McEvoy, P. M. (2007). Effectiveness of cognitive behavioural group therapy for social phobia in a community clinic: A benchmarking study. *Behaviour Research and Therapy, 45,* 3030–3040.
McFarlane, W. R. (2006). Family expressed emotion prior to onset of psychosis. In S. R. H. Beach, M. Z. Wamboldt, N. J. Kaslow, R. E. Heyman, M. B. First, et al., (Eds), *Relationalprocesses and DSM-V: Neuroscience, assessment,prevention, and treatment* (pp. 77–87). Washington, DC: American Psychiatric Association.
McGrath, P. (2002). Qualitative findings on the experience of end-of-life care for hematological malignancies. *American Journal of Hospice & Palliative Care, 19*(2), 103–111.
McGue, M., Bouchard, T. J., Iacono, W. G., & Lykken, D. T. (1993). Behavioral genetics of cognitive ability: A life-span perspective. In R. Plomin & G. McClearn (Eds.), *Nature, nurture, and psychology.* Washington, DC: American Psychological Association.
McGuire, B. E., Hogan, M. J., & Morrison, T. G. (2008). Dimensionality and reliability assessment of the Pain Patient Profile Questionnaire. *European Journal of Psychological Assessment, 24,* 22–26.
McKellar, P. (1972). Imagery from the standpoint of introspection. In P. W. Sheehan (Ed.), *The function and nature of imagery.* New York: Academic Press.
McKim, W. A. (2002). *Drugs and behavior: An introduction to behavioral pharmacology* (5th ed). Englewood Cliffs, NJ: Prentice Hall.
McKinney, C., Donnelly, R., & Renk, K. (2008). Perceived parenting, positive and negative perceptions of parents, and late adolescent emotional adjustment. *Child and Adolescent Mental Health, 13,* 66–73.
McNamara, J. M., Barta, Z., Fromhage, L., & Houston, A. I. (2008). The coevolution of choosiness and cooperation. *Nature, 451,* 189–201. 7896
McNaughton, B. L., Battaglia, F. P., Jensen, O., Moser, E. I., Moser, M. (2006). Path integration and the neural basis of the 'cognitive map'. *Nature Reviews Neuroscience.* 7(8), 663–678.
McNicholas, W. T., & Javaheri, S. (2007). Pathophysiologic mechanisms of cardiovascular disease in obstructive sleep apnea. *Sleep Medicine Clinics, 2,* 539–547.
McQuown, S. C., Belluzzi, J. D., & Leslie, F. M. (2007). Low dose nicotine treatment during early adolescence increases subsequent cocaine reward. *Neurotoxicology and Teratology, 29,* 66–73.
Meeus, W., & Raaijmakers, Q. (1989). Autoritätsgehorsam in Experimenten des Milgram-Typs: Eine Forschungsübersicht [Obedience to authority in Milgram-type studies: A research review]. *Zeitschrift für Sozialpsychologie, 20*(2), 70–85.
Mehrabian, A. (1968). A relationship of attitude to seated posture orientation and distance. *Journal of Personality and Social Psychology,* 10, 26–30.
Mehrabian, A. (1971). Silent messages. Belmont, CA: Wadsworth.
Mehrabian, A. (2007). Nonverbal communication. New Brunswick, NJ: Aldine Transaction.
Meltzer, G. (2000). Genetics and etiology of schizophrenia and bipolar disorder. *Biological Psychiatry, 47*(3), 171–178.
Meltzer, L. (2004). Resilience and learning disabilities: Research on internal and external protective dynamics. *Learning Disabilities Research & Practice, 19*(1), 1–2.
Meltzoff, A. N., & Moore, M. K. (1977). Imitation of facial and manual gestures by human neonates. *Science, 198,* 75–78.
Meltzoff, A. N., & Moore, M. K. (1985). Cognitive foundations and social functions of imitation and intermodal representation in infancy. In J. Mehler & R. Fox (Eds.), *Neonate cognition: Beyond the blooming buzzing confusion.* Hillsdale, NJ: Erlbaum.
Meltzoff, A. N., & Moore, M. K. (1994). Imitation, memory, and the representation of persons. *Infant Behavior and Development, 17,* 83–99.
Melzack, R. (1999). Pain and stress: A new perspective. In R. J. Gatchel & D. C. Turk (Eds.), *Psychosocial factors in pain: Critical perspectives.* New York: Guilford Press.
Melzack, R., & Wall, P. D. (1965). Pain mechanisms: A new theory. *Science, 150,* 971–979.
Menec, V. H. (2003). The relation between everyday activities and successful aging: A 6-year longitudinal study. *Journals of Gerontology B: Psychological Sciences and Social Sciences, 58,* 574–582.
Messinger, L.M. (2001). *Georgia O'Keeffe.* London: Thames & Hudson.
Metcalf, P., & Huntington, R. (1991). *Celebrations of death: The anthropology of mortuary ritual* (2nd ed.). Cambridge, England: Cambridge University Press.
Meyer, U., Nyffleler, M., Schwendener, S., Knuesel, I., Yee, B. K., & Feldon, J. (2008). Relative prenatal and postnatal maternal contributions to schizophrenia-related neurochemical dysfunction after in utero immune challenge. *Neuropsychopharmacology, 33,* 441–456.
Migueles, M, & Garcia-Bajos, E. (1999). Recall, recognition, and confidence patterns in eyewitness testimony. *Applied Cognitive Psychology, 13,* 257–268.
Mikulincer, M., & Goodman, G. S. (Eds.). (2006). *Dynamics of romantic love: Attachment, caregiving, and sex.* New York: Guilford Press.
Milgram, S. (1963). Behavioral study of obedience. *Journal of Abnormal and Social Psychology,* 67, 371–378.
Milgram, S. (1974). *Obedience to authority: An experimental view.* New York: Harper & Row.
Miller, G. A. (1956). The magical number seven, plus or minus two: Some limits on our capacity for processing information. *Psychological Review, 63,* 81–97.
Miller, J. G., & Bersoff, D. M. (1998). The role of liking in perceptions of the moral responsibility to help: A cultural perspective. *Journal of Experimental Social Psychology, 34,* 443–469.

Miller, L. K. (2005). What the savant syndrome can tell us about the nature and nurture of talent. *Journal for the Education of the Gifted, 28,* 361–373.

Millon, T. (2004). *Masters of the mind: Exploring the story of mental illness from ancient times to the new millennium.* Hoboken, NJ: Wiley.

Miltenberger, R. G. (2008). *Behavior modification: Principles and procedures* (4th ed.). Belmont, CA: Cengage.

Milton, J., & Wiseman, R. (1999). Does psi exist? Lack of replication of an anomalous process of information transfer. *Psychological Bulletin, 125,* 387–391.

Milton, J., & Wiseman, R. (2001). Does psi exist? Reply to Storm and Ertel 2000. *Psychological Bulletin, 127* (3), 434–438.

Mineka, S., & Oehlberg, K. (2008). The relevance of recent developments in classical conditioning to understanding the etiology and maintenance of anxiety disorders. *Acta Psychologica, 127,* 567–580.

Mingo, C., Herman, C. J., & Jasperse, M. (2000). Women's stories: Ethnic variations in women's attitudes and experiences of menopause, hysterectomy, and hormone replacement therapy. *Journal of Women's Health and Gender Based Medicine, 9,* S27–S38.

Mingroni, M. A. (2004). The secular rise in IQ. *Intelligence, 32,* 65–83.

Minuchin, S., Lee, W-Y., & Simon, G. M. (2007). *Mastering family therapy: Journeys of growth and transformation* (2nd ed.). Hoboken, NJ: Wiley.

Minzenberg, M. J., Poole, J. H., & Vinogradov, S. (2008). A neurocognitive model of borderline personality disorder: Effects of childhood sexual abuse and relationship to adult social attachment disturbance. *Development and Psychopathology, 20,* 341–368.

Miracle, T. S., Miracle, A. W., & Baumeister, R. F. (2006). *Human sexuality: Meeting your basic needs* (2nd ed.). Upper Saddle River, NJ: Pearson Education.

Mirescu, C., Peters, J. D., Noiman, L., & Gould, E. (2006). Sleep deprivation inhibits adult neurogenesis in the hippocampus by elevating glucocorticoids. *PNAS Proceedings of the National Academy of Sciences of the United States of America, 103*(50), 19170–19175.

Mischel, W, Shoda, Y., & Ayduk, O. (2008). *Introduction to personality: Toward an integrative science of the person* (8th ed.). Hoboken, NJ: Wiley.

Mita, T. H., Dermer, M., & Knight, J. (1977). Reversed facial images and the mere-exposure hypothesis. *Journal of Personality and Social Psychology, 35*(8), 597–601.

Mitchell, J. P., Dodson, C. S., & Schacter, D. L. (2005). fMRI evidence for the role of recollection in suppressing misattribution errors: The illusory truth effect. *Journal of Cognitive Neuroscience, 17,* 800–810.

Mitchell, R. (2003). Ideological reflections on the DSM-IV-R (or pay no attention to that man behind the curtain, Dorothy!). *Child & Youth Care Forum, 32*(5), 281–298.

Mittag, O., & Maurischat, C. (2004). Die Cook-Medley Hostility Scale (Ho-Skala) im Vergleich zu den Inhaltsskalen "Zynismus", "Ärger" sowie "Typ A" aus dem MMPI-2: Zur zukünftigen Operationalisierung von Feindseligkeit [A comparison of the Cook-Medley Hostility Scale (Ho-scale) and the content scales "cynicism," "anger," and "type A" out of the MMPI-2: On the future assessment of hostility]. *Zeitschrift für Medizinische Psychologie, 13*(1), 7–12.

Mograss, M. A., Guillem, F., & Godbout, R. (2008). Event-related potentials differentiates the processes involved in the effects of sleep on recognition memory. *Psychophysiology, 45,* 420–434.

Möhler, H., Rudolph, U., Boison, D., Singer, P., Feldon, J., & Yee, B. K. (2008). *Regulation of cognition and symptoms of psychosis: Focus on GABA-sub(A) receptors and glycine transporter 1. Pharmacology, Biochemistry & Behavior, 90,* 58–64.

Monastra, V. J. (2008). Electroencephalographic biofeedback in the treatment of ADHD. In V. J. Monastra (Ed.), *Unlocking the potential of patients with ADHD: A model for clinical practice.* Washington, DC: APA

Moneta, G. B., & Siu, C. M. Y. (2002). Trait intrinsic and extrinsic motivations, academic performance, and creativity in Hong Kong college students. *Journal of College Student Development, 43*(5), 664–683.

Monin, B. (2003). The warm glow heuristic: When liking leads to familiarity. *Journal of Personality and Social Psychology, 85*(6), 1035–1048.

Monks, C. P., Ortega-Ruiz, R., & Rodríguez-Hidalgo, A. J. (2008). Peer victimization in multicultural schools in Spain and England. *European Journal of Developmental Psychology, 5,* 507–535.

Montgomery, S. A. (2008). The under-recognized role of dopamine in the treatment of major depressive disorder. *International Clinical Psychopharmacology, 23,* 63–69.

Moore, M. M. (1998). The science of sexual signaling. In G. C. Brannigan, E. R. Allgeier, & A. R. Allgeier (Eds.), *The sex scientists* (pp. 61–75). New York: Longman.

Moore, S., Grunberg, L., & Greenberg, E. (2004). Repeated downsizing contact: The effects of similar and dissimilar layoff experiences on work and well-being outcomes. *Journal of Occupational Health Psychology, 9*(3), 247–257.

Morgenthaler, T. I., Lee-Chiong, T., Alessi, C., Friedman, L., Aurora, R. N., Boehlecke, B., Brown, T., Chesson Jr., A. L., Kapur, V., Maganti, R., Owens, J., Pancer, J., Swick, T. J., Zak, R., & Standards of Practice Committee of the AASM. (2007). Practice parameters for the clinical evaluation and treatment of circadian rhythm sleep disorders: An American academy of sleep medicine report. *Sleep: Journal of Sleep and Sleep Disorders Research, 30,* 1445–1459.

Morris, S. G. (2007). Influences on childrens' narrative coherence: Age, memory breadth, and verbal comprehension. *Dissertation Abstracts International: Section B: The Sciences and Engineering, 68*(6-B), 4157.

Morrison, C., & Westman, A. S. (2001). Women report being more likely than men to model their relationships after what they have seen on TV. *Psychological Reports, 89*(2), 252–254.

Morry, M. M. (2005). Relationship satisfaction as a predictor of similarity ratings: A test of the attraction-similarity hypothesis. *Journal of Social and Personal Relationships, 22,* 561–584.

Moss, D. (2004). Biofeedback. *Applied Psychophysiology & Biofeedback, 29*(1), 75–78.

Mueser, K. T., & Jeste, D. V. (Eds.). (2008). *Clinical handbook of schizophrenia.* New York: Guildford Press.

Mulckhuyse, M., van Zoest, W., & Theeuwes, J. (2008). Capture of the eyes by relevant and irrelevant onsets. *Experimental Brain Research, 186,* 225–235.

Munro, I., & Edward, K-L. (2008). Mental illness and substance use: An Australian perspective. *International Journal of Mental Health Nursing, 17,* 255–260.

Murakami, K., & Hayashi, T. (2002). Interaction between mind-heart and gene. *Journal of International Society of Life Information Science, 20*(1), 122–126.

Murphy, L. L. (2006). Endocannabinoids and endocrine function. In E. S. Onaivi, T. Sugiura, & V. Di Marzo (Eds.), *Endocannabinoids: The brain and body's marijuana and beyond* (pp. 467–474). Boca Raton, FL: CRC Press.

Murray, H. A. (1938). *Explorations in personality.* New York: Oxford University Press.

Myers, L. B., & Vetere, A. (2002). Adult romantic attachment styles and health-related measures. *Psychology, Health & Medicine, 7*(2), 175–180.

Myerson, J., Rank, M. R., Raines, F. Q., & Schnitzler, M. A. (1998). Race and general cognitive ability: The myth of diminishing returns to education. *Psychological Science, 9,* 139–142.

Nabi, R. L., Moyer-Gusé, E., & Byrne, S. (2007). All joking aside: A serious investigation into the persuasive effect of funny social issue messages. *Communication Monographs, 74,* 29–54.

Naglieri, J. A., & Ronning, M. E. (2000). Comparison of White, African American, Hispanic, and Asian children on the Naglieri Nonverbal Ability Test. *Psychological Assessment, 12*(3), 328–334.

Nahas, G. G., Frick, H. C., Lattimer, J. K., Latour, C., & Harvey, D. (2002). Pharmacokinetics of THC in brain and testis, male gametotoxicity and premature apoptosis of spermatozoa. *Human Psychopharmacology: Clinical & Experimental, 17*(2), 103–113.

Naisch, P. L. N. (2007). Time to explain the nature of hypnosis? *Contemporary Hypnosis, 23,* 33–46.

Nakamura, K. (2006). The history of psychotherapy in Japan. *International Medical Journal, 13,* 13–18.

Nash, M., & Barnier, A. (Eds.). (2008). *The Oxford handbook of hypnosis.* New York: Oxford University Press.

Näslund, E., & Hellström, P. M. (2007). Appetite signaling: From gut peptides and enteric nerves to brain. *Physiology & Behavior, 92,* 256–262.

Nathan, R., Rollinson, L., Harvey, K., & Hill, J. (2003). The Liverpool violence assessment: An investigator-based measure of serious violence. *Criminal Behaviour & Mental Health, 13*(2), 106–120.

National Institute of Mental Health (NIMH). (2001). *Facts about anxiety disorders.* [On-line]. Available: http://www.nimh.nih.gov/publicat/ad facts.cfm.

National Institute of Mental Health (NIMH). (2008). *Anxiety disorders.* [On-line]. Available: http://www.nimh.nih.gov/health/publications/anxiety-disorders/summary.shtml.

National Institute of Mental Health (NIMH). (2008). *When unwanted thoughts take over: Obsessive-compulsive disorder.* [On-line]. Available: http://www.nimh.nih.gov/health/publications/when-unwantedthoughts-take-over-obsessive-compulsivedisorder/complete-publication.shtml.

National Organization on Fetal Alcohol Syndrome. (2008). *Facts about FAS and FASD.* [On-line]. Available: http://www.nofas.org/family/facts.aspx

National Research Council. (2003). *The polygraph and lie detection.* [On-line]. Available: http://fermat.nap.edu/catalog/10420.html.

National Sleep Foundation. (2007). *Myths and facts about sleep.* [On-line]. Available: http://www. sleepfoundation.org/site/c.huI XKjM0IxF/b.2419251/k.2773/Myths__and_Facts__ About_Sleep.

Navarro, M. (2008). Who are we? New dialogue on mixed race. *New York Times.* [On-line]. Available: http://www.nytimes.com/2008/03/31/us/politics/31race.html.

Neale, J. M., Oltmanns, T. F., & Winters, K. C. (1983). Recent developments in the assessment and conceptualization of schizophrenia. *Behavioral Assessment, 5,* 33–54.

Neenan, M. (2008). Tackling procrastination: An REBT perspective for coaches. *Journal of Rational-Emotive & Cognitive Behavior Therapy, 26,* 53–62.

Neher, A. (1991). Maslow's theory of motivation: A critique. *Journal of Humanistic Psychology, 31,* 89–112.

Neisser, U. (1967). *Cognitive psychology.* New York: Appleton-Century-Crofts.

Neisser, U. (1998). Introduction: Rising test scores and what they mean. In U. Neisser (Ed.), *The rising curve: Long-term gains in IQ and related measures.* Washington, DC: American Psychological Association.

Nelson III, C. A., Zeanah, C. H., & Fox, N. A. (2007). The effects of early deprivation on brain-behavioral development: The Bucharest Early Intervention Project. In D. Romer & E. F. Walker (Eds.), *Adolescent psychopathology and the developing brain: Integrating brain and prevention science* (pp. 197–215). New York: Oxford University Press.

Nelson, R. J., & Chiavegatto, S. (2001). Molecular basis of aggression. *Trends in Neurosciences, 24*(12), 713–719.

Neria, Y., Bromet, E. J., Sievers, S., Lavelle, J., & Fochtmann, L. J. (2002). Trauma exposure and posttraumatic stress disorder in psychosis: Findings from a first-admission cohort. *Journal of Consulting & Clinical Psychology, 70*(1), 246–251.

Nesse, R. M. (2000). Is depression an adaptation? *Archives of General Psychiatry, 57,* 14–20.

Nesse, R. M., & Jackson, E. D. (2006). Evolution: Psychiatric nosology's missing biological foundation. *Clinical Neuropsychiatry: Journal of Treatment Evaluation, 3,* 121–131.

Neto, F., & Furnham, A. (2005). Gender-role portrayals in children's television advertisements. *International Journal of Adolescence & Youth, 12*(1–2), 69–90.

Neubauer, A. C., Grabner, R. H., Freudenthaler, H. H., Beckmann, J. F., & Guthke, J. (2004). Intelligence and individual differences in becoming neurally efficient. *Acta Psychologica, 116*(1), 55–74.

Neugarten, B. L., Havighurst, R. J., & Tobin, S. S. (1968). The measurement of life satisfaction. *Journal of Gerontology, 16,* 134–143.

Neuringer, A., Deiss, C., & Olson, G. (2000). Reinforced variability and operant learning. *Journal of Experimental Psychology: Animal Behavior Processes, 26*(1), 98–111.

Ng, S. M., Li, A. M., Lou, V. W. Q., Tso, I. F., Wan, P. Y. P., & Chan, D. F. Y. (2008). Incorporating family therapy into asthma group intervention: A randomized waitlist-controlled trial. *Family Process, 47,* 115–130.

Niaura, R., Todaro, J. F., Stroud, L., Spiro, A., Ward, K. D., & Weiss, S. (2002). Hostility, the metabolic syndrome, and incident cornary heart disease. *Health Psychology, 21*(6), 598–593.

Nicholson, A., Pikhart, H., Pajak, A., Malyutina, S., Kubinova, R., Peasey, A., Topor-Madry, R., Nikitin, Y., Capkova, N., Marmot, M., & Bobak, M. (2008). Socioeconomic status over the life-course and depressive symptoms in men and women in Eastern Europe. *Journal of Affective Disorders, 105,* 125–136.

Nickerson, R. (1998). Confirmation bias: A ubiquitous phenomenon in many guises. *Review of General Psychology, 2,* 175–220.

Nicotra, A., Critchley, H. D., Mathias, C. J., & Dolan, R. J. (2006). Emotional and autonomic consequences of spinal cord injury explored using functional brain imaging. *Brain: A Journal of Neurology, 129,* 718–728.

Nishimoto, R. (1988). A cross-cultural analysis of psychiatric symptom expression using Langer's twenty-two item index. *Journal of Sociology and Social Welfare, 15,* 45–62.

Nogueiras, R., & Tschöp, M. (2005). Separation of conjoined hormones yields appetite rivals. *Science, 310,* 985–986.

Nolen-Hoeksema, S., Larson, J., & Grayson, C. (2000). Explaining the gender difference in depressive symptoms. *Journal of Personality and Social Psychology, 77,* 1061–1072.

Norenzayan, A. (2006). Cultural variation in reasoning. In R. Viale, D. Andler, & L. Hirschfeld (Eds.), *Biological and cultural bases of human inference.* Mahwah, NJ: Erlbaum.

Nouchi, R., & Hyodo, M. (2007). The congruence between the emotional valences of recalled episodes and mood states influences the mood congruence effect. *Japanese Journal of Psychology, 78,* 25–32.

O'Farrell, T. J., Murphy, M., Alter, J., & Fals-Stewart, W. (2008). Brief family treatment intervention to promote continuing care among alcohol-dependent patients in inpatient detoxification: A randomized pilot study. *Journal of Substance Abuse Treatment, 34,* 363–369.

Oei, T. P. S., & Dingle, G. (2008). The effectiveness of group cognitive behaviour therapy for unipolar depressive disorders. *Journal of Affective Disorders, 107,* 5–21.

Oishi, K., Ohkura, N., Sei, H., Matsuda, J., & Ishida, N. (2007). CLOCK regulates the circadian rhythm of kaolin-induced writhing behavior in mice. *Neuroreport: For Rapid Communication of Neuroscience Research, 18,* 1925–1928.

Olds, J., & Milner, P. M. (1954). Positive reinforcement produced by electrical stimulation of septal area and other regions of rat brains. *Journal of Comparative and Physiological Psychology, 47,* 419–427.

Olfson, M., Marcus, S., Pincus, H. A., Zito, J. M., Thompson, J. W., & Zarin, D. A. (1998). Antidepressant prescribing practices of outpatient psychiatrists. *Archives of General Psychiatry, 55,* 310, 316.

Oliver, R. J. (2002). Tobacco abuse in pregnancy. *Journal of Prenatal Psychology & Health, 17*(2), 153–166.

Oppenheimer, D. M. (2004). Spontaneous discounting of availability in frequency judgment tasks. *Psychological Science, 15*(2), 100–105.

Orne, M. T. (2006). The nature of hypnosis. artifact and essence: An experimental study. *Dissertation Abstracts International: Section B: The Sciences and Engineering, 67*(2-B), 1207.

Orth-Gomér, K. (2007). Job strain and risk of recurrent coronary events. *Journal of the American Medical Association,* 298, 1693–1694.

Oshima, K. (2000). Ethnic jokes and social function in Hawaii. *Humor: International Journal of Humor Research, 13*(1), 41–57.

Osler, M., McGue, M., Lund, R., & Christensen, K. (2008). Marital status and twins' health and behavior: An analysis of middle-aged Danish twins. *Psychosomatic Medicine, 70,* 482–487.

Ostrov, J. M., & Keating, C. F. (2004). Gender differences in preschool aggression during free play and structured interactions: An observational study. *Social Development, 13*(2), 255–277.

Overmier, J. B., & Murison, R. (2000). Anxiety and helplessness in the face of stress predisposes, precipitates, and sustains gastric ulceration. *Behavioural Brain Research, 110*(1–2), 161–174.

Ozawa-de Silva, C. (2007). Demystifying Japanese therapy: An analysis of Naikan and the Ajase complex through Buddhist thought. *Ethos, 35,* 411–446.

Paice, E., Rutter, H., Wetherell, M., Winder, B., & McManus, I. C. (2002). Stressful incidents, stress and coping strategies in the pre-registration house officer year. *Medical Education, 36*(1), 56–65.

Palfai, T. P., Monti, P. M., Ostafin, B., & Hutchinson, K. (2000). Effects of nicotine deprivation on alcohol-related information processing and drinking behavior. *Journal of Abnormal Psychology, 109,* 96–105.

Palmer, S., & Gyllensten, K. (2008). How cognitive behavioural, rational emotive behavioural or multimodal coaching could prevent mental health problems, enhance performance, and reduce work related stress. *Journal of Rational-Emotive & Cognitive Behavior Therapy, 26,* 38–52.

Panksepp, J. (2005). Affective consciousness: Core emotional feelings in animals and humans. *Consciousness & Cognition: An International Journal, 14*(1), 30–80.

Papadelis, C, Chen, Z., Kourtidou-Papadeli, C., Bamidis, P. D, Chouvarda, I., Bekiaris, E., & Maglaveras, N. (2007). Monitoring sleepiness with on-board electrophysiological recordings for preventing sleep-deprived traffic accidents. *Clinical Neurophysiology, 118,* 1906–1922.

Paquet, F., Soucy, J. P., Stip, E., Lévesque, M., Elie, A., & Bédard, M. A. (2004). Comparison between olanzapine and haloperidol on procedural learning and the relationship with striatal D-sub-2 receptor occupancy in schizophrenia. *Journal of Neuropsychiatry & Clinical Neurosciences, 16*(1), 47–56.

Parker-Oliver, D. (2002). Redefining hope for the terminally ill. *American Journal of Hospice & Palliative Care, 19*(2), 115–120.

Parrott, A., Morinan, A., Moss, M., & Scholey, A. (2004). *Understanding drugs and behavior.* Hoboken, NJ: Wiley.

Pasternak, T., & Greenlee, M. W. (2005). Working memory in primate sensory systems. *Nature Reviews Neuroscience, 6*(2), 97–107.

Patterson, F., & Linden, E. (1981). *The education of Koko.* New York: Holt, Rinehart and Winston.

Patterson, J. M., Holm, K. E., & Gurney, J. G. (2004). The impact of childhood cancer on the family: A qualitative analysis of strains, resources, and coping behaviors. *Psycho-Oncology, 13*(6), 390–407.

Patterson, P. (2002). *Penny's journal: Koko wants to have a baby.* [On-line]. Available: http://www.koko.org/world/journal.phtml?offset58.

Patterson, T. G., & Joseph, S. (2007). Person-centered personality theory: Support from self determination theory and positive psychology. *Journal of Humanistic Psychology, 47,* 117–139.

Paul, D. B., & Blumenthal, A. L. (1989). On the trail of little Albert. *The Psychological Record, 39,* 547–553.

Pearlman, C. (2002). Electroconvulsive therapy in clinical psychopharmacology. *Journal of Clinical Psychopharmacology, 22*(4), 345–346.

Pearson, N. J., Johnson, I. M., & Nahin, R. L. (2006). Insomnia, trouble sleeping, and complementary and alternative medicine: Analysis of the 2002 National Health Interview Survey data. *Archives of Internal Medicine, 166,* 1775–1782.

Pedrazzoli, M., Pontes, J. C., Peirano, P., & Tufik, S. (2007). HLA-DQB1 genotyping in a family with narcolepsy-cataplexy. *Brain Research, 1165,* 1–4.

Peleg, G., Katzir, G., Peleg, O., Kamara, M., Brodsky, L., Hel-Or, H., et al. (2006). Hereditary family signature of facial expression. *Proceedings of the National Academy of Sciences, 103,* 15921–15926.

Pellegrino, J. E., & Pellegrino, L. (2008). Fetal alcohol syndrome and related disorders. In P. J. Accardo (Ed.), *Capute and Accardo's neurodevelopmental disabilities in infancy and childhood: Vol 1: Neurodevelopmental diagnosis and treatment* (3rd ed.) (pp. 269–284). Baltimore, MD: Paul H Brookes.

Penfield, W. (1947). Some observations in the cerebral cortex of man. *Proceedings of the Royal Society, 134,* 349.

Pérez-Mata, N., & Diges, M. (2007). False recollections and the congruence of suggested information. *Memory, 15,* 701–717.

Persson, J., Lind, J., Larsson, A., Ingvar, M., Sleegers, K., Van Broeckhoven, C., Adolfsson, R., Nilsson, L-G., & Nyberg, L. (2008). Altered deactivation in individuals with genetic risk for Alzheimer's disease. *Neuropsychologia, 46,* 1679–1687.

Pettigrew, T. F. (1998). Reactions towards the new minorities of Western Europe. *Annual Review of Sociology, 24,* 77–103.

Pham, T.M., Winblad, B., Granholm, A-C., & Mohammed, A. H. (2002). Environmental influences on brain neurotrophins in rats. *Pharmacology, Biochemistry & Behavior, 73*(1), 167–175.

Phillips, S. T., & Ziller, R. C. (1997). Toward a theory and measure of the nature of nonprejudice. *Journal of Personality and Social Psychology, 72,* 420–434.

Piaget, J. (1962). *Play, dreams and imitation in Childhood.* New York: Norton.

Picard, H., Amado, I., Mouchet-Mages, S., Olié, J-P., & Krebs, M-O. (2008). The role of the cerebellum in schizophrenia: An update of clinical, cognitive, and functional evidences. *Schizophrenia Bulletin, 34,* 155–172.

Pierce, J. P. (2007). Tobacco industry marketing, population-based tobacco control, and smoking behavior. *American Journal of Preventive Medicine, 33,* 327–334.
Plomin, R. (1990). The role of inheritance in behavior. *Science, 248,* 183–188.
Plomin, R. (1999). Genetics and general cognitive ability. *Nature, 402,* C25–C29.
Plomin, R., & Crabbe, J. (2000). DNA. *Psychological Bulletin, 126,* 806–828.
Plomin, R., DeFries, J. C., & Fulker, D. W. (2007). *Nature and nurture during infancy and early childhood.* New York: Cambridge University Press.
Plummer, D. C. (2001). The quest for modern manhood: Masculine stereotypes, peer culture and the social significance of homophobia. *Journal of Adolescence, 24*(1), 15–23.
Plutchik, R. (1984). Emotions: A general psychoevolutionary theory. In K. R. Scherer, & P. Ekman (Eds.), *Approaches to emotion.* Hillsdale, NJ: Erlbaum.
Plutchik, R. (1994). *The psychology and biology of emotion.* New York: HarperCollins.
Plutchik, R. (2000). *Emotions in the practice of psychotherapy: Clinical implications of affect theories.* Washington, DC: American Psychological Association.
Pomponio, A. T. (2002). *Psychological consequences of terror.* New York: Wiley.
Ponzi, A. (2008). Dynamical model of salience gated working memory, action selection and reinforcement based on basal ganglia and dopamine feedback. *Neural Networks, 21,* 322–330.
Pope Jr., H. G., Barry, S., Bodkin, J. A., & Hudson, J. (2007). "Scientific study of the dissociative disorders": Reply. *Psychotherapy and Psychosomatics, 76,* 401–403.
Popma, A., Vermeiren, R., Geluk, C. A. M. L., Rinne, T., van den Brink, W., Knol, D. L., Jansen, L. M. C., van Engeland, H., & Doreleijers, T. A. H. (2007). Cortisol moderates the relationship between testosterone and aggression in delinquent male adolescents. *Biological Psychiatry, 61,* 405–411.
Porzelius, L. K., Dinsmore, B. D., & Staffelbach, D. (2001). Eating disorders. In M. Hersen & V. B. Van Hasselt (Eds.), *Advanced abnormal psychology* (2nd ed.). Netherlands: Klewer Academic Publishers.
Posthuma, D., de Geus, E. J. C., & Boomsma, D. I. (2001). Perceptual speed and IQ are associated through common genetic factors. *Behavior Genetics, 31*(6), 593–602.
Posthuma, D., Neale, M. C., Boomsma, D. I., & de Geus, E. J. C. (2001). Are smarter brains running faster? Heritability of alpha peak frequency, IQ, and their interrelation. *Behavior Genetics, 31*(6), 567–579.
Powell, D. H. (1998), *The nine myths of aging: Maximizing the quality of later life.* San Francisco: Freeman.
Powell-Hopson, D., & Hopson, D. S. (1988). Implications of doll color preferences among Black preschool children and White preschool children. *Journal of Black Psychology, 14,* 57–63.
Prabhu, V., Sutton, C., & Sauser, W. (2008). Creativity and certain personality traits: Understanding the mediating effect of intrinsic motivation. *Creativity Research Journal, 20,* 53–66.
Pratt, M. W., Skoe, E. E., & Arnold, M. (2004). Care reasoning development and family socialisation patterns in later adolescence: A longitudinal analysis. *International Journal of Behavioral Development, 28*(2), 139–147.
Preuss, U. W., Zetzsche, T., Jäger, M., Groll, C., Frodl, T., Bottlender, R., Leinsinger, G., Hegerl, U., Hahn, K., Möller, H. J., & Meisenzahl, E. M. (2005). Thalamic volume in first-episode and chronic schizophrenic subjects: A volumetric MRI study. *Schizophrenia Research, 73*(1), 91–101.
Prigatano, G. P., & Gray, J. A. (2008). Predictors of performance on three developmentally sensitive neuropsychological tests in children with and without traumatic brain injury. *Brain Injury, 22,* 491–500.
Primavera, L. H., & Herron, W. G. (1996). The effect of viewing television violence on aggression. *International Journal of Instructional Media, 23,* 91–104.
Pring, L., Woolf, K., & Tadic, V. (2008). Melody and pitch processing in five musical savants with congenital blindness. *Perception, 37,* 290–307.
Prkachin, K. M. (2005). Effects of deliberate control on verbal and facial expressions of pain. *Pain, 114,* 328–338.
Prkachin, K. M., & Silverman, B. E. (2002). Hostility and facial expression in young men and women: Is social regulation more important than negative affect? *Health Psychology, 21*(1), 33–39.
Prouix, M. J. (2007). Bottom-up guidance in visual search for conjunctions. *Journal of Experimental Psychology, 33,* 48–56.
Pullum, G. K. (1991). *The great Eskimo vocabulary hoax and other irreverent essays on the study of language.* Chicago: University of Chicago Press.
Qualls, S. H. (2008). Caregiver family therapy. In K. Laidlaw & B. Knight (Eds.), *Handbook of emotional disorders in later life: Assessment and treatment.* New York: Oxford University Press.
Quinn, J., Barrowclough, C., & Tarrier, N. (2003). The Family Questionnaire (FQ): A scale for measuring symptom appraisal in relatives of schizophrenic patients. *Acta Psychiatrica Scandinavica, 108*(4), 290–296.
Quintanilla, Y. T. (2007). Achievement motivation strategies: An integrative achievement motivation program with first year seminar students. *Dissertation Abstracts International Section A: Humanities and Social Sciences, 68* (6-A), 2339.
Raevuori, A., Keski-Rahkonen, A., Hoek, H. W., Sihvola, E., Rissanen, A., & Kaprio, J. (2008). Lifetime anorexia nervosa in young men in the community: Five cases and their co-twins. *International Journal of Eating Disorders, 41,* 458–463.
Raine, A., & Yang, Y. (2006). The neuroanatomical bases of psychopathy: A review of brain imaging findings. In C. J. Patrick (Ed.), *Handbook of the psychopathy* (pp. 278–295). New York: Guilford Press.
Ran, M. (2007). Experimental research on the reliability of the testimony from eyewitnesses. *Psychological Science (China), 30,* 727–730.
Randi, J. (1997). *An encyclopedia of claims, frauds, and hoaxes of the occult and supernatural: James Randi's decidedly skeptical definitions of alternate realities.* New York: St Martin's Press.
Rattaz, C., Goubet, N., & Bullinger, A. (2005). The calming effect of a familiar odor on full-term newborns. *Journal of Developmental & Behavioral Pediatrics, 26,* 86–92.
Rauer, A. J. (2007). Identifying happy, healthy marriages for men, women, and children. *Dissertation Abstracts International: Section B: The Sciences and Engineering, 67*(10-B), 6098.
Rechtschaffen, A., & Siegel, J.M. (2000). Sleep and Dreaming. In E. R. Kandel, J. H. Schwartz, & T. M. Jessel (Eds.), *Principles of Neuroscience* (4th ed., pp. 936–947). New York: McGraw-Hill.
Reeve, J. (2005). *Understanding motivation and emotion* (4th ed.). Hoboken, NJ: Wiley.
Regan, P. (1998). What if you can't get what you want? Willingness to compromise ideal mate selection standards as a function of sex, mate value, and relationship context. *Personality and Social Psychology Bulletin, 24,* 1294–1303.
Regier, D. A., Narrow, W. E., Rae, D. S., Manderscheid, R. W., Locke, B. Z., & Goodwin, F. K. (1993). The de facto US mental and addictive disorders service system. *Archives of General Psychiatry, 50,* 85–93.
Reich, D. A. (2004). What you expect is not always what you get: The roles of extremity, optimism, and pessimism in the behavioral confirmation process. *Journal of Experimental Social Psychology, 40*(2), 199–215.

Reifman, A. (2000). Revisiting the Bell Curve. *Psychology, 11,* 21–29.

Renzetti, C., Curran, D., & Kennedy-Bergen, R. (2006). *Understanding diversity.* Boston: Allyn & Bacon/Longman.

Resing, W. C., & Nijland, M. I. (2002). Worden kinderen intelligenter? Een kwart eeuw onderzoek met de Leidse Diagnostische Test [Are children becoming more intelligent? Twenty-five years' research using the Leiden Diagnostic Test.] *Kind en Adolescent, 23*(1), 42–49.

Ressler, K., & Davis, M. (2003). Genetics of childhood disorders: L. Learning and memory, part 3: Fear conditioning. *Journal of the American Academy of Child & Adolescent Psychiatry, 42*(5), 612–615.

Rest, J., Narvaez, D., Bebeau, M., & Thoma, S. (1999). A neo-Kohlbergian approach: The DIT and schema theory. *Educational Psychology Review, 11*(4), 291–324.

Reulbach, U., Bleich, S., Biermann, T., Pfahlberg, A., & Sperling, W. (2007). Late-onset schizophrenia in child survivors of the Holocaust. *Journal of Nervous and Mental Disease, 195,* 315–319.

Revonsuo, A., (2006). *Inner presence: Consciousness as a biological phenomenon.* Cambridge, MA: MIT Press.

Reynolds, M. R., Keith, T. Z., Ridley, K. P., & Patel, P. G. (2008). Sex differences in latent general and broad cognitive abilities for children and youth: Evidence from higher-order MG-MACS and MIMIC models. *Intelligence, 36,* 236–260.

Rezayof, A., Alijanpour, S., Zarrindast, M-R., & Rassouli, Y. (2008). Ethanol state-dependent memory: Involvement of dorsal hippocampal muscarinic and nicotinic receptors. *Neurobiology of Learning and Memory,* 89, 441–447.

Rhodes, G., Halberstadt, J., & Brajkovich, G. (2001). Generalization of mere exposure effects to averaged composite faces. *Social Cognition, 19*(1), 57–70.

Riebe, D., Garber, C. E., Rossi, J. S., Greaney, M. L., Nigg, C. R., Lees, F. D., Burbank, P. M., & Clark, P. G. (2005). Physical activity, physical function, and stages of change in older adults. *American Journal of Health Behavior, 29,* 70–80.

Riemann, D., & Voderholzer, U. (2003). Primary insomnia: a risk factor to develop depression? *Journal of Affective Disorders, 76*(1–3), 255–259.

Rizolatti, G., Fadiga, L., Fogassi, L. & Gallese, V. (2002). From mirror neurons to imitation: Facts and speculations. In A. N. Meltzoff & W. Prinz (Eds.), *The imitative mind: Development, evolution, and brain bases.* Cambridge, MA: Cambridge University Press.

Rizolatti, G., Fogassi, L. & Gallese, V. (2006, November). In the mind. *Scientific American,* 54–61.

Roberge, P., Marchand, A., Reinharz, D., & Savard, P. (2008). Cognitive-behavioral treatment for panic disorder with agoraphobia: A randomized, controlled trial and cost-effectiveness analysis. *Behavior Modification, 32,* 333–351.

Roberts, M. (2006, March). Idol dreams. *Schizophrenia Digest,* 30–33.

Roberts, W. W., & Nagel, J. (1996). First-order projections activated by stimulation of hypothalamic sites eliciting attack and flight in rats. *Behavioral Neuroscience, 110,* 509–527.

Robinson, F. P. (1970). *Effective study* (4th ed.). New York: Harper & Row.

Robinson, R. J. (2008). Comorbidity of alcohol abuse and depression: Exploring the self-medication hypothesis. *Dissertation Abstracts International: Section B: The Sciences and Engineering, 68*(9-B), 6332.

Rodrigues, A., Assmar, E. M., & Jablonski, B. (2005). Social-psychology and the invasion of Iraq. *Revista de Psicología Social, 20,* 387–398.

Rogers, C. R. (1961). *On becoming a person.* Boston: Houghton Mifflin.

Rogers, C. R. (1980). *A way of being.* Boston: Houghton Mifflin.

Rogosch, F. A., & Cicchetti, D. (2004). Child maltreatment and emergent personality organization: Perspectives from the fivefactor model. *Journal of Abnormal Child Psychology, 32*(2), 123–145.

Romero, S. G., McFarland, D. J., Faust, R., Farrell, L., & Cacace, A. T. (2008). Electrophysiological markers of skill-related neuroplasticity. *Biological Psychology, 78,* 221–230.

Rönnberg, J., Rudner, M., & Ingvar, M. (2004). Neural correlates of working memory for sign language. *Cognitive Brain Research, 20*(2), 165–182.

Rosch, E. (1978). Principles of organization. In E. Rosch & H. L. Lloyd (Eds.), *Cognition and categorization.* Hillsdale, NJ: Erlbaum.

Rosch, E. H. (1973). Natural categories. *Cognitive Psychology, 4,* 328–350.

Rose, C. R., & Konnerth, A. (2002). Exciting glial oscillations. *Nature Neuroscience, 4,* 773–774.

Rosenthal, H. G. (2007). The Tiger Woods analogy: The seven-minute active listening solution. In L. L. Hecker & C. F Sori (Eds.), *The therapist's notebook: More homework, handouts, and activities for use in psychotherapy (Vol. 2) (pp. 277–280). Haworth practical practice in mental health.* New York: Haworth Press.

Rosenzweig, M. R., Bennett, E. L., & Diamond, M. C. (1972). Brain changes in response to experience. *Scientific American, 226,* 22–29.

Rossato, M., Pagano, C., & Vettor, R. (2008). The cannabinoid system and male reproductive functions. *Journal of Neuroendocrinology, 20,* 90–93.

Rossignol, S., Barrière, G., Frigon, A., Barthélemy, D., Bouyer, L., Provencher, J., Leblond, H., & Bernard, G. (2008). Plasticity of locomotor sensorimotor interactions after peripheral and/or spinal lesions. *Brain Research Reviews, 57,* 228–240.

Roth, M. D., Whittaker, K., Salehi, K., Tashkin, D. P., & Baldwin, G. C. (2004). Mechanisms for impaired effector function in alveolar macrophages from marijuana and cocaine smokers. *Journal of Neuroimmunology, 147*(1–2), 82–86.

Roth, M. L., Tripp, D. A. Harrison, M. H., Sullivan, M., & Carson, P. (2007). Demographic and psychosocial predictors of acute perioperative pain for total knee arthroplasty. *Pain Research & Management, 12,* 185–194.

Rothstein, J. B., Jensen, G., & Neuringer, A. (2008). Human choice among five alternatives when reinforcers decay. *Behavioural Processes, 78,* 231–239.

Rotter, J. B. (1954). *Social learning and clinical psychology.* Englewood Cliffs, NJ: Prentice Hall.

Rotter, J. B. (1990). Internal versus external control of reinforcement: A case history of a variable. *American Psychologist, 45,* 489–493.

Rowe, R., Maughan, B., Worthman, C. M., Costello, E. J., & Angold, A. (2004). Testosterone, antisocial behavior, and social dominance in boys: Pubertal development and biosocial interaction. *Biological Psychiatry, 55*(5), 546–552.

Rozencwajg, P., Cherfi, M., Ferrandez, A. M., Lautrey, J., Lemoine, C., & Loarer, E. (2005). Age related differences in the strategies used by middle aged adults to solve a block design task. *International Journal of Aging & Human Development, 60*(2), 159–182.

Rubin, Z. (1970). Measurement of romantic love. *Journal of Personality and Social Psychology, 16,* 265–273.

Rubinstein, E. (2008). Judicial perceptions of eyewitness testimony. *Dissertation Abstracts International: Section B: The Sciences and Engineering, 68*(8-B), 5592.

Ruffolo, J. S., Phillips, K. A., Menard, W., Fay, C., & Weisberg, R. B. (2006). Comorbidity of body dysmorphic disorder and eating disorders: Severity of psychopathology and body image disturbance. *International Journal of Eating Disorders, 39,* 11–19.

Rumbaugh, D. M., von Glasersfeld, E. C., Warner, H., Pisani, P., & Gill, T. V. (1974). Lana (chimpanzee) learning language: A progress report. *Brain & Language, 1*(2), 205–212.

Rushton, J. P., Bons, T. A., & Hur, Y-M. (2008). The genetics and evolution of the general factor of personality. *Journal of Research in Personality,* 42, 1173–1185.

Russo, N. F., & Tartaro, J. (2008). Women and mental health. In F. L. Denmark & M. A. Paludi (Eds.), *Psychology of women: A handbook of issues and theories* (2nd ed.) (pp. 440–483). *Women's psychology.* Westport, CT: Praeger/Greenwood.

Ruthig, J. C., Chipperfield, J. G., Perry, R. P. Newall, N. E., & Swift, A. (2007). Comparative risk and perceived control: Implications for psychological and physical well-being among older adults. *Journal of Social Psychology, 147,* 345–369.

Rutter, M. (2007). Gene-environment interdependence. *Developmental Science, 10,* 12–18.

Rutter, V. E. (2005). The case for divorce: Under what conditions is divorce beneficial and for whom? *Dissertation Abstracts International Section A: Humanities and Social Sciences, 65*(7-A), 2784.

Ryback, D., Ikemi, A., & Miki, Y. (2001). Japanese psychology in crisis: Thinking inside the (empty) box. *Journal of Humanistic Psychology, 41*(4), 124–136.

Rymer, R. (1993). *Genie: An abused child's first flight from silence.* New York: HarperCollins.

Sabet, K. A. (2007). The (often unheard) case against marijuana leniency. In M. Earleywine (Ed), *Pot politics: Marijuana and the costs of prohibition* (pp. 325–352). New York: Oxford University Press.

Sabini, J., & Silver, M. (1993). Critical thinking and obedience to authority. In J. Chaffee (Ed.), *Critical thinking* (2nd ed., pp. 367–376). Palo Alto, CA: Houghton Mifflin.

Sacchetti, B., Sacco, T., & Strata, P. (2007). Reversible inactivation of amygdala and cerebellum but not perirhinal cortex impairs reactivated fear memories. *European Journal of Neuroscience, 25*(9), 2875–2884.

Sachdev, P., Mondraty, N., Wen, W., & Gulliford, K. (2008). Brains of anorexia nervosa patients process self-images differently from non-self-images: An fMRI study. *Neuropsychologia, 46,* 2161–2168.

Sack, R. L., Auckley, D., Auger, R. R., Carskadon, M. A., Wright, Jr., K. P., Vitiello, M. V., & Zhdanova, I. V. (2007). Circadian rhythm sleep disorders: Part I, basic principles, shift work and jet lag disorders: An American academy of sleep medicine review. *Sleep: Journal of Sleep and Sleep Disorders Research, 30,* 1460–1483.

Salovey, P., & Mayer, J. D. (1990). Emotional intelligence. *Imagination, Cognition, and Personality, 9,* 185–211.

Salovey, P., Bedell, B. T., Detweiler, J. B., & Mayer, J. D. (2000). Current directions in emotional intelligence research. In M. Lewis & J. M. Haviland (Eds.), *Handbook of emotions* (2nd ed., pp. 504–520). New York: Guilford Press.

Saltus, R. (2000, June 22). Brain cells are coaxed into repair duty. *Boston Globe,* A18.

Salvatore, P., Ghidini, S., Zita, G., De Panfilis, C., Lambertino, S., Maggini, C., & Baldessarini, R. J. (2008). Circadian activity rhythm abnormalities in ill and recovered bipolar I disorder patients. *Bipolar Disorders, 10,* 256–265.

Sampselle, C. M., Harris, V. Harlow, S. D., & Sowers, M. F. (2002). Midlife development and menopause in African American and Caucasian women. *Health Care for Women International, 23*(4), 351–363.

Sanchez Jr., J. (2006). Life satisfaction factors impacting the older Cuban-American population. *Dissertation Abstracts International Section A: Humanities and Social Sciences, 67*(5-A), 1864.

Sangha, S., McComb, C., Scheibenstock, A., Johannes, C., & Lukowiak, K. (2002). The effects of continuous versus partial reinforcement schedules on associative learning, memory, and extinction in Lymnaea stagnalis. *Journal of Experimental Biology, 205,* 1171–1178.

Sangwan, S. (2001). Ecological factors as related to I.Q. of children. *Psycho-Lingua, 31*(2), 89–92.

Sapolsky, R. (2003). Taming stress. *Scientific American,* pp. 87–95.

Sarafino, E. P. (2008). *Health psychology: Biopsychosocial interactions* (6th ed.). Hoboken, NJ: Wiley.

Sassenberg, K., Moskowitz, G. B., Jacoby, J., & Hansen, N. (2007). The carry-over effect of competition: The impact of competition on prejudice towards uninvolved outgroups. *Journal of Experimental Social Psychology, 43,* 529–538.

Satcher, N. D. (2007). Social and moral reasoning of high school athletes and non-athletes. *Dissertation Abstracts International Section 7A: Humanities and Social Sciences, 68*(3-A), 928.

Sathyaprabha, T. N., Satishchandra, P., Pradhan, C., Sinha, S., Kaveri, B., Thennarasu, K., Murthy, B. T. C., & Raju, T. R. (2008). Modulation of cardiac autonomic balance with adjuvant yoga therapy in patients with refractory epilepsy. *Epilepsy & Behavior, 12,* 245–252.

Satler, C., Garrido, L. M., Sarmiento, E. P., Leme, S., Conde, C., & Tomaz, C. (2007). Emotional arousal enhances declarative memory in patients with Alzheimer's disease. *Acta Neurologica Scandinavica, 116,* 355–360.

Sato, S. M., Schulz, K. M., Sisk, C. L., & Wood, R. I. (2008). Adolescents and androgens, receptors and rewards. *Hormones and Behavior, 53,* 647–658.

Saudino, K. J. (1997). Moving beyond the heritability question: New directions in behavioral genetic studies of personality. *Current Directions in Psychological Science, 6,* 86–90.

Savage-Rumbaugh, E. S. (1990). Language acquisition in a nonhuman species: Implications for the innateness debate. *Developmental Psychobiology, 23,* 599–620.

Savic, I., Berglund, H., & Lindström, P. (2007). Brain response to putative pheromones in homosexual men. In G. Einstein (Ed.), *Sex and the brain* (pp. 731–738). Cambridge, MA: MIT Press.

Saxe, L., & Ben-Shakhar, G. (1999). Admissibility of polygraph tests: The application of scientific standards post-Daubert. *Psychology, Public Policy, & Law, 5*(1), 203–223.

Schachter, S., & Singer, J. E. (1962). Cognitive, social, and physiological determinants of emotional state. *Psychological Review, 69,* 379–399.

Schaefer, R. T. (2008). Power and power elite. In V. Parillo (Ed.), *Encyclopedia of social problems.* Thousand Oaks, CA: Sage.

Schaie, K. W. (1994). The life course of adult intellectual development. *American Psychologist, 49,* 304–313.

Schaie, K. W. (2008). A lifespan developmental perspective of psychological ageing. In K. Laidlaw & B. Knight (Eds.), *Handbook of emotional disorders in later life: Assessment and treatment.* New York: Oxford University Press.

Schiffer, F, Zaidel, E, Bogen, J, & Chasan-Taber, S. (1998). Different psychological status in the two hemispheres of two split-brain patients. *Neuropsychiatry, Neuropsychology, & Behavioral Neurology, 11,* 151–156.

Schmahl, C. G., Vermetten, E., Elzinga, B. M., & Bremner, J. D. (2004). A positron emission tomography study of memories of childhood abuse in borderline personality disorder. *Biological Psychiatry, 55*(7), 759–765.

Schmidt, U. (2004). Undue influence of weight on self-evaluation: A population-based twin study of gender differences. *International Journal of Eating Disorders, 35*(2), 133–135.

Schopp, L. H., Good, G. E., Mazurek, M. O., Barker, K. B., & Stucky, R. C. (2007). Masculine role variables and outcomes among men with spinal cord injury. *Disability and Rehabilitation: An International, Multidisciplinary Journal, 29*, 625–633.

Schuett, S., Heywood, C. A., Kentridge, R. W., & Zihl, J. (2008). The significance of visual information processing in reading: Insights from hemianopic dyslexia. *Neuropsychologia, 46*, 2445–2462.

Schülein, J. A. (2007). Science and psychoanalysis. *Scandinavian Psychoanalytic Review, 30*, 13–21.

Schunk, D. H. (2008). Attributions as motivators of self-regulated learning. In D. H. Schunk & B. J. Zimmerman (Eds.), *Motivation and self-regulated learning: Theory, research, and applications* (pp. 245–266). Mahwah, NJ: Erlbaum.

Scribner, S. (1977). Modes of thinking and ways of speaking: Culture and logic reconsidered. In P. N. Johnson-Laird & P. C. Wason (Eds.), *Thinking: Readings in cognitive science* (pp. 324–339). New York: Cambridge University Press.

Scully, J. A., Tosi, H., & Banning, K. (2000). Life event checklists: Revisiting the Social Readjustment Rating Scale after 30 years. *Educational & Psychological Measurement, 60*(6), 864–876.

Sebre, S., Sprugevica, I., Novotni, A., Bonevski, D., Pakalniskiene, V., Popescu, D., Turchina, T., Friedrich, W., & Lewis, O. (2004). Crosscultural comparisons of child-reported emotional and physical abuse: Rates, risk factors and psychosocial symptoms. *Child Abuse & Neglect, 28*(1), 113–127.

Segal, N. L., & Bouchard, T. J. (2000). *Entwined lives: Twins and what they tell us about human behavior.* New York: Plumsock.

Segerstrale, U. (2000). *Defenders of the truth: The battle for science in the sociobiology debate and beyond.* London: Oxford University Press.

Segerstrom, S., C., & Miller, G. E. (2004). Psychological stress and the human immune system: A meta-analytic study of 30 years of inquiry. *Psychological Bulletin, 130*(4), 601–630.

Seligman, M. E. P. (1975) *Helplessness: On depression, development, and death.* San Francisco: Freeman.

Seligman, M. E. P. (1994). *What you can change and what you can't.* New York: Alfred A. Knopf.

Seligman, M. E. P. (2003). The past and future of positive psychology. In C. L. M. Keyes & J. Daidt (Eds.), *Flourishing: Positive psychology and the life well-lived.* Washington, DC: American Psychological Association.

Seligman, M. E. P. (2007). Coaching and positive psychology. *Australian Psychologist, 42*, 266–267.

Selye, H. (1936). A syndrome produced by diverse nocuous agents. *Nature, 138*, 32.

Selye, H. (1974). *Stress without distress.* New York: Harper & Row.

Senko, C., Durik, A. M., & Harackiewicz, J. M. (2008). Historical perspectives and new directions in achievement goal theory: Understanding the effects of mastery and performance-approach goals. In J. Y. Shah & W. L. Gardner (Eds.), *Handbook of motivation science* (pp. 100–113). New York: Guilford Press.

Sequeira, A., Mamdani, F., Lalovic, A., Anguelova, M., Lesage, A., Seguin, M., Chawky, N., Desautels, A., & Turecki, G. (2004). Alpha 2A adrenergic receptor gene and suicide. *Psychiatry Research, 125*(2), 87–93.

Sergerdahl, P., Fields, W., & Savage-Rumbaugh, E. S. (2006). *Kanzai's primal language: The cultural initiation of primates into language.* New York: Palgrave Macmillan.

Shahim, S. (2008). Sex differences in relational aggression in preschool children in Iran. *Psychological Reports, 102*, 235–238.

Sharps, M. J., Hess, A. B., Casner, H., Ranes, B., & Jones, J. (2007). Eyewitness memory in context: Toward a systematic understanding of eyewitness evidence. *The Forensic Examiner, 16*, 20–27.

Shaver, P. R., & Mikulincer, M. (2005). Attachment theory and research: Resurrection of the psychodynamic approach to personality. *Journal of Research in Personality, 39*(1), 22–45.

Shea, D. J. (2008). Effects of sexual abuse by Catholic priests on adults victimized as children. *Sexual Addiction & Compulsivity, 15*, 250–268.

Shea, S. C. (1988). *Psychiatric interviewieng: The art of understanding.* Philadelphia, PA: Saunders.

Shear, K., Halmi, K. A., Widiger, T. A., & Boyce, C. (2007). Sociocultural factors and gender. In W. E. Narrow, M. B. First, P. J. Sirovatka, & D. A. Regier (Eds.), *Age and gender considerations in psychiatric diagnosis: A research agenda for DSM-V* (pp. 65–79). Arlington, VA: American Psychiatric Publishing.

Sheehy, G. (1976). *Passages. Predctable crises of adult life.* New York: Dutton.

Shepperd, J., Malone, W., & Sweeny, K. (2008). Exploring causes of the self-serving bias. *Social and Personality Psychology Compass, 2*, 895–908.

Sher, K. J., Grekin, E. R., & Williams, N. A. (2005). The development of alcohol use disorders. *Annual Review of Clinical Psychology, 1*, 493–523.

Sherif, M. (1966). *In common predicament: Social psychology of intergroup conflict and cooperation.* Boston: Houghton Mifflin.

Sherif, M. (1998). Experiments in group conflict. In J. M. Jenkins, K. Oatley, & N. L. Stein (Eds.), *Human emotions: A reader* (pp. 245–252). Malden, MA: Blackwell.

Shields, C. D. (2008). The relationship between goal orientation, parenting style, and self-handicapping in adolescents. *Dissertation Abstracts International Section A: Humanities and Social Sciences, 68*(10-A), 4200.

Sias, P. M., Heath, R. G., Perry, T., Silva, D., & Fix, B. (2004). Narratives of workplace friendship deterioration. *Journal of Social & Personal Relationships, 21*(3), 321–340.

Siccoli, M. M., Rölli-Baumeler, N., Acher-mann, P., & Bassetti, C. L. (2008). Correlation between sleep and cognitive functions after hemispheric ischaemic stroke. *European Journal of Neurology, 15*, 565–572.

Siegala, M., & Varley, R. (2008). If we could talk to the animals. *Behavioral & Brain Sciences, 31*, 146–147.

Siegel, J. M. (2000, January). Narcolepsy. *Scientific American*, pp. 76–81.

Siegel, J. M. (2008). Do all animals sleep? *Trends in Neurosciences, 31*, 208–213.

Siegel, J. P. (2007). The enduring crisis of divorce for children and their parents. In N. B. Boyd (Ed.), *Play therapy with children in crisis: Individual, group, and family treatment* (3rd ed.) (pp. 133–151). New York: Guilford Press.

Siever, L. J. (2008). Neurobiology of aggression and violence. *American Journal of Psychiatry, 165*, 429–442.

Sigall, H., & Johnson, M. (2006). The relationship between facial contact with a pillow and mood. *Journal of Applied Social Psychology, 36*, 505–526.

Silverman, W. K., Pina, A. A., & Viswesvaran, C. (2008). Evidence-based psychosocial treatments for phobic and anxiety disorders in children and adolescents. *Journal of Clinical Child and Adolescent Psychology, 37*, 105–130.

Silvestri, A. J., & Root, D. H. (2008). Effects of REM deprivation and an NMDA agonist on the extinction of conditioned fear. *Physiology & Behavior, 93*, 274–281.

Sim, T., & Wong, D. (2008). Working with Chinese families in adolescent drug treatment. *Journal of Social Work Practice, 22*, 103–118.

Singelis, T. M., Triandis, H. C., Bhawuk, D. S., & Gelfand, M. (1995). Horizontal and vertical dimensions of individualism and collectivism: A theoretical and measurement refinement. *Cross-Cultural Research, 29*, 240–275.

Skelhorn, J., Griksaitis, D., & Rowe, C. (2008). Colour biases are more than a question of taste. *Animal Behaviour, 75*, 827–835.

Skinner, B. F. (1948). Superstition in the pigeon. *Journal of Experimental Psychology, 38,* 168–172.

Skinner, B. F. (1953). *Science and human behavior.* New York: Macmillan.

Skinner, B. F. (1961). Diagramming schedules of reinforcement. *Journal of the Experimental Analysis of Behavior, 1,* 67–68.

Skinner, B. F. (1992). "Superstition" in the pigeon. *Journal of Experimental Psychology: General, 121*(3), 273–274.

Skinta, M. D. (2008). The effects of bullying and internalized homophobia on psychopathological symptom severity in a community sample of gay men. *Dissertation Abstracts International: Section B: The Sciences and Engineering, 68*(7-B), 4847.

Slobogin, C. (2006). *Minding justice: Laws that deprive people with mental disability of life and liberty.* Cambridge, MA: Harvard University Press.

Slováčková, B., & Slováček, L. (2007). Moral judgement competence and moral attitudes of medical students. *Nursing Ethics, 14,* 320–328.

Smart, D. W., & Smart, J. F. (1997). DSMIV and culturally sensitive diagnosis: Some observations for counselors. *Journal of Counseling and Development, 75,* 392–398.

Smith, E. E. (1995). Concepts and categorization. In E. E. Smith & D. N. Osherson (Eds.), *Thinking: An invitation to cognitive science* (2nd ed., Vol. 3, pp. 3–33). Cambridge: MIT Press.

Smith, K. D. (2007). Spinning straw into gold: Dynamics of Rumpelstiltskin style of leadership. *Dissertation Abstracts International Section A: Humanities and Social Sciences, 68*(5-A), 1760.

Smith, M. T., Huang, M. I., & Manber, R. (2005). Cognitive behavior therapy for chronic insomnia occurring within the context of medical and psychiatric disorders. *Clinical Psychology Review, 25,* 559–592.

Smith, P., Frank, J., Bondy, S., & Mustard, C. (2008). Do changes in job control predict differences in health status? Results from a longitudinal national survey of Canadians. *Psychosomatic Medicine, 70,* 85–91.

Smithson, M., & Baker, C. (2008). Risk orientation, loving, and liking in long-term romantic relationships. *Journal of Social and Personal Relationships, 25,* 87–103.

Snarey, J. R. (1985). Cross-cultural universality of social-moral development: A critical review of Kohlbergian research. *Psychological Bulletin, 97,* 202–233.

Snarey, J. R. (1995). In communitarian voice: The sociological expansion of Kohlbergian theory, research, and practice. In W. M. Kurtines & J. L. Gerwirtz (Eds.), *Moral development: An introduction* (pp. 109–134). Boston: Allyn & Bacon.

Snyder, C. R. (2003). "Me conform? No way": Classroom demonstrations for sensitizing students to their conformity. *Teaching of Psychology, 30*(1), 59–61.

Snyder, J. S., & Alain, C. (2007). Sequential auditory sense analysis is preserved in normal aging adults. *Cerebral Cortex, 17,* 501–512.

Solan, H. A., & Mozlin, R. (2001). Children in poverty: Impact on health, visual development, and school failure. *Issues in Interdisciplinary Care, 3*(4), 271–288.

Sollod, R. N., Monte, C. F., & Wilson, J. P. (2009). *Beneath the mask: An introduction to theories of personality* (8th ed.). Hoboken, NJ: Wiley.

Sotres-Bayón, F., & Pellicer, F. (2000). The role of the dopaminergic mesolimbic system in the affective component of chronic pain. *Salud Mental, 23*(1), 23–29.

Sowell, E. R., Mattson, S. N., Kan, E., Thompson, P. M., Riley, E. P., Edward, P., & Toga, A. W. (2008). Abnormal cortical thickness and brain-behavior correlation patterns in individuals with heavy prenatal alcohol exposure. *Cerebral Cortex, 18,* 136–144.

Spearman, C. (1923). *The nature of "intelligence" and the principles of cognition.* London: Macmillan.

Spence, S. A. (2003). Detecting lies and deceit: The psychology of lying and the implications for professional practice. *Cognitive Neuropsychiatry, 8*(1), 76–77.

Sperling, G. (1960). The information available in brief visual presentations. *Psychological Monographs, 74* (Whole No. 498).

Spiegel, D., & Maldonado, J. R. (1999). Dissociative disorders. In R. E. Hales, S. C. Yudofsky, & J. C. Talbott (Eds.), *American psychiatric press textbook of psychiatry.* Washington, DC: American Psychiatric Press.

Spitz, R. A., & Wolf, K. M. (1946). The smiling response: A contribution to the ontogenesis of social relations. *Genetic Psychology Monographs, 34,* 57–123.

Spokane, A. R., Meir, E. I., & Catalano, M. (2000). Person-environment congruence and Holland's theory: A review and reconsideration. *Journal of Vocational Behavior, 57*(2), 137–187.

Sprecher, S., & Regan, P. C. (2002). Liking some things (in some people) more than others: Partner preferences in romantic relationships and friendships. *Journal of Social & Personal Relationships, 19*(4), 463–481.

Squier, L. H., & Domhoff, G. W. (1998). The presentation of dreaming and dreams in introductory psychology textbooks: A critical examination. *Dreaming, 10,* 21–26.

Stafford, J., & Lynn, S. J. (2002). Cultural scripts, memories of childhood abuse, and multiple identities: A study of role-played enactments. *International Journal of Clinical & Experimental Hypnosis, 50*(1), 67–85.

Staley, C. (2008). Facial attractiveness and the sentencing of male defendants. *Dissertation Abstracts International: Section B: The Sciences and Engineering, 68*(8-B), 5639.

Steele, C. M. (2003). Through the back door to theory. *Psychological Inquiry, 14*(3–4), 314–317.

Steele, C. M., & Aronson, J. (1995). Stereotype threat and the intellectual test performance of African Americans. *Journal of Personality and Social Psychology, 69,* 797–811.

Steele, J., James, J. B., & Barnett, R. C. (2002). Learning in a man's world: Examining the perceptions of undergraduate women in male-dominated academic areas. *Psychology of Women Quarterly, 26*(1), 46–50.

Stein, D. J., & Matsunaga, H. (2006). Specific phobia: A disorder of fear conditioning and extinction. *CNS Spectrums, 11*(4), 248–251.

Steinberg, L. (2008). A social neuroscience perspective on adolescent risk-taking. *Developmental Review,* 28, 78–106.

Steinhart, P. (1986, March). Personal boundaries. *Audubon,* pp. 8–11.

Stelmack, R. M., Knott, V., & Beauchamp, C. M. (2003). Intelligence and neural transmission time: A brain stem auditory evoked potential analysis. *Personality & Individual Differences, 34*(1), 97–107.

Sternberg, R. J. (1985). *Beyond IQ: A triarchic theory of human intelligence.* New York: Cambridge University Press.

Sternberg, R. J. (1998). Principles of teaching for successful intelligence. *Educational Psychologist, 33,* 65–72.

Sternberg, R. J. (2005). The importance of converging operations in the study of human intelligence. *Cortex. 41*(2), 243–244.

Sternberg, R. J. (2007). Developing successful intelligence in all children: A potential solution to underachievement in ethnic minority children. In M. C. Wang & R. D. Taylor (Eds.), *Closing the achievement gap.* Philadelphia: Laboratory for Student Success at Temple University.

Sternberg, R. J. (2008). The triarchic theory of human intelligence. In N. Salkind (Ed.), *Handbook of multicultural assessment* (3rd ed.). New York: Jossey-Bass.

Sternberg, R. J. (2009). *Cognitive psychology* (5th ed.). Belmont, CA: Wadsworth.

Sternberg, R. J. (Ed.). (2004). *Definitions and conceptions of giftedness.* Thousand Oaks, CA: Corwin Press.

Sternberg, R. J., & Grigorenko, E. L. (2008). Ability testing across cultures. In L. Suzuki (Ed.), *Handbook of multicultural assessment* (3rd ed.). New York: Jossey-Bass.

Sternberg, R. J., & Hedlund, J. (2002). Practical intelligence, g, and work psychology. *Human Performance, 15*(1–2), 143–160.

Sternberg, R. J., & Lubart, T. I. (1992). Buy low and sell high: An investment approach to creativity. *Current Directions in Psychological Science, 1*(1), 1–5.

Sternberg, R. J., & Lubart, T. I. (1996). Investing in creativity. *American Psychologist, 51*(7), 677–688.

Stetter, F., & Kupper, S. (2002). Autogenic training: A meta-analysis of clinical outcome studies. *Applied Psychophysiology & Biofeedback, 27*(1), 45–98.

Stiles, W. B., Barkham, M., Mellor-Clark, J., & Connell, J. (2008). Effectiveness of cognitive-behavioural, person-centered, and psychodynamic therapies in UK primary-care routine practice: Replication in a larger sample. *Psychological Medicine, 39,* 677–688.

Stompe, T. G., Ortwein-Swoboda, K., Ritter, K., & Schanda, H. (2003). Old wine in new bottles? Stability and plasticity of the contents of schizophrenic delusions. *Psychopathology, 36*(1), 6–12.

Stone, K. L., & Redline, S. (2006). Sleep-related breathing disorders in the elderly. *Sleep Medicine Clinics, 1*(2), 247–262.

Stoner, J. A. (1961). *A comparison of individual and group decisions involving risk.* Unpublished master's thesis, School of Industrial Management, MIT, Cambridge, MA.

Strack, F., Martin, L. L., & Stepper, S. (1988). Inhibiting and facilitating conditions of the human smile: A nonobstrusive test of the facial feedback hypothesis. *Journal of Personality and Social Psychology, 54,* 768–777.

Stratton, G. M. (1896). Some preliminary experiments on vision without inversion of the retinal image. *Psychological Review, 3,* 611–617.

Straub, R. O. (2007). *Health psychology: A biopsychosocial approach* (2nd ed.). New York: Worth.

Subramanian, S., & Vollmer, R. R. (2002). Sympathetic activation fenfluramine depletes brown adipose tissue norepinephrine content in rats. *Pharmacology, Biochemistry & Behavior, 73*(3), 639–646.

Sue, D.W., & Sue, D. (2008). *Counseling the culturally diverse: Theory and practice* (5th ed.). Hoboken, NJ: Wiley.

Sullivan, M. J. L. (2008). Toward a biopsychomotor conceptualization of pain: Implications for research and intervention. *Clinical Journal of Pain, 24,* 281–290.

Sullivan, M. J. L., Tripp, D. A., & Santor, D. (1998). *Gender differences in pain and pain behavior: The role of catastrophizing.* Paper presented at the annual meeting of the American Psychological Association, San Francisco.

Sullivan, T. P., & Holt, L. J. (2008). PTSD symptom clusters are differentially related to substance use among community women exposed to intimate partner violence. *Journal of Traumatic Stress, 21,* 173–180.

Suzuki, W. A., & Amaral, D. G. (2004). Functional neuroanatomy of the medial temporal lobe memory system. *Cortex, 40*(1), 220–222.

Swami, V., & Furnham, A. (2008). *The psychology of physical attraction.* New York: Routledge/Taylor & Francis Group.

Swartz, K. L. & Margolis, S. (2004). *Depression and anxiety.* Johns Hopkins White Papers. Baltimore: Johns Hopkins Medical Institutions.

Swartz, K. L. (2008). Depression and anxiety. *Johns Hopkins White Papers.* Baltimore: Johns Hopkins Medical Institutions.

Szasz, T. (1960). The myth of mental illness. *American Psychologist, 15,* 113–118.

Szasz, T. (2000). Second commentary on "Aristotle's function argument." *Philosophy, Psychiatry, & Psychology, 7,* 3–16.

Szasz, T. (2004). The psychiatric protection order for the "battered mental patient." *British Medical Journal, 327*(7429), 1449–1451.

Takano, Y., & Sogon, S. (2008). Are Japanese more collectivistic than Americans? Examining conformity in in-groups and the reference-group effect. *Journal of Cross-Cultural Psychology, 39,* 237–250.

Tal-Or, N., & Papirman, Y. (2007). The fundamental attribution error in attributing fictional figures' characteristics to the actors. *Media Psychology, 9,* 331–345.

Talarico, J. M., & Rubin, D. C. (2007). Flashbulb memories are special after all; in phenomenology, not accuracy. *Applied Cognitive Psychology, 21,* 557–578.

Talley, A. E., & Bettencourt, B. A. (2008). Evaluations and aggression directed at a gay male target: The role of threat and antigay prejudice. *Journal of Applied Social Psychology, 38,* 647–683.

Tanaka, T., Yoshida, M., Yokoo, H., Tomita, M., & Tanaka, M. (1998). Expression of aggression attenuates both stress-induced gastric ulcer formation and increases in noradrenaline release in the rat amygdala assessed intracerebral microdialysis. *Pharmacology, Biochemistry & Behavior, 59*(1), 27–31.

Tandon, R., Keshavan, M. S., & Nasrallah, H. A. (2008). Schizophrenia, "just the facts" what we know in 2008. 2. Epidemiology and etiology. *Schizophrenia Research, 102,* 1–18.

Tang, Y.-P., Wang, H., Feng, R., Kyin, M., & Tsien, J. Z. (2001). Differential effects of enrichment on learning and memory function in NR2B transgenic mice. *Neuropharmacology, 41*(6), 779–790.

Tarbox, R. S. F., Ghezzi, P. M., & Wilson, G. (2006). The effects of token reinforcement on attending in a young child with autism. *Behavioral Interventions, 21,* 155–164.

Tarter, R. E., Vanyukov, M., Kirisci, L., Reynolds, M., & Clark, D. B. (2006). Predictors of marijuana use in adolescents before and after licit drug use: Examination of the gateway hypothesis. *American Journal of Psychiatry, 163,* 2134–2140.

Task Force of the National Advisory Council on Alcohol Abuse and Alcoholism, National Institute on Alcohol Abuse and Alcoholism (2002). *A call to action: Changing the culture of drinking at U.S. colleges.* Washington, DC: National Institutes of Health. [On-line]. Available: http://www.collegedrinkingprevention.gov/.NIAAACollegeMaterials/Task-Force/TaskForce_TOC.aspx.

Tatrow, K., Blanchard, E. B., & Silverman, D. J. (2003). Posttraumatic headache: An exploratory treatment study. *Applied Psychophysiology & Biofeedback, 28*(4), 267–279.

Taub, E. (2004). Harnessing brain plasticity through behavioral techniques to produce new treatments in neurorehabilitation. *American Psychologist, 59*(8), 692–704.

Taub, E., & Morris, D. (2002). Constraint-induced movement therapy to enhance recovery after stroke. *Current Artherosclerosis Reports, 3,* 279–286.

Taub, E., Gitendra, U., & Thomas, E. (2002). New treatments in neurorehabilitation founded on basic research. *Nature Reviews Neuroscience, 3,* 228–236.

Taub, E., Ramey, S. L., DeLuca, S., & Echols, K. (2004). Efficacy of constraint-induced movement therapy for children with cerebral palsy with asymmetric motor impairment. *Pediatrics, 113,* 305–312.

Taylor, D. J., Lichstein, K. L., & Durrence, H. H. (2003). Insomnia as a risk factor. *Behavioral Sleep Medicine, 1,* 227–247.

Taylor, J. Y., Caldwell, C. H., Baser, R. E., Faison, N., & Jackson, J. S. (2007). Prevalence of eating disorders among Blacks in the national survey of American life. *International Journal of Eating Disorders, 40*(Supl), S10-S14.

Taylor, R. C., Harris, N. A., Singleton, E. G., Moolchan, E. T., & Heishman, S. J. (2000).

Tobacco craving: Intensity-related effects of imagery scripts in drug abusers. *Experimental and Clinical Psychopharmacology, 8*(1), 75–87.

Taylor, S. E., & Sherman, D. K. (2008). Self-enhancement and self-affirmation: The consequences of positive self-thoughts for motivation and health. In J. Y. Shah & W. L. Gardner (Eds.), *Handbook of motivation science* (pp. 57–70). New York: Guilford Press.

Teasdale, T. W., & Owen, D. R. (2008). Secular declines in cognitive test scores: A reversal of the Flynn Effect. *Intelligence, 36,* 121–126.

Tebartz van Elst, L, Hesslinger, B., Thiel, T., Geiger, E., Haegele, K., Lemieux, L., Lieb, K., Bohus, M., Hennig, J., & Ebert, D. (2003). Frontolimbic brain abnormalities in patients with borderline personality disorder: A volumetric magnetic resonance imaging study. *Biological Psychiatry, 54*(2), 163–171.

Tecchio, F., Zappasodi, F., Pasqualetti, P., De Gennaro, L., Pellicciari, M. C., Ercolani, M., Squitti, R., & Rossini, P. M. (2008). Age dependence of primary motor cortex plasticity induced by paired associative stimulation. *Clinical Neurophysiology, 119,* 675–682.

Tellegen, A. (1985). Structures of mood and personality and their relevance to assessing anxiety with an emphasis on self-report. In A. H. Tuma & J. D. Maser (Eds.), *Anxiety and the anxiety disorders* (pp. 681–706). Hillsdale, NJ: Erlbaum.

Temcheff, C. E., Serbin, L. A., Martin-Storey, A., Stack, D. M., Hodgins, S., Ledingham, J., & Schwartzman, A. E. (2008). Continuity and pathways from aggression in childhood to family violence in adulthood: A 30-year longitudinal study. *Journal of Family Violence, 23,* 231–242.

Terman, L. M. (1916). *The measurement of intelligence.* Boston: Houghton Mifflin.

Terman, L. M. (1954). Scientists and nonscientists in a group of 800 gifted men. *Psychological Monographs, 68*(7), 1–44.

Terrace, H. S. (1979, November). How Nim Chimpsky changed my mind. *Psychology Today,* pp. 65–76.

Terry, S. W. (2003). *Learning and memory: Basic principles, processes, and procedures* (2nd ed.). Boston: Allyn & Bacon.

Tett, R. P., & Murphy, P. J. (2002). Personality and situations in co-worker preference: Similarity and complementarity in worker compatibility. *Journal of Business & Psychology, 17*(2), 223–243.

Thomas, S. E., Randall, P. K., Book, S. W., & Randall, C. L. (2008). The complex relationship between co-occurring social anxiety and alcohol use disorders: What effect does treating social anxiety have on drinking? *Alcoholism: Clinical and Experimental Research, 32,* 77–84.

Thompson, R. F. (2005). In search of memory traces. *Annual Review of Clinical Psychology, 56,* 1–23.

Thomson, E. (2007). *Mind in life: Biology, phenomenology, and the sciences of mind.* Cambridge, MA: Belknap Press.

Thorndike, E. L. (1898). Animal intelligence. *Psychological Review Monograph, 2*(8).

Thorndike, E. L. (1911). *Animal intelligence.* New York: Macmillan.

Thornhill, R., Gangestad, S. W., Miller, R., Scheyd, G., McCollough, J. K., & Franklin, M. (2003). Major histocompatibility complex genes, symmetry, and body scent attractiveness in men and women. *Behavioral Ecology, 14*(5), 668–678.

Thurstone, L. L. (1938). *Primary mental abilities.* Chicago: University of Chicago Press.

Tirodkar, M. A., & Jain, A. (2003). Food messages on African American television shows. *American Journal of Public Health, 93*(3), 439–441.

Tolin, D. F., Robison, J. T., Gaztambide, S., Horowitz, S., & Blank, K. (2007). Ataques de nervios and psychiatric disorders in older Puerto Rican primary care patients. *Journal of Cross-Cultural Psychology, 38,* 659–669.

Tolman, E. C., & Honzik, C. H. (1930). Introduction and removal of reward and maze performance in rats. *University of California Publications in Psychology, 4,* 257–275.

Tom, S. E. (2008). Menopause and women's health transitions through mid-life. *Dissertation Abstracts International Section A: Humanities and Social Sciences, 68*(8-A), 3605.

Torges, C. M., Stewart, A. J., & Duncan, L. E. (2008). Achieving ego-integrity: Personality development in late midlife. *Journal of Research in Personality, 42,* 1004–1019.

Torrey, E. F., & Yolken, R. H. (1998). Is household crowding a risk factor for schizophrenia and bipolar disorder? *Schizophrenia Bulletin, 24,* 321–324.

Trainor, B. C., Bird, I. M., & Marler, C. A. (2004). Opposing hormonal mechanisms of aggression revealed through short-lived testosterone manipulations and multiple winning experiences. *Hormones & Behavior, 45*(2), 115–121.

Tremblay, P. F., Graham, K., & Wells, S. (2008). Severity of physical aggression reported by university students: A test of the interaction between trait aggression and alcohol consumption. *Personality and Individual Differences, 45,* 3–9.

Triandis, H. C. (2007). Culture and psychology: A history of the study of their relationship. In S. Kitayama & D. Cohen (Eds), *Handbook of cultural psychology.* New York: Guilford.

Troll, S. J., Miller, J., & Atchley, R. C. (1979). *Families in later life.* Belmont, CA: Wadsworth.

Trull, T., Sher, K. J., Minks-Brown, C., Durbin, J., & Burr, R. (2000). Borderline personality disorder and substance use disorders: A review and integration. *Clinical Psychology Review, 20,* 235–253.

Tryon, W. W. (2008). Whatever happened to symptom substitution? *Clinical Psychology Review, 28,* 963–968.

Tseng, W. (2004). Culture and psychotherapy: Asian perspectives. *Journal of Mental Health, 13*(2), 151–161.

Tsien, J. Z. (2000, April). Building a brainier mouse. *Scientific American,* pp. 62–68.

Tsuang, M. T., Stone, W. S., & Faraone, S. V. (2001). Genes, environment and schizophrenia. *British Journal of Psychiatry, Suppl 40,* s18–24.

Tulving, E., & Thompson, D. M. (1973). Encoding specificity and retrieval processes in episodic memory. *Psychological Review, 80,* 352–373.

Tümkaya, S. (2007). Burnout and humor relationship among university lecturers. *Humor: International Journal of Humor Research, 20,* 73–92.

Tversky, A., & Kahneman, D. (1974). Judgment under uncertainty: Heuristics and biases. *Science, 185,* 1124–1131.

Tversky, A., & Kahneman, D. (1993). Probabilistic reasoning. In A. I. Goldman (Ed.), *Readings in philosophy and cognitive science.* Cambridge, MA: The MIT Press.

Twenge, J. M., Baumeister, R. F., Tice, D. M., & Stucke, T. S. (2001). If you can't join them, beat them: Effects of social exclusion on aggressive behavior. *Journal of Personality and Social Psychology, 81*(6), 1058–1069.

Uhlmann, E., & Swanson, J. (2004). Exposure to violent video games increases automatic aggressiveness. *Journal of Adolescence, 27*(1), 41–52.

Ulrich, R. E., Stachnik, T. J., & Stainton, N. R. (1963). Student acceptance of generalized personality interpretations. *Psychological Reports, 13,* 831–834.

Valeo, T. & Beyerstein, L. (2008). Curses! In D. Gordon (Ed.), *Your brain on Cubs: Inside the players and fans* (pp. 59–74, 139–140). Washington, DC: Dana Press.

Van de Carr, F. R., & Lehrer, M. (1997). *While you are expecting: Your own prenatal classroom.* New York: Humanics Publishing.

van der Sluis, S., Derom, C., Thiery, E., Bartels, M., Polderman, T. J. C., Verhulst, F. C., Jacobs, N., van Gestel, S., de Geus, E. J. C., Dolan, C. V.,

Boomsma, D. I., & Posthuma, D. (2008). Sex differences on the WISC-R in Belgium and the Netherlands. *Intelligence, 36,* 48–67.

van Lier, P., Boivin, M., Dionne, G., Vitaro, F., Brendgen, M., Koot, H., Tremblay, R. E., & Pérusse, D. (2007). Kindergarten children's genetic variabilities interact with friends' aggression to promote children's own aggression. *Journal of the American Academy of Child & Adolescent Psychiatry, 46,* 1080–1087.

van Stegeren, A. H. (2008). The role of the noradrenergic system in emotional memory. *Acta Psychologica, 127,* 532–541.

Vaughn, D. (1996). *The Challenger launch decision: Risky technology, culture, and deviance at NASA.* Chicago: University of Chicago Press.

Venkatesh, V., Morris, M. G., Sykes, T. A., & Ackerman, P. L. (2004). Individual reactions to new technologies in the workplace: The role of gender as a psychological construct. *Journal of Applied Social Psychology, 34*(3), 445–467.

Vertosick, F. T. (2000). *Why we hurt: The natural history of pain.* New York: Harcourt.

Vickers, J. C., Dickson, T. C., Adlard, P. A., Saunders, H. L., King, C. E., & McCormack, G. (2000). The cause of neuronal degeneration in Alzheimer's disease. *Progress in Neurobiology, 60*(2), 139–165.

Vidmar, N. (1997). Generic prejudice and the presumption of guilt in sex abuse trials. *Law and Human Behavior, 21*(1), 5–25.

Vijgen, S. M. C., van Baal, P. H. M., Hoogenveen, R. T., de Wit, G. A., & Feenstra, T. L. (2008). Cost-effectiveness analyses of health promotion programs: A case study of smoking prevention and cessation among Dutch students. *Health Education Research, 23,* 310–318.

Vogt, D. S., Rizvi, S. L., Shipherd, J. C., & Resick, P. A. (2008). Longitudinal investigation of reciprocal relationship between stress reactions and hardiness. *Personality and Social Psychology Bulletin, 34,* 61–73.

Völker, S. (2007). Infants' vocal engagement oriented towards mother versus stranger at 3 months and avoidant attachment behavior at 12 months. *International Journal of Behavioral Development, 31,* 88–95.

von Hippel, W., Brener, L., & von Hippel, C. (2008). Implicit prejudice toward injecting drug users predicts intentions to change jobs among drug and alcohol nurses. *Psychological Science, 19,* 7–11.

Vorria, P., Vairami, M., Gialaouzidis, M., Kotroni, E., Koutra, G., Markou, N., Marti, E., & Pantoleon, I. (2007). Romantic relationships, attachment syles, and experiences of childhood. *Hellenic Journal of Psychology, 4,* 281–309.

Wachtel, P. L. (2008). *Relational theory and the practice of psychotherapy.* New York: Guilford.

Wagner, U., Christ, O., & Pettigrew, T. F. (2008). Prejudice and group-related behaviors in Germany. *Journal of Social Issues, 64,* 403–416.

Wagner, U., Hallschmid, M., Verleger, R., & Born, J. (2003). Signs of REM sleep dependent enhancement of implicit face memory: A repetition priming study. *Biological Psychology, 62*(3), 197–210.

Wagstaff, G. G., Cole, J., Wheatcroft, J., Marshall, M., & Barsby, I. (2007). A componential approach to hypnotic memory facilitation: Focused meditation, context reinstatement and eye movements. *Contemporary Hypnosis, 24,* 97–108.

Wakfield, M., Flay, B., Nichter, M., & Giovino, G. (2003). Role of the media in influencing trajectories of youth smoking. *Addiction, 98*(Suppl 1), 79–103.

Walker, E., Kestler, L., Bollini, A., & Hochman, K. M. Schizophrenia: Etiology and course. (2004). *Annual Review of Psychology, 55,* 401–430.

Walker, F. R., Hinwood, M., Masters, L., Deilenberg, R. A., & Trevor, A. (2008). Individual differences predict susceptibility to conditioned fear arising from psychosocial trauma. *Journal of Psychiatric Research, 42,* 371–383.

Waller, M. R., & McLanahan, S. S. (2005). "His" and "her" marriage expectations: Determinants and consequences. *Journal of Marriage & Family, 67*(1), 53–67.

Waller, N. G., & Ross, C. A. (1997). The prevalence and biometric structure of pathological dissociation in the general population: Taxometric and behavior genetics findings. *Journal of Abnormal Psychology, 106,* 499–510.

Wallerstein, G. (2008). *The pleasure instinct: Why we crave adventure, chocolate, pheromones, and music.* New York: Wiley.

Walter, T. & Siebert, A. (1990). *Student success* (5th ed.). Fort Worth, TX: Holt, Rinehart & Winston.

Wang, Q. (2008). Emotion knowledge and autobiographical memory across the preschool years: A cross-cultural longitudinal investigation. *Cognition, 108,* 117–135.

Wardlaw, G. M., & Hampl, J. (2007). *Perspectives in nutrition* (7th ed.). New York: McGraw-Hill.

Warr, P., Butcher, V., & Robertson, I. (2004). Activity and psychological well-being in older people. *Aging & Mental Health, 8*(2), 172–183.

Wason, P. C. (1968). Reasoning about a rule. *Quarterly Journal of Experimental Psychology, 20*(3), 273–281.

Wass, T. S. (2008). Neuroanatomical and neurobehavioral effects of heavy prenatal alcohol exposure. In J. Brick (Ed.), *Handbook of the medical consequences of alcohol and drug abuse* (2nd ed.) (pp. 177–217). *The Haworth Press series in neuropharmacology.* New York: Haworth Press/Taylor and Francis Group.

Wasserman, E., Epstein, C., & Ziemann, U. (Eds.). (2008). *Oxford handbook of transcranial stimulation.* New York: Oxford University Press.

Waters, A. J., & Gobet, F. (2008). Mental imagery and chunks: Empirical and conventional findings. *Memory & Cognition, 36,* 505–517.

Watkins, L. R., & Maier, S. F. (2000). The pain of being sick: Implications of immune-to-brain communication for understanding pain. *Annual Review of Psychology, 51,* 29–57.

Watson, J. (1913). Psychology as the behaviorist views it. *Psychological Review, 20,* 158–177.

Watson, J. B., & Rayner, R. (1920). Conditioned emotional reactions. *Journal of Experimental Psychology, 3,* 1–14.

Waye, K. P., Bengtsson, J., Rylander, R., Hucklebridge, F., Evans, P., & Clow, A. (2002). Low frequency noise enhances cortisol among noise sensitive subjects during work performance. *Life Sciences, 70*(7), 745–758.

Weaver, M. F., & Schnoll, S. H. (2008). Hallucinogens and club drugs. In M. Galanter & H. D. Kleber (Eds.), *The American Psychiatric Publishing textbook of substance abuse treatment* (4th ed.) (pp. 191–200). Arlington, VA: American Psychiatric Publishing.

Weber, G. N. (2008). Using to numb the pain: Substance use and abuse among lesbian, gay and bisexual individuals. *Journal of Mental Health Counseling, 30,* 31–48.

Wechsler, D. (1944). *The measurement of adult intelligence* (3rd ed.). Baltimore: Williams & Wilkins.

Wechsler, D. (1977). *Manual for the Wechsler Intelligence Scale for Children* (Rev.). New York: Psychological Corporation.

Wechsler, H., Lee, J. E., Kuo, M., Sei-bring, M., Nelson, T. F., & Lee, H. (2002). Trends in college binge drinking during a period of increased prevention efforts: Findings from 4 Harvard School of Public Health College Alcohol Study Surveys, 1993–2001. *Journal of American College Health, 50,* 203–217.

Wegener, D. T., Clarl, J. K., & Petty, R. E. (2006). Not all stereotyping is created equal: Differential consequences of thoughtful versus nonthoughtful stereotyping. *Journal of Personality and Social Psychology, 90,* 42–59.

Wei, R. (2007). Effects of playing violent video games on Chinese adolescents' pro-violence attitudes, attitudes toward others, and aggressive behavior. *CyberPsychology & Behavior, 10,* 371–380.

Weiner, B. (1972). *Theories of motivation.* Chicago: Rand-McNally. REFERENCES 479 767964_References 2nd.qxd 12/21/06 4:38 PM Page 479.

Weiner, B. (1982). The emotional consequences of causal attributions. In M. S. Clark & S. T. Fiske (Eds.), *Affect and cognition.* Hillsdale, NJ: Erlbaum.

Weiner, B. (2006). *Social motivation, justice, and the moral emotions: An attributional approach.* Mahwah, NJ: Erlbaum.

Weiner, G. (2008). *Handbook of personality assessment.* Hoboken, NJ: Wiley.

Weiner, M. F. (2008). Perspective on race and ethnicity in Alzheimer's disease research. *Alzheimer's & Dementia, 4,* 233–238.

Weiss, A., Bates, T. C., & Luciano, M. (2008). Happiness is a personal(ity) thing: The genetics of personality and well-being in a representative sample. *Psychological Science, 19,* 205–210.

Weitlauf, J. C., Cervone, D., Smith, R. E., & Wright, P. M. (2001). Assessing generalization in perceived self-efficacy: Multidomain and global assessments of the effects of self-defense training for women. *Personality & Social Psychology Bulletin, 27*(12), 1683–1691.

Wenger, A., & Fowers, B. J. (2008). Positive illusions in parenting: Every child is above average. *Journal of Applied Social Psychology, 38,* 611–634.

Wentz, E., Mellström, D., Gillberg, I. C., Gillberg, C., & Råstam, M. (2007). Brief report: Decreased bone mineral density as a long-term complication of teenage-onset anorexia nervosa. *European Eating Disorders Review, 15,* 290–295.

Werner, J. S., & Wooten, B. R. (1979). Human infant color vision and color perception. *Infant Behavior and Development, 2*(3), 241–273.

Werner, K. H., Roberts, N. A., Rosen, H. J., Dean, D. L., Kramer, J. H., Weiner, M. W., Miller, B. L., & Levenson, R. W. (2007). Emotional reactivity and emotion recognition in frontotemporal lobar degeneration. *Neurology, 69,* 148–155.

Werth, J. L. Jr., Wright, K. S., Archambault, R. J., & Bardash, R. (2003). When does the "duty to protect" apply with a client who has anorexia nervosa? *Counseling Psychologist, 31*(4), 427–450.

West, D. A., & Lichtenstein B. (2006). Andrea Yates and the criminalization of the filicidal maternal body. *Feminist Criminology, 1,* 173–187.

Westen, D. (1998). Unconscious thought, feeling, and motivation: The end of a century-long debate. In R. F. Bornstein & J. M. Masling (Eds.), *Empirical perspectives on the psychoanalytic unconscious.* Washington, DC: American Psychological Association.

Whitbourne, S. K. (2009). *Adult development and aging: Biopsychosocial perspectives* (3rd ed.). Hoboken, NJ: Wiley.

White, L. K., & Rogers, S. J. (1997). Strong support but uneasy relationships: Coresidence and adult children's relationships with their parents. *Journal of Marriage and the Family, 59,* 62–76.

White, T., Andreasen, N. C., & Nopoulos, P. (2002). Brain volumes and surface morphology in monozygotic twins. *Cerebral Cortex, 12*(5), 486–493.

Whorf, B. L. (1956). Science and linguistics. In J. B. Carroll (Ed.), *Language, thought and reality.* Cambridge, MA: MIT Press.

Whyte, M. K., (1992, March–April). Choosing mates—the American way. *Society,* 71–77.

Wickramasekera II, I. (2008a). Review of how can we help witnesses to remember more? It's an eyes open and shut case. *American Journal of Clinical Hypnosis, 50,* 290–291.

Wieseler-Frank, J., Maier, S. F., & Watkins, L. R. (2005). Immune-to-brain communication dynamically modulates pain: Physiological and pathological consequences. *Brain, Behavior, & Immunity, 19*(2), 104–111.

Williams, A. L., Haber, D., Weaver, G. D., & Freeman, J. L. (1998). Altruistic activity: Does it make a difference in the senior center? *Activities, Adaptation and Aging, 22*(4), 31–39.

Williams, G., Cai, X. J., Elliott, J. C., & Harrold, J. A. (2004). Anabolic neuropeptides. *Physiology & Behavior, 81*(2), 211–222.

Williams, J. E., & Best, D. L. (1990). *Sex and psyche: Gender and self viewed cross-culturally.* Newbury Park, CA: Sage.

Williams, S. S. (2001). Sexual lying among college students in close and casual relationships. *Journal of Applied Social Psychology, 31*(11), 2322–2338.

Williamson, M. (1997). Circumcision an esthesia: A study of nursing implication for dorsal penile nerve block. *Pediatric Nursing, 23,* 59–63.

Willingham, D.B. (2001). *Cognition: The thinking animal.* Upper Saddle River, NJ: Prentice Hall.

Willis, M. S., Esqueda, C. W., & Schacht, R. N. (2008). Social perceptions of individuals missing upper front teeth. *Perceptual and Motor Skills, 106,* 423–435.

Wilson, E. O. (1975). *Sociobiology: The new synthesis.* Cambridge, MA: Harvard University Press.

Wilson, E. O. (1978). *On human nature.* Cambridge, MA: Harvard University Press.

Wilson, R. S., Gilley, D. W., Bennett, D. A., Beckett L. A., & Evans, D. A. (2000). Person-specific paths of cognitive decline in Alzheimer's disease and their relation to age. *Psychology & Aging, 15*(1), 18–28.

Wilson, S., & Nutt, D. (2008). *Sleep disorders.* Oxford: Oxford University Press.

Winterich, J. A. (2003). Sex, menopause, and culture: Sexual orientation and the meaning of menopause for women's sex lives. *Gender & Society, 17*(4), 627–642.

Wise, D., & Rosqvist, J. (2006). Explanatory style and well-being. In J. C. Thomas, D. L. Segal, & M. Hersen (Eds.), *Comprehensive Handbook of Personality and Psychopathology, Vol. 1: Personality and Everyday Functioning* (pp. 285–305). Hoboken, NJ: Wiley.

Wiste, A. K., Arango, V., Ellis, S. P., Mann, J. J., & Underwood, M. D. (2008). Norepinephrine and serotonin imbalance in the locus coeruleus in bipolar disorder. *Bipolar Disorders, 10,* 349–359.

Witelson, S. F., Kigar, D. L., & Harvey, T. (1999). The exceptional brain of Albert Einstein. *The Lancet, 353,* 2149–2153.

Wixted, J. T. (2004). The psychology and neuroscience of forgetting. *Annual Review of Psychology, 55,* 235–269.

Wolf, G (2008). Want to remember everything you'll ever learn? Surrender to thhis algorithm. Wired Magazine [On-line]. Available: http://www.wired.com/print/medtech/health/magazine/16-05/FF_Wozniak

Wolpe, J. & Plaud, J. J. (1997). Pavlov's contributions to behavior therapy. *American Psychologist, 52*(9), 966–972.

Wood, V. F., & Bell, P. A. (2008). Predicting interpersonal conflict resolution styles from personality characteristics. *Personality and Individual Differences, 45,* 126–131.

Wood, W., Christensen, P. N., Hebl, M. R., & Rothgerber, H. (1997). Conformity to sex-typed norms, affect, and the self-concept. *Journal of Personality and Social Psychology, 73*(3), 523–535.

Woodruff-Pak, D. S., & Disterhoft, J. F. (2008). Where is the trace in trace conditioning? *Trends in Neurosciences, 31,* 105–112.

Woollams, A. M., Taylor, J. R., Karayanidis, F., & Henson, R. N. (2008). Event-related potentials associated with masked priming of test cues reveal multiple potential contributions to recognition memory. *Journal of Cognitive Neuroscience, 20,* 1114–1129.

Workman, L., & Reader, W. (2008). *Evolutionary psychology: An introduction.* New York: Cambridge University Press.

World Health Organization (WHO). (2007). *Cultural diversity presents special challenges for men-*

tal health. [On-line]. Available: http://www.paho.org/English/DD/PIN/pr071010.htm.

World Health Organization (WHO). (2008). *Mental health and substance abuse.* [On-line]. Available: http://www.searo.who.int/en/section1174/section1199/section1567_6741.htm

World Health Organization. (1992). *ICD-10: International statistical classification of diseases and related health problems* (10th rev. ed). Arlington, VA: American Psychiatric Publishing.

Wright, J. H., & Beck, A. T. (1999). Cognitive therapies. In R. E. Hales, S. C. Yudofsky, & J. A. Talbott (Eds.), *American Psychiatric Press textbook of psychiatry.* Washington, DC: American Psychiatric Press.

Wyman, A. J., & Vyse, S. (2008). Science versus the stars: A double-blind test of the validity of the NEO Five Factor Inventory and computer-generated astrological natal charts. *Journal of General Psychology, 135,* 287–300.

Wynne, C. D. L. (2007). What the ape said. *Ethology, 113,* 411–413.

Yacoubian, G. S., Jr., Green, M. K., & Peters, R. J. (2003). Identifying the prevalence and correlates of ecstasy and other club drug (EOCD) use among high school seniors. *Journal of Ethnicity in Substance Abuse, 2*(2), 53–66.

Yamada, H. (1997). *Different games, different rules: Why Americans and Japanese misunderstand each other.* London: Oxford University Press.

Yang, Y. K., Yao, W. J., Yeh, T. L., Lee, I. H., Chen, P. S., Lu, R. B, & Chiu, N. T. (2008). Decreased dopamine transporter availability in male smokers—A dual isotope SPECT study. *Progress in Neuro-Psychopharmacology & Biological Psychiatry, 32,* 274–279.

Yarmey, A. D. (2004). Eyewitness recall and photo identification: A field experiment. *Psychology, Crime & Law, 10*(1), 53–68.

Yee, A. H., Fairchild, H. H., Weizmann, F., & Wyatt, G. E. (1993). Addressing psychology's problem with race. *American Psychologist, 48,* 1132–1140.

Yegneswaran, B., & Shapiro, C. (2007). Do sleep deprivation and alcohol have the same effects on psychomotor performance? *Journal of Psychosomatic Research, 63,* 569–572.

Yeh, S. J., & Lo, S. K. (2004). Living alone, social support, and feeling lonely among the elderly. *Social Behavior & Personality, 32*(2), 129–138.

Yoo, S-S., Hu, P. T., Gujar, N., Jolesz, F. A., & Walker, M. P. (2007). A deficit in the ability to form new human memories without sleep. *Nature Neuroscience, 10,* 385–392.

Young, J. D., & Taylor, E. (1998). Meditation as a voluntary hypometabolic state of biological estivation. *News in Physiological Science, 13,* 149–153.

Young, T. (1802). *Color vision.* Philosophical Transactions of the Royal Society, p. 12.

Zalaquett, C. P., Fuerth, K. M., Stein, C., Ivey, A. E., & Ivey, M. B. (2008). Reframing the DSM-IV-TR from a multicultural/social justice perspective. *Journal of Counseling & Development, 86,* 364–371.

Zanetti, L., Picciotto, M. R., & Zoli, M. (2007). Differential effects of nicotinic antagonists perfused into the nucleus accumbens or the ventral tegmental area on cocaine-induced dopamine release in the nucleus accumbens of mice. *Psychopharmacology, 190,* 189–199.

Zarrindast, M-R., Fazli-Tabaei, S., Khalilzadeh, A., Farahmanfar, M., & Yahyavi, S-Y. (2005). Cross state-dependent retrieval between histamine and lithium. *Physiology & Behavior, 86,* 154–163.

Zarrindast, M-R., Shendy, M. M., & Ahmadi, S. (2007). Nitric oxide modulates states dependency induced by lithum in an inhibitory avoidance task in mice. *Behavioural Pharmacology, 18,* 289–295.

Zeanah, C. H. (2000). Disturbances of attachment in young children adopted from institutions. *Journal of Developmental & Behavioral Pediatrics, 21*(3), 230–236.

Zhaoping, L., & Guyader, N. (2007). Interference with bottom-up feature detection by higher-level object recognition. *Current biology, 17,* 26–31.

Ziegert, Jonathan C., & Hanges, Paul J. (2005). Employment Discrimination: The Role of Implicit Attitudes, Motivation, and a Climate for Racial Bias. *Journal of Applied Psychology, 90*(3), 553–562.

Zillmer, E. A., Spiers, M. V., & Culbertson, W. (2008). *Principles of neuropsychology* (2nd ed.). Belmont, CA: Cengage.

Zimbardo, P. (2007). *The Lucifer effect: Understanding how good people turn evil.* New York: Random House.

Zimbardo, P. G. (1993). Stanford prison experiment: A 20-year retrospective. *Invited presentation at the meeting of the Western Psychological Association,* Phoenix, AZ.

Zimbardo, P. G. (2004). A situationist perspective on the psychology of evil: Understanding how good people are transformed into perpetrators. In A. G. Miller (Ed.), *The social psychology of good and evil* (pp. 21–50). New York: Guilford Press.

Zimbardo, P. G., Ebbeson, E. B., & Maslach, C. (1977). *Influencing attitudes and changing behavior.* Reading, MA: Addison-Wesley.

Zucker, K. J. (2008). Special issue: Biological research on sex-dimorphic behavior and sexual orientation. *Archives of Sexual Behavior, 37,* 1.

Zuckerman, L., &, Weiner, I. (2005). Maternal immune activation leads to behavioral and pharmacological changes in the adult offspring. *Journal of Psychiatric Research, 39*(3), 311–323.

Zuckerman, M. (1978, February). The search for high sensation. *Psychology Today,* 38–46.

Zuckerman, M. (1979). *Sensation seeking: Beyond the optimal level of arousal.* Hillsdale, NJ: Erlbaum.

Zuckerman, M. (1994). *Behavioral expressions and biosocial bases of sensation seeking.* New York: Cambridge University Press.

Zuckerman, M. (2004). The shaping of personality: Genes, environments, and chance encounters. *Journal of Personality Assessment, 82*(1), 11–22.

Zuckerman, M. (2008). Rose is a rose is a rose: Content and construct validity. *Personality and Individual Differences, 45,* 110–112.

Zweig, J. (2007). *Your money and your brain: How the new science of neuroeconomics can help make you rich.* New York: Simon & Schuster.

PHOTO CREDITS

Chapter 1

Pages 2–3: (inset top) Alaska Stock Images/NGS Image Sales, (inset bottom) Alaska Stock Images/NGS Image Sales, Katherine Feng/NGS Image Sales; Page 3: (top) SUPERSTOCK, (top center) Bettmann/Corbis Images, (center) Photodisc/Getty Images, (bottom center) Jeffrey Greenberg/Photo Researchers, (bottom) Fuse/Getty Images, Inc.; Page 5: (top) Henry Groskinsky/Time Life Pictures/Getty Images, (bottom) Stuart Pearce/Age Fotostock; Page 7: Bettmann/Corbis Images; Page 8: Nina Leen/Time Life Pictures/Getty Images; Page 10: (bottom) Tim Mantoani/Masterfile, (top) Tim Mantoani/Masterfile; Page 11: (bottom left) Psychology Archives—The University of Akron, (top left) Deborah Feingold/Archive Photos/Getty Images; Page 12: Cary Wolinsky/NGS Image Sales; Page 14: Photodisc/Getty Images; Page 18: (bottom left) © The New Yorker Collection 2004 David Sipress from cartoonbank.com. All Rights Reserved., (bottom right) Jeffrey Greenberg/Photo Researchers; Page 19: (left) Jeffrey Greenberg/Photo Researchers, Inc. From Damasio H., Grabowski T., Frank R., Galaburda A.M., Damasio A.R.: The return of Phineas Gage: Clues about the brain from a famous patient. Science, 264:1102–1105, 1994, (right) Department of Neurology & Image Analysis Facility, University; Page 20: Science Pictures Ltd./Photo Researchers; Page 21: (left) Chris Carroll/Corbis Images, (right) Digital Vision/Getty Images; Page 22: (top) Science Pictures Ltd./Photo Researchers, (top center) Photo Researchers, Inc., (center) iStockphoto, (bottom center) Gary D. Landsman/©Corbis, (bottom) ANS; Page 23: (top) Mehau Kulyk/Photo Researchers, (top center) N.I.H./Photo Researchers, (center) Scott Camazine/Photo Researchers, (bottom center) Science Photo Library/Photo Researchers, Inc., (bottom) ©AP/Wide World Photos; Page 25: Gordon Gahan/NGS Image Sales; Page 26: Fuse/Getty Images, Inc.; Page 27: (top) Gordon Gahan/NGS Image Sales; Page 27: Stuart Pearce/Age Fotostock, Nina Leen/Time Life Pictures/Getty Images; Page 28: (top) Cary Wolinsky/NGS Image Sales, (bottom right) Jeffrey Greenberg/Photo Researchers; Page 29: Fuse/Getty Images, Inc.; Page 30: Jonathan Selig/Getty Images; Page 31: (top right) Jeffrey Greenberg/Photo Researchers, (bottom right) Gordon Gahan/NGS Image Sales.

Chapter 2

Pages 32–33: AFP PHOTO/FILES/Paolo Cocco/NewsCom; Page 33: (top) Ira Block/NGS Image Sales, (top center) age fotostock/SUPERSTOCK, (bottom center) Courtesy Taub Therapy Clinc/UAB Media Relations®, (bottom) Associated Press; Page 35: W.E. Garrett/NG Images, (center) Ira Block/NGS Image Sales, (bottom) Tony Freeman/PhotoEdit; Page 36: David Young-Wolff/Stone/Getty Images, Inc; Page 37: Courtesy E.R. Lewis, Berkeley; Page 38: (right) Courtesy E.R. Lewis, Berkeley; Page 40: (center) Courtesy E.R. Lewis, Berkeley; Page 41: age fotostock/SUPERSTOCK; Page 45: Courtesy Taub Therapy Clinc/UAB Media Relations®; Page 48: (left) Mike Powell/The Image Bank/Getty Images, Inc, (right) ALASKA STOCK IMAGES/NGS Image Sales; Page 49: (top) Rob McEwan/Everett Collection, Inc., (bottom) Rick Rusing/Stone/Getty Images; Page 50: Getty Images News and Sport Services; Page 52: (center right) ALASKA STOCK IMAGES/NGS Image Sales, (right) Adrian Weinbrecht/Stone/Getty Images, (center left) Corbis Images, (left) Justin Guariglia/NGS Image Sales; Page 54: (left) Associated Press, (right) Associated Press; Page 59: Courtesy E.R. Lewis, Berkeley; Page 60: Courtesy Karen Huffman.

Chapter 3

Pages 62–63: Ira Block/NGS Image Sales; Page 63: (top) UpperCut Images/SUPERSTOCK, (top center) Digital Vision/SUPERSTOCK, (bottom center) Will & Deni McIntyre/Photo Researchers, Inc., (bottom) ©AP/Wide World Photos; Page 65: UpperCut Images/SUPERSTOCK; Page 66: Photofest Inc.; Page 68: Art Resource; Page 69: Eye of Science/Photo Researchers, Inc.; Page 70: Stuart Hughes/©Corbis; Page 71: Digital Vision/SUPERSTOCK; Page 72: Joel Sartore/NGS Image Sales; Page 74: Comstock/Superstock; Page 75: (top) Michael Siluk/The Image Works, (bottom) Michael Siluk/Index Stock; Page 76: Simon McComb Photography/Photonica/Getty Images; Page 77: Dennis MacDonald/PhotoEdit; Page 78: Will & Deni McIntyre/Photo Researchers, Inc.; Page 79: (top) ©AP/Wide World Photos, (bottom) ©AP/Wide World Photos; Page 80: (top) OJO Images/SUPERSTOCK, (top center) © Remi Ochlik/IP3/MAXPPP/ NewsCom, (bottom center) AFP PHOTO/FILES/Paolo Cocco/NewsCom, (bottom) NewsCom; Page 81: (top) NewsCom, (top center) age fotostock/SUPERSTOCK, (bottom center) Dynamic Graphics Value/SUPERSTOCK, (bottom) Corbis/SUPERSTOCK; Page 82: (left) Digital Vision/SUPERSTOCK, (right) Will & Deni McIntyre/Photo Researchers, Inc.; Page 83: ©AP/Wide World Photos; Page 84: Michael Newman/PhotoEdit; Page 85: (left) Digital Vision/SUPERSTOCK, (right) Will & Deni McIntyre/Photo Researchers, Inc.

Chapter 4

Pages 86–87: SUPERSTOCK; Page 87: (center) Corbis/SUPERSTOCK, (top) Chris Johns/NGS Image Sales, (top center) Rich Reid/NG Images, (bottom center) John Dominis/Time Life Pictures/Getty Images, (bottom) Medford Taylor/NGS Image Sales; Page 89: (top left) Chris Johns/NGS Image Sales, (bottom) Phanie/Photo Researchers, Inc.; Page 90: Terje Rakke/Riser/Getty Images, Inc; Page 91: Wolcott Henry/NGS Image Sales; Page 95: ThinkStock/SUPERSTOCK; Page 98: (left) Roy Toft/NG Image Collection, (center) John Dominis/Time Life Pictures/Getty Images, Bob Rosato/Sports Illustrated/Getty Images; Page 99: Gala/SUPERSTOCK; Page 100: (top left) Ryan McVay/Photodisc Green/Getty, (top) Medford Taylor/NGS Image Sales, (bottom) Cleo Photo/Alamy; Page 102: (top) Rykoff Collection/Corbis, (bottom) ©Cordon Art b.v.; Page 103: (top) Thomas Barwick/Riser/Getty Images, Inc., (bottom right) Mark Raycroft/Minden Pictures/Getty Images; Page 105: Ed George/NG Image Collection; Page 107: Associated Press; Page 109: (top left) Phanie/Photo Researchers, Inc., (bottom right) ThinkStock/SUPERSTOCK; Page 110: (top) John Dominis/Time Life Pictures/Getty Images; Page 111: Lindsey Hebberd/Corbis; Page 113: (top left) John Dominis/Time Life Pictures/Getty Images, (top) Bob Rosato/Sports Illustrated/Getty Images, (bottom left) Rykoff Collection/Corbis, (center right) Associated Press.

Chapter 5

Pages 114–115: Creatas/SUPERSTOCK; Page 115: (top) David Evans/NGS Image Sales, (center) Jim Varney/Science Photo Library/Photo Researchers, Inc., (bottom) Justin Guariglia®/NGS Image Sales; Page 117: (top) David Evans/NGS Image Sales, (just below top, left) Walter Hodges/Stone/Getty Images, (top center) Joel Sartore/NGS Image Sales, (bottom right) Photodisc/Getty Images, (bottom right) Maria Stenzel/NGS Image Sales; Page 120: (top) Hank Morgan/Photo Researchers; Page 123: Shinya Sasaki/Neovision/amana/Getty Images; Page 124: (center) Courtesy Stephen David Smith, D.M.D., (bottom) Photo Researchers, Inc.; Page 125: Urban Zone/Alamy; Page 126: Todd Gipstein/NGS Image Sales; Page 128: (top) Richard Nowitz/NGS Image Sales, (top center) Taylor S. Kennedy/NGS Image Sales, (center) Sam Abell/NGS Image Sales, (bottom center) Uwe Schmid/OKAPIA/Photo Researchers, (bottom) Joel Sartore/NGS

Image Sales; Page 129: Zefa/SUPERSTOCK; Page 130: (left) Uwe Schmid/OKAPIA/Photo Researchers, (right) Jim Varney/Science Photo Library/Photo Researchers, Inc.; Page 132: (top left) Courtesy DEA, (center left) Sinclair Stammers/Photo Researchers, Inc., (bottom center) AFP/Getty Images, (bottom right) Matt Cardy/Getty Images, (top) Evan Roberts/Aurora Photos, (center right) Tom & DeeAnn McCarthy/Corbis, (bottom left) Richard Nowitz/NGS Image Sales; Page 133: (left) Emme Lee/Photographer's Choice RF/Getty Images, (right) Terry Vine/Blend/Getty Images; Page 134: Justin Guariglia®/NGS Image Sales; Page 135: (left) Bill Bridges/Getty Images, (right) Michael Newman/PhotoEdit; Page 137: Urban Zone/Alamy; Page 138: (top left) Uwe Schmid/OKAPIA/Photo Researchers, (top) Jim Varney/Science Photo Library/Photo Researchers, Inc., (center top left) Richard Nowitz/NGS Image Sales, (center top right) Taylor S. Kennedy/NGS Image Sales, (center bottom right) Joel Sartore/NGS Image Sales, (bottom left) Emme Lee/Photographer's Choice RF/Getty Images, (bottom right) Terry Vine/Blend/Getty Images; Page 139: (bottom left) Dorling Kindersley/Getty Images, Inc., (bottom right) Neo Vision/Getty Images, Inc.; Page 141: (top) Taylor S. Kennedy/NGS Image Sales, (top left) Richard Nowitz/NGS Image Sales, (bottom right) Uwe Schmid/OKAPIA/Photo Researchers, (bottom left) Joel Sartore/NGS Image Sales.

Chapter 6

Pages 142–143: Associated Press, (inset) Randy Olsen/NG Image Collection; Page 143: (top) Courtesy Benjamin Harris, (top center) Kaku Kurita/Getty Images, Inc., (center) Courtesy Albert Bandura, (bottom center) PhotoDisc Green/Getty Images, (bottom) Joe Raedle/Getty Images News and Sport Services; Page 146: Courtesy Benjamin Harris; Page 147: (top right) ©The New Yorker Collection 1998 Sam Gross from the cartoonbank.com. All Rights Reserved; Page 150: Blend Images/SUPERSTOCK, ©AP/Wide World Photos; Page 152: (top) Gerd Ludwig/NGS Image Sales, (bottom) age fotostock/SUPERSTOCK; Page 153: Kaku Kurita/Getty Images, Inc., (bottom) DAJ/Getty Images; Page 155: Joel Sartore/NGS Image Sales; Page 156: (right) Courtesy American Philosophical Society; Page 157: age fotostock/SUPERSTOCK; Page 158: Courtesy Albert Bandura; Page 160: PhotoDisc Green/Getty Images, (right) Jodi Cobb/NG Image Collection; Page 162: ©Wildlife/Alamy; Page 163: Associated Press; Page 164: Jodi Cobb/NGS Image Sales; Page 165: (top) Index Source Limited/Index Stock, (left) Robert Goldstein/Science Photo Library/Photo Researchers, Inc., (bottom) Associated Press; Page 166: (left) Associated Press, (right) Joe Raedle/Getty Images News and Sport Services; Page 167: (top) Courtesy Benjamin Harris, (bottom left) Blend Images/SUPERSTOCK; Page 168: (right) Courtesy American Philosophical Society, (center left) ©Wildlife/Alamy, (right) Robert Goldstein/Science Photo Library/Photo Researchers, Inc.; Page 169: Chris Fitzgerald/CandidatePhotos/NewsCom; Page 170: Courtesy Benjamin Harris; Page 171: (left) Courtesy American Philosophical Society, (top right) Robert Goldstein/Science Photo Library/Photo Researchers, Inc., (bottom right) Joe Raedle/Getty Images News and Sport Services.

Chapter 7

Pages 172–173: Peter Leyden/iStockphoto; Page 173: (top) Dean Conger/NGS Image Sales, (top center) Dr. Bill Rudman/Australian Museum, (bottom) Tom Hussey/The Image Bank/Getty; Page 176: (left) Stephen St. John/NGS Image Sales, (right) Blue Jean Images/Getty Images; Page 177: Dean Conger/NGS Image Sales; Page 181: (top) Ted Tamburo/NG Image Collection, (center) Antonio M. Rosario/Photographers Choice/Getty Images; Page 183: Dr. Bill Rudman/Australian Museum; Page 185: Science Source/Photo Researchers, Science Source/Photo Researchers; Page 188: (left) ©AP/Wide World Photos, (right) ©AP/Wide World Photos; Page 189: Nat Farbman/Time & Life Pictures/Getty Images; Page 191: Tom Hussey/The Image Bank/Getty; Page 192: Photo by Jean Greenwald. Courtesy Elizabeth Loftus; Page 194: (top) Stacy Gold/NG Image Collection, (center left) Science Source/Photo Researchers, (center right) Science Source/Photo Researchers; Page 195: Nat Farbman/Time & Life/Getty Images; Page 196: Associated Press.

Chapter 8

Pages 198–199: Associated Press; Page 199: (top) Blend Images/SUPERSTOCK, (top center) Michael Nichols/NGS Image Sales, (bottom center) Associated Press, (bottom) REUTERS/Larry Downing/Landov; Page 200: Blend Images/SUPERSTOCK; Page 201: (left) Stephen St. John/NG Image Collection, (right) Gordon Wiltsie/NG Image Collection; Page 203: Hubcapcreatures/Photo Researchers; Page 205: Courtesy U. S. Department of the Interior, National Parks Service, Edison National Historic Site, WENN/NewsCom; Page 207: (right) Stacy Gold/NGS Image Sales, (left) David Young-Wolff/PhotoEdit; Page 208: (bottom) Thinkstock/SUPERSTOCK; Page 209: (top) iStockphoto, (center) iStockphoto, (bottom) iStockphoto; Page 210: (top) Courtesy Georgia State University Language Research Center, (bottom) Michael Nichols/NGS Image Sales; Page 212: (left) Peter Kramer/Getty Images News and Sport Services, (center) Harry How/Getty Images News and Sport Services, (right) Associated Press; Page 215: Harry How/Getty Images News and Sport Services; Page 216: Timothy Fadek/©Polaris Images; Page 217: Courtesy Richard J. Haier, University of California-Irvine; Page 218: PhotoDisc, Inc./Getty Images; Page 219: REUTERS/Larry Downing/Landov; Page 221: (top) Blend Images/SUPERSTOCK, (center) David Young-Wolff/PhotoEdit; Page 222: Timothy Fadek/©Polaris Images; Page 223: T.K. Wanstal/The Image Works; Page 225: REUTERS/Larry Downing/Landov.

Chapter 9

Pages 226–227: Alexandra Day/©Corbis-Bettmann; Page 227: (top) Paul Costello/Stone/Getty Images, (center) Tyrone Turner/NGS Image Sales, (bottom) Ellen Senisi/The Image Works; Page 229: Patricia McDonough/Photonica/Getty Images; Page 232: (left) Francis Leroy, Biocosmos/Photo Researchers, (right) David M. Phillips/Photo Researchers, Inc.; Page 233: (right) Biophoto Associates/Photo Researchers, (left) Petit Format/Nestle/Photo Researchers; Page 234: (left) Streissguth, A.P., & Little, R.E. (1994). Unit 5: Alcohol, Pregnancy, and the Fetal Alcohol Syndrome: Second Edition of the Project Cork Institute Medical School Cirriculum (slide lecture series) on Biomedical Education: Alcohol Use and Its Medical Consequences, produced by Dartmouth Medical School, (right) David H. Wells/Corbis; Page 236: Victor Englebert/Photo Researchers, Inc.; Page 237: Randy Olson/NG Image Collection; Page 238: (left) Tyrone Turner/NGS Image Sales, (right) W. Robert Moore/NGS Image Sales; Page 240: Willie B. Thomas/iStockphoto; Page 241: MPI/Getty Images; Page 244: (left) Doug Goodman//Photo Researchers, (right) Doug Goodman//Photo Researchers, Inc.; Page 245: Ellen B. Senisi/The Image Works, Ellen B. Senisi/The Image Works, Ellen B. Senisi/The Image Works; Page 247: Gordon Wilsie/NGS Image Sales; Page 248: A.N. Meltzoff & M.K. Moore, Imitation of facial and manual gestures by human neonates. Science, 1977, 198, 75–78; Page 249: Ellen Senisi/The Image Works, (bottom) Paul Costello/Stone/Getty Images; Page 250: (top) Victor Englebert/Photo Researchers, Inc., (bottom) Gordon Wilsie/NGS Image Sales;

Page 251: George F. Mobley/NG Image Collection; Page 253: (left) Doug Goodman//Photo Researchers, (right) Doug Goodman//Photo Researchers, Inc.

Chapter 10

Pages 254–255: image100/SUPERSTOCK; Page 255: (top) Nina Leen/Getty Images/Time Life Pictures, (center) Ranald Mackechnie/Stone/Getty Images, (bottom) Blend Images/SuperStock, Inc.; Page 256: (bottom) Nina Leen/Getty Images/Time Life Pictures; Page 257: (top) Nina Leen/Time Life Pictures/Getty Images, (bottom left) James L. Stanfield/NG Image Collection, (bottom right) Jodi Cobb/NGS Image Sales; Page 258: (left) Image Source/Getty Images, (right) Banana Stock/AgeFotostock; Page 259: (right) Andrew O Toole/Getty Images, (left) Rich Reid/NGS Image Sales; Page 261: Associated Press; Page 263: © The New Yorker Collection 2002 Alex Gregory from cartoonbank .com. All Rights Reserved; Page 264: (left) Gordon Wiltsie/NGS Image Sales, (center left) Rich Reid/NG Image Collection, (center right) Dynamic Graphics, Inc./Creatas, (right) PhotoDisc/Getty Images, Inc.; Page 265: (left) Joel Sartore/NGS Image Sales, (center left) Scott Barrow/SUPERSTOCK, (center right) Kate Thompson/NG Image Collection, (right) Pete Oxford/NGS Image Sales, (bottom left) James L. Stanfield/NG Image Collection; Page 267: Jon Feingersh/Iconica/Getty Images; Page 268: Associated Press; Page 269: (top) Comstock/SUPERSTOCK, (bottom) © Romell/Masterfile; Page 270: Ranald Mackechnie/Stone/Getty Images; Page 271: (top) Associated Press, (bottom) Michael Nichols/NGS Image Sales; Page 272: Paramount Pictures/Hulton Archive/Getty Images; Page 273: Annie Griffits Belt/NGS Image Sales; Page 274: (left) Zuma/NewsCom, (center) Richard Nowitz/NG Image Collection, (right) Jennifer Mitchell/Splash News/NewsCom; Page 275: Blend Images/SuperStock, Inc.; Page 277: (bottom) Susan Van Etten/PhotoEdit, (top) Associated Press; Page 278: (top) Banana Stock/AgeFotostock, (center) © Romell/Masterfile, (bottom) Annie Griffits Belt/NGS Image Sales; Page 279: David W. Hamilton/Getty Images; Page 280: (top left) Gordon Wiltsie/NGS Image Sales, (top) Rich Reid/NGS Image Sales, (bottom left) Dynamic Graphics, Inc./Creatas, (bottom right) PhotoDisc/Getty Images, Inc., (top left) Joel Sartore/NGS Image Sales, (top) Scott Barrow/SUPERSTOCK, (bottom left) Kate Thompson/NG Image Collection, (bottom right) Pete Oxford/NGS Image Sales; Page 281: Susan Van Etten/PhotoEdit.

Chapter 11

Pages 282–283: ©PHOTOPQR/Nice Matin/Eric Duliere/NewsCom; Page 283: (top) Paul Sutherland/NG Image Collection, (center) Don Smetzer/PhotoEdit; Page 284: (left) Paul Sutherland/NG Image Collection, (right) Associated Press; Page 285: Randy Olson/NGS Image Sales; Page 286: Cameron Lawson/NGS Image Sales; Page 287: Don Smetzer/PhotoEdit; Page 288: James A. Sugar/NG Image Collection; Page 290: Philip Teitelbaum; Page 291: Christian Thomas/Getty Images; Page 292: (left) Tony Freeman/PhotoEdit, (right) Associated Press; Page 293: Michael Newman/PhotoEdit; Page 295: Associated Press; Page 297: (top) William Albert Allard/NGS Image Sales, (bottom) Jeffery Meyers/Corbis Images; Page 299: (top) Upper Cut Images/SuperStock, Inc., (left) Courtesy Karen Huffman, (right) Courtesy Karen Huffman; Page 300: Digital Vision/Getty Images; Page 301: Mike Parry/NGS Image Sales; Page 302: Mark Owens/John Wiley & Sons, Inc., Mark Owens/John Wiley & Sons, Inc.; Page 305: (left) Jodi Cobb/NG Image Collection, (right) Joel Sartore/NGS Image Sales, Joel Sartore/NG Image Collection; Page 306: Jean-Leo Dugast/Panos Pictures, NG Maps; Page 307: Mark Henley/Panos Pictures, Giacomo Pirozzi/Panos Pictures; Page 308: Michael Abramson; Page 309: (top) Courtesy of Daniel Langleben, MD and Kosha Ruparel, MS, University of Pennsylvania, (bottom) Cameron Lawson/NGS Image Sales; Page 310: (top) Michael Newman/PhotoEdit, (bottom) Randy Olson/NGS Image Sales; Page 311: Tetra Images/SuperStock, Inc.; Page 313: Courtesy Karen Huffman.

Chapter 12

Pages 314–315: Colin Hawkins/The Image Bank/Getty Images; Page 315: (top) MASSIVE/Stone/Getty Images, (top center) Sam Bell/NGS Image Sales, (center) PhotoAlto/Getty Images, (bottom center) Taylor S. Kennedy/NGS Image Sales, (bottom) © Index Stock Imagery; Page 317: (top left) MASSIVE/Stone/Getty Images, (top) Jen Siska/Photonica/Getty Images, Inc., (bottom left) ThinkStock/SUPERSTOCK, (bottom right) © Larry Kolvoord/The Image Works; Page 318: Purestock/Superstock; Page 319: (left) Associated Press, (right) Associated Press; Page 321: © The New Yorker Collection 1979 Dana Fradon from cartoonbank.com. All Rights Reserved; Page 322: Karen Kasmauski/NGS Image Sales; Page 324: Sam Bell/NGS Image Sales; Page 325: (top left) Roger Wood/Corbis, (top) Charles & Josette Lenars/Corbis, (bottom) Corbis Images; Page 326: PEANUTS reprinted by permission of United Features Syndicate, Inc.; Page 327: © Masterfile; Page 328: PhotoAlto/Getty Images; Page 329: David Pluth/NG Image Collection; Page 331: © The New Yorker Collection 1997 Mike Twohy from cartoonbank.com. All Rights Reserved; Page 333: Taylor S. Kennedy/NGS Image Sales; Page 334: Joel Sartore/NGS Image Sales; Page 336: (left) © Index Stock Imagery, (right) Courtesy Harvard University Press; Page 337: (top) © Larry Kolvoord/The Image Works, (bottom) Karen Kasmauski/NGS Image Sales; Page 338: (left) © Masterfile, (top) Taylor S. Kennedy/NGS Image Sales, (bottom right) Joel Sartore/NGS Image Sales; Page 339: William Thomas Gain/Getty Images; Page 341: © Index Stock Imagery.

Chapter 13

Pages 342–343: age fotostock/SUPERSTOCK; Page 343: (top) Digital Vision/SuperStock, Inc., (second from the top) Mark Clarke/Photo Researchers, (top center) Big Cheese Photo/SuperStock, Inc., (bottom center) Frazier Harrison/Getty Images, Inc., (second from bottom) Frank Siteman/Index Stock, (bottom) David Alan Harvey/NG Image Collection; Page 345: (top left) zefa/Corbis Images, (top) Digital Vision/SuperStock, Inc., (bottom left) ThinkStock, LLC/Index Stock, (bottom right) Digital Vision/Getty Images; Page 346: Lego/Getty Images; Page 347: (top) Associated Press; Page 349: (top) Corbis Images, (top center) Bobby Model/NGS Image Sales, (bottom center) Blair Seitz/Photo Researchers, David Young-Wolff/PhotoEdit; Page 351: Mark Clarke/Photo Researchers; Page 352: Digital Vision/SUPERSTOCK; Page 355: Big Cheese Photo/SuperStock, Inc.; Page 357: (top) Frazier Harrison/Getty Images, (bottom) Associated Press; Page 359: Photo Researchers; Page 360: ©Dreamworks/Photofest; Page 361: Frank Siteman/Index Stock; Page 362: Associated Press; Page 363: Associated Press; Page 365: Pixland/Index Stock; Page 366: (left) David Alan Harvey/NG Image Collection, (right) Benelux/Zefa/Corbis; Page 368: (top) Lego/Getty Images; Page 371: Jodi Cobb/NG Image Collection; Page 368: (bottom) Mark Clarke/Photo Researchers; Page 369: (top) Britt Erlanson/Getty Images, (center) Photo Researchers, (bottom) Frank Siteman/Index Stock; Page 370: Pixland/Index Stock.

Chapter 14

Pages 374–375: Mauritius//SUPERSTOCK; Page 375: (top) Don Hammond/AgeFotostock, (top center) Kevin Curtis/Photo Researchers, (bottom center) Will McIntyre/Photo Researchers, (bottom) Zia

Soleil/Iconica/Getty Images, Inc.; Page 378: Stockbyte/SUPERSTOCK; Page 381: ©John Wiley & Sons; Page 382: (top) Comstock/SUPERSTOCK, (bottom) Don Hammond/AgeFotostock; Page 383: David Young Wolff/PhotoEdit; Page 384: PhotoDisc/Getty Images, Inc.; Page 385: (top) Sidney Harris, (bottom) Kevin Curtis/Photo Researchers; Page 386: James King-Holmes/Photo Researchers; Page 387: Stephen Mallon/The Image Bank/Getty Images; Page 389: ©The New Yorker Collection 1993 Lee Lorenz from cartoonbank.com. All Rights Reserved; Page 391: Will McIntyre/Photo Researchers; Page 392: Cordelia Molloy/Photo Researchers; Page 394: Jedd Cadge/The Image Bank/Getty Images; Page 395: Sky Bonillo/PhotoEdit; Page 396: (left) Zia Soleil/Getty Images, Inc., (right) James Shaffer/PhotoEdit; Page 397: (top) David Young Wolff/PhotoEdit, (bottom) Kevin Curtis/Photo Researchers; Page 398: (top) Will McIntyre/Photo Researchers, (bottom) Sky Bonillo/PhotoEdit; Page 399: Photos 12/Alamy; Page 401: Will McIntyre/Photo Researchers.

Chapter 15

Pages 402–403: Alex Wong/Getty Images, Inc.; Page 403: (top) Associated Press, (top center) Alaska Stock Images/NGS Image Sales, (bottom center) Associated Press, (bottom) Bob Daemmrich/PhotoEdit; Page 405: (top) Gerd Ludwig/NG Image Collection, (bottom) Associated Press; Page 407: (bottom) GlowImages/AgeFotostock; Page 408: (left) Justin Guariglia/NGS Image Sales, (right) Getty Images News and Sport Services; Page 409: (left) Associated Press, (right) Anthony Njuguna/Reuters/Corbis; Page 410: (left) Corbis/SUPERSTOCK, (center) Arat Wolfe/Stone/Getty Images, (right) William Albert Allard/NG Image Collection; Page 411: (left) Bernhard Kuhmsted/Retna, (right) Bernhard Kuhmsted/Retna; Page 412: (left) Sisse Brimberg & Cotton Coulson/NG Images, (center) Annie Griffiths Belt/NGS Image Sales, (right) Alaska Stock Images/NGS Image Sales; Page 414: Associated Press; Page 415: Associated Press; Page 416: ©1965 by Stanley Milgram. From the film OBEDIENCE, distributed by the New York University Film Library; Page 418: Philip G. Zimbardo, Inc.; Page 419: Associated Press; Page 420: Joel Satore/NGS Image Sales; Page 421: Jodi Cobb/NG Image Collection; Page 422: ©AP/Wide World Photos; Page 423: (top) SUPERSTOCK, (bottom) Marcelo Santos/Getty Images, Inc.; Page 425: Bob Daemmrich/PhotoEdit; Page 426: ©Reuters/Corbis; Page 427: (top) Gerd Ludwig/NG Image Collection, (center) GlowImages/AgeFotostock, (bottom) ©1965 by Stanley Milgram. From the film OBEDIENCE, distributed by the New York University Film Library; Page 428: Bob Daemmrich/PhotoEdit; Page 429: (top) Najlah Feanny-Hicks/Corbis.

Appendix A

Page 433: Associated Press.

TEXT, TABLE, AND ILLUSTRATION CREDITS

Chapter 1

Figure 1.2: From *2009 Graduate Study in Psychology.* Copyright © 2009 by the American Psychological Association. Adapted with permission. Pie chart" depicting "Doctorates Awarded in Psychology, by Subfield, 2006–07", adapted from *2009 Graduate Study in Psychology.* No further reproduction or distribution is permitted without written permission from the American Psychological Association. Figure 1.7, Study Organizer 1.2-drawings (page 15): From Huffman, Karen. *Psychology in Action, 9e.* Reprinted with permission of John Wiley & Sons, Inc.

Chapter 2

Figure 2.1: From Huffman, Karen. *Psychology in Action, 9e.* Reprinted with permission of John Wiley & Sons, Inc. Figure 2.5: From Tortora and Derrickson; *Principles of Anatomy and Physiology, 12e.* Reprinted with permission of John Wiley & Sons, Inc. Figure 2.6, What a Scientist Sees A-D (page 41), Figure 2.7, Figure 2.9, Figure 2.11, Applying Psychology-drawings A-C (page 47), Figure 2.12, Figure 2.13, Figure 2.15, Figure 2.16, Figure 2.17, Figure 2.19, Figure 2.20, Figure 2.21: From Huffman, Karen. *Psychology in Action, 9e.* Reprinted with permission of John Wiley & Sons, Inc.

Chapter 3

Figure 3.1: From Huffman, Karen. *Psychology in Action, 9e.* Reprinted with permission of John Wiley & Sons, Inc. Table 3.1: Reprinted from the Journal of Psychosomatic Research, Vol. III; Holmes and Rahe: "The Social Readjustment Rating Scale." 213–218, with permission from Elsevier. Figure 3.5, Figure 3.12: From Huffman, Karen. *Psychology in Action, 9e.* Reprinted with permission of John Wiley & Sons, Inc.

Chapter 4

Figure 4.1, Figure 4.3, Figure 4.6, Figure 4.7, Figure 4.10, What a Psychologist Sees drawings of coins and Ames room (page 103): From Huffman, Karen. *Psychology in Action, 9e.* Reprinted with permission of John Wiley & Sons, Inc.

Chapter 5

Figure 5.2: From Huffman, Karen. *Psychology in Action, 9e.* Reprinted with permission of John Wiley & Sons, Inc. Applying Psychology-Are You Sleep-Deprived? quiz (page 119): From POWER SLEEP by James B. Maas and M L Wherry, copyright © 1997 by James Maas. Used by permission of Villard Books, a division of Random House, Inc. Figure 5.9, What a Psychologist Sees, drawings of the brain (page 134): From Huffman, Karen. *Psychology in Action, 9e.* Reprinted with permission of John Wiley & Sons, Inc.

Chapter 6

Figure 6.1, Figure 6.4, Figure 6.13: From Huffman, Karen. *Psychology in Action, 9e.* Reprinted with permission of John Wiley & Sons, Inc.

Chapter 7

Figure 7.1, Figure 7.2, Figure 7.3, Figure 7.4, Figure 7.5, Study Organizer 7.1-drawings (page 179), Figure 7.6c, Psychological Science-drawing (page 184), Figure 7.8: From Huffman, Karen. *Psychology in Action, 9e.* Reprinted with permission of John Wiley & Sons, Inc. Psychological Science-graph, (page 186): Bahrick, H. P., Bahrick, P. O., & Wittlinger, R. P. (1974). Long-term memory: Those unforgettable high-school days. *Psychology Today,* 8, 50–56. Reprinted with permission from Psychology Today Magazine, Copyright © 1978 Sussex Publishers, LLC. Figure 7.9, Applying Psychology-drawings A-C (page 193): From Huffman, Karen. *Psychology in Action, 9e.* Reprinted with permission of John Wiley & Sons, Inc.

Chapter 8

Figure 8.3, Figure 8.5, Applying Psychology-drawing (page 204), Figure 8.7: From Huffman, Karen. *Psychology in Action, 9e.* Reprinted with permission of John Wiley & Sons, Inc. Table 8.4: From Sternberg: *Beyond IQ.* ©1985 Cambridge University Press. Reprinted with permission of the Cambridge University Press. Figure 8.12, Figure 8.13, What a Psychologist Sees-graph (page 218), Psychological Science-drawing (page 219), Solution to the nine-dot problem (page 220), Coin problem solution (page 220): From Huffman, Karen. *Psychology in Action, 9e.* Reprinted with permission of John Wiley & Sons, Inc.

Chapter 9

Figure 9.1: From Huffman, Karen. *Psychology in Action, 9e.* Reprinted with permission of John Wiley & Sons, Inc. Figure 9.2: From Schaie: *The life course of Adult intellectual abilities, American Psychologist* - American Psychological Association. Reprinted with permission. Figure 9.4a: From Huffman, Karen. *Psychology in Action, 9e.* Reprinted with permission of John Wiley & Sons, Inc. Figure 9.6: Reproduced with permission from *Pediatrics,* Vol 89, Issue 1, pages 91–97, Copyright 1992. Figure 9.8a, Graph in What a Psychologist Sees (page 237): From Huffman, Karen. *Psychology in Action, 9e.* Reprinted with permission of John Wiley & Sons, Inc. Figure 9.10: Reproduced from Tanner J.M. Whitehouse, R.N. and Takaislu, M. "Male/female growth spurt." Archives of Diseases in childhood, 41, 454–471, 1996 with permission from BMJ Publishing Group Ltd. Figure 9.11: From Huffman, Karen. *Psychology in Action, 9e.* Reprinted with permission of John Wiley & Sons, Inc. Figure 9.13: From the chart of Elderly Achievers in, "The Brain - A User's Manual." 1982; Reprinted with permission by Diagram Visual Information. Figure 9.14, Applying Psychology-table (page 246): From Huffman, Karen. *Psychology in Action, 9e.* Reprinted with permission of John Wiley & Sons, Inc.

Chapter 10

Figure 10.2: From Huffman, Karen. *Psychology in Action, 9e.* Reprinted with permission of John Wiley & Sons, Inc. Figure 10.4: Kohlberg's Stages of Moral Development: Original Sketch: Adapted from Kohlberg, L. "Stage and Sequence: The Cognitive Developmental Approach to Socialization." in D.A. Goslin, *The Handbook of Socialization Theory and Research.* Chicago: Rand McNally, 1969. P 376 (Table 6.2). Table 10.3: Adapted and reproduced with special permission of the Publisher, Psychological Assessment Resources, Inc., 16204 North Florida Avenue, Lutz, Florida, 33549, from the Dictionary of Holland Occupational Codes, Third Edition, by Gary D. Gottfredson, Ph.D., and John L. Holland, Ph.D., Copyright 1982, 1989, 1996. Further reproduction is prohibited without permission from PAR, Inc. Figure 10.7a: From Miracle, Tina S., Miracle, Andrew W., and Baumeister, R.F., Study Guide: *Human sexuality: Meeting your basic needs,* 2nd edition. © 2006. Reprinted by permission of Pearson Education, Inc. Upper Saddle River, NJ. Figure 10.13: *The Social Context of Emotional Experience,* Annual Review of Geriatrics & Gerontology, From Laura L. Carstensen, James J. Gross, and Helen H. Fung, Reproduced with the permission of Springer Publishing Company, LLC, New York, NY 10036. Psychological Science-graph (page 276): From Huffman, Karen. *Psychology in Action, 9e.* Reprinted with permission of John Wiley & Sons, Inc.

Chapter 11

Figure 11.2a, Figure 11.3: From Huffman, Karen. *Psychology in Action, 9e.* Reprinted with permission of John Wiley & Sons, Inc. Applying Psychology-list of questions (page 286): From Zuckerman, M. (1978, February). The search for high sensation. *Psychology Today,* 38–46. Reprinted with permission from Psychology Today Magazine. Copyright © 1978 Sussex Publishers, LLC. Figure 11.5: From Maslow, A.H./Frager, R.D./Fadiman, J., MOTIVATION AND PERSONALITY, © 1987. Adapted by permission of Pearson Education, Inc., Upper Saddle River, New Jersey. Figure 11.7a: From Huffman, Karen. *Psychology in Action, 9e.* Reprinted with permission of John Wiley & Sons, Inc. Figure 11.12b: From Masters, W. H., & Johnson, V. E. (1966). *Human sexual response.* Boston: Little, Brown. Reprinted with permission of Geraldine B. Masters. Psychological Science-graph: Extrinsic versus intrinsic rewards (page 297): Copyright © 1973 by the American Psychological Association. Adapted with permission. Lepper, Mark R; Greene, David; Nisbett, Richard E. "Undermining children's intrinsic interest with intrinsic reward: A Test of the 'Overjustification' Hypothesis." *Journal of Personality and Social Psychology.* No further reproduction or distribution is permitted without written permission from the American Psychological Association. Figure 11.14, Figure 11.17, Figure 11.19: From Huffman, Karen. *Psychology in Action, 9e.* Reprinted with permission of John Wiley & Sons, Inc. Figure 11.20 Plutchik, EMOTION: PSYCHOEVOLUTIONARY SYNTHESIS, "Plutchick's Wheel of Emotion", © 1979 Individual Dynamics Inc Reproduced by permission of Pearson Education, Inc. Figure 11.23c, What a Psychologist Sees: (page 308–9): From Huffman, Karen. *Psychology in Action, 9e.* Reprinted with permission of John Wiley & Sons, Inc.

Chapter 12

Applying Psychology-"Mate selection around the world" (page 318): From Buss et al., "*International Preferences in Selecting Mates,*" Journal of Cross-Cultural Psychology, 21, pp. 5–47 © 1990. Reprinted by permission of SAGE Publications, Inc. Figure 12.3, Figure 12.5, Figure 12.7, What a Psychologist Sees-drawing (page 327), Figure 12.13, Psychological Science-pie chart (page 333): From Huffman, Karen. *Psychology in Action, 9e.* Reprinted with permission of John Wiley & Sons, Inc.

Chapter 13

Figure 13.1: From Huffman, Karen. *Psychology in Action, 9e.* Reprinted with permission of John Wiley & Sons, Inc. Study Organizer 13.1 (page 349), Figure 13.2, Figure 13.3: From the Diagnostic and Statistical Manual of Mental Disorders, fourth edition text revision, Washington, DC, © 2000 American Psychiatric Association. Reprinted with permission from the Diagnostic and Statistical Manual of Mental Disorders, Text Revision, Fourth Edition, (Copyright 2000). American Psychiatric Association. Figure 13.6, Figure 13.7: From Huffman, Karen. *Psychology in Action, 9e.* Reprinted with permission of John Wiley & Sons, Inc. Figure 13.11: "Graph: 'Genetics and schizophreinia' from the book, SCHIZOPHRENIA GENESIS by I. I. Gottesman. Copyright © 1991 by W. H. Freeman and Company. Reprinted by arrangement with Henry Holt and Company, LLC. Figure 13.19: From Huffman, Karen. *Psychology in Action, 9e.* Reprinted with permission of John Wiley & Sons, Inc.

Chapter 14

Figure 14.2, Figure 14.3: From Huffman, Karen. *Psychology in Action, 9e.* Reprinted with permission of John Wiley & Sons, Inc. Figure 14.4: From Beck, Judith S. *Cognitive Therapy: Basics and Beyond* © 1993. Dysfunctional Thought Record, Figure 9.1, page 126. Reprinted with permission of Guilford Publications, Inc. Figure 14.11: From Huffman, Karen. *Psychology in Action, 9e.* Reprinted with permission of John Wiley & Sons, Inc.

Chapter 15

Figure 15.3, Figure 15.4, Figure 15.5: From Huffman, Karen. *Psychology in Action, 9e.* Reprinted with permission of John Wiley & Sons, Inc. Table 15.1: From Rubin, Z. (1970). "Measurement of romantic love," *Journal of Personality and Social Psychology, 16,* 265–273. Copyright © 1970 by Zick Rubin. Reprinted by permission of the author. Graph in What a Psychologist Sees (page 412) Figure 15.9, What a Psychologist Sees (page 416) Figure 15.16: From Huffman, Karen. *Psychology in Action, 9e.* Reprinted with permission of John Wiley & Sons, Inc.

NAME INDEX

I

J

K

S

T

SUBJECT INDEX

D

E

H

I

J

K

L

M